Modern Optical Spectroscopy

William W. Parson

Modern Optical Spectroscopy

With Exercises and Examples from Biophysics and Biochemistry

Second Edition

Springer

William W. Parson
University of Washington
Seattle
Washington
USA

ISBN 978-3-662-50021-7 ISBN 978-3-662-46777-0 (eBook)
DOI 10.1007/978-3-662-46777-0

Springer Heidelberg New York Dordrecht London

Softcover reprint of the hardcover 2nd edition 2015

Printed on acid-free paper

Springer-Verlag GmbH Berlin Heidelberg is part of Springer Science+Business Media (www.springer.com)

Preface to the Second Edition

Applications of optical spectroscopy in chemistry, biochemistry, and biophysics continue to flourish. This revised edition of Modern Optical Spectroscopy includes expanded discussions of quantum optics, metal-ligand charge-transfer transitions, entropy changes during photoexcitation, electron transfer from excited molecules, normal-mode calculations, vibrational Stark effects, studies of fast processes by resonance energy transfer in single molecules, and two-dimensional electronic and vibrational spectroscopy. I have added new figures where I thought they would help and modified some of the original figures for greater clarity. The references have been updated and moved from the end of the book to the chapters where they are cited. Each chapter also has a set of exercises for students.

I greatly appreciate the constructive comments from readers calling my attention to errors or points that needed clarification in the first edition. Discussions with Bill Hazelton, Bob Knox, Ross McKenzie, Nagarajan, and Steve Boxer were particularly helpful. I also thank my Springer editors Jutta Lindenborn and Sabine Schwarz for their excellent suggestions and my wife Polly for her continuing patience and encouragement.

Seattle, WA
April 2015

William W. Parson

Preface to the First Edition

This book began as lecture notes for a course on optical spectroscopy that I taught for graduate students in biochemistry, chemistry, and our interdisciplinary programs in molecular biophysics and biomolecular structure and design. I started expanding the notes partly to try to illuminate the stream of new experimental information on photosynthetic antennas and reaction centers, but mostly just for fun. I hope that readers will find the results not only useful, but also as stimulating as I have.

One of my goals has been to write in a way that will be accessible to readers with little prior training in quantum mechanics. But any contemporary discussion of how light interacts with molecules must begin with quantum mechanics, just as experimental observations on blackbody radiation, interference, and the photoelectric effect form the springboard for almost any introduction to quantum mechanics. To make the reasoning as transparent as possible, I have tried to adopt a consistent theoretical approach, minimize jargon, and explain any terms or mathematical methods that might be unfamiliar. I have provided numerous figures to relate spectroscopic properties to molecular structure, dynamics, and electronic and vibrational wavefunctions. I also describe classical pictures in many cases and indicate where these either have continued to be useful or have been supplanted by quantum mechanical treatments. Readers with experience in quantum mechanics should be able to skip quickly through many of the explanations, but will find that the discussion of topics such as density matrices and wavepackets often progresses well beyond the level of a typical 1-year course in quantum mechanics. I have tried to take each topic far enough to provide a solid stepping-stone to current theoretical and experimental work in the area.

Although much of the book focuses on physical theory, I have emphasized aspects of optical spectroscopy that are especially pertinent to molecular biophysics, and I have drawn most of the examples from this area. The book therefore covers topics that receive little attention in most general books on molecular spectroscopy, including exciton interactions, resonance energy transfer, single molecule spectroscopy, high-resolution fluorescence microscopy, femtosecond pump–probe spectroscopy, and photon echoes. It says less than is customary about atomic spectroscopy and about rotational and vibrational spectroscopy of small molecules. These choices reflect my personal interests and the realization that I had to stop somewhere, and I can only apologize to readers whose selections would have been different. I apologize also for using work from my own laboratory in many of the illustrations when other excellent illustrations of the same points are available in the literature. This was just a matter of convenience.

I could not have written this book without the patient encouragement of my wife Polly. I also have enjoyed many thought-provoking discussions with Arieh Warshel, Nagarajan, Martin Gouterman, and numerous other colleagues and students, particularly including Rhett Alden, Edouard Alphandéry, Hiro Arata, Donner Babcock, Mike Becker, Bob Blankenship, Steve Boxer, Jacques Breton, Jim Callis, Patrik Callis, Rod Clayton, Richard Cogdell, Tom Ebrey, Tom Engel, Graham Fleming, Eric Heller, Dewey Holten, Ethan Johnson, Amanda Jonsson, Chris Kirmaier, David Klug, Bob Knox, Rich Mathies, Eric Merkley, Don Middendorf, Tom Moore, Jim Norris, Oleg Prezhdo, Phil Reid, Bruce Robinson, Karen Rutherford, Ken Sauer, Dustin Schaefer, Craig Schenck, Peter Schellenberg, Avigdor Scherz, Mickey Schurr, Gerry Small, Rienk van Grondelle, Maurice Windsor, and Neal Woodbury. Patrik Callis kindly provided the atomic coefficients used in Chaps. 4 and 5 for the molecular orbitals of 3-methylindole. Any errors, however, are entirely mine. I will appreciate receiving any corrections or suggestions for improvements.

Seattle, WA William W. Parson
October 2006

Contents

List of Boxes

Introduction 1

1.1 Overview

Because of their extraordinary sensitivity and speed, optical spectroscopic techniques are well suited for addressing a broad range of questions in molecular and cellular biophysics. Photomultipliers sensitive enough to detect a single photon make it possible to measure the fluorescence from individual molecules, and lasers providing light pulses with widths of less than 10^{-14} s can be used to probe molecular behavior on the time scale of nuclear motions. Spectroscopic properties such as absorbance, fluorescence and linear and circular dichroism can report on the identities, concentrations, energies, conformations or dynamics of molecules and can be sensitive to small changes in molecular structure or surroundings. Resonance energy transfer provides a way to probe intermolecular distances. Because they usually are not destructive, spectrophotometric techniques can be used with samples that must be recovered after an experiment. They also can provide analytical methods that avoid the need for radioisotopes or hazardous reagents. When combined with genetic engineering and microscopy, they provide windows to the locations, dynamics and turnover of particular molecules in living cells.

In addition to describing applications of optical spectroscopy in biophysics and biochemistry, this book is about light and how light interacts with matter. These are topics that have puzzled and astonished people for thousands of years, and continue to do so today. To understand how molecules respond to light we first must inquire into why molecules exist in well-defined states and how they change from one state to another. Thinking about these questions underwent a series of revolutions with the development of quantum mechanics, and today quantum mechanics forms the scaffold for almost any investigation of molecular properties. Although most of the molecules that interest biophysicists are far too large and complex to be treated exactly by quantum mechanical techniques, their properties often can be rationalized by quantum mechanical principles that have been refined on simpler systems. We'll discuss these principles in Chap. 2. For now, the most salient points are just that a molecule can exist in a variety of states depending on how its

W.W. Parson, *Modern Optical Spectroscopy*, DOI 10.1007/978-3-662-46777-0_1

electrons are distributed among a set of molecular orbitals, and that each of these quantum states is associated with a definite energy. For a molecule with $2n$ electrons, the electronic state with the lowest total energy usually is obtained when there are two electrons with antiparallel spins in each of the n lowest orbitals and all the higher orbitals are empty. This is the *ground state*. In the absence of external perturbations, a molecule placed in the ground state will remain there indefinitely.

Chapter 3 will discuss the nature of light, beginning with a classical description of an oscillating electromagnetic field. Exposing a molecule to such a field causes the potential energies of the electrons to fluctuate with time, so that the original molecular orbitals no longer limit the possibilities. The result of this can be that an electron moves from one of the occupied molecular orbitals to an unoccupied orbital with a higher energy. Two main requirements must be met in order for such a transition to occur. First, the electromagnetic field must oscillate at the right frequency. The required frequency (ν) is

$$\nu = \Delta E / h, \tag{1.1}$$

where ΔE is the difference between the energies of the ground and excited states and h is Planck's constant (6.63×10^{-34} J s, 4.12×10^{-15} eV s, or 3.34×10^{-11} cm^{-1} s). This expression is in accord with our experience that a given type of molecule, or a molecule in a particular environment, absorbs light of some colors and not of others. In Chap. 4 we will see that the frequency rule emerges straightforwardly from the classical electromagnetic theory of light, as long as we treat the absorbing molecule quantum mechanically. It is not necessary at this point to use a quantum mechanical picture of light.

The second requirement is perhaps less familiar than the first, and has to do with the shapes of the two molecular orbitals and the disposition of the orbitals in space relative to the polarization of the oscillating electrical field. The two orbitals must have different geometrical symmetries and must be oriented in an appropriate way with respect to the field. This requirement rationalizes the observation that absorption bands of various molecules vary widely in strength. It also explains why the absorbance of an anisotropic sample depends on the polarization of the light beam.

The molecular property that determines both the strength of an absorption band and the optimal polarization of the light is a vector called the *transition dipole*, which can be calculated from the molecular orbitals of the ground and excited state. The square of the magnitude of the transition dipole is termed the *dipole strength*, and is proportional to the strength of absorption. Chapter 4 develops these notions more fully and examines how they arise from the principles of quantum mechanics. This provides the theoretical groundwork for discussing how measurements of the wavelength, strength, or polarization of electronic absorption bands can provide information on molecular structure and dynamics. In Chaps. 10 and 11 we extend the quantum mechanical treatment of absorption to large ensembles of molecules that interact with their surroundings in a variety of ways. Various types of vibrational spectroscopy are discussed in Chaps. 6, 11 and 12.

A molecule that has been excited by light can decay back to the ground state by several possible paths. One possibility is to reemit energy as fluorescence. Although spontaneous fluorescence is not simply the reverse of absorption, it shares the same requirements for energy matching and appropriate orbital symmetry. Again, the frequency of the emitted radiation is proportional to the energy difference between the excited and ground states and the polarization of the radiation depends the orientation of the excited molecule, although both the orientation and the energy of the excited molecule usually change in the interval between absorption and emission. As we will see in Chap. 7, the same requirements underlie another mechanism by which an excited molecule can decay, the transfer of energy to a neighboring molecule. The relationship between fluorescence and absorption is developed in Chap. 5, where the need for a quantum theory of light finally comes to the front.

1.2 The Beer-Lambert Law

A beam of light passing through a solution of absorbing molecules transfers energy to the molecules as it proceeds, and thus decreases progressively in intensity. The decrease in the intensity, or irradiance (I) over the course of a small volume element is proportional to the irradiance of the light entering the element, the concentration of absorbers (C), and the length of the path through the element (dx):

$$\frac{dI}{dx} = -\varepsilon' I C. \tag{1.2}$$

The proportionality constant (ε') depends on the wavelength of the light and on the absorber's structure, orientation and environment. Integrating Eq. (1.2) shows that if light with irradiance I_o is incident on a cell of thickness l, the irradiance of the transmitted light will be:

$$I = I_o \exp(-\varepsilon' C l) = I_o 10^{-\varepsilon C l} \equiv I_o 10^{-A}. \tag{1.3}$$

Here A is the *absorbance* or *optical density* of the sample ($A = \varepsilon\, C\, l$) and ε is called the *molar extinction coefficient* or *molar absorption coefficient* [$\varepsilon = \varepsilon'/\ln(10) = \varepsilon'/2.303$]. The absorbance is a dimensionless quantity, so if C is given in units of molarity (1 M = 1 mol/l) and c in cm, ε must have dimensions of $M^{-1}\ cm^{-1}$.

Equations (1.1) and (1.2) are statements of *Beer's law*, or more accurately, the *Beer-Lambert law*. Johann Lambert, a physicist, mathematician and astronomer born in 1728, observed that the fraction of the light that is transmitted (I/I_o) is independent of I_o. Wilhelm Beer, a banker and astronomer who lived from 1797 to 1850, noted the exponential dependence on C.

In the classical electromagnetic theory of light, the oscillation frequency ν is related to the wavelength (λ), the velocity of light in a vacuum (c), and the refractive index of the medium (n) by the expression

$$\nu = c/n\lambda. \tag{1.4}$$

Light with a single wavelength, or more realistically, with a narrow band of wavelengths is called *monochromatic*.

The light intensity, or *irradiance* (I) in Eqs. (1.2) and (1.3) represents the flux of radiant energy per unit cross-sectional area of the beam (joules per second per square cm or watts per square cm). We usually are concerned with the radiation in a particular frequency interval ($\Delta\nu$), so that I has units of joule per frequency interval per second per square cm. For a light beam with a cross-sectional area of 1 cm^2, the amplitude of the signal that might be recorded by a photomultiplier or other detector is proportional to $I(\nu)\Delta\nu$. In the quantum theory of light that we'll discuss briefly in Sect. 1.6 and at greater depth in Chap. 3, itensities often are expressed in terms of the flux of photons rather than energy [photons per frequency interval per second per square cm]. A beam with an irradiance of 1 W cm^{-2} has a photon flux of $5.05 \times (\lambda/\mathrm{nm}) \times 10^{15}$ photons cm^{-2}.

The dependence of the absorbance on the frequency of light can be displayed by plotting A or ε as a function of the frequency (ν), the wavelength (λ), or the *wavenumber* ($\overline{\nu}$). The wavenumber is simply the reciprocal of the wavelength in a vacuum: $\overline{\nu} = 1/\lambda = \nu/c$, and has units of cm^{-1}. Sometimes the percent of the incident light that is absorbed or transmitted is plotted. The percent absorbed is $100 \times (I_o - I)/I_o = 100 \times (1 - 10^{-A})$, which is proportional to A if $A \ll 1$.

1.3 Regions of the Electromagnetic Spectrum

The regions of the electromagnetic spectrum that will be most pertinent to our discussion involve wavelengths between 10^{-9} and 10^{-2} cm. Visible light fills only the small part of this range between 3×10^{-5} and 8×10^{-5} cm (Fig. 1.1). Transitions of bonding electrons occur mainly in this region and the neighboring ultraviolet (UV) region; vibrational transitions occur in the infrared (IR). Rotational transitions are measurable in the far-IR region in small molecules, but in macromolecules these transitions are too congested to resolve. Radiation in the X-ray region can cause transitions in which 1 *s* or other core electrons are excited to atomic 3*d* or 4*f* shells or are dislodged completely from a molecule. These transitions can report on the oxidation and coordination states of metal atoms in metalloproteins.

The inherent sensitivity of absorption measurements in different regions of the electromagnetic spectrum decreases with increasing wavelength because, in the idealized case of a molecule that absorbs and emits radiation at a single frequency, it depends on the difference between the populations of molecules in the ground and excited states. If the two populations are the same, radiation at the resonance frequency will cause upward and downward transitions at the same rate, giving a net absorbance of zero. At thermal equilibrium, the fractional difference in the populations is given by

$$\frac{(N_g - N_e)}{(N_g + N_e)} = \frac{\{1 - (S_e/S_g)\exp(-\Delta E/k_B T)\}}{\{1 + (S_e/S_g)\exp(-\Delta E/k_B T)\}} \approx \frac{\{1 - \exp(-h\nu/k_B T)\}}{\{1 + \exp(-h\nu/k_B T)\}}, \quad (1.5)$$

where S_e/S_g is an entropic (degeneracy) factor (1 for a system with only two states), k_B is the Boltzmann constant (0.69502 cm^{-1}/K) and T is the temperature. At room temperature ($k_B T \approx 200\ \mathrm{cm}^{-1}$), $(N_g\text{-}N_e)/(N_g+N_e)$ is essentially 1 for an electronic transition with $\lambda = 500$ nm [$h\nu = \overline{\nu} = 2 \times 10^4\ \mathrm{cm}^{-1}$, $\exp(-h\nu/k_B T) \approx \exp(-100) \approx 10^{-43}$], compared to only 1 part in 10^4 for a proton magnetic transition in a 600 Mhz spectrometer [$\lambda = 50$ cm, $\overline{\nu} = 0.02\ \mathrm{cm}^{-1}$, $\exp(-h\nu/k_B T) \approx e^{-0.00010} \approx 0.99990$]. The greater specificity of NMR of course often compensates for the lower sensitivity.

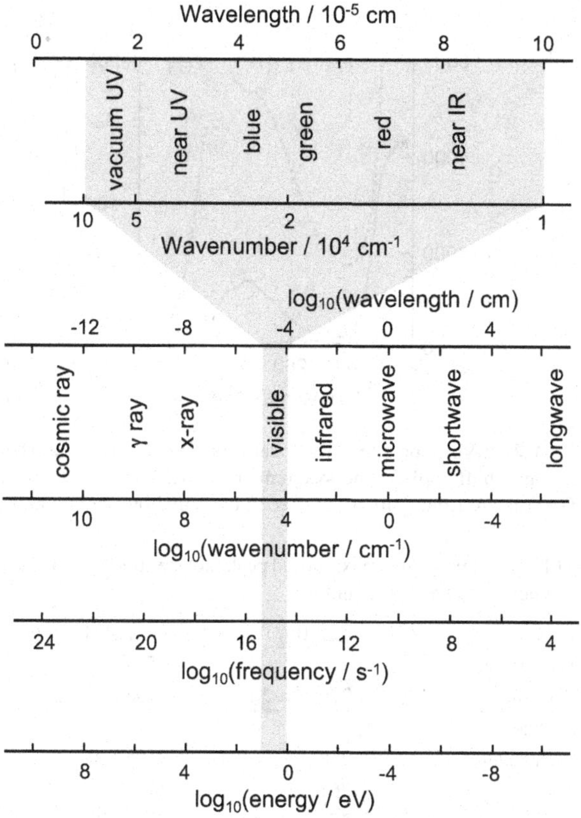

Fig. 1.1 Regions of the electromagnetic spectrum. The upper spectrum shows the visible, UV and near-IR regions on a linear wavelength scale. More extended spectra are shown on logarithmic wavelength, wavenumber, frequency and energy scales

1.4 Absorption Spectra of Proteins and Nucleic Acids

Most of the absorbance of proteins in the near-UV region between 250 and 300 nm is due to the aromatic amino acids, particularly tyrosine [1, 2]. Figure 1.2 shows the absorption spectra of tyrosine, phenylalanine and tryptophan in solution, and Table 1.1 gives the absorption maxima (λ_{max}) and the peak molar extinction coefficients (ε_{max}). Although tryptophan has a larger extinction coefficient than phenylalanine or tyrosine, it makes only a minor contribution to the absorbance of most proteins because of its lower abundance. Cystine disulfide groups have a weak absorption band in the region of 260 nm, which shifts to longer wavelengths as the C–S–S–C dihedral angle is twisted away from 90° [3]. At shorter wavelengths,

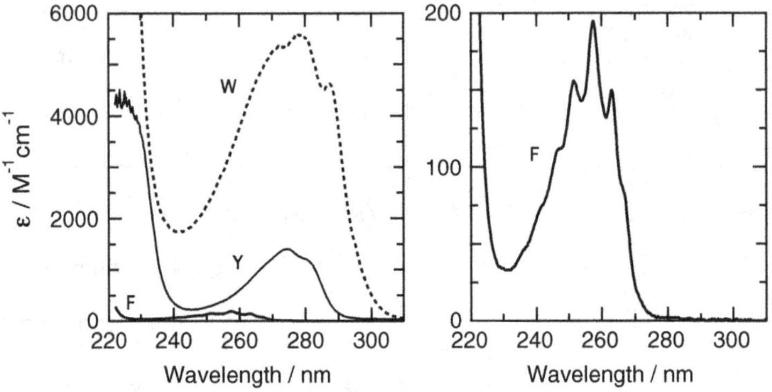

Fig. 1.2 Absorption spectra of phenylalanine (*F*), tyrosine (*Y*) and tryptophan (*W*) in 0.1 M phosphate buffer, pH 7. The spectrum of phenylalanine is shown on an expanded scale on the *right*. (The data are from a library of spectra measured by Lindsey and coworkers [34, 35])

Table 1.1 Absorption maxima and peak molar extinction coefficients of amino acids, purines and pyrimidines in aqueous solution

Absorber	λ_{max} (nm)[a]	ε_{peak} (M^{-1} cm^{-1})[a]	ε_{280} (M^{-1} cm^{-1})[b]
Tryptophan	278	5,580	5,500
Tyrosine	274	1,405	1,490
Cystine	–	–	125
Phenylalanine	258	195	
Adenine	261	13,400	
Guanine	273	13,150	
Uracil	258	8,200	
Thymine	264	7,900	
Cytosine	265	4,480	

[a]0.1 M phosphate buffer, pH 7.0. From Fasman [33]
[b]Best fit of values for 80 folded proteins in water. From Pace et al. [2]

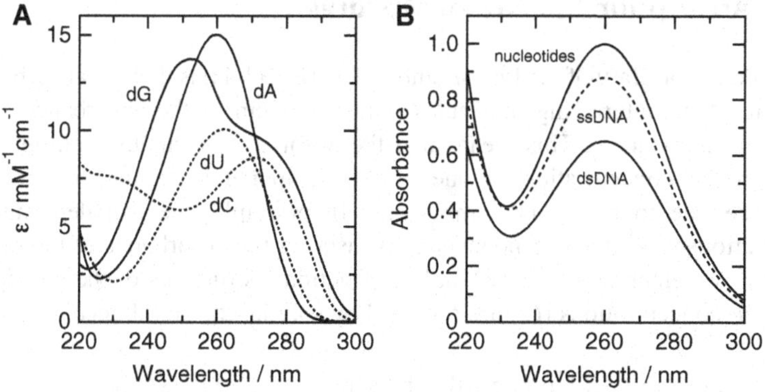

Fig. 1.3 (**A**) Absorption spectra of 2′-deoxyadenosine (dA), 2′-deoxyguanosine (dG), 2′-deoxyuridine (dU) and 2′-deoxycytidine (dC) at pH 7.1. (**B**) Absorption spectra of *Escherichia coli* DNA at 25° (double-stranded DNA, *ds-DNA*) and 82° (single-stranded DNA, *ss-DNA*), and at 25° after enzymatic digestion (*nucleotides*). *E. coli* DNA is double-stranded at 25° and single-stranded at 82°; enzymatic digestion yields the component nucleotides (the data in **B** are from Voet et al. [36])

most of the absorbance of proteins comes from the peptide backbone. The –C(O)–N(H)- group has absorption bands near 190 and 215 nm with peak extinction coefficients of about 7,000 and 100 M^{-1} cm^{-1}, respectively. The stronger band represents a $\pi - \pi^*$ transition, in which an electron is excited from a π (bonding) molecular orbital to a π^* (antibonding) orbital; the weaker band comes from an $n - \pi^*$ transition, in which a non-bonding electron of the oxygen atom is promoted to a π^* orbital.

In addition to containing phenylalanine, tyrosine and tryptophan, many proteins bind small molecules that absorb in the UV or visible regions of the spectrum. NADH, for example, absorbs at 340 nm; flavins, in the region of 400 nm. Hemes have strong absorption bands between 410 and 450 and weaker bands between 550 and 600 nm.

The common purine and pyrimidine bases all absorb in the region of 260 nm, and the absorption spectra of the nucleosides and nucleotides are similar to those of the free bases (Fig. 1.3A and Table 1.1). However, the absorbance of double-stranded DNA in the region of 260 nm is 30–40% smaller than the sum of the absorbances of the individual bases (Fig. 1.3B). Single-stranded DNA gives an intermediate absorbance. As we'll discuss in Chap. 8, this *hypochromism* results from electronic coupling of the individual nucleotides in the nucleic acid. The excited states of the oligomer include contributions from multiple nucleotides, with the result that some of the absorption strength moves from the near-UV region to shorter wavelengths. Polypeptides show similar effects: the absorbance of an α-helix near 200 nm is lower than that of a random coil or β-sheet.

1.5 Absorption Spectra of Mixtures

An important corollary of the Beer-Lambert law (Eq. (1.1)) is that the absorbance of a mixture of noninteracting molecules is just the sum of the absorbances of the individual components. This means that the absorbance change resulting from a change in the concentration of one of the components is independent of the absorbance due to the other components. In principle, we can determine the concentrations of all the components by measuring the absorbance of the solution at a set of wavelengths where the molar extinction coefficients of the components differ. The concentrations (C_i) are obtained by solving the simultaneous equations

$$\begin{aligned} &\varepsilon_{1(\lambda 1)}C_1 + \varepsilon_{2(\lambda 1)}C_2 + \varepsilon_{3(\lambda 1)}C_3 + \cdots = A_{\lambda 1}/l \\ &\varepsilon_{1(\lambda 2)}C_1 + \varepsilon_{2(\lambda 2)}C_2 + \varepsilon_{3(\lambda 2)}C_3 + \cdots = A_{\lambda 2}/l \\ &\varepsilon_{1(\lambda 3)}C_1 + \varepsilon_{2(\lambda 3)}C_2 + \varepsilon_{3(\lambda 3)}C_3 + \cdots = A_{\lambda 3}/l \\ &\cdots, \end{aligned} \tag{1.6}$$

where $\varepsilon_{i(\lambda a)}$ and $A_{\lambda a}$ are the molar extinction coefficient of component i and the absorbance of the solution at wavelength λ_a, and l again is the optical path length. (A method for solving such a set of equations is given in Box 8.1.) The concentrations are completely determined when the number of measurement wavelengths is the same as the number of components, as long as the extinction coefficients of the components differ significantly at each wavelength. Measurements at additional wavelengths can be used to increase the reliability of the results. The best way to calculate the concentrations then probably is to use singular-value decomposition [4].

Although two chemically distinct molecules usually have characteristically different absorption spectra, their extinction coefficients may be identical at one or more wavelengths. In the notation used above, $\varepsilon_{i(\lambda a)}$ and $\varepsilon_{j(\lambda a)}$ (the extinction coefficients of components i and j at wavelength a) may be the same (Fig. 1.4A). Such a wavelength is termed an *isosbestic point*. If we change the ratio of the concentrations C_i and C_j in a solution, the absorbance at the isosbestic points will remain constant (Fig. 1.4B). If the solution contains a third component (k) we might have $\varepsilon_{i(\lambda b)} = \varepsilon_{k(\lambda b)}$ and $\varepsilon_{j(\lambda c)} = \varepsilon_{k(\lambda c)}$ at two other wavelengths (b and c), but it would be unlikely for all three components to have identical extinction coefficients at any wavelength. Thus if we have a solution containing an unknown number of components and the absorption spectrum of the solution changes as a function of a parameter such as pH, temperature or time, the observation of an isosbestic point indicates that the absorbance changes probably reflect changes in the ratio of only two components. The reliability of this conclusion increases if there are two isosbestic points.

Protein concentrations often are estimated from the absorbance at 280 nm, where the only amino acids that absorb significantly are tryptophan, tyrosine and cystine (Table 1.1). The molar extinction coefficient of a protein at this wavelength is given by $\varepsilon_{280} \approx 5{,}500 \times W + 1{,}490 \times Y + 125 \times CC\ \mathrm{M^{-1}\,cm^{-1}}$, where W, Y and CC are the numbers of tryptophan, tyrosine and cystine residues in the protein [2].

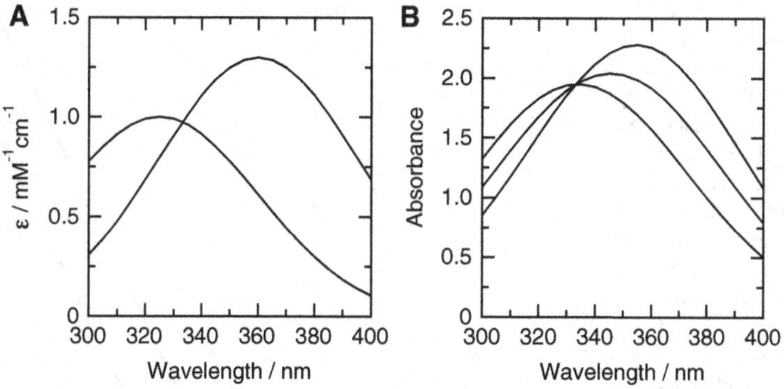

Fig. 1.4 Absorption spectra of two species that have the same extinction coefficient at 333 nm (**A**) and of mixtures of the same two components in molar ratios of 1:3, 1:1 and 3:1 (**B**). The total absorbance at the isosbestic point (333 nm) is constant

1.6 The Photoelectric Effect

Ejection of electrons from a solid surface exposed to radiation is called the *photoelectric effect*. The electrons that are released are called *photoelectrons*. In a landmark paper, Einstein [5] pointed out three key features of the effect: (1) photoelectrons are released only if the frequency of the radiation (ν) exceeds a certain minimum (the *work function*, ν_o) that is characteristic of the solid material (Fig. 1.5); (2) the kinetic energy of the departing photoelectrons is proportional to ($\nu-\nu_o$); and (3) the process appears to occur instantaneously, even at low light intensities. These observations suggested that electromagnetic radiation has a particulate nature, and that each particle bears a definite amount of energy that is proportional to ν (Eq. 1.1). G.N. Lewis [6] coined the term *photon* for such a particle from the Greek word "phos" for "light." The first two observations mentioned above can be explained by noting that dislodging an electron from a solid requires a certain minimum amount of energy that depends on the composition of the solid and is analogous to the ionization energy of a molecule. If the photon's energy exceeds this threshold, the photon is absorbed and all its energy is transferred to the solid and the departing electron. The instantaneous nature of photoejection (observation 3) indicates that the process involves an all-or-nothing absorption of an individual photon, rather than a gradual accumulation of energy over a period of time. The all-or-nothing nature of the effect is difficult to explain by the classical electromagnetic theory of light, in which the energy depends on the square of the electromagnetic field and varies continuously as a function of the distance from the light source. In the quantum picture of light suggested by Einstein, the square of the electromagnetic field strength is a measure of the number of photons per unit volume. The photon density looks like a continuous variable at high light intensities, but varies discontinuously at very low intensities.

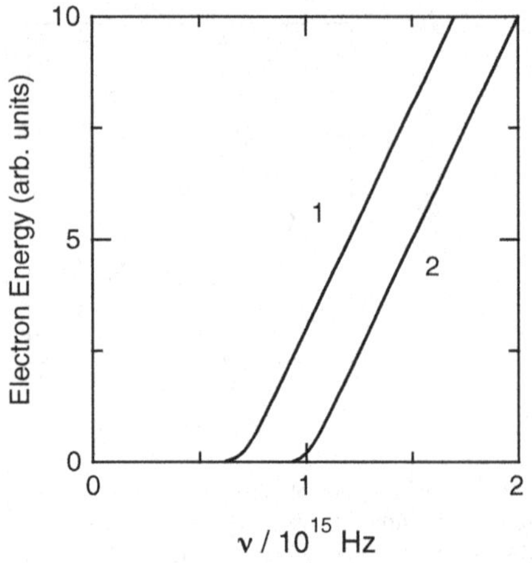

Fig. 1.5 Some materials release electrons when they are struck by electromagnetic radiation with a frequency (ν) greater than a threshold value that depends on the material, as shown here schematically for two different materials (curves 1 and 2). Above the threshold, the kinetic energy of the photoelectrons increases linearly with ν. Inhomogeneity in the material can cause a slight curvature of the plots near the threshold

Einstein, who was 25 at the time, introduced the theories of special relativity and Brownian motion the same year. He received the physics Nobel prize for his explanation of the photoelectric effect, but later expressed serious doubts about the existence of photons. The central problem was that, because the photoelectric effect involves interaction of light with a solid material, it shows only that the energy levels of the material are quantized. Convincing evidence that light also obeys the laws of quantum mechanics came only some 80 years later from experiments on the interactions of two photons in short pulses of light [7–9]. We'll discuss these experiments in Sect. 3.5.

1.7 Techniques for Measuring Absorbance

Light intensities can be measured with a *photomultiplier*, which is a vacuum tube with a negatively charged plate or surface (the *cathode* or *photocathode*) that releases photoelectrons when it absorbs photons (Fig. 1.6). The cathode is coated with a material that responds to light over a broad range of frequencies, and as explained in the previous section, the departing photoelectron acquires a kinetic energy equal to the difference between the photon energy at a particular frequency ($h\nu$) and the minimum energy needed to dislodge the electron ($h\nu_o$). The photoelectron is accelerated by an electric field in the tube and is drawn to a second coated plate (a *dynode*). Here the kinetic energy released upon impact dislodges several new electrons. These electrons are accelerated toward a second dynode, where they dislodge additional electrons. Repeated amplification steps over a series of 6–14 stages can give an overall current amplification of 10^6–10^8, depending on the design of the tube and the voltage applied to the chain of dynodes. At the final dynode (the *anode*) the current is passed to an amplifier and recorded either as a continuous signal or as a pulse that is registered digitally.

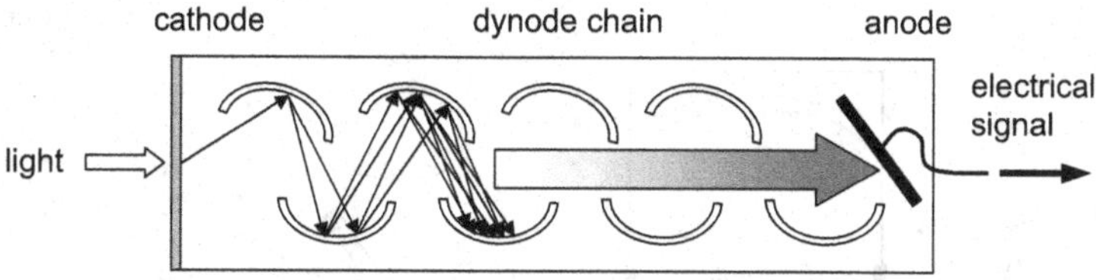

Fig. 1.6 Schematic design of a photomultiplier tube. Pins at the base of the tube provide a negative electrical potential at the cathode relative to the anode, and intermediate potentials along the chain of dynodes

Photomultipliers are remarkably linear over many orders of magnitude of light intensity. However, they are best suited for very low light intensities. They typically have *quantum efficiencies* on the order of 0.25, which means that about 25% of the photons striking the cathode give rise to current pulses at the anodes. The response saturates at high light intensities because depletion of electrons raises the electric potential on the cathode. Cathodes coated with GaAs and related materials have higher quantum efficiencies and cover exceptionally broad spectral ranges, but are less robust than most other commonly used materials.

The time resolution of a photomultiplier is limited mainly by the variations in the paths that electrons take in reaching the anode. Because of the spread in transit times, the anode pulse resulting from the absorption of a single photon typically has a width on the order of 10^{-9}–10^{-8} s. The spread of transit times is smaller in *microchannel plate* photomultipliers, which work on the same principles as ordinary photomultipliers except that the electronic amplification steps occur along the walls of small capillaries. The anode pulse width in a microchannel plate detector can be as short as 2×10^{-11} s.

Light intensities also can be measured with *photodiodes* made from silicon, germanium or other semiconductors. These devices have junctions between regions (*N* and *P*) that are doped with other elements to give an excess of either electrons or holes, respectively. In the dark, electrons diffuse from the *N* to the *P* region and holes diffuse in the opposite direction, creating an electric field across the junction. Absorption of light separates additional electron–hole pairs, which diffuse in the directions dictated by the electric field and create a flow of current across the device. Silicon photodiodes have quantum efficiencies of about 80% and function well at comparatively high light intensities. Their time resolution usually is on the order of 10^{-9}–10^{-8} s, but can be about 10^{-11} s in diodes with very small active areas.

A spectrophotometer for measuring absorption spectra typically includes a continuous light source, a monochromator for dispersing the white light and selecting a narrow band of wavelengths, and a chopper for separating the light into two beams (Fig. 1.7). One beam passes through the specimen of interest; the other, through a reference (blank) cuvette. The intensities of the two beams are measured with a photomultiplier or other detector, and used to calculate the

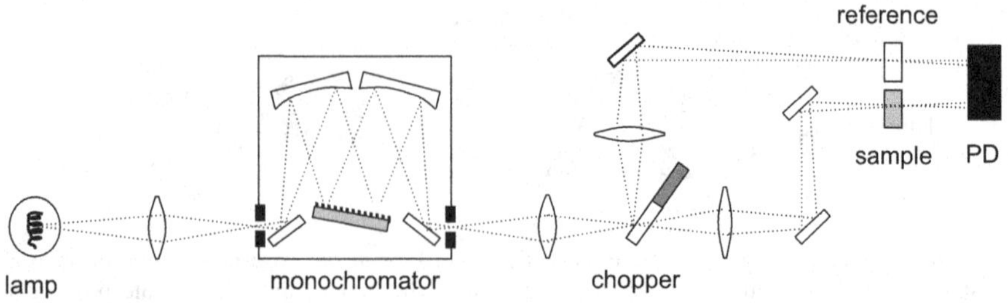

Fig. 1.7 Elements of a spectrophotometer. The light chopper typically is a spinning mirror with sectors that alternately transmit and reflect the beam, so that the photodetector alternately registers the intensities of light beams passing through the sample or the reference cuvette. The monochromator sketched here consists of a rotatable diffraction grating (*shaded*), two curved mirrors, two planar mirrors, and entrance and exit slits for adjusting the spectral resolution. *PD* photodetector (photomultiplier or photodiode)

absorbance of the sample as a function of wavelength ($\Delta A = \log_{10}(I_o/I_s) - \log_{10}(I_o/I_r) = \log_{10}(I_r/I_s)$, where I_o, I_s and I_r are the incident light intensity and the intensities of the light transmitted through the specimen and reference, respectively). The measurement of the reference signal allows the instrument to discount the absorbance due to the solvent and the walls of the cuvette. Judicious choice of the reference also can minimize errors resulting from the loss of light by scattering from turbid samples.

The conventional spectrophotometer sketched in Fig. 1.7 has several limitations. First, the light reaching the detector at any given time covers only a narrow band of wavelengths. Because a grating, prism or mirror in the monochromator must be rotated to move this window across the spectral region of interest acquisition of an absorption spectrum typically takes several minutes, during which time the sample may change. In addition, narrowing the entrance and exit slits of the monochromator to improve the spectral resolution decreases the amount of light reaching the photomultiplier, which makes the signal noisier. Although the signal/noise ratio can be improved by averaging the signal over longer periods of time, this further slows acquisition of the spectrum. These limitations are surmounted to some extent in instruments that use photodiode arrays to detect light at many different wavelengths simultaneously.

The difficulties mentioned above also can be overcome by a Fourier-transform technique that is especially useful for IR spectroscopy. In a Fourier-transform IR (FTIR) spectrometer, radiation from an IR source is split into two beams that ultimately are recombined and focused on the detector (Fig. 1.8). The detector senses light covering a broad band of wavelengths. If the two beams traverse different optical path lengths, their contributions to the radiation field at the detector will interfere constructively or destructively depending on the wavelength and the path difference, ΔL. The length of one of the paths in the spectrometer is modulated by moving a mirror forward and backward over a distance of several cm, causing the signal registered by the detector to oscillate. If the radiation were monochromatic, a plot of the signal intensity *versus* ΔL (an *interferogram*) would be a

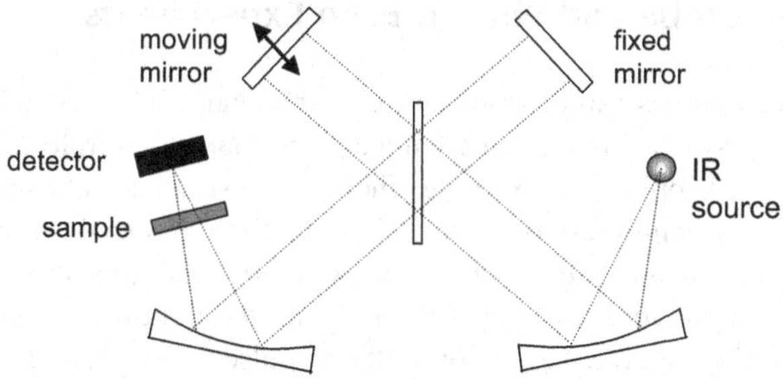

Fig. 1.8 Elements of an FTIR spectrometer. Radiation from the IR source is collimated by a spherical mirror and then split into two beams by a partially silvered mirror (*center*). The beams are reflected by planar mirrors and are recombined after they pass through the sample. Moving one of the planar mirrors back and forth as indicated by the *double arrow* changes the length of one of the optical paths

sinusoidal function with a period determined by the wavelength of the light; with broad-band radiation, the interferogram includes superimposed oscillations with many different periods. The spectrum of the radiation can be calculated by taking a Fourier transform of the interferogram. (See Appendix A3 for an introduction to Fourier transforms.) When a specimen that absorbs radiation at a particular wavelength is placed in the beam, the corresponding oscillations in the interferogram are attenuated. The absorption spectrum is calculated as $A(\overline{\nu}) = \log_{10}(S_r(\overline{\nu})/S_s(\overline{\nu}))$, where $S_s(\overline{\nu})$ and $S_r(\overline{\nu})$ are the Fourier transforms of interferograms obtained with and without the sample in the beam.

FTIR spectrometers have another advantage in addition to faster data acquisition and an improved signal/noise ratio: the wavelength scale can be calibrated very accurately by replacing the IR source by a laser with a sharp emission line. A He-Ne laser, which emits at 6,328 Å, is often used for this purpose. The accurate calibration makes it possible to measure small shifts in spectra measured with different samples, such as materials enriched in ^{13}C, ^{15}N or ^{18}O. In addition, any stray light that reaches the detector without passing through the interferometer causes relatively little error in an FTIR spectrometer because the signal resulting from this light is not correlated with ΔL. This feature makes FTIR spectrometers well suited for measuring absorption spectra of samples that transmit only a small fraction of the incident light.

Sections 1.8, 1.10 and 11.5 describe several other ways of measuring absorption spectra when more conventional approaches are inadequate. Absorption spectra also can be obtained indirectly by measuring fluorescence (Sect. 1.11 and Chap. 5) or other processes that result from the excitation. In a classical example of this approach, Otto Warburg discovered that cytochrome-*c* oxidase was a hemoprotein by inhibiting respiration with carbon monoxide and measuring photochemical reversal of the inhibition by light of various wavelengths. The resulting "action spectrum" gave the absorption spectrum of the heme bound to the protein.

1.8 Pump-probe and Photon-Echo Experiments

Optical delay techniques can be used to probe absorbance changes with very high time resolution in systems that react chemically or relax structurally when they are excited with light. Biophysical examples include the structural changes that occur in myoglobin or cytochrome-*c* oxidase following photodissociation of bound CO, the isomerization and subsequent structural excursions induced by illumination of rhodopsin, bacteriorhodopsin or phytochrome, and the light-driven electron-transfer reactions of photosynthesis. A short "pump" pulse excites the reactive system, and the absorbance is examined as a function of time by measuring the transmitted intensity of a second ("probe") pulse. To control the timing of the two pulses, one of the pulses is sent over a path whose length is adjusted by moving a mirror on a translation stage (Fig. 1.9). Because light moves through air at 3×10^{10} cm s^{-1}, changing the path length by 1 cm changes the delay time by 0.33×10^{-10} s. The time resolution in this technique is determined mainly by the widths of the excitation and probe pulses, which can be less than 10^{-14} s. If the probe pulse includes light over a broad range of frequencies and a monochromator is used to disperse the transmitted probe pulse onto an array of photodiodes, a full spectrum of the absorbance changes can be captured after each excitation. However, such measurements usually are averaged over many pairs of pump and probe pulses to increase the signal/noise ratio.

Similar experimental techniques make it possible to study the temporal coherence (synchronization) of members of an ensemble of molecules. If many of the molecules interact with a pulse of light during the same short interval of time, the ensemble initially will have some "memory" of this event, but this memory will

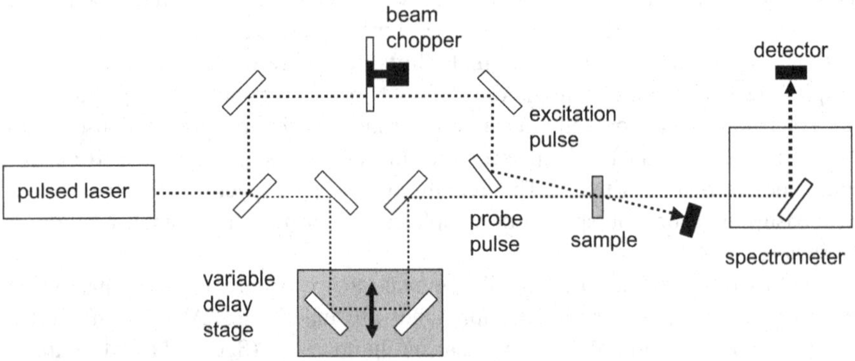

Fig. 1.9 Absorbance changes can be measured with high time resolution by exciting a sample with a short "pump" pulse from a laser and measuring the transmission of a "probe" pulse that passes through the sample after an adjustable delay. The pump beam usually is interrupted periodically by a chopper, and the difference between the amplitudes of the transmitted probe beam with and without the pump light is averaged over a large number of pulses. The apparatus often includes additional lasers or optical devices for generating probe beams with various wavelengths or polarizations

fade with time as a result of random thermal interactions of the molecules with their surroundings. The molecules might, for example, undergo particular vibrational motions in unison at short times after the pulse, but vibrate with random phases at later times. The rate at which the ensemble loses such coherence provides information on the dynamics and strength of the interactions with the surroundings. In Chap. 11, we'll discuss *photon-echo* experiments, in which a series of short pulses are used to create coherence and then partially regenerate it at later times.

1.9 Linear and Circular Dichroism

We mentioned above that absorption of light by a molecule requires that the oscillating electromagnetic field be oriented in a particular way relative to the molecular axes. The absorbance is proportional to $\cos^2\theta$, where θ is the angle between the electric field and the molecule's transition dipole. *Linear dichroism* is the dependence of absorption strength on the linear polarization of the light beam relative to a macroscopic "laboratory" axis. Molecules in solution usually do not exhibit linear dichroism because the individual molecules are oriented randomly relative to the laboratory axes. However, proteins and other macromolecules often can be oriented by flow through a fine tube, or by compression or stretching of samples embedded in gels such as polyacrylamide or polyvinylalcohol. Membranes can be oriented by magnetic fields or by drying in multiple layers on a glass slide. The linear dichroism of such samples can be measured with a conventional spectrophotometer equipped with a polarizing filter that is oriented alternately parallel and perpendicular to the orientation axis of the sample.

Isotropic (disordered) samples often can be made to exhibit a transient linear dichroism (*induced dichroism*) by excitation with a polarized flash of light, because the flash selectively excites those molecules that happen to be oriented in a particular way at the time of the flash. The decay kinetics of the induced dichroism provides information on the rotational dynamics of the molecule. Such measurements are useful for exploring the disposition and rotational mobility of small chromophores (light-absorbing molecules or groups) bound to macromolecules or embedded in membranes, and for dissecting absorption spectra of complex systems containing multiple interacting pigments. We'll discuss linear dichroism further in Chap. 4.

A beam of light also can be circularly polarized, which means that the orientation of the electric field at a given position along the beam rotates with time. The rotation frequency is the same as the classical oscillation frequency of the field (ν), but the direction of the rotation can be either clockwise or counterclockwise from the perspective of an observer looking into the oncoming beam. These directions correspond to left- and right-handed screws, and are called "left" and "right" circular polarization, respectively. Many natural materials exhibit differences between their absorbance of left and right circularly polarized light. This is *circular dichroism* (CD). The effect typically is small (about 1 part in 10^4), but can be measured by using an electro-optic modulator to switch the measuring beam rapidly back and forth between right and left circular polarization. A phase- and frequency-sensitive amplifier is used to extract the small oscillatory component of the light beam transmitted through the sample.

Because it represents the difference between two absorption strengths, CD can be either positive or negative. It differs in this respect from the ordinary absorbance of a sample, which is always positive. Its magnitude is expressed most directly by the difference between the molar extinction coefficients for left and right circularly polarized light ($\Delta\varepsilon = \varepsilon_{left} - \varepsilon_{right}$ in units of M^{-1} cm^{-1}). However, for historical reasons circular dichroism often is expressed in angular units (*ellipticity* or *molar ellipticity*), which relate to the elliptical polarization that is generated when a beam of linearly polarized light is partially absorbed.

In order to exhibit CD, the absorbing molecule must be distinguishable from its mirror image. Proteins and nucleic acids generally meet this criterion. As mentioned above in connection with hypochromism, their UV absorption bands represent coupled transitions of multiple chromophores (peptide bonds or purine and pyrimidine bases), which are arranged stereospecifically with respect to each other. The arrangement of the peptide bonds in a right-handed α-helix, for example, is distinguishable from that in a left-handed α-helix. Such oligomers can have relatively strong CD even if the individual units have little or none [10]. Circular dichroism thus provides a convenient and sensitive way to probe secondary structure in proteins, nucleic acids, and other multimolecular complexes [11–13]. Figure 1.10 shows typical CD spectra of polypeptides in α-helical and β conformations. The α-helix has a positive band near 195 nm and a characteristic pair of negative bands near 210 and 220 nm; the β-sheet has a weaker positive band near 200 nm and a solitary negative band around 215 nm. In Chap. 9, we will see that CD arises from interactions of the absorber with both the magnetic and the electric fields of the incident radiation and we'll discuss how its magnitude depends on the geometry of the system.

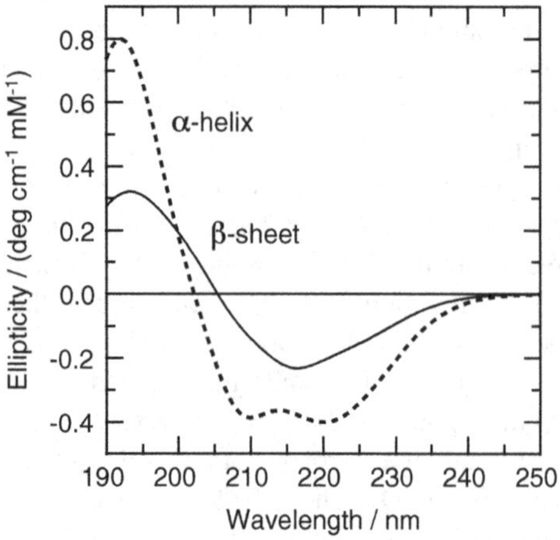

Fig. 1.10 Typical circular dichroism spectra of a polypeptide in α-helical and antiparallel β-sheet conformations

1.10 Distortions of Absorption Spectra by Light Scattering or Nonuniform Distributions of the Absorbing Molecules

Absorption spectra of suspensions of cells, organelles, or large molecular complexes can be distorted severely by light scattering. Scattering results from interference between optical rays that pass through or by the suspended particles, and it becomes increasingly important when the physical dimensions of the particles approach the wavelength of the light. Its hallmark is a baseline that departs from zero outside the true absorption bands of the material and that rises with decreasing wavelength. Light that is scattered by more than a certain angle with respect to the incident beam misses the detector and registers as an apparent absorbance. The effect is greatest if the spectrophotometer has a long distance between the sample and the detector. There are several ways to minimize this distortion. One is to use an *integrating sphere*, a device designed to make the probability of reaching the detector relatively independent of the angle at which light leaves the sample. This can be achieved by placing the specimen in a chamber with white walls and positioning the detector behind a baffle so that light reaches the detector only after bouncing off the walls many times (Fig. 1.11).

An integrating sphere also can be made by lining the walls with a large number of photodiodes. In some cases light scattering can be decreased by fragmenting the offending particles or by increasing the refractive index of the solvent, for example by the addition of albumen. Mathematical corrections for scattering can be made if the absorption spectrum is measured with the sample placed at several different distances from the detector [14, 15].

Absorption spectra of turbid materials and even of specimens as dense as intact leaves or lobster shells also can be measured by *photoacoustic spectroscopy*. A photoacoustic spectrometer measures the heat that is dissipated when a sample is excited by light and then decays nonradiatively back to the ground state. In a typical design, release of heat causes a gas or liquid surrounding the sample to expand and the expansion is detected by a microphone. Such measurements can be used to

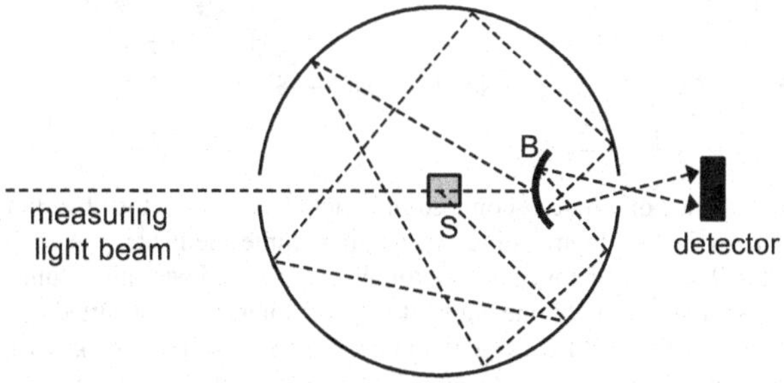

Fig. 1.11 An integrating sphere. The *square in the middle* represents a cuvette containing a turbid sample (*S*). Transmitted and scattered light reach the detector after many bounces off the white walls of the chamber. A baffle (B) blocks the direct path to the detector. In one design [37], the light passes through separate integrating spheres before and after the sample

study intermolecular energy transfer and volume changes that follow the absorption of light, in addition to the absorption itself [16–20].

A different type of distortion occurs if the absorbing molecules are not dispersed uniformly in the solution, but rather are sequestered in microscopic domains such as cells or organelles. Figure 1.12 illustrates the problem. Because the light intensity decreases as a beam of light traverses a domain that contains absorbers, molecules at the rear of the domain are screened by molecules toward the front and have less opportunity to contribute to the total absorbance. If light beams passing through different regions of the sample encounter significantly different numbers of absorbers, the measured absorption band will be flattened compared to the band that would be observed for a uniform solution. Mathematical corrections for this effect can be applied if the size and shape of the microscopic domains is known [14, 15, 21, 22]. Alternatively, a comparison of the spectra before and after the absorbing molecules are dispersed with detergents can be used to estimate the size of the domains. However, the latter procedure assumes that dispersing the molecules does not affect their intrinsic absorbance. In Chap. 8 we'll show that molecular interactions can make the spectroscopic properties of an oligomer intrinsically very different from the properties of the individual subunits.

It is interesting to note in passing that frogs, octupi and some other animals use the spectral flattening effect to change their appearance. They appear dark when pigmented cells in their skin spread out to cover the surface more or less uniformly, and pale when the cells clump together.

Deviations from Beer's law also can arise from a concentration-dependent equilibrium between different chemical species, or from changes in the refractive index of the solution with concentration. Instrumental non-linearities or noise may be significant if the concentration of the absorber is too high or too low.

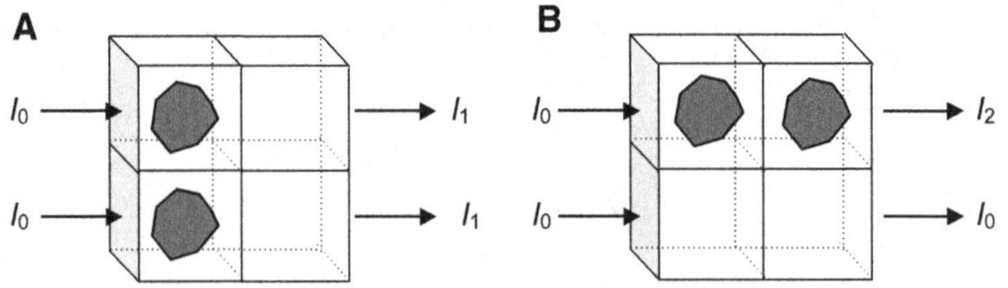

Fig. 1.12 The absorbance of a macroscopic sample depends on how the absorbing materials are distributed. The *boxes* in this figure represent small volume elements; the *shaded objects* are absorbing molecules. The overall concentration of absorbers is 0.5 per unit volume in both panels. If the light intensity incident on a volume element is I_o, the intensity transmitted by the element can be written $I_1 = \kappa I_o$, where $0 < \kappa < 1$ if the element contains an absorber and $\kappa = 1$ if it does not. In the arrangement shown in panel (**A**), light beams that pass through the specimen in different places encounter the same number of absorbers. The light transmitted per unit area is $I_1 = \kappa I_o$. In (**B**), the two beams encounter different number of absorbers. Here the light transmitted per unit area is $(I_2 + I_o)/2 = (\kappa^2 I_o + I_o)/2$. The difference ($B$–$A$) is $\kappa^2 I_o + I_o - 2\kappa I_o)/2 = I_o(\kappa - 1)^2/2$, which is greater than zero

1.11 Fluorescence

When they are excited in the near UV, the tryptophan residues of many proteins fluoresce in the region of 340 nm. The emission peak varies from about 308 to 355 nm depending on the environment in the protein. The fluorescence emission spectra of tryptophan, tyrosine and phenylalanine in solution are shown in Fig. 1.13. Although tyrosine fluoresces strongly in solution, the fluorescence from tyrosine and phenylalanine in proteins usually is very weak, in part because the excitation energy can transfer to tryptophans. Most purines and pyrimidines also do not fluoresce strongly, but the unusual nucleotide Y base found in yeast t-RNA^{PHE} is highly fluorescent. Flavins and pyridine nucleotides have characteristic fluorescence that provides sensitive assays of their oxidation states: the reduced pyridine nucleotides (NADH and NADPH) fluoresce around 450 nm, while the oxidized forms are nonfluorescent; the oxidized flavin coenzymes (FMN and FAD) fluoresce while their reduced forms do not. A large variety of fluorescent dyes have been used to label proteins or nucleic acids. These include fluorescein, rhodamine, dansyl chloride, naphthylamine, and ethenoATP. Proteins also can be labeled by fusing them genetically with *green fluorescent protein*, a protein with a remarkable, built-in chromophore formed by spontaneous cyclization and oxidation of a Ser–Tyr–Gly sequence.

Fluorescence excitation and emission spectra like those shown in Fig. 1.13 usually are measured by using two monochromators, one between the excitation light source and the sample and the other between the sample and the photodetector (Fig. 1.14). For a molecule with a single chromophore, the excitation spectrum usually is similar to a spectrum of $(1-T)$, where T is the fraction of the incident light that is transmitted ($T = I/I_o = 10^{-A}$). The emission spectrum is shifted to longer wavelengths because the excited molecule transfers a portion of its energy to the surroundings as heat before it fluoresces. We'll discuss this relaxation in Chaps. 5 and 10. Accurate measurements of a fluorescence excitation or emission spectrum require the sample to be sufficiently dilute so that only a negligible fraction of the incident or emitted light is absorbed. An emission spectrum also requires corrections for the wavelength-dependence of the sensitivity of the photomultiplier and other components of the fluorometer; these corrections can be made by measuring the apparent emission spectrum of a lamp with a known temperature (Chap. 3).

The *fluorescence yield* is the fraction of the excited molecules that decay by emitting fluorescence. This is measured most easily by comparing the integrated intensity of the emission spectrum with that obtained from a sample whose fluorescence yield is known. Standards with yields close to 100% are available for such measurements [23]. Fluorescence yields also can be determined by comparing the fluorescence with the intensity of the light that is scattered by a turbid sample [24, 25].

Fluorescence yields often are much less than unity because collisions with O_2 or other *quenchers* return the excited molecule rapidly to the ground state by nonradiative pathways. Measurements of fluorescence yields thus can provide

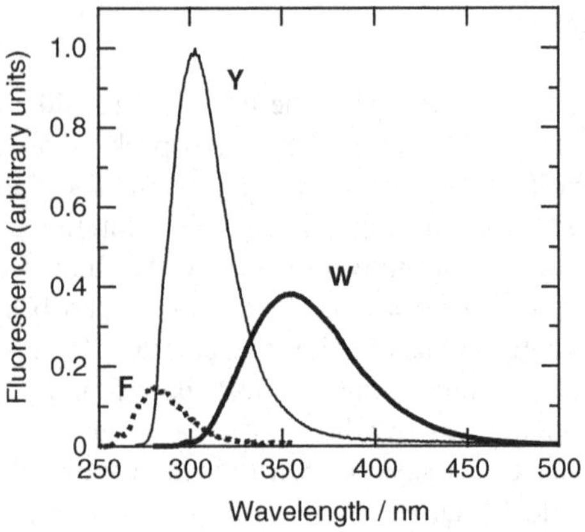

Fig. 1.13 Fluorescence emission spectra of tryptophan (*W*), tyrosine (*Y*) and phenylalanine (*F*) in aqueous solution. Tryptophan was excited at 270 nm, tyrosine at 260 nm, and phenylalanine at 240 nm. The spectra are from Du et al. [34] and Dixon et al. [35] and are scaled so that the areas under the curves are proportional to the fluorescence quantum yields. The measured quantum yields are 0.12 for tryptophan, 0.13 for tyrosine, and 0.022 for phenylalanine [23, 38]

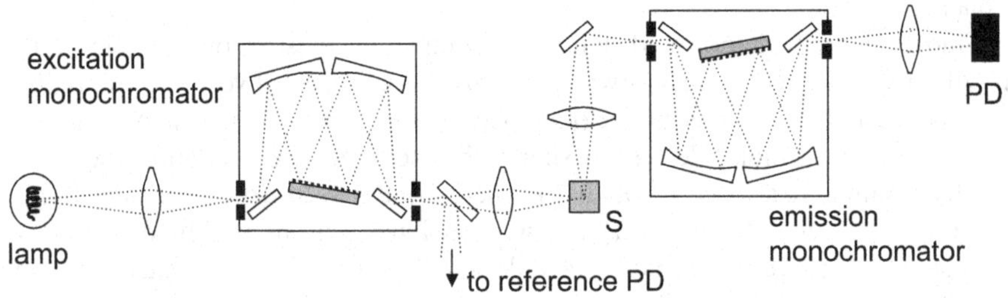

Fig. 1.14 Components of a fluorometer. The emission usually is collected at 90° with respect to the excitation to minimize spurious signals from transmitted or scattered excitation light. A portion of the excitation beam is split off and measured by a reference photodetector so that the excitation intensity can be held constant as the wavelength is varied. *S* sample, *PD* photodetector

information on whether a chromophore attached to a macromolecule is accessible to quenchers dissolved in the surrounding solution or attached to a different site on the molecule. Fluorescence also can be quenched by intramolecular electron transfer, and by *intersystem crossing* of the excited molecule from a *singlet* to a *triplet state*. Intersystem crossing involves a change in the relationship between the spins of the electrons in the partly-occupied molecular orbitals.

If a sample is excited with a short pulse of light the time course of the fluorescence intensity often can be described by an exponential function:

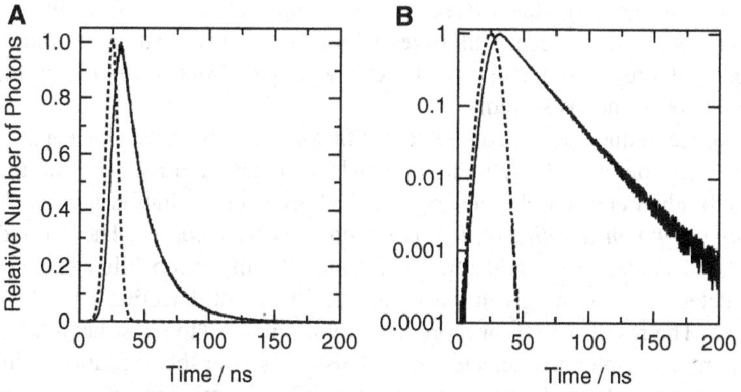

Fig. 1.15 The time course of fluorescence (*solid curve*) from a sample with a fluorescence lifetime of 20 ns, plotted on arbitrary linear (**A**) and logarithmic (**B**) scales. The *dotted curve* is the excitation pulse. The fluorescence signal is a convolution of the excitation pulse with the exponential decay function. (The signal at any given time includes fluorescence from molecules that were excited at a variety of earlier times.) The noise in the signal is proportional to the square root of the amplitude

$$F(t) = k_r[M^*(t)] = k_r[M^*(0)]\exp(-t/\tau) = F(0)\exp(-t/\tau). \quad (1.7)$$

Here $F(t)$ is the relative probability that the sample emits a photon during the short interval of time between t and $t + dt$ after the excitation; $[M^*(t)]$ is the concentration of molecules that remain in the excited state at time t; and k_r is a first-order rate constant for emission of light. The time constant τ is called the *fluorescence lifetime*, and is the average length of time that the molecule remains in the excited state. Tryptophan, for example, typically has a fluorescence lifetime of about 5×10^{-9} s. The reciprocal of the fluorescence lifetime ($1/\tau$) is the overall rate constant for decay of the excited molecule back to the ground state by any mechanism, including the quenching processes mentioned above in addition to fluorescence. Figure 1.15 illustrates the time course of fluorescence from a sample with a fluorescence lifetime of 20 ns.

In general, the excited molecules in an inhomogeneous sample interact with their surroundings in different ways and the measured fluorescence is given more accurately by a sum of exponential terms:

$$F(t) = \sum_i F_i(0)\exp(-t/\tau_i), \quad (1.8)$$

where $F_i(0)$ and τ_i are the initial amplitude and decay time constant of component i. Several procedures can be used to fit such expressions to experimental data, taking into account the width of the excitation flash and the finite response time of the detection instrumentation [26–28].

In a homogeneous sample, the fluorescence lifetime is proportional to the fluorescence yield: quenching processes that cause the excited molecule to decay rapidly to the ground state decrease the lifetime and the yield in parallel. Measurements of fluorescence lifetimes thus can give information similar to that provided by fluorescence yields, but also can be used to probe the dynamics of more complex, heterogeneous systems.

Several techniques arey used to measure fluorescence lifetimes. One approach is to excite a sample repeatedly with short pulses of light and measure the times at which individual emitted photons are detected after the excitation flashes. This is called *single photon counting* or *time-correlated photon counting*. The emitted light is attenuated so that, on average, there is a relatively small probability that a photon will be detected after any given flash, and the chance of detecting two photons is negligible. The results of 10^5 or more excitations are used to construct a histogram of the numbers of photons detected at various times after the excitation. This plot reflects the time dependence of the probability of emission from the excited molecules. The signal-to-noise ratio in a given bin of the histogram is proportional to the total number of photons counted in the bin (Fig. 1.15). If a sufficiently large number of photons are counted, the time resolution is limited by the photomultiplier and the associated electronics and can be on the order of 5×10^{-11} s.

Fluorescence decay kinetics also can be measured by exciting the sample with continuous light whose intensity is modulated sinusoidally at a frequency (ω) on the order of $1/\tau$, where τ again is the fluorescence lifetime. The fluorescence oscillates sinusoidally at the same frequency, but the amplitude and phase of its oscillations relative to the oscillations of the excitation light depend on the product of ω and τ (Fig. 1.16 and Appendix A4). If $\omega\tau$ is much less than 1, the fluorescence amplitude tracks the excitation intensity closely; if $\omega\tau$ is larger, the oscillations are delayed in phase and damped (demodulated) relative to the excitation [28–30]. Fluorescence with multiexponential decay kinetics can be analyzed by measuring the fluorescence modulation amplitude or phase shift with several different frequencies of modulated excitation.

A third technique, *fluorescence upconversion*, is to focus the emitted light into a material with nonlinear optical properties, such as a crystal of KH_2PO_4. If a separate short pulse of light is focused into the same crystal so that the two light beams overlap temporally and spatially, the crystal emits light at a new frequency that is the sum of the frequencies of the fluorescence and the probe pulse. The time dependence of the fluorescence intensity is obtained by varying the timing of the probe pulse relative to the pulse that excites the fluorescence, in the manner described above for pump-probe absorbance measurements. As with absorbance measurements, the time resolution of fluorescence upconversion can be on the order of 10^{-14} s. This technique is well suited for studying relaxations that cause the emission spectrum of an excited molecule to evolve rapidly with time.

To measure *fluorescence polarization*, or *anisotropy*, a sample is excited with polarized light and the emission is measured through a polarizer aligned either parallel or perpendicular to the excitation polarization (Fig. 1.17). If emission occurs from the same state that is generated by excitation, and the excited molecule does not rotate between these two events, then the fluorescence polarized parallel to

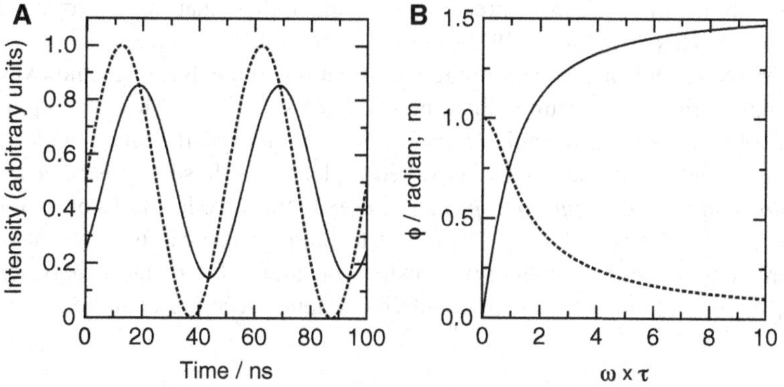

Fig. 1.16 (**A**) Fluorescence (*solid curve*) from a molecule that is excited with sinusoidally modulated light (*dotted curve*). If the fluorescence decays exponentially with single time constant τ, the phase shift (ϕ) and the relative modulation of the fluorescence amplitude (m) are related to τ and the angular frequency of the modulation (ω) by $\phi = \arctan(\omega\tau)$ and $m = (1 + \omega^2\tau^2)^{-1/2}$ (Appendix A4). The *curves* shown here are calculated for $\tau = 8$ ns, $\omega = 1.257 \times 10^8$ rad/s (20 MHz) and 100% modulation of the excitation light ($\phi = 0.788$ rad, $m = 0.705$). (**B**) Phase shift (ϕ, *solid curve*) and relative modulation (m, *dotted curve*) of the fluorescence of a molecule that decays with a single exponential time constant, plotted as a function of the product $\omega\tau$. The relationships among ϕ, m, τ and ω become more complicated if the fluorescence decays with multiexponential kinetics (Appendix A4)

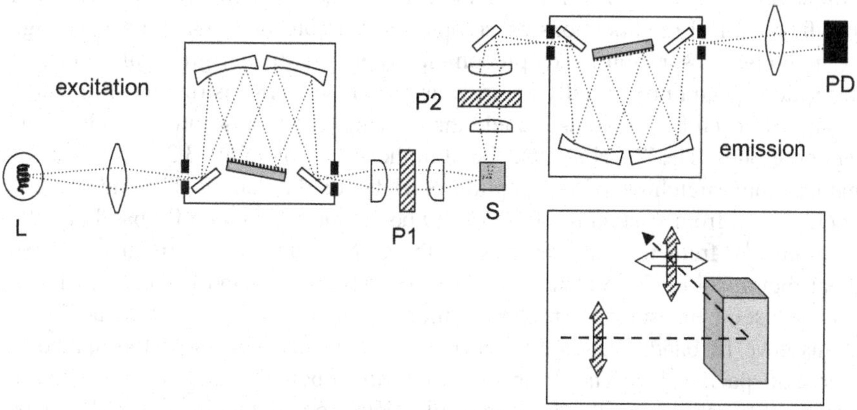

Fig. 1.17 Apparatus for measuring fluorescence anisotropy. In the main drawing (a *top view*), a polarizing filter or prisim ($P1$) polarizes the excitation light so that the electric vector is normal to the plane of the paper. The intensity of the fluorescence is measured through a second polarizer ($P2$), which is oriented either parallel or perpendicular to P1. The drawing at the *lower right* shows a perspective view of the sample and the polarizers. To check for bias of the detection monochromator in favor of a particular polarization, measurements also are made with polarizer P1 rotated by 90°. The fluorescence signal then should be same with either orientation of P2 because both orientations are perpendicular to P1. *L* lamp, *S* sample, *PD* photodetector

the excitation will be about three times stronger than that with perpendicular polarization. This difference will disappear as the molecule rotates.

Fluorescence anisotropy can provide information on the dynamics and extent of molecular motions that occur on the time scale of 10^{-9}–10^{-7} s. Slower motions can be studied by measuring the anisotropy of *phosphorescence*, the much longer-lived emission of light from molecules in excited triplet states. In some cases, *near-field* or *confocal fluorescence microscopy* can be used to track the locations of individual fluorescent molecules with time. These last techniques also can be combined with measurements of resonance energy transfer to examine the distance between the energy donor and the acceptor in an individual donor-acceptor complex.

1.12 IR and Raman Spectroscopy

In a semiclassical picture, the bond in a heteronuclear diatomic molecule such as CO vibrates at a characteristic frequency that increases with the order of the bond and decreases with the reduced mass of the two atoms. Quantum mechanically, the molecule has a series of allowed nuclear wavefunctions with a ladder of approximately equally spaced energies. Electromagnetic radiation with an appropriate frequency in the IR region can excite the molecule from one of these states to the next. The situation is more complicated in a polyatomic molecule. As we discuss in Chap. 6, the molecule still has a set of discrete vibrational modes, but these generally involve more than just two atoms and so cannot be assigned uniquely to individual bonds. The IR absorption spectrum now includes multiple bands that provide a complex fingerprint of the molecular structure. However, the effects of specific chemical modifications or isotopic substitutions often allow one to assign the individual absorption bands predominantly to a particular group of atoms. IR absorption spectra now usually are measured with an FTIR instrument (Fig. 1.8).

The IR spectra of proteins include three characteristic absorption bands of the peptide groups. The "N–H stretch" band in the region of 3,280–3,300 cm^{-1} stems mainly from stretching of the peptide N–H bond; the "amide I" band (1,620–1,660 cm^{-1}), from stretching of the C = O bond; and the "amide II" band (1,520–1,550 cm^{-1}), from in-plane bending of the C–N–H angle. The frequencies and linear dichroism of these bands are different for α-helices and β-sheets, and thus provide useful measures of protein conformation. In addition, amino acid side chains have IR bands that can be used to probe events such as proton uptake or release at specific sites. The IR absorption bands of bound ligands such as CO can provide information on the location and orientation of the ligand, and can be measured with high time resolution by pump-probe techniques.

Like IR spectroscopy, *Raman spectroscopy* reflects transitions between different vibrational states of a molecule. However, rather than a simple upward transition driven by the absorption of a photon, Raman scattering is a two-photon process in which one photon is absorbed and a photon with a different energy is emitted almost simultaneously, leaving the molecule in a different vibrational state. It resembles Raleigh light scattering, which involves absorption and emission of photons with identical energies. Raman spectrometers resemble fluorometers (Fig. 1.14), but usually employ a laser to generate a sharply defined excitation frequency. Two monochromators often are used in series in the emission arm of the instrument. A Raman emission spectrum differs from fluorescence in having sharp lines at frequencies that differ from the excitation frequency (ν_{ex}) by $\pm \upsilon_i$, where the υ_i are the molecule's vibrational frequencies. The lines at $\nu_{ex} - \upsilon_i$ represent net retention of a quantum of vibrational energy ($h\upsilon_i$) by the molecule (*Stokes scattering*), while those at $\nu_{ex} + \upsilon_i$ represent net release of vibrational energy (*anti-Stokes scattering*). Anti-Stokes scattering usually is much weaker because it occurs only from molecules that are in excited vibrational states before the excitation.

Probably the most useful form of Raman spectroscopy in molecular biophysics has been *resonance Raman* spectroscopy, in which the frequency of the excitation light is tuned to fall within an electronic absorption band. The electronic resonance has two benefits: it greatly increases the strength of the Raman scattering, and it can make the measurement specific for a chromophore with a particular electronic absorption spectrum. Resonance Raman scattering thus lends itself to probing the states of a ligand such as retinal or heme bound to a protein with little interference from the protein atoms. We consider Raman scattering and other two-photon processes in Chap. 12.

1.13 Lasers

Modern spectroscopic techniques depend heavily on lasers as light sources. Lasers can provide either short pulses of light covering broad bands of frequencies, or continuous beams with extremely narrow bandwidths, and they can be focused to extremely small spots. Figure 1.18 shows the main components of a laser designed to provide ultrashort pulses. In this system, a continuous beam of green light from another source such as an Ar^+ gas laser is focused on a thin crystal of sapphire (Al_2O_3) doped with Ti^{3+}. The pump light raises Ti atoms in the crystal to an excited state, from which they relax rapidly to a metastable state at lower energy. (In some lasers, the metastable state is simply a vibrationally relaxed level of the excited electronic state generated by the pump light; in others it is a different electronic state. The main requirements are that atoms or molecules in the metastable state do not absorb light strongly at the pump frequency, that they fluoresce at a longer wavelength, and that their decay by radiationless processes be comparatively slow.) With continued pumping, the population of the metastable state grows until it exceeds the population of the ground state. However, transitions from the metastable state to the ground state can be stimulated by light with frequency $\nu = \Delta E/h$, where ΔE is the energy difference between these two states. This *stimulated emission* is just the reverse of absorption. The excited atom or molecule emits light with the same frequency, direction, polarization and phase as the stimulating radiation. In the laser, photons emitted along a particular axis are reflected by a series of mirrors and returned to the crystal, where they stimulate other atoms to decay in the same manner. Emission that begins spontaneously thus can lead to an

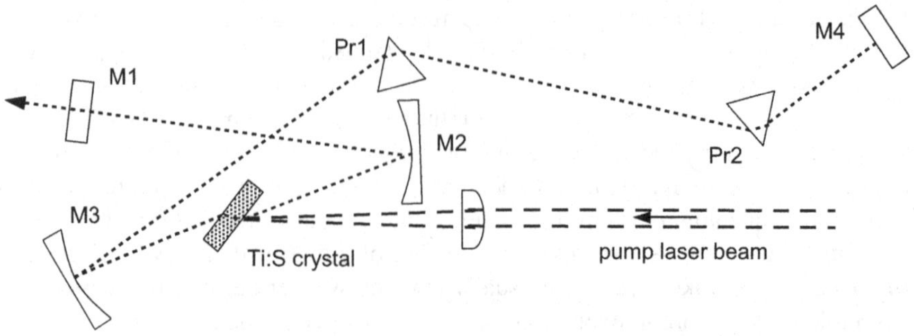

Fig. 1.18 The main components of a pulsed Ti:sapphire (Ti:S) laser. Light from a continuous "pump" laser (typically 532 nm) is focused on a Ti:sapphire crystal. The crystal fluoresces in the region of 800 nm. Light emitted along a particular axis (*dotted line*) is collimated by curved mirrors *M*2 and *M*3 and reflected back to the crystal by mirrors *M*1 and *M*4. Prisms *Pr*1 and *Pr*2 compensate for the wavelength dependence of the refractive index of the crystal. The pump beam usually enters the laser through M2; it is displaced here for clarity

explosive collapse of the population in the metastable state and release of a pulse of light. The pulse emerges through the mirror at one end of the optical cavity (M1 in Fig. 1.18), which is coated to transmit about 10% of the light and reflect the remainder.

The width of the pulses emitted by such a laser depends on the spectral bandwidth: to provide ultrashort pulses, the lasing medium must be capable of emitting light over a broad band of frequencies. Ti:sapphire and organic dyes of the rhodamine family are ideal in this regard because their fluorescence emission spectra cover several hundred nanometers. (Ti:sapphire emits in the region of 800 nm and rhodamines in the region of 500–700 nm, depending on the molecular structure.) However, photons with different wavelengths suffer different optical delays as they pass through the lasing medium because the refractive index depends on the wavelength. In the Ti:sapphire laser diagrammed in Fig. 1.18, a pair of prisms correct for this dispersion by extending the optical cavity length for light with longer wavelengths. Pulses with widths of less than 10 fs can be obtained by choosing the prism material judiciously and adjusting the spacing of the prisms and the amount of the prism material in the path [31, 32]. To obtain a continuous laser beam, the pair of prisms can be replaced by a single, rotatable prism or other optical element that selects a light of a particular wavelength.

1.14 Nomenclature

In the following chapters, our aim often will be to develop mathematical expressions that show how a spectroscopic property depends on molecular structure or environmental variables. Such expressions often can be stated most concisely in terms of vectors and matrices. Explanations of these mathematical tools are given in Appendices A1 and A2. We will indicate a vector by a bold-face letter in italics ($\boldsymbol{R}$), or for vectors in 3-dimensional physical space, sometimes as (R_x,R_y,R_z), where R_x, R_y, and R_z are the components parallel to the x-, y- and z- axes of a Cartesian coordinate system. A vector with unit length parallel to one of the axes of the coordinate system will be designated by a letter with a caret (^) on top. In particular, $\hat{x}$, $\hat{y}$ and $\hat{z}$ denote unit vectors parallel to the x-, y- and z- axes of a Cartesian coordinate system. Matrices will be represented by un-italicized letters in bold face.

We also will make frequent use of *operators*. An operator is just a recipe for a mathematical procedure that uses a given variable or function as input. We'll designate an operator by a letter with a tilde (~) on top, and indicate the input by a letter or other symbol immediately after the operator. For example, if operator $\tilde{A}$ is "add 3, take the square and divide by 2," then $\tilde{\mathrm{A}}x = (x + 3)^2/2$. The input and output can be a scalar, vector, or matrix depending on the nature of the operator. Some of the operators that work on matrices are defined in Appendix A2.

Exercises

1. Complete the following table of oscillation frequencies, wavenumbers and photon energies of monochromatic light with wavelengths of 300 and 700 nm and the molar excitation energies of molecules that absorb at these wavelengths.

Wavelength (nm)	Frequency (s^{-1})	Wavenumber (cm^{-1})	Photon Energy (J $photon^{-1}$)	Excitation Energy (kJ mol^{-1})
300				
700				

2. What fractions of the incident light at a given wavelength are absorbed by samples that have absorbances of (*a*) 0.1, (*b*) 0.5, and (*c*) 2.0?
3. Molecule A has molar extinction coefficients of 10,000 M^{-1} cm^{-1} at 300 nm and 20,000 M^{-1} cm^{-1} at 400 nm. The molar extinction coefficients of molecule B at these wavelengths are 15,000 and 12,000 M^{-1} cm^{-1}, respectively. If a solution containing only a mixture of A and B has absorbances of 0.5 at 300 nm and 0.8 at 400 nm, what are the concentrations of A and B?
4. The minimum energy that a photon must transfer to an electron in order to free the electron from the surface of a particular metal is called the "photoelectric work function." The work function for Cs is approximately 2.1 eV. What is the maximum wavelength of light that would free an electron from a Cs surface?
5. The fluorescence from a sample of interest decays with multiphasic kinetics extending over several time domains. What would be the relative merits and limitations of measuring these kinetics by (*a*) time-correlated photon counting, (*b*) fluorescence phase-shift and amplitude modulation, and (*c*) fluorescence upconversion?

6. Quenching of protein fluorescence by a water-soluble agent such as I^- can be used to probe the exposure of a tryptophan residue to the solvent. If such a quencher decreases the yield of fluorescence from a homogeneous sample by 90%, how would the fluorescence lifetime change?
7. What are the technical advantages and limitations of a Fourier transform spectrometer compared to a conventional spectrophotometer?

References

1. Wetlaufer, D.B.: Ultraviolet spectra of proteins and amino acids. Adv. Prot. Chem. **17**, 303–391 (1962)
2. Pace, C.N., Vajdos, F., Lee, L., Grimsley, G., Gray, T.: How to measure and predict the molar absorption coefficient of a protein. Protein Sci. **4**, 2411–2423 (1995)
3. Boyd, D.B.: Conformational dependence of electronic energy levels in disulfides. J. Am. Chem. Soc. **94**, 8799–8804 (1972)
4. Press, W.H., Flannery, B.P., Teukolsky, S.A., Vetterling, W.T.: Numerical Recipes in Fortran 77: The Art of Scientific Computing. Cambridge University Press, Cambridge (1989)
5. Einstein, A.: Uber einen die Erzeugung und Verwandlung des Lichtes betreffenden heuristischen Gesichtspunkt. Ann. der Phys. **17**, 132–146 (1905)
6. Lewis, G.N.: The conservation of photons. Nature **118**, 874–875 (1926)
7. Hong, C.K., Ou, Z.Y., Mandel, L.: Measurement of subpicosecond time intervals between two photons by interference. Phys. Rev. Lett. **59**, 2044–2046 (1987)
8. Kwiat, P.G., Steinberg, A.M., Chiao, R.Y.: Observation of a "quantum eraser": a revival of coherence in a two-photon interference experiment. Phys. Rev. A **45**, 7729–7739 (1992)
9. Pittman, T.B., Strekalov, D.V., Migdall, A., Rubin, M.H., Sergienko, A.V., et al.: Can two-photon interference be considered the interference of two photons? Phys. Rev. Lett. **77**, 1917–1920 (1996)
10. Tinoco Jr., I.: Hypochromism in polynucleotides. J. Am. Chem. Soc. **4784**, 5047 (1961). Erratum J. Am. Chem. Soc. 4784: 5047 (1961)
11. Greenfield, N., Fasman, G.D.: Computed circular dichroism spectra for the evaluation of protein conformation. Biochemistry **8**, 4108–4116 (1969)
12. Johnson, W.C., Tinoco, I.: Circular dichroism of polypeptide solutions in vacuum ultraviolet. J. Am. Chem. Soc. **94**, 4389–4390 (1972)
13. Brahms, S., Spach, G., Brack, A.: Identification of β, β-turns and unordered conformations in polypeptide chains by vacuum UV circular dichroism. Proc. Natl. Acad. Sci. **74**, 3208–3212 (1977)
14. Latimer, P., Eubanks, C.A.H.: Absorption spectrophotometry of turbid suspensions: a method of correcting for large systematic distortions. Arch. Biochem. Biophys. **98**, 274–285 (1962)
15. Naqvi, K.R., Melo, T.B., Raju, B.B., Javorfi, T., Garab, G.: Comparison of the absorption spectra of trimers and aggregates of chlorophyll a/b light-harvesting complex LHC II. Spectrochim. Acta A **53**, 1925–1936 (1997)
16. Ort, D.R., Parson, W.W.: Flash-induced volume changes of bacteriorhodopsin-containing membrane fragments and their relationship to proton movements and absorbance transients. J. Biol. Chem. **253**, 6158–6164 (1978)
17. Arata, H., Parson, W.W.: Enthalpy and volume changes accompanying electron transfer from P-870 to quinones in Rhodopseudomonas sphaeroides reaction centers. Biochim. Biophys. Acta **636**, 70–81 (1981)
18. Braslavsky, S.E., Heibel, G.E.: Time-resolved photothermal and photoacoustic methods applied to photoinduced processes in solution. Chem. Rev. **92**, 1381–1410 (1992)

19. Feitelson, J., Mauzerall, D.: Photoacoustic evaluation of volume and entropy changes in energy and electron transfer. Triplet state porphyrin with oxygen and naphthoquinone-2-sulfonate. J. Phys. Chem. **100**, 7698–7703 (1996)
20. Sun, K., Mauzerall, D.: Fast photoinduced electron transfer from polyalkyl- to polyfluoro-metalloporphyrins in lipid bilayer membranes. J. Phys. Chem. B **102**, 6440–6447 (1998)
21. Duysens, L.N.M.: The flattening of the absorption spectrum of suspensions, as compared to that of solutions. Biochim. Biophys. Acta **19**, 1–12 (1956)
22. Pulles, M.P.J., Van Gorkom, H.J., Verschoor, G.A.M.: Primary reactions of photosystem II at low pH. 2. Light-induced changes of absorbance and electron spin resonance in spinach chloroplasts. Biochim. Biophys. Acta **440**, 98–106 (1976)
23. Chen, R.F.: Measurements of absolute values in biochemical fluorescence spectroscopy. J. Res. Natl. Bur. Stand. **76A**, 593–606 (1972)
24. Weber, G., Teale, F.W.J.: Determination of the absolute quantum yield of fluorescent solutions. Trans. Farad. Soc. **53**, 646–655 (1957)
25. Wang, R.T., Clayton, R.K.: Absolute yield of bacteriochlorophyll fluorescence in vivo. Photochem. Photobiol. **13**, 215–224 (1971)
26. Beechem, J.M.: Global analysis of biochemical and biophysical data. Meth. Enzymol. **210**, 37–54 (1992)
27. Brochon, J.C.: Maximum entropy method of data analysis in time-resolved spectroscopy. Meth. Enzymol. **240**, 262–311 (1994)
28. Lakowicz, J.R.: Principles of Fluorescence Spectroscopy, 3rd edn. Springer, New York (2006)
29. Birks, J.B., Dyson, D.J.: The relationship between absorption intensity and fluorescence lifetime of a molecule. Proc. R. Soc. Lond. Ser. A **275**, 135–148 (1963)
30. Gratton, E., Jameson, D.M., Hall, R.D.: Multifrequency phase and modulation fluorometry. Annu. Rev. Biophys. Bioeng. **13**, 105–124 (1984)
31. Christov, I.P., Stoev, V.D., Murnane, M.M., Kapteyn, H.C.: Sub-10-fs operation of Kerr-lens mode-locked lasers. Opt. Lett. **21**, 1493–1495 (1996)
32. Rundquist, A., Durfee, C., Chang, Z., Taft, G., Zeek, E., et al.: Ultrafast laser and amplifier sources. Appl. Phys. B. Lasers and Optics **65**, 161–174 (1997)
33. Fasman, G.D.: Handbook of Biochemistry and Molecular Biology. Proteins I, 3rd edn. CRC Press, Cleveland (1976)
34. Du, H., Fuh, R.A., Li, J., Corkan, A., Lindsey, J.S.: PhotochemCAD: A computer-aided design and research tool in photochemistry. Photochem. Photobiol. **68**, 141–142 (1998)
35. Dixon, J.M., Taniguchi, M., Lindsey, J.S.: PhotochemCAD 2: a refined program with accompanying spectral databases for photochemical calculations. Photochem. Photobiol. **81**, 212–213 (2005)
36. Voet, D., Gratzer, W.B., Cox, R.A., Doty, P.: Absorption spectra of nucleotides, polynucleotides, and nucleic acids in the far ultraviolet. Biopolymers **1**, 193–205 (1963)
37. Kramer, D.M., Sacksteder, C.A.: A diffused-optics flash kinetic spectrophotometer (DOFS) for measurements of absorbance changes in intact plants in the steady-state. Photosynth. Res. **56**, 103–112 (1998)
38. Chen, R.F.: Fluorescence quantum yields of tryptophan and tyrosine. Anal. Lett. **1**, 35–42 (1967)

Basic Concepts of Quantum Mechanics

2

2.1 Wavefunctions, Operators and Expectation Values

2.1.1 Wavefunctions

In this chapter we discuss the basic principles of quantum mechanics that underlie optical spectroscopy. More comprehensive treatments are available in the classic texts by Dirac [1] and Pauling and Wilson [2], a collection of historical papers edited by van der Waerden [3], and numerous more recent texts [4–9]. Atkins [10] is a useful source of leading references and concise discussions of the main ideas.

The most basic notion of quantum mechanics is that all the properties of a system that depend on position and time hinge on a mathematical function called the system's *wavefunction*. The wavefunction of an electron, for example, contains all the information needed to specify the probability that the electron is located in a given region of space at a particular time. If we designate a position by vector $\boldsymbol{r}$, the probability P that a system with wavefunction $\Psi(\boldsymbol{r},t)$ is located in a small volume element $d\sigma$ around $\boldsymbol{r}$ at time t is

$$P(\boldsymbol{r},t)d\sigma = \Psi^*(\boldsymbol{r},t)\Psi(\boldsymbol{r},t)d\sigma = |\Psi(\boldsymbol{r},t)|^2 d\sigma. \tag{2.1}$$

Here $\Psi^*(\boldsymbol{r},t)$ is the complex conjugate of $\Psi(\boldsymbol{r},t)$, which means that if the wavefunction contains any imaginary numbers (as it generally does), i is replaced everywhere by $-i$. The product $\Psi^*\Psi$ is always a real number, as we would expect for a measurable property such as the probability of finding an electron at a particular place.

The interpretation of $\Psi^*\Psi$ as a probability per unit volume, or probability density, was developed by Max Born [11]. Although universally accepted today, Born's interpretation was controversial at the time. Some of the major contributors to the field held that the wavefunction of a system *is* the system in a more direct sense, while others disputed the idea that probabilities had to replace the causal laws of classical physics. Reichenbach [12], Pais [13] and Jammer [14] give

W.W. Parson, *Modern Optical Spectroscopy*, DOI 10.1007/978-3-662-46777-0_2

interesting accounts of how these views evolved as the profound implications of quantum mechanics came into focus.

If we accept Born's interpretation, the total probability that the system exists somewhere is obtained by integrating Eq. (2.1) over all space. In the widely used *bra-ket notation* introduced by Paul Dirac [1], this integral is denoted by $\langle \Psi | \Psi \rangle$:

$$\langle \Psi | \Psi \rangle \equiv \int \Psi^* \Psi d\sigma = \int P(\mathbf{r}, t) d\sigma. \tag{2.2}$$

If the system exists, $\langle \Psi | \Psi \rangle$ must be 1; if it doesn't exist, $\langle \Psi | \Psi \rangle = 0$. Note that the asterisk denoting the complex conjugate of the wavefunction on the left is omitted in the symbol $\langle \Psi |$ to simplify the nomenclature, but is implied.

In a Cartesian coordinate system, the volume elements in Eqs. (2.1) and (2.2) can be written $d\sigma = dxdydz$. For a one-dimensional system, the position vector $\boldsymbol{r}$ reduces to a single coordinate (x) and Eq. (2.2) becomes simply

$$\int P(x, t) d\sigma = \langle \Psi(x, t) | \Psi(x, t) \rangle = \int_{-\infty}^{\infty} \Psi^*(x, t) \Psi(x, t)\, dx. \tag{2.3}$$

If the system contains more than one particle, we have to integrate over the coordinates of all the particles.

Born's interpretation of the wavefunction puts restrictions on the mathematical functions that can be physically meaningful wavefunctions. First, Ψ must be a single-valued function of position. There should be only one value for the probability of finding a system at any given point. Second, the integral of $\Psi^* \Psi d\sigma$ over any region of space must be finite; the probability of finding the system in any particular volume element should not go to infinity. Third, the integral $\langle \Psi | \Psi \rangle$ must exist and must be finite. In addition, it seems reasonable to assume that Ψ is a continuous function of the spatial coordinates, and that the first derivatives of Ψ with respect to these coordinates also are continuous except possibly at boundaries of the system. These last restrictions guarantee that second derivatives with respect to the spatial coordinates exist.

2.1.2 Operators and Expectation Values

The second fundamental idea of quantum mechanics is that for any measurable physical property A of a system there is a particular mathematical *operator*, $\tilde{A}$, that can be used with the wavefunction to obtain an expression for A as a function of time, or at least for the most probable result of measuring this property. An operator is simply an instruction to do something, such as to multiply the amplitude of Ψ by a constant. The expression for A is obtained in the following way: (1) perform operation $\tilde{A}$ on wavefunction Ψ at a particular position and time, (2) multiply by the value of Ψ^* at the same position and time, and (3) integrate the result of these first two steps over all possible positions. In bra-ket notation, this three-step procedure is denoted by $\langle \Psi | \tilde{A} | \Psi \rangle$:

$$A = \left\langle \Psi \middle| \tilde{A} \middle| \Psi \right\rangle \equiv \int \Psi^* \tilde{A} \Psi d\sigma. \tag{2.4}$$

The calculated value of the property (A) is called the *expectation value*. If the wavefunction depends on the positions of several particles, the spatial integral denoted by $\langle \Psi | \tilde{A} | \Psi \rangle$ represents a multiple integral over all the possible positions of all the particles.

Equation (2.4) is a remarkably general assertion, considering that it claims to apply to any measurable property of an arbitrary system. Note, however, that we are discussing an individual system. If we measure property A in an ensemble of many systems, each with its own wavefunction, the result is not necessarily a simple average of the expectation values for the individual systems. We'll return to this point in Chap. 10.

In addition to the question of how to deal with ensembles of many systems, there are two obvious problems in trying to use Eq. (2.4). We have to know the wavefunction Ψ, and we have to know what operator corresponds to the property of interest. Let's first consider how to select the operator.

In the description we will use, the operator for position ($\tilde{r}$) is simply multiplication by the position vector, $\boldsymbol{r}$. So to find the expected x, y and z coordinates of an electron with wavefunction Ψ we just evaluate the integrals $\langle \Psi | x | \Psi \rangle$, $\langle \Psi | y | \Psi \rangle$ and $\langle \Psi | z | \Psi \rangle$. This amounts to integrating over all the positions where the electron could be found, weighting the contribution of each position by the probability function $\Psi^* \Psi$.

The operator for momentum is more complicated. In classical mechanics, the linear momentum of a particle traveling along the x axis has a magnitude of $p = mw$, where m and w are the particle's mass and velocity. The quantum mechanical operator for momentum in the x direction is $\tilde{p}_x = (\hbar/i)\partial/\partial x$, where $\hbar$ ("h-bar") is Planck's constant (h) divided by 2π. The recipe for finding the momentum according to Eq. (2.4), therefore, is to differentiate Ψ with respect to x, multiply by $\hbar/i$, multiply by Ψ^*, and integrate the result over all space: $p_x = \langle \Psi | \tilde{p}_x | \Psi \rangle = \langle \Psi | (\hbar/i) \partial \Psi | \partial x \rangle$. Because $\hbar/i$, though an imaginary number, is just a constant, this is the same as $p_x = (\hbar/i) \langle \Psi | \partial \Psi / \partial x \rangle$.

In three dimensions, momentum is a vector ($\boldsymbol{p}$) with x, y and z components p_x, p_y and p_z. We can indicate this by writing $\tilde{p}$ as $(\hbar/i)\tilde{\nabla}$, where $\tilde{\nabla}$ is the *gradient operator*. The gradient operator acts on a scalar function such as Ψ to generate a vector whose x, y and z components are the derivatives of Ψ with respect to x, y and z, respectively:

$$\tilde{p}\Psi = (\hbar/i)\tilde{\nabla}\Psi = (\hbar/i)(\partial\Psi/\partial x, \partial\Psi/\partial y, \partial\Psi/\partial z) \tag{2.5}$$

(Appendix A.2). The momentum operator $(\hbar/i)\tilde{\nabla}$ also can be written as $-i\hbar\tilde{\nabla}$ because $i^{-1} = -i$. To see this equality, multiply i^{-1} by unity in the form of i/i.

It may seem disconcerting that $\tilde{\mathrm{p}}$ involves imaginary numbers, because the momentum of a free particle is a real, measurable quantity. The quantum mechanical momentum given by $\langle \Psi | \tilde{\mathrm{p}} | \Psi \rangle$, however, also proves to be a real number (Box 2.1). The formula for $\tilde{\mathrm{p}}$ emerged from the realization by Max Born, Werner Heisenberg, Paul Dirac and others in the period 1925–1927 that the momentum of a bound particle such as an electron in an atom cannot be specified precisely as a function of the particle's position.

Box 2.1. Operators for Observable Properties must be Hermitian

An operator $\tilde{\mathrm{A}}$ for a measurable physical quantity must have the property that

$$\left\langle \Psi_b \middle| \tilde{\mathrm{A}} \middle| \Psi_a \right\rangle = \left\langle \Psi_a \middle| \tilde{\mathrm{A}} \middle| \Psi_b \right\rangle^* \tag{B2.1.1}$$

for any pair of wavefunctions Ψ_a and Ψ_b that are eigenfunctions of the operator. Operators with this property are called *Hermitian* after the French mathematician Charles Hermite. If $\tilde{\mathrm{A}}$ is Hermitian, then

$$\left\langle \Psi_a \middle| \tilde{\mathrm{A}} \middle| \Psi_a \right\rangle = \left\langle \Psi_a \middle| \tilde{\mathrm{A}} \middle| \Psi_a \right\rangle^* \tag{B2.1.2}$$

for any eigenfunction Ψ_a of $\tilde{\mathrm{A}}$. This implies that $\langle \Psi_a | \tilde{\mathrm{A}} | \Psi_a \rangle$ is real, because the complex conjugate of a number can be equal to the number itself only if the number has no imaginary part. The operators for position, momentum and energy all are Hermitian (see [4] for a proof). However, if $\Psi_a \neq \Psi_b$, the integral $\langle \Psi_b | \tilde{\mathrm{A}} | \Psi_a \rangle$ may or may not be real, depending on the nature of the operator. We'll encounter imaginary integrals of this type in Chap. 9 when we consider circular dichroism.

Classical physics put no theoretical restrictions on the precision to which we can specify the position and momentum of a particle. For a particle moving with a given velocity along a known path, it seemed possible to express the momentum precisely as a function of position and time. However, on the very small scales of distance and energy that apply to individual electrons, momentum and position turn out to be interdependent, so that specifying the value of one of them automatically introduces uncertainty in the value of the other. This comes about because the result of combining the position and momentum operators depends on the order in which the two operations are performed: $(\tilde{\mathrm{r}}\tilde{\mathrm{p}}\psi - \tilde{\mathrm{p}}\tilde{\mathrm{r}}\psi)$ is not zero, as (***rp–pr***) is in classical mechanics, but $i\hbar\ \psi$. The fact that the position and momentum operators do not obey the law of commutation that holds for multiplication of pure numbers is expressed by saying that the two operators do not *commute* (see Box 2.2). It was the recognition of this fundamental breakdown of classical mechanics that led Heisenberg to suggest that position and momentum should not be treated simply as numbers, but rather as operators or matrices.

Box 2.2 Commutators and Formulations of the Position, Momentum and Hamiltonian Operators

Heisenberg's commutation relationship for the position and momentum operators,

$$\widetilde{\mathrm{r}}\widetilde{\mathrm{p}} - \widetilde{\mathrm{p}}\widetilde{\mathrm{r}} = i\hbar, \tag{B2.2.1}$$

can be written succinctly by defining the *commutator* of two operators $\tilde{\mathrm{A}}$ and $\widetilde{\mathrm{B}}$ as

$$\left[\widetilde{\mathrm{A}}, \widetilde{\mathrm{B}}\right] = \widetilde{\mathrm{A}}\widetilde{\mathrm{B}} - \widetilde{\mathrm{B}}\widetilde{\mathrm{A}}. \tag{B2.2.2}$$

With this notation, Heisenberg's expression reads

$$[\widetilde{\mathrm{r}}, \widetilde{\mathrm{p}}] = i\hbar. \tag{B2.2.3}$$

Equation (B2.2.3) is often said to be the most fundamental equation of quantum mechanics. The prescription for the momentum operator, Eq. (2.5), is a solution to this equation if we take the x component of the position operator to be multiplication by x. To see this, just substitute $\widetilde{\mathrm{r}}_x = x$ and $\widetilde{\mathrm{p}}_x = (\hbar/i)\partial/\partial x$ in Eq. (B2.2.3) and evaluate the results when the commutator operates on an arbitrary function ψ:

$$\begin{aligned}[x, \widetilde{\mathrm{p}}_x]\psi &= x\widetilde{\mathrm{p}}_x\psi - \widetilde{\mathrm{p}}_x(x\psi) = x\frac{\hbar}{i}\frac{\partial\psi}{\partial x} - \frac{\hbar}{i}\frac{\partial(x\psi)}{\partial x} \\ &= x\frac{\hbar}{i}\frac{\partial\psi}{\partial x} - \frac{\hbar}{i}\left(\frac{x\partial\psi}{\partial x} + \psi\right) = -\frac{\hbar}{i}\psi = i\hbar\psi.\end{aligned} \tag{B2.2.4}$$

In order for the observables corresponding to two operators $\tilde{\mathrm{A}}$ and $\widetilde{\mathrm{B}}$ to have precisely defined values simultaneously, it is necessary and sufficient that the operators commute, i.e. that $\left[\widetilde{\mathrm{A}}, \widetilde{\mathrm{B}}\right] = 0$. The operators for the x, y and z components of position, for example, commute with each other, so the three components of a position vector can all be specified simultaneously to arbitrary precision. The x component of the position also can be specified simultaneously with the y or z component of the momentum (p_y or p_z), but not with p_x.

One other basic point to note concerning commutators is that

(continued)

Box 2.2 (continued)

$$\left[\tilde{\mathrm{B}},\tilde{\mathrm{A}}\right] = -\left[\tilde{\mathrm{A}},\tilde{\mathrm{B}}\right]. \tag{B2.2.5}$$

Although we cannot specify both the energy and the exact position of a particle simultaneously, we can determine the expectation values of both properties with arbitrary precision. This is because the expectation value of the commutator of the Hamiltonian and position operators is zero, even though the commutator itself is not zero. If a system is in a state with wavefunction ψ_n, the expectation value of the commutator $\left[\tilde{\mathrm{H}},\tilde{\mathrm{r}}\right]$ is:

$$\left\langle \psi_n \middle| \left[\tilde{\mathrm{H}},\tilde{\mathrm{r}}\right] \middle| \psi_n \right\rangle = \left\langle \psi_n \middle| \tilde{\mathrm{H}}\tilde{\mathrm{r}} \middle| \psi_n \right\rangle - \left\langle \psi_n \middle| \tilde{\mathrm{r}}\tilde{\mathrm{H}} \middle| \psi_n \right\rangle. \tag{B2.2.6}$$

To evaluate the integrals on the right side of this expression, first expand ψ_n formally in terms of a set of normalized, orthogonal eigenfunctions of $\tilde{\mathrm{H}}$ for the system:

$$\psi_n = \sum_i C_i \psi_i, \tag{B2.2.7}$$

with $C_i = 1$ for $i = n$ and zero otherwise. Here we are anticipating some of the results from Sect. 2.2. By saying that ψ_i is an eigenfunction of $\tilde{\mathrm{H}}$, we mean that $\tilde{\mathrm{H}}\psi_i = E_i\psi_i$, where E_i is the energy of wavefunction ψ_i and is a constant, independent of both position and time. Specifying that the eigenfunctions are normalized and orthogonal means that $\langle\psi_i|\psi_i\rangle = 1$ and $\langle\psi_i|\psi_j\rangle = 0$ if $j \neq i$.

Now use the matrix representation

$$\left\langle \psi_i \middle| \tilde{\mathrm{A}} \middle| \psi_j \right\rangle = A_{ij}, \tag{B2.2.8}$$

and the procedure for matrix multiplication,

$$\left\langle \psi_i \middle| \tilde{\mathrm{A}}\tilde{\mathrm{B}} \middle| \psi_j \right\rangle = \sum_k A_{ik} B_{ki} \tag{B2.2.9}$$

(Appendix A.2). The result is that the integrals on the right side of Eq. (B2.2.6) are just products of the expectation values of the energy and the position:

(continued)

Box 2.2 (continued)

$$\left\langle \psi_n \middle| \tilde{\mathrm{r}}\tilde{\mathrm{H}} \middle| \psi_n \right\rangle = \sum_k \langle \psi_n | \tilde{\mathrm{r}} | \psi_k \rangle \left\langle \psi_k \middle| \tilde{\mathrm{H}} \middle| \psi_n \right\rangle$$
$$= \sum_k \{ (r_k \langle \psi_n | \psi_k \rangle) (E_n \langle \psi_k | \psi_n \rangle) \} = r_n E_n, \qquad \text{(B2.2.10a)}$$

and

$$\left\langle \psi_n \middle| \tilde{\mathrm{H}}\tilde{\mathrm{r}} \middle| \psi_n \right\rangle = \sum_k \left\langle \psi_n \middle| \tilde{\mathrm{H}} \middle| \psi_k \right\rangle \langle \psi_k | \tilde{\mathrm{r}} | \psi_n \rangle$$
$$= \sum_k \{ (E_n \langle \psi_k | \psi_n \rangle) (r_k \langle \psi_n | \psi_k \rangle) \} = E_n r_n. \qquad \text{(B2.2.10b)}$$

All the terms with $k \neq n$ drop out because the eigenfunctions are orthogonal. Because $r_n E_n = E_n r_n$, $\left\langle \psi_n \middle| \left[\tilde{\mathrm{H}}, \tilde{\mathrm{r}} \right] \middle| \psi_n \right\rangle$ must be zero.

In an alternative formulation called the "momentum representation," the momentum operator is taken to be simply multiplication by the classical momentum and the position operator is $i\hbar$ times the derivative with respect to momentum. The "position representation" described above is more widely used, but all the predictions concerning observable quantities are the same.

Given the operators for position and momentum, it is relatively straightforward to find the appropriate operator for any other dynamical property of a system, that is, for any property that depends on position and time. The operator is obtained by starting with the classical equation for the property and replacing the classical variables for position and momentum by the corresponding quantum mechanical operators.

The operator for the total energy of a system is called the *Hamiltonian* operator $\left(\tilde{\mathrm{H}}\right)$ after the nineteenth century Anglo-Irish physicist and mathematician W.R. Hamilton, who developed a general scheme for writing the equations of motion of a dynamical system in terms of coordinates and momenta. Hamilton wrote the classical energy of a system as the sum of the kinetic energy (T) and the potential energy (V), and $\tilde{\mathrm{H}}$ is written as the analogous sum operators for kinetic $\left(\tilde{\mathrm{T}}\right)$ and potential energy ($\tilde{\mathrm{V}}$):

$$\tilde{\mathrm{H}} = \tilde{\mathrm{T}} + \tilde{\mathrm{V}}. \qquad (2.6a)$$

The expectation value of $\tilde{\mathrm{H}}$ for a system thus is the total energy (E):

$$E = \left\langle \Psi \middle| \widetilde{\mathrm{H}} \middle| \Psi \right\rangle. \tag{2.6b}$$

In accordance with the prescription given above, the kinetic energy operator is obtained from the classical expression for kinetic energy $\left(T = |\boldsymbol{p}|^2/2m\right)$ by substituting $\widetilde{\mathrm{p}}$ for $\boldsymbol{p}$. The quantum mechanical operator that corresponds to $|\boldsymbol{p}|^2$ requires performing the operation designated by $\widetilde{\mathrm{p}}$ a second time, rather than simply squaring the result of a single operation. In one dimension, performing $\widetilde{\mathrm{p}}$ twice generates $(\hbar/i)^2$ times the second derivative with respect to position. In three dimensions, the result can be written $(\hbar/i)^2\widetilde{\nabla}^2$, or $-\hbar^2\widetilde{\nabla}^2$, where

$$\widetilde{\nabla}^2\Psi = \left(\partial^2\Psi/\partial x^2 + \partial^2\Psi/\partial y^2 + \partial^2\Psi/\partial z^2\right). \tag{2.7}$$

The kinetic energy operator is, therefore,

$$\widetilde{\mathrm{T}} = \widetilde{\mathrm{p}}^2/2m = -\left(\hbar^2/2m\right)\widetilde{\nabla}^2 \tag{2.8a}$$

$$= -\left(\hbar^2/2m\right)\left(\partial^2/\partial x^2 + \partial^2/\partial y^2 + \partial^2/\partial z^2\right). \tag{2.8b}$$

The operator $\widetilde{\nabla}^2$ is called the *Laplacian* operator and is read "del-squared."

The potential energy operator $\tilde{\mathrm{V}}$ is obtained similarly from the classical expression for the potential energy of the system by substituting the quantum mechanical operators for position and momentum wherever necessary. The starting classical expression depends on the system. In an atom or molecule, the potential energy of an electron depends primarily on interactions with the nuclei and other electrons. If a molecule is placed in an electric or magnetic field, the potential energy also depends on the interactions of the electrons and nuclei with the field.

2.2 The Time-Dependent and Time-Independent Schrödinger Equations

Our next problem is to find the wavefunction $\Psi(\boldsymbol{r},t)$. In 1926, Erwin Schrödinger proposed that the wavefunction obeys the differential equation

$$\widetilde{\mathrm{H}}\Psi = i\hbar\partial\Psi/\partial t. \tag{2.9}$$

This is the *time-dependent Schrödinger equation*. Schrödinger, who lived from 1887 to 1961 and was trained as a mathematician, developed this elegantly simple expression by considering the transition from geometrical optics to wave optics and seeking a parallel transition in mechanics. He was stimulated by Louis deBroglie's proposal that a particle such as an electron has an associated wavelength that is inversely proportional to the particle's momentum (Box 2.3). However, the insight that led to the Schrödinger equation does not constitute a proof from first principles, and most theorists today take Eq. (2.9) as a basic postulate in the spirit of Eqs. (2.1) and (2.3). Its ultimate justification rests on the fact that the Schrödinger equation accounts for an extraordinarily broad range of experimental observations and leads to predictions that, so far, have unfailingly proved to be correct.

Box 2.3. The Origin of the Time-Dependent Schrödinger Equation
In his ground-breaking paper on the photoelectric effect (Sect. 1.6), Einstein [15] noted that if light consists of discrete particles ("photons" in today's vernacular), then each such particle should have a momentum that is inversely proportional to the wavelength (λ). This relationship follows from the theory of relativity, which requires that a particle with energy E and velocity w have momentum Ew/c^2. A photon with energy $h\nu$ and velocity c should, therefore, have momentum

$$p = h\nu c/c^2 = h\nu/c = h/\lambda = h\overline{\nu}, \tag{B2.3.1}$$

where $\overline{\nu}$ is the wavenumber ($1/\lambda$). Convincing experimental support for this reasoning came in 1923, when A. Compton measured the scattering of X-ray photons by crystalline materials.

It occurred to de Broglie [16] that the same reasoning might hold in reverse. If light with wavelength λ has momentum p, then perhaps a particle with momentum p has an associated wave with wavelength

$$\lambda = p/h. \tag{B2.3.2}$$

The wave-like properties of electrons were confirmed four years later by Davison and Germer's measurements of electron diffraction by crystals and P.G. Thompson's measurements of diffraction by gold foil.

Consider a one-dimensional, non-relativistic particle with mass m moving in a potential field V. Suppose further, and this is critical to the argument, that V is independent of position. The relation between the energy of and momentum of the particle then is

$$E = \frac{1}{2m}p^2 + V. \tag{B2.3.3}$$

Inserting the relationships $E = h\nu$ and $p = h\overline{\nu}$ into this expression gives

$$h\nu = \frac{h^2}{2m}\overline{\nu}^2 + V. \tag{B2.3.4}$$

(continued)

Box 2.3 (continued)
A relationship between the energy and momentum of a particle (Eq. B2.3.3) thus can be transformed into a relationship between the frequency and wavenumber of a wave Eq. (B2.3.4).

Now suppose we write the wave as

$$\Psi = A\exp\{2\pi i(\overline{\nu}x - \nu t)\}, \tag{B2.3.5}$$

where A is a constant. This is a general expression for a monochromatic plane wave moving in the x direction with velocity $\nu/\overline{\nu}$. The derivatives of Ψ with respect to position and time then are:

$$\frac{1}{2\pi i}\frac{\partial \Psi}{\partial x} = \overline{\nu}\Psi, \quad \frac{1}{(2\pi i)^2}\frac{\partial^2 \Psi}{\partial x^2} = \overline{\nu}^2\Psi \quad \text{and} \quad -\frac{1}{2\pi i}\frac{\partial \Psi}{\partial t} = \nu\Psi. \tag{B2.3.6}$$

These expressions can be used to change Eq. (B2.3.4) into a differential equation. Multiplying each term of (B2.3.4) by Ψ and substituting in the derivatives gives:

$$-\frac{h}{2\pi i}\frac{\partial \Psi}{\partial t} = \frac{h^2}{2m(2\pi i)^2}\frac{\partial^2 \Psi}{\partial t^2} + V\Psi,$$

or

$$i\hbar\frac{\partial \Psi}{\partial t} = -\frac{\hbar^2}{2m}\frac{\partial^2 \Psi}{\partial t^2} + V\Psi. \tag{B2.3.7}$$

This is the time-dependent Schrödinger equation for a one-dimensional system in a constant potential.

Extending these arguments to three dimensions is straightforward. But there is no *a priori* justification for extending Eq. (B2.3.7) to the more general situation in which the potential varies with position, because in that case the simple wave expression (Eq. B2.3.5) no longer holds. Schrödinger, however, saw that the general equation

$$i\hbar\frac{\partial \Psi}{\partial t} = -\frac{\hbar^2}{2m}\frac{\partial^2 \Psi}{\partial x^2} + V(x,y,z)\Psi \tag{B2.3.8}$$

has solutions of the form

$$\Psi = \psi(x,y,z)\exp(-2\pi i Et/\hbar), \tag{B2.3.9}$$

and that these solutions have exactly the properties needed to account for the quantized energy levels of the hydrogen atom [17, 18].

Further discussion of the origin of the Schrödinger equation and philosophical interpretations of quantum mechanics can be found in [12, 14, 19].

To see how the Schrödinger equation can be used to find Ψ, let's first suppose that $\widetilde{\mathrm{H}}$ depends on the position of a particle but is independent of time: $\widetilde{\mathrm{H}} = \widetilde{\mathrm{H}}_o(\boldsymbol{r})$. The wavefunction then can be written as a product of two functions, one (ψ) that depends only on position and one (ϕ) that depends only on time:

$$\Psi(r,t) = \psi(r)\phi(t). \tag{2.10}$$

To prove this assertion, we need only show that all the wavefunctions given by Eq. (2.10) satisfy the Schrödinger equation as long as $\widetilde{\mathrm{H}}$ remains independent of time. Replacing $\widetilde{\mathrm{H}}$ by $\widetilde{\mathrm{H}}_o$ and Ψ by $\psi(\boldsymbol{r})\phi(t)$ on the left-hand side of Eq. (2.9) gives:

$$\widetilde{\mathrm{H}}\Psi \to \widetilde{\mathrm{H}}_o\{\psi(\boldsymbol{r})\phi(t)\} = \phi(t)\widetilde{\mathrm{H}}_o\psi(\boldsymbol{r}). \tag{2.11}$$

(Multiplication by $\phi(t)$ can be done after $\widetilde{\mathrm{H}}_o$ operates on $\psi(\boldsymbol{r})$ because $\phi(t)$ is just a constant as far as the time-independent operator $\widetilde{\mathrm{H}}_o$ is concerned.) Making the same substitution for Ψ on the right side gives

$$i\hbar\partial\Psi/\partial t \to i\hbar\psi\left(\vec{r}\right)\partial\phi(t)/\partial t. \tag{2.12}$$

The wavefunction given by Eq. (2.10) thus will satisfy the Schrödinger equation if

$$\phi(t)\widetilde{\mathrm{H}}_o\psi(\boldsymbol{r}) = i\hbar\psi(\boldsymbol{r})\partial\phi(t)/\partial t, \tag{2.13a}$$

or

$$\frac{1}{\psi(\boldsymbol{r})}\widetilde{\mathrm{H}}_o\psi(\boldsymbol{r}) = i\hbar\frac{1}{\phi(t)}\partial\phi(t)/\partial t. \tag{2.13b}$$

Note that while the left-hand side of Eq. (2.13b) is a function only of position, the right side is a function only of time. The two sides can be equivalent only if both sides are equal to a constant that does not depend on either time or position. This allows us to solve the two sides of the equation independently. First, consider the left side. If we call the constant E, then we have $[1/\psi(\boldsymbol{r})]\ \widetilde{\mathrm{H}}_o\psi(\boldsymbol{r}) = E$. Multiplying by $\psi(\boldsymbol{r})$ gives the *time-independent Schrödinger equation*:

$$\widetilde{\mathrm{H}}_o\psi(\boldsymbol{r}) = E\psi(\boldsymbol{r}). \tag{2.14}$$

This expression says that when operator $\widetilde{\mathrm{H}}_o$ works on function ψ, the result is simply a constant (E) times the original function. An equation of this type is called an *eigenfunction equation* or *eigenvalue equation*, and in general, it has a set of solutions, called *eigenfunctions*, $\psi_k(\boldsymbol{r})$. If these solutions meet Born's restrictions on physically meaningful wavefunctions, we can interpret them as representing various possible states of the system, each of which corresponds to a particular distribution of the system in space. For an electron in an atom or a molecule, the acceptable eigenfunctions are the set of atomic or molecular *orbitals*. Each eigenfunction represents an *eigenstate* of the system and is associated with a particular value of the constant E (E_k), which is called an *eigenvalue*.

Now consider the right-hand side of Eq. (2.13b), which also must equal E: $i\hbar\,[1/\phi(t)]\partial\phi(t)/\partial t = E$. This expression has one solution (ϕ_k) for each allowable value of E (E_k):

$$\phi_k(t) = \exp\{-i(E_k t/\hbar + \zeta)\}, \tag{2.15}$$

where ζ is an arbitrary constant. Because $\exp(i\alpha) = \cos(\alpha) + i\,\sin(\alpha)$, Eq. (2.15) represents a function with real and imaginary parts, both of which oscillate in time at a frequency of $E_k/2\pi\hbar$, or E_k/h. The constant term ζ in the exponential is a phase shift that depends on our choice of zero time. As long as we are considering only a single particle, ζ will not affect any measurable properties of the system and we can simply set it to zero. (We'll return to this point for systems containing many particles in Chap. 10.)

Combining Eqs. (2.10) and (2.15) and setting $\zeta = 0$ gives a complete wavefunction for state k:

$$\Psi_k(\boldsymbol{r}, t) = \psi_k(\boldsymbol{r})\exp(-iE_k t/\hbar). \tag{2.16}$$

The complex conjugate of Ψ_k is $\psi_k^* \exp(iE_k t/\hbar)$, so the probability function P (Eqs. 2.1 and 2.2) is $\psi_k^*\exp(iE_k t/\hbar)\psi_k\exp(-iE_k t/\hbar) = \psi_k^*\psi_k = |\psi_k|^2$. The time dependence of Ψ thus does not appear in the probability density. The time dependence can, however, be critical when wavefunctions of systems with different energies are combined and when we consider transitions of a system from one state to another.

The meaning of the eigenvalues of the Schrödinger equation emerges if we use Eq. (2.4) to find the expectation value of the energy for a particular state, Ψ_k:

$$\left\langle \Psi_k \middle| \tilde{\mathrm{H}} \middle| \Psi_k \right\rangle = \langle \psi_k | E_k \psi_k \rangle \{\exp(iE_k t/\hbar)\exp(-iE_k t/\hbar)\} = E_k \langle \psi_k | \psi_k \rangle = E_k. \tag{2.17}$$

Eigenvalue E_k of the Hamiltonian operator thus is the energy of the system in state k. Note that the energy is independent of time; it has to be, because we assumed that $\tilde{\mathrm{H}}$ is independent of time.

Equation (2.17) sheds some light on our basic postulate concerning the expectation value of an observable property (Eq. (2.4)). If we know that a system has wavefunction Ψ_k and that this wavefunction is an eigenfunction of operator $\tilde{\mathrm{A}}$ with eigenvalue A_k (i.e., that $\langle \Psi_k | \Psi_k \rangle = 1$ and $\tilde{\mathrm{A}}\Psi_k = A_k \Psi_k$), then barring experimental errors, each measurement of the property must give the value A_k. This is just what Eq. (2.4) states for such a situation:

$$\left\langle \Psi_k \middle| \tilde{\mathrm{A}} \middle| \Psi_k \right\rangle = \langle \Psi_k | A_k \Psi_k \rangle = A_k \langle \Psi_k | \Psi_k \rangle = A_k. \tag{2.18}$$

On the other hand, if Ψ_k is *not* an eigenfunction of operator $\tilde{\mathrm{A}}$ (i.e., $\tilde{\mathrm{A}}\Psi_k \neq A_k\Psi_k$ for any value of A_k), then we cannot predict the result of an individual measurement of the property. We can, however, still predict the average result of many measurements, and Eq. (2.4) asserts that this statistical average is given by $\left\langle \Psi_k \middle| \tilde{\mathrm{A}} \middle| \Psi_k \right\rangle$.

The time-independent Schrödinger equation will have at least one solution for any finite value of E. However, if the potential energy function $V(\boldsymbol{r})$ confines a particle to a definite region of space, only certain values of E are consistent with Born's restrictions on physically meaningful wavefunctions. All other values of E either make ψ go to infinity somewhere or have some other unacceptable property such as a discontinuous first derivative with respect to position. This means that the possible energies for a bound particle are *quantized*, as illustrated in Fig. 2.1. The solutions to the Schrödinger equation for a free particle are not quantized in this way; such a particle can have any energy above the threshold needed to set it free (Fig. 2.1).

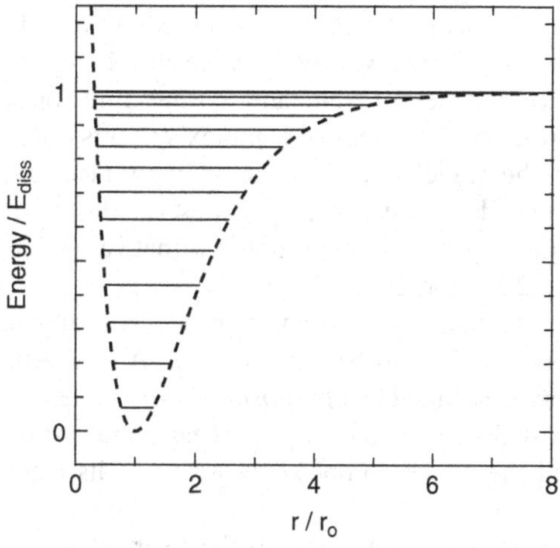

Fig. 2.1 The energy of a bound particle is quantized. The *horizontal lines* indicate the eigenvalues of the total energy of a system whose potential energy depends on position (r) as indicated by the *dashed curve*. The potential function shown is the Morse potential, $V = E_{diss}[1.0 - \exp(-a(r - r_o))]^2$, where r_o is the equilibrium value of the coordinate and a is a factor that determines the asymmetry (anharmonicity) of the potential well. Here $a = 0.035/r_o$. For energies less than or equal to E_{diss}, the eigenvalues are, to a very good approximation, $E_n = h\upsilon_o[(n + 1/2) - (h\upsilon_o/4E_{diss})(n + 1/2)^2]$, where n is an integer and υ_o depends on a, E_{diss} and the reduced mass of the system [35, 36]. If the total energy exceeds E_{diss}, the energy is not quantized. The particle then can escape from the well on the right-hand side, with its surplus energy taking the form of kinetic energy. Some of the eigenfunctions of this potential are shown in Fig. 6.5

2.2.1 Superposition States

The eigenfunctions of the Hamiltonian operator can be shown to form a *complete set of orthogonal functions*. By "orthogonal" we mean that the product of any two different members of such a set (Ψ_i and Ψ_k), when integrated over all space, is zero:

$$\langle \Psi_i | \Psi_k \rangle = 0. \tag{2.19}$$

The functions sin(x) and sin($2x$), for example, are orthogonal.

A general property of eigenvalue equations such as the time-independent Schrödinger equation is that if ψ_i is a solution, then so is the product of ψ_i with any constant. This means that we can multiply any non-zero eigenfunction by a suitable factor so that

$$\langle \Psi_i | \Psi_i \rangle = 1. \tag{2.20}$$

Such an eigenfunction is said to be *normalized*, and eigenfunctions that satisfy both Eqs. (2.19) and (2.20) are called *orthonormal*.

A complete set of functions of a given variable has the property that a linear combination of its members can be used to construct any well-behaved function of that variable, even if the target function is not a member of the set. "Well-behaved" in this context means a function that is finite everywhere in a defined interval and has finite first and second derivatives everywhere in this region. Fourier analysis uses this property of the set of sine and cosine functions (Appendix A.3). In quantum mechanics, wavefunctions of complex systems often are approximated by combinations of the wavefunctions of simpler or idealized systems. Wavefunctions of time-dependent systems can be described similarly by combining wavefunctions of stationary systems in proportions that vary with time. We'll return to this point in Sects. 2.3.6 and 2.5.

If Ψ_i and Ψ_j are eigenfunctions of an operator, then any linear combination $C_i\Psi_i + C_j\Psi_j$ also is an eigenfunction of that operator. A state represented by such a combined wavefunction is called a *superposition state*. One of the basic tenets of quantum mechanics is that we should use a linear combination of this nature to describe a system whenever we do not know which of its eigenstates the system is in.

The idea of a superposition state gets to the heart of the difference between classical and quantum mechanics. Suppose that a particular experiment always gives result x_1 when it is performed on systems known to be in state 1, and a different result, x_2, for systems in state 2. We then have $\langle \Psi_1 | \tilde{\mathrm{x}} | \Psi_1 \rangle = x_1$ and $\langle \Psi_2 | \tilde{\mathrm{x}} | \Psi_2 \rangle = x_2$, where Ψ_1 and Ψ_2 are the wavefunctions of the two states and $\tilde{\mathrm{x}}$ is the operator for the property examined in the experiment. Now suppose further that a given system has probability $|C_1|^2$ of being in state 1 and probability $|C_2|^2$ of being in state 2. Then, according to classical mechanics, the average result of a large number of measurements on the system would be the weighted sum

$|C_1|^2 x_1 + |C_2|^2 x_2$. However, the individual probabilities do not add in this way for system in a quantum mechanical superposition state. Instead, the expectation value is $\langle C_1\Psi_1 + C_2\Psi_2|\tilde{x}|C_1\Psi_1 + C_2\Psi_2\rangle$. The classical and quantum mechanical predictions can be very different, because in addition to $|C_1|^2 x_1 + |C_2|^2 x_2$, the latter contains an "interference" term, $C_1^* C_2\langle\Psi_1|\tilde{x}|\Psi_2\rangle + C_2^* C_1\langle\Psi_2|\tilde{x}|\Psi_1\rangle$.

One well known illustration of quantum interference is the pattern of bright and dark bands cast on a screen by light that is diffracted at a pair of slits. When Thomas Young first described these interference "fringes" in 1804, they seemed to comport well with the wave theory of light and to be incompatible with the notion that light consists of particles. They are, however, consistent with the particle picture if we accept the idea that a photon passing through the slits exists in a superposition state involving both slits until it strikes the screen where it is detected. Because the same pattern of fringes is seen even when the light intensity is reduced so that no more than one photon is in transit through the apparatus at any given time, the interference evidently reflects a property of the wavefunction of an individual photon rather than interactions between photons. Similar manifestations of interference have been described for electrons, neutrons, He atoms and even $^9Be^+$ ions [20–22]. Additional observations of this type on photons are discussed in Sect. 3.5.

In most experiments, we do not deal with single photons or other particles, but rather with an ensemble containing a very large number of particles. Such an ensemble generally cannot be described by a single wavefunction, because the wavefunctions of the individual systems in the ensemble typically have random phases with respect to each other. An ensemble of systems whose individual wavefunctions have random, uncorrelated phases is said to be in a *mixed state*. Young himself observed that diffraction fringes were not seen if the light passing through the two slits came from separate sources. We now can generalize this observation to mean that the interference term in the expectation value for an observable disappears if it is averaged over a large number of uncorrelated systems. We'll discuss the quantum mechanics of mixed states in more detail in Chap. 10.

2.3 Spatial Wavefunctions

The Schrödinger equation can be solved exactly for the electron wavefunctions of the hydrogen atom and some other relatively simple systems. We will now discuss some of the main results. More detailed derivations are available in the texts on quantum mechanics cited in Sect. 2.1.

2.3.1 A Free Particle

For a free particle, the potential energy is constant and we can choose it to be zero. The time-independent Schrödinger equation in one dimension then becomes simply

$$-(\hbar^2/2m)\partial^2\psi/\partial x^2 = E\psi, \tag{2.21}$$

where m and E are the mass and total energy of the particle and x is the dimensional coordinate (Eqs. 2.8a, b and 2.14). Equation (2.21) has a solution for any positive value of E:

$$\psi(x) = A\exp\left(i\sqrt{2mE}\,x/\hbar\right), \tag{2.22}$$

where A is an arbitrary amplitude with dimensions of $\mathrm{cm}^{-1/2}$. The energy of a free particle thus is not quantized.

2.3.2 A Particle in a Box

Now consider a particle confined in a one-dimensional rectangular box with infinitely high walls. Let the box extend from $x=0$ to $x=l$, and suppose again that the potential energy ($V(x)$) inside the box is zero. The Hamiltonian operator is

$$\widetilde{\mathrm{H}} = -(\hbar^2/2m)\partial^2/\partial x^2 + V(x), \tag{2.23}$$

where $V(x)=0$ for x between 0 and l, $V(x)=\infty$ outside this region, and m is the mass of the particle. The time-independent Schrödinger equation for this Hamiltonian has a set of solutions for the region in the box:

$$\psi_n(x) = (2/l)^{1/2}\sin(n\pi x/l) \tag{2.24}$$

where n is any integer. The energies associated with these wavefunctions are

$$E_n = n^2h^2/8ml^2. \tag{2.25}$$

Each eigenfunction (ψ_n) thus is determined by a particular value of an integer *quantum number* (n), and the energies (E_n) increase quadratically with this number. Figure 2.2 shows the first five eigenfunctions, their energies, and the corresponding probability density functions ($|\psi_n|^2$). Outside the box, the wavefunctions must be zero because there is no chance of finding the particle in a region where its potential energy would be infinite. To avoid discontinuities, the wavefunctions must go to zero at both ends of the box, and it is this boundary condition that forces n to be an integer.

There are several important points to note here. First, the energies are quantized. Second, excluding the trivial solution $n=0$, which corresponds to an empty box, the lowest possible energy is not zero as it would be for a classical particle at rest in a box, but rather $h^2/8ml^2$. The smaller the box, the higher the energy. Finally, the number of *nodes* (surfaces where the wavefunction crosses zero, or in a

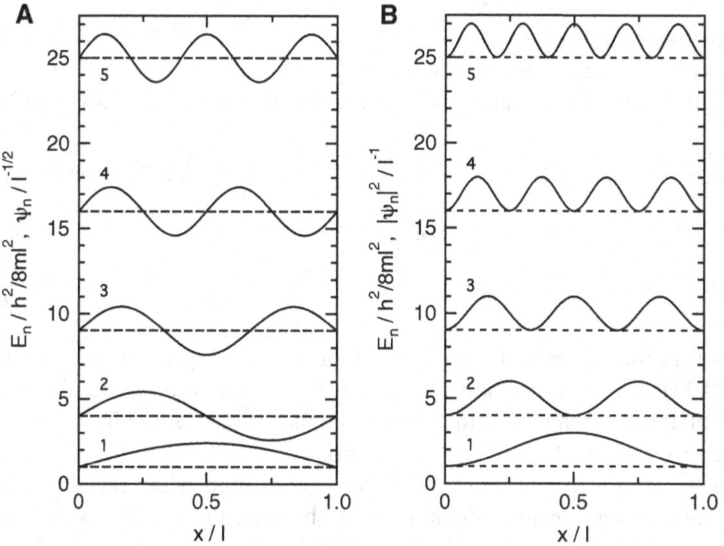

Fig. 2.2 Eigenfunctions (**A**) and probability densities (**B**) of a particle in a one-dimensional rectangular box of length l and infinitely high walls. The *dashed lines* indicate the energies of the first five eigenstates ($n = 1, 2, \ldots 5$), and the eigenfunctions and probability densities (*solid curves*) are displaced vertically to align them with the corresponding energies

one-dimensional system, points where this occurs) increases linearly with n, so that the probability distribution becomes more uniform with increasing n (Fig. 2.2). The momentum of the particle is discussed in Box 2.4.

Box 2.4. Linear Momentum

Because we took the potential energy inside the box in Fig. 2.2 to be zero, the energies given by Eq. (2.22) are entirely kinetic energy. From classical physics, a particle with kinetic energy E_n and mass m should have a linear momentum $\boldsymbol{p}$ with magnitude

$$p = |\boldsymbol{p}| = \sqrt{2mE_n} = nh/2l. \qquad \text{(B2.4.1)}$$

(*B*2.4.1)

Referring to Fig. 2.2A, we see that wavefunction ψ_n is equivalent to a *standing wave* with a wavelength λ_n of $2l/n$. We therefore also could write the momentum ($nh/2l$) as h/λ_n or $h\overline{\nu}$, where $\overline{\nu}$ is the wavenumber ($1/\lambda$). This is consistent with deBroglie's expression linking the momentum of a free particle to the wavenumber of an associated wave (Eq. B2.3.1). But the

(continued)

Box 2.4 (continued)
expectation value of the momentum of the particle in a box with infinitely high walls is not $h\overline{\nu}$; it's zero. We can see this by using Eqs. (2.4) and (2.5):

$$\begin{aligned}\langle\psi_n|\widetilde{\mathrm{p}}|\psi_n\rangle &= \langle\psi_n|(\hbar/i)\partial\psi_n/\partial x\rangle = (\hbar/i)\Big\langle(2/l)^{1/2}\sin(n\pi x/l)\Big|(2/l)^{1/2}\partial\sin(n\pi x/l)/\partial x\Big\rangle \\ &= (\hbar/i)(2/l)(n\pi/l)\int_0^l \sin(n\pi x/l)\cos(n\pi x/l)dx = 0.\end{aligned} \tag{B2.4.2}$$

This result makes sense if we view each of the standing waves described by Eq. (2.22) as a superposition of two waves moving in opposite directions. Individual measurements of the momentum then might give either positive or negative values but would average to zero.

From a more formal perspective, we cannot predict the outcome of an individual measurement of the momentum because the wavefunctions given by Eq. (2.22) are not eigenfunctions of the momentum operator. This is apparent because the action of the momentum operator on wavefunction ψ_n gives

$$\widetilde{\mathrm{p}}\psi_n = (\hbar/i)\partial\psi_n/\partial x = (\hbar/i)\left(2^{1/2}n\pi/l^{3/2}\right)\cos(n\pi x/l), \tag{B2.4.3}$$

which is not equal to a constant times ψ_n. However, the expectation value obtained by Eq. (B2.4.2) still gives the correct result for the momentum of a standing wave (zero).

The Schrödinger equation does have solutions that are eigenfunctions of the momentum operator for an electron moving through an unbounded region of constant potential. These can be written

$$\psi_\pm = A\exp\{2\pi i(\overline{\nu}x \pm \nu t)\}, \tag{B2.4.4}$$

where A is a constant and the plus and minus signs are for particles moving in the two directions. Applying the momentum operator to these wavefunctions gives

$$\widetilde{\mathrm{p}}\psi_\pm = (\hbar/i)\partial\psi_\pm/\partial x = (\hbar/i)A(2\pi i\overline{\nu})\exp\{2\pi i(\overline{\nu}x \pm \nu t)\} = h\overline{\nu}\psi_\pm, \tag{B2.4.5}$$

which leads to the correct expectation value of the momentum of a free particle ($h\overline{\nu}$).

Although the probability of finding the particle at a given position varies with the position, the eigenfunctions given by Eq. (2.24) extend over the full length of the box. Wavefunctions for a particle that is more localized in space can be constructed from linear combinations of these eigenfunctions. A superposition state formed in this way from vibrational wavefunctions is called a *wavepacket*. The combination $\psi_1 - \psi_3 + \psi_5 - \psi_7$, for example, gives a wavepacket whose amplitude peaks strongly at the center of the box ($x = l/2$), where the individual spatial wavefunctions interfere constructively. At positions far from the center, the wavefunctions interfere destructively and the summed amplitude is small. Because the time-dependent factors in the complete wavefunctions $[\exp(-iE_n t/\hbar)]$ oscillate at different frequencies, the wavepacket will not remain fixed in position, but will move and change shape with time. Wavepackets provide a way of representing an atom or macroscopic particle whose energy eigenvalues are uncertain relative to the separation between the eigenvalues. The spacing between the energies for a particle in a one-dimensional box is inversely proportional to both the mass of the particle and l^2 (Eq. 2.25), and will be very small for any macroscopic particle with a substantial mass or size.

If the potential energy outside the square box described in Sect. 2.3.2 is not infinite, there is some probability of finding the particle here even if the potential energy is greater than the total energy of the system. For a one-dimensional square box with infinitely thick walls of height V, the wavefunctions in the region $x > l$ are given by

$$\psi_n = A_n \exp\{-(x - l)\zeta\}, \tag{2.26}$$

where $\zeta = [2m(V - E_n)]^{1/2}\hbar^{-1}$ and A_n is a constant set by the boundary conditions. Inside the box, ψ_n has a sinusoidal form similar to that of a wavefunction for a box with infinitely high walls (Eq. 2.24), but its amplitude goes to A_n rather than zero at the boundary ($x = l$). Both the amplitudes and the slopes of the wavefunctions are continuous across the boundary.

A quantum mechanical particle thus can *tunnel* into a potential-energy wall, although the probability of finding it here decreases exponentially with the distance into the wall. A classical particle could not penetrate the wall because the condition $V > E_n$ requires the kinetic energy to be negative, which is not possible in classical physics.

2.3.3 The Harmonic Oscillator

The potential well that restrains a chemical bond near its mean length is described reasonably well by the quadratic expression

$$V(x) = \frac{1}{2}kx^2, \tag{2.27}$$

where x is the difference between the length of the bond and the mean length, and k is a force constant. If the bond is stretched or compressed, a restoring force proportional to the distortion ($F = -kx$) acts to return the bond to its mean length, provided that the distortion is not too large. Such a parabolic potential well is described as *harmonic*. At larger departures from the equilibrium length the potential well becomes increasingly *anharmonic*, rising more steeply for compression than for stretching, as illustrated in Fig. 2.1.

A classical particle of mass m in a harmonic potential well oscillates about its equilibrium position with a frequency given by

$$\upsilon = \frac{1}{2\pi}\left(\frac{k}{m}\right)^{1/2}. \tag{2.28}$$

Such a system is called a *harmonic oscillator*. Equation (2.28) also applies to the frequency of the classical vibrations of a pair of bonded atoms if we replace m by the reduced mass of the pair [$m_r = m_1 m_2/(m_1 + m_2)$, where m_1 and m_2 are the masses of the individual atoms]. The angular frequency ω is $2\pi\upsilon = (k/m)^{1/2}$.

The sum of the kinetic and potential energies of a classical harmonic oscillator is

$$E_{classical} = \frac{1}{2m_r}|\boldsymbol{p}|^2 + \frac{1}{2}kx^2, \tag{2.29}$$

which can have any non-negative value including zero. The quantum mechanical picture again is significantly different. The eigenvalues of the Hamiltonian operator for a one-dimensional harmonic oscillator are an evenly spaced ladder of energies, starting not at zero but at $(1/2)h\upsilon$:

$$E_n = (n + 1/2)h\upsilon, \tag{2.30}$$

with $n = 0, 1, 2, \ldots$. The corresponding wavefunctions are

$$\chi_n(x) = N_n H_n(u)\exp(-u^2/2). \tag{2.31}$$

Here N_n is a normalization factor, $H_n(u)$ is a *Hermite polynomial*, and u is a dimensionless positional coordinate obtained by dividing the Cartesian coordinate x by $(\hbar/2\pi m_r\upsilon)^{1/2}$:

$$u = x/(\hbar/2\pi m_r \upsilon)^{1/2}. \tag{2.32}$$

The Hermite polynomials and the normalization factor N_n are given in Box 2.5 and the first six wavefunctions are shown in Fig. 2.3.

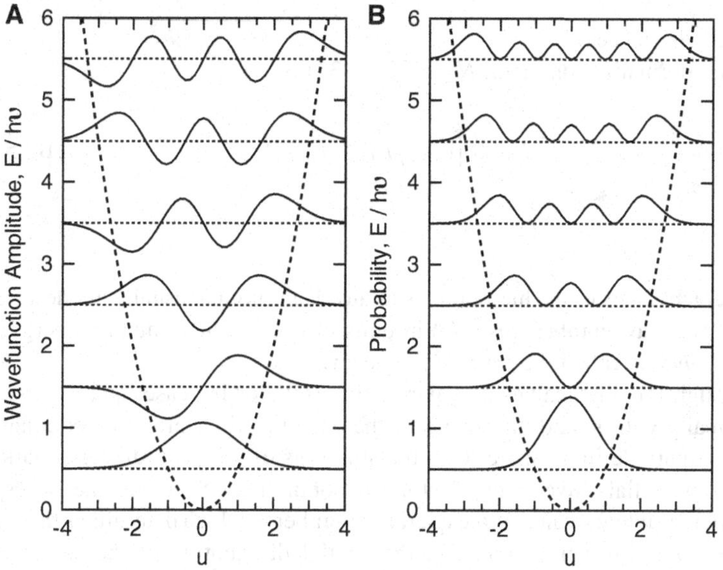

Fig. 2.3 Wavefunctions (**A**) and probability densities (**B**) of a harmonic oscillator. The dimensionless quantity plotted on the abscissa (u) is the distance of the particle from its equilibrium position divided by $(\hbar/2\pi m_r \upsilon)^{1/2}$, where m_r is the reduced mass and υ is the classical oscillation frequency. The *dashed curves* represent the potential energy, and the *dotted lines* indicate the first six eigenvalues of the total energy, E_n, in units of $h\upsilon$. The wavefunctions are shifted vertically to align them with the energies

Box 2.5. Hermite Polynomials

The Hermite polynomials are defined by the expression

$$H_n(u) = (-1)^n \exp(u^2) \frac{d^n}{du^n} \exp(-u^2). \tag{B2.5.1}$$

They can be generated by starting with Eq. (B2.5.1) to find $H_0 = 1$ and $H_1 = 2u$, and then using the recursion formula

$$H_{n+1}(u) = 2uH_n(u) - 2nH_{n-1}(u). \tag{B2.5.2}$$

The first six Hermite polynomials are:

$$\begin{aligned} &H_0 = 1, && H_3 = 8u^3 - 12u, \\ &H_1 = 2u, && H_4 = 16u^4 - 48u^2 + 12, \\ &H_2 = 4u^2 - 2, && H_5 = 32u^5 - 160u^3 + 120u. \end{aligned} \tag{B2.5.3}$$

(continued)

Box 2.5 (continued)

The normalization factor N_n in (Eq. 2.31) is

$$N_n = \left[(2\pi \upsilon/\hbar)^{1/2}/(2^n n!)\right]^{1/2}. \tag{B2.5.4}$$

The eigenvalues of the harmonic oscillator Hamiltonian usually are described in terms of the wavenumber ($\omega = \upsilon/c$) in units of cm^{-1}. The minimum energy, (1/2) $h\upsilon$ or $(1/2)\hbar\omega$, is called the *zero-point energy*.

Although the eigenvalues of a harmonic oscillator increase linearly instead of quadratically with n, and the shapes of the wavefunctions are more complex than those of a particle in a square well, the solutions of the Schrödinger equation for these two potentials have several features in common. Each eigenvalue corresponds to a particular integer value of the quantum number n, which determines the number of nodes in the wavefunction, and the spatial distribution of the wavefunction becomes more uniform as n increases. As in the case of a box with finite walls, a quantum mechanical harmonic oscillator has a definite probability of being outside the region bounded by the potential energy curve (Fig. 2.3B). Finally, as mentioned above and discussed in more detail in Chap. 11, a wavepacket for a particle at a particular position can be constructed from a linear combination of harmonic-oscillator wavefunctions. The position of such a wavepacket oscillates in the potential well at the classical oscillation frequency υ.

2.3.4 Atomic Orbitals

Spatial wavefunctions for the electron in a hydrogen atom, or more generally, for a single electron with charge $-e$ bound to a nucleus of charge $+Ze$, can be written as products of two functions, $R_{n,l}(r)$ and $Y_{l,m}(\theta, \phi)$, where the variables r, θ, and ϕ specify positions in polar coordinates relative to the nucleus and an arbitrary z axis:

$$\psi_{nlm} = R_{n,l}(r)Y_{l,m}(\theta,\phi). \tag{2.33}$$

(See Fig. 4.4 for an explanation of polar coordinates.) The subscripts n, l and m in these functions denote integer quantum numbers with the following possible values:

principal quantum number: $n = 1, 2, 3, \ldots,$

angular momentum (azimuthal) quantum number: $l = 0, 1, 2, \ldots, n-1,$

magnetic quantum number: $m = -l, \ldots, 0, \ldots, l.$

Table 2.1 Hydrogen-atom wavefunctions

Orbital	n	l	m	$R_{n,l}(r)$ [a]	$Y_{l,m}(\theta,\phi)$ [b]
$1s$	1	0	0	$(Z/a_o)^{3/2}2\exp(-\rho)$	$(2\sqrt{\pi})^{-1}$
$2s$	2	0	0	$\frac{(Z/a_o)^{3/2}}{2\sqrt{2}}(2-\rho)\exp(-\rho/2)$	$(2\sqrt{\pi})^{-1}$
$2p_z$	2	1	0	$\frac{(Za_o)^{3/2}}{2\sqrt{6}}\rho\exp(-\rho/2)$	$(3/4\pi)^{1/2}\cos\theta$
$2p_-$	2	1	−1	$\frac{(Za_o)^{3/2}}{2\sqrt{6}}\rho\exp(-\rho/2)$	$(3/8\pi)^{1/2}\sin\theta\{\cos\phi - i\sin\phi)$ $= (3/8\pi)^{1/2}\sin\theta\exp(-i\phi)$
$2p_+$	2	1	1	$\frac{(Za_o)^{3/2}}{2\sqrt{6}}\rho\exp(-\rho/2)$	$(3/8\pi)^{1/2}\sin\theta\{\cos\phi + i\sin\phi)$ $= (3/8\pi)^{1/2}\sin\theta\exp(i\phi)$

[a] $\rho = Zr/a_o$, where Z is the nuclear charge, r is the distance from the nucleus and a_o is the Bohr radius (0.529 Å)

[b] θ and ϕ are the angles with respect to the z and x axes in polar coordinates (Fig. 4.4)

The energy of the orbital depends mainly on the principal quantum number (n) and is given by

$$E_n = -16\pi^2 Z^2 m_r e^4/n^2h^2, \tag{2.34}$$

where m_r is the reduced mass of the electron and the nucleus. (This expression uses the cgs system of units, which is discussed in Sect. 3.1.1.) The angular momentum or azimuthal quantum number (l) determines the electron's angular momentum, while the magnetic quantum number (m) determines the component of the angular momentum along a specified axis and relates to a splitting of the energy levels in a magnetic field (Sect. 9.5).

The wavefunctions for the first few hydrogen-atom orbitals are given in Table 2.1, and Figs. 2.4, 2.5, and 2.6 show their shapes. Atomic orbitals with $l = 0$, 1, 2 and 3 are conventionally labeled s, p, d and f. The s wavefunctions peak at the nucleus and are spherically symmetrical (Figs. 2.4 and 2.5). The $1s$ wavefunction has the same sign everywhere, whereas $2s$ changes sign at $r = 2\hbar^2/m_r e^2 Z$, or $2a_o/Z$, where a_o (the *Bohr radius*) is $\hbar^2/m_r e^2$ (0.5292 Å). Because the volume element ($d\sigma$) between between spherical shells with radii r and $r + dr$ is (for small values of dr) $4\pi r^2 dr$, the probability of finding an s electron at distance r from the nucleus (the *radial distribution function*) depends on $4\pi r^2\psi(r)^2 dr$. The radial distribution function for the $1s$ wavefunction peaks at $r = a_o$ (Fig. 2.4E, F).

The p orbitals have nodal planes that pass through the nucleus (Figs. 2.5 and 2.6). The choice of a coordinate system for describing the orientations of the orbitals is arbitrary unless the atom is in a magnetic field, in which case the z axis is taken to be the direction of the field. The $2p$ orbital with $m = 0$ is oriented along this axis and is called $2p_z$. The $2p$ orbitals with $m = \pm 1$ ($2p_+$ and $2p_-$) are complex functions that are maximal in the plane normal to the z axis and rotate in opposite directions around the z axis with time. However, they can be combined to give two real

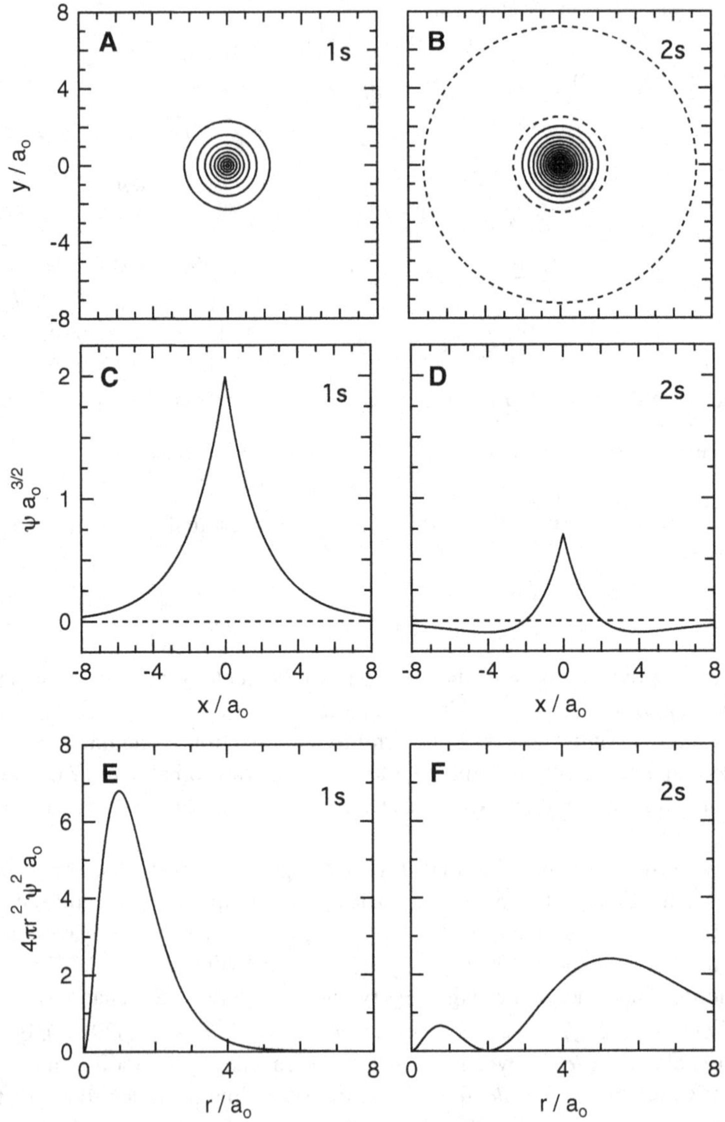

Fig. 2.4 The hydrogen-atom 1*s* and 2*s* orbitals. (**A**, **B**) Contour plots (lines of constant amplitude) of the wavefunctions in the plane of the nucleus, with *solid lines* for positive amplitudes and *dashed lines* for negative amplitudes. The Cartesian coordinates x and y are expressed as dimensionless multiples of the Bohr radius ($a_o = 0.529$ Å). The contour intervals for the amplitude are $0.2a_o^{2/3}$ in (**A**) and $0.05a_o^{2/3}$ in (**B**). (**C**, **D**) The amplitudes of the wavefunctions as functions of the x coordinate. (**E**, **F**) The radial distribution functions

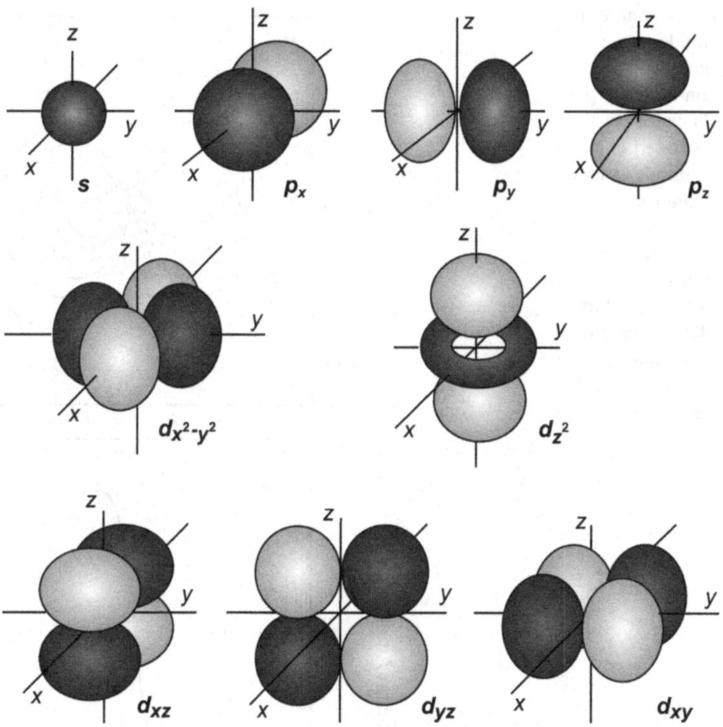

Fig. 2.5 Perspective drawings of the "boundary surfaces" (regions where an electron is most likely to be found) of the 1*s*, 2*p* and 3*d* wavefunctions of the hydrogen atom. The wavefunctions have a constant sign in the regions with *dark shading*, and the opposite sign in the regions with *light shading*. These drawings are only approximately to scale. Figure 4.30 shows similar drawings of the 4*d* wavefunctions

functions that are oriented along definite *x*- and *y*-axes and have no net rotational motion ($2p_x$ and $2p_y$). The $2p_x$ and $2p_y$ wavefunctions then are identical to $2p_z$ except for their orientations in space. This is essentially the same as constructing standing waves from wavefunctions for electrons moving in opposite directions, as described in Box 2.4. If we use the scaled coordinates $z = \rho\cos\theta$, $x = \rho\sin\theta\cos\phi$ and $y = \rho\sin\theta\sin\phi$, with $\rho = Zr/a_o$, the three 2*p* wavefunctions can be written:

$$2p_z = (Z/a_o)^{5/2}(32\pi)^{-1/2} z\exp(-\rho/2), \tag{2.35a}$$

$$2p_x = \frac{1}{\sqrt{2}}\{2p_- + 2p_+\} = (Z/a_o)^{5/2}(32\pi)^{-1/2} x\exp(-\rho/2), \tag{2.35b}$$

and

$$2p_y = \frac{i}{\sqrt{2}}\{2p_- - 2p_+\} = (Z/a_o)^{5/2}(32\pi)^{-1/2} y\exp(-\rho/2). \tag{2.35c}$$

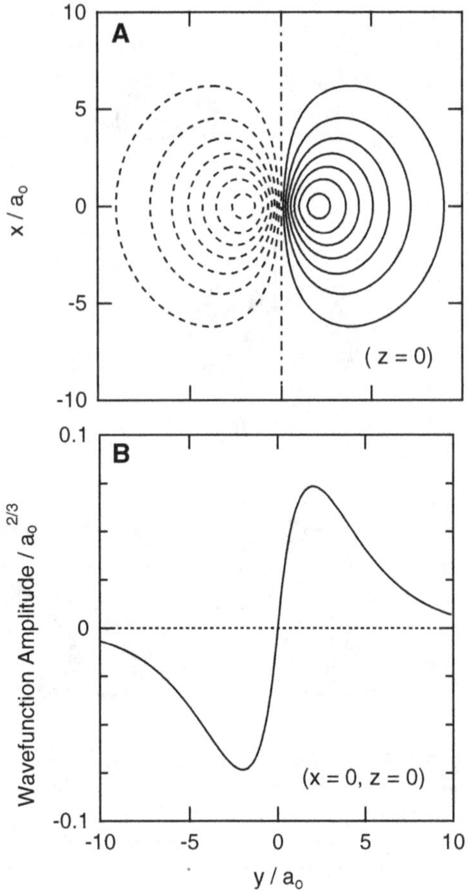

Fig. 2.6 (**A**) A contour plot of the amplitude of the hydrogen-atom $2p_y$ wavefunction in the *xy* plane, with *solid lines* for positive amplitudes, *dashed lines* for negative amplitudes and a *dot-dashed line* for zero. Distances are given as dimensionless multiples of the Bohr radius. The contour intervals for the amplitude are $0.01a_o^{2/3}$. (**B**) The amplitude of the $2p_y$ wavefunction in the *xy* plane as a function of position along the *y*-axis

The 3*d* orbitals with $m = \pm 1$ or ± 2 also are complex functions that can be combined to give a set of real wavefunctions. The boundary surfaces of the real wavefunctions are shown in Fig. 2.5.

2.3.5 Molecular Orbitals

The Schrödinger equation has not been solved exactly for electrons in molecules larger than the H_2^+ ion; the interactions of multiple electrons become too complex to handle. However, the eigenfunctions of the Hamiltonian operator provide a complete set of functions, and as mentioned in Sect. 2.2.1, a linear combination of such functions can be used to construct any well-behaved function of the same coordinates. This suggests the possibility of representing a molecular electronic wavefunction by a linear combination of hydrogen atomic orbitals centered at the nuclear positions. In principle, we should include the entire set of atomic orbitals for each nucleus, but smaller sets often provide useful approximations. For example, the π orbitals of a molecule with N conjugated atoms can be written as

$$\psi_k \approx \sum_{n=1}^{N} C_n^k \psi_{2z(n)}, \tag{2.36}$$

where $\psi_{2z(n)}$ is an atomic $2p_z$ orbital centered on atom n, and coefficient $C_{\mathrm{n}}^{\mathrm{k}}$ indicates the contribution that $\psi_{2z(n)}$ makes to molecular orbital k. If the atomic wave functions are orthogonal and normalized ($\langle \psi_{2z(n)} | \psi_{2z(n)} \rangle = 1$ for all n, and $\langle \psi_{2z(n)} | \psi_{2z(m)} \rangle = 0$ for $m \neq n$) the molecular wavefunctions can be normalized by scaling the coefficients so that

$$\sum_{n=1}^{N} |C_n^k|^2 = 1. \tag{2.37}$$

The highest occupied molecular orbital (HOMO) of ethylene, a bonding π orbital, can be approximated reasonably well by a symmetric combination of carbon $2p_z$ orbitals centered on carbon atoms 1 and 2 and oriented so that their z-axes are parallel:

$$\begin{aligned} \psi_\pi(\mathbf{r}) &\approx 2^{-1/2}\psi_{2pz(1)} + 2^{-1/2}\psi_{2pz(1)} \\ &= 2^{-1/2}\psi_{2pz}(\mathbf{r}-\mathbf{r}_1) + 2^{-1/2}\psi_{2pz}(\mathbf{r}-\mathbf{r}_2), \end{aligned} \tag{2.38}$$

where $\mathbf{r}_1$ and $\mathbf{r}_2$ denote the positions of the two carbon atoms (Fig. 2.7). The contributions from $\psi_{2z(1)}$ and $\psi_{2z(2)}$ combine constructively in the region between the two atoms, leading to a build-up of electron density in this region. The molecular orbital resembles the $2p_z$ atomic orbitals in having a node in the xy plane.

The lowest unoccupied molecular orbital (LUMO) of ethylene, an antibonding (π*) orbital, can be represented as a similar linear combination but with opposite sign:

$$\begin{aligned} \psi_\pi(\mathbf{r}) &\approx 2^{-1/2}\psi_{2pz(1)} - 2^{-1/2}\psi_{2pz(1)} \\ &= 2^{-1/2}\psi_{2pz}(\mathbf{r}-\mathbf{r}_1) - 2^{-1/2}\psi_{2pz}(\mathbf{r}-\mathbf{r}_2). \end{aligned} \tag{2.39}$$

In this case, $\psi_{2z(1)}$ and $\psi_{2z(2)}$ interfere destructively in the region between the two carbons, and the molecular wavefunction has a node in the xz plane as well as the xy plane (Fig. 2.7C). The combinations described by Eqs. (2.38) and (2.39) are called *symmetric* and *antisymmetric*, respectively.

Techniques for finding the best coefficients C_i^k for more complex molecules have been refined by comparing calculated molecular energies, dipole moments and other properties with experimental measurements, and a variety of software packages for such calculations are available [6, 23–30]. The *basis wavefunctions* employed in these descriptions usually are not the atomic orbitals obtained by solving the

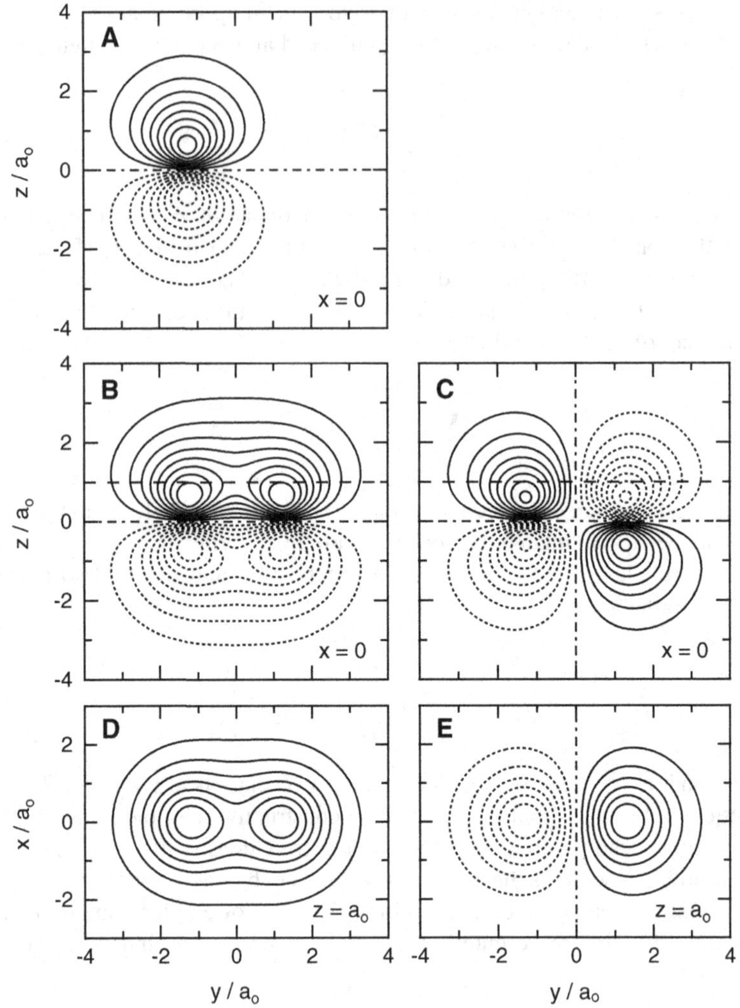

Fig. 2.7 Combining atomic 2*p* wavefunctions to form π and π^* molecular orbitals. (**A**) Contour plot of a $2p_z$ wavefunction of an individual carbon atom. Coordinates are given as multiples of the Bohr radius (a_o), and the carbon $2p_z$ wavefunction is represented as a Slater-type $2p_z$ orbital (Eq. 2.40) with $\zeta = 3.071/\text{Å}$ ($1.625/a_o$). The plot shows lines of constant amplitude in the *yz* plane, with *solid lines* for positive amplitudes, *dotted lines* for negative amplitudes, and a *dot-dashed line* for zero. (**B**) Contour plot of the amplitude in the *yz* plane for the bonding (π) molecular orbital formed by symmetric combination of $2p_z$ wavefunctions of two carbon atoms separated by $2.51a_o$ (1.33 Å) in the *y* direction. (**C**) Same as (**B**), but for the antibonding (π^*) molecular orbital created by antisymmetric combination of the atomic wavefunctions. (**D**, **E**) Same as (**B**) and (**C**), respectively, but showing the amplitudes of the wavefunctions in the plane parallel to the *xy* plane and a_o above *xy* (horizontal *dashed lines* at $z = a_o$ in **B**, **C**). A map of the amplitudes a_o below the plane of the ring would be the same except for an interchange of positive and negative signs. The contour intervals for the amplitude are $0.1a_o^{2/3}$ in all five panels

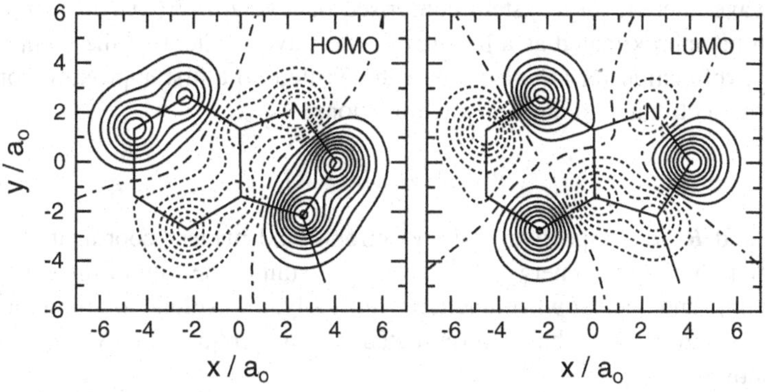

Fig. 2.8 Contour plots of the wavefunction amplitudes for the highest occupied molecular orbital (*HOMO*) and lowest unoccupied molecular orbital (*LUMO*) of 3-methylindole. Positive amplitudes are indicated by *solid lines*, negative amplitudes by *dotted lines*, and zero by *dot-dashed lines*. The plane of the map is parallel to the plane of the indole ring and is above the ring by a_o as in Fig. 2.7, panels D and E. The contour intervals for the amplitude are $0.05a_o^{3/2}$. Small contributions from the carbon and hydrogen atoms of the methyl group are neglected. The *straight black lines* indicate the carbon and nitrogen skeleton of the molecule. The atomic coefficients for the molecular orbitals were obtained as described by Callis [37–39]. Slater-type atomic orbitals (Eq. 2.40) with with $\zeta = 3.071/\text{Å}$ ($1.625/a_o$) and $3.685/\text{Å}$ ($1.949/a_o$) were used to represent C and N, respectively

Schrödinger equation for the hydrogen atom, but rather idealized wavefunctions with mathematical forms that are easier to manipulate. They include parameters that are adjusted semiempirically to model a particular type of atom and to adjust for overlap with neighboring atoms. A standard form is the *Slater-type orbital*,

$$\psi_{nlm} = Nr^{n^*-1}\exp\{-\zeta r\}Y_{l,m}, \tag{2.40}$$

where N is a normalization factor, n^* and ζ are parameters related to the principal quantum number (n) and the effective nuclear charge, and $Y_{l,m}$ is a spherical harmonic function of the polar coordinates θ and ϕ. The Slater $2p_z$ orbitals of carbon, nitrogen and oxygen, for example, take the form $(\zeta^5/\pi)^{1/2}r{\cdot}\cos(\theta)\ \exp(-\zeta r)$ with $\zeta = 3.071$, 3.685 and 4.299 Å^{-1}, respectively. Gaussian functions also are commonly used because, although they differ substantially in shape from the hydrogen-atom wavefunctions, they have particularly convenient mathematical properties. For example, the product of two Gaussians centered at positions $\boldsymbol{r}_1$ and $\boldsymbol{r}_2$ is another Gaussian centered midway between these points. Current programs use linear combinations of three or more Gaussians to replace each Slater-type orbital. See [31] for further information on these and other semiempirical orbitals.

Figure 2.8 shows contour plots of calculated amplitudes of the wavefunctions for the HOMO and LUMO of 3-methylindole, which is a good model of the sidechain of tryptophan. Note that the HOMO has two nodal curves in the plane of the drawing, while the LUMO has three; the LUMO thus has less bonding character.

The wavefunction for a system comprised of several *independent* components often can be approximated as a *product* of the wavefunctions of the components. Thus a wavefunction for a molecule can be written, to a first approximation, as a product of electronic and nuclear wavefunctions:

$$\Psi(\boldsymbol{r}, \boldsymbol{R}) \approx \psi(\boldsymbol{r})\chi(\boldsymbol{R}), \tag{2.41}$$

where $\boldsymbol{r}$ and $\boldsymbol{R}$ are, respectively, the electronic and nuclear coordinates. In this approximation, the total energy of the system is simply the sum of the energies of the electronic and nuclear wavefunctions. Similarly, a wavefunction for a molecule with N electrons can be approximated as a product of N one-electron wavefunctions.

2.3.6 Wavefunctions for Large Systems

Extending the ideas described in the previous section, linear combinations of molecular wavefunctions can be used to generate approximate wavefunctions for systems containing more than one molecule. For example, a wavefunction representing an excited state of an oligonucleotide can be described as a linear combination of wavefunctions for the excited states of the individual nucleotides.

The basis wavefunctions used in such constructions are said to be *diabatic*, which means that they are not eigenfunctions of the complete Hamiltonian; they do not consider all the intermolecular interactions that contribute to the energy of the actual system. A measurement of the energy must give one of the eigenvalues associated with the *adiabatic* wavefunctions of the full Hamiltonian, which usually will not be an eigenvalue of the Hamiltonian for any one of the basis functions. The energies of the full system are, however, given approximately by

$$E_k \approx \sum_{n=1}^{N} |C_n^k|^2 E_n, \tag{2.42}$$

where E_n is the energy of basis function ψ_n and $\mathrm{C_n^k}$ is the corresponding coefficient for state k of the larger system (Eq. 2.36). The accuracy of this approximation depends on the choice of the basis functions, the number of terms that are included in the sum, and the reliability of the coefficients.

Although the individual basis wavefunctions are not eigenfunctions of the full Hamiltonian, it is possible in principle to find linear combinations of these wavefunctions that do give such eigenfunctions, at least to the extent that the basis functions are a complete, orthonormal set. Equation (2.42) then becomes exact. The coefficients and eigenvalues are obtained by solving the simultaneous linear equations

$$\begin{aligned}
&C_1^k H_{11} + C_2^k H_{12} + C_3^k H_{13} + \cdots + C_1^k H_{1n} = C_1^k E_k \\
&C_2^k H_{21} + C_2^k H_{22} + C_2^k H_{23} + \cdots + C_2^k H_{2n} = C_2^k E_k \\
&C_3^k H_{31} + C_3^k H_{23} + C_3^k H_{33} + \cdots + C_3^k H_{3n} = C_3^k E_k \, . \\
&\vdots \\
&C_n^k H_{1n} + C_n^k H_{2n} + C_n^k H_{3n} + \cdots + C_n^k H_{nn} = C_n^k E_k
\end{aligned} \tag{2.43}$$

Here $H_{jk} = \left\langle \psi_j | \widetilde{\mathrm{H}} | \psi_k \right\rangle$, and $\widetilde{\mathrm{H}}$ is the complete Hamiltonian of the system, including terms that act on the basis functions for the individual molecules $\left(\widetilde{\mathrm{H}}_{kk}\right)$ and terms that "couple" or "mix" two basis functions ($\widetilde{\mathrm{H}}_{jk}$ with $j \neq k$). We'll go through the derivation of Eqs. (2.43) for a system of two molecules in Chap. 8

Equations (2.43) can be written compactly as a matrix equation by using the rules for matrix multiplication (Appendix 2):

$$\mathbf{H} \cdot C_k = E_k C_k, \tag{2.44}$$

where **H** denotes a matrix of the Hamiltonian integrals (H_{jk}),

$$\mathbf{H} = \begin{bmatrix} H_{11} & H_{12} & \cdots & H_{1n} \\ H_{21} & H_{22} & \cdots & H_{2n} \\ & \vdots & & \\ H_{n1} & H_{n2} & \cdots & H_{nn} \end{bmatrix}, \tag{2.45a}$$

and $\boldsymbol{C}_k$ is a column vector of coefficients:

$$C_k = \begin{bmatrix} C_1^k \\ C_2^k \\ \vdots \\ C_n^k \end{bmatrix}. \tag{2.45b}$$

In general, there will be n eigenvalues of the energy (E_k) that satisfy Eq. (2.44), each with its own eigenvector $\boldsymbol{C}_k$. The eigenvectors can be arranged in a square matrix **C**, in which each column corresponds to a particular eigenvalue:

$$\mathbf{C} = \begin{bmatrix} C_1^1 & C_1^2 & \cdots & C_1^n \\ C_2^1 & C_2^2 & \cdots & C_{12}^n \\ & \vdots & & \\ C_n^1 & C_n^2 & \cdots & C_n^n \end{bmatrix}. \tag{2.46a}$$

If we then find $\mathbf{C}^{-1}$ (the inverse of **C**), the product $\mathbf{C}^{-1} \cdot \mathbf{H} \cdot \mathbf{C}$ turns out to be a diagonal matrix with the eigenvalues on the diagonal (see Appendix 2):

$$\mathbf{C}^{-1} \cdot \mathbf{H} \cdot \mathbf{C} = \begin{bmatrix} E_1 & 0 & \cdots & 0 \\ 0 & E_2 & \cdots & 0 \\ & \vdots & & \\ 0 & 0 & \cdots & E_n \end{bmatrix}. \tag{2.46b}$$

The problem of finding the eigenvalues and eigenfunctions therefore is to *diagonalize* the Hamiltonian matrix **H** by finding another matrix **C** and its inverse such that the product $\mathbf{C}^{-1}{\cdot}\mathbf{H}{\cdot}\mathbf{C}$ is diagonal. The diagonal elements of $\mathbf{C}^{-1}{\cdot}\mathbf{H}{\cdot}\mathbf{C}$ are the eigenvalues of the adiabatic states, and column k of **C** is the set of coefficients corresponding to eigenvalue E_k.

The Hamiltonian matrix is always Hermitian, and for all the cases that will concern us is symmetric (Appendix A.2). Its eigenvectors ($\boldsymbol{C}_k$) are, therefore, always real. In addition, there is always an orthonormal set of eigenvectors ($\boldsymbol{C}_i{\cdot}\boldsymbol{C}_j = 0$ for $i \neq j$, and $\boldsymbol{C}_i{\cdot}\boldsymbol{C}_i = 1$) [32].

2.4 Spin Wavefunctions and Singlet and Triplet States

Electrons, protons and other nuclei have an intrinsic angular momentum or "spin" that is characterized by two *spin quantum numbers*, s and m_s. The magnitude of the angular momentum is $[s + (s + 1)]^{1/2}\hbar$. For an individual electron, $s = 1/2$, so the angular momentum has a magnitude of $(3^{1/2}/2)\hbar$. The component of the angular momentum parallel to a single prescribed axis (z), also is quantized and is given by $m_s\hbar$, where $m_s = s,\ s - 1,\ \ldots,\ -s$. For $s = 1/2$, m_s is limited to $\pm 1/2$, making the momentum along the z axis $\pm\hbar/2$. The magnitude of the spin and the component of the spin in this direction do not commute with the components of the momentum in orthogonal directions, so these components are indeterminate. The angular momentum vector for an electron therefore can lie anywhere on a cone whose half-angle (θ) with respect to the z axis is given by $\cos(\theta) = m_s/(s(s+1))^{1/2}$ ($\theta \approx 54.7°$; Fig. 2.9).

The two possible values of m_s for an electron can be described formally by two *spin wavefunctions*, α and β, which we can view as "spin up" ($m_s = +1/2$) and "spin down" ($m_s = -1/2$), respectively. These functions are orthogonal and normalized so that $\langle\alpha|\alpha\rangle = \langle\beta|\beta\rangle = 1$ and $\langle\alpha|\beta\rangle = \langle\beta|\alpha\rangle = 0$. The variable of integration here is not a spatial coordinate, but rather a spin variable that represents the orientation of the angular momentum.

The notion of electronic spin was first proposed by Uhlenbeck and Goudsmit in 1925 to account for the splitting of some of the lines seen in atomic spectra. They and others showed that the electronic spin also accounted for the anomalous effects of magnetic fields (Zeeman effects) on the spectra of many atoms. However, it was necessary to postulate that the magnetic moment associated with electronic spin is not simply the product of the angular momentum and $e/2mc$, as is true of orbital magnetic moments, but rather twice this value. The extra factor of 2 is called the *Landé g factor*. When Dirac [33] reformulated quantum mechanics to be consistent

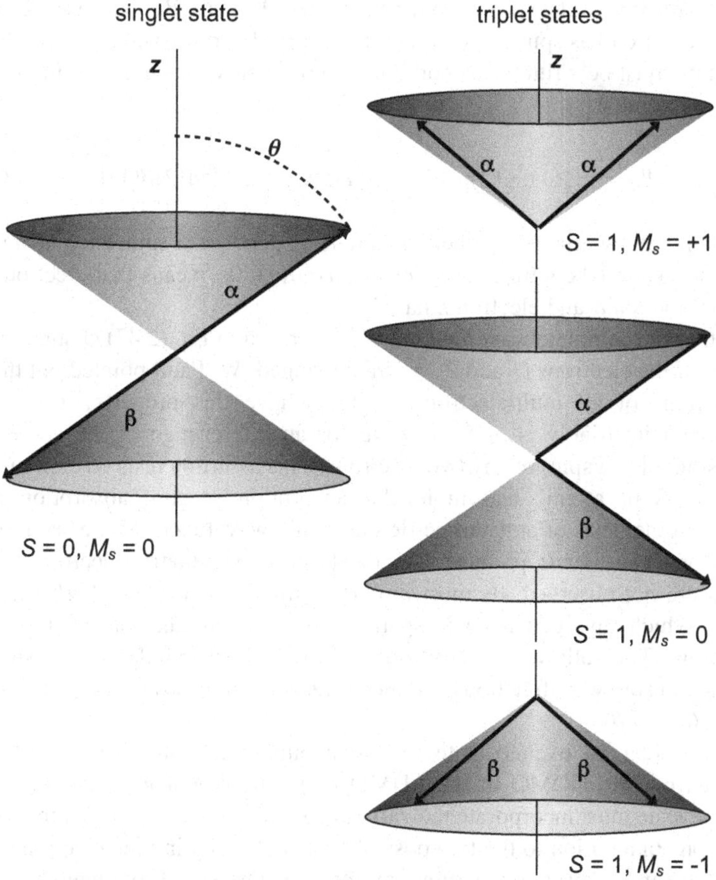

Fig. 2.9 A vectorial representation of the electronic spins in the four possible spin states of a system with two coupled electrons. Each *arrow* represents a spin with magnitude $3^{1/2}\hbar/2$ constrained to the surface of a cone with half-angle $\theta = \cos^{-1}(\pm 3^{-1/2})$ relative to the z axis (54.7° for spin α and 125.3° for β). In the singlet state (*left*), the two vectors are antiparallel, giving a total spin quantum number (S) of zero. In the triplet states (*right*) $S = 1$ and $M_s = 1, 0$ or -1. This requires the two individual spin vectors to be arranged so that their resultant (not shown) has magnitude $2^{1/2}\hbar$ and lies on a cone with half-angle 45°, 90° or 135° from z for $M_s = 1, 0$ or -1, respectively

with special relativity, both the intrinsic angular momentum and the anomalous factor of 2 emerged automatically without any *ad hoc* postulates. Dirac shared the Physics Nobel Prize with Schrödinger in 1933.

In a system of two interacting electrons, the total spin (S) is quantized and can be either 1 or 0 depending on whether the individual spins are the same (e.g., both β) or different (one α, the other β). The component of the total spin along a prescribed axis is $M_s\hbar$ with $M_s = S, S-1, \ldots, -S$, i.e., 0 if $S = 0$ and either 1, 0 or -1 if $S = 1$. For most organic molecules in their ground state, the HOMO contains two electrons with different antiparallel spins, making both S and M_s zero. But because we cannot tell which electron has spin α and which has β, the electronic wavefunction for the ground state must be written as a combination of expressions representing the two possible assignments:

$$\Psi_a = [\psi_h(1)\psi_h(2)]\left[2^{-1/2}\alpha(1)\beta(2) - 2^{-1/2}\alpha(2)\beta(1)\right]. \tag{2.47}$$

Here ψ_h denotes a spatial wavefunction that is independent of spin, and the numbers in parentheses are labels for the two electrons; $\alpha(j)\beta(k)$ means that electron j has spin wavefunction α and electron k has β.

Note that the complete wavefunction Ψ_a as written in Eq. (2.47) changes sign if the labels of the electrons (1 and 2) are interchanged. W. Pauli pointed out that the wavefunctions of all multielectronic systems have this property. The overall wavefunction invariably is *antisymmetric* for an interchange of the coordinates (both positional and spin) of any two electrons. This assertion rests on experimental measurements of atomic and molecular absorption spectra: absorption bands predicted on the basis of antisymmetric electronic wavefunctions are seen experimentally, whereas bands predicted on the basis of symmetric electronic wavefunctions are not observed. Its most important implication is the *Pauli exclusion principle*, which says that a given spatial wavefunction can hold no more than two electrons. This follows if an electron can be described completely by specifying its spatial and spin wavefunctions and electrons have only two possible spin wavefunctions (α and β).

Now consider an excited state in which either electron 1 or electron 2 is promoted from the HOMO to the LUMO (ψ_l). The complete wavefunction for the excited state must incorporate the various possible assignments of spin α or β to the electrons in addition to the two possible ways of assigning the electrons to the two orbitals, and again the wavefunction must change sign if we interchange the coordinates of the electrons. If there is no change in the net spin of the system during the excitation, as is usually the case, the spin part of the wavefunction remains the same as for the ground state so that both S and M_s remain zero. The spatial part is more complicated:

$${}^1\Psi_b = \left[2^{-1/2}\psi_h(1)\psi_l(2) + 2^{-1/2}\psi_h(2)\psi_l(1)\right]\left[2^{-1/2}\alpha(1)\beta(2) - 2^{-1/2}\alpha(2)\beta(1)\right]. \tag{2.48}$$

The choice of a + sign for the combination of the spatial wavefunctions in the first brackets satisfies the requirement that the overall wavefunction be antisymmetric for an exchange of the two electrons. If antisymmetric combinations were used in both brackets, the overall product of the spatial and spin wavefunctions would be symmetric, in conflict with experiment. The state described by such a wavefunction is referred to as a *singlet* state because when the spatial wavefunction is symmetric there is only one possible combination of spin wavefunctions (the one written in

Eq. 2.48). The ground state (Eq. 2.47) also is a singlet state. Singlet states often are indicated by a superscript "1" as in Eq. (2.48).

If, instead, we choose the antisymmetric combination for the spatial part of the wavefunction in the excited state, there are three possible symmetric combinations of spin wavefunctions that make the overall wavefunction antisymmetric:

$$ {}^{3}\Psi_b^{+1} = \left[2^{-1/2}\psi_h(1)\psi_l(2) - 2^{-1/2}\psi_h(2)\psi_l(1)\right]\ [\alpha(1)\alpha(2)] \tag{2.49a} $$

$$ {}^{3}\Psi_b^{0} = \left[2^{-1/2}\psi_h(1)\psi_l(2) - 2^{-1/2}\psi_h(2)\psi_l(1)\right]\left[2^{-1/2}\alpha(1)\beta(2) + 2^{-1/2}\alpha(2)\beta(1)\right] \tag{2.49b} $$

$$ {}^{3}\Psi_b^{-1} = \left[2^{-1/2}\psi_h(1)\psi_l(2) - 2^{-1/2}\psi_h(2)\psi_l(1)\right]\ [\beta(1)\beta(2)]. \tag{2.49c} $$

These are the three excited *triplet* states corresponding to $S = 1$ and $M_s = 1$, 0 and -1, respectively.

Figure 2.9 shows a vectorial representation of the angular momenta of the two electrons in the singlet and triplet states when the vertical (z) axis is defined by a magnetic field. The vectors are drawn to satisfy the quantization of the spin and the z-component of the spin simultaneously for both the individual electrons and the combined system. Although the x and y components of the individual spins are still indeterminate, the angle between the two spins is fixed. In the triplet state with $M_s = 0$, the resultant vector representing the total spin is in the xy plane; in the states with $M_s = \pm 1$, it lies on a cone with a half-angle of 45° or 135° with respect to the z axis.

The triplet state ${}^{3}\Psi_b^{0}$ invariably has a lower energy than the corresponding singlet state, ${}^{1}\Psi_b$. This statement, which is often called *Hundt's rule*, is a consequence of the different spatial wavefunctions, not the spin wavefunctions. The motions of the two electrons are correlated in a way that tends to keep the electrons farther apart in the triplet wavefunction, decreasing their repulsive interactions. The difference between the two energies, called the *singlet-triplet splitting*, is given by $2K_{hl}$, where K_{hl} is the *exchange integral*:

$$ K_{hl} = \left\langle \psi_l(1)\psi_h(2)\left|\frac{e^2}{r_{12}}\right|\psi_h(1)\psi_l(2)\right\rangle. \tag{2.50} $$

K_{hl} is always positive [34].

A special situation arises when the two orbitals are equivalent in the sense that they have the same energy and can be transformed into each other by a symmetry operation such as rotation. The triplet state then can be lower in energy than a singlet state in which both electrons reside in one of the orbitals. This is the case for O_2, for which the ground state is a triplet and the lowest excited state is a singlet.

The z axis that defines the orientation of an electronic spin often is determined uniquely by an external magnetic field. Because of their different magnetic moments, the three triplet states split apart in energy in the presence of a magnetic field, with ${}^3\Psi_b^{+1}$ moving up in energy and ${}^3\Psi_b^{-1}$ moving down; ${}^3\Psi_b^0$ is not affected. Although the triplet states are degenerate in the absence of a magnetic field, orbital motions of the electrons can create local magnetic fields that lift this degeneracy. This *zero-field splitting* can be measured by imposing an oscillating microwave field to induce transitions between the triplet states. The magnetic dipoles of the three zero-field triplet states generally can be associated with x, y and z structural axes of the molecule.

Wavefunctions for more than two electrons can be approximated as linear combinations of all the allowable products of electronic and spin wavefunctions for the individual electrons. Combinations that satisfy the Pauli exclusion principle can be written conveniently as determinants:

$$\Psi = (1/N!)^{1/2} \begin{vmatrix} \psi_a(1)\alpha(1) & \psi_a(2)\alpha(2) & \psi_a(3)\alpha(3) & \cdots & \psi_a(N)\alpha(N) \\ \psi_a(1)\beta(1) & \psi_a(2)\beta(2) & \psi_a(3)\beta(3) & \cdots & \psi_a(N)\beta(N) \\ \psi_b(1)\alpha(1) & \psi_b(2)\alpha(2) & \psi_b(3)\alpha(3) & \cdots & \psi_b(N)\alpha(N) \\ \psi_b(1)\beta(1) & \psi_b(2)\beta(2) & \psi_b(3)\beta(3) & \cdots & \psi_b(N)\beta(N) \\ \vdots & \vdots & \vdots & & \vdots \\ \psi_N(1)\beta(1) & \psi_N(2)\beta(2) & \psi_N(3)\beta(3) & \cdots & \psi_N(N)\beta(N) \end{vmatrix}, \tag{2.51}$$

where the ψ_i are the individual one-electron spatial wavefunctions and N is the total number of electrons. Equation (2.47), for two electrons and a single spatial wavefunction, consists of the 2×2 block at the top left corner of Eq. (2.51). This formulation guarantees that the overall wavefunction will be antisymmetric for interchange of any two electrons, because the value of a determinant always changes sign if any two columns or rows are interchanged. Such determinants are called *Slater determinants* after J. C. Slater, who developed the procedure. They often are denoted compactly by omitting the normalization factor $(1/N!)^{1/2}$, listing only the diagonal terms of the determinant, representing the combinations $\psi_j(k)\alpha(k)$ and $\psi_j(k)\beta(k)$ by $\psi_j(k)$ and $\overline{\psi}_j(k)$, respectively, and omitting the indices for the electrons:

$$\Psi = \left|\psi_a\overline{\psi_a}\psi_b\overline{\psi_b}\cdots\overline{\psi_N}\right|. \tag{2.52}$$

In the nomenclature of Eq. (2.48), the Slater determinant for the HOMO and LUMO "spin-orbital" wavefunctions of the ground, excited singlet and excited triplet states (Eqs. 2.47–2.49) would be written:

$$\Psi_a = \left|\psi_h\overline{\psi_h}\right|, \tag{2.53a}$$

$$^1\Psi_b = 2^{-1/2}\left(\left|\psi_h\overline{\psi_l}\right| + \left|\psi_l\overline{\psi_h}\right|\right), \tag{2.53b}$$

$$^3\Psi_b^{+1} = 2^{-1/2}\left(\left|\psi_h\psi_l\right| - \left|\psi_l\psi_h\right|\right), \tag{2.53c}$$

$$^3\Psi_b^{0} = 2^{-1/2}\left(\left|\psi_h\overline{\psi_l}\right| - \left|\psi_l\overline{\psi_h}\right|\right), \tag{2.53d}$$

and

$$^3\Psi_b^{-1} = 2^{-1/2}\left(\left|\overline{\psi_h}\ \overline{\psi_l}\right| - \left|\overline{\psi_l}\ \overline{\psi_h}\right|\right). \tag{2.53e}$$

Transitions between singlet and triplet states will be discussed in Sect. 4.9.

Protons resemble electrons in having spin quantum numbers $s = 1/2$ and $m_s = \pm 1/2$. Wavefunctions of systems containing multiple protons, or any other particles with half-integer spin (*fermions*), also resemble electronic wavefunctions in being antisymmetric for interchange of any two identical particles. Wavefunctions for systems containing multiple particles with integer or zero spin (*bosons*), by contrast, must be symmetric for interchange of two identical particles. Deuterons, which have spin quantum numbers $s = 1$ and $m_s = -1, 0$ of 1, fall in this second group. Because interchanging bosons leaves their combined wavefunction unchanged, any number of bosons can occupy the same wavefunction. Fermions and bosons obey different statistics that become increasingly distinct at low temperatures (Box 2.6).

Box 2.6. Boltzmann, Fermi-Dirac and Bose-Einstein Statistics

As Eqs. (2.47–2.52) illustrate, wavefunctions of a system comprised of multiple non-interacting components can be written as linear combinations of products of the form

$$\Psi = \Psi_a(1)\Psi_b(2)\Psi_c(3)\cdots. \tag{B2.6.1}$$

The *Boltzmann distribution law* says that, if the individual components are distinguishable and the number of components is very large, the probability of finding a given component in a state with wavefunction Ψ_m at temperature T is

$$P_m = Z^{-1} g_m \exp[-E_m/k_B T]. \tag{B2.6.2}$$

Here E_m is the energy of state m, g_m is the *degeneracy*, or *multiplicity*, of the state (the number of substates with the same energy), k_B is the Boltzmann constant (1.3709×10^{-16} erg), and Z (the *partition function*) is a temperature-dependent factor that normalizes the sum of the probabilities over all the states:

(continued)

Box 2.6 (continued)

$$Z = \sum_{m} g_m \exp[-E_m/k_B T]. \tag{B2.6.3}$$

Electronic states with the same spatial wavefunction but different spins (α or β) can be described as having a multiplicity (g_m) of 2 if the energy difference between the two spin sublevels is negligible. Alternatively, the spin states can be enumerated separately, each with $g_m = 1$.

Writing combined wavefunctions of the form of Eq. (B2.6.1) becomes problematic if the individual components of the system are indistinguishable. As we have discussed, wavefunctions that are symmetric for interchange of two identical particles are not available to fermions, while wavefunctions that are antisymmetric for such interchanges are not available to bosons. One consequence of this is that systems of fermions or bosons follow different distribution laws. Fermions obey the *Fermi-Dirac distribution*,

$$P_m = \{A_F g_m \exp[E_m/k_B T] + N\}^{-1}, \tag{B2.6.4}$$

where N is the number of particles and A_F is defined to make the total probability 1. Bosons follow the *Bose-Einstein distribution*,

$$P_m = \{A_B g_m \exp[E_m/k_B T] - N\}^{-1}, \tag{B2.6.5}$$

with A_B again defined to give a total probability of 1. The Boltzmann, Fermi-Dirac and Bose-Einstein distribution laws can be derived by defining the entropy of a distribution as $S = k_B \ln \Omega$, where Ω is the number of different ways that the particles could be assigned to various states, and then maximizing the entropy while keeping the total energy of the system constant. The formulas for Ω depend on whether the particles are distinguishable and on whether or not more than one particle can occupy the same state (Sect. 4.15 and Fig. 4.31).

Figure 2.10 illustrates Boltzmann, Fermi-Dirac and Bose-Einstein distributions of four particles among five states with equally spaced energies. Each state is assumed to have a multiplicity of 2. The populations of the five states are plotted as functions of $k_B T/E$, where E is the energy difference between adjacent states. The populations all converge on 0.2 at high temperature, where a given particle has nearly equal probabilities of being in any of the five states. However, the three distributions differ notably at low temperature. Here the Fermi-Dirac distribution puts two particles in each of the states with

(continued)

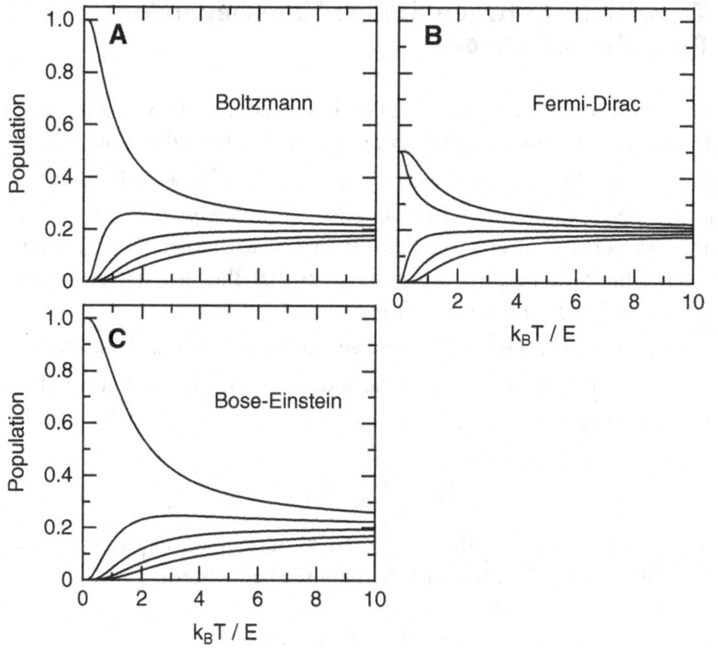

Fig. 2.10 The Boltzmann (**A**), Fermi-Dirac (**B**) and Bose-Einstein (**C**) distributions of four particles among five states with energies of 0, E, $2E$, $3E$, and $4E$ in arbitrary energy units. Each state is assumed to have a multiplicity of 2. The populations of the five states are plotted as functions of k_BT/E. In each panel, the *upper-most curve* is the population in the state with the lowest energy, and the *bottom curve* is for the state with the highest energy

Box 2.6 (continued)

the two lowest energies, making the probability of finding a given particle in a given one of these states 0.5 (Fig. 2.10B). This corresponds to the Pauli exclusion principle that a spatial wavefunction can hold no more than two electrons, which must have different spins. The Boltzmann and Bose-Einstein distributions, in contrast, both put all four particles in the state with the lowest energy (Fig. 2.10A, C). The Bose-Einstein distribution differs from the Boltzmann distribution in changing more gradually with temperature.

The Bose–Einstein distribution law was derived by S.N. Bose in 1924 to describe a photon gas. Einstein extended it to material gasses. Fermi developed the Fermi-Dirac distribution law in 1926 by exploring the Pauli exclusion principle, and Dirac obtained it independently in the same year by considering antisymmetric wavefunctions.

2.5 Transitions Between States: Time-Dependent Perturbation Theory

The wavefunctions obtained by solving the time-independent Schrödinger equation (Eq. 2.14) describe *stationary states*. A system that is placed in one of these states evidently will stay there forever as long as $\widetilde{\mathrm{H}} = \widetilde{\mathrm{H}}_\mathrm{o}$, because the energy of the system is independent of time. But suppose $\widetilde{\mathrm{H}}$ changes with time. We could, for example, switch on an electric field or bring two molecules together so that they interact. This will perturb the system so that the original solutions to the Schrödinger equation are no longer entirely valid.

If the change in $\widetilde{\mathrm{H}}$ is relatively small, we can write the total Hamiltonian of the perturbed system as a sum of the time-independent $\widetilde{\mathrm{H}}_\mathrm{o}$ and a smaller, time-dependent term, $\widetilde{\mathrm{H}}'(t)$:

$$\widetilde{\mathrm{H}} = \widetilde{\mathrm{H}}_\mathrm{o} + \widetilde{\mathrm{H}}'(t). \tag{2.54}$$

To find the wavefunction of the perturbed system, let's express it as a linear combination of the eigenfunctions of the unperturbed system:

$$\Psi = C_a\Psi_a + C_b\Psi_b + \cdots . \tag{2.55}$$

where the coefficients C_k are functions of time. The value of $|C_k|^2$ at a given time represents the extent to which Ψ resembles the wavefunction of basis state k (Ψ_k). This approach makes use of the fact that the original eigenfunctions form a complete set of functions, as discussed in Sect. 2.2.1.

Suppose we know that the molecule is in state Ψ_a before we introduce the perturbation $\widetilde{\mathrm{H}}'$. How rapidly does the wavefunction begin to resemble that of some other basis state, say Ψ_b? The answer should lie in the time-dependent Schrödinger equation (Eq. 2.9). Using Eqs. (2.54) and (2.55), we can expand the left-hand side of the Schrödinger equation to:

$$\begin{aligned}\left[\widetilde{\mathrm{H}}_o + \widetilde{\mathrm{H}}'(t)\right][C_a(t)\Psi_a + C_b(t)\Psi_b + \cdots] &= \widetilde{\mathrm{H}}_o[C_a\Psi_a + C_b\Psi_b + \cdots] + \widetilde{\mathrm{H}}'[C_a\Psi_a + C_b\Psi_b + \cdots]\\ &= C_a\widetilde{\mathrm{H}}_o\Psi_a + C_b\widetilde{\mathrm{H}}_o\Psi_b + \cdots + C_a\widetilde{\mathrm{H}}'\Psi_a + C_b\widetilde{\mathrm{H}}'\Psi_b + \cdots .\end{aligned} \tag{2.56}$$

The right-hand side of the Schrödinger equation can be expanded similarly to

$$i\hbar[\Psi_a\partial C_a/\partial t + \Psi_b\partial C_b/\partial t + \cdots + C_a\partial\Psi_a/\partial t + C_b\partial\Psi_b/\partial t + \cdots]. \tag{2.57}$$

We have assumed that $\widetilde{\mathrm{H}}C_k\Psi = C_k\widetilde{\mathrm{H}}\Psi$, which means that the operator $\widetilde{\mathrm{H}}$ and multiplication by C_k commute; the result of performing the two operations is independent of the order in which they are performed. As discussed in Box 2.2, this assumption must be valid if the energy of the system and the values of the coefficients can be known simultaneously.

For the unperturbed system, we know that $\widetilde{\mathrm{H}}_o\Psi_a = i\hbar\partial\Psi_a/\partial t$ and $\widetilde{\mathrm{H}}_o\Psi_b = i\hbar\partial\Psi_b/\partial t$, because each eigenfunction satisfied the Schrödinger equation before we changed $\widetilde{\mathrm{H}}$. Canceling the corresponding terms on opposite sides of the Schrödinger equation (e.g., subtracting $C_a\widetilde{\mathrm{H}}_o\Psi_a$ from Eq. (2.56) and $C_a i\hbar\partial\Psi_a/\partial t$ from Eq. (2.57)) leaves:

$$C_a\widetilde{\mathrm{H}}'\Psi_a + C_b\mathrm{H}'\Psi_b + \cdots = i\hbar[\Psi_a\partial C_a/\partial t + \Psi_b\partial C_b/\partial t + \cdots]. \quad (2.58)$$

We can simplify this equation by multiplying each term by $\Psi_b{}^*$ and integrating over all space, because this allows us to use the orthogonality relationships (Eqs. (2.19–2.20)) to set many of the integrals to 0 or 1:

$$C_a\left\langle\Psi_b\middle|\widetilde{\mathrm{H}}'\middle|\Psi_a\right\rangle + C_b\left\langle\Psi_b\middle|\widetilde{\mathrm{H}}'\middle|\Psi_b\right\rangle + \cdots = i\hbar\left[\langle\Psi_b|\Psi_a\rangle\partial C_a/\partial t + \langle\Psi_b|\Psi_b\rangle\partial C_b/\partial t + \cdots\right]$$
$$= i\hbar\partial C_b/\partial t. \quad (2.59)$$

If we know that the system is in state a at a particular time, then C_a must be 1, and C_b and all the other coefficients must be zero. So all but one of the terms on the left-hand side of Eq. (2.59) drop out. This leaves us with $\left\langle\Psi_b\middle|\widetilde{\mathrm{H}}'\middle|\Psi_a\right\rangle = i\hbar\partial C_b/\partial t$, or

$$\partial C_b/\partial t = (1/i\hbar)\left\langle\Psi_b\middle|\widetilde{\mathrm{H}}'\middle|\Psi_a\right\rangle = (-i/\hbar)\left\langle\Psi_b\middle|\widetilde{\mathrm{H}}'\middle|\Psi_a\right\rangle. \quad (2.60)$$

Equation (2.56) tells us how coefficient C_b increases with time at early times when there is still a high probability that the system is still in state a. But the wavefunctions Ψ_a and $\Psi_b{}^*$ in the equation are themselves functions of time. From the general solution to the time-dependent Schrödinger equation (Eq. 2.16) we can separate the spatial and time-dependent parts of these wavefunctions as follows:

$$\Psi_a = \psi_a(\mathbf{r})\ \exp(-iE_at/\hbar) \quad (2.61a)$$

and

$$\Psi_b{}^* = \psi_b{}^*(\mathbf{r})\ \exp(iE_bt/\hbar). \quad (2.61b)$$

Inserting these relationships into Eq. (2.60) yields the following result for the growth of C_b with time:

$$\partial C_b/\partial t = -(i/\hbar)\exp(iE_bt/\hbar)\exp(-iE_at/\hbar)\left\langle\psi_b\middle|\widetilde{\mathrm{H}}'\middle|\psi_a\right\rangle$$
$$= -(i/\hbar)\exp[i(E_b - E_a)t/\hbar]\left\langle\psi_b\middle|\widetilde{\mathrm{H}}'\middle|\psi_a\right\rangle = -(i/\hbar)\exp[i(E_b - E_a)t/\hbar]H'_{ba}. \quad (2.62)$$

Equation (2.62) factors $\partial C_b/\partial t$ into an oscillatory component that hinges on the difference between the energies of states a and b ($\exp[i(E_b - E_a)t/\hbar]$) and an integral that depends on the time-dependent perturbation $\left(\tilde{\mathrm{H}}'\right)$ and the spatial wavefunctions ψ_a and ψ_b. The integral $\left\langle \psi_b \middle| \tilde{\mathrm{H}}' \middle| \psi_a \right\rangle$, or H'_{ba}, is called a *matrix element* of $\tilde{\mathrm{H}}'$. The terminology is the same as that used in connection with Eq. (2.45a), except that here $\tilde{\mathrm{H}}'$ is just the perturbation term in the Hamiltonian rather than the complete Hamiltonian. A similar matrix element, $H'_{kj} = \left\langle \psi_k \middle| \tilde{\mathrm{H}}' \middle| \psi_j \right\rangle$, could be written to describe the build up of the coefficient C_k for any other state of the system.

To obtain the value of C_b after a short interval of time, τ, we need to integrate Eq. (2.62) from 0 to τ:

$$C_b(\tau) = \int_0^\tau (\partial C_b/\partial t)dt. \tag{2.63}$$

During the time that the perturbation is applied, the system cannot be said to be in either state a or state b because these are no longer eigenstates of the Hamiltonian. So what physical interpretation should we place on the coefficients C_a and C_b for these states in Eq. (2.56)? Suppose that at time τ we perform a measurement that has different expectation values for states a and b in the unperturbed system. If the operator corresponding to the measurement is $\tilde{\mathrm{A}}$, the expectation value for observations on the perturbed system will be

$$\begin{aligned} A = \left\langle \Psi \middle| \tilde{\mathrm{A}} \middle| \Psi \right\rangle &= \left\langle \sum_j C_j \psi_j \middle| \tilde{\mathrm{A}} \middle| \sum_k C_k \psi_k \right\rangle = \sum_j \sum_k C_j C_k \left\langle \psi_j \middle| \tilde{\mathrm{A}} \middle| \psi_k \right\rangle \\ &= \sum_j \sum_k C_j C_k \langle \psi_j | A_k \psi_k \rangle = \sum_j \sum_k C_j C_k A_k \langle \psi_j | \psi_k \rangle = \sum_k |C_k|^2 A_k, \end{aligned} \tag{2.64}$$

where A_k is the expectation value for observations on an unperturbed system in state k. Equation (2.64) is a generalization of Eq. (2.46a), which gives the expectation value of the energy. The magnitude of $|C_b(\tau)|^2$ thus tells us the extent to which an arbitrary measurement on the perturbed system at time τ will resemble A_b, the result of the same measurement on a system known to be in state b.

Now suppose we turn off the perturbation abruptly after we make a measurement on the perturbed system. With the perturbation removed, the basis states again become eigenstates of the system, and evolution of the coefficients C_a and C_b comes to a halt. It thus seems reasonable to view $|C_b(\tau)|^2$ as the statistical probability that the system has evolved into state b at time τ. If this probability increases linearly with time for small values of τ so that $|C_b(\tau)|^2 = \kappa\tau$ where κ is a constant, we can identify κ as the *rate constant* for the transition.

To evaluate $\partial C_b/\partial t$ for any particular case, we need to specify the time-dependence of $\widetilde{\mathrm{H}}'$ more explicitly. We'll do this in Chap. 4 for an oscillating electromagnetic field. In Chaps. 5–8 we'll consider several types of perturbations introduced by bringing two molecules together so that they interact, and in Chaps. 10 and 11 we'll consider the effects of randomly fluctuating interactions between a system and its surroundings. But even without going into the nature of $\widetilde{\mathrm{H}}'$, we can see that if the energies of the reactant and product states (E_a and E_b) are very different, the factor $\exp[i(E_b - E_a)t/\hbar]$ in Eq. (2.62) will oscillate rapidly with time and will average to zero. On the other hand, if the two energies are the same so that the argument of the exponent is zero, and if H'_{ba} is constant, C_b will increase linearly with time. Transitions from state a to state b therefore occur at a significant rate only if E_b is the same as or close to E_a. This is the condition for resonance between the two states, and is the quantum mechanical expression of the classical principle that the transition must conserve energy. For transitions that involve absorption or emission of light, the energy of one of the states includes the energy of the photon that is absorbed or emitted.

The general conclusion that transitions from state a to state b depend on the matrix element $\left\langle \psi_b \middle| \widetilde{\mathrm{H}}' \middle| \psi_a \right\rangle$ merits a few additional comments. Note that H'_{ba} is an off-diagonal matrix element of the time-dependent perturbation to the Hamiltonian. In Sect. 2.3.6, we discussed how linear combinations of basis wavefunctions can be used to construct wavefunctions for more complex systems. Finding the sets of coefficients that give eigenfunctions of the complete Hamiltonian, and so give stationary states of the system, requires diagonalizing the Hamiltonian. A time-dependent perturbation thus can drive transitions between diabatic states a and b, which are non-stationary in the presence of the perturbation, but it cannot drive transitions between the linear combinations of these states that make the Hamiltonian diagonal.

2.6 Lifetimes of States and the Uncertainty Principle

As discussed in Sect. 2.1.2, the position and momentum operators do not commute; the combined action of the two operators gives different results, depending on which operator is used first:

$$[\widetilde{\mathrm{r}}, \widetilde{\mathrm{p}}]\psi = \widetilde{\mathrm{r}}\widetilde{\mathrm{p}}\psi - \widetilde{\mathrm{p}}\widetilde{\mathrm{r}}\psi = i\hbar\psi. \tag{2.65}$$

With some algebra, it follows from Eq. (2.65) that the product of the uncertainties (root-mean-square deviations) in the expectation values for position and momentum must be $\geq \hbar/2$ [4]. This is a statement of Heisenberg's *uncertainty principle.*

The potential energy of a particle can be specified precisely as a function of position. However, the Hamiltonian operator $\widetilde{\mathrm{H}}$ also includes a term for kinetic energy. Because kinetic energy depends on momentum, $\widetilde{\mathrm{H}}$ does not commute with $\widetilde{\mathrm{r}}$.

We therefore cannot specify both the energy and the position of a particle simultaneously with arbitrary precision. Similarly, because the dipole moment of a molecule depends on the positions of all the electrons and nuclei, we cannot specify the dipole moment together with the energy to arbitrary precision. If we know that a system is in a state with a particular energy, a measurement of the dipole moment will give a real result, but we cannot be sure exactly what result will be obtained on any given measurement. However, the average result of many such measurements is given by the expectation value, which is an integral over all possible positions. The expectation values of both the energy and the dipole moment can, therefore, be stated precisely, at least in principle (Box 2.2).

There is no uncertainty principle comparable to the one for momentum and position that links the energy of a state with the state's lifetime. Indeed, there is no quantum mechanical operator for the lifetime of a state. There is, nevertheless, a relationship between the lifetime and our ability to assign the state a definite energy. One way to view this relationship is to recall that the full wavefunction for a system with energy E_a is an oscillating function of time, and that the oscillation frequency is proportional to the energy (Eq. 2.16):

$$\Psi(\boldsymbol{r},t) = \psi(\boldsymbol{r})\exp(-iE_a t/\hbar). \tag{2.66}$$

According to this expression, if E_a is constant the probability of finding the system in the state is independent of time ($P = \Psi^*\Psi = \psi^*\psi$). Conversely, if a system remains in one state indefinitely we can specify its oscillation frequency (E_a/h), and thus its energy, with arbitrarily high precision. But if the particle can make a transition to another state the probability density for the initial state clearly must decrease with time.

Suppose the probability of finding the system in the initial state decays to zero by first-order kinetics with a time constant T:

$$P(t) = \langle\Psi(\boldsymbol{r},t)|\Psi(\boldsymbol{r},t)\rangle = \langle\Psi(\boldsymbol{r},0)|\Psi(\boldsymbol{r},0)\rangle\exp(-t/T), \tag{2.67}$$

where $\Psi(\boldsymbol{r},0)$ is the amplitude of the wavefunction at zero time. Equations (2.66) and (2.67) then require the wavefunction to be an oscillatory function that decreases in amplitude with a time constant of $2T$:

$$\begin{aligned}\Psi(\boldsymbol{r},t) &= \psi(\boldsymbol{r})\exp(-iE_a t/\hbar)\exp(-t/2T)\\ &= \psi(\boldsymbol{r})\exp\{-[(iE_a/\hbar) + (1/2T)]t\}.\end{aligned} \tag{2.68}$$

The time dependence of such a function is illustrated in Fig. 2.11.

We can equate the time-dependent function in Eq. (2.68) to a superposition of many oscillating functions, all of the form $\exp(-Et/\hbar)$ but with a range of energies:

$$\exp\{-[(iE_a/\hbar) + (1/2T)]t\} = \int_{-\infty}^{\infty} G(E)\exp(-iEt/\hbar)dE. \tag{2.69}$$

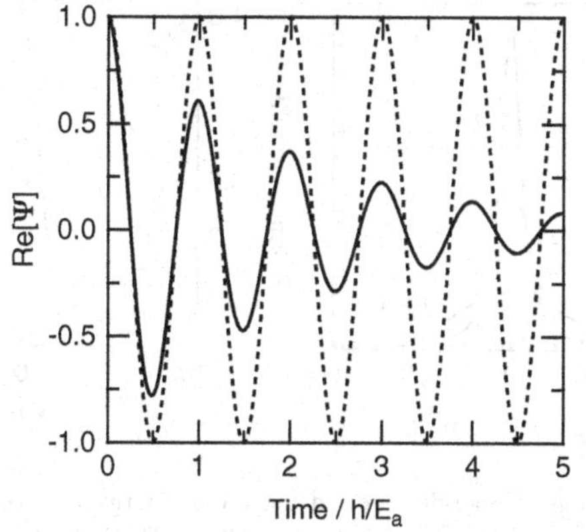

Fig. 2.11 Wavefunctions for particles with different lifetimes. The *dotted curve* is the real part of an undamped wavefunction $\psi_o \exp(-iE_a t/\hbar)$ with $\psi_o = 1$; this represents a particle with an infinite lifetime. The energy (E_a) is defined precisely. The *solid curve* is the real part of the wavefunction $\psi_o \exp(-iE_a t/\hbar)\exp(-t/2T)$ with a damping time constant $2T$ that here is set equal to $2\,h/E_a$; this represents a particle with an energy of E_a but a finite lifetime of $T = h/E_a$

Inspection of Eq. (2.69) shows that the distribution function $G(E)$ is the Fourier transform of the time-dependent part of Ψ (Appendix A.3). In general, for such an equality to hold, $G(E)$ must be a complex quantity. The real part of $G(E)dE$, Re[$G(E)$]dE, can be interpreted as the probability that the energy of the system is in the interval between $E - dE/2$ and $E + dE/2$, which can be normalized so that

$$\int_{-\infty}^{\infty} \mathrm{Re}[G(E)]\,dE = 1. \tag{2.70}$$

The imaginary part of the Fourier transform, Im[$G(E)$], relates to the phases of the different oscillation frequencies. The phases must be such that the oscillations interfere constructively at $t = 0$, where $|\Psi|$ is maximal. As time increases, the interference must become predominantly destructive so that $|\Psi|$ decays to zero.

The solution to Eq. (2.69) is that Re[$G(E)$] is a *Lorentzian function*:

$$\mathrm{Re}[G(E)] = \left(\frac{1}{\pi}\right)\frac{\hbar/2T}{(E - E_a)^2 + (\hbar/2T)^2}. \tag{2.71}$$

This function peaks at $E = E_a$, but has broad wings stretching out on either side (Fig. 2.12). It falls to half its maximum amplitude when $E = E_a \pm \hbar/2T$, and its full width at the half-maximal points (FWHM) is $\hbar/T$. ($\hbar$ is 5.308×10^{-12} cm^{-1} s, where 1 cm^{-1} = 1.240×10^{-4} eV = 2.844 cal/mol.) A Lorentzian has wider wings than a Gaussian with the same integrated area and FWHM (Fig. 2.12B).

If we interpret the FWHM of the Lorentzian as representing an uncertainty δE in the energy caused by the finite lifetime of the state, we see that

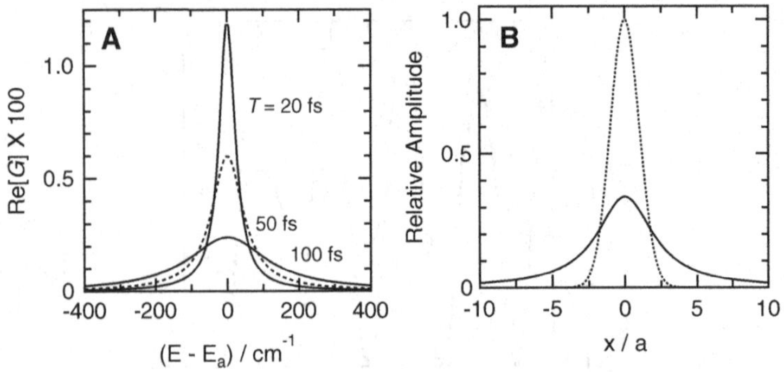

Fig. 2.12 (**A**) Distribution of energies around the mean energy (E_a) of a system that decays exponentially with a time constant (T) of 100, 50 or 20 fs. The Lorentzian distribution function (Eq. 2.71) is normalized to keep the area under the curve constant. (**B**) Comparison of Lorentzian (*solid line*) and Gaussian (*dotted line*) functions with the same integrated area and the same full width at half-maximum amplitude (FWHM). The Gaussian $a^{-1}(2p)^{-1/2}\exp(-x^2/2a^2)$ has a peak height of $a^{-1}(2\pi)^{-1/2}$ and a FWHM of $(8{\cdot}\ln 2)^{1/2}a$. The Lorentzian $\pi^{-1}a/(x^2+a^2)$ has a peak height of $a^{-1}\pi^{-1}$ and a FWHM of a

$$\delta E \approx \hbar/T. \tag{2.72}$$

This uncertainty, or *lifetime broadening*, puts a lower limit on the width of an absorption line associated with exciting a molecule into a transient state.

Equation (2.71) is used to describe the shapes of magnetic resonance absorption lines in terms of the "transverse" relaxation time constant (T_2). We'll use it similarly in Chaps. 4 and 10 to describe the shapes of optical absorption bands.

Exercises

1. (*a*) Let ψ_1 be the complete, normalized wavefunction of an enzyme, and ψ_2 the complete, normalized wavefunction of the substrate. What wavefunction would you use as a first approximation for the enzyme-substrate complex? Your wavefunction should be consistent with the fact that the enzyme and the substrate both exist simultaneously as well as individually, and should be normalized. (*b*) Why is your wavefunction for the combined system only an approximation?
2. Given a complete set of orthonormalized eigenfunctions of the Hamiltonian operator, $\psi_1, \psi_2, \ldots, \psi_n$, you could describe any arbitrary wavefunction as a linear combination of the form $\Psi = \sum_i C_i\psi_i$. (*a*) Show that Ψ is normalized if $\sum_i C_i^* C_i = 1$. (*b*) When might it might be appropriate to describe a system by such a linear combination? (*c*) Why would this be a poor choice for the enzyme-substrate complex considered in problem 1?

3. Using the treatment described in problem 2, you find that only two eigenfunctions (ψ_1 and ψ_2) make significant contributions to Ψ. Suppose the system is twice as likely to be in state ψ_1 as to be in ψ_2. (*a*) Find the values of C_1 and C_2, assuming that both values are real. (*b*) If the energies of ψ_1 and ψ_2 are E_1 and E_2, what is the energy of Ψ?
4. (*a*) Consider two spatial wavefunctions for a free, one-dimensional particle of mass m, $\psi_1 = A\exp\left(i\sqrt{2mE_1}x/\hbar\right)$ and $\psi_2 = A\exp\left(i\sqrt{2mE_2}x/\hbar\right)$. Show that the momentum operator $(\tilde{\mathrm{p}} = -i\hbar\partial/\partial x)$ conforms to the relationship $\langle\psi_2|\tilde{\mathrm{p}}|\psi_1\rangle = \langle\psi_1|\tilde{\mathrm{p}}|\psi_2\rangle^*$ for these wavefunctions. (*b*) The relationship $\left[\tilde{\mathrm{B}}, \tilde{\mathrm{A}}\right] = -\left[\tilde{\mathrm{A}}, \tilde{\mathrm{B}}\right]$ for any two operators $\tilde{\mathrm{A}}$ and $\tilde{\mathrm{B}}$ follows simply from the definition of the commutator $\left[\tilde{\mathrm{A}}, \tilde{\mathrm{B}}\right]$ (Box 2.2). Show that it holds in particular for the one-dimensional position and momentum operators by evaluating $[\tilde{\mathrm{x}}, \tilde{\mathrm{p}}]\psi$ and $[\tilde{\mathrm{p}}, \tilde{\mathrm{x}}]\psi$ explicitly for an arbitrary wavefunction ψ.
5. Assuming that the energies of the atomic orbitals of carbon increase in the same order as those of hydrogen and that Hund's rule holds, what are the electron configurations of (*a*) the ground state and (*b*) the first excited state of atomic carbon?
6. Find the expectation values of the position of an electron in the first two non-trivial eigenstates ($n = 1$ and 2) of a particle in a one-dimensional rectangular box with infinitely high walls.
7. (*a*) Write a Slater determinant for a singlet-state wavefunction of a system of four electrons. (*b*) Expand the determinant to write out the combination of spatial and spin wavefunctions that it represents. (*c*) Using whichever of the two representations you prefer, show that the wavefunction is antisymmetric for interchange of two electrons.
8. Consider two eigenstates of a one-dimensional system, with singlet wavefunctions $\psi_a(x,t)$ and $\psi_b(x,t)$. Show that, according to first-order perturbation theory, a perturbation $\tilde{\mathrm{H}}'$ to the Hamiltonian can cause transitions between the two states only if $\tilde{\mathrm{H}}'$ is a function of position (x).

References

1. Dirac, P.M.: The Principles of Quantum Mechanics. Oxford University Press, Oxford (1930)
2. Pauling, L., Wilson, E.B.: Introduction to Quantum Mechanics. McGraw-Hill, New York (1935)
3. van der Waerden, B.L. (ed.): Sources of Quantum Mechanics. Dover, New York (1968)
4. Atkins, P.W.: Molecular Quantum Mechanics, 2nd edn. Oxford Univ. Press, Oxford (1983)
5. Levine, I.N.: Quantum Chemistry. Prentice-Hall, Englewood Cliffs, NJ (2000)
6. Szabo, A., Ostlund, N.S.: Modern Quantum Chemistry: Introduction to Advanced Electronic Structure Theory. Macmillan, New York (1982)

7. Jensen, F.: Introduction to Computational Chemistry. Wiley, New York (1999)
8. Simons, J., Nichols, J.: Quantum Mechanics in Chemistry. Oxford University Press, New York (1997)
9. Engel, T.: Quantum Chemistry and Spectroscopy. Benjamin Cummings, San Francisco (2006)
10. Atkins, P.W.: Quanta: A Handbook of Concepts, p. 434. Oxford University Press, Oxford (1991)
11. Born, M.: The quantum mechanics of the impact process. Z. Phys. **37**, 863–867 (1926)
12. Reichenbach, H.: Philosophic Foundations of Quantum Mechanics, p. 182. University of California Press, Berkeley & Los Angeles (1944)
13. Pais, A.: Max Born's statistical interpretation of quantum mechanics. Science **218**, 1193–1198 (1982)
14. Jammer, M.: The Philosophy of Quantum Mechanics: The Interpretation of Quantum Mechanics in Historical Perspective. Wiley, New York (1974)
15. Einstein, A.: Uber einen die Erzeugung und Verwandlung des Lichtes betreffenden heuristischen Gesichtspunkt. Ann. der Phys. **17**, 132–146 (1905)
16. de Broglie, L.: Radiations—ondes et quanta. Comptes rendus **177**, 507–510 (1923)
17. Schrödinger, E.: Quantisierung als eigenwertproblem. Ann. der Phys. **79**, 489–527 (1926)
18. Schrödinger, E.: Collected Papers on Wave Mechanics. Blackie & Son, London (1928)
19. Jammer, M.: The Conceptual Development of Quantum Mechanics. McGraw-Hill, New York (1966)
20. Marton, L., Simpson, J.A., Suddeth, J.A.: Electron beam interferometer. Phys. Rev. **90**, 490–491 (1953)
21. Carnal, O., Mlynek, J.: Young's double-slit experiment with atoms: a simple atom interferometer. Phys. Rev. Lett. **66**, 2689–2692 (1991)
22. Monroe, C., Meekhof, D.M., King, B.E., Wineland, D.J.: A "Schrödinger cat" superposition state of an atom. Science **272**, 1131–1136 (1996)
23. Pople, J.A.: Nobel lecture: quantum chemical models. Rev. Mod. Phys. **71**, 1267–1274 (1999)
24. Pople, J.A., Beveridge, D.L.: Approximate Molecular Orbital Theory. McGraw-Hill, New York (1970)
25. Angeli, C.: DALTON, a molecular electronic structure program, Release 2.0 (2005). See http://www.kjemi.uio.no/software/dalton/dalton.html. (2005)
26. Parr, R.G.: Density-functional theory of atoms and molecules. Clarendon, Oxford (1989)
27. Ayscough, P.B.: Library of physical chemistry software, vol. 2. Oxford University Press & W. H. Freeman, New York (1990)
28. Kong, J., White, C.A., Krylov, A., Sherrill, D., Adamson, R.D., et al.: Q-chem 2.0: a high-performance ab initio electronic structure program package. J. Comp. Chem. **21**, 1532–1548 (2000)
29. Becke, A.D.: Perspective: fifty years of density-functional theory in chemical physics. J. Chem. Phys. **140**, 18A301 (2014)
30. Burke, K., Werschnik, J., Gross, E.K.U.: Time-dependent density functional theory: past, present, and future. J. Chem. Phys. **123**, 62206–62209 (2005)
31. McGlynn, S.P., Vanquickenborne, L.C., Kinoshita, M., Carroll, D.G.: Introduction to Applied Quantum Chemistry. Holt, Reinhardt & Winston, New York (1972)
32. Press, W.H., Flannery, B.P., Teukolsky, S.A., Vetterling, W.T.: Numerical Recipes in Fortran 77: The Art of Scientific Computing. Cambridge University Press, Cambridge (1989)
33. Dirac, P.M.: The quantum theory of the electron. Part II. Proc. Roy. Soc. **A118**, 351–361 (1928)
34. Roothaan, C.C.J.: New developments in molecular orbital theory. Rev. Mod. Phys. **23**, 69–89 (1951)
35. Morse, P.M.: Diatomic molecules according to the wave mechanics. II. Vibrational levels. Phys. Rev. **34**, 57–64 (1929)

36. ter Haar, D.: The vibrational levels of an anharmonic oscillator. Phys. Rev. **70**, 222–223 (1946)
37. Callis, P.R.: Molecular orbital theory of the ^1L_b and ^1L_a states of indole. J. Chem. Phys. **95**, 4230–4240 (1991)
38. Slater, L.S., Callis, P.R.: Molecular orbital theory of the ^1L_a and ^1L_b states of indole. 2. An ab initio study. J. Phys. Chem. **99**, 4230–4240 (1995)
39. Callis, P.R.: ^1L_a and ^1L_b transitions of tryptophan: applications of theory and experimental observations to fluorescence of proteins. Meth. Enzymol. **278**, 113–150 (1997)

Light

3

3.1 Electromagnetic Fields

In this chapter we consider classical and quantum mechanical descriptions of electromagnetic radiation. We develop expressions for the energy density and irradiance of light passing through a homogeneous medium, and we discuss the Planck black-body radiation law and linear and circular polarization. Readers anxious to get on to the interactions of light with matter can skip ahead to Chap. 4 and return to the present chapter as the need arises.

3.1.1 Electrostatic Forces and Fields

The classical picture of light as an oscillating electromagnetic field provides a reasonably satisfactory basis for discussing the spectroscopic properties of molecules, provided that we take the quantum mechanical nature of matter into account. To develop this picture, let's start by reviewing some of the principles of classical electrostatics.

Charged particles exert forces that conventionally are described in terms of electric and magnetic fields. Consider two particles with charges q_1 and q_2 located at positions $\boldsymbol{r}_1$ and $\boldsymbol{r}_2$ in a vacuum. According to Coulomb's law, the *electrostatic force* acting on particle 1 is

$$\boldsymbol{F} = \frac{q_1 q_2}{|\boldsymbol{r}_{12}|^2}\hat{\boldsymbol{r}}_{12}, \tag{3.1}$$

where $\boldsymbol{r}_{12} = \boldsymbol{r}_1 - \boldsymbol{r}_2$ and $\hat{\boldsymbol{r}}_{12}$ is a unit vector parallel to $\boldsymbol{r}_{12}$. $\boldsymbol{F}$ is directed along $\boldsymbol{r}_{12}$ if the two charges have the same sign, and in the opposite direction if the signs are different. The *electric field* $\boldsymbol{E}$ at any given position is defined as the electrostatic force on an infinitesimally small, positive "test" charge at this position. For two particles in a vacuum, the field at $\boldsymbol{r}_1$ is simply the derivative of $\boldsymbol{F}$ with respect to q_1:

W.W. Parson, *Modern Optical Spectroscopy*, DOI 10.1007/978-3-662-46777-0_3

$$\boldsymbol{E}(\boldsymbol{r}_1) = \lim_{q_1 \to 0} \frac{\partial \boldsymbol{F}(\boldsymbol{r}_1)}{\partial q_1} = \frac{q_2}{|\boldsymbol{r}_{12}|^2} \hat{r}_{12}. \tag{3.2}$$

Fields are additive: if the system contains additional charged particles the field at $\boldsymbol{r}_1$ is the sum of the fields from all the other particles.

The *magnetic field* ($\boldsymbol{B}$) at position $\boldsymbol{r}_1$ is defined similarly as the magnetic force on an infinitesimally small magnetic pole m_1. Magnetic fields are generated by moving electrical charges, and conversely, a changing magnetic field generates an electrical field that can cause an electrical charge to move.

Equations (3.1) and (3.2) are written in the *electrostatic* or *cgs* system of units, in which charge is given in electrostatic units (esu), distance in cm, and force in dynes. The electron charge e is -4.803×10^{-10} esu. The electrostatic unit of charge is also called the *statcoulomb* or *franklin*. In the *MKS* units adopted by the System International, distance is expressed in meters, charge in coulombs ($1\ \text{C} = 3 \times 10^9$ esu; $e = -1.602 \times 10^{-19}$ C), and force in newtons ($1\ \text{N} = 10^5$ dynes). In MKS units, the force between two charged particles is

$$\boldsymbol{F} = \frac{1}{4\pi\varepsilon_o} \frac{q_1 q_2}{|\boldsymbol{r}_{12}|^2} \hat{r}_{12}, \tag{3.3}$$

where ε_o is a constant called the *permitivity of free space* ($8.854 \times 10^{-12}\ \text{C}^2\ \text{N}^{-1}\ \text{m}^{-2}$). In the cgs system ε_o is equal to $1/4\pi$ so that the proportionality constants in Eqs. (3.1) and (3.2) are unity. Because this simplifies the equations of electromagnetism, the cgs system continues to be widely used. Appendix A4 gives a table of equivalent units in the two systems.

3.1.2 Electrostatic Potentials

What is the energy of electrostatic interaction of two charges in a vacuum? Suppose we put particle 2 at the origin of the coordinate system and hold it there while we bring particle 1 in from infinity. To move particle 1 at a constant velocity (i.e., without using any extra force to accelerate it), we must apply a force $\boldsymbol{F}_{app}$ that is always equal and opposite to the electrostatic force on the particle. The electrostatic energy (E_{elec}) is obtained by integrating the dot product $\boldsymbol{F}_{app}(\boldsymbol{r})\cdot d\boldsymbol{r}$, where $\boldsymbol{r}$ represents the variable position of particle 1 and $d\boldsymbol{r}$ is an incremental change in position during the approach. Because the final energy must be independent of the path, we can assume for simplicity that particle 1 moves in a straight line directly toward 2, so that $d\boldsymbol{r}$ and $\boldsymbol{F}_{app}$ are always parallel to $\boldsymbol{r}$. We then have (using cgs units again):

$$E_{elec} = \int_{\infty}^{r_{12}} \boldsymbol{F}_{app} \cdot d\boldsymbol{r} = -\int_{\infty}^{r_{12}} \boldsymbol{F}(\boldsymbol{r}) \cdot d\boldsymbol{r} = -\int_{\infty}^{r_{12}} \frac{q_1 q_2}{|\boldsymbol{r}|^2} \hat{r} \cdot d\boldsymbol{r} = \frac{q_1 q_2}{|\boldsymbol{r}_{12}|}. \tag{3.4}$$

The scalar *electrostatic potential* V_{elec} at $\boldsymbol{r}_1$ is defined as the electrostatic energy of a positive test charge at this position. This is just the derivative of E_{elec} with respect to the charge at $\boldsymbol{r}_1$. In a vacuum, the potential at $\boldsymbol{r}_1$ created by a charge at $\boldsymbol{r}_2$ is

$$V_{elec}(\boldsymbol{r}_1) = \frac{\partial E_{elec}(\boldsymbol{r}_1)}{\partial q_1} = \frac{q_2}{|\boldsymbol{r}_{12}|}. \tag{3.5}$$

The electrostatic energy of a pair of charges in a vacuum is simply the product of charge q_1 and the potential at $\boldsymbol{r}_1$:

$$E_{elec} = q_1 V_{elec}(\boldsymbol{r}_1). \tag{3.6}$$

In a system with more than two charges, the total electrostatic energy is given similarly by

$$E_{elec} = \frac{1}{2}\sum_i q_i V_{elec}(\boldsymbol{r}_i) = \frac{1}{2}\sum_i q_i \sum_{j \neq i} \frac{q_j}{|\boldsymbol{r}_{ij}|}, \tag{3.7}$$

where $V_{elec}(\boldsymbol{r}_i)$ is the electrostatic potential at $\boldsymbol{r}_i$ resulting from the fields from all the other charges; the factor of 1/2 prevents counting the pairwise interactions twice. Figure 3.1 shows a contour plot of the electrostatic potential resulting from a pair of positive and negative charges.

Note that Eq. (3.7) still refers to a set of stationary charges. Introducing a charge at $\boldsymbol{r}_i$ will change the potential at this point if the field from the new charge causes other charged particles to move.

In the cgs system, potentials have units of statvolts (ergs per esu of charge). In the MKS system, the potential difference between two points is one volt if one joule of work is required to move one coulomb of charge between the points. 1 V = 1 J/C = (10^7 ergs)/(3 × 10^9 esu) = 3 × 10^{-2} erg/esu = 3 × 10^{-2} (dyne-cm)/esu.

The electric field at a given point, which we defined above in terms of forces (Eq. 3.2), also can be defined as

$$\boldsymbol{E}(\boldsymbol{r}) = -\tilde{\nabla} V_{elec}(\boldsymbol{r}), \tag{3.8}$$

where $\tilde{\nabla} V_{elec}$ is the *gradient* of the electrostatic potential at that point. The gradient of a scalar function V is a vector whose components are the derivatives of V with respect to the coordinates:

$$\tilde{\nabla} V = (\partial V/\partial x, \partial V/\partial y, \partial V/\partial z) = \hat{x}\,\partial V/\partial x + \hat{y}\,\partial V/\partial y + \hat{z}\,\partial V/\partial z \tag{3.9}$$

(Eq. 2.5). Thus, the electric field at $\boldsymbol{r}_1$ generated by a charged particle at $\boldsymbol{r}_2$, is

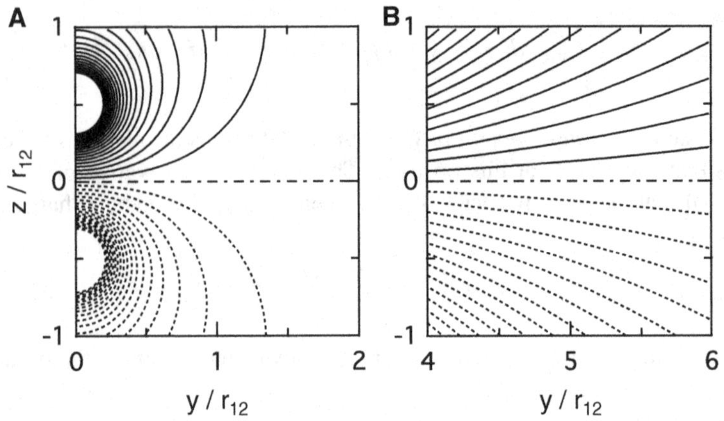

Fig. 3.1 Contour plots of the electric potential (V_{elec}) generated an electric dipole oriented along the z axis, as functions of position in the yz plane. The dipole consists of a unit positive charge at (y, z) = (0, $r_{12}/2$) and a unit negative charge at (0, $-r_{12}/2$). *Solid lines* represent positive potentials; *dotted lines*, negative potentials. The contour intervals are 0.2 e/r_{12} in **A** and 0.001 e/d_{12} in **B**; lines for $|V_{elec}| > 4e/r_{12}$ are omitted for clarity. The electric field vectors (not shown) are oriented normal to the contour lines of the potential, pointing in the direction of more positive potential. Their magnitudes are inversely proportional to the distances between the contour lines

$$\begin{aligned}\boldsymbol{E}(\boldsymbol{r}) &= -\tilde{\nabla} V_{elec}(\boldsymbol{r}) = -\tilde{\nabla}\left(\frac{q_2}{|\boldsymbol{r}_{12}|}\right) = -q_2\left(\frac{\partial\left(|\boldsymbol{r}_{12}|^{-1}\right)}{\partial x_1}, \frac{\partial\left(|\boldsymbol{r}_{12}|^{-1}\right)}{\partial y_1}, \frac{\partial\left(|\boldsymbol{r}_{12}|^{-1}\right)}{\partial z_1}\right)\\ &= q_2\frac{(x_1 - x_2,\ y_1 - y_2,\ z_1 - z_2)}{\left[(x_1 - x_2)^2 + (y_1 - y_2)^2 + (z_1 - z_2)^2\right]^{3/2}} = \frac{q_2}{|\boldsymbol{r}_{12}|^2}\hat{r}_{12},\end{aligned} \tag{3.10}$$

where $|\boldsymbol{r}_{12}| = [(x_1\text{-}x_2)^2 + (y_1\text{-}y_2)^2 + (z_1\text{-}z_2)^2]^{1/2}$ and $\hat{r}_{12} = (x_1 - x_2, y_1 - y_2, z_1 - z_2)/|\boldsymbol{r}_{12}|$. This is the same as Eq. (3.2).

Equation (3.8) implies that the line integral of the field over any path between two points $\boldsymbol{r}_1$ and $\boldsymbol{r}_2$ is just the difference between the potentials at the two points:

$$\int_{\boldsymbol{r}_1}^{\boldsymbol{r}_2} \boldsymbol{E}\cdot d\boldsymbol{r} = -[V(\boldsymbol{r}_2) - V(\boldsymbol{r}_1)] = V(\boldsymbol{r}_1) - V(\boldsymbol{r}_2). \tag{3.11}$$

This expression is similar to Eq. (3.4), in which we integrated the electrostatic force acting on a charged particle as another particle came in from a large distance. Here we integrate the component of the field that is parallel to the path element $d\boldsymbol{r}$ at each point along the path. Again, the result is independent of the path. Taking Eq. (3.11) one step further, we see that the line integral of the field over any closed path must be zero:

$$\oint \boldsymbol{E} \cdot d\boldsymbol{r} = 0. \tag{3.12}$$

We'll use this result later in this chapter to see what happens to the electric field when light enters a refracting medium.

3.1.3 Electromagnetic Radiation

The electric and magnetic fields ($\boldsymbol{E}$ and $\boldsymbol{B}$) generated by a pair of positive and negative charges (an *electric dipole*) are simply the sum of the fields from the individual charges. If the orientation of the dipole oscillates with time, the fields in the vicinity will oscillate at the same frequency. It is found experimentally, however, that the fields at various positions do not all change in phase: the oscillations at larger distances from the dipole lag behind those at shorter distances, with the result that the oscillating fields spread out in waves. The oscillating components of $\boldsymbol{E}$ and $\boldsymbol{B}$ at a given position are perpendicular to each other, and at large distances from the dipole, they also are perpendicular to the position vector ($\boldsymbol{r}$) relative to the center of the dipole (Fig. 3.1). They fall off in magnitude with $1/r$, and with the sine of the angle (θ) between $\boldsymbol{r}$ and the dipole axis. Such a coupled set of oscillating electric and magnetic fields together constitute an *electromagnetic radiation field.*

The strength of electromagnetic radiation often is expressed as the *irradiance*, which is a measure of the amount of energy flowing across a specified plane per unit area and time. The irradiance at a given position is proportional to the square of the magnitude of the electric field strength, $|\boldsymbol{E}|^2$ (Sect. 3.1.4). At large distances, the irradiance from an oscillating dipole therefore decreases with the square of the distance from the source and is proportional to $\sin^2(\theta)$, as shown in Fig. 3.2. The irradiance is symmetrical around the axis of the oscillations.

A spreading radiation field like that illustrated in Figs. 3.1 and 3.2 can be collimated by a lens or mirror to generate a *plane wave* that propagates in a single direction with constant irradiance. The electric and magnetic fields in such a wave oscillate sinusoidally along the propagation axis as illustrated in Fig. 3.3, but are independent of position normal to this axis. Polarizing devices can be used to restrict the orientation of the electric and magnetic fields to a particular axis in the plane. The plane wave illustrated in Fig. 3.3 is said to be *linearly polarized* because the electric field vector is always parallel to a fixed axis. Because $\boldsymbol{E}$ is confined to the plane normal to the axis of propagation, the wave also can be described as *plane polarized.* An *unpolarized* light beam propagating in the y direction consists of electric and magnetic fields oscillating in the xz plane at all angles with respect to the z axis.

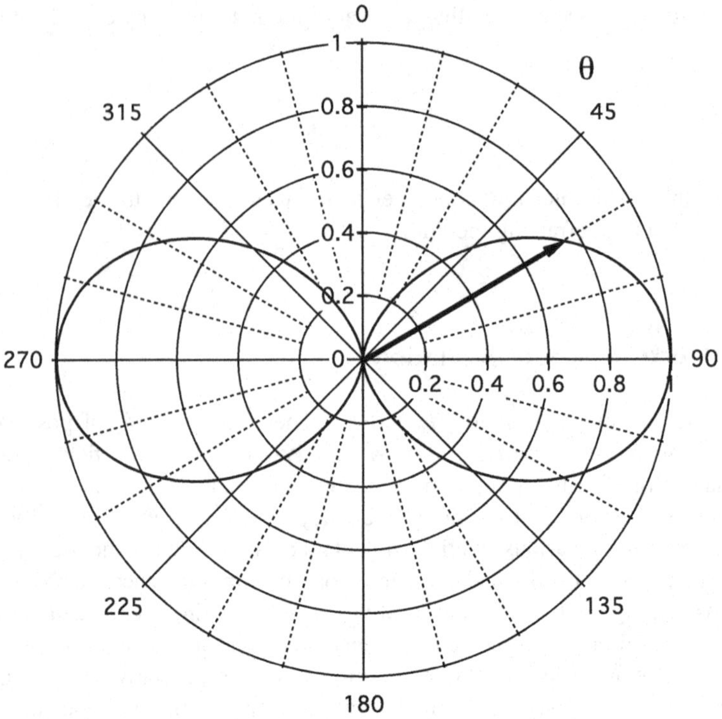

Fig. 3.2 Irradiance of electromagnetic radiation from an electric charge that oscillates in position along a vertical axis. In this polar plot, the angular coordinate is the radiation angle (θ) in degrees relative to the oscillation axis. The radial distance of the curve from the origin gives the relative irradiance of the wave propagating in the corresponding direction, which is proportional to $\sin^2(\theta)$. For example, the irradiance of the wave propagating at 60° (*arrow*) is 75% that of the wave propagating at 90°

The properties of electromagnetic fields are described empirically by four coupled equations that were set forth by J.C. Maxwell in 1865 (Box 3.1). These very general equations apply to both static and oscillating fields, and they encapsulate the salient features of electromagnetic radiation. In words, they state that:

1. Both $\boldsymbol{E}$ and $\boldsymbol{B}$ are always perpendicular to the direction of propagation of the radiation (i.e., the waves are *transverse*).
2. $\boldsymbol{E}$ and $\boldsymbol{B}$ are perpendicular to each other.
3. $\boldsymbol{E}$ and $\boldsymbol{B}$ oscillate in phase.
4. if we look in the direction of propagation, a rotation from the direction of $\boldsymbol{E}$ to the direction of $\boldsymbol{B}$ is clockwise.

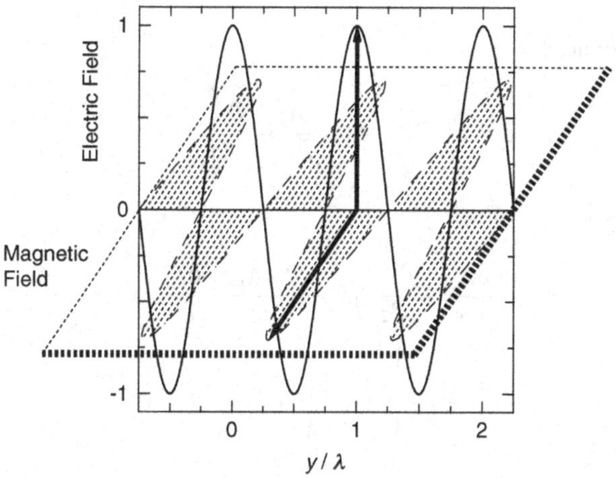

Fig. 3.3 Electric and magnetic fields at a given time in a linearly polarized plane wave propagating in the y-direction, as a function of position along the propagation axis. The *solid curve* is the component of the electric field parallel to the polarization axis, relative to the maximum amplitude ($2|\boldsymbol{E}_o|$); λ is the wavelength. The *dashed curve with shading to the propagation axis* is a perspective view of the magnetic field, which is perpendicular to both the propagation axis and the electric field. The directions of the fields at $y/\lambda = 1$ are indicated by *arrows*

Box 3.1 Maxwell's Equations and the Vector Potential

Maxwell's equations describe experimentally observed relationships between the electric and magnetic fields ($\boldsymbol{E}$ and $\boldsymbol{B}$) and the densities of charge and current in the medium. The charge density (ρ_q) at a given point is defined so that the total charge in a small volume element $d\sigma$ including the point is $q = \rho_q d\sigma$. If the charge moves with a velocity $\boldsymbol{v}$, the current density ($\boldsymbol{J}$) at the point is $\boldsymbol{J} = q\boldsymbol{v}$. In cgs units, Maxwell's equations read

$$\mathrm{div}\boldsymbol{E} = \frac{4\pi\rho_q}{\varepsilon}, \tag{B3.1.1}$$

$$\mathrm{div}\boldsymbol{B} = 0, \tag{B3.1.2}$$

$$\mathrm{curl}\boldsymbol{E} = -\frac{1}{c}\frac{\partial \boldsymbol{B}}{\partial t}, \tag{B3.1.3}$$

and

$$\mathrm{curl}\boldsymbol{B} = \frac{4\pi}{c}\boldsymbol{J} + \frac{\varepsilon}{c}\frac{\partial \boldsymbol{E}}{\partial t}, \tag{B3.1.4}$$

where c and ε are constants, and the vector operators div and curl are defined as follows:

(continued)

Box 3.1 (continued)

$$\mathrm{div}\boldsymbol{A} = \tilde{\nabla} \cdot \boldsymbol{A} = \frac{\partial A_x}{\partial x} + \frac{\partial A_y}{\partial y} + \frac{\partial A_z}{\partial z} \tag{B3.1.5}$$

and

$$\begin{aligned}\mathrm{curl}\boldsymbol{A} = \tilde{\nabla} \times \boldsymbol{A} &= \left(\frac{\partial A_z}{\partial y} - \frac{\partial A_y}{\partial z}\right)\hat{x} + \left(\frac{\partial A_x}{\partial z} - \frac{\partial A_z}{\partial x}\right)\hat{y} + \left(\frac{\partial A_y}{\partial x} - \frac{\partial A_x}{\partial y}\right)\hat{z} \\ &= \begin{vmatrix} \hat{x} & \hat{y} & \hat{z} \\ \partial/\partial x & \partial/\partial y & \partial/\partial z \\ A_x & A_y & A_z \end{vmatrix}.\end{aligned} \tag{B3.1.6}$$

(Appendix A1.)

The constant ε in Eqs. (B3.1.1) and (B3.1.4) is the *dielectric constant* of the medium, which is defined as the ratio of the energy density (energy per unit volume) associated with an electric field in a medium to that for the same field in a vacuum. As we'll discuss later in this chapter, the difference between the energy densities in a condensed medium and a vacuum reflects polarization of the medium by the field.

In free space, or more generally, in a uniform, isotropic, nonconducting medium with no free charges, ρ_q and $\boldsymbol{J}$ are zero and ε is independent of position and orientation, and Eqs. (B3.1.1) and (B3.1.4) simplify to $\mathrm{div}\boldsymbol{E} = 0$ and $\mathrm{curl}\boldsymbol{B} = (\varepsilon/c)\partial\boldsymbol{E}/\partial t$. $\boldsymbol{E}$ and $\boldsymbol{B}$ then can be eliminated from two of Maxwell's equations to give:

$$\tilde{\nabla}^2\boldsymbol{E} = \frac{\varepsilon}{c^2}\frac{\partial^2\boldsymbol{E}}{\partial t^2} \tag{B3.1.7}$$

and

$$\tilde{\nabla}^2\boldsymbol{B} = \frac{\varepsilon}{c^2}\frac{\partial^2\boldsymbol{B}}{\partial t^2}, \tag{B3.1.8}$$

where the action of the Laplacian operator $\tilde{\nabla}^2$ on a vector $\boldsymbol{A}$ is defined as

(continued)

Box 3.1 (continued)

$$\tilde{\nabla}^2 \boldsymbol{A} = \left(\frac{\partial^2 A_x}{\partial x^2} + \frac{\partial^2 A_x}{\partial y^2} + \frac{\partial^2 A_x}{\partial z^2}\right)\hat{x} + \left(\frac{\partial^2 A_y}{\partial x^2} + \frac{\partial^2 A_y}{\partial y^2} + \frac{\partial^2 A_y}{\partial z^2}\right)\hat{y}$$
$$+ \left(\frac{\partial^2 A_z}{\partial x^2} + \frac{\partial^2 A_z}{\partial y^2} + \frac{\partial^2 A_z}{\partial z^2}\right)\hat{z}. \tag{B3.1.9}$$

Equations (B3.1.7) and (B3.1.8) are classical, three-dimensional wave equations for waves that move through space with velocity

$$u = c/\sqrt{\varepsilon}. \tag{B3.1.10}$$

Since $\varepsilon = 1$ in a vacuum, the constant c that appears in Eqs. (B3.1.7) and (B3.1.8) must be the speed at which electromagnetic waves travel in a vacuum. This was an unanticipated result when Maxwell discovered it. He had obtained the value of the constant from experimental data on the magnetic field generated by a steady current, and there had been no reason to think that it had anything to do with light. The realization that c was, to within a very small experimental error, the same as the measured speed of light led Maxwell [1] to suggest that light consists of electromagnetic waves.

For a plane wave propagating in the y direction with $\boldsymbol{E}$ polarized parallel to the z axis, Eq. (B3.1.7) reduces to

$$\frac{\partial^2 E_z}{\partial y^2} = \frac{\varepsilon}{c^2}\frac{\partial^2 E_z}{\partial t^2}. \tag{B3.1.11}$$

Solutions to Maxwell's equations also can be obtained in terms of a *vector potential* $\boldsymbol{V}$ and a *scalar potential* ϕ, which are related to the electric and magnetic fields by the expressions

$$\boldsymbol{E} = -\frac{1}{c}\frac{\partial \boldsymbol{V}}{\partial t} - \tilde{\nabla}\phi \tag{B3.1.12}$$

and

$$\boldsymbol{B} = \mathrm{curl}\,\boldsymbol{V}. \tag{B3.1.13}$$

This description has the advantage that only four parameters are needed to specify the electromagnetic fields (the magnitude of ϕ and the three components of $\boldsymbol{V}$), instead of the six components of $\boldsymbol{E}$ and $\boldsymbol{B}$. The description is not unique because adding any arbitrary function of time to ϕ does not

(continued)

Box 3.1 (continued)
affect the values of the physical observables $\boldsymbol{E}$ and $\boldsymbol{B}$, which makes it possible to simplify the description further. If ϕ is chosen so that

$$\operatorname{div}\boldsymbol{V} + \frac{1}{c}\frac{\partial \phi}{\partial t} = 0, \tag{B3.1.14}$$

the scalar potential drops out and an electromagnetic radiation field can be represented in terms of the vector potential alone [2]. This choice of ϕ is called the *Lorentz guage*. An alternative choice called the *Coulomb guage* is often used for static systems.

Using the Lorentz guage, $\boldsymbol{V}$ for a uniform, isotropic, nonconducting medium with no free charges is determined by the equations

$$\tilde{\nabla}^2 \boldsymbol{V} = \frac{\varepsilon}{c^2}\frac{\partial^2 \boldsymbol{V}}{\partial t^2} \tag{B3.1.15}$$

and

$$\operatorname{div}\boldsymbol{V} = 0, \tag{B3.1.16}$$

while $\boldsymbol{E}$ is parallel to $\boldsymbol{V}$ and is given by

$$\boldsymbol{E} = -\frac{1}{c}\frac{\partial \boldsymbol{V}}{\partial t}. \tag{B3.1.17}$$

Equation (B3.1.13) still holds for $\boldsymbol{B}$.

See [1, 3] for Maxwell's own description of electromagnetism, and [2, 4–6] for additional discussion.

For our purposes, we will not need to use Maxwell's equations themselves; we can focus on a solution to these equations for a particular situation such as the plane wave of monochromatic, polarized light illustrated in Fig. 3.3. In a uniform, homogeneous, nonconducting medium with no free charges, Maxwell's equations for $\boldsymbol{E}$ in a one-dimensional plane wave reduce to

$$\frac{\partial^2 \boldsymbol{E}}{\partial y^2} = \frac{\varepsilon}{c^2}\frac{\partial^2 \boldsymbol{E}}{\partial t^2}, \tag{3.13}$$

where c is the velocity of light in a vacuum and ε is the dielectric constant of the medium (Box 3.1). An identical expression holds for the magnetic field. Eq. (3.13) is a classical wave equation for a wave that moves with velocity

$$u = c/\sqrt{\varepsilon}. \tag{3.14}$$

Solutions to Eq. (3.13) can be written in exponential notation as

$$\boldsymbol{E} = \boldsymbol{E}_o\{\exp[2\pi i(vt - y/\lambda + \delta)] + \exp[-2\pi i(vt - y/\lambda + \delta)]\}, \tag{3.15a}$$

or, by using the identity $\exp(i\theta) = \cos(\theta) + i\,\sin(\theta)$ and the relationships $\cos(-\theta) = \cos(\theta)$ and $\sin(-\theta) = -\sin(\theta)$, as

$$\boldsymbol{E} = 2\boldsymbol{E}_o \cos[2\pi(vt - y/\lambda + \delta)]. \tag{3.15b}$$

In Eqs. (3.15a) and (3.15b), $\boldsymbol{E}_o$ is a constant vector that expresses the magnitude and polarization of the field (parallel to the z axis for the wave shown in Fig. 3.3), v is the frequency of the oscillations, λ is the wavelength and δ is phase shift that depends on an arbitrary choice of zero time. The frequency and wavelength are linked by the expression

$$\lambda = u/v, \tag{3.16}$$

More generally, we can describe the electric field at point $\boldsymbol{r}$ in a plane wave of monochromatic, linearly polarized light propagating in an arbitrary direction $(\hat{k})$ by

$$\boldsymbol{E}(\boldsymbol{r}, t) = \boldsymbol{E}_o\{\exp[2\pi i(vt - \boldsymbol{k}\cdot\boldsymbol{r} + \delta)] + \exp[-2\pi i(vt - \boldsymbol{k}\cdot\boldsymbol{r} + \delta)]\}, \tag{3.17}$$

where $\boldsymbol{k}$, the *wavevector*, is a vector with magnitude $1/\lambda$ pointing in direction $\hat{k}$. Note that each of the exponential terms in Eq. (3.17) could be written as a product of a factor that depends only on time, another factor that depends only on position, and a third factor that depends only on the phase shift. We'll return to this point in Sect. 3.4.

As discussed in Box 3.1, convenient solutions to Maxwell's equations also can be obtained in terms of a *vector potential* $\boldsymbol{V}$ instead of electric and magnetic fields. Using the same formalism as Eq. (3.17) but omitting the phase shift for simplicity, the vector potential for a plane wave of monochromatic, linearly polarized light can be written

$$\boldsymbol{V}(\boldsymbol{r}, t) = \boldsymbol{V}_o\{\exp[2\pi i(vt - \boldsymbol{k}\cdot\boldsymbol{r})] + \exp[-2\pi i(vt - \boldsymbol{k}\cdot\boldsymbol{r})]\}. \tag{3.18}$$

We will use this expression in Sect. 3.4 when we consider the quantum mechanical theory of electromagnetic radiation.

The velocity of light in a vacuum, 2.9979×10^{10} cm/s, has been denoted almost universally by c since the early 1900s, probably for *celeritas*, the Latin word for speed. The first accurate measurements of the velocity of light in air were made by A. Fizeau in 1849 and L. Foucault in 1850. Fizeau passed a beam of light through a gap between teeth at the edge of a spinning disk, reflected the light back to the disk with a distant mirror, and increased the speed of the disk until the returning light passed through the next gap. Foucault used a system of rotating mirrors. Today, c is taken to be an exactly defined number rather than a measured quantity and is used to define the length of the meter.

If monochromatic light moves from a vacuum into a non-absorbing medium with a refractive index n, the frequency ν remains the same but the velocity and wavelength decrease to c/n and λ/n. Equation (3.14) indicates that the refractive index, defined as c/u, can be equated with $\varepsilon^{1/2}$:

$$n \equiv c/u = \sqrt{\varepsilon}. \tag{3.19}$$

Most solvents have values of n between 1.2 and 1.6 for visible light. The refractive index of most materials increases with ν, and such media are said to have positive *dispersion*. As we'll discuss in Sects. 3.1.4 and 3.5, Eqs. (3.14) and (3.19) do not necessarily hold in regions of the spectrum where n varies significantly with the wavelength. At frequencies where the medium absorbs light, the refractive index can vary strongly with ν and the velocity at which energy moves through the medium is not necessarily given simply by c/n, particularly if the light includes a broad band of frequencies.

We will be interested mainly in the time-dependent oscillations of the electrical and magnetic fields in small, fixed regions of space. Because molecular dimensions typically are much smaller than the wavelength of visible light, the amplitude of the electrical field at a particular time will be nearly the same everywhere in the molecule. We also will restrict ourselves initially to phenomena that relate to averages of the field over many cycles of the oscillation and do not depend on coherent superposition of light beams with fixed phase relationships. With these restrictions, we can neglect the dependence of $\mathbf{E}$ on position and the phase shift, and write

$$\mathbf{E}(t) = \mathbf{E}_o[\exp(2\pi i\nu t) + \exp(-2\pi i\nu t)] = 2\mathbf{E}_o \cos(2\pi\nu t). \tag{3.20}$$

We will have to use a more complete expression that includes changes of the fields with position in a molecule when we discuss circular dichroism. We'll need to include phase shifts when we consider the light emitted by ensembles of many molecules.

3.1.4 Energy Density and Irradiance

Because electromagnetic radiation fields cause charged particles to move, they clearly can transmit energy. To evaluate the rate at which absorbing molecules take up energy from a beam of light, we will need to know how much energy the radiation field contains and how rapidly this energy flows from one place to another. We usually will be interested in the energy of the fields in a specified spectral region with frequencies between ν and $\nu + d\nu$. The amount of energy per unit volume in such a spectral interval can be expressed as $\rho(\nu)d\nu$, where $\rho(\nu)$ is the *energy density* of the field. The *irradiance*, $I(\nu)d\nu$, is the amount of energy in a specified spectral interval that crosses a given plane per unit area and time. In a homogeneous, nonabsorbing medium, the irradiance is

$$I(\nu)d\nu = u(\nu)\rho(\nu)d\nu, \tag{3.21}$$

where u, as before, is the velocity of light in the medium.

Several different measures are used to describe the strengths of light sources. The *radiant intensity* is the energy per unit time that a source radiates into a unit solid angle in a given direction. It usually is expressed in units of watts per steradian. *Luminance*, a measure of the amount of visible light leaving or passing through a surface of unit area, requires correcting the irradiance for the spectrum of sensitivity of the human eye, which peaks near 555 nm. Luminance is given in units of *candela* (cd) per square meter, or *nits*. For arcane historical reasons having to do with the apparent brightness of a hot bar of platinum, one candela is defined as the luminous intensity of a source that emits 540-nm monochromatic light with a radiant intensity of 1/683 (0.001464) watt per steradian. The total luminous flux from a source into a given solid angle , the product of the luminance and the solid angle, is expressed in *lumens*. One *lux* is one lumen per square meter. Bright sunlight has an illuminance on the order of 5×10^4 to 1×10^5 lux, and an luminance of 3×10^3 to 6×10^3 cd m^{-2}.

From Maxwell's equations one can show that the energy density of electromagnetic radiation depends on the square of the electric and magnetic field strengths [4–6]. For radiation in a vacuum, the relationship is

$$\rho(\nu) = \overline{\left[|E(\nu)|^2 + |B(\nu)|^2\right]}\rho_\nu(\nu)/8\pi, \tag{3.22}$$

where the bar above the quantity in brackets means an average over the spatial region of interest and $\rho_\nu(\nu)d\nu$ is the number of *modes* (frequencies) of oscillation in the small interval between ν and $\nu + d\nu$. An oscillation mode for electromagnetic radiation is analogous to a standing wave for an electron in a box (Eq. 2.23a). But the drawing in Fig. 3.3, which represents the electric field for such an individual mode (monochromatic light), is an idealization. In practice, electromagnetic radiation is never strictly monochromatic: it always includes fields oscillating over a range of frequencies. We'll discuss the nature of this distribution in Sect. 3.6.

Equation (3.22) is written in cgs units, which are particularly convenient here because the electric and magnetic fields in a vacuum have the same magnitude:

$$|\boldsymbol{B}| = |\boldsymbol{E}|. \tag{3.23}$$

(In MKS units, $|\boldsymbol{B}| = |\boldsymbol{E}|/c$.) The energy density of a radiation field in a vacuum is, therefore,

$$\rho(\nu) = \overline{|E(\nu)|^2}\rho_\nu(\nu)/4\pi. \tag{3.24}$$

If we now use Eq. (3.15b) to express the dependence of $\boldsymbol{E}$ on time and position, we have

$$\overline{|\boldsymbol{E}|^2} = \overline{[2\boldsymbol{E}_o \cos(2\pi\nu t - 2\pi y/\lambda + \delta)]^2} = 4|\boldsymbol{E}_o|^2\overline{\cos^2(2\pi\nu t - 2\pi y/\lambda + \delta)}. \quad (3.25)$$

Since the fields in a plane wave are by definition independent of position perpendicular to the propagation axis (y), the average denoted by the bar in Eq. (3.25) requires only averaging over a distance in the y direction. If this distance is much longer than λ (or is an integer multiple of λ), the average of $\cos^2(2\pi\nu t - 2\pi y/\lambda + \delta)$ is 1/2, and Eq. (3.25) simplifies to

$$\overline{|\boldsymbol{E}|^2} = 2|\boldsymbol{E}_o|^2. \quad (3.26)$$

So, for a plane wave of light in a vacuum:

$$\rho(\nu) = |\boldsymbol{E}_o|^2\rho_\nu(\nu)/2\pi \quad (3.27)$$

and

$$I(\nu) = c|\boldsymbol{E}_o|^2\rho_\nu(\nu)/2\pi. \quad (3.28)$$

If a beam of light in a vacuum strikes the surface of a refractive medium, part of the beam is reflected while another part enters the medium. We can use Eq. (3.28) to relate the irradiances of the incident and reflected light to the amplitudes of the corresponding fields, because the fields on this side of the interface are in a vacuum. But we need a comparable expression that relates the transmitted irradiance to the amplitude of the field in the medium, and for this we must consider the effect of the field on the medium.

As light passes through the medium, the electric field causes electrons in the material to move, setting up electric dipoles that generate an oscillating *polarization field* (***P***). In an isotropic, nonabsorbing and nonconducting medium ***P*** is proportional to ***E*** and can be written

$$\boldsymbol{P} = \chi_e\boldsymbol{E}. \quad (3.29)$$

The proportionality constant χ_e is called the *electric susceptibility* of the medium, and materials in which ***P*** and ***E*** are related linearly in this way are called *linear* optical materials. The field in the medium at any given time and position (***E***) can be viewed as the resultant of the polarization field and the *electric displacement* (***D***), which is the field that hypothetically would be present in the absence of the polarization. In a vacuum, ***P*** is zero and $\boldsymbol{E} = \boldsymbol{D}$. In cgs units, the field in a linear medium is given by

$$\begin{aligned} \boldsymbol{E} &= \boldsymbol{D} - 4\pi\boldsymbol{P} = \boldsymbol{D} - 4\pi\chi_e\boldsymbol{E} \\ &= \boldsymbol{D}/(1 + 4\pi\chi_e) = \boldsymbol{D}/\varepsilon, \end{aligned} \quad (3.30)$$

where, as in Maxwell's equations (Box 3.1), ε is the dielectric constant of the medium. Rearranging these relationships gives $\varepsilon = 1 + 4\pi\chi_e$.

We will be interested mainly in electromagnetic fields that oscillate with frequencies on the order of 10^{15} Hz, which is too rapid for nuclear motions to follow. The polarization described by $\boldsymbol{P}$ therefore reflects only the rapidly oscillating induced dipoles created by electronic motions, and the corresponding dielectric constant in Eq. (3.30) is called the *high-frequency* or *optical dielectric constant*. For a nonabsorbing medium, the electric susceptibility is independent of the oscillation frequency, and the high-frequency dielectric constant is equal to the square of the refractive index. The magnetic field of light passing through a refractive medium is affected analogously by induced magnetic dipoles, but in nonconducting materials this is a much smaller effect and usually is neglible.

In a linear, nonabsorbing medium, the relationship between the amplitudes of the magnetic and electric fields in cgs units becomes [4–6]

$$|\boldsymbol{B}| = \sqrt{\varepsilon}|\boldsymbol{E}|, \tag{3.31}$$

and the energy density of electromagnetic radiation is

$$\begin{aligned}\rho(\nu) &= \overline{\left[\boldsymbol{E}(\nu)\cdot\boldsymbol{D}(\nu) + |\boldsymbol{B}(\nu)|^2\right]}\rho_\nu(\nu)/8\pi \\ &= \overline{\left[\boldsymbol{E}(\nu)\cdot\varepsilon\boldsymbol{E}(\nu) + |\sqrt{\varepsilon}\boldsymbol{E}(\nu)|^2\right]}\rho_\nu(\nu)/8\pi \\ &= \varepsilon\overline{|\boldsymbol{E}(\nu)|^2}\rho_\nu(\nu)/4\pi = \varepsilon|\boldsymbol{E}_o|^2\rho_\nu(\nu)/2\pi\end{aligned} \tag{3.32}$$

We have assumed again that we are interested in the average energy over a region that is large relative to the wavelength of the radiation.

Equation (3.32) indicates that, for equal field strengths, the energy density in a refracting medium is ε times that in a vacuum. The additional energy resides in the polarization of the medium. But we are not quite through. To find the irradiance in the medium, we also need to know the velocity at which energy moves through the medium. This *energy velocity*, or *group velocity* (u) is not necessarily simply c/n because waves with different frequencies will travel at different rates if n varies with ν. The group velocity describes the speed at which a packet of waves with similar frequencies travels as a whole (Sect. 3.5). It is related generally to c, n and ε by [7, 8]

$$u = cn/\varepsilon, \tag{3.33}$$

which reduces to $u = c/n$ (Eq. 3.19) if $\varepsilon = n^2$, as it does if n is independent of ν.

Combining Eqs. (3.33) with (3.21) and (3.32) gives the irradiance in the medium:

$$I(\nu) = \frac{c}{n}\rho(\nu) = cn|\boldsymbol{E}_o|^2\rho_\nu(\nu)/2\pi. \tag{3.34}$$

This is an important result for our purposes because it relates the irradiance of a light beam in a condensed medium to the refractive index and the amplitude of the electric field. We'll need this relationship in Chap. 4 in order to connect the strength

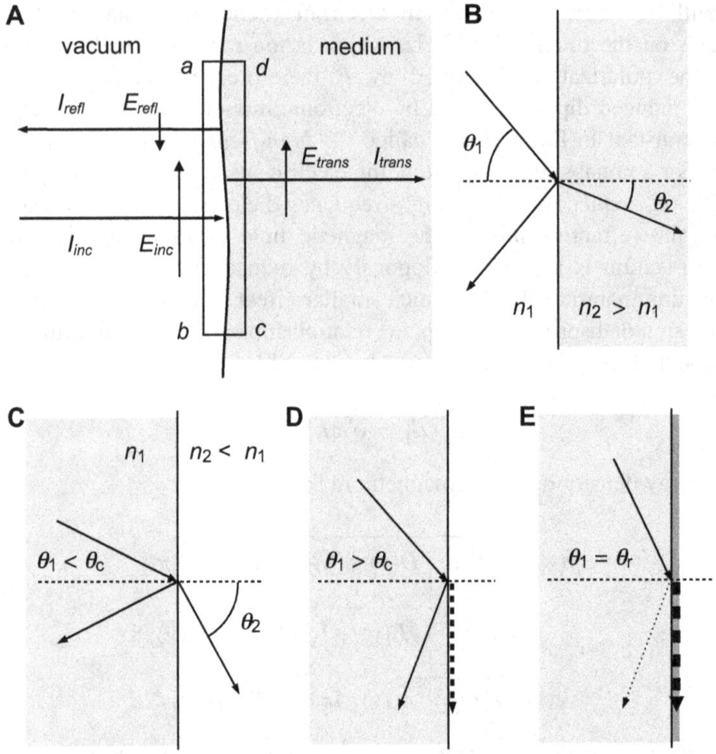

Fig. 3.4 (A) When a light beam propagating in a vacuum enters a refractive medium, the incident irradiance (I_{in}) must equal the sum of the transmitted and reflected irradiances (I_{trans} and I_{refl}). In addition, the electric field of the transmitted (refracted) beam in the plane of the interface ($\boldsymbol{E}_{trans}$) must equal the sum of the electric fields of the incident and reflected beams ($\boldsymbol{E}_{inc}+\boldsymbol{E}_{refl}$) because the path integral of the field over any closed loop must be zero (Eq. 3.12). Here, the angle of incidence is normal to the surface and the light is linearly polarized so that the electric field is parallel to edges *ab* and *cd* of rectangle *abcd* (vectors $\boldsymbol{b}-\boldsymbol{a}$ and $\boldsymbol{d}-\boldsymbol{c}$), and perpendicular to edges *bc* and *ad*. The path integral of the field over the loop $a\rightarrow b\rightarrow c\rightarrow d\rightarrow a$ is $(\boldsymbol{E}_{inc}+\boldsymbol{E}_{refl})\cdot(\boldsymbol{b}-\boldsymbol{a})+\boldsymbol{E}_{trans}\cdot(\boldsymbol{d}-\boldsymbol{c})=(\boldsymbol{E}_{inc}+\boldsymbol{E}_{refl}-\boldsymbol{E}_{trans})\cdot(\boldsymbol{b}-\boldsymbol{a})$. (**B–D**) If a beam propagating in a medium with refractive index n_1 encounters an interface with a medium with refractive index n_2, the angle of the refracted beam relative to the normal (θ_2) is related to the angle of incidence (θ_1) by $n_2\sin(\theta_2)=n_1\sin(\theta_1)$. The refracted beam is bent toward the normal if $n_2>n_1$ (**B**) and away from the normal if $n_2<n_1$ (**C**). Total internal reflection occurs if $n_2<n_1$ and $\theta_2\geq\theta_c$, where $\theta_c=\arcsin(n_2/n_1)$ (**D**). In this situation, an evanescent wave (*dotted arrow*) propagates along the interface and penetrates a short distance into the second medium. (**E**) If the surface of the medium with the higher refractive index is coated with a semitransparent layer of silver (*gray line*) and the angle of incidence at the interface with the second medium matches a resonance angle θ_r of about 60°, total internal reflection creates surface plasmons in the metal coating, greatly enhancing the evanescent field

of an electronic absorption band to the electronic structure of a molecule. As discussed in Box 3.2, Eq. (3.34) also can be used to find the fractions of a light beam that are reflected and transmitted at a surface.

Box 3.2 Reflection, Transmission, Evanescent Radiation and Surface Plasmons
What happens to the electric field and irradiance when a beam of light moves from a vacuum into a refractive, but non-absorbing medium? Let the irradiance of the incident beam be I_{inc}. At the interface, some of the radiation is transmitted, giving an irradiance I_{trans} that continues forward in the medium, while a portion with irradiance I_{refl} is reflected. I_{trans} must equal $I_{inc} - I_{refl}$ to balance the flux of energy across the interface (Fig. 3.4A):

$$I_{trans} = I_{inc} - I_{refl}. \tag{B3.2.1}$$

The fraction of the incident irradiance that is transmitted depends on the angle of incidence and the refractive index of the medium (n). Suppose the incident beam is normal to the surface so that the electric and magnetic fields are in the plane of the surface. For fields in the plane of the interface, the instantaneous electric field on the medium side of the interface ($\boldsymbol{E}_{trans}$) must be equal to the field on the vacuum side, which is the sum of the fields of the incident and reflected beams ($\boldsymbol{E}_{inc}$ and $\boldsymbol{E}_{refl}$):

$$\boldsymbol{E}_{trans} = \boldsymbol{E}_{inc} + \boldsymbol{E}_{refl} \tag{B3.2.2}$$

This follows from the fact that the path integral of the field around any closed path is zero (Eq. (3.12) and Fig. 3.4A).

Using Eq. (3.32), we can replace Eq. (B3.2.1) by a second relationship between the fields:

$$n|\boldsymbol{E}_{trans}|^2 = |\boldsymbol{E}_{inc}|^2 - |\boldsymbol{E}_{refl}|^2. \tag{B3.2.3}$$

Eliminating $\boldsymbol{E}_{refl}$ from Eqs. (B3.2.2) and (B3.2.3) then gives

$$\boldsymbol{E}_{trans} = \frac{2}{n+1}\boldsymbol{E}_{inc}. \tag{B3.2.4}$$

And finally, using Eq. (3.32) again,

$$I_{trans} = cn|\boldsymbol{E}_{trans}|^2 \rho_\nu(\nu)/2\pi = \frac{4n}{(n+1)^2} I_{inc}. \tag{B3.2.5}$$

For a typical refractive index of 1.5, Eqs. (B3.2.4 and B3.2.5) give $|\boldsymbol{E}_{trans}| = 0.8|\boldsymbol{E}_{vac}|$ and $I_{trans} = 0.96 I_{vac}$.

The same approach can be used for other angles of incidence to generate Snell's law, which relates the angle of the refracted beam to the refractive index [6]. In general, when light passes from a nonabsorbing medium with

(continued)

Box 3.2 (continued)
refractive index n_1 to a second nonabsorbing medium with refractive index n_2,

$$n_2 \sin\theta_2 = n_1 \sin\theta_1, \tag{B3.2.6}$$

where θ_1 and θ_2 are the angles of the incident and refracted beams relative to an axis normal to the surface. If $n_2 > n_1$, the refracted beam is bent toward the normal (Fig. 3.4B); if $n_2 < n_1$, it is bent away (Fig. 3.4C). For $n_2 < n_1$, the refracted beam becomes parallel to the interface ($\theta_2 = 90°$) when the angle of incidence reaches the "critical angle" θ_c defined by $\sin\theta_c = n_2/n_1$. Values of $\theta_1 > \theta_c$ give *total internal reflection*: all the incident radiation is reflected at the interface and no beam continues forward through the second medium (Fig. 3.4D). The critical angle is about 61.1° at a glass/water interface ($n_1 = 1.52$ and $n_2 = 1.33$) and 41.1° at a glass/air interface.

Because the electric and magnetic fields must be continuous across the interface, the radiation must penetrate a finite distance into the second medium even in the case of total internal reflection. Constructive interference of the incident and reflected fields in this situation creates a wave of *evanescent* ("vanishing") *radiation* that propagates parallel to the interface but drops off quickly in amplitude beyond the interface. The fall-off of the field with distance (z) in the second medium is given by $\mathbf{E} = \mathbf{E}_o \exp(-z/d)$ with $d = \left(n_1^2 \sin^2\theta_1 - n_2^2\right)^{-1/2} \lambda_1/4\pi$, which typically gives a penetration depth on the order of 500 to 1,000 nm [9, 10]. For angles of incidence slightly greater than θ_c, $\mathbf{E}_o \approx \mathbf{E}_{inc} + \mathbf{E}_{refl} \approx 2\mathbf{E}_{inc}$. The intensity of the evanescent radiation at $z = 0$ is therefore about four times that of the incident radiation. The intensity decreases gradually as the angle of incidence is raised above θ_c.

The existence of evanescent radiation can be demonstrated from the effects of objects in the second medium close to the interface. For example, if a third medium with a higher refractive index is placed near the interface between the first and second materials, radiation can tunnel through the barrier imposed by the second medium. This process, called *attenuated total internal reflection*, is essentially the same as the tunnelling of an electron between two potential wells separated by a region of higher potential (Sect. 2.3.2). Isaac Newton is said to have discovered the phenomenon when he placed a convex lens against the internally-reflecting face of a prism: the spot of light entering the lens was larger than the point where the two glass surfaces actually touched. As we'll discuss in Chap. 5, evanescent radiation also can excite fluorescence from molecules situated close to the interface. However, an absorbing medium does not obey Snell's law because, for plane waves entering the medium at other than normal incidence, the amplitude of the surviving light is not constant across a wavefront of constant phase.

(continued)

Box 3.2 (continued)

A striking phenomenon called *surface plasmon resonance* can occur at a glass-water interface if the glass is coated with a partially transmitting layer of gold or silver (Fig. 3.4E). As the angle of incidence is increased above θ_c, total internal reflection of the incident light first occurs just as at an uncoated surface. But at an angle of about 56° (the exact value depends on the wavelength, the metal coating, and the refractive indices of the two media), the intensity of the reflected beam drops almost to zero and evanescent radiation with an intensity as high as 50 times the intensity of the incident radiation can be detected on the aqueous side of the interface [11–14]. The strong evanescent radiation reflects movements of electron clouds ("plasmon surface polaritons" or "surface plasmons" for short) in the conduction band of the metal, and the loss of the reflected beam at the resonance angle results from destructive interference of the reflected and surface waves. Interactions of the evanescent field with molecules in the aqueous solution close to the interface can be detected by their fluorescence or strongly enhanced Raman scattering, or by their effects on the resonance angle. Intense fields from surface plasmons also can be generated in colloidal gold or silver, metal-coated particles that are small relative to the wavelength of the radiation and in microscopically patterned surfaces made by lithography [15, 16].

3.1.5 The Complex Electric Susceptibility and Refractive Index

In an absorbing medium, the electric susceptibility χ_e has an imaginary component that can give the refractive index a strong dependence on the frequency. The classical theory of this dependence on frequency (dispersion) is described in Box 3.3. In this theory, the ordinary index of refraction (n) is related to the real part of the complex index of refraction (n_c), and absorption is related to the imaginary part. Figure 3.5 shows the predicted real and imaginary parts of n_c in the region of a weak absorption band that is well removed from any other bands. In qualitative agreement with experiment, the theory predicts that n will increase with frequency except in the region surrounding the absorption maximum. The inversion of the slope near the absorption peak is known as *anomalous dispersion*. The classical theory also reproduces the Lorentzian shape of a homogenous absorption line.

Box 3.3 The Classical Theory of Dielectric Dispersion

In classical physics, dielectric media were modeled by considering an electron bound to a mean position by a force (k) that increased linearly with the

(continued)

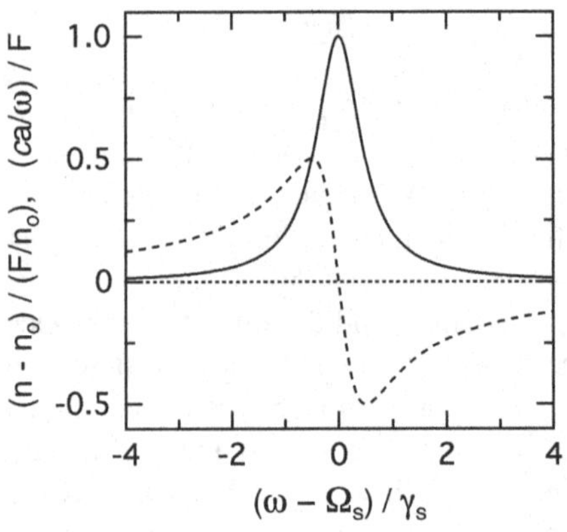

Fig. 3.5 Plots of the absorption (ca/ω, *solid curve*) and refractive index ($n - n_o$, *dashed curve*) as functions of angular frequency (ω) in the region of an absorption band centered at frequency Ω_s, as predicted by the classical theory of dielectric dispersion (Eqs. B3.3.16 and B3.3.17). Frequencies are plotted relative to the damping constant γ_s; ca/ω, relative to the factor $F = 2\pi Ne2f_s/m_e\Omega_s$; and $(n - n_o)$, relative to the factor F/n_o

Box 3.3 (continued)
displacement (x) from this position. The classical equation of motion for such an electron is

$$m_e \frac{d^2x}{dt^2} + g\frac{dx}{dt} + kx = 0, \tag{B3.3.1}$$

where m_e is the electron mass and g is a damping constant. The damping factor is included to account for loss of energy as heat, which in the classical theory is assumed to depend on the electron's velocity. The solution to Eq. (B3.3.1) is $x = \exp(i\omega_o t - \gamma t/2)$ with $\gamma = g/m_e$ and $\omega_o = (k/m_e - \gamma^2/4)^{1/2}$, or $\omega_o \approx (k/m_e)^{1/2}$ if γ is small. The displacement thus executes damped oscillations at frequency ω_o, decaying to zero at long times.

An external electromagnetic field $\boldsymbol{E}(t)$ oscillating at angular frequency ω perturbs the positions of the electrons, setting up an oscillating polarization $\boldsymbol{P}(t)$ that contributes to the total field. The equation of motion for the electron under these conditions becomes

$$m_e \frac{d^2x}{dt^2} + g\frac{dx}{dt} + kx = e\boldsymbol{E}(t) + \frac{4\pi}{3}e\boldsymbol{P}(t), \tag{B3.3.2}$$

where e is the electron charge. The polarization at any given time is proportional to the electron displacement and charge, and also to the number of electrons per unit volume (N):

(continued)

Box 3.3 (continued)

$$\boldsymbol{P}(t) = x(t)eN. \tag{B3.3.3}$$

Combining Eqs. (B3.3.2) and (B3.2.3) gives a differential equation for $\boldsymbol{P}(t)$ with the following solution when γ is small [4]:

$$\boldsymbol{P}(t) = \left(\frac{Ne^2}{3m_e}\right) \frac{\boldsymbol{E}(t)}{(\omega_o)^2 - \omega^2 - (4\pi Ne^2/3m_e) + i\gamma\omega}. \tag{B3.3.4}$$

According to this expression, $\boldsymbol{P}$ oscillates at the same frequency as $\boldsymbol{E}$, but with an amplitude that is a complex function of the frequency. The complex susceptibility (χ_e) is defined as the ratio of the polarization to the field:

$$\chi_e(\omega) = \boldsymbol{P}(t)/\boldsymbol{E}(t) = \left(\frac{Ne^2}{3m_e}\right) \frac{1}{(\omega_o)^2 - \omega^2 - (4\pi Ne^2/3m_e) + i\gamma\omega}. \tag{B3.3.5}$$

The oscillating electron considered above is said to have an *oscillator strength* of unity. The classical theory assumes that each molecule in a dielectric medium could have a set of S electrons with various oscillation frequencies (ω_s), damping factors (γ_s) and oscillator strengths (O_s) between 0 and 1 [4]. If we define

$$(\Omega_s)^2 = (\omega_s)^2 - \left(\frac{4\pi N_m e^2}{3m_e}\right) O_s, \tag{B3.3.6}$$

where N_m is the number of molecules per unit volume, the complex susceptibility becomes

$$\chi_e = \left(\frac{N_m e^2}{m_e}\right) \sum_{s=1}^{S} \frac{O_s}{(\Omega_s)^2 - \omega^2 + i\gamma_s\omega}. \tag{B3.3.7}$$

Equation (B3.3.7) is the fundamental equation of classical dispersion theory [4]. Because χ_e is related to the high-frequency dielectric constant by Eq. (3.30), and the high-frequency dielectric constant is the square of the refractive index (Eq. 3.30), it appears that the dielectric constant and refractive index also should be treated as complex numbers. To indicate this, we'll rewrite Eqs. (3.19) and (3.30) using ε_c and n_c to distinguish the complex dielectric constant and refractive index from, ε and n, the more familiar, real quantities that apply to non-absorbing media:

(continued)

Box 3.3 (continued)

$$(n_c)^2 = \varepsilon_c = 1 + 4\pi\chi_e. \tag{B3.3.8}$$

The meaning of the complex refractive index will become clearer if we consider what happens when a plane wave of monochromatic light passes from a vacuum into an absorbing, but non-scattering dielectric medium. If the propagation axis (y) is normal to the surface, the electric field at the interface can be written

$$\boldsymbol{E} = \boldsymbol{E}_o\exp[i(\omega t - \kappa y)] + \boldsymbol{E}_o\exp[-i(\omega t - \kappa y)], \tag{B3.3.9}$$

where ω is the angular frequency, $\kappa = n\omega/c$, and n is the ordinary refractive index. According to Lambert's law (Eq. 1.3), the intensity of the light will decrease exponentially with position as the ray moves through the medium. If the absorbance is A, the amplitude of the electric field falls off as exp(-ay), where $a = A\ln(10)/2$. The field in the medium thus is

$$\begin{aligned}\boldsymbol{E} &= \boldsymbol{E}_o\exp[i(\omega t - \kappa y) - ay] + \boldsymbol{E}_o\exp[-i(\omega t - \kappa y) - ay] \\ &= \boldsymbol{E}_o\exp\left\{i\omega\left[t - \frac{n}{c}\left(1 - i\frac{a}{\kappa}\right)y\right]\right\} + \boldsymbol{E}_o\exp\left\{-i\omega\left[t - \frac{n}{c}\left(1 - i\frac{a}{\kappa}\right)y\right]\right\} \\ &= \boldsymbol{E}_o\exp\left\{i\omega\left[t - \frac{n_c}{c}y\right]\right\} + \boldsymbol{E}_o\exp\left\{-i\omega\left[t - \frac{n_c{}^*}{c}y\right]\right\},\end{aligned} \tag{B3.3.10}$$

where

$$n_c = n - i(n/\kappa)a = n - i(c/\omega)a. \tag{B3.3.11}$$

Equation (B3.3.11) shows that the imaginary part of n_c is proportional to a, and thus to the absorbance, whereas the real part of n_c pertains to refraction. The real part of n_c cannot, however, be used in Snell's law to calculate the angle of refraction for light entering an absorbing medium, because Snell's law does not hold in the presence of absorption (Box 3.2).

Let's examine the behavior of the complex refractive index in the region of an absorption band representing an individual oscillator with natural frequency ω_s and damping constant γ_s. If ω_s is well removed from the frequencies of the other oscillators in the medium, we can rewrite Eqs. (B3.3.7) and (B3.3.8) as

$$(n_c)^2 = (n_o)^2 + 4\pi\left(\frac{Ne^2}{m_e}\right)\frac{\boldsymbol{O}_s}{(\Omega_s)^2 - \omega^2 + i\gamma_s\omega}, \tag{B3.3.12}$$

where n_o is the contribution of all the other oscillators to the real part of n_c.

(continued)

Box 3.3 (continued)
Absorption due to these other oscillators is assumed to be negligible over the frequency range of interest. Making the approximations $\omega \approx \Omega_s$ and $\Omega_s^2 - \omega^2 = (\Omega_s + \omega)(\Omega_s - \omega) \approx 2\Omega_s(\Omega_s - \omega)$, we obtain

$$(n_c)^2 = (n_o)^2 + \left(\frac{4\pi Ne^2\mathbf{O}_s}{m_e\Omega_s}\right)\frac{1}{2(\Omega_s - \omega) + i\gamma_s\omega}$$
$$= (n_o)^2 + \left(\frac{4\pi Ne^2\mathbf{O}_s}{m_e\Omega_s}\right)\left[\frac{2(\Omega_s - \omega)}{4(\Omega_s - \omega)^2 + (\gamma_s)^2}\right. \quad \text{(B3.3.13)}$$
$$\left. - i\frac{\gamma_s}{4(\Omega_s - \omega)^2 + (\gamma_s)^2}\right].$$

From Eq. (B3.3.11), we also have

$$(n_c)^2 = n^2 - (ca/\omega)^2 - 2i(ca/\omega). \quad \text{(B3.3.14)}$$

Equating the real and imaginary parts of Eqs. (B3.3.13) and (B3.3.14) gives

$$n^2 - (ca/\omega)^2 - (n_o)^2 = \left(\frac{8\pi Ne^2\mathbf{O}_s}{m_e\Omega_s}\right)\left[\frac{\Omega_s - \omega}{4(\Omega_s - \omega)^2 + (\gamma_s)^2}\right] \quad \text{(B3.3.15)}$$

and

$$ca/\omega = \left(\frac{2\pi Ne^2\mathbf{O}_s}{m_e\Omega_s}\right)\left[\frac{\gamma_s}{4(\Omega_s - \omega)^2 + (\gamma_s)^2}\right]. \quad \text{(B3.3.16)}$$

If $(n - n_o)$ and ca/ω are small relative to n_o, the left-hand side of Eq. (B3.3.15) is approximately equal to $2n_o(n - n_o)$, and

$$n \approx n_o + \left(\frac{4\pi Ne^2\mathbf{O}_s}{n_o m_e\Omega_s}\right)\left[\frac{\Omega_s - \omega}{4(\Omega_s - \omega)^2 + (\gamma_s)^2}\right]. \quad \text{(B3.3.17)}$$

Figure 3.5 shows plots of ca/ω and $n - n_o$ as given by Eqs. (B3.3.16) and (B3.3.17). The contribution of the oscillator to the refractive index changes sign at $\omega = \Omega_s$, where the absorption peaks.

In spectral regions that are far from any absorption bands, the term $i\gamma_s\omega_s$ in Eq. (3.27) drops out and the refractive index becomes purely real. The predicted frequency dependence of the refractive index then becomes

(continued)

Box 3.3 (continued)

$$n^2 - 1 = \left(\frac{4\pi N_m e^2}{m_e}\right)\sum_{s=1}^{S}\frac{\boldsymbol{O}_s}{(\Omega_s)^2 - \omega^2}. \tag{B3.3.18}$$

The quantum theory of electric susceptibility is discussed in Box 12.1.

The theory outlined here considers a linear dielectric, in which the polarization of the medium (***P***) is directly proportional to the radiation field (***E***). This linearity breaks down at high field strengths, revealing components of ***P*** that depend on the square or higher powers of ***E***. Since $\cos^2\omega = [1 + \cos(2\omega)]/2$ and $\cos^3\omega = [3\cos\omega + \cos(3\omega)]/4$, these components can give rise to absorption or emission of light at various multiples of the fundamental frequency. Studies of spectroscopic phenomena that reflect second-, third- and even fifth-order polarizations have blossomed with the development of pulsed lasers, which can provide extremely strong electromagnetic fields. We'll discuss some of these experiments from a quantum mechanical approach in Chap. 11. A quantum mechanical description of polarizability is given in Box 12.1.

3.1.6 Local-Field Correction Factors

Now consider an absorbing molecule dissolved in the linear medium we have been discussing. If the molecular polarizability is different from the polarizability of the medium, the local electric field "inside" the molecule ($\boldsymbol{E}_{loc}$) will differ from the field in the medium ($\boldsymbol{E}_{med}$). The ratio of the two fields ($|\boldsymbol{E}_{loc}|/|\boldsymbol{E}_{med}|$), or *local-field correction* factor (f), depends on the shape and polarizability of the molecule and the refractive index of the medium. One model for this effect is an empty spherical cavity embedded in a homogeneous medium with dielectric constant ε. For high-frequency fields ($\varepsilon = n^2$), the electric field in such a cavity is given by

$$\boldsymbol{E}_{cav} = \left(\frac{3n^2}{2n^2 + 1}\right)\boldsymbol{E}_{med} = f_{cav}\boldsymbol{E}_{med} \tag{3.35}$$

[17]. Although a spherical cavity is clearly a very simplistic model for a molecule, models of this type are useful in quantum mechanical theories that include explicit treatments of a molecule's electronic structure. The macroscopic dielectric constant ε or n^2 can be used to describe the electronic polarization of the surrounding medium, while the intramolecular electrons are treated microscopically.

Equation (3.35) neglects the *reaction field* due to polarization of the medium by the molecule itself. The reaction field results partly from interactions of the medium with oscillating dipoles that are induced in the molecule by the electromagnetic radiation (Fig. 3.6). (Again, we are concerned only with electronic induced dipoles

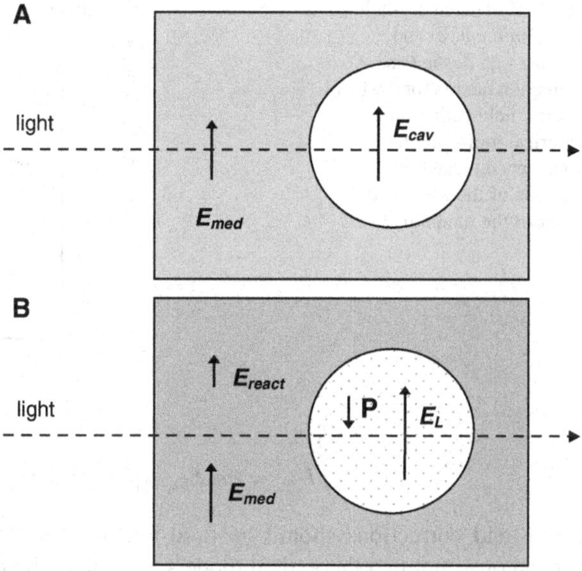

Fig. 3.6 The effective electric field acting on a molecule in a polarizable medium (*shaded rectangles*) is $E_{loc} = fE_{med}$, where E_{med} is the field in the medium and f is the local-field correction factor. In the cavity-field model (**A**) E_{loc} is the field that would be present if the molecule were replaced by an empty cavity (E_{cav}); in the Lorentz model (**B**) E_{loc} is the sum of E_{cav} and the reaction field (E_{react}) resulting from polarization of the medium by induced dipoles within the molecule (*P*)

that can follow the high-frequency oscillation of electromagnetic radiation. If the medium contains molecules that can rotate or bend so as to realign their permanent dipole moments, the reaction field also includes a static component.) The high-frequency field acting on a spherical molecule can be taken to be the sum of the reaction field and the cavity field described by Eq. (3.35b). An approximate expression for this total field, due to H.A. Lorentz [18], is

$$\mathbf{E}_L = \left(\frac{n^2+2}{3}\right)\mathbf{E}_{med} = f_L\mathbf{E}_{med}. \tag{3.36}$$

The factor $(n^2 + 2)/3$ is called the *Lorentz correction.* Liptay [19] gives expressions that include the molecular radius, dipole moment and polarizability explicitly. More elaborate expressions for f also have been derived for cylindrical or elipsoidal cavities that are closer to actual molecular shapes [20, 21].

Figure 3.7 shows the local-field correction factors given by Eqs. (3.35) and (3.36). The Lorentz correction is somewhat larger and may tend to overestimate the contribution of the reaction field, because the cavity-field expression agrees better with experiment in some cases (Fig. 4.5).

With the local-field correction factor, the relationships between the energy density and irradiance in the medium ($\rho(\nu)$ and $I(\nu)$) and the amplitude of the local field ($|\mathbf{E}_{loc(o)}|$) become:

$$\rho(\nu) = n^2\left|\mathbf{E}_{loc(o)}\right|^2\rho_\nu(\nu)/2\pi f^2 \tag{3.37}$$

and

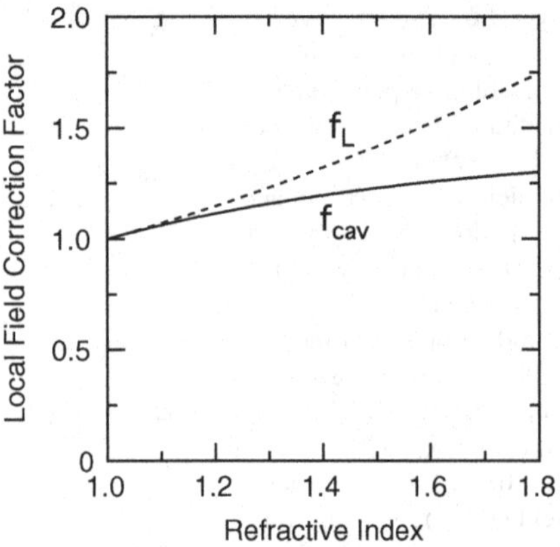

Fig. 3.7 The cavity-field (f_{cav}, *solid curve*) and Lorentz (f_L, *dashed curve*) correction factors for the local electric field acting on a spherical molecule in a homogeneous medium, as a function of the refractive index of the medium

$$I(\nu) = cn\left|\boldsymbol{E}_{loc(o)}\right|^2 \rho_\nu(\nu)/2\pi f^2. \tag{3.38}$$

Local-field corrections should be used with caution, bearing in mind that they depend on simplified theoretical models and cannot be measured directly.

3.2 The Black-Body Radiation Law

It has long been known that the radiation emitted by a heated object shifts to higher frequencies as the temperature is increased. When we discuss fluorescence in Chap. 5, we will need to consider the electromagnetic radiation fields inside a closed box whose walls are at a given temperature. Lord Rayleigh derived an expression for the energy distribution of this *black-body radiation* in 1900 by considering the number of possible modes of oscillation (standing waves) with frequencies between ν and $\nu + d\nu$ in a cube of volume V. Considering the two possible polarizations of the radiation (Sect. 3.3), taking the refractive index (n) of the medium inside the cube into account, and including a correction pointed out by Jeans, the number of oscillation modes in the frequency interval $d\nu$ is

$$\rho_\nu(\nu)V d\nu = \left(8\pi n^3 \nu^2 V/c^3\right)d\nu \tag{3.39}$$

[22]. Following classical statistical mechanics, Rayleigh assumed that each mode would have an average energy of $k_B T$, independent of the frequency. Since ρ_ν increases quadratically with ν (Eq. 3.39), this analysis led to the alarming conclusion that the *energy density* (the product of ρ_ν and the average energy per mode) goes to infinity at high frequencies. Experimentally, the energy density was found to increase with frequency in accord with Rayleigh's prediction at low frequencies, but then to pass through a maximum and decrease to zero.

Max Planck saw that the observed dependence of the energy density on frequency could be reproduced by introducing the *ad hoc* hypothesis that the material in the walls of the box emit or absorb energy only in integral multiples of $h\nu$, where h is a constant. He assumed further that, if the material is at thermal equilibrium at temperature T, the probability of emitting an amount of energy, E_j, is proportional to $\exp(-E_j/k_BT)$ where k_B is the Boltzmann constant (Box 2.6). With these assumptions, the average energy of an oscillation mode with frequency ν beomes

$$\overline{E} = \left(\sum_j E_j \exp(-E_j/k_BT)\right) / \left(\sum_j \exp(-E_j/k_BT)\right) \tag{3.40a}$$

$$= \left(\sum_j jh\nu \exp(-jh\nu/k_BT)\right) / \left(\sum_j \exp(jh\nu/k_BT)\right). \tag{3.40b}$$

The sums can be evaluated by letting $x = \exp(-h\nu/k_BT)$, using the expansion $(1-x)^{-1} = 1 + x + x^2 + \ldots = \sum_j x^j$, and noting that $\sum_j jx^j = xd\left(\sum_j x^j\right)/dx$. This yields

$$\overline{E} = \frac{h\nu}{\exp(h\nu/k_BT) - 1}. \tag{3.40c}$$

(Note that $\bar{E}$ denotes a thermal average of the energy, a scalar quantity, not the electric field vector $\boldsymbol{E}$.) Finally, multiplying Planck's expression for $\bar{E}$ (Eq. 3.40c) by ρ_ν (Eq. 3.39) gives

$$\rho(\nu) = \overline{E}(\nu)\rho_\nu = \left(8\pi h n^3 \nu^3/c^3\right)/[\exp(h\nu/k_BT) - 1], \tag{3.41a}$$

or expressed as irradiance,

$$I(\nu) = \rho(\nu)c/n = \left(8\pi h n^3 \nu^3/c^2\right)/[\exp(h\nu/k_BT) - 1]. \tag{3.41b}$$

Figure 3.8 shows plots of $\rho(\nu)$ as functions of the wavelength and wavenumber of the radiation. The predictions are in accord with the measured energy density of black-body radiation at all accessible frequencies and temperatures. In particular, Eq. (3.41a) accounts for the observations that the energy density of blackbody radiation increases with the fourth power of T (the *Stefan-Boltzmann law*) and that the wavelength of peak energy density is inversely proportional to T (the *Wien displacement law*). Although the derivation outlined above invokes a Boltzmann distribution, the same result can be obtained by using the Bose-Einstein distribution for a photon gas at thermal equilibrium [23].

Planck's theory did not require the radiation field itself to be quantized, and Planck did not conclude that it is [24]. Because the radiation inside a black-body box is emitted and absorbed by the walls of the box, only the energy levels of the

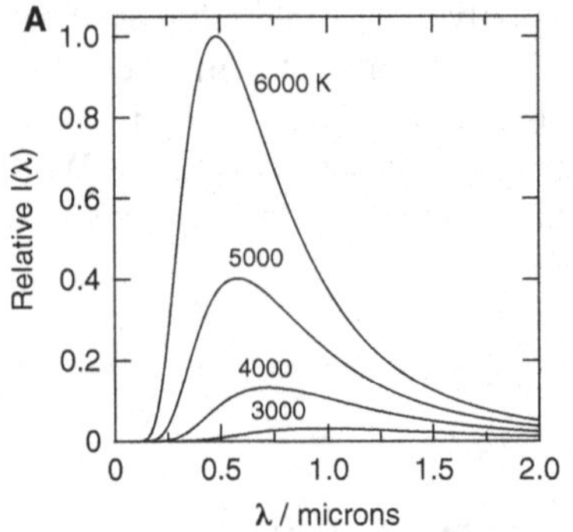

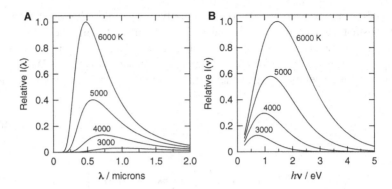

Fig. 3.8 The intensity (I) of the black-body radiation emitted by an object at a temperature of 3,000, 4,000, 5,000 or 6,000 K, plotted as a function of wavelength (λ, *left*) and photon energy ($h\nu$, *right*). The surface temperature of the sun is about 6,600 K, and a tungsten-halogen lamp filament typically has an effective temperature of about 3,000 K

material comprising the walls must be quantized. The theory is consistent with quantization of the radiation field as well, but does not demand it.

In addition to its pivotal contribution in the development of quantum theory, the black-body radiation law has practical applications in spectroscopy. Because it describes the spectrum of the light emitted by an incandescent lamp with a filament at a known temperature, Eq. (3.41) can be used to calibrate the frequency-dependence of a photodetector or monochromator. However, the actual emission spectrum of a lamp can depart somewhat from Eq. (3.41), depending on the material used for the filament [25].

3.3 Linear and Circular Polarization

In the quantum theory described in the following section, the eigenfunctions of the Schrödinger equation for a radiation field have angular momentum quantum numbers $s = 1$ and $m_s = \pm 1$. The two possible values of m_s correspond to left ($m_s = +1$) and right ($m_s = -1$) *circularly polarized light*. Figure 3.9 illustrates this property. In this depiction, the electric field vector $\boldsymbol{E}$ has a constant magnitude but its orientation rotates with time at frequency ν. The component of $\boldsymbol{E}$ parallel to any given axis normal to the propagation axis oscillates at the same frequency, and oscillates along the the propagation axis with wavelength λ.

For radiation propagating in the y direction, the time dependence of the rotating field can be written (neglecting an arbitrary phase shift):

$$\boldsymbol{E}_{\pm} = 2E_o[\cos(2\pi\nu t)\hat{z} \pm \sin(2\pi\nu t)\hat{x}][\cos(2\pi y/\lambda)\hat{z} \pm \sin(2\pi y/\lambda)\hat{x}], \quad (3.42a)$$

$$\boldsymbol{B}_{\pm} = 2E_o[\cos(2\pi\nu t)\hat{x} \mp \sin(2\pi\nu t)\hat{z}][\cos(2\pi y/\lambda)\hat{x} \mp \sin(2\pi y/\lambda)\hat{z}], \quad (3.42b)$$

where E_o is a scalar amplitude and the + and − subscripts refer to left and right

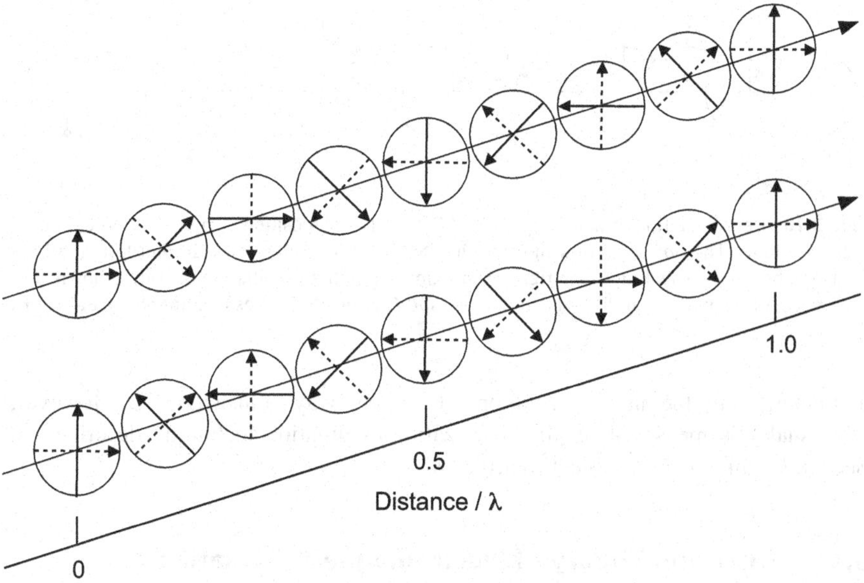

Fig. 3.9 Electric and magnetic fields in right (*top*) and left (*bottom*) circularly polarized light. In this illustration, both beams propagate diagonally upward from left to right. The *solid arrows* in the disks indicate the orientations of the electric field (***E***) at a given time as a function of position along the propagation axis; the *dashed arrows* show the orientations of the magnetic field (***B***). The field vectors both rotate so that their tips describe right- or left-handed corkscrews, making one full turn in a distance corresponding to the wavelength of the light (λ)

circular polarization, respectively. These expressions are solutions to the general wave equation that satisfies Maxwell's equations for a non-conducting medium with no free charges (Eqs. (B3.1.7 and B3.1.8) in Box 3.1). The different algebraic combinations of the z and x components in Eqs. (3.42a) and (3.42b) keep the magnetic field pependicular to the electric field for both polarizations.

A linearly polarized beam of light can be treated as a coherent superposition of left and right circularly polarized light, as shown in Fig. 3.10. Changing the phase of one of the circularly polarized components relative to the other rotates the plane of the linear polarization. Unpolarized light consists of a mixture of photons with left and right circular polarization and with electric fields rotating at all possible phase angles, or equivalently, a mixture of linearly polarized light with all possible orientation angles.

If linearly polarized light passes through a sample that preferentially absorbs one of the circularly polarized components, the transmitted beam will emerge with elliptical polarization. The *ellipticity* is defined as the arctangent of the ratio I_{minor}/I_{major}, where I_{minor} and I_{major} are the light intensities measured through polarizers parallel to the minor and major axes of the elipse. Ellipticity can be measured by using a quarter-wave plate to convert the elliptically polarized light back to a linearly polarized beam that is rotated in alignment relative to the original beam,

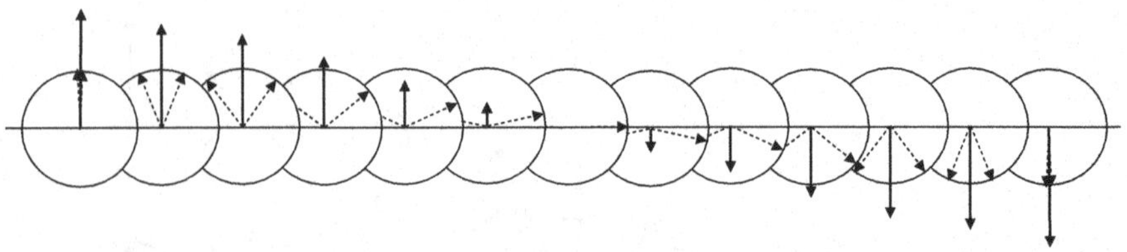

Fig. 3.10 Representation of linearly polarized light as a superposition of right and left circularly polarized light. The *dotted arrows* indicate the electric field vectors of right and left circularly polarized beams propagating to the right. Approximately half an oscillation period is shown. The vector sum of the two circularly polarized fields (*solid arrows*) oscillates in amplitude parallel to a fixed axis

and determining the angle of rotation. This is one way to measure CD. However, CD usually is measured by the polarization-modulation technique described in Sect. 1.8, which is less subject to artifacts.

3.4 Quantum Theory of Electromagnetic Radiation

The Schrödinger equation was first applied to electromagnetic radiation in 1927 by Paul Dirac [26]. The notion of a quantized radiation field that emerged from this work reconciled some of the apparent contradictions between earlier wave and particle theories of light, and as we will see in Chap. 5, led to a consistent explanation of the "spontaneous" fluorescence of excited molecules.

To develop the quantum theory of electromagnetic radiation, it is convenient to describe a radiation field in terms of the vector potential $\boldsymbol{V}$ that is introduced in Box 3.1. Consider the vector potential associated with a plane wave of light propagating in the y direction and polarized in the z direction, and suppose the radiation is confined within a cube with edge L. As discussed in Sect. 3.2, the radiation with frequencies in a specified interval is restricted to a finite number of oscillation modes, each with a discrete wavelength $\lambda_j = L/2\pi n_j$ where n_j is a positive integer. The total energy of the radiation is the sum of the energies of these individual modes, and the total vector potential evidently is a similar sum of the individual vector potentials:

$$\boldsymbol{V} = \sum_j \boldsymbol{V}_j. \tag{3.43}$$

According to Eq. (3.18), the contribution to $\boldsymbol{V}$ from mode j can be written

$$\boldsymbol{V}_j = \left(\frac{4\pi c^2}{L^3}\right)^{1/2} \hat{z}\left[\exp(-i\omega_j t)\exp(iy/\lambda_j) + \exp(i\omega_j t)\exp(-iy/\lambda_j)\right] \tag{3.44a}$$

$$= q_j(t)\boldsymbol{A}_j(y) + q_j{}^*(t)\boldsymbol{A}_j{}^*(y), \tag{3.44b}$$

where

$$q_j = \exp(-i\omega_j t), \tag{3.45a}$$

and

$$A_j = \left(\frac{4\pi c^2}{L^3}\right)^{1/2} \hat{z}\exp(iy/\lambda_j), \tag{3.45b}$$

and $\omega_j = 2\pi\nu_j = c/\lambda_j$. The vector fields V_j, which are real quantities, are written here in terms of products of two complex functions and their complex conjugates. The first function (q_j) is a scalar that depends only on time; the second (A_j) is a vector function of position. The position-dependent factors are normalized so that $\langle A_j|A_j\rangle = 4\pi c^2$, while the factors for different modes are orthogonal: $\langle A_i|A_j\rangle = 0$ for $i \neq j$. Equations (3.43–3.45b) hold for progressive waves as well as for standing waves, although the restriction on the number of possible modes applies only to standing waves.

To put the energy of the radiation field in Hamiltonian form, we now define two real variables,

$$Q_j(t) = q_j(t) + q_j{}^*(t), \tag{3.46}$$

and

$$P_j(t) = \frac{\partial Q_j}{\partial t} = -i\omega_j\{q_j(t) - q_j{}^*(t)\}. \tag{3.47}$$

From Eqs. (B3.1.17), (3.24), (3.44a,b) and (3.45a,b), the contribution of mode j to the energy of the field, integrated over the volume of the cube, then takes the form [2, 4]

$$E_j = \frac{1}{2}\left\{P_j^2 + \omega_j^2 Q_j^2\right\}. \tag{3.48}$$

A little algegra will show that Q_j obeys a classical wave equation homologous to Eqs. (3.13) and (B3.1.7), and that Q_j and P_j have the formal properties of a time-dependent position (Q_j) and its conjugate momentum (P_j) in Hamiltonian's classical equations of motion:

$$\frac{\partial E_j}{\partial P_j} = \frac{\partial Q_j}{\partial t}, \tag{3.49a}$$

and

$$\frac{\partial E_j}{\partial Q_j} = \frac{\partial P_j}{\partial t}. \tag{3.49b}$$

In addition, Dirac noted that Eq. (3.48) is identical to the classical expression for the energy of a harmonic oscillator with unit mass (Eq. 2.28). The first term in the braces corresponds formally to the kinetic energy of the oscillator; the second, to the potential energy. It follows that if we replace P_j and Q_j by momentum and position operators $\widetilde{P}_j$ and $\widetilde{Q}_j$, respectively, the eigenstates of the Schrödinger equation for electromagnetic radiation will be the same as those for harmonic oscillators. In particular, each oscillation mode will have a ladder of states with wavefunctions $\chi_{j(n_j)}$ and energies

$$E_{j(n_j)} = (n_j + 1/2)h\nu_j, \tag{3.50}$$

where $n_j = 0, 1, 2 \ldots.$

The transformation of the time-dependent function P_j into a momentum operator is consistent with Einstein's description of light in terms of particles (photons), each of which has momentum $h\overline{\nu}$ (Sect. 1.6 and Box 2.3). We can interpret the quantum number n_j in Eq. (3.50) either as the particular excited state occupied by oscillator j, or as the number of photons with frequency ν_j. The oscillating electric and magnetic fields associated with a photon can still be described by Eqs. (3.44) and (3.45) if the amplitude factor E_o is scaled appropriately. However, we will be less concerned with the spatial properties of photon wavefunctions themselves than with the matrix elements of the position operator $\widetilde{Q}$. These matrix elements play a central role in the quantum theory of absorption and emission, as we'll discuss in Chap. 5.

The total energy of a radiation field is the sum of the energies of its individual modes, and according to Eq. (3.50), the energy of each mode increases with the number of photons in the mode. But, like the harmonic oscillator, an electromagnetic radiation field has a *zero-point* or *vacuum energy* when every oscillator is in its lowest level ($n_j = 0$). Because a radiation wave in free space could have an infinite number of different oscillation frequencies, the total zero-point energy of the universe appears to be infinite, which may seem a nonsensical result. One way to escape this dilemma would be to argue that the universe is bounded, so that there is no such thing as completely free space. In this picture, the zero-point energy of the universe becomes an unknown, but finite constant. However, we can arbitrarily set the zero-point energy of each oscillation mode to zero by simply subtracting a constant ($h\nu_j/2$) from the Hamiltonian of Eq. (3.48), making the energy associated with each mode

$$E_{j(n_j)} = n_j h\nu_j. \tag{3.51}$$

This is common practice for other types of energy, which can be expressed with respect to any convienent reference. The relativistic rest-energy mc^2, for example, usually is omitted in discussions of the nonrelativistic energy of a particle. See [2, 4, 27] for further discussion of this point.

The zero-point eigenstate is a critical feature of the quantum theory. It suggests, surprisingly, that a radiation field might interact with a molecule even if the number of photons in the field is zero! In Chap. 5, we will discuss how this interaction gives rise to fluorescence. We also will see there that most transitions between different states of a radiation field change the energy of the field by $\pm h\nu_j$, and result in creation or disappearance of a single photon. The identification of the quantum states of radiation field with those of harmonic oscillators makes it possible to evaluate the matrix elements for such transitions.

Experimental evidence for the existence of the vacuum radiation field has come from observations of the *Casimir effect*. First predicted by the Dutch physicist H. Casimir in 1948, this is an attractive force between reflective objects in a vacuum. Consider two polished square plates with parallel faces separated by distance L. Standing waves in the gap between the plates must have wavelengths of $L/2n$ where n is a positive integer. Because the number of such possible radiation modes is proportional to L, the total energy of the vacuum field decreases when the plates move closer together. The attractive force has been measured for objects of a variety of shapes and found to agree well with predictions [28–30].

The fact that they have integer spin ($m_s = \pm 1$) implies that photons obey Bose-Einstein statistics (Box 2.6). This means that any number of photons can have the same energy ($h\nu$) and spatial properties, and conversely, that an individual radiation mode can have any number of photons. Accumulation of many photons in a single radiation mode makes possible the coherent radiation emitted by lasers.

In addition to accounting for the quantization of radiation, Dirac introduced a relativistic theory of electrons. But even with these advances, the quantum theory at this stage left fundamental questions unresolved, including the mechanism by which charged particles interact at a distance in a vacuum. What does it mean to say that an electron gives rise to an electromagnetic field? An answer to this question came from work by R. Feynman, J. Schwinger and S.-I.Tomonaga in the 1950s. In the theory of *quantum electrodynamics* that emerged from these studies, charged particles interact by exchanging photons, and the charge of a particle is a measure of its tendency to absorb or emit photons. However, photons that move from one particle to another are termed "virtual" photons because they cannot be intercepted and measured directly.

Contrary to the principles of classical optics, the theory of quantum electrodynamics asserts that photons do not necessarily travel in straight lines. To find the probability that a photon will move from point A to point B, we must sum the amplitudes of wavefunctions for all possible paths between the two points, including even round-about routes via distant galaxies and paths in which the photon splits transiently into an electron-positron pair. Although there are an infinite number of paths between any two points, destructive interferences cancel most of the contributions from all the indirect paths, leaving only small (but sometimes significant) corrections to the laws of classical optics and electrostatics. This is because small differences between the indirect routes have large effects on the overall lengths of the paths, causing phase shifts of the oscillations that photons travelling by these routes contribute at the destination. Feynman [31] has provided a readable introduction to the theory of quantum electrodynamics. For more complete treatments see [32–34].

3.5 Superposition States and Interference Effects in Quantum Optics

Sections 2.2.1 and 2.3.2 introduced the idea of a superpostion state whose wavefunction is a linear combination of two or more eigenfunctions with fixed phases. We asserted there that a system must be described by such a linear combination if it cannot be assigned uniquely to an individual eigenstate. The interference terms in expectation values for superposition states lead to some of the most intriguing aspects of quantum optics, including the fringes in Young's classical double-slit experiments. Figure 3.11A shows an experiment that illustrates the point well [35–41]. The apparatus is called a Mach-Zender interferometer. Photons enter the interferometer at the upper left and are detected by a pair of photon counters, D1 and D2. The light intensity is low enough so that no more than one photon is in the apparatus at any given time. BS1 is a beamsplitter that, on average, transmits 50% of the photons and reflects the other 50%. Mirrors M1 and M2 reflect all the photons reaching them to a second beamsplitter (BS2), which again transmits 50% and reflects 50%. If beam-splitter BS2 is removed, a photon is detected by detector D1 half the time and by D2 the other half, just as we would expect. If either of the two paths between BS1 and BS2 is blocked, half the photons get through, and again half of these are detected at D1 and half at D2. That also seems expected. But if both paths are open, all the photons are detected at D1, and none at D2!

To account for this surprising result, note that when there are two possible paths of equal length between BS1 and BS2, we have no way of knowing which path a given photon follows. According to the prescriptions of quantum electrodynamics (Sect. 3.4), we therefore must write the wavefunction for a photon reaching D1 or D2 as a sum of the two possibilities. Suppose that each of the beamsplitters transmits photons with probability $|C_T|^2$ and reflects photons with probability $|C_R|^2$. The wavefunction for photons reaching D1 then is the weighted sum:

$$\Psi_{D1} = C_R C_T \Psi + C_T C_R \Psi = (C_R C_T + C_T C_R)\Psi, \tag{3.52a}$$

where $\Psi(r,t)$ is the wavefunction for an individual photon (Fig. 3.11A). The wavefunction for photons reaching D2 is, similarly,

$$\Psi_{D2} = C_T C_T \Psi + C_R C_R \Psi = (C_T C_T + C_R C_R)\Psi. \tag{3.52b}$$

We have omitted the coefficient for reflection at mirrors M1 and M2 since this has a

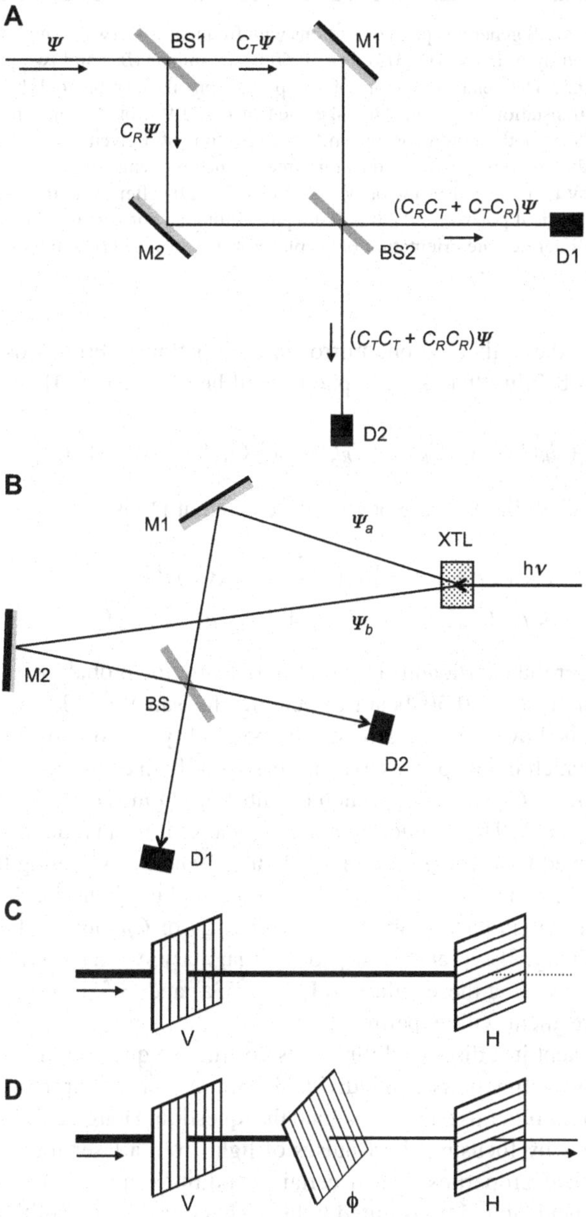

Fig. 3.11 (**A**) Single-photon interference in a Mach-Zender interferometer with equal arms. *BS*1 and *BS*2 are beam-splitters with coefficients C_T and C_R for transmission and reflection, respectively ($|C_T|^2 = |C_R|^2 = 1/2$); *M*1 and *M*2 are mirrors (100% reflecting); and *D*1 and *D*2 are photon-counting detectors. The dependence of photon wavefunction Ψ on time and the distance along the optical path is not indicated explicitly. If BS2 is removed, or if either path is blocked before BS2, photons are detected at D1 and D2 with equal probability; but when BS2 is present and both paths are open, photons are detected only at D1. (**B**) Two-photon quantum interference. Short pulses of light with frequency ν are focused into a crystal with nonlinear optical properties (*XTL*). This

Fig. 3.11 (continued) generates pairs of photons with frequency $\nu/2$ (wavefunctions Ψ_a and Ψ_b), which are reflected by mirrors ($M1$, $M2$) to a 50:50 beam-splitter (BS) and two photon-counting detectors ($D1$, $D2$). The relative lengths of the paths from the crystal to D1 and D2 can be controlled by a translation stage like that sketched in Fig. 1.9 (not shown here). If the beam-splitter either reflects both photons or transmits both photons of a given pair, D1 and D2 should each get one of the photons; however, quantum interference prevents this unless the photons are distinguishable by their arrival times or polarizations. (**C, D**) After an unpolarized light beam passes through a vertical polarizer (V), it will not pass through a horizontal polarizer (H). But if a polarizer with an intermediate orientation (ϕ) is placed between V and H, some of the light passes through both this polarizer and H

magnitude of 1.0 and there is one mirror in each pathway. From Eqs. (3.52a) and (3.52b), the probability that a given photon will be detected at D1 is

$$P_{D1} = \langle (C_R C_T + C_T C_R)\Psi | (C_R C_T + C_T C_R)\Psi \rangle = 4|C_R|^2|C_T|^2, \qquad (3.53a)$$

whereas the probability that the photon is detected at D2 is

$$\begin{aligned} P_{D2} &= \langle (C_T C_T + C_R C_R)\Psi | (C_T C_T + C_R C_R)\Psi \rangle \\ &= |C_T|^2|C_T|^2 + |C_R|^2|C_R|^2 + (C_T^* C_R)^2 + (C_R^* C_T)^2. \end{aligned} \qquad (3.53b)$$

Now consider the coefficients C_T and C_R. Because the probabilities of reflection and transmission at a 50:50 beamsplitter are the same, we know that $|C_T|^2 = |C_R|^2 = 1/2$. This, however, leaves open the possibility that one of the coefficients is imaginary, which in fact proves to be necessary. If both coefficients were real, we would have $C_T = \pm C_R = \pm 2^{-1/2}$, which on substitution in Eqs. (3.53a) and (3.53b) gives $P_{D1} = P_{D2} = 1$. That cannot be correct because it means that a single photon would be detected with 100% certainty at both D1 and D2, violating the conservation of energy. Trying $C_T = \pm C_R = \pm i2^{-1/2}$ (i.e., making both coefficients imaginary) gives the same unacceptable result. Either C_T or C_R, but not both, therefore must be imaginary. If we choose C_R to be imaginary, we have $C_T = \pm 2^{-1/2}$ and $C_R = \pm i2^{-1/2}$. Inserting these values in Eqs. (3.53a) and (3.53b) gives $P_{D1} = 1$ and $P_{D2} = 0$ in agreement with experiment.

The experiment just discussed illustrates destructive quantum mechanical interference in the wavefunction of an individual photon. Can similar interference occur between different photons? To investigate this question, Hong et al. [42] generated pairs of photons by focusing short pulses of light from a laser into a crystal with non-linear optical properties. When certain conditions are met for matching the phases of the incident and transmitted light, such a crystal can "split" photons with incident frequency ν into two photons with frequency $\nu/2$. Hong et al. sent the two photons along separate paths to a 50:50 beamsplitter and on to a pair of detectors (D1 and D2) as shown in Fig. 3.11B. At the beamsplitter, there are four possibilites. Both photons might be transmitted; both might be reflected; photon a might be transmitted and photon b reflected; or photon a might be reflected and b transmitted. In the first two cases, detectors D1 and D2 each would receive a photon; in the third and fourth cases, one detector would get two photons and the other none. If the paths from the crystal to D2 and D2 are of different lengths, the four possible outcomes have equal probabilities. But when the paths are adjusted to be the same, a surprising thing happens: the two photons of each pair always go to the same detector. To see how this occurs, note that in cases 3 and 4 (one photon transmitted and the other reflected), the overall wavefunction acting on the detectors can be written as a product of the individual waveforms (Ψ_a and Ψ_b) with the applicable coefficients for transmission and reflection ($C_R\Psi_a \cdot C_T\Psi_b$ or $C_T\Psi_a \cdot C_R\Psi_b$). Inserting the values of C_T and C_R gives a non-zero amplitude. On the other hand, if the detectors can not distinguish between the two photons by their arrival times, frequencies or polarizations, we have no way to distinguish between cases 1 and 2 (both photons transmitted or both photons reflected). We therefore must write the overall waveform for these cases as a sum of $C_T\Psi_a \cdot C_T\Psi_b$ and $C_R\Psi_a \cdot C_R\Psi_b$. The amplitude of this sum evaluates to zero:

$$\begin{aligned}\Psi &= 2^{-1/2}(C_T\Psi_a C_T\Psi_b + C_R\Psi_a C_R\Psi_b)\\ &= 2^{-1/2}\left(2^{-1/2}2^{-1/2} + i2^{-1/2}i2^{-1/2}\right)\Psi_a\Psi_b.\end{aligned} \tag{3.54}$$

Two-photon quantum interference sometimes is ascribed to local interference of the photons at the beamsplitter. However, Kwiat *et al.* [43] and Pittman *et al.* [44] demonstrated experimentally that the critical factor is the whether the wavefunctions are distinguishable at the detectors. The photons do not need to be at the beamsplitter simultaneously.

Linearly polarized light provides another instructive illustration of superposition states in quantum optics. As discussed in Sect. 3.3, unpolarized light can be viewed as a mixture of photons with all possible linear polarizations. Light that is polarized at an angle θ with respect to an arbitrary "vertical" axis can be viewed as a coherent superposition of vertically and horizontally polarized light with coefficients $C_V = \cos(\theta)$ and $C_H = \sin(\theta)$:

$$\Psi_\theta = C_V\Psi_V + C_H\Psi_H = \cos(\theta)\Psi_V + \sin(\theta)\Psi_H. \tag{3.55}$$

The probability that a photon with this polarization will pass through a vertical polarizer is $\cos^2(\theta)$. After passage through such polarizer, the light is completely polarized in the vertical direction, and the probability that it will pass through a horizontal polarizer is zero (Fig. 3.11C). But if a polarizer with a different orientation, ϕ, is placed between the vertical and horizontal polarizers, photons will go through this second polarizer with probability $\cos^2(\phi)$, and (assuming that $0 < \cos^2(\phi) < 1$) some of those photons now will pass through the horizontal polarizer (Fig. 3.11D). The interpretation is that the vertically polarized light consists of a coherent superposition of light polarized parallel and perpendicular to the second polarizer, both of which have a non-zero projection on the horizontal axis. The intensity of the light of passing the horizontal polarizer peaks at $\phi = 45°$, where it is 1/8 relative to the intensity reaching the second polarizer.

3.6 Distribution of Frequencies in Short Pulses of Light

The idealized wave described by Eq. (3.15a) or (3.15b) continues indefinitely for all time and all values of y. Any real beam of light must start and stop at some point, and so cannot be described completely by this expression. It can, however, be described by a linear combination of idealized waves with a distribution of frequencies. Such a combination of waves is called a *wave group* or *wavepacket*. The details of the distribution function depend on the width and shape of the pulse. This description is essentially the same as using a linear combination of wavefunctions for a localized particle in a box (Sect. 2.2.2) or a harmonic potential well (Sect. 2.2.1 and Chap. 11).

The short pulses produced by mode-locked Ti:sapphire or dye lasers typically have Gaussian or $\mathrm{sech}^2(t)$ shapes. If all the oscillations are in phase at $t = 0$, when the electric field of the light peaks, the dependence of the field strength on time in a Gaussian pulse can be written

$$\boldsymbol{E}(t) = \left(2\pi\ \tau^2\right)^{-1/2} \exp\left(-t^2/2\tau^2\right) \cos\left(\omega_o t\right), \tag{3.56}$$

where τ is a time constant and ω_o is the center angular frequency of the light (radians per second, or 2π times the center frequency in herz). The Gaussian function $\exp(-t^2/2\tau^2)$ has a full width at half-maximal amplitude (FWHM) of $(8 \cdot \ln 2)^{1/2}\tau$, or 2.355τ. The factor $(2\pi\tau^2)^{-1/2}$ normalizes the area under the curve to 1. The measured intensity of the pulse, being proportional to $|\boldsymbol{E}(t)|^2$ or $\exp(-t^2/\tau^2)$, would be narrower by a factor of $2^{1/2}$. Its FWHM thus is $2(\ln 2)^{1/2}\tau$, or 1.665τ.

We can equate the time-dependent function on the right side of Eq. (3.56) to a frequency-dependent function

$$|\boldsymbol{E}(t)| = \frac{1}{\sqrt{2\pi}} \int_{-\infty}^{\infty} G(\omega) \exp(i\omega t) d\omega, \tag{3.57}$$

where $G(\omega)$ is the distribution of angular frequencies in the pulse. This means that the temporal shape of the field pulse, $|\boldsymbol{E}(t)|$, is the Fourier transform of the frequency distribution $G(\omega)$, and conversely, that $G(\omega)$ is the Fourier transform of $|\boldsymbol{E}(t)|$ (Appendix A3). For the Gaussian shape function $\exp(-t^2/2\tau^2)$, the solution to $G(\omega)$ is

$$G(\omega) = \left(\frac{\tau}{\sqrt{2\pi}}\right) \exp\left\{-(\omega - \omega_o)^2 \tau^2/2\right\}. \tag{3.58}$$

The field thus includes a Gaussian distribution of angular frequencies around ω_o, with a FWHM of $(8 \cdot \ln 2)^{1/2}/\tau$, or $2.35/\tau$ radians/s. In frequency units, the FWHM is $(8 \cdot \ln 2)^{1/2}/2\pi\tau$, or $0.375/\tau$ Hz. The shorter the pulse, the broader the spread of frequencies. The measured spectrum of the intensity again is narrower by a factor of $2^{1/2}$, and has a FWHM of $0.265/\tau$ Hz.

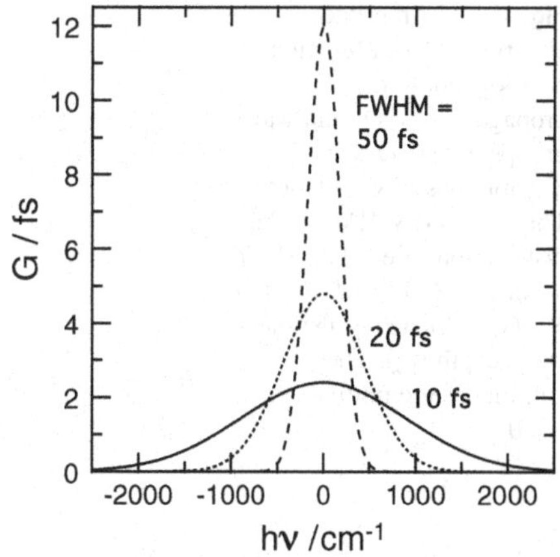

Fig. 3.12 Spectra of the electric fields in homogeneously broadened Gaussian pulses with widths (*FWHM*) of 10, 20 and 50 fs ($\tau = 6$, 12 and 30 fs)

A Gaussian pulse with $\tau = 6$ fs (a measured FWHM of 10 fs, the order of magnitude of the shortest pulses that can be generated by current Ti:sapphire lasers) includes a span of about 6.25×10^{13} Hz, which corresponds to an energy ($h\nu$) band of 2.0×10^3 cm^{-1}. If the spectrum is centered at 800 nm (12,500 cm^{-1}), its FWHM is 274 nm. Figure 3.12 shows the energy distribution function for such a pulse and for pulses with FWHM's of 20 and 50 fs.

For a Gaussian pulse, the product of the measured temporal width (1.665τ) and the frequency width ($0.265/\tau$) is fixed at 0.441. Fleming [45] gives corresponding expressions for pulses with other shapes. For a square pulse, the product of the measured temporal and frequency widths is 0.886; for a pulse in which the intensity is proportional to $\text{sech}^2(t)$, the product is 0.315. These expressions assume that the frequency distribution arises solely from the finite length of the pulse. Such a pulse is said to be *transform-limited*. Light from an incoherent source such as a xenon flash lamp contains a distribution of frequencies that are unrelated to the length of the pulse because atoms or ions with many different energies contribute to the emission.

Figure 3.13 shows the electric field at a fixed time as a function of position in wave groups with three different distributions of wavelengths. If the distribution is very narrow relative to the mean wavelength (λ_o), the wave group resembles a pure sine wave over many periods of the oscillation. With broader distributions, the envelope of the oscillations is more bunched, and the envelope moves through space with a group velocity u that can be written [4]

$$u = \frac{c}{n}\left\{1 + \frac{\lambda}{n}\frac{dn}{d\lambda}\right\}. \tag{3.59}$$

For visible light, this is within 5% of c/n in most liquids. The envelope smears out with time as the individual oscillations get increasingly out of phase.

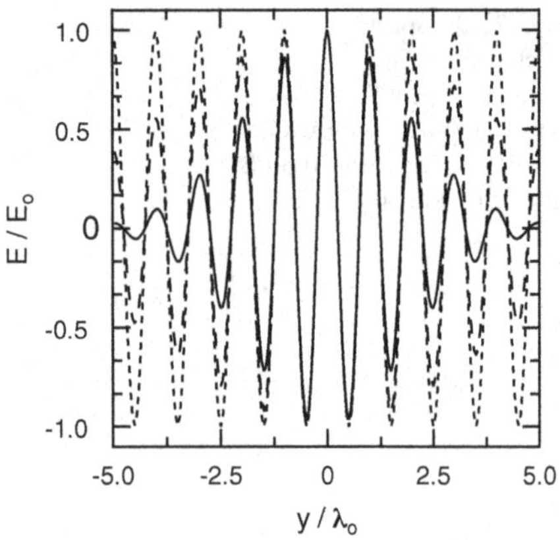

Fig. 3.13 Amplitude of the electric field as a function of position along the propagation axis (y) in wave groups with Gaussian distributions of wavelengths. The widths (FWHM) of the distributions are 0.1% (*short dashes*), 2% (*long dashes*) or 5% (*solid curve*) of the mean wavelength (λ_o). The amplitudes are normalized at $y = 0$

Exercises

1. Consider three particles with the following charges (esu) and coordinates (cm): $q_1 = 0.5$, $\boldsymbol{r}_1 = (1.0, 0.0, 0.5)$; $q_2 = -0.6$, $\boldsymbol{r}_2 = (-0.5, 0.5, 0.0)$; $q_3 = -0.4$, $\boldsymbol{r}_3 = (1.5, 1.0, -0.5)$. Calculate the following electrostatic quantities and give the results in both CGS and MKS units: (a) the electrostatic field at particle 1 from particles 2 and 3; (b) the potential at particle 1; (c) the energy of electrostatic interaction of particle 1 with with field; (*d*) the electrostatic force acting on particle; and (e) the total electrostatic energy of the system.
2. (a) How long does it take a photon to travel from the sun to the earth, assuming that the photon takes the most direct path (mean distance, 1.496×10^{11} m) and neglecting effects of the earth's atmosphere? (b) Suppose a 500-nm photon travels simultaneously two paths with different lengths. What difference in length would make the radiation arrive 180° out of phase, so that the resultant field strength is zero?
3. Consider a plane wave of light passing through a medium with a refractive index of 1.2. (a) Neglecting local-field corrections, what is the magnitude ($|\boldsymbol{E}_0|$) of the oscillating electric field at frequency ν if the irradiance $I(\nu)d\nu$ is 1 W cm^{-2}? (b) What is the total energy density ($\rho(\nu)d\nu$) in frequency interval $d\nu$? (c) Including the cavity-field correction, calculate the magnitude of the field acting on a molecule in a spherical cavity embedded in the medium. (d) Calculate the magnitude of the field acting on a molecule using the Lorentz correction. (e) What component of the field is included (approximately) in the Lorentz correction but not in the cavity-field correction.
4. What is the energy density of radiation at 500 nm emitted by a black-body source at the surface temperature of the sun (6,600 K)?
5. (a) What are the frequencies (s^{-1}) of the first six modes of a radiation field in a rectangular box with dimensions $1{,}000 \times 1{,}500 \times 2{,}000$ Å^3? ("First" here means the modes with lowest energies.) (*b*) What is the total energy of these six modes if there are no photons in any of the modes. (Indicate the

convention you use for zero energy.) (*c*) Using the same convention for zero energy, what is the total energy of the six modes if there is one photon in the first mode, two in the second, and none in the higher modes?

6. A light beam propagating along the x axis of a Cartesian coordinate system is passed through a polarizer oriented in either the y or the z direction. The intensity of the light measured with the y orientation of the polarizer is 1.3 times that measured with z. What is the ellipticity of the light? (*b*) What ellipticity corresponds to complete circular polarization?
7. What are the spectral bandwidths (FWHM of the frequency distribution) of pulses of light that have a Gaussian temporal shapes with widths of (a) 1 fs and (b) 1 ps?

References

1. Maxwell, J.C.: A dynamical theory of the electromagnetic field. Philos. Trans. R. Soc. **155**, 459–512 (1865)
2. Hameka, H.: Advanced Quantum Chemistry. Addison-Wesley, Reading, MA (1965)
3. Maxwell, J.C.: A Treatise on Electricity and Magnetism. Clarendon, Oxford (1873)
4. Ditchburn, R.W.: Light, 3rd edn. Academic, New York (1976)
5. Schatz, G.C., Ratner, M.A.: Quantum Mechanics in Chemistry, p. 325. Prentice-Hall, Englewood Cliffs, NJ (1993)
6. Griffiths, D.J.: Introduction to Electrodynamics, 3rd edn. Prentice-Hall, Upper Saddle River, NJ (1999)
7. Brillouin, L.: Wave Propagation and Group Velocity. Academic, New York (1960)
8. Knox, R.S.: Refractive index dependence of the Förster resonance excitation transfer rate. J. Phys. Chem. B **106**, 5289–5293 (2002)
9. de Fornel, F.: Evanescent Waves: From Newtonian Optics to Atomic Optics. Springer, Berlin (2001)
10. Bekefi, G., Barrett, A.H.: Electromagnetic Vibrations, Waves, and Radiation. MIT Press, Cambridge, MA (1987)
11. Liebermann, T., Knoll, W.: Surface-plasmon field-enhanced fluorescence spectroscopy. Colloids Surf. A Physiochem. Eng. Asp **10**, 115–130 (2000)
12. Moscovits, M.: Surface-enhanced spectroscopy. Rev. Mod. Phys. **57**, 783–826 (1985)
13. Knoll, W.: Interfaces and thin films as seen by bound electromagnetic waves. Annu. Rev. Phys. Chem. **49**, 565–634 (1998)
14. Aslan, K., Lakowicz, J.R., Geddes, C.D.: Plasmon light scattering in biology and medicine: new sensing approaches, visions and perspectives. Curr. Opin. Chem. Biol. **9**, 538–544 (2005)
15. Haynes, C.L., Van Duyne, R.P.: Plasmon-sampled surface-enhanced Raman excitation spectroscopy. J. Phys. Chem. B **107**, 7426–7433 (2003)
16. Wang, Z.J., Pan, S.L., Krauss, T.D., Du, H., Rothberg, L.J.: The structural basis for giant enhancement enabling single-molecule Raman scattering. Proc. Natl. Acad. Sci. U.S.A. **100**, 8638–8643 (2003)
17. Böttcher, C.J.F.: Theory of Electric Polarization, 2nd edn. Elsevier, Amsterdam (1973)
18. Lorentz, H.A.: The Theory of Electrons. Dover, New York (1952)
19. Liptay, W.: Dipole moments of molecules in excited states and the effect of external electric fields on the optical absorption of molecules in solution. In: Sinanoglu, O. (ed.) Modern Quantum Chemistry Part III: Action of Light and Organic Crystals. Academic, New York (1965)

20. Chen, F.P., Hansom, D.M., Fox, D.: Origin of Stark shifts and splittings in molecular crystal spectra. 1. Effective molecular polarizability and local electric field. Durene and Naphthalene. J. Chem. Phys. **63**, 3878–3885 (1975)
21. Myers, A.B., Birge, R.R.: The effect of solvent environment on molecular electronic oscillator strengths. J. Chem. Phys. **73**, 5314–5321 (1980)
22. Atkins, P.W.: Molecular Quantum Mechanics, 2nd edn. Oxford Univ. Press, Oxford (1983)
23. Landau, L.D., Lifshitz, E.M.: Statistical Physics. Addison-Wesley, Reading, MA (1958)
24. Planck, M.: The Theory of Heat Radiation, 2nd edn. (Engl transl by M. Masius). Dover, New York (1959).
25. Touloukian, Y.S., DeWitt, D.P.: Thermophysical Properties of Matter. Vol. 7. Thermal Radiative Properties: Metallic Elements and Alloys. IFI/Plenum, New York (1970)
26. Dirac, P.M.: The Principles of Quantum Mechanics. Oxford University Press, Oxford (1930)
27. Heitler, W.: Quantum Theory of Radiation. Oxford University Press, Oxford (1954)
28. Bordag, M., Mohideen, U., Mostepanenko, V.M.: New developments in the Casimir effect. Phys. Rep. **353**, 1–205 (2001)
29. Bressi, G., Carugno, G., Onofrio, R., Ruoso, G.: Measurement of the Casimir force between parallel metallic surfaces. Phys. Rev. Lett. 88: Art. No. 041804 (2002).
30. Lamoreaux, S.K.: Demonstration of the Casimir force in the 0.6 to 6 μm range. Phys. Rev. Lett. **78**, 5–8 (1997)
31. Feynman, R.P.: QED: The Strange Theory of Light and Matter. Princeton University Press, Princeton, NJ (1985)
32. Feynman, R.P., Hibbs, A.R.: Quantum Mechanics and Path Integrals. McGraw-Hill, New York (1965)
33. Craig, D.P., Thirunamachandran, T.: Molecular Quantum Electrodynamics: An Introduction to Radiation-Molecule Interactions. Academic, London (1984)
34. Peskin, M.E., Schroeder, D.V.: An Introduction to Quantum Field Theory. Perseus Press, Reading, MA (1995)
35. Rioux, F.: Illustrating the superposition principle with single-photon interference. Chem. Educator **10**, 424–426 (2005)
36. Scarani, V., Suarez, A.: Introducing quantum mechanics: one-particle interferences. Am. J. Phys. **66**, 718–721 (1998)
37. Zeilinger, A.: General properties of lossless beam splitters in interferometry. Am. J. Phys. **49**, 882–883 (1981)
38. Glauber, R.J.: Dirac's famous dictum on interference: one photon or two? Am. J. Phys. **63**, 12 (1995)
39. Monroe, C., Meekhof, D.M., King, B.E., Wineland, D.J.: A "Schrödinger cat" superposition state of an atom. Science **272**, 1131–1136 (1996)
40. Marton, L., Simpson, J.A., Suddeth, J.A.: Electron beam interferometer. Phys. Rev. **90**, 490–491 (1953)
41. Carnal, O., Mlynek, J.: Young's double-slit experiment with atoms: a simple atom interferometer. Phys. Rev. Lett. **66**, 2689–2692 (1991)
42. Hong, C.K., Ou, Z.Y., Mandel, L.: Measurement of subpicosecond time intervals between two photons by interference. Phys. Rev. Lett. **59**, 2044–2046 (1987)
43. Kwiat, P.G., Steinberg, A.M., Chiao, R.Y.: Observation of a "quantum eraser": a revival of coherence in a two-photon interference experiment. Phys. Rev. A **45**, 7729–7739 (1992)
44. Pittman, T.B., Strekalov, D.V., Migdall, A., Rubin, M.H., Sergienko, A.V., et al.: Can two-photon interference be considered the interference of two photons? Phys. Rev. Lett. **77**, 1917–1920 (1996)
45. Fleming, G.R.: Chemical Applications of Ultrafast Spectroscopy. Oxford University Press, New York (1986)

Electronic Absorption

4

4.1 Interactions of Electrons with Oscillating Electric Fields

This chapter begins with a discussion of how the oscillating electric field of light can raise a molecule to an excited electronic state. We then explore the factors that determine the wavelength, strength, linear dichroism, and shapes of molecular absorption bands. Our approach is to treat the molecule quantum mechanically with time-dependent perturbation theory (Chap. 2) but to consider light, the perturbation, as a purely classical oscillating electric field. Because many of the phenomena associated with absorption of light can be explained well by this semiclassical approach, we defer considering the quantum nature of light until Chap. 5. Interactions with the magnetic field of light will be discussed in Chap. 9.

Let's start by considering the interaction of an electron with the oscillating electrical field ($\boldsymbol{E}$) of linearly polarized light as in Eq. (3.15a):

$$\boldsymbol{E}(t) = \boldsymbol{E}_o(t)[\exp(2\pi i\nu t) + \exp(-2\pi i\nu t)]. \tag{4.1}$$

The oscillating field adds a time-dependent term to the Hamiltonian operator for the electron. To an approximation that often proves acceptable, we can write the perturbation as the dot product of $\boldsymbol{E}$ with the *dipole operator*, $\tilde{\mu}$:

$$\tilde{\mathrm{H}}'(t) = -\boldsymbol{E}(t) \cdot \tilde{\upmu}. \tag{4.2}$$

The dipole operator for an electron is simply

$$\tilde{\upmu} = e\tilde{\mathrm{r}} = e\boldsymbol{r}, \tag{4.3}$$

where e is the electron charge (-4.803×10^{-10} esu in the cgs system or -1.602×10^{-19} C in MKS units), $\tilde{\mathrm{r}}$ is the position operator, and $\boldsymbol{r}$ is the position of the electron. Thus Eq. (4.2) also can be written

W.W. Parson, *Modern Optical Spectroscopy*, DOI 10.1007/978-3-662-46777-0_4

$$\tilde{\mathrm{H}}'(t) = -e\boldsymbol{E}(t)\cdot\boldsymbol{r} = -e|\boldsymbol{E}_o|[\exp(2\pi i\nu t) + \exp(-2\pi i\nu t)]|\boldsymbol{r}|\cos\theta, \tag{4.4}$$

where θ is the angle between $\boldsymbol{E}_o$ and $\boldsymbol{r}$.

Moving a classical particle with charge e by a small distance $\boldsymbol{dr}$ in an electric field changes the potential energy of the particle by $dV = -e\boldsymbol{E}(\boldsymbol{r})\cdot d\mathbf{r}$ (Box 4.1). So if the field is independent of position, moving the particle from the origin of the coordinate system to position $\boldsymbol{r}$ changes the classical energy by $-e\boldsymbol{E}\cdot\boldsymbol{r}$. Quantum mechanically, if an electron is described by wavefunction Ψ, interaction with a uniform field changes the potential energy of the electron by $-e\langle\Psi|\boldsymbol{E}\cdot\boldsymbol{r}|\Psi\rangle$, assuming that the field does not alter the wavefunction itself.

Box 4.1 Energy of a Dipole in an External Electric Field

In Sect. 3.1 we discussed the energy of a charged particle in the electric field from another charge. The same considerations apply to a set of charged particles in an external electric field, such as the field between the plates of a capacitor. Consider a pair of parallel, oppositely charged plates separated by a small gap. The field in the region between the plates ($\boldsymbol{E}$) is normal to the plates, points from the positive plate to the negative, and (if we are sufficiently far from the edges of the plates) is independent of position. The electrostatic potential (V_{elec}) therefore increases linearly from the negative plate to the positive. Let's put the origin of our coordinate system at the center of the negative plate and express $V_{elec}(\boldsymbol{r})$ relative to the potential here. The field from the plates then changes the electrostatic energy of particle i by

$$E_{q_i,field} = q_i V_{elec}(\boldsymbol{r}_i) = -q_i\int_0^{r_i}\boldsymbol{E}\cdot d\boldsymbol{r} = -q_i\boldsymbol{E}\cdot\boldsymbol{r}_i, \tag{B4.1.1}$$

where q_i and $\boldsymbol{r}_i$ are the charge and position of the particle.

Summing over all the charged particles between the plates gives the total energy of interaction of the particles with the external field:

$$\begin{aligned} E_{Q,field} &= -\sum_i q_i\boldsymbol{E}\cdot\boldsymbol{r}_i = -\boldsymbol{E}\cdot\left\{\boldsymbol{R}Q + \sum_i q_i\boldsymbol{r}_i{}^o\right\} \\ &= -\boldsymbol{E}\cdot\{\boldsymbol{R}Q + \boldsymbol{\mu}\}. \end{aligned} \tag{B4.1.2}$$

Here Q is the net charge of the system (Σq_i); $\boldsymbol{R}$ is the center of charge defined as

(continued)

Box 4.1 (continued)

$$\boldsymbol{R} = \frac{1}{Q}\sum_i \boldsymbol{r}_i q_i; \tag{B4.1.3}$$

$\boldsymbol{r}_i^o$ is the position of charge i with respect to the center of charge ($\boldsymbol{r}_i^o = \boldsymbol{r}_i - \boldsymbol{R}$); and $\boldsymbol{\mu}$ is the *electric dipole*, or *electric dipole moment* of the system of charges:

$$\boldsymbol{\mu} = \sum_i q_i \boldsymbol{r}_i^o. \tag{B4.1.4}$$

In the convention used here, the dipole of a pair of charges with opposite signs points from the negative to the positive charge.

We'll show later in this chapter that the term $\boldsymbol{R}Q$ in Eq. (B4.1.2) drops out of the interactions of a molecule with the oscillating field of light, so that the choice of the coordinate system is immaterial for these interactions. $\boldsymbol{R}Q$ also drops out for static fields if we use the center of charge as the origin of the coordinate system, or if the net charge (Q) is zero. In these situations, we can write $\boldsymbol{\mu}$ simply as

$$\boldsymbol{\mu} = \sum_i q_i \boldsymbol{r}_i. \tag{B4.1.5}$$

Equation (4.4) makes several approximations. In addition to treating light classically, we have neglected the magnetic component of the radiation field. This is often an acceptable approximation because the effects of the electric field usually are much greater than those of the magnetic field. We will return to this point in Chap. 5 when we discuss the quantum theory of absorption and emission, and again in Chap. 9 when we take up circular dichroism. We also have assumed that $\boldsymbol{E}_o$ is independent of the position of the electron within the molecular orbital. This also is a reasonable approximation for most molecular chromophores, which typically are small relative to the wavelength of visible light (~ 5,000 Å). We could, however, consider the variation in the field with position by expanding the field strength in a Taylor series about the origin of the coordinate system. For light polarized in the z direction, this gives

$$\tilde{\mathrm{H}}'(x,y,z,t) = -ez\left\{ |\boldsymbol{E}(t)|_{x,y=0} + \left[x\left(\frac{\partial|\boldsymbol{E}(t)|}{\partial x}\right)_{x,y=0} + y\left(\frac{\partial|\boldsymbol{E}(t)|}{\partial y}\right)_{x,y=0} \right] + \cdots \right\} - \cdots. \tag{4.5}$$

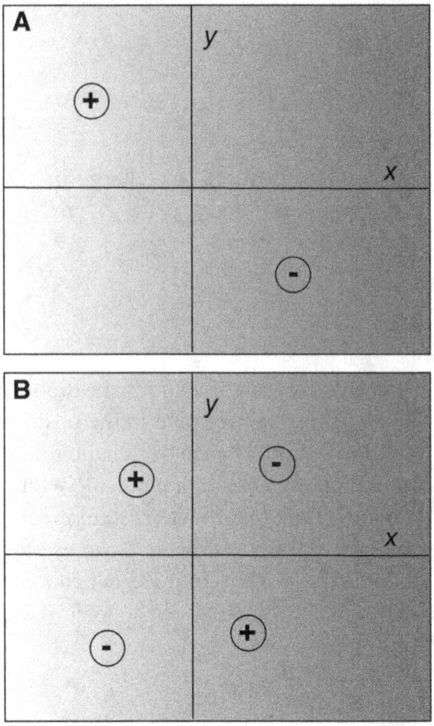

Fig. 4.1 Systems of charges with no net charge but with either a dipole moment (**A**) or a quadrupole moment but no dipole moment (**B**). The origin of the coordinate system is the center of charge in each case. The energy of interaction of the system with a constant external electric field depends on the dipole moment, and so is zero in (**B**). The quadrupole moment becomes important if the magnitude of the external electric field ($|\boldsymbol{E}|$) varies with position. The *shaded* backgrounds here represent a field that increases in strength with position in the x direction

The leading term in the expansion represents the dipole interaction; the subsequent terms represent quadrupolar, octupolar and higher-order interactions that usually are smaller in magnitude (Box 4.2 and Fig. 4.1).

Box 4.2 Multipole Expansion of the Energy of a Set of Charges in a Variable External Field

Figures 4.1A and B show two sets of charges in a field ($\boldsymbol{E}$) that points in the y direction and increases in strength with position in the x direction. For simplicity, suppose that all the charges are in the xy plane and the field strength is independent of the y coordinate. The energy of the interactions of the particles with the field then can be written

$$E_{Q,field} = -\sum_i q_i \int_0^{y_i} \boldsymbol{E}(x_i) \cdot dy, \tag{B4.2.1}$$

where q_i and (x_i, y_i) are the charge and position of particle i. One way to evaluate this sum is to choose the center of charge as the origin of the

(continued)

Box 4.2 (continued)
coordinate system and expand $\boldsymbol{E}$ as a Taylor series around this point. For the systems shown in Fig. 4.1, this gives:

$$E_{Q,field} = -(\hat{E}\cdot\hat{y})\left\{|\boldsymbol{E}|\sum_i y_i q_i + \left(\frac{\partial|\boldsymbol{E}|}{\partial x}\right)\sum_i x_i y_i q_i + \cdots\right\}, \quad \text{(B4.2.2)}$$

where $\hat{E}$ and $\hat{y}$ are unit vectors parallel to the field and the y axis, and $\boldsymbol{E}$ and its derivatives are evaluated at the center of charge. The factor $\hat{E}\cdot\hat{y}$ is just +1 in this illustration. The first term in the brackets in Eq. (B4.2.2) is non-zero for the set of charges shown in Fig. 4.1*A*, which have an electric dipole with a component in the y direction. This term vanishes for the set of charges in Fig. 4.1*B*, where the contribution from one pair of positive and negative charges cancels the contribution from the other. The second term in the brackets vanishes if the field is constant, but not if the field strength changes with x.

Equation (B4.2.2) can be written in a more general way by using matrices and matrix operators:

$$\begin{aligned} E_{Q,field} &= -\sum_i q_i \boldsymbol{E}(\boldsymbol{r}_i)\cdot\boldsymbol{r}_i = -\boldsymbol{E}\cdot\boldsymbol{R}Q - \boldsymbol{E}\cdot\boldsymbol{\mu} - Tr\left[\tilde{\nabla}\boldsymbol{E}\cdot\boldsymbol{\Theta}\right] + \cdots \\ &= V_{elec}(\boldsymbol{R})Q - \boldsymbol{E}\cdot\boldsymbol{\mu} - Tr\left[\tilde{\nabla}\boldsymbol{E}\cdot\boldsymbol{\Theta}\right] + \cdots. \end{aligned} \quad \text{(B4.2.3)}$$

Here $\boldsymbol{R}$ again is the center of charge, and $\boldsymbol{E}$ and its derivatives are evaluated at this point; Q is the total charge of the system; $V_{elec}(\boldsymbol{R})$ is the electrostatic potential at $\boldsymbol{R}$; μ is the electric dipole calculated with respect to $\boldsymbol{R}$ as in Eq. (B4.1.4) (or with respect to any coordinate system if Q is zero); Θ is a matrix called the *electric quadrupole moment* of the system of charges (see below); $\tilde{\nabla}\boldsymbol{E}$ is the gradient of $\boldsymbol{E}$; and $Tr\left[\tilde{\nabla}\boldsymbol{E}\cdot\Theta\right]$ means the trace of the matrix product $\tilde{\nabla}\boldsymbol{E}\cdot\Theta$. (See Appendix A.2 for definitions of the gradient of a vector, the product of two matrices, and the trace of a matrix.)

Equation (B4.2.3) is a *multipole expansion* of the interaction of a set of charges with an external field. The first term on the right side ($V_{elec}(\boldsymbol{R})$Q) is the interaction of the net charge of the system with the potential at the center of charge. The second term ($-\boldsymbol{E}\cdot\boldsymbol{\mu}$) describes the interaction of the dipole moment with the external field, and the third $\left(-Tr\left[\tilde{\nabla}\boldsymbol{E}\cdot\Theta\right]\right)$ describes the interaction of the quadrupole moment with the gradient of the field. The elipsis represents terms for the electric octupole and higher-order moments

(continued)

Box 4.2 (continued)
interacting with progressively higher derivatives of $\boldsymbol{E}$. In most of the situations that arise in optical spectroscopy, these higher-order terms are very small relative to the terms given in Eq. (B4.2.3), and even the quadrupole term usually is negligible compared to the dipole term.

The matrix elements of the quadrupole moment of a system of charges are defined as

$$\Theta_{\alpha,\beta} = \sum_i q_i r_{\alpha(i)} r_{\beta(i)}, \tag{B4.2.4}$$

where $r_{\alpha(i)}$ and $r_{\beta(i)}$ denote the x, y or z coordinate of charge i with respect to the center of charge. For example, letting α and β be 1, 2 or 3 for x, y or z, respectively, $\Theta_{1,2} = \sum q_i x_i y_i$ and $\Theta_{3,1} = \sum q_i z_i x_i$. Note that $\Theta_{\alpha,\beta} = \Theta_{\beta,\alpha}$, so Θ is symmetric.

With the recipes for matrix operations given Appendix A.2, the quadrupole term in Eq. (B4.2.3) becomes

$$\begin{aligned} Tr\left[\tilde{\nabla}\boldsymbol{E}\cdot\boldsymbol{\Theta}\right] &= \sum_i \sum_\alpha \sum_\beta r_{\alpha(i)} r_{\beta(i)} \partial E_\alpha / \partial r_\beta \\ &= \sum_i \left\{ x_i x_i \frac{\partial E_x}{\partial x} + y_i x_i \frac{\partial E_y}{\partial x} + z_i x_i \frac{\partial E_z}{\partial x} + x_i y_i \frac{\partial E_x}{\partial y} + y_i y_i \frac{\partial E_y}{\partial y} + z_i y_i \frac{\partial E_z}{\partial y} + \right. \\ &\qquad \left. + x_i z_i \frac{\partial E_x}{\partial z} + y_i z_i \frac{\partial E_y}{\partial z} + z_i z_i \frac{\partial E_z}{\partial z} \right\}. \end{aligned} \tag{B4.2.5}$$

If the field is oriented along the y axis and its magnitude depends only on x, as in Fig. 4.1, this expression reduces to

$$Tr\left[\tilde{\nabla}\boldsymbol{E}\cdot\boldsymbol{\Theta}\right] = \sum_i q_i y_i x_i \partial|\boldsymbol{E}|/\partial x, \tag{B4.2.6}$$

which is the same as the second term in the braces on the right side of Eq. (B4.2.2).

4.2 The Rates of Absorption and Stimulated Emission

Suppose that before we turn on the light our electron is in a state described by wavefunction Ψ_a. In the presence of the oscillating radiation field, this and the other solutions to the Schrödinger equation for the unperturbed system become unsatisfactory; they no longer represent *stationary* states. However, we can represent the wavefunction of the electron in the presence of the field by a linear combination of the original wavefunctions, $C_a\Psi_a + C_b\Psi_b + \ldots$, where the coefficients C_k are functions of time (Eq. 2.55). As long as the system is still in Ψ_a, $C_a = 1$ and all the other coefficients are zero; but if the perturbation is sufficiently strong C_a will decrease with time while C_b or one or more of the other coefficients increases. We can find the expected rate of growth of C_b by incorporating Eqs. (4.1) and (4.2) into Eq. (2.62):

$$\partial C_b/\partial t = (i/\hbar)\exp[i(E_b - E_a)t/\hbar][\exp(2\pi i\nu t) + \exp(-2\pi i\nu t)]\mathbf{E}_o \cdot \langle\psi_b|\tilde{\mu}|\psi_a\rangle \tag{4.6a}$$

$$= (i/\hbar)\{\exp[i(E_b - E_a + h\nu)t/\hbar] + \exp[i(E_b - E_a - h\nu)t/\hbar]\}\mathbf{E}_o \cdot \langle\psi_b|\tilde{\mu}|\psi_a\rangle, \tag{4.6b}$$

where E_a and E_b are the energies of states a and b.

The probability that the electron has made a transition from Ψ_a to Ψ_b by time τ is obtained by integrating Eq. (4.6b) from time $t = 0$ to τ (Eq. 2.55) and then evaluating $|C_b(\tau)|^2$. Integrating Eq. (4.6b) is straightforward, and gives the following result:

$$C_b(\tau) = \left(\frac{\exp[i(E_b - E_a + h\nu)\tau/\hbar] - 1}{E_b - E_a + h\nu} + \frac{\exp[i(E_b - E_a - h\nu)\tau/\hbar] - 1}{E_b - E_a - h\nu}\right)\mathbf{E}_0 \cdot \langle\psi_b|\tilde{\mu}|\psi_a\rangle. \tag{4.7}$$

Note that the two fractions in the large parentheses in Eq. (4.7) differ only in the sign of the term $h\nu$. Suppose that $E_b > E_a$, which means that Ψ_b lies above Ψ_a in energy. The denominator in the second term in the braces then becomes zero when $E_b - E_a = h\nu$. The numerator of this term is a complex number, but its magnitude also goes to zero when $E_b - E_a = h\nu$, and the ratio of the numerator to the denominator becomes $i\tau/\hbar$ (Box 4.3). On the other hand if $E_b < E_a$ (i.e., if Ψ_b lies below Ψ_a), then the ratio of the numerator to the denominator in the first term in the braces becomes $i\tau/\hbar$ when $E_a - E_b = h\nu$. If $|E_b - E_a|$ is very different from $h\nu$, both terms will be very small. (The amplitudes of the numerators cannot exceed 2 for any values of E_a, E_b or $h\nu$, whereas the denominators usually are large.) So something special evidently happens if $h\nu$ is close to the energy difference between the two states. We'll see shortly that the second term in the braces in Eq. (4.7) accounts for *absorption* of light when $E_b - E_a = h\nu$, and that the first term accounts for *induced* or *stimulated emission* of light when $E_a - E_b = h\nu$. Stimulated emission, a downward electronic transition in which light is given off, is just the reverse of absorption.

Box 4.3 The Behavior of the Function [exp(*iy*) – 1]/*y* as *y* goes to 0

To examine the behavior of Eq. (4.7) when $h\nu \approx |E_b - E_a|$, let $y = (E_b - E_a - h\nu)\tau/\hbar$. The term for absorption then is

$$\frac{\exp[i(E_b - E_a - h\nu)\tau/\hbar] - 1}{E_b - E_a - h\nu} = \left(\frac{\exp(iy) - 1}{y}\right)\frac{\tau}{\hbar}$$
$$= \left(\frac{1 + iy - y^2/2! + \ldots - 1}{y}\right)\frac{\tau}{\hbar}, \quad \text{(B4.3.1)}$$

which goes to $i\tau/\hbar$ as y approaches zero. Alternatively, we could write

$$\left(\frac{\exp(iy) - 1}{y}\right)\frac{\tau}{\hbar} = \left(\frac{\cos(y) + i\sin(y) - 1}{y}\right)\frac{\tau}{\hbar}, \quad \text{(B4.3.2)}$$

which also goes to $i\tau/\hbar$ as y goes to zero. The fact that $i\tau/\hbar$ is imaginary has no particular significance here, because we are interested in $|C_b(\tau)|^2$.

Equation (4.7) describes the effects of light with a single frequency (ν). As we discussed in Chap. 3, light always includes multiple oscillation modes spaning a range of frequencies. The rates of excitation caused by these individual modes are additive. To obtain the total rate of excitation, we therefore must integrate $|C_b(\tau)|^2$ over all the frequencies included in the radiation. The integral will be very small unless the region of integration includes a frequency for which $h\nu = |E_b - E_a|$. This means that we can safely take the integral from $\nu = 0$ to ∞, which is convenient because the result then appears in standard tables (Box 4.4). Also, as explained above, we only need to consider either the term for absorption or that for stimulated emission, depending on whether E_b is greater than or less than E_a. Integrating the term for absorption gives:

$$\int_0^\infty C_b^*(\tau,\nu)C_b(\tau,\nu)\rho_\nu d\nu$$
$$= \int_0^\infty \left[(\boldsymbol{E}_o \cdot \langle\psi_b|\tilde{\mu}|\psi_a\rangle)^2 \left(\frac{\{\exp[-i(E_b - E_a - h\nu)\tau/\hbar] - 1\}\{\exp[i(E_b - E_a - h\nu)\tau/\hbar] - 1\}}{(E_b - E_a - h\nu)^2}\right)\rho_\nu(\nu)\right]d\nu \quad (4.8a)$$

$$= (\boldsymbol{E}_o \cdot \langle\psi_b|\tilde{\mu}|\psi_a\rangle)^2 \rho_\nu(\nu_o)\ \tau/\hbar^2 \quad (4.8b)$$

$$= (\boldsymbol{E}_o \cdot \boldsymbol{\mu}_{ba})^2 \rho_\nu(\nu_o)\ \tau/\hbar^2, \quad (4.8c)$$

where $\rho_\nu(\nu)d\nu$ is the number of modes of oscillation in the frequency interval between ν and $\nu + d\nu$, $\nu_o = (E_b - E_a)/h$, and $\boldsymbol{\mu}_{ba} \equiv \langle\psi_b|\tilde{\mu}|\psi_a\rangle$. We have factored

$(\boldsymbol{E}_o \cdot \boldsymbol{\mu}_{ba})^2 \rho_v(v)$ out of the integral in Eq. (4.8*a*) on the assumption that the field is essentially independent of v over the small frequency interval where hv is close to $E_b - E_a$. The factor $\rho_v(v_o)$ in the final expression therefore pertains to this interval. Additional details of the derivation are in Box 4.4. Equation (4.8*c*) is a special case of a general expression that is often called the *golden rule* of quantum mechanics, which we met in Chap. 2 and will encounter again in a variety of contexts.

Box 4.4 The Function $\sin^2 x/x^2$ and Its Integral

To evaluate Eq. (4.8a), first pull the term $(\boldsymbol{E}_o \cdot \boldsymbol{\mu}_{ba})^2 \rho_v(v)$ out of the integral and let $s = (E_b - E_a - hv)\tau/\hbar$ as in Box 4.3. With this substitution, $ds = -2\pi\tau dv$ and the limits of integration are from $s = \infty$ to $s = -\infty$. The integral then can be evaluated as follows:

$$\int_{\infty}^{-\infty} \left\{ \frac{[\exp(-is) - 1][\exp(is) - 1]}{s^2(\hbar/\tau)^2(-2\pi\tau)} \right\} ds = \left(\frac{-\tau}{\pi\hbar^2}\right) \int_{\infty}^{-\infty} \left\{ \frac{1 - \cos(s)}{s^2} \right\} ds$$
$$= \left(\frac{\tau}{\pi\hbar^2}\right) \int_{-\infty}^{\infty} \frac{2\sin^2(s/2)}{s^2} ds = \left(\frac{\tau}{\pi\hbar^2}\right) \int_{-\infty}^{\infty} \left(\frac{\sin(s/2)}{(s/2)}\right)^2 d(s/2) \tag{B4.4.1}$$

$$= (\tau/\pi/\hbar^2)\pi = \tau/\hbar^2. \tag{B4.4.2}$$

Figure 4.2 shows the function $\sin^2 x/x^2$ that appears in Eq. (B4.4.1). This function, sometimes called $\mathrm{sinc}^2 x$, has a value of 1 at $x = 0$, and drops off rapidly on either side of zero. But could the spectrum of the light be so sharp that it would cover only a small fraction of the region where $\sin^2(s/2)/(s/2)^2$ is significantly different from zero? Note that s includes a product of an energy difference $(E_b - E_a - hv)$ and time (τ), and remember from Chaps. 2 and 3 that such products must cover a minimum range of approximately h. The spread of $s/2$ therefore must be at least on the order of $h/2\hbar$, or π, which would include a substantial part of the integral.

This analysis has led us to the resonance condition for absorption of light: $hv = E_b - E_a$. We also have obtained a very general expression for the rate at which a molecule will be excited from state a to state b when the resonance condition is met (Eq. 4.8c). Integrating the term for stimulated emission in Eq. (4.7) gives exactly the same result except that $\rho_v(v_o)$ refers to the frequency where $hv = E_a - E_b$. You may be surprised to see that we did not have to introduce the notion of photons with quantized energy (hv) in order to obtain the resonance condition. Our description of light was completely classical. Although the requirement for hv to match $|E_b - E_a|$ is a quantum mechanical result, it emerges in this treatment as a consequence of the quantization of the states of the absorber, not the

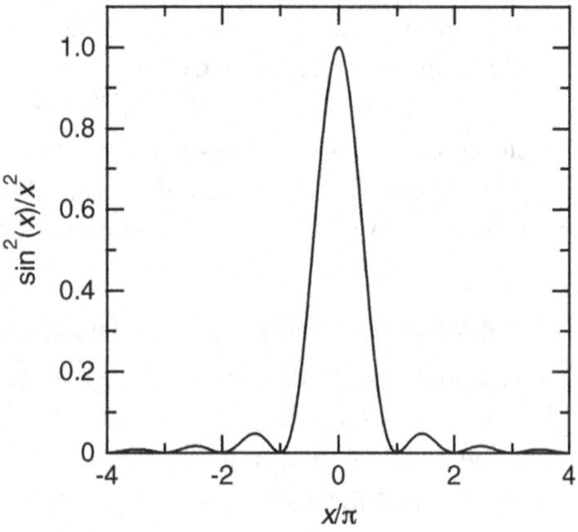

Fig. 4.2 The function $\sin^2(x)/x^2$. In Eq. (B4.4.1), $x = s/2 = (E_b - E_a - h\nu)\tau/2\hbar$

quantization of light. However, we will see in Chap. 5 that the same result is obtained by a fully quantum mechanical treatment that includes a quantized radiation field.

Equation (4.7) has the curious feature that, when $\left|E_b - E_a\right| \approx h\nu$, $C_b(\tau)$ is proportional to $-i\tau/\hbar$ (Box 4.3). This means that the probability density $\left|C_b(\tau)\right|^2$ is proportional to τ^2, at least for short times when the system still is most likely to be in state Ψ_a. In other words, the probability that the system has made a transition to state Ψ_b increases quadratically with time! By contrast, Eqs. (4.8b) and (4.8c) say that the probability that the system has made the transition grows linearly with time, which seems more in keeping with everyday observations. The quadratic time dependence predicted by Eq. (4.7) results from considering light with a single frequency, or equivalently, from considering a system with single, sharply defined value of $E_b - E_a$. Integrating over a distribution of frequencies gives Eq. (4.8). We could have obtained the same linear dependence on τ by considering a very large number of molecules with a distribution of energy gaps clustered around $h\nu$, or by considering a single molecule for which $E_b - E_a$ fluctuates rapidly with time. In Chap. 11 we will see that the dynamics of absorption in fact actually *are* expected to be nonlinear on very short time scales, although the extent of this nonlinearity depends on the fluctuating interactions of the system with the surroundings.

4.3 Transition Dipoles and Dipole Strengths

The matrix element $\left\langle\psi_b\middle|\tilde{\mu}\middle|\psi_a\right\rangle$ that we denote as $\boldsymbol{\mu}_{ba}$ in Eq. (4.8c) is called a *transition dipole*. Transition dipoles are vectors whose magnitudes have units of charge times distance. Note that $\boldsymbol{\mu}_{ba}$ differs from $\left\langle\psi_a\middle|\tilde{\mu}\middle|\psi_a\right\rangle$, or $\boldsymbol{\mu}_{aa}$, which is the contribution that an electron in orbital ψ_a makes to the *permanent* dipole of the molecule. The total electric dipole of the molecule is given by a sum of terms corresponding to $\boldsymbol{\mu}_{aa}$ for all the wavefunctions of all the charged particles in the molecule, including both electrons and nuclei. As we'll discuss in Sect. 4.10, the change in the permanent dipole when the system is excited from ψ_a to ψ_b ($\boldsymbol{\mu}_{bb} - \boldsymbol{\mu}_{aa}$) bears on how interactions with the surroundings affect the energy difference between the excited state and the ground state. The transition dipole, on the other hand, determines the *strength* of the absorption band associated with the excitation. The transition dipole can be related to the oscillatory component of the dipole in a superposition of the ground and excited states (Box 4.5).

Box 4.5 The Oscillating Electric Dipole of a Superposition State

Classically, energy can be transfered between a molecule and an oscillating electromagnetic field only if the molecule has an electric or magnetic dipole that oscillates in time at a frequency close to the oscillation frequency of the field; otherwise, the interactions of the molecule with the field will average to zero. A molecule in a state described by a superposition of its gound and excited states (Ψ_a and Ψ_b) can have such an oscillating dipole even though the individual states do not. Figure 4.3 illustrates this point for the first two eigenfunctions of an electron in a one-dimensional box. The dotted and dashed curves in panel *A* show the amplitudes of Ψ_a and Ψ_b at time $t=0$ (Eq. 2.23 and Fig. 2.2). The solid curve shows the sum ($\Psi_a + \Psi_b$) multiplied by the normalization factor $2^{-1/2}$. Because the time-dependent parts of Ψ_a and Ψ_b $[\exp(-iE_at/\hbar)$ and $\exp(-iE_bt/\hbar)$, where E_a and E_b are the energies of the pure states] are both unity at zero time, the superpostion state at this time is simply the sum of the spatial parts of the wavefunctions $(\psi_a + \psi_b)$. The corresponding probability functions are shown in Fig. 4.3*C*. The pure states have no dipole moment because the electron density ($e|\psi(x)|^2$) at any point where $x > 0$ is balanced by the electron density at a corresponding point with $x < 0$, making the integrals $\int e|\psi(x)|^2 x dx$ zero. This symmetry is broken in the superposition state, which has a higher electron density for negative values of x (Fig. 4.3*C*, solid curve).

To see that the dipole moment of the superposition state oscillates with time, consider the same states at time $t = (1/2)h/(E_b - E_a)$. For the particle in a box, the energy of the second eigenstate is four times that of the first, $E_b = 4E_a$ (Eq. 2.24), so $(1/2)h/(E_b - E_a) = (1/6)h/E_a$. The individual wavefunctions at this time have both real and imaginary parts. For the lower-energy state, we find by using the relationship $\exp(-i\theta) = \cos(\theta) - i\sin(\theta)$ that

$$\Psi_a(x,t) = \psi_a(x)\exp(-iE_at/\hbar) = \psi_a(x)\exp(-iE_ah/6E_a\hbar) = \psi_a(x)\exp(-i\pi/3)$$
$$= \psi_a(x)[\cos(\pi/3) - i\sin(\pi/3)]. \tag{B4.5.1}$$

For the higher-energy state, which oscillates four times more rapidly,

(continued)

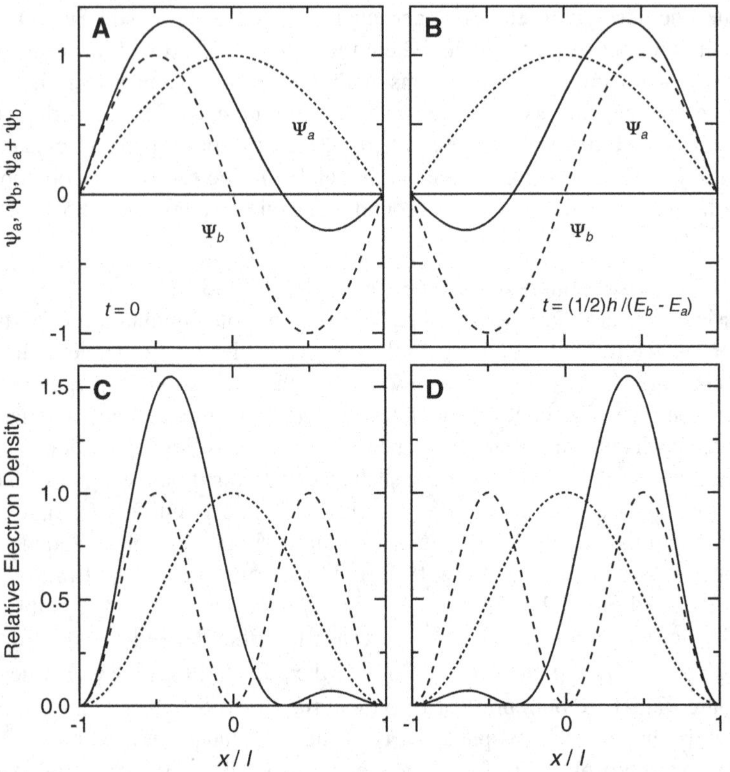

Fig. 4.3 Wavefunction amplitudes (**A**, **B**) and probability densities (**C**, **D**) for two pure states and a superposition state. The *dotted and dashed curves* are for the first two eigenstates of an electron in a one-dimensional box of unit length, as given by Eq. (2.23) with $n = 1$ or 2 (Ψ_a and Ψ_b, respectively). The *solid curves* are for the superposition $2^{-1/2}(\Psi_a + \Psi_b)$. (**A**, **C**) The signed amplitudes of the wavefunctions and the probability densities at time $t = 0$, when all the wavefunctions are real. (**B**, **D**) The corresponding functions at time $t = (1/2)h/(E_b - E_a)$, where E_a and E_b are the energies of the pure states

Box 4.5 (continued)

$$\Psi_b(x,t) = \psi_b(x)\exp(-iE_bt/\hbar) = \psi_b(x)\exp(-i4\pi/3)$$
$$= \psi_b(x)\{\cos(4\pi/3) - i\sin(4\pi/3)\} = -\psi_b(x)\{\cos(\pi/3) - i\sin(\pi/3)\}. \tag{B4.5.2}$$

Because $\cos(4\pi/3) = -\cos(\pi/3)$ and $\sin(4\pi/3) = -\sin(\pi/3)$, ψ_b has changed sign relative to ψ_a. The wavefunction of the superposition state at $t = (1/2)h/(E_b - E_a)$ is, therefore,

(continued)

Box 4.5 (continued)

$$2^{-1/2}\{\Psi_a(x,t)+\Psi_b(x,t)\}=2^{-1/2}\{\psi_a(x)-\psi_b(x)\}\{\cos(\pi/3)-i\sin(\pi/3)\}. \tag{B4.5.3}$$

This differs from the superposition state at zero time in that it depends on the difference between $\psi_a(x)$ and $\psi_b(x)$ instead of the sum (Fig. 4.3*B*). Inspection of the electron density function in Fig. 4.3*D* shows that the electric dipole of the superposition state has reversed direction relative to the orientation at $t=0$.

The wavefunction of the superposition state returns to its initial shape at $t=h/(E_b-E_a)$, when the phases of the time-dependent parts of Ψ_a and Ψ_b are $2\pi/3$ and $8\pi/3$, respectively. The spatial part of $\Psi_a+\Psi_b$ thus oscillates between symmetric and antisymmetric combinations of ψ_a and ψ_b ($\psi_a+\psi_b$ and $\psi_a-\psi_b$) with a period of $h/(E_b-E_a)$, and the electric dipole oscillates in concert.

We can relate the amplitude of the oscillating dipole of the superposition state to the transition dipole ($\boldsymbol{\mu}_{ba}$) as follows. For a superposition state $C_a\Psi_a+C_b\Psi_b$ with $\Psi_k=\psi_k\exp(-iE_kt/\hbar)$, the expectation value of the dipole is

$$\langle C_a\Psi_a+C_b\Psi_b|\tilde{\mu}|C_a\Psi_a+C_b\Psi_b\rangle$$

$$=|C_a|^2\langle\Psi_a|\tilde{\mu}|\Psi_a\rangle+|C_b|^2\langle\Psi_b|\tilde{\mu}|\Psi_b\rangle+C_a^*C_b\langle\Psi_a|\tilde{\mu}|\Psi_b\rangle+C_b^*C_a\langle\Psi_b|\tilde{\mu}|\Psi_a\rangle$$

$$=|C_a|^2\boldsymbol{\mu}_{aa}+|C_b|^2\boldsymbol{\mu}_{bb}+C_a{}^*C_b\boldsymbol{\mu}_{ab}\exp[i(E_a-E_b)t/\hbar]+C_b{}^*C_a\boldsymbol{\mu}_{ba}\exp[i(E_b-E_a)t/\hbar] \tag{B4.5.4a}$$

$$=|C_a|^2\boldsymbol{\mu}_{aa}+|C_b|^2\boldsymbol{\mu}_{bb}+2\mathrm{Re}\{C_bC_a\boldsymbol{\mu}_{ba}\exp[i(E_b-E_a)t/\hbar]\}. \tag{B4.5.4b}$$

We have used the equality $\theta+\theta^*=2\mathrm{Re}(\theta)$ where $\mathrm{Re}(\theta)$ is the real part of the complex number θ. Equation (B4.5.4*b*) shows that the dipole moment of the superposition state includes a component that oscillates sinusoidally with a period of $h/|E_b-E_a|$, and that the amplitude of this component is proportional to $\boldsymbol{\mu}_{ba}$.

Now suppose that a molecule in the superposition state is exposed to the oscillating electric field of light ($\boldsymbol{E}$). If the frequency of the light is very different from that of the oscillating molecular dipole ($|E_b-E_a|/h$), the interactions of the molecule with the radiation field will average to zero. On the other hand, if the two frequencies match and are in phase, the interaction energy will be proportional to $\boldsymbol{\mu}_{ba}\cdot\boldsymbol{E}$, and in general will be non-zero. The oscillating dipole of the superposition state thus appears to rationalize the dependence of the absorption of light on both the resonance condition and $\boldsymbol{\mu}_{ba}$. However, this argument has the problem that, in addition to being proportional to $\boldsymbol{\mu}_{ba}$, the oscillating dipole of the superposition state depends

(continued)

Box 4.5 (continued)
on the product of the coefficients C_a and C_b (Eq. B4.5.4*b*). If we know the system is in the ground state, then $C_b = 0$, and the amplitude of the oscillating dipole is zero. The perturbation treatment presented in Sect. 4.2 does not encounter this dilemma, and indeed predicts that the rate of absorption will be maximal when $C_a = 1$ and $C_b = 0$. We'll resolve this apparent contradiction in Chap. 10, when we discuss polarization of a medium by an electromagnetic radiation field.

The magnitudes of both permanent and transition dipole moments commonly are expressed in units of *debyes* (D) after Peter Debye, who received the Chemistry Nobel Prize in 1936 for showing how dipole moments can be measured and related to molecular structure. One debye is 10^{-18} esu cm in the cgs system and 3.336×10^{-30} C m in MKS units. Because the charge of an electron is -4.803×10^{-10} esu, and 1 Å $= 10^{-8}$ cm, the dipole moment associated with a pair of positive and negative elementary charges separated by 1 Å is 4.803 debyes.

According to Eq. (4.8c), the absorption strength is proportional to the square of the magnitude of $\boldsymbol{\mu}_{ba}$, which is called the *dipole strength*: (D_{ba}):

$$D_{ba} = |\boldsymbol{\mu}_{ba}|^2 = |\langle \psi_b | \tilde{\mu} | \psi_a \rangle|^2. \tag{4.9}$$

Dipole strength is a scalar with units of debye^2.

Suppose that a sample is illuminated with a light beam of irradiance I, with I defined as in Chap. 3 so that $I\Delta\nu$ is the flux of energy (e.g., in joules s^{-1}) in the frequency interval $\Delta\nu$ crossing a plane with an area of 1 cm^2. According to Eqs. (1.1) and (1.2), the light will decrease in intensity by $IC\varepsilon l\ln(10)$, where C is the concentration of the absorbing molecules (M), l is the sample's thickness (cm), and ε is the molar extinction coefficient in the frequency interval represented by $\Delta\nu$ (M^{-1} cm^{-1}). The rate at which a sample with a 1 cm^2 area absorbs energy in frequency interval $\Delta\nu$ is, therefore,

$$dE/dt = I\Delta\nu C\varepsilon l \ \ln(10). \tag{4.10}$$

Assume for now that ε has a constant value across the frequency interval $\Delta\nu$ and is zero everywhere else. Then Eq. (4.10) must account for all the absorption of light by the sample. Equation (4.8c), on the other hand, indicates that the rate at which molecules are excited is $(\boldsymbol{E}_o \cdot \boldsymbol{\mu}_{ba})^2 N_g \rho_\nu(\nu)/\hbar^2$ molecules per second, where N_g is the number of molecules in the ground state in the illuminated region. By combining these two expressions we can relate the molar extinction coefficient, an experimentally measureable quantity, to the transition dipole ($\boldsymbol{\mu}_{ba}$) and the dipole strength ($|\boldsymbol{\mu}_{ba}|^2$).

If the light beam has a cross-sectional area of 1 cm^2, the volume of the illuminated region of interest is l cm^3 and the total number of molecules in this volume is $N = 10^{-3} lCN_A$ where N_A is Avogadro's number. We usually can use N in place of the number of molecules in the ground state (N_g) because, in most measurements with continuous light sources, the light intensity is low enough and the decay of the excited state fast enough so that depletion of the ground-state population is negligible. Equation (4.10) thus can be rewritten as $dE/dt = I\Delta\nu 10^3 \ln(10)\varepsilon N/N_A$.

The dot product $\boldsymbol{E}_o \cdot \boldsymbol{\mu}_{ba}$ in Eq. (4.8*c*) depends on the cosine of the angle between the electric field vector of the light ($\boldsymbol{E}_o$) and the molecular transition dipole vector ($\boldsymbol{\mu}_{ba}$), and this angle usually varies from molecule to molecule in the sample. To find the average value of $(\boldsymbol{E}_o \cdot \boldsymbol{\mu}_{ba})^2$, imagine a Cartesian coordinate system in which the z axis is parallel to $\boldsymbol{E}_o$. The x and y axes can be chosen arbitrarily as long as they are perpendicular to z and to each other. In this coordinate system, the vector $\boldsymbol{\mu}_{ba}$ for an individual molecule can be written as (μ_x, μ_y, μ_z), where $\mu_z = \boldsymbol{E}_o \cdot \boldsymbol{\mu}_{ba}/|\boldsymbol{E}_o|$ and $\mu_x^2 + \mu_y^2 + \mu_z^2 = |\boldsymbol{\mu}_{ba}|^2$. If the sample is isotropic (i.e., if the absorbing molecules have no preferred orientation), then the average values of $\mu_x{}^2$, $\mu_y{}^2$ and $\mu_z{}^2$ must all be the same, and so must be $|\boldsymbol{\mu}_{ba}|^2/3$. The average value of $(\boldsymbol{E}_o \cdot \boldsymbol{\mu}_{ba})^2$ for an isotropic sample is, therefore,

$$\overline{(\boldsymbol{E}_o \cdot \boldsymbol{\mu}_{ba})^2} = (1/3)|\boldsymbol{E}_o|^2 |\boldsymbol{\mu}_{ba}|^2 \tag{4.11}$$

This result also can be obtaind by a more general approach that is described in Box 4.6.

Box 4.6 The Mean-Squared Energy of Interaction of an External Field with Dipoles in an Isotropic System

The average value of $(\boldsymbol{E}_o \cdot \boldsymbol{\mu}_{ba})^2$ is $|\boldsymbol{E}_o|^2 |\boldsymbol{\mu}_{ba}|^2 \overline{\cos^2\theta}$, where θ is the angle between $\boldsymbol{E}_o$ and μ_{ba} for an individual absorbing molecule and $\overline{\cos^2\theta}$ means the average of $\cos^2\theta$ over all the molecules in the system. The average value of $\cos^2\theta$ in an isotropic system can be obtained by representing θ as the angle of a vector $\boldsymbol{r}$ with respect to the z-axis in polar coordinates and integrating $\cos^2\theta$ over the surface of a sphere (Fig. 4.4).

If we let r be the length of $\boldsymbol{r}$, evaluating the integral gives:

$$\overline{\cos^2\theta} = \left(r^2 \int_0^{2\pi} d\phi \int_0^{\pi} \cos^2\theta \sin\theta d\theta \right) \Big/ \left(r^2 \int_0^{2\pi} d\phi \int_0^{\pi} \sin\theta d\theta \right)$$
$$= (4\pi/3)/4\pi = 1/3. \tag{B4.6.1}$$

The denominator in this expression is just the integral over the same surface without the weighting of the integrand by $\cos^2\theta$. Though more cumbersome than the justification of Eq. (4.11) given in the text, this analysis illustrates a more general way of dealing with related problems that arise in connection with fluorescence polarization (Chaps. 5 and 10).

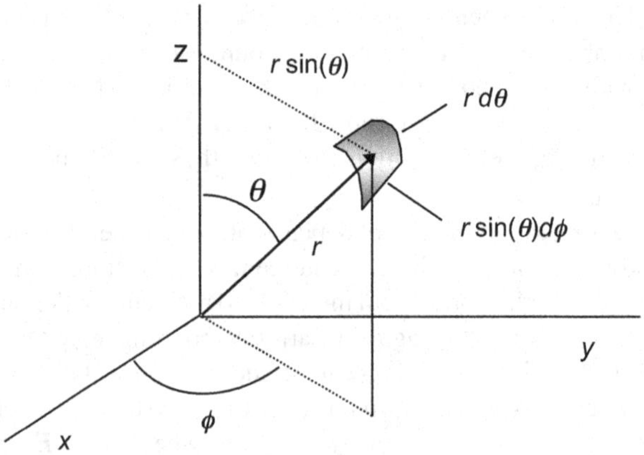

Fig. 4.4 The average value of $\cos^2\theta$ can be obtained by integrating over the surface of a sphere. The *arrow* represents a vector with length r parallel to the transition dipole of a particular molecule; the z axis is the polarization axis of the light. The area of a small element on the surface of the sphere is $r^2 \sin\theta d\phi d\theta$. The polar coordinates used here can be converted to Cartesian coordinates by the transformation $z = r\cos(\theta)$, $x = r\sin(\theta)\cos(\phi)$, $y = r\sin(\theta)\sin(\phi)$

If we now use Eq. (3.34) to relate $|\mathbf{E}_o|^2$ to the irradiance (I), still considering an isotropic sample, Eq. (4.11) becomes:

$$\overline{(E_o \cdot \boldsymbol{\mu}_{ba})^2} \rho_v(v) = \left(2\pi f^2/3cn\right)|\mu_{ba}|^2 I = \left(2\pi f^2/3cn\right) D_{ba} I, \tag{4.12}$$

where c is the speed of light in a vacuum, n is the refractive index of the solution and f is the local-field correction factor. Thus the excitation rate is:

$$-dN_g/dt = 10^{-3} ICN_A \left(2\pi f^2/3cn\hbar^2\right) D_{ba} I \text{ molecules s}^{-1}\text{ cm}^{-2}. \tag{4.13}$$

Because each excitation increases the energy of a molecule by $E_b - E_a$, or hv, the rate at which energy is transferred from the radiation field to the sample must be

$$dE/dt = 10^{-3} hvICN_A \left(2\pi f^2/3cn\hbar^2\right) D_{ba} I. \tag{4.14}$$

Finally, by equating the two expressions for dE/dt (Eqs. 4.10 and 4.14) we obtain

$$D_{ba} = \left(\frac{3{,}000 \ln(10) nhc}{8\pi^3 f^2 N_A}\right) \frac{\varepsilon}{v} \Delta v. \tag{4.15}$$

In deriving Eq. (4.15) we have assumed that all the transitions expected on the basis of dipole strength D_{ba} occur within a small frequency interval Δv over which ε is constant. This is fine for atomic transitions, but not for molecules. As we will discuss in Sect. 4.10, molecular absorption bands are broadened because a variety of nuclear transitions can accompany the electronic excitation. To include all of these transitions, we must relate D_{ba} to an integral over the absorption band:

$$D_{ba} = \left(\frac{3{,}000\ln(10)hc}{8\pi^3 N_A}\right)\int \frac{n\varepsilon}{f^2 \nu} d\nu \approx 9.186 \times 10^{-3}\left(\frac{n}{f^2}\right)\int \frac{\varepsilon}{\nu} d\nu \frac{\mathrm{D}^2}{\mathrm{M}^{-1}\mathrm{cm}^{-1}}, \quad (4.16a)$$

or

$$\int \frac{\varepsilon}{\nu} d\nu \approx \left(\frac{f^2}{n}\right)\left(\frac{4\pi^3 N_A}{3{,}000\ln(10)\hbar c}\right)|\boldsymbol{\mu}_{ba}|^2 = 108.86\left(\frac{f^2}{n}\right)|\boldsymbol{\mu}_{ba}|^2 \frac{\mathrm{M}^{-1}\mathrm{cm}^{-1}}{\mathrm{D}^2}. \quad (4.16b)$$

The values of the physical constants and conversion factors in Eqs. (4.16*a*) and (4.16*b*) are given in Box 4.7. The integral $\int(\varepsilon/\nu)d\nu$ is the same as $\int(\varepsilon/\lambda)d\lambda$, or $\int\varepsilon\ d\ln\lambda$, where λ is the wavelength; the units of ν or λ are immaterial because they cancel out in the integral.

Box 4.7 Physical Constants and Conversion Factors for Absorption of Light

The values of the physical constants in Eqs. (4.16*a*) and (4.16*b*) are:

$N_A = 6.0222 \times 10^{23}$ molecules mol^{-1}
$\hbar = 1.0546 \times 10^{-27}$ erg s $= 6.5821 \times 10^{-27}$ eV s
$c = 2.9979 \times 10^{10}$ cm s^{-1}
$\ln 10 = 2.30259$
1 D $= 10^{-18}$ esu cm
1 dyn $= 1$ esu^2 cm^{-2}
$4\pi^2 = 39.4784$
1 erg $= 1$ dyn cm $= 1$ esu^2 cm^{-1}.

If $\boldsymbol{\mu}_{ba}$ is given in debyes, then

$$\left(\frac{4\pi^3 N_A}{3{,}000\ln 10\hbar c}\right)|\boldsymbol{\mu}_{ba}|^2$$
$$= \frac{39.4784 \times \left(6.0222 \times 10^{23}\frac{\text{molecules}}{\text{mol}}\right) \times \left(|\boldsymbol{\mu}_{ba}|^2 \frac{\mathrm{D}^2}{\text{molecule}}\right) \times \left(10^{-36}\frac{\text{esu}^2\text{cm}^2}{\mathrm{D}^2}\right)}{3 \times \left(10^3 \frac{\text{cm}^3}{1}\right) \times 2.30259 \times (1.0546 \times 10^{-27}\text{erg s}) \times \left(2.9979 \times 10^{10}\frac{\text{cm}}{\text{s}}\right) \times \left(1\frac{\text{esu}^2}{\text{erg cm}}\right)}$$
$$= 108.86\ \text{mol}^{-1}\text{cm}^{-1}\text{l} = 108.86\ \text{M}^{-1}\text{cm}^{-1}.$$

If we use the cavity-field expression for f (Eq. 3.35) and $n = 1.33$ (the refractive index of water), then the factor (f^2/n) is $9n^3/(2n^2+1)^2 = 1.028$.

We have assumed that the refractive index (n) and the local-field correction factor (f) are essentially constant over the spectral region of the absorption band, so that the ratio n/f^2 can be extracted from the integral in Eq. (4.16*a*). As discussed in Chap. 3, f depends on the shape and polarizability of the molecule, and usually cannot be measured independently. If the Lorentz expression (Eq. 3.36) is used for f, as some authors recommend [1, 2], Eq. (4.16a) becomes

$$D_{ba} = 9.186 \times 10^{-3} \frac{9n}{\left(n^2+2\right)^2} \int \frac{\varepsilon}{\nu} d\nu \ \mathrm{D}^2. \tag{4.17a}$$

Using the cavity-field expression (Eq. 3.35) gives

$$D_{ba} = 9.186 \times 10^{-3} \frac{\left(2n^2+1\right)^2}{9n^3} \int \frac{\varepsilon}{\nu} d\nu \quad \text{debye}^2. \tag{4.17b}$$

Myers and Birge [3] give additional expressions for n/f^2 that depend on the shape of the chromophore.

Figure 4.5 illustrates how the treatment of the local-field correction factor in Eq. (4.16a) affects the dipole strength calculated for bacteriochlorophyll-a from the measured absorption spectrum [4]. For this molecule, using the Lorentz correction or just setting $f = 1$ leads to values for D_{ba} that change systematically with n, whereas the cavity-field expression gives values that are nearly independent of n. Excluding specific solvent-solute interactions such as hydrogen bonding, which could affect the molecular orbitals, D_{ba} should be an intrinsic molecular property that is relatively insensitive to the solvent. The cavity-field expression thus works reasonably well for bacteriochlorophyll, although the actual shape of the molecule

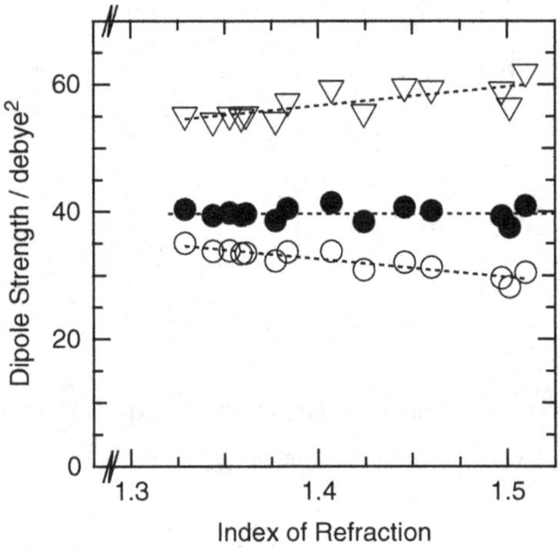

Fig. 4.5 Dipole strength of the long-wavelength absorption band of bacteriochlorophyll-a, calculated by Eq. (4.16a) from absorption spectra measured in solvents with various refractive indices. Three treatments of the local-field correction factor (f) were used: *down triangles*, $f = 1.0$ (no correction); *filled circles*, f is the cavity-field factor; *empty circles*, f is the Lorentz factor. The *dashed lines* are least-squares fits to the data. Spectra measured by Connolly et al. [148] were converted to dipole strengths as described by Alden et al. [4] and Knox and Spring [5]

is hardly spherical. The dependence of D_{ba} on n for chlorophyll-*a* is essentially the same as that for bacteriochlorophyll-*a* [5].

The strength of an absorption band sometimes is expressed in terms of the *oscillator strength*, a dimensionless quantity defined as

$$\begin{aligned} O_{ba} &= \frac{8\pi^2 m_e \nu}{3e^2 h} D_{ba} = \frac{2.303 \times 10^3 m_e c}{\pi e^2 N_A} \int \left(\frac{n}{f^2} \right) \varepsilon d\nu \\ &\approx 1.44 \times 10^{-19} \int \left(\frac{n}{f^2} \right) \varepsilon d\nu, \end{aligned} \tag{4.18}$$

where m_e is the electron mass (Box 3.3). Here the units of ν do matter; the numerical factors given in Eq. (4.18) are for ν in s^{-1}. The oscillator strength relates the rate of absorption of energy to the rate predicted for a classical electric dipole oscillating at the same frequency (ν). It is on the order of 1 for the strongest possible electronic absorption band of a single chromophore. According to the *Kuhn-Thomas* sum rule the sum of the oscillator strengths for all the absorption bands of a molecule is equal to the total number of electrons in the molecule; however, this rule usually is of little practical value because many of the absorption bands at high energies are not measureable.

The strength of an absorption band also can be expressed as the *absorption cross section* (σ), which (for ε in units of M^{-1} cm^{-1}) is given by $10^{-3}\ln(10)\varepsilon/N_a$, or $3.82 \times 10^{-21}\varepsilon$ in units of cm^2. If the incident light intensity is I photons cm^{-2} s^{-1} and the excited molecules return to the ground state rapidly relative to the rate of excitation, a molecule with absorption cross-section σ will be excited $I\sigma$ times per second. This result is independent of the concentration of absorbing molecules in the sample, although I drops off more rapidly with depth in the sample if the concentration is increased.

4.4 Calculating Transition Dipoles for π Molecular Orbitals

Theoretical dipole strengths for molecular transitions can be calculated by using linear combinations of atomic orbitals to represent the molecular wavefunctions of the excited and unexcited system. Discussing an example of such a calculation will help to bring out the vectorial nature of transition dipoles. Consider a molecule in which the highest normally occupied molecular orbital (HOMO) and the lowest normally unoccupied orbital (LUMO) are both π orbitals. We can describe these orbitals by writing, as in Eq. (2.35),

$$\psi_h \approx \sum_t C_t^h p_t \quad \text{and} \quad \psi_l \approx \sum_t C_t^l p_t, \tag{4.19}$$

where superscripts h and l stand for HOMO and LUMO, the sums run over the conjugated atoms of the π system, and p_t represents an atomic $2p_z$ orbital on atom t.

In the ground state, ψ_h usually contains two electrons. If we use the notation $\psi_k(j)$ to indicate that electron j is in orbital ψ_k, we can express the wavefunction for the ground state as a product:

$$\Psi_a = \psi_h(1)\psi_h(2). \tag{4.20}$$

We have factored out all the filled orbitals below the HOMO and have omitted them in Eq. (4.20) on the simplifying assumption that the electrons in these orbitals are not affected by the movement of one of the outer electrons from the HOMO to the LUMO. This clearly represents only a first approximation to the actual rearrangement of the electrons that accompanies the excitation.

In the excited state, either electron 1 or electron 2 could be promoted from the HOMO to the LUMO. Since we cannot distinguish the individual electrons, the wavefunction for the excited state must combine the various possible ways that the electrons could be assigned to the two orbitals:

$$\Psi_b = 2^{-1/2}\psi_h(1)\psi_l(2) + 2^{-1/2}\psi_h(2)\psi_l(1). \tag{4.21}$$

Again, we have factored out all the orbitals below the HOMO. For now we also neglect the spins of the two electrons and assume that no change of spin occurs during the excitation. The wavefunctions written for both the ground and the excited state pertain to *singlet* states, in which the spins of electrons 1 and 2 are antiparallel (Sect. 2.4). We will return to this point in Sect. 4.9.

By using Eqs. (4.20) and (4.21) for Ψ_a and Ψ_b, the transition dipole can be reduced to a sum of terms involving the atomic coordinates and the molecular orbital coefficients C_t^h and C_t^l for the HOMO and LUMO:

$$\boldsymbol{\mu}_{ba} \equiv \langle \Psi_b | \tilde{\mu} | \Psi_a \rangle = \langle \Psi_b | \tilde{\mu}(1) + \tilde{\mu}(2) | \Psi_a \rangle \tag{4.22a}$$

$$= \left\langle 2^{-1/2}(\psi_h(1)\psi_l(2) + \psi_h(2)\psi_l(1)) | \tilde{\mu}(1) + \tilde{\mu}(2) | \psi_h(1)\psi_h(2) \right\rangle \tag{4.22b}$$

$$\begin{aligned} &= 2^{-1/2}\langle \psi_l(1)|\tilde{\mu}(1)|\psi_h(1)\rangle \langle \psi_h(2)|\psi_h(2)\rangle \\ &\quad + 2^{-1/2}\langle \psi_l(2)|\tilde{\mu}(2)|\psi_h(2)\rangle \langle \psi_h(1)|\psi_h(1)\rangle \end{aligned} \tag{4.22c}$$

$$= \sqrt{2}\langle \psi_l(k)|\tilde{\mu}(k)|\psi_h(k)\rangle \approx \sqrt{2}\left\langle \sum_s C_s^l p_s |\tilde{\mu}| \sum_t C_t^h p_t \right\rangle \tag{4.22d}$$

$$= \sqrt{2}\ e \sum_s \sum_t C_s^l C_t^h \langle p_s|\tilde{\mathbf{r}}|p_t\rangle \approx \sqrt{2}\ e \sum_t C_s^l C_t^h \mathbf{r}_t, \tag{4.22e}$$

where $\mathbf{r}_i$ is the position of atom i. In this derivation we have separated the dipole operator $\tilde{\mu}$ into two parts that make identical contributions to the overall transition dipole. One part, $\tilde{\mu}(1)$, acts only on electron 1 while $\tilde{\mu}(2)$ acts only on electron 2;

$\langle\psi_l(2)|\tilde{\mu}(1)|\psi_h(2)\rangle$ and $\langle\psi_l(1)|\tilde{\mu}(2)|\psi_h(1)\rangle$ are zero. The final step of the derivation uses the fact that $\langle p_t|\tilde{r}|p_t\rangle = r_t$ and the approximation $|\langle p_s|\tilde{r}|p_t\rangle| \approx 0$ for $s \neq t$.

As an example consider ethylene, for which the HOMO and LUMO can be described approximately as symmetric and antisymmetric combinations of carbon $2p$ orbitals: $\Psi_a = 2^{-1/2}(p_1 + p_2)$ and $\Psi_b = 2^{-1/2}(p_1 - p_2)$, respectively (Fig. 2.7). The corresponding absorption band occurs at 175 nm. Equation (4.22e) gives a transition dipole of $(2^{1/2}/2)e(r_1 - r_2) = (2^{1/2}/2)e\boldsymbol{r}_{12}$, where $\boldsymbol{r}_{12}$ is the vector from carbon 2 to carbon 1. The transition dipole vector thus is aligned along the C=C bond. The calculated dipole strength is $D_{ba} = |\boldsymbol{\mu}_{ba}|^2 = (e^2/2)|\boldsymbol{r}_{12}|^2$, or $11.53|\boldsymbol{r}_{12}|^2$ debye^2 if $\boldsymbol{r}_{12}$ is given in Å. From this result, it might appear that the dipole strength would increase indefinitely with the square of the C=C bond length. But if the bond is stretched much beyond the length of a typical C=C double bond, the description of the HOMO and LUMO as symmetric and antisymmetric combinations of the two atomic p_z orbitals begins to break down. In the limit of a large interatomic distance, the orbitals are no longer shared by the two carbons, but instead are localized entirely at one site or the other. Equation (4.22*e*) then gives a dipole strength of zero because either C_i^h or C_i^l is zero in each of the products that enters into the sum.

Note that although the transition dipole calculated by Eq. (4.22*e*) has a definite direction, flipping it over by 180° would have no effect on the the absorption spectrum because the extinction coefficient depends on the dipole strength $(|\boldsymbol{\mu}_{ba}|^2)$ rather than on $\boldsymbol{\mu}_{ba}$ itself. This makes sense, considering that the electric field of light oscillates rapidly in sign. However, we later will consider transitions that are best described by linear combinations of excitations in which an electron moves between any of several different pairs of molecular orbitals, rather than simply from the HOMO to the LUMO (Sect. 4.7 and Chap. 8). Because the overall transition dipole in this situation is a vector combination of the weighted transition dipoles for the individual excitations, the relative signs of the individual contributions are important.

4.5 Molecular Symmetry and Forbidden and Allowed Transitions

We have seen that the transition dipole of ethylene is aligned along the bond between the two carbon atoms. Because the strength of absorption depends on the dot product of $\boldsymbol{\mu}_{ba}$ and the electrical field of the light (Eq. 4.8*c*), absorption of light polarized parallel to the C=C bond is said to be *allowed*, whereas absorption of light polarized perpendicular to the bond is *forbidden*. It often is easy to determine whether or not an absorption band is allowed, and if so for what polarization of light, by examining the symmetries of the molecular orbitals involved in the transition.

Note first that the transition dipole depends on the integral over all space of a product of the position vector and the amplitudes of the two wavefunctions:

$$\boldsymbol{\mu}_{ba} = \langle \Psi_b|\tilde{\mu}|\Psi_a\rangle = e\langle \Psi_b|\boldsymbol{r}|\Psi_a\rangle \equiv e\int \Psi_b^* \boldsymbol{r} \Psi_a d\sigma. \tag{4.23}$$

At a given point $\boldsymbol{r}$, each of these three quantities could have either a positive or a negative sign depending on our choice of the origin for the coordinate system. However, the magnitude of a molecular transition dipole does not depend on any particular choice of coordinate system. This should be clear in the case of ethylene, where Eq. (4.22) shows that $|\boldsymbol{\mu}_{ba}|$ depends only on the length of the carbon–carbon bond ($|\boldsymbol{r}_{12}|$) and is oriented along this bond. More generally, if we shift the origin by adding any constant vector $\boldsymbol{R}$ to $\boldsymbol{r}$, Eq. (4.23) becomes

$$\boldsymbol{\mu}_{ba} = e\langle \Psi_b|\boldsymbol{r} + \boldsymbol{R}|\Psi_a\rangle = e\langle \Psi_b|\boldsymbol{r}|\Psi_a\rangle + e\langle \Psi_b|\boldsymbol{R}|\Psi_a\rangle \tag{4.24a}$$

$$= e\langle \Psi_b|\boldsymbol{r}|\Psi_a\rangle + e\boldsymbol{R}\langle \Psi_b|\Psi_a\rangle = e\langle \Psi_b|\boldsymbol{r}|\Psi_a\rangle, \tag{4.24b}$$

which is the same as before. The term $e\boldsymbol{R}\langle \Psi_b|\Psi_a\rangle$ is zero as long as Ψ_a and Ψ_b are orthogonal. Similarly, rotating the coordinate system modifies the x, y and z components of μ_{ba} but does not change the magnitude or orientation of the vector.

Returning to ethylene, let's put the origin of the coordinate system midway between the two carbons and align the the C=C bond on y axis, making the z axes of the individual atoms perpendicular to the bond as shown in Fig. 4.6 (see also Fig. 2.7). Like the atomic p_z orbitals (Fig. 2.6B), the HOMO (Ψ_a) then has equal magnitudes but opposite signs on the two sides of the xy plane; it is said to be an *odd* or *antisymmetric* function of the z coordinate (Fig. 4.6A). In other words, $\Psi_a(x, y, z) = -\Psi_a(x, y, -z)$ for any fixed values of x and y. The same is true of the LUMO: $\Psi_b(x, y, z) = -\Psi_b(x, y, -z)$ (Fig. 4.6B). The product of the HOMO and the LUMO, on the other hand, has the same sign on the two sides of the xy plane, and thus is an *even* or *symmetric* function of z: $\Psi_b(x, y, z)\Psi_a(x, y, z) = \Psi_b(x, y, -z)\Psi_a(x, y, -z)$ (Fig. 4.6C). The product $\Psi_b\Psi_a$ also is an even function of x, but an odd function of y (Fig. 4.6D).

To evaluate the x, y or z component of the transition dipole, we have to multiply $\Psi_b\Psi_a$ by, respectively, x, y or z and integrate the result over all space. Since z has opposite signs on the two sides of the xy plane, whereas $\Psi_b\Psi_a$ has the same sign, the quantity $z\Psi_b\Psi_a$ is an odd function of z and will vanish if we integrate it along any line parallel to the z axis:

$$\int_{-\infty}^{\infty} z\Psi_b(x, y, z)\Psi_a(x, y, z)\, dz = 0. \tag{4.25}$$

The z component of $\boldsymbol{\mu}_{ba}$ therefore is zero. The same is true for the x component. By contrast, the quantity $y\Psi_b\Psi_a$ is an even function of y and will give either a positive result or zero if it is integrated along any line parallel to y (Fig. 4.6*C*, *D*). $y\Psi_b\Psi_a$ also is an even function of x and an even function of z, so its integral over all space must be nonzero. The transition dipole $\langle \Psi_b|\tilde{\mu}|\Psi_a\rangle$ thus has a nonzero y component.

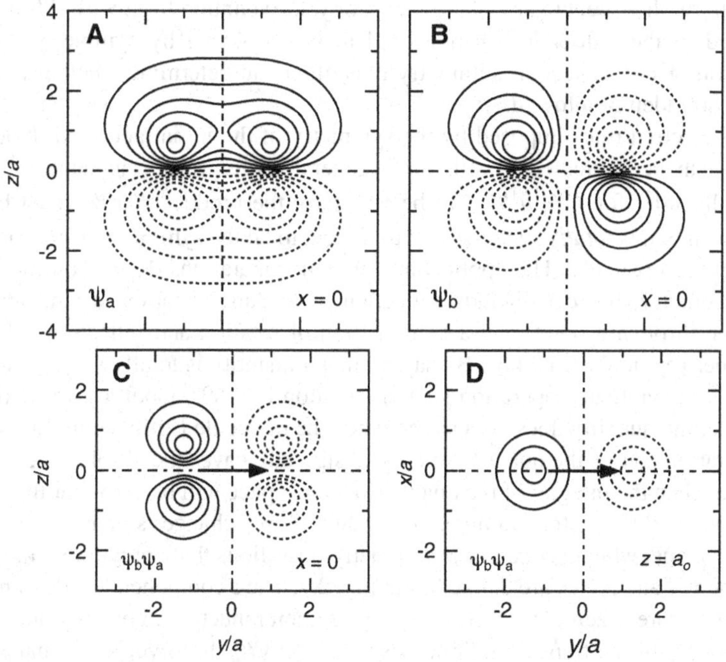

Fig. 4.6 Orbital symmetry in the first $\pi \rightarrow \pi^*$ electronic transition of ethylene. Panels **A**, **B** Contour plots of the amplitudes of the HOMO (π, **A**) and LUMO (π^*, **B**) wavefunctions. **C**, **D** Contour plots of the product of the two wavefunctions. The C=C bond is aligned with the *y* axis, and the atomic *z* axes are parallel to the molecular *z* axis. In **A–C**, the plane of the drawing coincides with the *yz* plane; in **D**, the plane of the drawing is parallel to the molecular *xy* plane and is above this plane by the Bohr radius ($a_o = 0.529$ Å). The wavefunctions are constructed as in Fig. 2.7. *Solid curves* represent positive amplitudes; *dotted curves*, negative. Distances are plotted as dimensionless multiples of a_o, and the *contour intervals* are $0.05a_o^{3/2}$ in **A** and **B** and $0.02a_o^{3/2}$ in **C** and **D**. The *arrows* in **C** and **D** show the transition dipole in units of eÅ$/a_o$ as calculated by Eq. (4.22*e*)

A particle in a one-dimensional box provides another simple illustration of these principles. Inspection of Fig. 2.2A shows that the wavefunctions for $n = 1, 3, 5 \ldots$ are all symmetric functions of the distance from the center of the box (Δx) whereas those for even $n = 2, 4, \ldots$ are all antisymmetric. The product of the wavefunctions for $n = 1$ (ψ_1) and $n = 2$ (ψ_2) thus has the same symmetry as Δx. The quantity $\Delta x \psi_1 \psi_2$ therefore will give a non-zero result if we integrate it over all values of Δx, which means that excitation from ψ_1 to ψ_2 has a non-zero transition dipole oriented along *x*. The same is true for excitation from ψ_1 to any of the higher states with even values of *n*, but not for transitions to states with odd values of *n*. In this system, the *selection rule* for absorption is simply that *n* must change from odd to even or vice versa.

To generalize the foregoing results, we can say that the *j* component ($j = x, y$ or z) of a transition dipole $\langle \Psi_b | \tilde{\mu} | \Psi_a \rangle$ will be zero if the product $j\Psi_b\Psi_a$ has odd reflection symmetry with respect to any plane (*xy*, *xz* or *yz*). Excitation from Ψ_a to Ψ_b by light polarized in the *j* direction then is said to be *forbidden* by symmetry. Simple considerations of molecular symmetry thus often can determine whether a transition is forbidden or allowed.

The selection rules imposed by the symmetry of the initial and final molecular orbitals can be expressed in still more general terms in the language of group theory. In order for $\langle \Psi_b | \tilde{\mu} | \Psi_a \rangle$ to be non-zero, the product $\Psi_b \mathbf{r} \Psi_a$ must have a component that is *totally symmetric* with respect to all the symmetry operations that apply to the molecule. The applicable symmetry operations depend on the molecular geometry, but can include reflection in a plane, rotation around an axis, inversion through a point, and a combination of rotation and reflection called an "improper rotation" (Box 4.8). By saying that a quantity is totally symmetric with respect to a symmetry operation such as rotation by 180° about a given axis, we mean that the quantity does not change when the molecule is rotated in this way. If the operation causes the quantity to change sign but leaves the absolute magnitude the same, then the integral of the quantity over all space will be zero. For the $\pi - \pi^*$ transition of ethylene, for example, the product $\Psi_b z \Psi_a$ changes sign upon reflection in the *xy* plane, which is one of the symmetry operations that apply to the molecule when the *x*, *y* and *z* axes are defined as in Fig. 2.7. The *z* component of the transition dipole therefore is zero. $\Psi_b x \Psi_a$ does the same upon reflection in the *yz* plane, so the *x* component of the transition dipole also zero. $\Psi_b y \Psi_a$, however, is unchanged by such a reflection or any of the other applicable symmetry operations (reflection in the *xz* or *yz* plane, rotation by 180° around the *x*, *y* or *z* axis, or inversion of the structure through the origin), so the *y* component of the transition dipole is non-zero. But that is all we can say based simply on the symmetry of the molecule. To find the actual magnitude of the *y* component of $\boldsymbol{\mu}_{ba}$, we have to evaluate the integral.

Box 4.8 Using Group Theory to Determine Whether a Transition is Forbidden by Symmetry

Molecular structures and orbitals can be classified into various *point groups* according to the *symmetry elements* they contain. Symmetry elements are lines, planes or points with respect to which various *symmetry operations* such as rotation can be performed without changing the structure. The symmetry operations that concern us here are as follows:

Rotation $(\tilde{\mathrm{C}}_n)$. If rotation by $2\pi/n$ radians ($1/n$ of a full rotation) about a particular axis generates a structure that (barring isotopic labeling) is indistinguishable from the original structure, the molecule is said to have a C_n *axis of rotational symmetry*.

Reflection $(\tilde{\sigma})$. The *xy* plane is said to be a *plane of reflection symmetry* or a *mirror plane* (σ) if moving each atom from its original position (*x*, *y*, *z*) to

(continued)

Box 4.8 (continued)

$(x, y, -z)$ gives an identical structure. A mirror plane often is designated as σ_h if it is normal to a principle rotation axis, or σ_v if it contains this axis.

Inversion $(\tilde{i})$. A molecule has a *point of inversion symmetry* (i) if moving each atom in a straight line through the point to the same distance on the opposite side of the molecule gives the same structure. If we use the point of inversion symmetry as the origin of the coordinate system, the inversion operation moves each atom from its original position (x, y, z) to $(-x, -y, -z)$.

Improper rotation $(\tilde{S}_n)$. Improper rotation is a rotation by $2\pi/n$ followed by reflection through a plane perpendicular to the rotation axis. An *axis of improper rotation* (S_n) is an axis about which this operation leaves the structure of a molecule unchanged.

Identity $(\tilde{E})$. The identity operator leaves all the particles in a molecule where it finds them, which is to say that it does nothing. It is, nevertheless, essential to group theory. All molecules have the identity symmetry element (E).

Translation, though an important symmetry operation in crystallography, is not included here because we are concerned with the symmetry of an individual molecule. A molecule whose center of mass has been shifted is, in principle, distinguishable from the original molecule.

Point groups are sets of symmetry elements corresponding to symmetry operators that obey four general rules of group theory:

1. Each group must include the identity operator $\tilde{E}$.
2. For each operator $\tilde{A}$ in the group, the group must include an *inverse* operator $\tilde{A}^{-1}$ with the properties that $\tilde{A}^{-1} \cdot \tilde{A} = \tilde{A} \cdot \tilde{A}^{-1} = \tilde{E}$.
3. If operators $\tilde{A}$ and $\tilde{B}$ are members of a group, then the products $\tilde{A} \cdot \tilde{B}$ and $\tilde{B} \cdot \tilde{A}$ also must be in the group. (As in Chap. 2, the operator written on the right in such a product acts first, followed by the one on the left. The order may or may not matter, depending on the operators and the point group. A symmetry operator necessarily commutes with its inverse and with $\tilde{E}$, but not necessarily with the other operators.)
4. Multiplication of the symmetry operators must be associative. This means that $\tilde{A} \cdot (\tilde{B} \cdot \tilde{C}) = (\tilde{A} \cdot \tilde{B}) \cdot \tilde{C}$ for any three operators in the group.

Let's look at a few examples. Ethylene has three perpendicular axes of twofold rotational symmetry [$C_2(x)$, $C_2(y)$ and $C_2(z)$], three planes of reflection symmetry [$\sigma(xz)$, $\sigma(yz)$ and $\sigma(xy)$], and a center of inversion symmetry (i) (Fig. 4.7A). When combined with the identity element E, these symmetry

(continued)

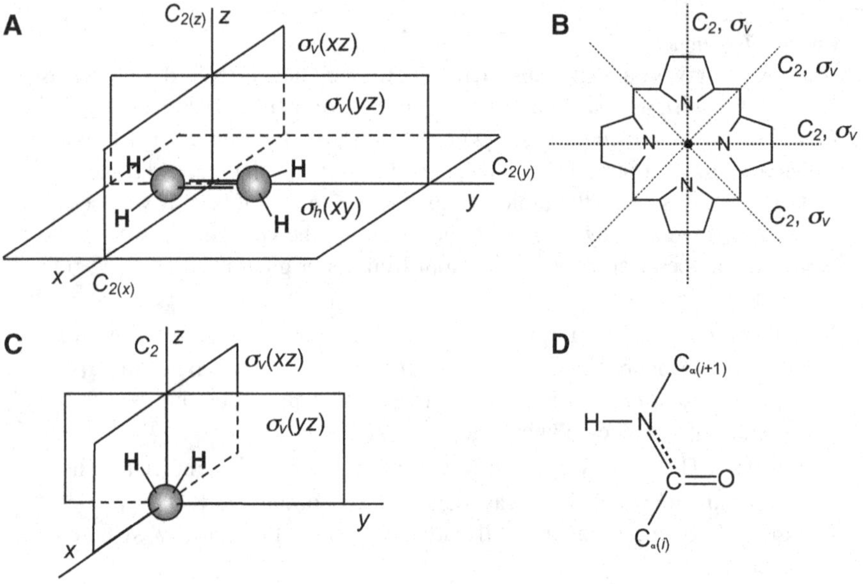

Fig. 4.7 Symmetry elements in ethylene (**A**), porphyrin (**B**), water (**C**), and a peptide (**D**). In **A**, ethylene is drawn in the *xy* plane, the *x*, *y* and *z* axes are all axes of twofold rotational symmetry (C_2), and the *xy*, *xz* and *yz* axes are planes of mirror symmetry. If we take *z* to be the "principal" axis of rotational symmetry, the mirror planes that contain this axis (*xz* and *yz*) are called "vertical" mirror planes (σ_h) and the mirror plane normal to *z* (xy) is called a "horizontal" plane of mirror symmetry (σ_h). In **B**, porphyrin is viewed along an axis (*z*, *filled circle*) normal to the plane of the macrocycle. The *z* axis is an axis of fourfold rotational symmetry (C_4) and is the principal symmetry axis. There are four C_2 axes in the *xy* plane (*dotted lines*), four vertical planes of mirror symmetry (σ_v), one horizontal plane of mirror symmetry (*xy*), and a point of inversion symmetry at the center. Water, drawn in the *yz* plane in **C**, has one C_2 axis (*z*) and two vertical planes of mirror symmetry (*xz* and *yz*). The peptide bond (**D**) has a plane of mirror symmetry (the plane of the drawing), but no other symmetry elements

Box 4.8 (continued)

elements obey the general rules for a group. They are called the D_{2h} point group.

The conjugated atoms of porphin have a C_4 axis (*z*), four C_2 axes in the *xy* plane, four planes of reflection symmetry containing the *z* axis (σ_v), a center of inversion symmetry, and identity (Fig. 4.7B). These elements form the D_{4h} point group.

Water has a C_2 axis that passes through the oxygen atom and bisects the H–O–H angle, and two perpendicular reflection planes that contain the C_2 axis, and again, identity (Fig. 4.7C). It is in point group C_{2v}.

(continued)

Table 4.1 Products of symmetry operations for the D_{2h} point group

	E	$C_2(z)$	$C_2(y)$	$C_2(x)$	i	$\sigma_h(xy)$	$\sigma_v(xz)$	$\sigma_v(yz)$
E	E	$C_2(z)$	$C_2(y)$	$C_2(x)$	i	$\sigma_h(xy)$	$\sigma_v(xz)$	$\sigma_v(yz)$
$C_2(z)$	$C_2(z)$	E	$C_2(x)$	$C_2(y)$	$\sigma_h(xy)$	i	$\sigma_v(yz)$	$\sigma_v(xz)$
$C_2(y)$	$C_2(y)$	$C_2(x)$	E	$C_2(z)$	$\sigma_v(xz)$	$\sigma_v(xz)$	i	$\sigma_h(xy)$
$C_2(x)$	$C_2(x)$	$C_2(y)$	$C_2(z)$	E	$\sigma_v(yz)$	$\sigma_v(yz)$	$\sigma_h(xy)$	i
i	i	$\sigma_h(xy)$	$\sigma_v(xz)$	$\sigma_v(yz)$	E	$C_2(z)$	$C_2(y)$	$C_2(x)$
$\sigma_h(xy)$	$\sigma_h(xy)$	i	$\sigma_v(yz)$	$\sigma_v(xz)$	$C_2(z)$	E	$C_2(x)$	$C_2(y)$
$\sigma_v(xz)$	$\sigma_v(xz)$	$\sigma_v(yz)$	i	$\sigma_h(xy)$	$C_2(y)$	$C_2(x)$	E	$C_2(z)$
$\sigma_v(yz)$	$\sigma_v(yz)$	$\sigma_v(xz)$	$\sigma_h(xy)$	i	$C_2(x)$	$C_2(y)$	$C_2(z)$	E

Box 4.8 (continued)

The backbone of a peptide has only one symmetry element other than identity, a mirror plane that includes the central N, C, O and C_α atoms (Fig. 4.7D). This puts it in point group C_s.

For a general procedure for finding the point group for a molecule see [6] or [7].

As shown in Table 4.1 for the D_{2h} point group, the products of the operations in a point group can be collected in a multiplication table. The entries in such a table are the results of performing the symmetry operation for the symmetry element given in row 1, followed by the operation for the element given in column 1. In the D_{2h} point group, the operators all commute with each other and each of the operators is its own inverse. For example, rotating around a C_2 axis by $2\pi/2$ and then rotating by an additional $2\pi/2$ around the same axis returns all the atoms to their original positions, so $\tilde{\mathrm{C}}_2 \cdot \tilde{\mathrm{C}}_2 = \tilde{\mathrm{E}}$. More generally, the inverse of $\tilde{\mathrm{C}}_n$ in any point group is $\left(\tilde{\mathrm{C}}_n\right)^{n-1}$. To work out the other products, it is helpful to make projection drawings of the type shown in Fig. 4.8, where the small filled circles represent points above the xy plane, and the empty circles represent points below this plane.

Symmetry operators and their products can be represented conveniently by matrices. Suppose an atom in the D_{2h} point group has coordinates $\boldsymbol{r} = (x, y, z)$ relative to the molecule's center of mass. If we represent the coordinates by a column vector and use the expression for multiplication of a vector by a matrix (Eq. A2.5 in Appendix A.2), the actions of the symmetry operators on the atom's location can be written:

$$\tilde{\mathrm{E}} \cdot \boldsymbol{r} = \begin{bmatrix} 1 & 0 & 0 \\ 0 & 1 & 0 \\ 0 & 0 & 1 \end{bmatrix} \begin{pmatrix} x \\ y \\ z \end{pmatrix} = \begin{pmatrix} x \\ y \\ z \end{pmatrix} \qquad \text{(B4.8.1}a\text{)}$$

(continued)

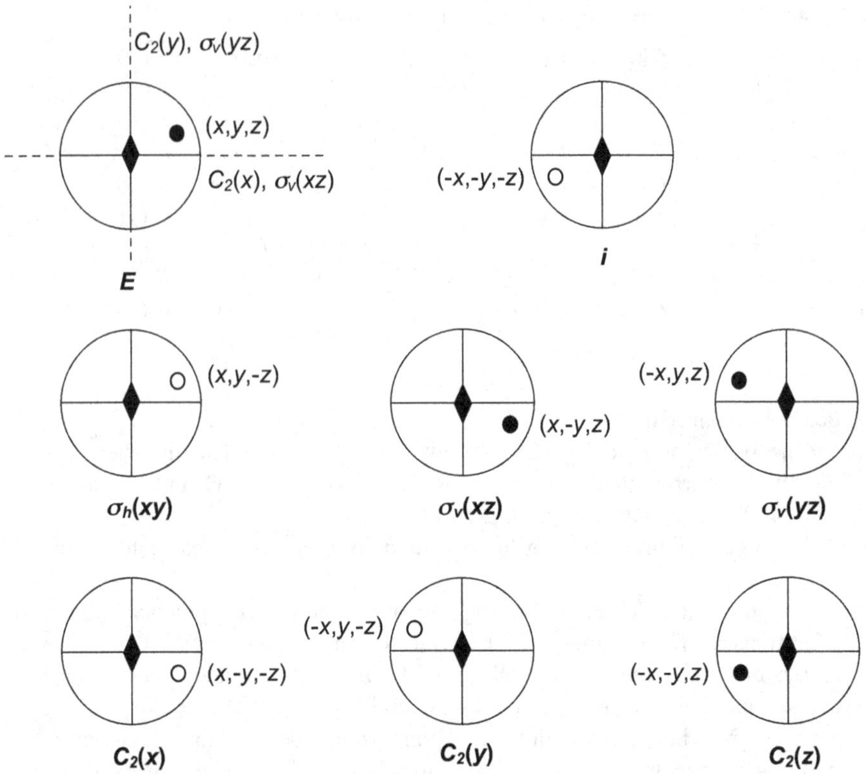

Fig. 4.8 Projection drawings of the effects of symmetry operations in point group D_{2h}. The *large circles* represent regions of space containing the atoms in the group; *horizontal* and *vertical lines* indicate the x and y axes; the z axis is normal to the plane of the paper. The *small filled circles* represent points above the xy plane; the *small empty circles*, points below the plane. The symmetry operations in point group D_{2h} move an atom from its initial position (x,y,z) to the indicated positions

Box 4.8 (continued)

$$\tilde{\mathrm{I}} \cdot \boldsymbol{r} = \begin{bmatrix} -1 & 0 & 0 \\ 0 & -1 & 0 \\ 0 & 0 & -1 \end{bmatrix} \begin{pmatrix} x \\ y \\ z \end{pmatrix} = \begin{pmatrix} -x \\ -y \\ -z \end{pmatrix} \tag{B4.8.1b}$$

$$\tilde{\mathrm{C}}_2(z) \cdot \boldsymbol{r} = \begin{bmatrix} \cos(2\pi/2) & \sin(2\pi/2) & 0 \\ -\sin(2\pi/2) & \cos(2\pi/2) & 0 \\ 0 & 0 & 1 \end{bmatrix} \begin{pmatrix} x \\ y \\ z \end{pmatrix}$$

$$= \begin{bmatrix} -1 & 0 & 0 \\ 0 & -1 & 0 \\ 0 & 0 & 1 \end{bmatrix} \begin{pmatrix} x \\ y \\ z \end{pmatrix} = \begin{pmatrix} -x \\ -y \\ z \end{pmatrix} \tag{B4.8.1c}$$

(continued)

Box 4.8 (continued)

$$\tilde{\mathrm{C}}_2(x)\cdot\boldsymbol{r}=\begin{bmatrix}1 & 0 & 0\\ \cos(2\pi/2) & 0 & \sin(2\pi/2)\\ -\sin(2\pi/2) & 0 & \cos(2\pi/2)\end{bmatrix}\begin{pmatrix}x\\ y\\ z\end{pmatrix}$$

$$=\begin{bmatrix}1 & 0 & 0\\ -1 & 0 & 0\\ 0 & 0 & -1\end{bmatrix}\begin{pmatrix}x\\ y\\ z\end{pmatrix}=\begin{pmatrix}x\\ -y\\ -z\end{pmatrix} \quad \text{(B4.8.1}d\text{)}$$

$$\tilde{\mathrm{C}}_2(y)\cdot\boldsymbol{r}=\begin{bmatrix}\cos(2\pi/2) & 0 & \sin(2\pi/2)\\ 0 & 1 & 0\\ -\sin(2\pi/2) & 0 & \cos(2\pi/2)\end{bmatrix}\begin{pmatrix}x\\ y\\ z\end{pmatrix}$$

$$=\begin{bmatrix}-1 & 0 & 0\\ 0 & 1 & 0\\ 0 & 0 & -1\end{bmatrix}\begin{pmatrix}x\\ y\\ z\end{pmatrix}=\begin{pmatrix}-x\\ y\\ -z\end{pmatrix} \quad \text{(B4.8.1}e\text{)}$$

$$\tilde{\sigma}(xy)\cdot\boldsymbol{r}=\begin{bmatrix}1 & 0 & 0\\ 0 & 1 & 0\\ 0 & 0 & -1\end{bmatrix}\begin{pmatrix}x\\ y\\ z\end{pmatrix}=\begin{pmatrix}x\\ y\\ -z\end{pmatrix} \quad \text{(B4.8.1}f\text{)}$$

$$\tilde{\sigma}(xz)\cdot\boldsymbol{r}=\begin{bmatrix}1 & 0 & 0\\ 0 & -1 & 0\\ 0 & 0 & 1\end{bmatrix}\begin{pmatrix}x\\ y\\ z\end{pmatrix}=\begin{pmatrix}x\\ -y\\ z\end{pmatrix} \quad \text{(B4.8.1}g\text{)}$$

and

$$\tilde{\sigma}(yz)\cdot\boldsymbol{r}=\begin{bmatrix}-1 & 0 & 0\\ 0 & 1 & 0\\ 0 & 0 & 1\end{bmatrix}\begin{pmatrix}x\\ y\\ z\end{pmatrix}=\begin{pmatrix}-x\\ y\\ z\end{pmatrix}. \quad \text{(B4.8.1}h\text{)}$$

By using the expression for the product of two matrices (Eq. A2.5) and the fact that matrix multiplication is associative, it is straightforward to show that these matrices have the same multiplication table as the operators. The products $\tilde{\mathrm{C}}_2(z)\cdot\tilde{\mathrm{C}}_2(z)$ and $\tilde{\mathrm{i}}\cdot\tilde{\mathrm{C}}_2(z)$, for example, are

(continued)

Box 4.8 (continued)

$$\tilde{\mathrm{C}}_2(z)\cdot\tilde{\mathrm{C}}_2(z)\cdot\boldsymbol{r}=\begin{bmatrix}-1&0&0\\0&-1&0\\0&0&1\end{bmatrix}\begin{bmatrix}-1&0&0\\0&-1&0\\0&0&1\end{bmatrix}\begin{pmatrix}x\\y\\z\end{pmatrix}$$
$$=\begin{bmatrix}1&0&0\\0&1&0\\0&0&1\end{bmatrix}\begin{pmatrix}x\\y\\z\end{pmatrix}=\begin{pmatrix}x\\y\\z\end{pmatrix}\qquad\text{(B4.8.2}a\text{)}$$

and

$$\tilde{\mathrm{I}}\cdot\tilde{\mathrm{C}}_2(z)\cdot\boldsymbol{r}=\begin{bmatrix}-1&0&0\\0&-1&0\\0&0&-1\end{bmatrix}\begin{bmatrix}-1&0&0\\0&-1&0\\0&0&1\end{bmatrix}\begin{pmatrix}x\\y\\z\end{pmatrix}$$
$$=\begin{bmatrix}1&0&0\\0&1&0\\0&0&-1\end{bmatrix}\begin{pmatrix}x\\y\\z\end{pmatrix}=\begin{pmatrix}x\\y\\-z\end{pmatrix}.\qquad\text{(B4.8.2}b\text{)}$$

You can verify these results by referring to Fig. 4.8. Multiplying any of the matrices by the matrix representing Ẽ (the *identity matrix*) leaves the first matrix unchanged, as it should:

$$\tilde{\mathrm{E}}\cdot\tilde{\mathrm{C}}_2(z)\cdot\boldsymbol{r}=\begin{bmatrix}1&0&0\\0&1&0\\0&0&1\end{bmatrix}\begin{bmatrix}-1&0&0\\0&-1&0\\0&0&1\end{bmatrix}\begin{pmatrix}x\\y\\z\end{pmatrix}$$
$$=\begin{bmatrix}-1&0&0\\0&-1&0\\0&0&1\end{bmatrix}\begin{pmatrix}x\\y\\z\end{pmatrix}=\begin{pmatrix}-x\\-y\\z\end{pmatrix}.\qquad\text{(B4.8.3)}$$

Since (*a*) the products of the matrices have the same multiplication table as the operators in the point group, (*b*) matrix multiplications are associative, (*c*) each of the products is the same as one of the original matrrices, and (*d*) every matrix with a non-zero determinant has an inverse (Appendix A.2), the set of matrices representing the symmetry operators in a point group (Eqs. B4.10.1a–B4.10.1h) meet the criteria stated above for a group. The matrix group thus provides a *representation* of the point group, just as the individual matrices provide representations of the individual symmetry operators.

The vector (*x*,*y*,*z*) that we used in this example is said to form a *basis* for a representation of the D_{2h} point group. There are an infinite number of possible choices for such a basis, and the matrices representing the symmetry operators depend on our choice. We could, for example, use the coordinates

(continued)

Box 4.8 (continued)
of the six atoms of ethylene, in which case we would need an 18 × 18 matrix to represent each of the operators. Other possibilities would be a set of bond lengths and angles, the molecular orbitals of ethylene, or other functions of the coordinates. But the choices can be reduced systematically to a small set of representations that are mathematically orthogonal to each other. The number of these *irreducible representations* depends on how many different classes of symmetry operations make up the point group. We will not go into this in detail here other than to say that two operators $\tilde{X}$ and $\tilde{Y}$ are in the same class, and are said to be *conjugate* operators, if and only if $\tilde{Y} = \tilde{Z}^{-1}\tilde{X}\tilde{Z}$ where $\tilde{Z}$ is some other operator. The procedure of transforming $\tilde{X}$ to $\tilde{Y}$ by forming the product $\tilde{Z}^{-1}\tilde{X}\tilde{Z}$ is called a *similarity transformation*. In the D_{2h} point group, each of the eight symmetry operators is of a different class, so there are eight different irreducible representations. Also, each of the irreducible representations for the operators in this point group is a single quantity (i.e., a one-dimensional representation, or a 1 × 1 matrix). Any more complicated (*reducible*) representation can be written as a linear combination of these eight irreducible representations, just as a vector can be constructed from Cartesian x, y and z components.

Information on the irreducible representations of the various point groups is presented customarily in tables that are called *character tables* because they give the *character* of the irreducible representation of each symmetry operation in a point group. The character of a representation is the trace (the sum of the diagonal elements) of the matrix that represents that operation. Character tables for the D_{2h}, C_{2v} and C_{4v} point groups are presented in Tables 4.2, 4.3 and 4.4. The symmetry elements of the point group are displayed in the top row of each table, and the conventional names, called *Mulliken symbols*, of the irreducible representations are given in the first column. Letters *A* and *B* are used as the Mulliken symbols for

(continued)

Table 4.2 Character table for the D_{2h} point group

D_{2h}	E	$C_2(z)$	$C_2(y)$	$C_2(x)$	i	$\sigma_h(xy)$	$\sigma_v(xz)$	$\sigma_v(yz)$	Functions
A_g	1	1	1	1	1	1	1	1	x^2, y^2, z^2
B_{1g}	1	1	−1	−1	1	1	−1	−1	xy
B_{2g}	1	−1	1	−1	1	−1	1	−1	xz
B_{3g}	1	−1	−1	1	1	−1	−1	1	yz
A_u	1	1	1	1	−1	−1	−1	−1	xyz
B_{1u}	1	1	−1	−1	−1	−1	1	1	z
B_{2u}	1	−1	1	−1	−1	1	−1	1	y
B_{3u}	1	−1	−1	1	−1	1	1	−1	x

Table 4.3 Character table for the C_{2v} point group

C_{2v}	E	C_2	$\sigma_v(xz)$	$\sigma'_v(yz)$	Functions
A_1	1	1	1	1	z, x^2, y^2, z^2
A_2	1	1	−1	−1	xy, xyz
B_1	1	−1	1	−1	x, xz
B_2	1	−1	−1	1	y, yz

Table 4.4 Character table for the C_{4v} point group

C_{4v}	E	$2C_4$	C_2	$2\sigma_v$	$2\sigma_d$	Functions
A_1	1	1	1	1	1	z, x^2+y^2, z^2
A_2	1	1	1	−1	−1	
B_1	1	−1	1	1	−1	x^2-y^2
B_2	1	−1	1	−1	1	xy, xyz
E	2	0	−2	0	0	x, y, xz, yz

Box 4.8 (continued)

one-dimensional representations like those of the D_{2h} and C_{2v} point groups. E denotes a two-dimensional matrix (regretably inviting confusion with the symbol for the identity symmetry element), and T denotes a three-dimensional matrix. Representations of types A and B differ in being, respectively, symmetric and antisymmetric with respect to a C_n axis. Subscripts g and u stand for the German terms *gerade* and *ungerade*, and indicate whether the representation is even (g) or odd (u) with respect to a center of inversion. The numbers 1, 2 and 3 in the subscripts distinguish different representations of the same general type. The three columns on the right side of the character table give some one-, two- and three-dimensional functions that have the same symmetry as the various irreducible representations. These are referred to as *basis functions* for the irreducible representations.

Looking at the character table for the D_{2h} point group (Table 4.2) we see that, because all the irreducible representations of this group are one-dimensional, their characters are either +1 or −1. Each symmetry operation therefore either leaves the representation unchanged, in which case the character is +1, or changes the sign of the representation, making the character −1. A_g is the *totally symmetric* representation. Like the quadratic functions x^2, y^2 and z^2, it is unaffected by any of the symmetry operations of the D_{2h} point group. A_u also is unaffected by rotation about the x, y or z axis but changes sign on inversion or reflection across the xy, xz or yz plane. It thus has the same symmetry properties as the product xyz in this point group. B_{1u}, B_{2u} and B_{3u} have the same symmetry as the coordinates z, y and x, each of

(continued)

Box 4.8 (continued)
which changes sign on rotation around two of the three C_2 axes, inversion, or reflection across one plane.

The behaviors of the basis functions on the right side of the character table for the D_{2h} point group are described by saying that x^2, y^2 and z^2 *transform* as A_g in this point group, xy transforms as B_{1g}, z transforms as B_{1u}, and so forth. Turning to the C_{2v} and C_{4v} point groups (Tables 4.3 and 4.4), note that here z transforms as the totally symmetric irreducible representation, A_1. Rotation around the C_2 axis does not change the z coordinate in these point groups, because the single C_2 axis coincides with the z axis (Fig. 4.7C). Reflection across the xz or yz plane also leaves the z coordinate unchanged. There is no center of inversion in the C_{2v} or C_{4v} point group, so none of the Mulliken symbols has a g or u subscript.

The entries $2C_4$, $2\sigma_v$ and $2\sigma_d$ in the top row of the character table for the C_{4v} point group mean that this point group has two independent symmetry operations in each of the C_4, σ_v and σ_d classes. The C_4 class includes both the C_4 operation itself, and the inverse of this operation, C_4^{-1}, which is the same as C_4^3. The basis functions for the two-dimensional irreducible representation (E) in the last row are pairs of coordinate values (x, y) or pairs of products of these values. The character 2 here means that the identity symmetry preserves both values, as it should, and the character –2 indicates that the C_2 operation changes the sign of both values.

Character tables for virtually all the point groups that are encountered in chemistry are available [6, 7]. The full tables also include information on how the symmetry operations affect the direction of molecular rotation around the x, y or z axis, which is pertinent to the rotational spectroscopy of small molecules.

Because molecular orbitals must recognize the symmetry of a molecule, a symmetry operation must either preserve the value of the wavefunction or simply change the sign of the wavefunction. The wavefunction therefore provides a basis for a representation of the molecule's point group. Inspection of the HOMO and LUMO molecular orbitals of ethylene (Fig. 4.6) shows that the HOMO transforms as the irreducible representation B_{1u} and z in the D_{2h} point group (Table 4.2), whereas the LUMO transforms as B_{3g} and yz.

Perhaps the most important feature of character tables for our purposes is that we can determine the symmetry of a product of two irreducible representations simply by looking at the products of the characters of these representations in the same point group. For example, the product of the HOMO and LUMO of ethylene transforms as B_{2u} in D_{2h}, as you can see by comparing the products of the characters of B_{3g} and B_{1u} with the corresponding characters of B_{2u}. Alternatively, we can just note that the product of B_{3g} and B_{1u} transforms as the product of z and yz, or yz^2, which

(continued)

Box 4.8 (continued)
is equivalent to y. Further, the product of these two wavefunctions and y transforms as the square of B_{2u}, or y^2, which is the same as the totally symmetric representation, A_g. The electronic transition from the HOMO to the LUMO is, therefore, allowed upon excitation with radiation polarized along the y axis.

In general, excitation of a centrosymmetric system is forbidden if the initial and final orbitals both have the same symmetry for inversion (either both g or both u). This principle is known as *Laporte's rule*.

To illustrate the vectorial nature of transition dipoles for a larger molecule, Fig. 4.9 shows the two highest occupied and two lowest unoccupied molecular orbitals of bacteriochlorophyll-*a*. These four wavefunctions are labeled $\psi_1 - \psi_4$ in order of increasing energy. The products of the wavefunctions for the four possible excitations ($\psi_1 \rightarrow \psi_3$, $\psi_1 \rightarrow \psi_4$, $\psi_2 \rightarrow \psi_3$, and $\psi_2 \rightarrow \psi_4$) are shown in Fig. 4.10. The conjugated atoms of bacteriochlorophyll-*a* form an approximately planar π system, and the wavefunctions and their products all have a plane of reflection symmetry that coincides with the plane of the macrocyclic ring. Because the wavefunctions have opposite signs on the two sides of this plane (z), their products are even functions of z. The z component of the transition dipole for excitation from ψ_1 or ψ_2 to ψ_3 or ψ_4 therefore will be zero: all the transition dipoles lie in the plane of the π-system. Inspection of Fig. 4.10 shows further that the products $\psi_1\psi_4$ and $\psi_2\psi_3$ both are approximately odd functions of the y (vertical) coordinate in the figure, and approximately even functions of the x (horizontal) coordinate. The transition dipoles for $\psi_1 \rightarrow \psi_4$ and $\psi_2 \rightarrow \psi_3$ are, therefore, oriented approximately parallel to the y axis. The other two products, $\psi_1\psi_3$ and $\psi_2\psi_4$, are approximately odd functions of x and approximately even functions of y, so the transition dipoles for $\psi_1 \rightarrow \psi_3$ and $\psi_2 \rightarrow \psi_4$ must be be approximately parallel to the x axis. The calculated transition dipoles confirm these qualitative predictions (Fig. 4.10). The fact that the transition dipoles for the $\psi_1 \rightarrow \psi_3$ and $\psi_2 \rightarrow \psi_4$ excitations point in opposite directions refelects arbitrary choices of the signs of the wavefunctions and has no particular significance.

Figure 4.11 shows similar calculations of the transition dipoles for the two highest occupied and lowest empty molecular orbitals of 3-methylindole, a model of the side chain of tryptophan. Again, the transition dipoles for excitation from one of these orbitals to another must lie in the plane of the π system. The transition dipole calculated for the $\psi_2 \rightarrow \psi_3$ and $\psi_1 \rightarrow \psi_4$ excitations are oriented approximately 30° from the x axis in the figure (Fig. 4.11E, F). That calculated for the $\psi_1 \rightarrow \psi_3$ and $\psi_2 \rightarrow \psi_4$ excitations is approximately 120° from the x axis (Fig. 4.11G, H).

Transitions that are formally forbidden by symmetry with the dipole operator sometimes are weakly allowed by the quadrupole or octupole terms in Eq. (4.5). Forbidden transitions also can be promoted by vibrational motions that perturb the symmetry or change the mixture of electronic configurations contributing to the

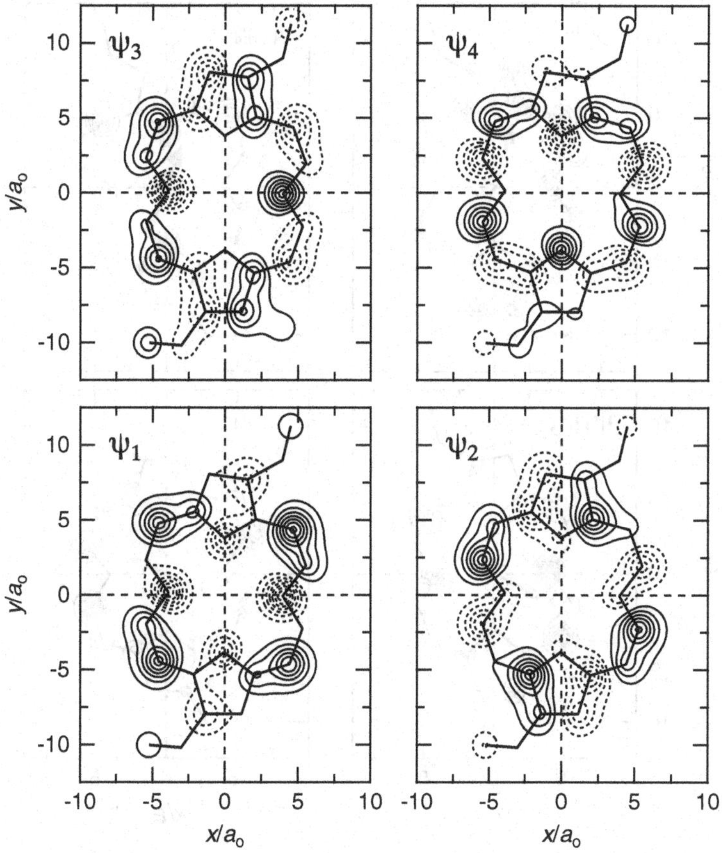

Fig. 4.9 Contour plots of the two highest occupied orbitals (ψ_1 and ψ_2) and the two lowest unoccupied molecular orbitals (ψ_3 and ψ_4) of bacteriochlorophyll-*a*. The plane of each drawing is parallel to the plane of the macrocyclic ring and is above the ring by the Bohr radius, a_o (Fig. 2.7, panels D, E). *Solid curves* represent positive amplitudes; *dotted curves*, negative amplitudes. The contours for zero amplitude are omitted for clarity. Distances are given as multiples of a_o, and the contour intervals are $0.02a_o^{3/2}$. The skeleton of the π system is shown with *heavy lines*. The coefficients for the atomic p_z orbitals were obtained with the program QCFF/PI [25, 149]

excited state. This is called *vibronic coupling*. Finally, some transitions with small electric transition dipoles can be driven by the magnetic field of light. We'll return to this point in Chap. 9. Figure 9.4 of that chapter also illustrates how the electric transition dipoles for the first four excitations of *trans*-butadiene depend on the symmetries of the molecular orbitals.

The spin of the photon imposes an additional orbital selection rule on absorption. To conserve angular momentum when a photon is absorbed, a change in electronic orbital angular momentum must balance the angular momentum provided by the photon. In atomic absorption, this means that the azimuthal quantum number l must change by either -1 or $+1$, depending on whether the photon has left ($m_s = +1$) or

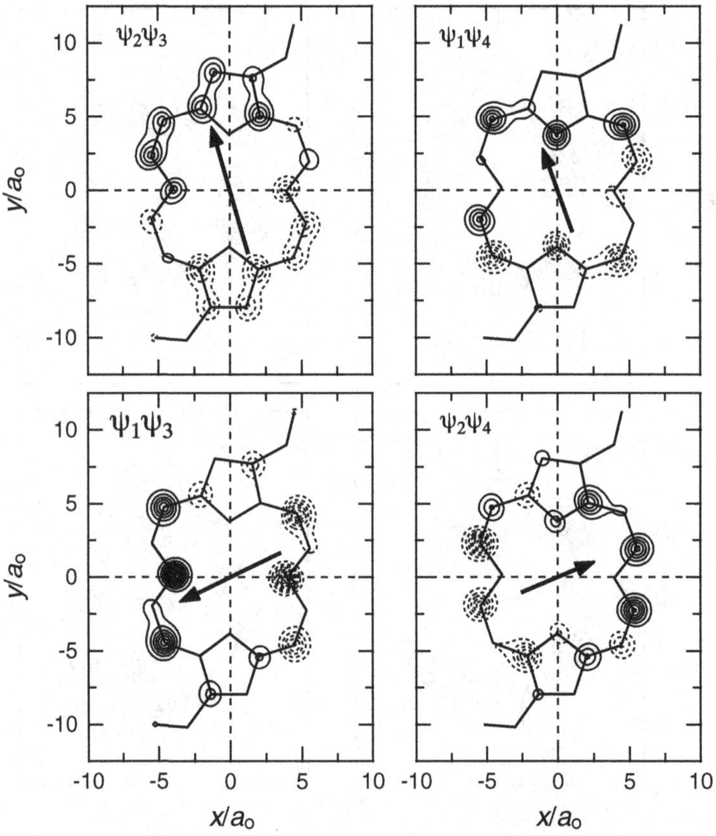

Fig. 4.10 Contour plots of products of the four molecular wavefunctions of bacteriochlorophyll-*a* shown in Fig. 4.9. The planes of the drawings and the line types for positive and negative amplitudes are as in Fig. 4.9. The contour intervals are $0.002a_o^{\,3}$. The *arrows* show the transition dipoles calculated by Eq. (4.22e), with the length in units of eÅ/5a_o

right ($m_s = -1$) circular polarization (Sect. 3.3). This forbids excitation from an s orbital to an f orbital, but allows excitation from s to p. Absorption of linearly polarized light does not impart angular momentum to the absorbing electron, because a linearly polarized photon is in a superposition of states with left and right circular polarization. Selection rules that depend on the electronic spins of the ground- and excited-state orbitals will be discussed in Sect. 4.9.

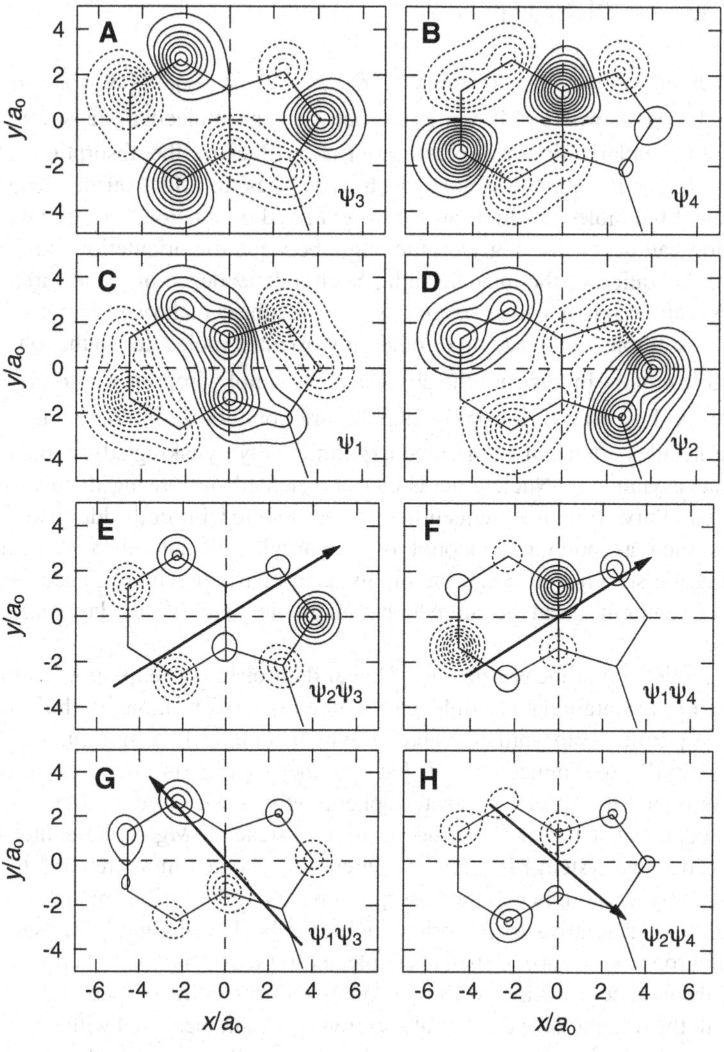

Fig. 4.11 Contour plots of the two lowest unoccupied molecular orbitals (ψ_3 and ψ_4, **A**, **B**) and the two highest occupied orbitals (ψ_1 and ψ_2, **C**, **D**) of 3-methylindole, and products of these wavefunctions ($\psi_2\psi_3$, $\psi_1\psi_4$, $\psi_1\psi_3$ and $\psi_2\psi_4$, **E**–**H**). Positive amplitudes are indicated with *solid lines*, negative amplitudes with *dotted lines*. Small contributions from the methyl group are neglected. The contour intervals are $0.02a_o^{3/2}$ in **A**–**D** and $0.005a_o^{3}$ in **E**–**H**. The *arrows* in **E**–**H** show the transition dipoles calculated by Eq. (4.22e), with the length in units of eÅ/2.5a_o. The atomic coefficients for the orbitals were obtained as described by Callis [36, 37, 150]

4.6 Linear Dichroism

As discussed above, the quantity $(\mathbf{E}_o \cdot \boldsymbol{\mu}_{ba})^2$ in Eq. (4.8c) is equal to $|\mathbf{E}_o|^2 |\boldsymbol{\mu}_{ba}|^2 \cos^2\theta$, where θ is the angle between the molecular transition dipole ($\boldsymbol{\mu}_{ba}$) and the polarization axis of the light ($\mathbf{E}_o$). Because the absorption strength depends on $\cos^2\theta$, rotating a molecule by 180° has no effect on the absorption. However, if the molecules in a sample have a fixed orientation, the strength of the absorption can depend strongly on the angle between the orientation axis and the axis of polarization of the incident light. Such a dependence of the absorbance on the polarization axis is called *linear dichroism*. Linear dichroism is not seen if a sample is isotropic, i.e., made up of randomly oriented molecules so that θ takes on all possible values. In this case the absorbance is simply proportional to $(1/3)|\mathbf{E}_o|^2 |\boldsymbol{\mu}_{ba}|^2$ (Box 4.6). But anisotropic materials are common in biology, and purified macromolecules often can be oriented experimentally by taking advantage of their molecular asymmetry. Nucleic acids can be oriented by flowing them through a narrow capillary. Proteins frequently can be oriented by embedding them in a polymer such as polyvinyl alcohol or polyacrylamide and then stretching or squeezing the specimen to align the highly asymmetric polymer molecules. Phospholipid membranes can be aligned by magnetic fields or by layering on flat surfaces.

One application of measurements of linear dichroism is to explore the structures of complexes containing multiple chromophores. An example is the "reaction center" of purple photosynthetic bacteria, which contains four molecules of bacteriochlorophyll, two molecules of bacteriopheophytin, and several additional pigments bound to a protein. (Bacteriopheophytin is the same as bacteriochlorophyll except that it has two hydrogen atoms instead of Mg at the center of the macrocyclic ring system.) If reaction centers are oriented in a stretched film or a squeezed polyacrylamide gel, the absorption bands of the various pigments exhibit linear dichroism relative to the orientation axis [8–13]. Figure 4.12B shows the linear dichroism spectrum of such a sample, expressed as $A_\perp - A_\parallel$, where $A_\perp$ and $A_\parallel$ are the absorbance measured with light polarized, respectively, perpendicular and parallel to the orientation axis. The absorption spectrum measured with unpolarized light is shown in Fig. 4.12A. The bands between 830 and 1,000 nm represent transitions of the bacteriochlorophylls, while the bands at 790 and 805 nm are assigned to the bacteriopheophytins. Note that the linear dichroism of the bacteriopheophytin bands is negative, whereas that of the bacteriochlorophyll bands is positive, indicating that the two types of pigments are oriented so that their transition dipoles are approximately perpendicular. We'll return to the absorption spectrum of reaction centers in Sect. 4.7 and again in Chap. 8.

A classic application of linear dichroism to study molecular orientations and motions in a complex biological system was R. Cone's study of *induced dichroism* in retinal rod outer segements [14]. Rhodopsin, the light-sensitive pigment-protein complex of the retina, contains 11-*cis*-retinal attached covalently to a protin (opsin) by a Schiff base linkage (Fig. 4.13A, B). Its transition dipole is oriented

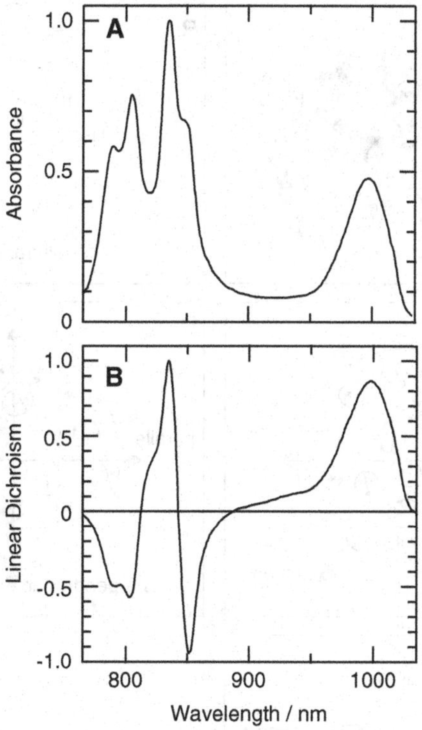

Fig. 4.12 Absorption (**A**) and linear dichroism (**B**) spectra of photosynthetic reaction centers of *Blastochloris* (formerly called *Rhodopseudomonas*) *viridis* at 10 K [12, 13]. Aggregates of the pigment–protein complex were embedded in a polyacrylamide gel and oriented by uniaxial squeezing. The spectra are normalized relative to the positive peak near 830 nm. Linear dichroism is expressed as $A_{\perp}$–$A_{\parallel}$, where $A_{\perp}$ and $A_{\parallel}$ are the absorbance measured with light polarized perpendicular and parallel to the compression axis. Transition dipoles that lie within about 35° of the plane of the aggregate's largest cross-section give positive linear dichroism, whereas transition dipoles that are closer to normal to this plane give negative linear dichroism. The absorption bands in the spectra shown here represent mixed Q_y transitions of the reaction center's four molecules of bacteriochlorophyll-*b* and two molecules of bacteriopheophytin-*b* (Sect. 4.7, Chap. 8). See [12, 13] for spectra extending to shorter wavelengths

approximately along the long axis of the retinylidine chromophore. Rhodopsin is an integral membrane protein and resides in flattened membrane vesicles ("disks") that are stacked in the outer segements of rod cells. Previous investigators had shown that when rod cells were illuminated from the side, light polarized parallel to the planes of the disk membranes was absorbed much more strongly than light polarized normal to the membranes. However, if the cells were illuminated end-on, there was no preference for any particular polarization in the plane of the membrane. These measurements indicated that the rhodopsin molecules are aligned so that the transition dipole of the chromophore in each molecule is approximately parallel to the plane of the membrane, but that the transition dipoles point in random directions within this plane (Fig. 4.13C).

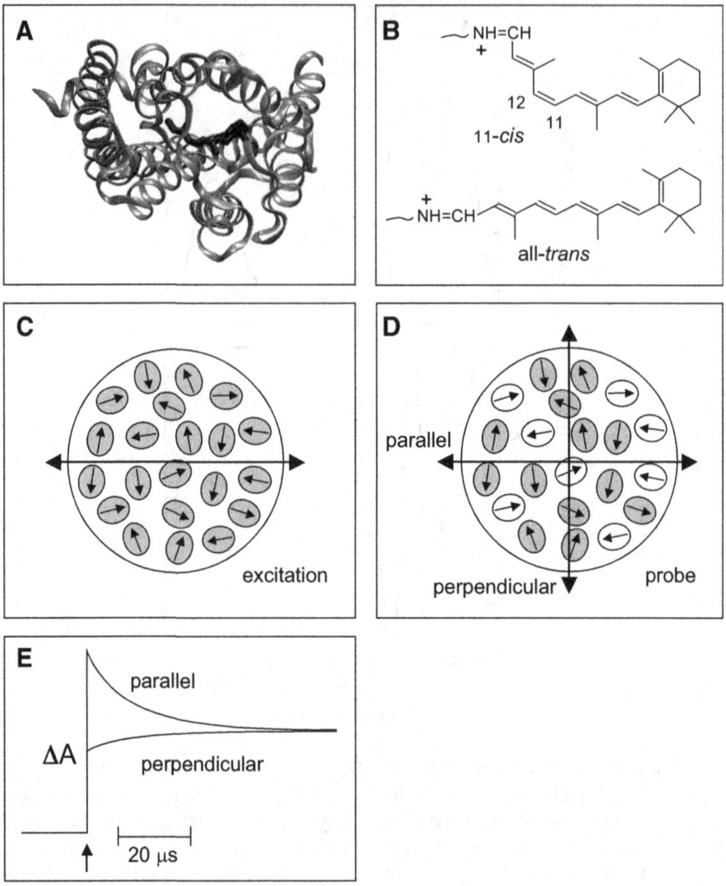

Fig. 4.13 (**A**) Crystal structure of bovine rhodopsin viewed from a perspective approximately normal to the membrane. The polypeptide backbone is represented by a ribbon model (*gray*) and the retinylidine chromophore by a licorice model (*black*). The coordinates are from Protein Data Bank file 1f88.pdb [69]. Some parts of the protein that protrude from the phosopholipid bilayer of the membrane are omitted for clarity. (**B**) The 11-*cis*-retinylidine chromophore is attached to a lysine residue by a protonated Schiff base linkage. Excitation results in isomerization around the 11–12 bond to give an all-*trans* structure. (**C, D**) Schematic depictions of a field of rhodopsin molecules in a rod cell disk membrane, viewed normal to the membrane. The *short arrows* in the *shaded ovals* represent the transition dipoles of individual rhodopsin molecules. (Each disk in a human retina contains approximately 1,000 rhodopsins.) The transition dipoles lie approximately in the plane of the membrane, but have no preferred orientation in this plane. A polarized excitation flash (*horizontal double-headed arrow* in **C**) selectively excites molecules that are oriented with their transition dipoles parallel to the polarization axis, causing some of them to isomerize and changing their absorption spectrum (*empty ovals* in **D**). (**E**) Smoothed records of the absorbance changes at 580 nm as a function of time, measured with "probe" light polarized either parallel or perpendicular to the excitation [14]. The *vertical arrow* indicates the time of the flash. The absorbance change initially depends on the polarization, but this dependence disappears as rhodopsin molecules rotate in the membrane

When rhodopsin is excited with light the 11-*cis*-retinyl chromophore is isomerized to all-*trans*, initiating conformational changes in the protein that ultimately result in vision (Fig. 4.13B). If rod cells are illuminated end-on with a weak flash of polarized light, the light is absorbed selectively by rhodopsin molecules that happen to be oriented with their transition dipoles parallel to the polarization axis (Fig. 4.13C, D). Molecules whose transition dipoles are oriented at an angle θ with respect to the polarization axis are excited with a probability that falls off with $\cos^2\theta$. On the order of 2/3 of the molecules that are excited undergo isomeration to an all-*trans* structure, and then evolve through a series of metastable states that can be distinguished by changes in their optical absorption spectrum. Cone [14] measured the absorbance changes associated with these transformations, again using polarized light that passed through the rod cells end-on. In the absence of the excitation flash, the absorbance measured with the probe beam was, as stated above, indendepent of the polarization of the probe. However, the absorbance changes resulting from the excitation flash were very different, depending on whether the measuring light was polarized parallel or perpendicular to the excitation light (Fig. 4.13E). The polarized excitation flash, by preferentially exciting molecules with a certain orientation, thus created a linear dichroism that could be probed at a later time. The difference between the signals measured with parallel and perpendicular polarizations of the probe beam decayed with a time constant of about 20 μs. Cone interpreted the decay of the induced dichroism as reflecting rotations of rhodopsin in the plane of the membrane. These experiments provided the first quantitative measurements of the fluidity of a biological membrane. The rotation dynamics are still timely, however, because measurements by atomic force microscopy indicate that rhodopsin may exist as dimers and paracrystalline arrays under some conditions [15].

In another application, Junge et al. [16, 17] measured the rate of rotation of the γ subunit of the chloroplast ATP synthase when the enzyme hydrolyzes ATP. Rotation of the γ subunit relative to the α and β subunits appears to couple transmembrane movement of protons to the synthesis or breakdown of ATP.

Another example is the use of linear dichroism to examine the orientation of CO bound to myoglobin. Compared to free heme, myoglobin discriminates strongly against the binding of CO relative to O_2. This discrimination initially was ascribed to steric factors that prevent the diatomic molecules from sitting along an axis normal to the heme; it was suggested that the molecular orbitals of O_2 were more suitable for an orientation off the normal than those of CO. Lim et al. [18] examined the orientation of bound CO by exciting carboxymyoglobin with polarized laser pulses at a wavelength absorbed by the heme. Excitation of the heme causes photodissociation of the bound CO, which remains associated with the protein in a pocket close to the heme. This results in an absorption decrease in an infrared band that reflects CO attached to the Fe, and in the appearance of a new band reflecting CO in the looser pocket. Polarized infrared probe pulses following the excitation pulse thus can be used to determine the orientations of CO molecules in the two sites relative to the transition dipole of the heme. Lim et al. found that the C–O bond of CO attached to the Fe is approximately normal to the heme plane, indicating that other factors must be responsible for the preference of myoglobin for O_2.

4.7 Configuration Interactions

Although many molecular absorption bands can be ascribed predominantly to transitions between the HOMO and LUMO, it often is necessary to take other transitions into account. One reason for this is that the HOMO and LUMO wavefunctions pertain to an unexcited molecule that has two electrons in each of the lower-lying orbitals, including the HOMO itself. The interactions among the electrons are somewhat different if there is an unpaired electron in each of the HOMO and LUMO. In addition, the wavefunctions themselves are approximations of varying reliability. Better descriptions of the excited state often can be obtained by considering the excitation as a linear combination of transitions from several of the occupied orbitals to several unoccupied orbitals. Each such orbital transition is termed a *configuration*, and the mixing of several configurations in an excitation is called *configuration interaction*. In general, two transitions will mix most strongly if they have similar energies and involve similar changes in the symmetry of the molecular orbitals.

Equation (4.22e) can be expanded straightforwardly to include a sum over the various configurations that contribute to an excitation:

$$\boldsymbol{\mu}_{ba} \approx \sqrt{2}\, e \sum_{j,k} A_{j,k}^{a,b} \sum_t C_t^j C_t^k \mathbf{r}_t, \tag{4.26}$$

where $A_{j,k}^{a,b}$ is the coefficient for the configuration $\psi_j \rightarrow \psi_k$ in the overall excitation from state a to state b. Box 4.9 describes a procedure for finding these coefficients.

Box 4.9 Evaluating Configuration-Interaction Coefficients

The procedure for finding the configuration-interaction (CI) coefficient $A_{j,k}^{a,b}$ involves constructing a matrix in which the diagonal elements are the energies of the individual transitions [19–24]. For excited singlet states of π molecular orbitals, the off-diagonal matrix elements that couple two configurations, $\psi_{j1} \rightarrow \psi_{k1}$ and $\psi_{j2} \rightarrow \psi_{k2}$ with $j1 \neq j2$ and $k1 \neq k2$, take the form

$$\begin{aligned} \langle \psi_{j1\rightarrow k1} | \tilde{\mathrm{H}} | \psi_{j2\rightarrow k2} \rangle &= \left\langle \psi_{j1\rightarrow k1} \left| \sum_s \sum_t e^2/r_{s,t} \right| \psi_{j2\rightarrow k2} \right\rangle \\ &\approx \sum_s \sum_t \left(2C_s^{j1} C_s^{k1} C_t^{j2} C_t^{k2} - C_s^{j1} C_s^{j2} C_t^{k1} C_t^{k2} \right) \gamma_{s,t}. \end{aligned} \tag{B4.8.1}$$

Here C_t^j represents the contribution of atom t to wavefunction ψ_j, as in Eqs. (4.19–4.22e) and (2.42), $r_{s,t}$ is the distance between atoms s and t, and $\gamma_{s,t}$ is a semiempirical function of this distance. A typical expression for $\gamma_{s,t}$,

(continued)

Box 4.9 (continued)
obtained by maximizing the agreement between the calculated and observed spectroscopic properties of a large number of molecules, is

$$\gamma_{s,t} = A\exp(-Br_{s,t}) + C/(D + r_{s,t}), \quad \text{(B4.8.2)}$$

with $A = 3.77 \times 10^4$, $B = 0.232/\text{Å}$, $C = 1.17 \times 10^5$ Å, and $D = 2.82$ [25]. The CI coefficients are obtained by diagonalizing the matrix as we describe in Sect. 2.3.6.

The restrictions that orbital symmetry imposes on configuration interactions can be analyzed in the same manner as the selection rules for the individual transition dipoles. As Eq. (B4.8.1) indicates, coupling of two transitions, $\psi_{j1} \rightarrow \psi_{k1}$ and $\psi_{j2} \rightarrow \psi_{k2}$, depends on the product of the four wavefunctions (ψ_{j1}, ψ_{k1}, ψ_{j2} and ψ_{k2}). In the language of group theory (Box 4.8) we can say that, if the product of ψ_{j1} and ψ_{k1} and the product of ψ_{j2} and ψ_{k2} both transform as x, the product of all four wavefunctions will transform as x^2 and will be totally symmetric. Mixing of the two transitions then will be allowed by symmetry. On the other hand, if the product of ψ_{j1} and ψ_{k1} transforms as x, say, while that of ψ_{j2} and ψ_{k2} transforms as y, the overall product will transform as xy and will integrate to zero. Inspection of Figs. 4.10 and 4.11 shows that if we denote the top two filled molecular orbitals and the first two empty orbitals of bacteriochlorophyll or 3-methylindole as $\psi_1 - \psi_4$ in order of increasing energy, the product $\psi_1\psi_3$ has the same symmetry as $\psi_2\psi_4$, whereas $\psi_2\psi_3$ has the same symmetry as $\psi_1\psi_4$. The $\psi_1 \rightarrow \psi_3$ transition thus should mix with $\psi_2 \rightarrow \psi_4$ and the $\psi_2 \rightarrow \psi_3$ transition should mix with $\psi_1 \rightarrow \psi_4$.

Porphyrins and their chlorin and bacteriochlorin derivatives provide numerous illustrations of the importance of configuration interactions [26]. In porphin (Fig. 4.14, *left*), the highly symmetrical parent compound, the two highest filled molecular orbitals (ψ_1 and ψ_2 in order of increasing energy) are nearly isoenergetic, as are the two lowest unoccupied orbitals (ψ_3 and ψ_4). The $\psi_1 \rightarrow \psi_4$ and $\psi_2 \rightarrow \psi_3$ transitions have the same symmetry and energy and thus mix strongly, as explained in Box 4.9; the $\psi_1 \rightarrow \psi_3$ transition mixes similarly with $\psi_2 \rightarrow \psi_4$. Because of the near degeneracy of the transition energies the two CI coefficients ($A^{a,b}_{j,k}$) are simply $\pm 2^{-1/2}$ in each case, but the signs of the coefficients can be either the same or opposite. The four orbitals thus give rise to four different excited states, which are commonly referred to as the B_y, B_x, Q_x and Q_y states. B_y and Q_y consist of the configurations $2^{-1/2}(\psi_1 \rightarrow \psi_4) \pm 2^{-1/2}(\psi_2 \rightarrow \psi_3)$; B_x and Q_x, of $2^{-1/2}(\psi_1 \rightarrow \psi_3) \pm 2^{-1/2}(\psi_2 \rightarrow \psi_4)$. However, the transition dipoles cancel each other almost exactly in two of the combinations and reinforce each other in the other two. The result is that the two lowest-energy absorption bands (Q_x and Q_y)

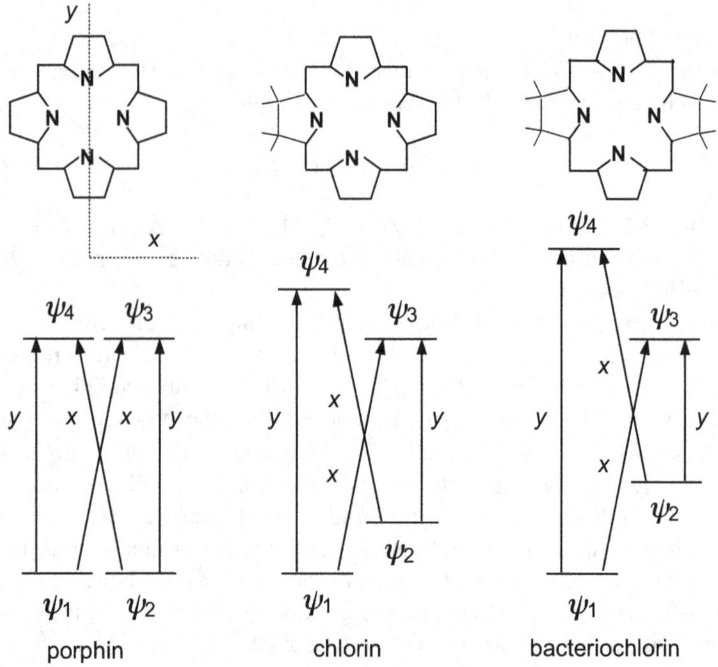

Fig. 4.14 Structures, energy diagrams, and excitations of porphin, chlorin and bacteriochlorin. The *horizontal bars* are schematic representations of the energies of the highest two occupied molecular orbitals (ψ_1 and ψ_2) and the first two unoccupied orbitals (ψ_3 and ψ_4), relative to the energy of ψ_1 in each molecule. *Arrows* indicate excitations from one of the ocupied orbitals to an empty orbital. x and y correspond to the molecular axes shown as *dotted lines* with the porphin structure, and convey the symmetry of the product of the initial and final wavefunctions for each configuration, e.g., $\psi_1 \rightarrow \psi_4$. Configurations with the same symmetry mix in the excited states, and the transition dipole for the mixed excitation is oriented approximately along the x or y molecular axis. Reduction of one or both of the tetrapyrrole rings in chlorin and bacteriochlorin moves the lowest-energy excited state to progressively lower energy and increases its dipole strength

occur at essentially the same energy and are very weak relative to the two higher-energy bands (B_x and B_y).

In chlorin (Fig. 4.14, *center*), one of the four pyrrole rings is partially reduced, removing two carbons from the π system. This perturbs the symmetry of the molecule and moves ψ_2 and ψ_4 up in energy relative to ψ_1 and ψ_3. As a result, the lowest-energy absorption band moves to a lower energy and gains dipole strength, whereas the highest-energy band moves to higher energy and loses dipole strength. The trend continues in the bacteriochlorins, in which two of the pyrrole rings are reduced (Fig. 4.14, *right*). Hemes, which are symmetrical iron porphyrins, thus absorb blue light strongly and absorb yellow or red light only weakly (Fig. 4.15A), while bacteriochlorophylls absorb intensely in the red or near-IR (Fig. 4.15B). This four-orbital model rationalizes a large body of experimental observations on the spectroscopic properties of metalloporphyrins, chlorophylls, bacteriochlorophylls and related molecules [26–28], and has been used to analyze

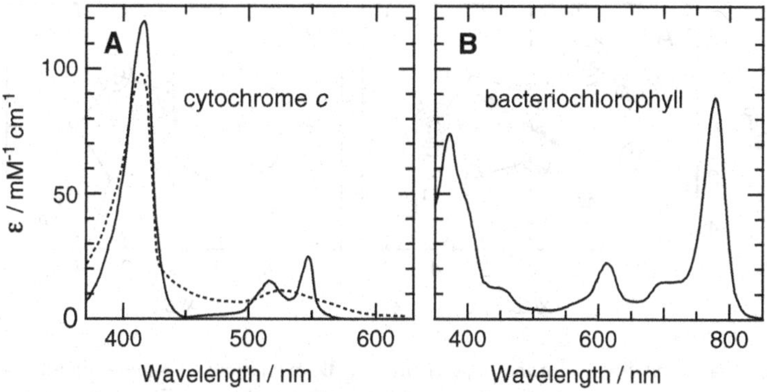

Fig. 4.15 Absorption spectra of cytochrome *c* with the bound heme in its reduced (*solid curve*) and oxidized (*dotted curve*) forms (**A**), and of bacteriochlorophyll *a* in methanol (**B**)

the spectroscopic properties of photosynthetic bacterial reaction centers and antenna complexes [4, 29, 30]. In the spectrum of *Bl. viridis* reaction centers shown in Fig. 4.12A, the absorption bands between 750 and 1,050 nm reflect Q_y transitions of the bacteriopheophytins and bacteriochlorophylls, which are mixed by exciton interactions as discussed in Chap. 8. The corresponding Q_x bands (not shown in the figure) are in the regions of 530–545 and 600 nm.

The indole side chain of tryptophan provides another example. Its absorption spectrum again involves significant contributions from the orbitals lying just below the HOMO and just above the LUMO (ψ_1 and ψ_4), in addition to the HOMO and LUMO themselves (ψ_2 and ψ_3). Transitions among the four orbitals give rise to two overlapping absorption bands in the region of 280 nm that are commonly called the ^1L_a and ^1L_b bands, and two higher-energy bands (^1B_a and ^1B_b) near 195 and 221 nm [31–37]. The ^1L_a excitation has a somewhat higher energy and greater dipole strength than ^1L_b, and, as we will see in Sect. 4.12, results in a larger change in the permanent dipole moment. Neglecting small contributions from higher-energy configurations, the excited singlet states associated with the ^1L_a and ^1L_b excitations can be described reasonably well by the combinations $^1L_a \approx 0.917(\psi_2 \rightarrow \psi_3) - 0.340(\psi_1 \rightarrow \psi_4)$ and $^1L_b \approx 0.732(\psi_1 \rightarrow \psi_3) + 0.634(\psi_2 \rightarrow \psi_4)$. Figure 4.16 shows the two transition dipoles calculated by Eq. (4.26).

If an improved description of an excited electronic state is wanted for other purposes such as calculations of exciton interactions (Chap. 8), the CI coefficients can be adjusted empirically to maximize the agreement between the calculated and observed transition energy or dipole strength. In one application of this idea [30], the CI coefficients for the Q_x and Q_y absorption bands of bacteriochloropyll and bacteriopheophytin were adjusted so that the dipole strengths calculated using the transition gradient operator (Sect. 4.8) matched the measured dipole strengths. The discrepancy between these dipole strengths and the values calculated with the dipole operator then was used to correct calculated energies of dipole-dipole interactions among the bacteriochlorophylls and bacteriopheophytins in photosynthetic bacterial reaction centers.

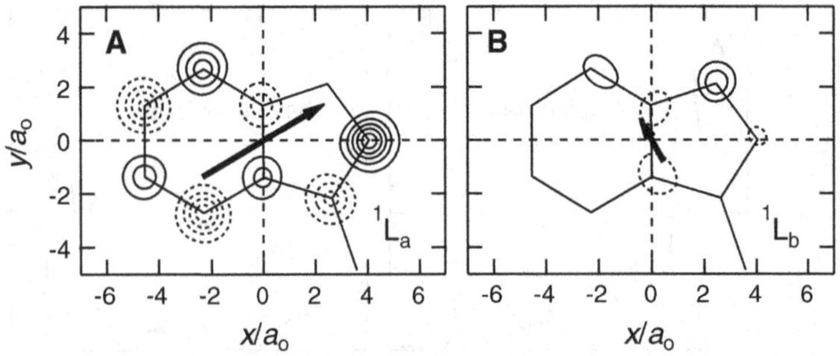

Fig. 4.16 Transition dipoles for the ^1L_a (**A**) and ^1L_b (**B**) excitations of 3-methylindole calculated from linear combinations of products of the molecular orbitals shown in Fig. 4.11. Contour plots of the functions $0.917\psi_2\psi_3 - 0.340\psi_1\psi_4$ (**A**) and $0.732\psi_1\psi_3 + 0.634\psi_2\psi_4$ (**B**) are shown with *solid lines* for positive amplitudes, *dotted lines* for negative amplitudes, and *contour intervals* of $0.005a_o^3$. The *arrows* show the transition dipoles calculated by Eq. (4.26), with the lengths in units of eÅ$/5a_o$

4.8 Calculating Electric Transition Dipoles with the Gradient Operator

When contributions from transitions involving the top two or three filled orbitals and the first few unoccupied orbitals are considered, dipole strengths calculated by using Eq. (4.26) typically agree with experimentally measured dipole strengths to within a factor of 2 or 3, which means that the magnitude of the transition dipole is correct to within about ±50%. Better agreement sometimes can be obtained by using the gradient operator, $\tilde{\nabla} = (\partial/\partial x, \partial/\partial y, \partial/\partial z)$, instead of $\tilde{\mu}$. Matrix elements of the gradient and dipole operators are related by the expression

$$\left\langle \Psi_b \middle| \tilde{\nabla} \middle| \Psi_a \right\rangle = \frac{-(E_b - E_a)m_e}{\hbar^2 e} \langle \Psi_b | \tilde{\mu} | \Psi_a \rangle, \qquad (4.27)$$

where m_e is the electron mass (Box 4.10). Thus, if the energy difference $E_b - E_a$ is known, we can obtain $\langle \Psi_b | \tilde{\mu} | \Psi_a \rangle$ from $\left\langle \Psi_b \middle| \tilde{\nabla} \middle| \Psi_a \right\rangle$ and vice versa. Transition dipoles calculated with $\tilde{\mu}$ and $\tilde{\nabla}$ should be identical if the molecular orbitals are exact, but with approximate orbitals the two methods usually give somewhat different results. The dipole strengths calculated with the dipole operator often are too large, while those obtained with the gradient operator agree better with experiment [28, 30, 38].

Box 4.10 The Relationship Between Matrix Elements of the Electric Dipole and Gradient Operators

Equation (4.27) can be derived by relating the gradient operator $\tilde{\nabla}$ to the commutator of the Hamiltonian and dipole operators $([\tilde{\mathrm{H}}, \tilde{\mu}])$. (See Box 2.2 for an introduction to commutators.) The Hamiltonian operator $\tilde{\mathrm{H}}$ includes terms for both potential energy ($\tilde{\mathrm{V}}$) and kinetic energy $(\tilde{\mathrm{T}})$; however, we only need to consider $\tilde{\mathrm{T}}$ because $\tilde{\mathrm{V}}$ commutes with the position operator $([\tilde{\mathrm{V}}, \tilde{\mathrm{r}}] = 0)$ and $\tilde{\mu}$ is simply $e\tilde{\mathrm{r}}$. For a one-dimensional system, in which $\tilde{\nabla}$ is just $\partial/\partial x$, the commutator of $\tilde{\mathrm{H}}$ and $\tilde{\mu}$ is

$$[\tilde{\mathrm{H}}, \tilde{\mu}] = [\tilde{\mathrm{T}}, e\tilde{x}] = -(\hbar^2 e/2m)\left[\partial^2/\partial x^2, x\right] \quad \text{(B4.10.1}a\text{)}$$

$$= -(\hbar^2 e/2m)\left[(d^2/dx^2)x - x(d^2/dx^2)\right] \quad \text{(B4.10.1}b\text{)}$$

$$= -(\hbar^2 e/2m)\left[2(d/dx) + x(d^2/dx^2) - x(d^2/dx^2)\right] \quad \text{(B4.10.1}c\text{)}$$

$$= -(\hbar^2 e/m)(d/dx) = -(\hbar^2 e/m)\tilde{\nabla}. \quad \text{(B4.10.1}d\text{)}$$

Generalizing to three dimensions, and treating the commutator as an operator gives

$$\langle \Psi_b | [\tilde{\mathrm{H}}, \tilde{\mu}] | \Psi_a \rangle = -(\hbar^2 e/m)\left\langle \Psi_b \middle| \tilde{\nabla} \middle| \Psi_a \right\rangle. \quad \text{(B4.10.2)}$$

We can relate the matrix element on the left side of Eq. (B4.10.2) to the transition dipole ($\boldsymbol{\mu}_{ba}$) by expanding Ψ_b and Ψ_a formally in the basis of all the eigenfunctions of $\tilde{\mathrm{H}}$ (Ψ_k) and using the procedure for matrix multiplication (Appendix A.2). This gives

$$\langle \Psi_b | [\tilde{\mathrm{H}}, \tilde{\mu}] | \Psi_a \rangle \equiv \langle \Psi_b | \tilde{\mathrm{H}}\tilde{\mu} | \Psi_a \rangle - \langle \Psi_b | \tilde{\mu}\tilde{\mathrm{H}} | \Psi_a \rangle \quad \text{(B4.10.3}a\text{)}$$

$$= \sum_k \langle \Psi_b | \tilde{\mathrm{H}} | \Psi_k \rangle \langle \Psi_k | \tilde{\mu} | \Psi_a \rangle - \sum_k \langle \Psi_b | \tilde{\mu} | \Psi_k \rangle \langle \Psi_k | \tilde{\mathrm{H}} | \Psi_a \rangle \quad \text{(B4.10.3}b\text{)}$$

$$= E_b \langle \Psi_b | \tilde{\mu} | \Psi_a \rangle - \langle \Psi_b | \tilde{\mu} | \Psi_a \rangle E_a = (E_b - E_a)\boldsymbol{\mu}_{ba}, \quad \text{(B4.10.3}c\text{)}$$

where E_b and E_a are the energies of states a and b. Equation (B4.10.3b) reduces to Eq. (B4.10.3c) because, in the absence of additional perturbations, the only non-zero Hamiltonian matrix elements involving Ψ_b or Ψ_a are $\langle \Psi_b | \tilde{\mathrm{H}} | \Psi_b \rangle$ and $\langle \Psi_a | \tilde{\mathrm{H}} | \Psi_a \rangle$. Equating the right-hand sides of Eqs. (B4.10.2) and (B4.10.3c) gives Eq. (4.27).

In the quantum theory of absorption that we discuss in Chap. 5, the transition gradient matrix element $\left(\left\langle \Psi_b \middle| \tilde{\nabla} \middle| \Psi_a \right\rangle\right)$ arises directly, rather than just as an alternative way of calculating $\langle \Psi_b|\tilde{\mu}|\Psi_a\rangle$. Matrix elements of $\tilde{\nabla}$ also play a fundamental role in the theory of circular dichroism (Chap. 9).

Figure 4.17 illustrates the functions that enter into the transition gradient matrix element for the HOMO → LUMO excitation of ethylene. Contour plots of the HOMO and LUMO (ψ_a and ψ_b) are reproduced from Fig. 4.6B in panels A and B

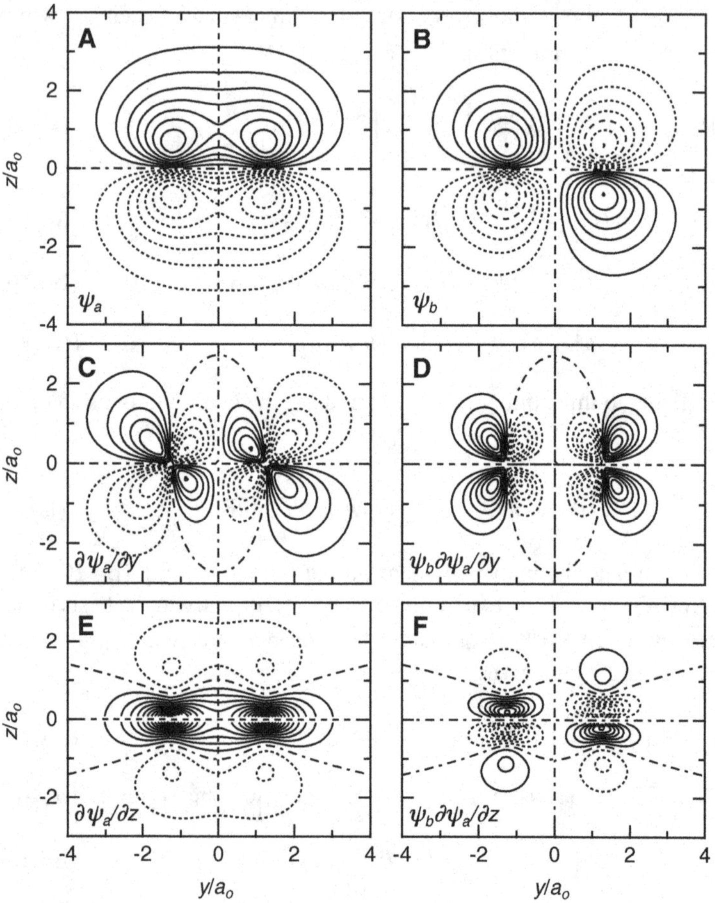

Fig. 4.17 Components of the transition matrix element of the gradient operator for excitation of ethylene. (**A**, **B**) Contour plots of the amplitudes of the HOMO (ψ_a) and LUMO (ψ_b) molecular orbitals in the *yz* plane. The C=C double bond lies on the *y* axis. (**C**, **E**) The derivatives of ψ_a with respect to *y* and *z*, respectively. (**D**, **F**) The products of these derivatives with ψ_b. The *y* and *z* components of $\left\langle \psi_b \middle| \tilde{\nabla} \middle| \psi_a \right\rangle$ are obtained by integrating $\psi_b \partial\psi_a/\partial y$ and $\psi_b \partial\psi_a/\partial z$, respectively, over all space. The symmetry of the wavefunctions is such that this integral is zero for $\psi_b \partial\psi_a/\partial z$ and (not shown) $\psi_b \partial\psi_a/\partial x$, but non-zero for $\psi_b \partial\psi_a/\partial y$

for reference. As before, the C=C bond is aligned along the y axis and ψ_a and ψ_b are constructed from atomic $2p_z$ orbitals. Panel C shows a contour plot of the derivative $\partial\psi_a/\partial y$, and D shows the result of multiplying this derivative by ψ_b. Integrating the product $\psi_b\partial\psi_a/\partial y$ over all space gives the y component of $\left\langle\psi_b\middle|\tilde{\nabla}\middle|\psi_a\right\rangle$. By inspecting Fig. 4.17D you can see that $\psi_b\partial\psi_a/\partial y$ is an even function of y, so except for points in the xy plane, where ψ_a and ψ_b both go through zero, integration along any line parallel to the y axis will give a non-zero result. (Although the contour plots in the figure show only the amplitudes in the yz plane, the corresponding plots for any other plane parallel to yz would be similar because ψ_a and ψ_b are both even functions of x.) By contrast, the function $\psi_b\partial\psi_a/\partial z$ (Fig. 4.17F) is an odd function of both y and z, which means that integrating this product over all space will give zero. This is true also of $\psi_b\partial\psi_a/\partial x$ (not shown). The vector $\left\langle\psi_b\middle|\tilde{\nabla}\middle|\psi_a\right\rangle$ evidently is oriented along the C=C bond, which is just what we found above for $\langle\psi_b|\tilde{\mu}|\psi_a\rangle$.

Calculating the matrix element $\left\langle\psi_b\middle|\tilde{\nabla}\middle|\psi_a\right\rangle$ for a transition between two π molecular orbitals is somewhat more cumbersome than calculating $\langle\psi_b|\tilde{\mu}|\psi_a\rangle$, but is still relatively straightforward if the molecular orbitals are constructed of linear combinations of Slater-type atomic orbitals. The transition matrix element for excitation to an excited singlet state then takes the same form as Eq. (4.22e):

$$\left\langle\Psi_b\middle|\tilde{\nabla}\middle|\Psi_a\right\rangle = \sqrt{2}\sum_s\sum_t C_s^b C_t^a \left\langle p_s\middle|\tilde{\nabla}\middle|p_t\right\rangle, \tag{4.28}$$

where C_s^b and C_t^a are the expansion coefficients for atomic $2p_z$ orbitals centered on atoms s and t (p_s and p_t) in molecular orbitals Ψ_b and Ψ_a, respectively, and $\left\langle p_s\middle|\tilde{\nabla}\middle|p_t\right\rangle$ is the matrix element of $\tilde{\nabla}$ for the two atomic orbitals. Box 4.11 outlines a general procedure for evaluating $\left\langle p_s\middle|\tilde{\nabla}\middle|p_t\right\rangle$ that allows the atomic orbitals to have any orientation with respect to each other. Equation (4.28) can be simplified slightly by noting that $\left\langle p_t\middle|\tilde{\nabla}\middle|p_t\right\rangle$ is zero and $\left\langle p_t\middle|\tilde{\nabla}\middle|p_s\right\rangle = -\left\langle p_s\middle|\tilde{\nabla}\middle|p_t\right\rangle$, as you can verify by studying Fig. 4.17. With these substitutions, the sum over atoms s and t becomes

$$\left\langle\Psi_b\middle|\tilde{\nabla}\middle|\Psi_a\right\rangle = \sqrt{2}\sum_{s>t}\sum_t 2\left(C_s^a C_t^b - C_s^b C_t^a\right)\left\langle p_s\middle|\tilde{\nabla}\middle|p_t\right\rangle, \tag{4.29}$$

which sometimes can be approximated well by a sum over just the pairs of bonded atoms [28, 30, 38–40].

Box 4.11 Matrix Elements of the Gradient Operator for Atomic 2*p* Orbitals

The integral $\left\langle p_s \middle| \tilde{\nabla} \middle| p_t \right\rangle$ in Eqs. (4.28) and (4.29) is the matrix element of the gradient operator ($\tilde{\nabla}$) for atomic 2*p* orbitals on atoms *s* and *t*. To allow the local *z* axis of atom *s* to have any orientation relative to that of atom *t*, the atomic matrix elements can be written

$$\begin{aligned}\left\langle p_s \middle| \tilde{\nabla} \middle| p_t \right\rangle \approx & \left(\eta_{x',s}\eta_{y',t} + \eta_{y',s}\eta_{x',t}\right)\nabla_{xy}\hat{\imath} \\ & + \left[\eta_{y',s}\eta_{y',t}\nabla_\sigma + \left(\eta_{x',s}\eta_{x',t} + \eta_{z',s}\eta_{z',t}\right)\nabla_\pi\right]\hat{\jmath} \\ & + \left(\eta_{z',s}\eta_{y',t} + \eta_{y',s}\eta_{z',t}\right)\nabla_{zy}\hat{k}.\end{aligned} \tag{B4.11.1}$$

Here $\eta_{x',t}$, $\eta_{y',t}$ and $\eta_{z',t}$ are direction cosines of the atomic *z* axis of orbital *t* with respect to a Cartesian coordinate system (x',y',z') defined so that atom *t* is at the origin and the y' axis points along the line from atom *t* to atom *s* (for example, $\eta_{y',t}$ is the cosine of the angle between y' and the local *z* axis of orbital *t*); $\hat{\imath}$, $\hat{\jmath}$ and $\hat{k}$ are unit vectors parallel to the x', y' and z' axes; and ∇_σ, ∇_π and ∇_{zy} are the matrix elements of $\tilde{\nabla}$ for pairs of Slater 2*p* orbitals in three canonical orientations. ∇_σ is for the end-on orientation shown in Fig. 4.18A; ∇_σ, for side-by-side orientation with parallel *z* axes along z' or x' (Fig. 4.18B); and ∇_{zy}, for one orbital displaced along y' and rotated by 90° around an axis parallel to x'(Fig. 4.18C). ∇_{xz} is 0, and ∇_{xy} is the same as ∇_{zy}. Inspection of Eq. (B4.11.1) and Fig. 4.18 shows that if the *z* axes of atoms *t* and *s* are parallel to each other and perpendicular to y', as is approximately the case for π-electron systems, $\left\langle p_s \middle| \tilde{\nabla} \middle| p_t \right\rangle$ points from atom *t* to atom *s* and has magnitude ∇_π.

The Slater $2p_z$ orbital, in polar coordinates centered on atom *s* (r_s,θ_s,ϕ_s), is

$$p_s = \left(\zeta_s^5/\pi\right)^{1/2} r_s \cos\theta_s \exp(-\zeta_s r_s/2), \tag{B4.11.2}$$

where θ_s is the angle with respect to the *z* axis and $\zeta_s = 3.071$, 3.685 and 4.299 Å^{-1} for C, N and O, respectively. The dependences of ∇_σ, ∇_π and ∇_{zy} on the interatomic distance (*R*) and the Slater orbital parameters for the two atoms (ζ_s and ζ_t) can be evaluated as follows, using expressions described originally by Mulliken et al. [41] and Král [42] (see [4] and [43]):

First define the two functions

(continued)

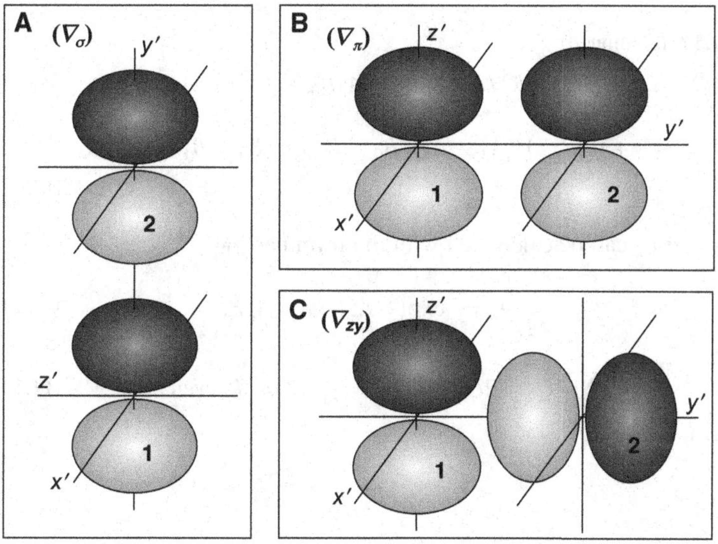

Fig. 4.18 Canonical orientations of $2p_z$ orbitals of two carbon atoms. The *shaded regions* represent boundary surfaces of the wavefunctions as in Fig. 2.5. The $x'y'z'$ Cartesian coordinate system is centered on atom 1 with the y' axis aligned along the interatomic vector. The transition gradient matrix elements ∇_σ, ∇_π and ∇_{zy} are for the orientations shown in **A**, **B** and **C**, respectively. Matrix elements for an arbitrary orientation can be expressed as linear combinations of these canonical matrix elements

Box 4.11 (continued)

$$A_k = \int_1^\infty w^k \exp(-Pw)dw \quad \text{and} \quad B_k = \int_{-1}^{1} w^k \exp(-Qw)dw, \qquad \text{(B4.11.3)}$$

where $P = (\zeta_s + \zeta_t)R/2$ and $Q = (\zeta_s - \zeta_t)R/2$. Then,

$$\nabla_\sigma = (\zeta_s\zeta_t)^{5/2}(R^4/8)\{A_0B_2 - A_2B_0 + A_1B_3 - A_3B_1 + (\zeta_t R/2)[A_1(B_0 - B_2) + B_1(A_0 - A_2) + A_3(B_4 - B_2) + B_3(A_4 - A_2)]\}, \qquad \text{(B4.11.4)}$$

$$\nabla_\pi = (\zeta_s\zeta_t)^{5/2}(\zeta_t R^5/32)[(B_1 - B_3)(A_0 - 2A_2 + A_4) + (A_1 - A_3)(B_0 - 2B_2 + B_4)], \qquad \text{(B4.11.5)}$$

and

(continued)

Box 4.11 (continued)

$$\nabla_{zy} = (\zeta_s\zeta_t)^{5/2}(R^4/8)\{A_0B_2 - A_2B_0 + A_1B_3 - A_3B_1 + (\zeta_t R/4)[(A_3 - A_1)(B_0 - B_4) + (B_3 - B_1)(A_0 - A_4)]\}. \quad \text{(B4.11.6)}$$

A_k and B_k can be calculated with the formulas [44]:

$$A_k = [\exp(-s) + kA_{k-1}]/P, \quad \text{(B4.11.7)}$$

$$B_k = 2\sum_{i=0}^{3}[Q^{2i}/(2i)!(k+2i+1)] \quad for\ k\ even, \quad \text{(B4.11.8)}$$

and

$$B_k = -2\sum_{i=0}^{3}[Q^{2i+1}/(2i+1)!(k+2i+2)] \quad for\ k\ odd. \quad \text{(B4.11.9)}$$

See [28, 30, 39, 40] for other semiempirical expressions for $\langle p_s|\tilde{\nabla}|p_t\rangle$.

Figure 4.19 illustrates the use of this approach to calculate transition matrix elements for *trans*-butadiene, which provides a useful model for carotenoids and retinals. Panels *A–D* show the two highest filled molecular orbitals and the two lowest unoccupied orbitals (ψ_1 to ψ_4 in order of increasing energy). The vector diagram in panel *E* shows the direction and relative magnitude of the matrix element of $\tilde{\nabla}$ for each pair of bonded atoms, $\langle p_s|\tilde{\nabla}|p_t\rangle$, weighted by the coefficients for that pair of atoms in wavefunctions ψ_2 and ψ_3 ($C_s^bC_t^a$). Combining the vectors for each pair of atoms gives an overall transition gradient matrix element $\left(\sum_s\sum_t C_s^bC_t^a\langle p_s|\tilde{\nabla}|p_t\rangle\right)$ oriented on the long axis of the molecule. (Contributions from pairs of non-bonded atoms do not affect the overall matrix element significantly.) Excitation of an electron from ψ_2 and ψ_3, which is the lowest-energy configuration in the excited singlet state, therefore has a non-zero transition dipole with this orientation.

Figure 4.19H is a similar vector diagram for the highest-energy configuration in an excited singlet state of the four wavefunctions ($\psi_1 \rightarrow \psi_4$). The transition-gradient dipole for this excitation has a smaller magnitude and a different orientation than that for $\psi_2 \rightarrow \psi_3$ because the $C_s^bC_t^a\langle p_s|\tilde{\nabla}|p_t\rangle$ vector for the central two atoms partly cancels the contributions from the outer pairs of atoms.

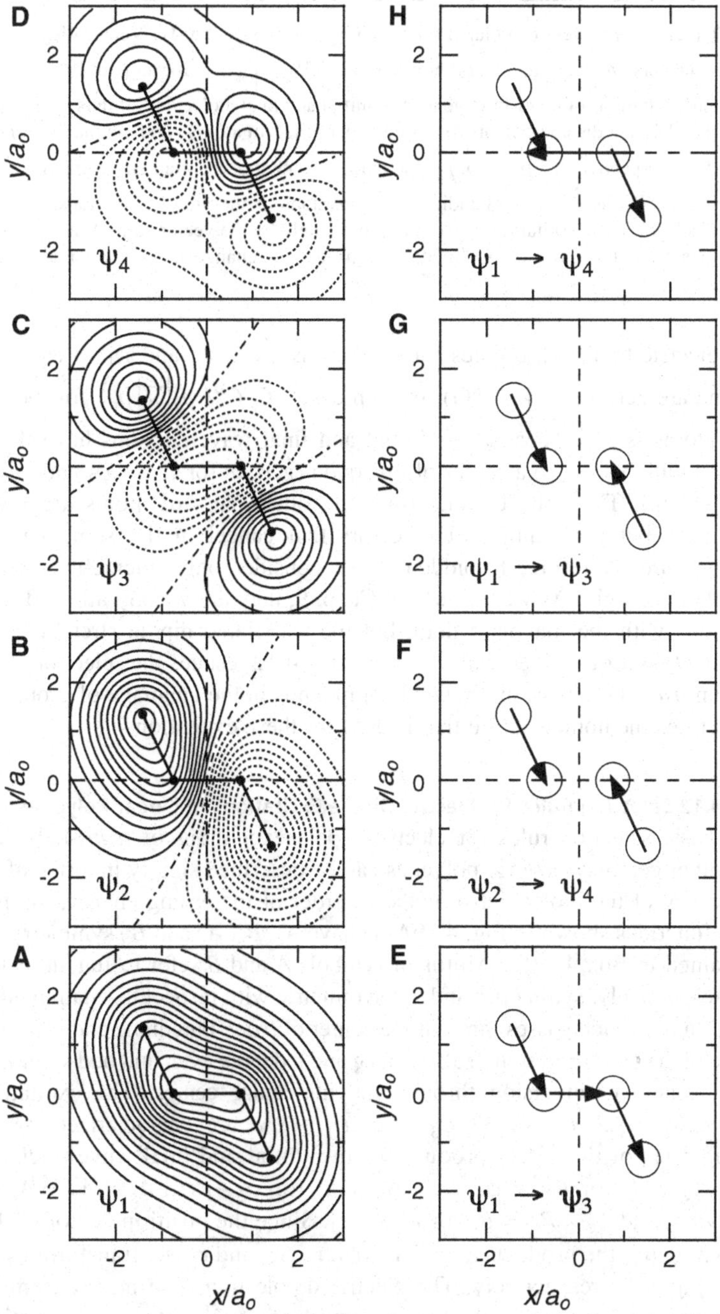

Fig. 4.19 (**A–D**) Contour plots of the two highest occupied and the two lowest unoccupied molecular orbitals in the ground state of *trans*-butadiene ($\psi_1 - \psi_4$ in order of increasing energy). The *small filled circles* indicate the positions of the atoms. The *contour lines* give the amplitudes of the wavefunctions in a plane parallel to the plane of the π system, and above that plane by the Bohr radius (a_o). *Solid lines* indicate positive amplitudes; *dotted lines*, negative. The wavefunctions

Fig. 4.19 (continued) were calculated with QCFF-PI [25, 149]. (**E–H**) Vector diagrams of the directions and relative magnitudes of the products $C_s^b C_t^a \langle p_s | \tilde{\nabla} | p_t \rangle$ for pairs of bonded atoms in the first four excitations of *trans*-butadiene. The initial and final molecular orbitals are indicated in each panel and C_t^a are the coefficients for atomic $2p_z$ orbitals of atoms *s* and *t* in the final and initial wavefunction, respectively; $\langle p_s | \tilde{\nabla} | p_t \rangle$ is the matrix element of the gradient operator for the two atomic orbitals. The *empty circles* indicate the positions of the atoms. The transition gradient matrix element for each excitation is given approximately by the vector sum of the *arrows*. (The contributions from the pairs of nonbonded atoms do not change the overall matrix element significantly)

The electric transition dipoles for excitations $\psi_2 \rightarrow \psi_4$ and $\psi_1 \rightarrow \psi_3$ of *trans*-butadiene are zero (Fig. 4.19F,G). In both cases, $C_s^b C_t^a \langle p_s | \tilde{\nabla} | p_t \rangle$ for the central pair of atoms is zero because the initial and final wavefunctions have the same symmetry with respect to a center of inversion (*gerade* for $\psi_1 \rightarrow \psi_3$ and *ungerade* for $\psi_2 \rightarrow \psi_4$). The contributions from the other pairs of atoms are not zero individually, but give antiparallel vectors that cancel in the sum. These two excitations are, therefore, forbidden if we consider only interactions with the electric field of light. As we will see in Chap. 9, they are weakly allowed through interactions with the magnetic field, but the associated dipole strength is much lower than that for $\psi_2 \rightarrow \psi_3$ and $\psi_1 \rightarrow \psi_4$. Box 4.12 restates the selection rules for excitation *trans*-butadiene and related compounds in the language of group theory and describes the nomenclature that is used for these excitations.

Box 4.12 Selection Rules for Electric-Dipole Excitations of Linear Polyenes

The basic selection rules for electric-dipole excitations of *trans*-butadiene and other centrosymmetric polyenes can be described simply in terms of the symmetry of the molecular orbitals. In order of increasing energy, the four wavefunctions shown in Fig. 4.19A–D have A_g, B_u, A_g and B_u symmetry. As explained in Box 4.8, the Mulliken symbols *A* and *B* refer to functions that are, respectively, symmetric and antisymmetric with respect to rotation about the C_2 axis, which passes through the center of *trans*-butadiene and is normal to the plane of the π system; subscripts *g* and *u* denote even and odd symmetry with respect to inversion through the molecular center. The excitations $\psi_2 \rightarrow \psi_3$, $\psi_2 \rightarrow \psi_4$, $\psi_1 \rightarrow \psi_3$ and $\psi_1 \rightarrow \psi_4$ are characterized by the symmetries of the direct products of the initial and final wavefunctions: $B_u \times A_g = B_u$ for $\psi_2 \rightarrow \psi_3$, $B_u \times B_u = A_g$ for $\psi_2 \rightarrow \psi_4$, $A_g \times A_g = A_g$ for $\psi_1 \rightarrow \psi_3$, and $A_g \times B_u = B_u$ for $\psi_1 \rightarrow \psi_4$. Since the position vector ($\mathbf{r}$) has B_u symmetry, the products $\mathbf{r}\psi_2\psi_3$, $\mathbf{r}\psi_2\psi_4$, $\mathbf{r}\psi_1\psi_3$ and $\mathbf{r}\psi_1\psi_4$ transform as A_g, B_u, B_u and A_g, respectively. The electric dipole transition matrix elements are, therefore, nonzero for the first and fourth excitations (the two excitations with B_u symmetry), and zero for the second and third (the two with A_g

(continued)

Box 4.12 (continued)

symmetry). Similar considerations apply to longer polyenes including carotenoids, although twisting and bending distortions leave these molecules only approximately centrosymmetric.

Note that the vectors in Fig. 4.19E–H pertain to the pure configurations $\psi_2 \rightarrow \psi_3$, $\psi_2 \rightarrow \psi_4$, $\psi_1 \rightarrow \psi_3$ and $\psi_1 \rightarrow \psi_4$. An accurate description of the actual excited states of *trans*-butadiene and other such polyenes requires extensive configuration interactions [45]. The $\psi_2 \rightarrow \psi_4$ excitation mixes strongly with $\psi_1 \rightarrow \psi_3$ and other higher-energy configurations, with the result that an excited state with this symmetry moves below the first B_u state in energy [45–47]. The lowest-energy allowed absorption band therefore reflects excitation to the second excited state rather than the first, which has significant consequences for the functions of carotenoids as energy donors and acceptors in photosynthesis (Sect. 7.4). The "forbidden" first excited state of several polyenes has been detected by Raman spectroscopy [46], two-photon spectroscopy (Chap. 12 and [48]), and time-resolved measurements of absorption and fluorescence [47, 49–51].

In the nomenclature that is commonly used, the ground state of a centrosymmetric polyene is labeled $1^1A_g^-$; the first excited singlet state with A_g symmetry is $2^1A_g^-$, and the first with B_u symmetry (the second excited singlet state in order of energy) is $1^1B_u^+$. The first number indicates the position of the state in order of increasing energy among states with the same symmetry. The first superscript identifes the spin multiplicity (singlet or triplet), and the superscript "+" or "−" indicates whether the state is predominantly covalent (−) or ionic (+). Because the same selection rules also hold as a first approximation in larger polyenes that are not strictly centrosymmetric, this nomenclature also is used for retinals and complex carotenoids.

4.9 Transition Dipoles for Excitations to Singlet and Triplet States

As we discussed in Chap. 2 (Sect. 2.4), electrons have an intrinsic angular momentum or "spin" that is characterized by spin quantum numbers $s = 1/2$ and $m_s = \pm 1/2$. The different values of m_s can be described by two spin wavefunctions, α ("spin up") for $m_s = +1/2$ and β ("spin down") for $m_s = -1/2$. For systems with more than one electron, wavefunctions that include the electronic spins must be written in a way so that the complete wavefunction changes sign if we interchange any two electrons. We skipped over this point quickly when we derived expressions for the transition dipole for forming an excited singlet state (Eqs. 4.22a–4.22e), so let's check whether we get the same expressions if we write the spin wavefunctions explicitly. We'll also examine the transition dipoles for excitation to triplet states.

Using the notation of Eqs. (2.47), the transition dipole for excitation from a singlet ground state to an excited singlet state of a system with two electrons takes the form

$$\boldsymbol{\mu}_{ba} = \langle {}^1\Psi_b|\tilde{\mu}(1)+\tilde{\mu}(2)|\Psi_a\rangle = \langle\{2^{-1/2}[\psi_h(1)\psi_l(2)+\psi_h(2)\psi_l(1)]2^{-1/2}[\alpha(1)\beta(2) \\ -\alpha(2)\beta(1)]\}|\tilde{\mu}(1)+\tilde{\mu}(2)|\left\{\psi_h(1)\psi_h(2)2^{-1/2}[\alpha(1)\beta(2)-\alpha(2)\beta(1)]\right\}\rangle. \quad (4.30)$$

Because the electric dipole operator does not act on the spin wavefunctions, integrals such as $\langle\alpha(1)|\alpha(1)\rangle$ and $\langle\alpha(1)|\beta(1)\rangle$ can be factored out of the overall integral. By doing this and making the same approximations we made above in Eqs. (4.22a–4.22c), we obtain

$$\boldsymbol{\mu}_{ba} = 2^{-3/2}[\langle\psi_l(1)|\tilde{\mu}(1)|\psi_h(1)\rangle\langle\psi_h(2)|\psi_h(2)\rangle + \langle\psi_l(2)|\tilde{\mu}(2)|\psi_h(2)\rangle\langle\psi_h(1)|\psi_h(1)\rangle]\times \\ [\langle\alpha(1)|\alpha(1)\rangle\langle\beta(2)|\beta(2)\rangle - \langle\alpha(1)|\beta(1)\rangle\langle\beta(2)|\alpha(2)\rangle - \langle\alpha(2)|\beta(2)\rangle\langle\beta(1)|\alpha(1)\rangle \\ +\langle\alpha(2)|\alpha(2)\rangle\langle\beta(1)|\beta(1)\rangle]. \quad (4.31)$$

The factors $\langle\psi_h(1)|\psi_h(1)\rangle$ and $\langle\psi_h(2)|\psi_h(2)\rangle$ in Eq. (4.31) are both unity. The spin integrals also can be evaluated immediately because the spin wavefunctions are orthogonal and normalized: $\langle\alpha(1)|\alpha(1)\rangle = \langle\beta(1)|\beta(1)\rangle = 1$, and $\langle\alpha(1)|\beta(1)\rangle = \langle\beta(1)|\alpha(1)\rangle = 0$. The terms in the second square brackets thus are $1\times 1 - 0\times 0 - 0\times 0 + 1\times 1 = 2$, so

$$\boldsymbol{\mu}_{ba} = \sqrt{2}\langle\psi_l(k)|\tilde{\mu}(k)|\psi_h(k)\rangle \approx \sqrt{2}e\sum_i C_i^l C_i^h \boldsymbol{r}_i. \quad (4.32)$$

This is the same result as Eq. (4.22e).

The transition dipoles for transitions from the singlet ground state to the excited triplet states are very different from the transition dipole for forming the excited singlet state. The transition dipoles for forming the triplet states all evaluate to zero because terms with opposite signs cancel or because each term includes an integral of the form $\langle\alpha(1)|\beta(1)\rangle$. For ${}^3\Psi_b^0$, we have:

$$\langle {}^3\Psi_b^0|\tilde{\mu}(1)+\tilde{\mu}(2)|\Psi_a\rangle = \langle\{2^{-1}[\psi_h(1)\psi_l(2)-\psi_h(2)\psi_l(1)]\,[\alpha(1)\beta(2)+\alpha(2)\beta(1)]\} \\ \times|\tilde{\mu}(1)+\tilde{\mu}(2)|\left\{\psi_h(1)\psi_h(2)2^{-1/2}[\alpha(1)\beta(2)-\alpha(2)\beta(1)]\right\}\rangle \quad (4.33a)$$

$$= 2^{-3/2}[\langle\psi_l(2)|\tilde{\mu}(2)|\psi_h(2)\rangle\langle\psi_h(1)\psi_h(1)|\rangle - \langle\psi_l(1)|\tilde{\mu}(1)|\psi_h(1)\rangle\langle\psi_h(2)\psi_h(2)|\rangle] \\ \times[\langle\alpha(1)|\alpha(1)\rangle\langle\beta(2)|\beta(2)\rangle - \langle\alpha(1)|\alpha(2)\rangle\langle\beta(2)|\beta(1)\rangle + \langle\alpha(2)|\alpha(1)\rangle\langle\beta(1)|\beta(1)\rangle \\ - \langle\alpha(2)|\alpha(2)\rangle\langle\beta(1)|\beta(1)\rangle]. \quad (4.33b)$$

The products in the first line of Eq. (4.33b) reduce to $[\langle\psi_l(2)|\tilde{\mu}(2)|\psi_h(2)\rangle - \langle\psi_l(1)|\tilde{\mu}(1)|\psi_h(1)\rangle]$, which is zero because the transition dipole does not depend on how we label the electron. The sum of products of spin integrals in the second and third lines evaluates to $1\times1-0\times0+0\times0-1\times1$, which also is 0.

Similarly, for ${}^3\Psi_b^{+1}$:

$$\left\langle {}^3\Psi_b^{+1}|\tilde{\mu}(1)+\tilde{\mu}(2)|\Psi_a\right\rangle = \left\langle\{2^{-1/2}[\psi_h(1)\psi_l(2)-\psi_h(2)\psi_l(1)]\,[\alpha(1)\alpha(2)\}\right.$$
$$\left.\times|\tilde{\mu}(1)+\tilde{\mu}(2)|\{\psi_h(1)\psi_h(2)2^{-1/2}[\alpha(1)\beta(2)-\alpha(2)\beta(1)]\}\right\rangle \quad (4.34a)$$

$$=2^{-1}[\langle\psi_l(2)|\tilde{\mu}(2)|\psi_h(2)\rangle\langle\psi_h(1)\psi_h(1)\rangle - \langle\psi_l(1)|\tilde{\mu}(1)|\psi_h(1)\rangle\langle\psi_h(2)|\psi_h(2)\rangle]$$
$$\times[\langle\alpha(1)|\alpha(1)\rangle\langle\alpha(2)|\beta(2)\rangle - \langle\alpha(1)|\beta(1)\rangle\langle\alpha(2)|\alpha(2)\rangle]. \quad (4.34b)$$

Here again, both the first and the second lines of the final expression are zero. Evaluating the transition dipole for ${}^3\Psi_b^{-1}$ similarly gives the same result.

Excitations from the ground state to an excited triplet state are, therefore, formally forbidden. In practice, weak optical transitions between singlet and triplet states sometimes are observable. Triplet states also can be created by *intersystem crossing* from excited singlet states, as we'll discuss in Chap. 5. This process results mainly from coupling of the magnetic dipoles associated with electronic spin and orbital electronic motion.

4.10 The Born-Oppenheimer Approximation, Franck-Condon Factors, and the Shapes of Electronic Absorption Bands

So far, we have focused on the effects of light on electrons. The complete wavefunction for a molecule must describe the nuclei also. But because nuclei have very large masses compared to electrons, it is reasonable for some purposes to view them as being more or less fixed in position. The Hamiltonian operator for the electrons then will include slowly changing fields from the nuclei, while the Hamiltonian for the nuclei includes the nuclear kinetic energies and averaged fields from the surrounding clouds of rapidly-moving electrons. Using the electronic Hamiltonian in the Schrödinger equation leads to a set of electronic wavefunctions $\psi_i(\boldsymbol{r},\boldsymbol{R})$ that depend on both the electron coordinates ($\boldsymbol{r}$) and the coordinates of the nuclei ($\boldsymbol{R}$). Using the nuclear wavefunction in the Schrödinger equation provides a set of vibrational-rotational nuclear wavefunctions $\chi_{n(i)}(\boldsymbol{R})$ for each electronic state. The solutions to the Schrödinger equation for the full Hamiltonian, which includes the motions of both the nuclei and the electrons, can be written as a linear combination of products of these partial wavefunctions:

$$\Psi(\boldsymbol{r},\boldsymbol{R}) = \sum_i\sum_n \psi_i(\boldsymbol{r},\boldsymbol{R})\chi_{n(i)}(\boldsymbol{R}). \quad (4.35)$$

This description is most useful when the double sum on the right-hand side of Eq. (4.35) is dominated by a single term, because we then can express the complete wavefunction as a simple product of an electronic wavefunction and a nuclear wavefunction:

$$\Psi(\boldsymbol{r},\boldsymbol{R}) \approx \psi_i(\boldsymbol{r},\boldsymbol{R})\chi_{n(i)}(\boldsymbol{R}). \tag{4.36}$$

This is called the *Born-Oppenheimer approximation*.

The Born-Oppenheimer approximation proves to be reasonably satisfactory under a wide range of conditions. This is of fundamental importance in molecular spectroscopy because it allows us to assign transitions as primarily either electronic, vibrational or rotational in nature. In addition, it leads to tidy explanations of how electronic transitions depend on nuclear wavefunctions and temperature. A more complete discussion of the basis of the Born-Oppenheimer approximation and of the situations in which it breaks down can be found in [52, 53].

For a diatomic molecule, the potential energy term in the nuclear Hamiltonian is approximately a quadratic function of the distance between the two nuclei, with a minimum at the mean bond length. Such a Hamiltonian gives a set of vibrational wavefunctions with equally spaced energies (Eq. 2.29 and Fig. 2.3). The energy levels are $E_n = (n + 1/2)h\upsilon$, where $n = 0, 1, 2, 3 \ldots$, and υ is the classical bond vibration frequency.

A combination of vibrational and electronic wavefunctions χ_n and ψ_i is referred to as a *vibronic* state or level. Consider a transition from a particular vibronic level of the ground electronic state, $\Psi_{a,n} = \psi_a(r,R)\chi_n(R)$, to a vibronic level of an excited state, $\Psi_{b,m} = \psi_b(r,R)\chi_m(R)$. The transition could involve a change in vibrational wavefunction from χ_n to χ_m in addition to the change in electronic wavefunction from ψ_a to ψ_b. We can analyze the matrix element for this process by writing the dipole operator as a sum of separate operators for the electrons and the nuclei:

$$\tilde{\mu} = \tilde{\mu}_{el} + \tilde{\mu}_{nuc} = \sum_i e\boldsymbol{r}_i + \sum_j z\boldsymbol{R}_j, \tag{4.37}$$

where $\boldsymbol{r}_i$ is the position of electron i, and $\boldsymbol{R}_j$ and z_j are the position and charge of nucleus j. For one electron and one nucleus, the transition dipole then is:

$$\begin{aligned}\boldsymbol{\mu}_{ba,mn} &= \langle\psi_b(\boldsymbol{r},\boldsymbol{R})\chi_m(\boldsymbol{R})|\tilde{\mu}_{el}|\psi_a(\boldsymbol{r},\boldsymbol{R})\chi_n(\boldsymbol{R})\rangle + \langle\psi_b(\boldsymbol{r},\boldsymbol{R})\chi_m(\boldsymbol{R})|\tilde{\mu}_{nuc}|\psi_a(\boldsymbol{r},\boldsymbol{R})\chi_n(\boldsymbol{R})\rangle \\ &= e\int\chi_m^*(\boldsymbol{R})\chi_n(\boldsymbol{R})d\boldsymbol{R}\int\psi_b^*(\boldsymbol{r},\boldsymbol{R})\ \psi_a(\boldsymbol{r},\boldsymbol{R})\boldsymbol{r}d\boldsymbol{r} \\ &+ z\int\chi_m^*(\boldsymbol{R})\chi_n(\boldsymbol{R})\boldsymbol{R}d\boldsymbol{R}\int\psi_b^*(\boldsymbol{r},\boldsymbol{R})\ \psi_a(\boldsymbol{r},\boldsymbol{R})d\boldsymbol{r}.\end{aligned} \tag{4.38}$$

The integral $\int\psi_b^*(\boldsymbol{r},\boldsymbol{R})\psi_a(\boldsymbol{r},\boldsymbol{R})d\boldsymbol{r}$ on the right side of Eq. (4.38) is just $\langle\psi_b|\psi_a\rangle$ for a particular value of $\boldsymbol{R}$, which is zero if the electronic wavefunctions are orthogonal for all $\boldsymbol{R}$. Therefore,

$$\begin{aligned}\boldsymbol{\mu}_{ba,mn} &= \langle \psi_b(\boldsymbol{r},\boldsymbol{R})\chi_m(\boldsymbol{R})|\tilde{\mu}_{el}|\psi_a(\boldsymbol{r},\boldsymbol{R})\chi_n(\boldsymbol{R})\rangle \\ &= e\int \chi_m^*(\boldsymbol{R})\chi_n(\boldsymbol{R})d\boldsymbol{R}\int \psi_b^*(\boldsymbol{r},\boldsymbol{R})\psi_a(\boldsymbol{r},\boldsymbol{R})\boldsymbol{r}d\boldsymbol{r}.\end{aligned} \tag{4.39}$$

The double integral in Eq. (4.39) cannot be factored rigorously into a product of the form $e\langle\chi_m(\boldsymbol{R})|\chi_n(\boldsymbol{R})\rangle\,\langle\psi_b(\boldsymbol{r})|\boldsymbol{r}|\psi_a(\boldsymbol{r})\rangle$ because the electronic wavefunctions depend on $\boldsymbol{R}$ in addition to $\boldsymbol{r}$. We can, however, write

$$\begin{aligned}\langle \psi_b(\boldsymbol{r},\boldsymbol{R})\chi_m(\boldsymbol{R})|\tilde{\mu}_{el}|\psi_a(\boldsymbol{r},\boldsymbol{R})\chi_n(\boldsymbol{R})\rangle &= \langle\chi_m(\boldsymbol{R})|\chi_n(\boldsymbol{R})\rangle\langle\psi_b(\boldsymbol{r},\boldsymbol{R})|\tilde{\mu}_{el}|\psi_a(\boldsymbol{r},\boldsymbol{R})\rangle \\ &= \langle\chi_m(\boldsymbol{R})|\chi_n(\boldsymbol{R})\,\boldsymbol{U}_{ba}(\boldsymbol{R})\rangle,\end{aligned} \tag{4.40}$$

where $\boldsymbol{U}_{ba}(\boldsymbol{R})$ is an electronic transition dipole that is a function of $\boldsymbol{R}$. If $\boldsymbol{U}_{ba}(\boldsymbol{R})$ does not vary greatly over the range of $\boldsymbol{R}$ where both χ_m and χ_n have substantial amplitudes, then

$$\boldsymbol{\mu}_{ba,mn} \approx \langle\chi_m(\boldsymbol{R})|\chi_n(\boldsymbol{R})\rangle\overline{\boldsymbol{U}}_{ba}, \tag{4.41}$$

where $\overline{\boldsymbol{U}}_{ba}$ denotes an average of $\boldsymbol{U}_{ba}(\boldsymbol{R})$ over the nuclear coordinates of the initial and final vibrational states. To a good approximation, the overall transition dipole $\boldsymbol{\mu}_{ba,mn}$ thus depends on the product of a *nuclear overlap integral*, $\langle\chi_m|\chi_n\rangle$, and an electronic transition dipole ($\boldsymbol{\mu}_{ba}$) that is averaged over the nuclear coordinates. This is called the *Condon approximation* [54].

The contribution that a particular vibronic transition makes to the dipole strength depends on $|\boldsymbol{\mu}_{ba,mn}|^2$, and thus on the square of the nuclear overlap integral, $|\langle\chi_m|\chi_n\rangle|^2$. The square of such a nuclear overlap integral is called a *Franck-Condon factor*. Franck-Condon factors provide quantitative quantum mechanical expressions of the classical notion that nuclei, being much heavier than electrons, do not move significantly on the short time scale of electronic transitions [54]. For a particular vibronic transition to occur, the Franck-Condon factor must be non-zero.

The different vibrational wavefunctions for a given electronic state are orthogonal to each other. So if the same set of vibrational wavefunctions apply to the ground and excited electronic states, the nuclear overlap integral $\langle\chi_m|\chi_n\rangle$ is 1 for $m = n$ and zero for $m \neq n$. If the vibrational potential is harmonic, the energies of the allowed vibronic transitions ($\Psi_{a,n} \rightarrow \Psi_{b,n}$) will be the same for all n and the absorption spectrum will consist of a single line at the frequency set by the electronic energy difference, $E_a - E_b$, as shown in Fig. 4.20A.

In most cases, the vibrational wavefunctions differ somewhat in the ground and excited electronic states because the electron distributions in the molecule are different. This makes $\langle\chi_m|\chi_n\rangle$ non-zero for $m \neq n$, and less than 1 for $m = n$, allowing transitions between different vibrational levels to occur in concert with the electronic transition. The absorption spectrum therefore includes lines at multiple frequencies corresponding to various vibronic transitions (Fig. 4.20B). At low temperatures, most of the molecules will be at the lowest vibrational level of the

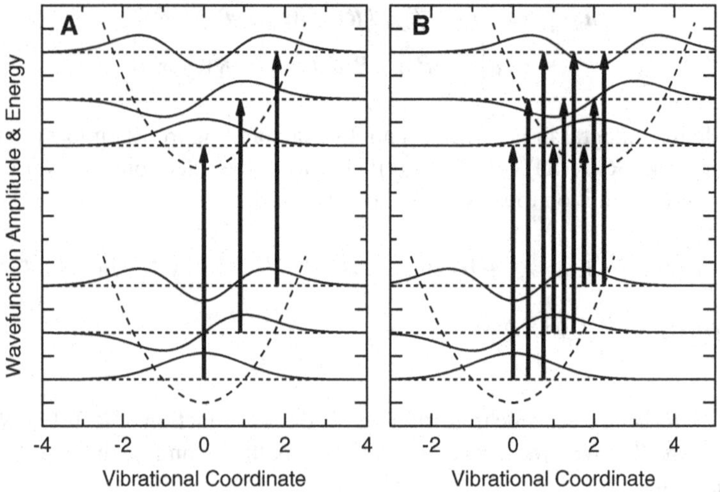

Fig. 4.20 If the vibrational potential energy surfaces are the same in the ground and excited electronic states (**A**), the Franck-Condon factors are nonzero only for vibronic transitions between corresponding vibrational levels, and these transitions all have the same energy; the absorption spectrum consists of a single line at this energy. If the minimum of the potential energy surface is displaced along the nuclear coordinate in the excited state, as in (**B**), Franck-Condon factors for multiple vibronic transitions are nonzero and the spectrum includes lines at multiple energies

ground state (the zero-point level) and the absorption line with the lowest energy will be the (0–0) transition. At elevated temperatures, higher vibrational levels will be populated, giving rise to absorption lines at energies below the 0–0 transition energy.

Approximate Franck-Condon factors for vibronic transitions can be obtained by using the wavefunctions of the harmonic oscillator (Eq. 2.30 and Fig. 2.3). Consider the vibronic transitions of ethylene. Because the HOMO is a bonding orbital and the LUMO is antibonding, the equilibrium length of the C=C bond is slightly longer when the molecule is in the excited state than it is in the gound state, but the change in the bond vibration frequency (υ) is relatively small. The vibrational potential energy wells for the two states have approximately the same shape but are displaced along the horizontal coordinate (bond length) as represented in Fig. 4.20B. It is convenient to express the change in bond length ($b_e - b_g$) in terms of the dimensionless quantity Δ defined by the expression

$$\Delta = 2\pi\sqrt{m_r \upsilon / h}\left(b_e - b_g\right), \tag{4.42}$$

where m_r is the reduced mass of the vibrating atoms. If we define the *coupling strength*

$$S = \frac{1}{2}\Delta^2, \tag{4.43}$$

then the Franck-Condon factor for a transition from the lowest vibrational level of the ground state (χ_0) to level m of the excited state (χ_m) can be written as:

$$|\langle\chi_m|\chi_0\rangle|^2 = \frac{S^m \exp(-S)}{m!}. \tag{4.44}$$

Corresponding expressions can be derived for the Franck-Condon factors for transitions that start in higher vibrational levels of the ground state (Box 4.13). The coupling strength S is sometimes called the *Huang-Rhys factor*.

Box 4.13 Recursion Formulas for Vibrational Overlap Integrals

Overlap integrals for harmonic-oscillator wavefunctions can be calculated using recursion formulas derived by Manneback [55]. Manneback treated the general case that the two vibrational states have different vibrational frequencies in addition to a displacement. Here we give only the results when the frequencies are the same. Let the dimensionless displacement of wavefunction χ_m with respect to χ_n be Δ, and define the *Huang-Rhys factor* or *coupling strength*, S, as $\Delta^2/2$. The overlap integral for the two zero-point wavefunctions then is:

$$\langle\chi_0|\chi_0\rangle = \exp(-S/2). \tag{B4.13.1}$$

Note that the vibrational wavefunctions in the bra and ket portions of this expression implicitly pertain to different electronic states. We have dropped the indices a and b for the electronic states to simplify the notation.

Overlap integrals for other combinations of vibrational levels can be built from $\langle\chi_0|\chi_0\rangle$ by the recursion formulas:

$$\langle\chi_{m+1}|\chi_n\rangle = (m+1)^{-1/2}\left\{n^{1/2}\langle\chi_m|\chi_{n-1}\rangle - S^{1/2}\langle\chi_m|\chi_n\rangle\right\} \tag{B4.13.2a}$$

and

$$\langle\chi_m|\chi_{n+1}\rangle = (n+1)^{-1/2}\left\{m^{1/2}\langle\chi_{m-1}|\chi_n\rangle + S^{1/2}\langle\chi_m|\chi_n\rangle\right\}, \tag{B4.13.2b}$$

with $\langle\chi_m|\chi_{-1}\rangle = \langle\chi_{-1}|\chi_m\rangle = 0$. For the overlap of the lowest vibrational level of the ground state with level m of the excited state, these formulas give

$$\langle\chi_m|\chi_0\rangle = \exp(-S/2)\,(-1)^m S^{m/2}/(m!)^{1/2}. \tag{B4.13.3}$$

The Franck-Condon factors are the squares of the overlap integrals:

$$|\langle\chi_m|\chi_0\rangle|^2 = \exp(-S)S^m/m!. \tag{B4.13.4}$$

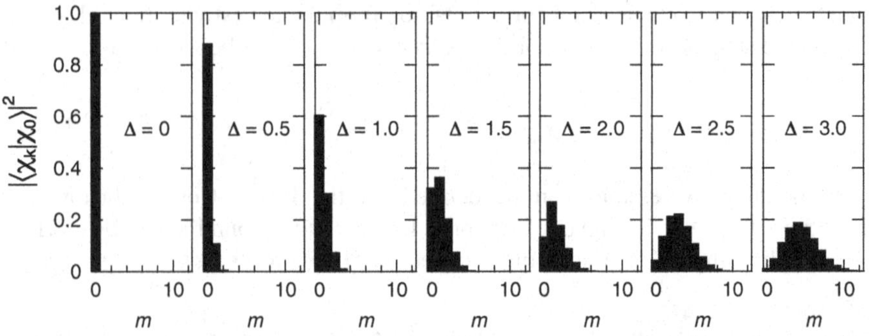

Fig. 4.21 Franck-Condon factors for vibronic transitions from the lowest vibrational level ($n = 0$) of a ground electronic state to various vibrational levels of an excited electronic state, as functions of the dimensionless displacement (Δ), for a system with a single, harmonic vibrational mode. The abscissa in each panel is the vibrational quantum number (m) of the excited state

Figure 4.21 shows how the Franck-Condon factors for transitions from the lowest vibrational level of the ground state (χ_0) to various levels of the excited state change as a function of Δ according to Eq. (4.44). If $\Delta = 0$, only $|\langle\chi_0|\chi_0\rangle|^2$ is nonzero. As $|\Delta|$ increases, $|\langle\chi_0|\chi_0\rangle|^2$ shrinks and the Franck-Condon factors for higher-energy vibronic transitions grow, with the sum of all the Franck-Condon factors remaining constant at 1.0. If $|\Delta| > 1$, the absorption spectrum peaks at an energy approximately $Sh\upsilon$ above the 0–0 energy.

As we'll discuss in Chap. 6, a nonlinear molecule with N atoms has $3N - 6$ vibrational modes, each involving movements of at least two, and sometimes many atoms. The overall vibrational wavefunction can be written as a product of wavefunctions for these individual modes, and the overall Franck-Condon factor for a given vibronic transition is the product of the Franck-Condon factors for all the modes. When the molecule is raised to an excited electronic state some of its vibrational modes will be affected but others may not be. The coupling factor (S) provides a measure of these effects. Vibrational modes for which S is large are strongly *coupled* to the excitation, and ladders of lines corresponding to vibronic transitions of each of these modes will feature most prominently in the absorption spectrum.

In the ground state, molecules are distributed among the different vibrational states depending on the energies of these states. At thermal equilibrium, the relative population of level n_k of vibrational mode k is given by the Boltzmann expression (Eq. B.2.6.3):

$$B_k = \frac{1}{Z_k}\exp(-n_k h\upsilon_k / k_B T), \tag{4.45}$$

where k_B is the Boltzmann constant, T the temperature, and Z_k the vibrational partition function for the mode:

$$Z_k = \sum_{n=0}^{\infty} \exp(-n_k h\upsilon_k / k_B T) = [1 - \exp(-h\upsilon_k / k_B T)]^{-1}. \tag{4.46}$$

The absorption strength at a given frequency depends on the sum of the Boltzmann-weighted Franck-Condon factors for all the vibronic transitions in which the change in the total energy (electronic plus vibrational) matches the photon energy $h\nu$ (Box 4.14).

Box 4.14 Thermally Weighted Franck-Condon Factors

Vibronic spatial wavefunctions for the ground and excited states of a molecule with N harmonic vibrational modes can be written as products of wavefunctions of the individual modes:

$$\Psi_a = \psi_a \prod_{k=1}^{N} \chi_{k,n_k}^a \quad \text{and} \quad \Psi_b = \psi_b \prod_{k=1}^{N} \chi_{k,m_k}^b, \tag{B4.14.1}$$

where ψ_a and ψ_b are electronic wavefunctions and χ_{k,n_k}^a denotes the n_k^{th} vibrational wavefunction of mode k in electronic state a. If the vibrational frequencies (υ_k) do not change significantly when the molecule is excited, the strength of absorption at frequency ν and temperature T can be related to a sum of weighted Franck-Condon factors:

$$W(\nu, T) = \frac{1}{Z} \sum_{\boldsymbol{n}} \exp\left(-E_{a,\boldsymbol{n}}^{vib} / k_B T\right) \left\{ \sum_{\boldsymbol{m}} \delta(h\nu - E_{\boldsymbol{m},\boldsymbol{n}}) \prod_{k=1}^{N} \left| \left\langle \chi_{k,m_k}^b \middle| \chi_{k,n_k}^a \right\rangle \right|^2 \right\}. \tag{B4.14.2}$$

Here the bold-face subscripts $\boldsymbol{m}$ and $\boldsymbol{n}$ denote vectorial representations of the vibrational levels of all the modes in the two electronic states: $\boldsymbol{m} = (m_1, m_2, \ldots, m_N)$ and $\boldsymbol{n} = (n_1, n_2, \ldots, n_N)$. The other terms are defined as follows:

$$E_{a,\boldsymbol{n}}^{vib} = \sum_{k=1}^{N} (n_k + 1/2)\ h\upsilon_k, \tag{B4.14.3a}$$

$$E_{b,\boldsymbol{m}}^{vib} = \sum_{k=1}^{N} (m_k + 1/2)\ h\upsilon_k, \tag{B4.14.3b}$$

$$E_{\boldsymbol{m},\boldsymbol{n}} = \left(E_b^{elec} + E_{b,\boldsymbol{m}}^{vib}\right) - \left(E_a^{elec} + E_{a,\boldsymbol{n}}^{vib}\right), \tag{B4.14.3c}$$

(continued)

Box 4.14 (continued)

$$Z = \sum_n \exp\left(-E^{vib}_{a,n}/k_BT\right) \qquad \text{(B4.14.4}a\text{)}$$

$$= \prod_k Z_k = \prod_k \left[1 - \exp(-h\upsilon_k/k_BT)\right]^{-1}, \qquad \text{(B4.14.4}b\text{)}$$

and E^{elec}_a and E^{elec}_b are the electronic energies of the two states. The Kroneker delta function $\delta(h\nu - E_{m,n})$ is 1 if the excitation energy $h\nu$ is equal to the total energy difference given by Eq. (B4.14.3c), and zero otherwise. In the Condon approximation, the absorption strength at frequency ν depends on $W(\nu, T)|\boldsymbol{\mu}_{ba}|^2$, where $\boldsymbol{\mu}_{ba}$ is the electronic transition dipole averaged over the nuclear coordinates as in Eq. (4.41).

The sum over $\boldsymbol{n}$ in Eq. (B4.14.2) runs over all possible vibrational levels of the ground state. A given level represents a particular distribution of energy among the N vibrational modes, and its vibrational energy $\left(E^{vib}_{a,n}\right)$ is the sum of the vibrational energies of all the modes (Eq. B4.14.3a). Each level is weighted by the Boltzmann factor $\exp\left(-E^{vib}_{a,n}/k_BT\right)/Z$, where Z is the complete vibrational partition function for the ground state. The vibrational partition function for a system with multiple vibrational modes is the product of the partition functions of all the individual modes (Z_k in Eq. B4.14.4b), as you can see by writing out the sum in Eq. (B4.14.4a) for a system with two or three modes. The sum over $\boldsymbol{m}$ in Eq. (B4.14.2) considers all possible vibrational levels of the excited state, but the delta function preserves only the levels for which $E_{m,n} = h\nu$. If a level meets this resonance condition, the Franck-Condon factor for the corresponding vibronic transition is the product of the Franck-Condon factors for all the individual vibrational modes.

The function $W(\nu,T)$ defined by Eq. (B4.14.2) gives a set of lines at frequencies (ν_{mn}) where $h\nu = E_{m,n}$. As we discuss below and in Chaps. 10 and 11, each of these absorption lines typically has a Lorentzian or Gaussian shape with a width that depends on the lifetime of the excited vibronic state. To incorporate this effect, the delta function in Eq. (B4.14.2) can be replaced by a line-shape function $w_{m,n}(\nu - \nu_{mn})$ for each combination of $\boldsymbol{m}$ and $\boldsymbol{n}$ vectors that meets the resonance condition.

At low temperatures, molecules in the ground state are confined largely to the zero-point levels of their vibrational modes. The spectrum described by Eq. (B4.14.2) then simplifies to

(continued)

Box 4.14 (continued)

$$W(\nu, T=0) = \sum_{k=1}^{N}\sum_{m}\delta(h\nu - E_{m,0})\prod_{k=1}^{N}\left|\left\langle \chi_{k,m_k}^{b} \middle| \chi_{k,0}^{a}\right\rangle\right|^2. \tag{B4.14.5}$$

By writing out the Franck-Condon factors and introducing the line-shape functions $w_{m,0}(\nu - \nu_{m0})$, we can recast this expression in the form

$$W(\nu, T=0) = \exp(-S_t)\sum_{k=1}^{N}\sum_{m=0}^{\infty}((S_k)^m/m!)w_{m,0}, \tag{B4.14.6}$$

where S_k is the Huang-Rhys factor (coupling strength) for mode k and S_t is the sum of the Huang-Rhys factors for all the modes that are coupled to the electronic transition.

The foregoing expressions consider only the spectrum of a homogeneous system in which all the molecules have identical values of E_a^{elec} and E_b^{elec}. In an inhomogeneous system, the molecules will have a distribution of electronic energy differences, and the spectrum will be a convolution of this *site-distribution function* with Eq. (B4.14.2) or (B4.14.6).

For further discussion of the Franck-Condon factors, line-shape functions and site-distribution functions for bacteriochlorophyll-*a* and related molecules see [56, 57].

Because the eigenfunctions of the harmonic oscillator form a complete set, the Franck-Condon factors for excitation from any given vibrational level of the ground electronic state to all the vibrational levels of an excited electronic state must sum to 1. $|\langle\chi_0|\chi_0\rangle|^2$ therefore gives the ratio of the strength of the 0–0 transition to the total dipole strength. This ratio is called the *Debye-Waller factor*. From Eq. (4.44), the Debye-Waller factor is $\exp(-S/2)$.

The analysis of Franck-Condon factors described in Boxes 4.13 and 4.14 assumes that the chromophore's vibrational modes are essentially the same in the excited and ground electronic states, differing only in the location of the energy minimum on the vibrational coordinate and a possible shift in the vibrational frequency. Breakdowns of this assumption are referred to as *Duschinsky effects*. They can be treated in some cases by representing the vibrational modes for one state as a linear combination of those for the other [58–60].

The width of an absorption band for an individual vibronic transition depends on how long the excited molecule remains in the state created by the excitation. According to Eq. (2.70), the spectrum for excitation to a state that decays exponentially with time should be a Lorentzian function of frequency. The shorter the lifetime of the excited state, the broader the Lorentzian (Fig. 2.12). A variety of processes can cause an excited molecule to evolve with time, and thus can broaden the absorption line. The molecule might, for example, decay to another vibrational state by redistributing energy among its internal vibrational modes or by releasing energy to the surroundings. Molecules in higher vibrational levels have a larger number of possible relaxation pathways that are thermodynamically favorable, and thus relax more rapidly than molecules in lower levels. In addition, the energies of the individual molecules in a sample will fluctuate as a result of randomly changing interactions with the surroundings. These fluctuations cause the time-dependent parts of the wavefunctions of the molecules to get out of phase, a process termed *pure dephasing*. In Chap. 10 we will see that the composite time constant (T_2) that determines the width of a Lorentzian vibronic absorption band depends on both the equilibration time constant for true decay processes (T_1) and the time constant for pure dephasing (T_2*). The width of the Lorentzian at half-maximal amplitude is $\hbar/T_2$.

If T_2 is long, as it can be for molecules in the gas phase and for some molecules chilled to low temperatures in inert matrices, the absorption line can be very sharp. The width of such of an absorption line for an individual molecule in a fixed environment, or for an ensemble of identical molecules with the same solvational energies, is termed the *homogeneous* line width. An absorption band representing molecules that interact with their surroundings in a variety of ways is said to be *inhomogeneously broadened* and its width is termed the *inhomogeneous* line width. The spectrum generated by a family of Lorentzians with a Gaussian distribution of center energies is called a *Voigt spectrum*.

4.11 Spectroscopic Hole Burning

The homogeneous absorption lines that underlie an inhomogeneous spectrum can be probed experimentally by *hole-burning* spectroscopy at low temperatures [61–63]. In *photochemical* hole-burning, the excited molecule evolves into a triplet state or another long-lived product, leaving a hole in the absorption spectrum at the frequency of the original excitation. In *nonphotochemical* hole-burning, the excited molecule decays back to the original ground state with conversion of the excitation energy to heat. The thermal energy that is released causes rearrangements of the molecule's immediate surroundings, shifting the absorption spectrum and again leaving a hole at the excitation frequency.

At room temperature, non-photochemical spectral holes usually are filled in by fluctuations of the surroundings on the picosecond time scale. This process, termed *spectral diffusion*, can be studied by picosecond pump-probe techniques. At temperatures below 4 K, non-photochemical spectral holes can persist almost indefinitely and can be measured with a conventional spectrophotometer. The shape of the hole depends on the lifetime of the excited state and the coupling of the electronic excitation to vibrational modes of the solvent, both of which depend in turn on the excitation wavelength. Excitation on the far-red edge of the absorption band populates mainly the lowest vibrational level of the excited state, which has a relatively long lifetime, and the resulting *zero-phonon hole* is correspondingly sharp (Fig. 4.22A). The zero-phonon hole typically is accompanied by one or more *phonon side bands* that reflect vibrational excitation of the solvent in concert with electronic excitation of the chromophore. The side bands are broader than the zero-

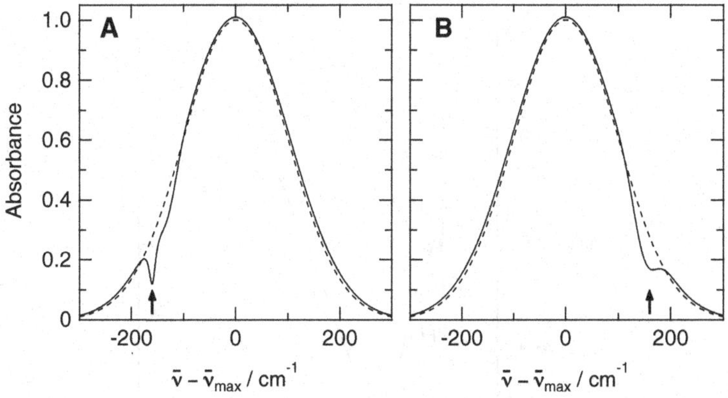

Fig. 4.22 Non-photochemical hole-burning. An inhomogenous absorption spectrum (*dotted curves*) usually is an envelope of spectra for chromophores in many different local environments. If an inhomogeneous sample is irradiated with light covering a narrow band of frequencies (*vertical arrows*), the spectra of molecules that absorb here can be shifted to higher or lower frequencies, leaving a hole in the inhomogeneous spectrum (*solid curves*). If the sample is excited on the red (low-energy) edge of the absorption band, the hole often consists of a sharp zero-phonon hole with one or more broad phonon side-bands at somewhat higer energies (**A**). The width of the zero-phonon hole provides information on the lifetime of the excited electronic state, while the phonon side bands report on solvent vibrations that are coupled to the electronic excitation. Excitation on the blue side of the absorption band usually gives a broader, unstructured hole (**B**)

phonon hole because the excited solvent molecules relax rapidly by transfering the excess vibrational energy to the surroundings. In addition, the phonon side bands sometimes represent a variety of vibrational modes, or a quasi-continuum of closely lying vibrational states. Excitation on the blue side of the absorption band populates higher-energy vibrational levels of both the chromophore and the solvent, which usually decay rapidly and give a broad, unstructured hole (Fig. 4.22B).

Studies of antenna complexes from photosynthetic bacteria by Small and coworkers [64–66] provide a good illustration of nonphotochemical hole-burning. These complexes have extensive manifolds of excited electronic states that lie close together in energy. Excitation on the long-wavelength edge of the main absorption band populates mainly the lowest vibrational level of the lowest excited electronic state, which decays with a time constant on the order of 10 ps. Holes burnt in this region of the spectrum have correspondingly narrow widths of about 3 cm^{-1}. Excitation at shorter wavelengths populates higher excited electronic states, which evidently decay to the lowest state in 0.01 to 0.1 ps. Holes burnt near the center of the absorption band therefore have widths on the order of 200 cm^{-1}.

In similar studies of photosynthetic reaction centers, the width of the zero-phonon hole for the reactive bacteriochlorophyll dimer was related to the time constant for electron transfer to a neighboring molecule [56, 67, 68]. Figure 4.23 shows a typical hole spectrum (the difference between absorption spectra measured with the excitation laser on and off) for a sample of reaction centers that was excited at 10,912 cm^{-1} at 5 K. The holes in this experiment resulted from the

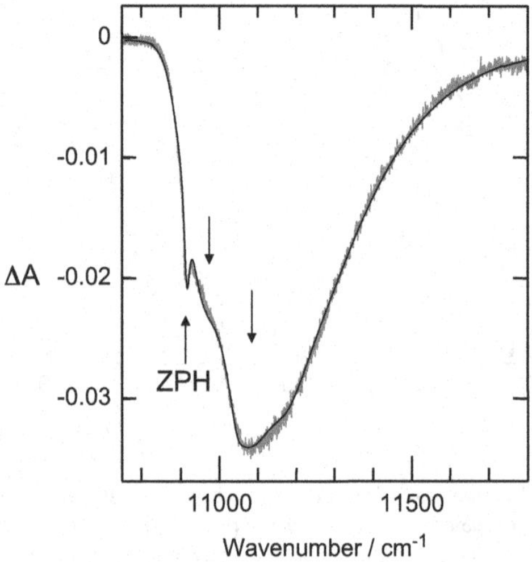

Fig. 4.23 The spectrum of a photochemical hole burned in the long-wavelength absorption band of a sample of photosynthetic bacterial reaction centers at 5 K [68]. The *gray curve* is the difference between absorption spectra measured with the excitation laser on and off. The excitation frequency was 10,912 cm^{-1}. Note the sharp zero-phonon hole (ZPH, *upward arrow*) at 10,980 cm^{-1}. The *downward arrows* indicate the centers of two discrete vibrational (phonon) bands that are linked to the zero-phonon transition. The *solid curve* is a theoretical hole spectrum calculated as described in the text

photochemical electron-transfer reaction followed by conversion of the bacteriochlorophyll dimer to an excited triplet state. They are broadened by strong vibronic coupling to motions of the protein surrounding the bacteriochlorophylls. The holes generated by burning in the red edge of the spectrum reveal a zero-phonon line with a width of about 6 cm^{-1}, which corresponds to an electron-transfer time constant of about 1 ps. They also have several discrete phonon side bands that can be assigned to two characteristic vibrational modes. In the spectrum shown in Fig. 4.23 the zero-phonon hole is at 10,980 cm^{-1}, and the prominent vibrational modes have frequencies of approximately 30 and 130 cm^{-1}. Both the hole spectrum and the original absorption spectrum can be fit well by using Eq. (B4.14.6) with these two vibrational modes, a Gaussian distribution of zero-phonon transition energies, and relatively simple Lorentzian and Gaussian functions for the shapes of the zero-, one- and two-phonon absorption lines [56, 57, 68].

Each of the homogeneous lines in a vibronic absorption spectrum actually consists of a family of transitions between various rotational states of the molecule. The rotational fine structure in the spectrum can be seen for small molecules in the gas phase, but for large molecules the rotational lines are too close together to be resolved.

4.12 Effects of the Surroundings on Molecular Transition Energies

Interactions with the surroundings can shift the energy of an absorption band to either higher or lower energies, depending on the nature of the chromophore and the solvent. Such shifts are called *solvatochromic effects*. Consider, for example, an n–π* transition, in which an electron is excited from a nonbonding orbital of an oxygen atom to an antibonding molecular orbital distributed between O and C atoms (Sect. 9.1). In the ground state, electrons in the nonbonding orbital can be stabilized by hydrogen-bonding or dielectric effects of the solvent. In the excited state, these favorable interactions are disrupted. Although solvent molecules will tend to reorient in response to the new distribution of electrons in the chromophore, this reorientation is too slow to occur during the excitation itself. An n–π* transition therefore shifts to higher energy in more polar or H-bonding solvents relative to less polar solvents. A shift of an absorption band in this direction is called a "blue shift." The energies of π–π* transitions are less sensitive to the polarity of the solvent but still depend on the solvent's high-frequency polarizability, which as we noted in Chap. 3, increases quadratically with the refractive index. Increasing the refractive index usually decreases the transition energy of a π–π* transition, causing a "red shift" of the absorption band. The terms "blue" and "red"often are used in this context without regard to the position of the absorption band relative to the spectrum of visible light. For example, a shift of an IR band to lower energies is generally called a red shift even though the band moves away from the region of the visible spectrum that we perceive as red.

The visual pigments provide dramatic illustrations of how minor changes in protein structure can shift the absorption spectrum of a bound chromophore. As in many other vertebrates, the human retina contains three types of cone cells whose pigments (cone-opsins) absorb in different regions of the visible spectrum. Cone-opsin from human "blue" cones absorbs maximally near 414 nm, while those from "green" and "red" cones have absorption maxima near 530 and 560 nm, respectively (Fig. 4.24). The visual pigments from other organisms have absorption maxima ranging from 355 to 575 nm. Yet all these pigments resemble rhodopsin, the pigment from retinal rod cells, in containing a protonated Schiff base of 11-*cis*-retinal. (The chromophores in some other organisms are based on retinal A_2, which has one more conjugated double bond than retinal A_1 and can push the absorption maximum as far to the red as 620 nm.) The proteins from vertebrate rods and cones have homologous amino acid sequences and, although a crystal structure currently is available only for bovine rhodopsin [69], probably have very similar three-dimensional structures [70, 71]. If the proteins are denatured by acid, the absorption bands all move to the region of 440 nm and resemble the spectrum of the protonated Schiff base of 11-*cis*-retinal in methanol. Resonance Raman measurements have shown that the vibrational structure of the chromophore does not vary greatly among the different proteins, indicating that the spectral shifts probably result mainly from electrostatic interactions with the surrounding protein rather than from changes in the conformation of the chromophore [72, 73]. Studies of proteins

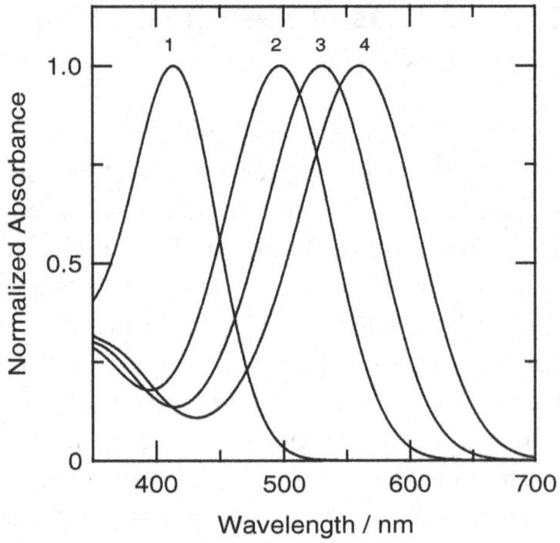

Fig. 4.24 Normalized absorption spectra of rhodopsin from human retinal rod cells (*curve 2*), and cone-opsins from human "blue", "green" and "red" cone cells (*curves 1, 3 and 4*, respectively). The spectra are drawn as described by Stavenga et al. [151]

containing conformationally constrained analogs of 11-*cis*-retinal also have lent support to this view [74, 75].

Correlations of the absorption spectra with the amino acid sequences of the proteins from a variety of organisms, together with studies of the effects of site-directed mutations, indicate that the shifts of the spectra of the visual pigments reflect changes in a small number of polar or polarizable amino acid residues near the chromophore [73, 76–79] (Fig. 4.25). The 30-nm shift of the human red cone pigment relative to the green pigment can be attributed entirely to changes of seven residues, including the replacement of several Ala residues by Ser and Thr. The more polarizable side chains of Ser and Thr probably facilitate delocalization of the positive charge from the N atom of the retinyl Schiff base toward the β-ionone ring in the excited state. The red and the green cone pigments also have a bound Cl^- ion that contributes to the red-shift of the spectrum relative to rhodopsin and the blue cone pigment. Variations in the position of a Glu carboxylate group that serves as a counterion for the protonated Schiff base also may contribute to the spectral shifts, although the changes in the distance from the counterion to the proton probably are relatively small [72–74, 80].

DNA photolyases, which use the energy of blue light to split pyrimidine dimers formed by UV irradiation of DNA, provide other examples of large and variable shifts in the absorption spectrum of a bound chromophore. These enzymes contain a bound pterin (methylenetetrahydrofolate, MTHF) or deazaflavin, which serves to absorb light and transfer energy to a flavin radical in the active site [81]. The absorption maximum of MTHF occurs at 360 nm in solution, but ranges from 377 to 415 nm in the enzymes from different organisms [82].

Solvatochromic effects on the transition energies of molecules in solution often can be related phenomenologically to the solvent's dielectric constant and refractive index. The analysis is similar to that used for local-field correction factors (Sect. 3.1.5). Polar solvent molecules around the chromophore will be ordered in response to the chromophore's ground-state dipole moment ($\boldsymbol{\mu}_{aa}$), and the oriented

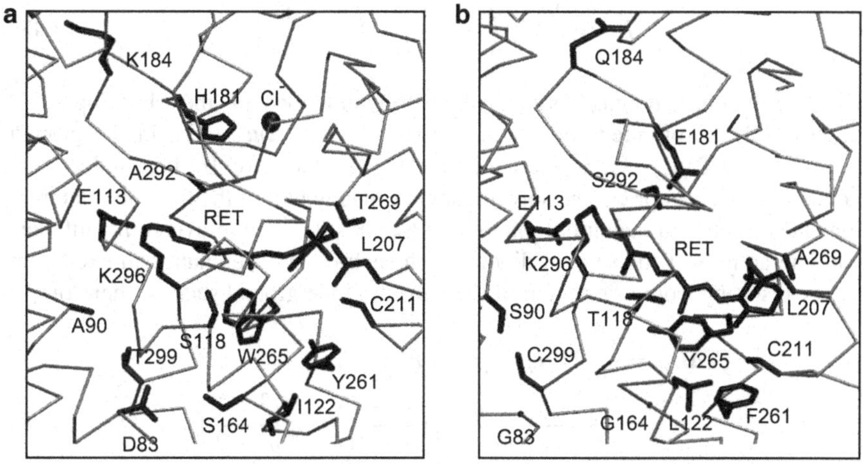

Fig. 4.25 Models of the region surrounding the retinyl chromophore in the visual pigments from human "blue" (**a**) and "red" (**b**) cone cells. The main absorption band is shifted to longer wavelengths by almost 150 nm in the red pigment. The chromophore (RET), the Lys residue that forms the protonated Schiff base (K296 in the rhodopsin numbering scheme), the Glu residue that acts as a counter ion (E113), another nearby Glu (E181), and other residues that contribute to regulation of the color of the human cone pigments (residues 83, 90, 118, 122, 164, 184, 265, 269, 292 and 299) are shown as licorice models in *black*. The protein backbone is shown in *gray*. The models were constructed by homology with bovine rhodopsin [70], followed by addition of water (and, for the red pigment, Cl^- ion) and minimization of the energy by short molecular dynamics trajectories

solvent molecules provide a reaction field that acts back on the chromophore. A simple model of the system is a dipole at the center of a sphere of radius R embedded in a homogeneous medium with dielectric constant ε_s. The reaction field felt by such a dipole is given approximately by [23]:

$$\boldsymbol{E} = \frac{2\boldsymbol{\mu}_{aa}}{R^3}\left(\frac{\varepsilon_s - 1}{\varepsilon_s + 2}\right). \tag{4.47}$$

Now suppose that excitation of the chromophore changes its dipole moment to $\boldsymbol{\mu}_{bb}$. Although the solvent molecules cannot reorient instantaneously in response, the dielectric constant ε_s includes electronic polarization of the solvent in addition to orientational polarization, and changes in electronic polarization can occur essentially instantaneously in response to changing electric fields. The high-frequency component of the dielectric constant is the square of the refractive index (n) (Sects. 3.1.4 and 3.1.5). If we subtract the part of the reaction field that is attributable to electronic polarization, the part due to orientation of the solvent ($\boldsymbol{E}_{or}$) can be written

$$\boldsymbol{E}_{or} = \frac{2\boldsymbol{\mu}_{aa}}{R^3}\left[\left(\frac{\varepsilon_s - 1}{\varepsilon_s + 2}\right) - \left(\frac{n^2 - 1}{n^2 + 2}\right)\right]. \tag{4.48}$$

The solvation energies associated with interactions of the chromophore's dipoles with the oriented solvent molecules are $-(1/2)\boldsymbol{\mu}_{aa} \cdot \boldsymbol{E}_{or}$ in the ground state and $-(1/2)\boldsymbol{\mu}_{bb} \cdot \boldsymbol{E}_{or}$ in the excited state. The factor of 1/2 here reflects the fact that, in order to orient the solvent, approximately half of the favorable interaction energy between the chromophore and $\boldsymbol{E}_{or}$ must be used to overcome unfavorable interactions of the solvent dipoles with each other. The change in excitation energy resulting from orientation of the solvent in the ground state is, therefore,

$$\begin{aligned}\Delta E_{or} &= (1/2)(\boldsymbol{\mu}_{aa} - \boldsymbol{\mu}_{bb}) \cdot \boldsymbol{E}_{or} \\ &= \frac{(\boldsymbol{\mu}_{aa} - \boldsymbol{\mu}_{bb}) \cdot \boldsymbol{\mu}_{aa}}{R^3}\left[\left(\frac{\varepsilon_s - 1}{\varepsilon_s + 2}\right) - \left(\frac{n^2 - 1}{n^2 + 2}\right)\right].\end{aligned} \tag{4.49}$$

This expression shows that interactions with a polar solvent can either increase or decrease the transition energy, depending on the sign of $(\boldsymbol{\mu}_{aa} - \boldsymbol{\mu}_{bb}) \cdot \boldsymbol{\mu}_{aa}$. The change in energy ($\Delta E_{or}$) is expected to be small if the excitation involves little change in dipole moment ($\boldsymbol{\mu}_{aa} \approx \boldsymbol{\mu}_{bb}$), if the chromophore is nonpolar in the ground state ($\boldsymbol{\mu}_{aa} \approx 0$), or if the solvent is nonpolar ($\varepsilon_s \approx n^2$).

Equations (4.48) and (4.49) do not include the effects of electronic polarizability. The solvent can be polarized electronically by both the permanent and transition dipoles of the chromophore. Quantum mechanically, this *inductive polarization* can be viewed as mixing of the excited and ground electronic states of the solvent under the perturbation caused by electric fields from the solute (Box 12.1). The chromophore experiences a similar inductive polarization by fields from the solvent. In nonpolar solvents, $\pi - \pi^*$ absorption bands of nonpolar molecules typically decrease in energy with increasing refractive index, and the decrease is approximately linear in the function $(n^2 - 1)/(n^2 + 2)$. Some authors use the function $(n^2 - 1)/(2n^2 + 1)$, which gives very similar results (see [23] for a review of early work and [83] for a more recent study). Figure 4.26 illustrates this shift for the long-wavelength absorption band of bacteriochlorophyll-*a*. Extrapolation to $n = 1$ gives a "vacuum" transition energy on the order of 1,000 cm^{-1} above the energy measured for bacteriochlorophyll in solution.

The energy of interactions of a molecule with its surroundings can be treated in a more microscopic way by using Eqs. (4.19–4.21) to describe the ground and excited-state wavefunctions, ψ_a and ψ_b. The solvation energy of the molecule in the ground state is

$$E_a^{solv} \approx -2e\sum_{i=1}^{N} \left(C_i^a\right)^2 V_i + E_{core}^{solv}, \tag{4.50}$$

where e is the electronic charge, V_i is the electric potential at the position of atom i, C_i^a is the atomic expansion coefficient for atom i in ψ_a, N is the total number of

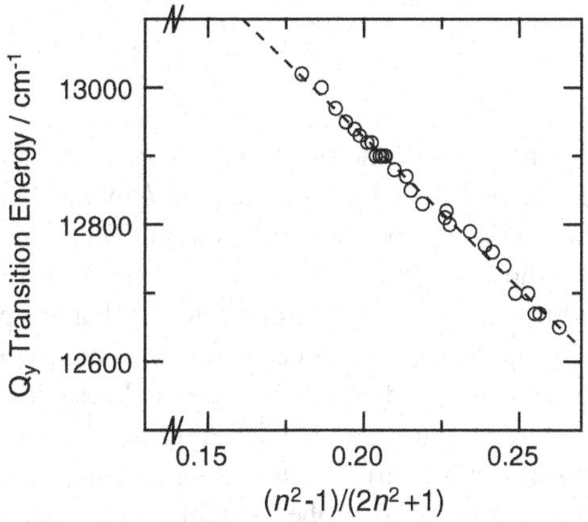

Fig. 4.26 Dependence of the transition energy of the long-wavelength (Q_y) absorption band of bacteriochlorophyll-*a* on the refractive index (n) in nonpolar solvents. Experimental data from Limantara et al. [83] are replotted as a function of $(n^2-1)/(2n^2+1)$. Extrapolating to $n = 1$ (0 on the abscissa) gives 13,810 cm^{-1} for the transition energy in a vacuum. A similar plot of the data vs $(n^2-1)/(n^2+2)$ gives 13,600 cm^{-1}

atoms that participate in the wavefunction, and E_{core}^{solv} represents the effects of the surroundings on the nuclei and electrons in the orbitals other than ψ_a. The contribution of ψ_a to the electronic charge on atom i is proportional to $(C_i^a)^2$, and the factor of 2 before the sum reflects the assumption that there are two electrons in ψ_a in the ground state. Similarly, the solvation energy in the excited state is:

$$E_b^{solv} \approx -e\sum_{i=1}^{N}\left[\left(C_i^a\right)^2 + \left(C_i^b\right)^2\right]V_i + E_{core}^{solv}, \tag{4.51}$$

where C_i^b is the coefficient for atom i in ψ_b. The change in solvation energy upon excitation is the difference between the two solvation energies:

$$E_b^{solv} - E_a^{solv} \approx -e\sum_{i=1}^{N}\left[\left(C_i^b\right)^2 - \left(C_i^a\right)^2\right]V_i. \tag{4.52}$$

Note that these expressions consider only a single configuration, excitation from ψ_a to ψ_b. If several configurations contribute to the absorption band, the relative contribution of a given configuration to the changes in electronic charge is proportional to the square of the coefficient for this configuration in the overall excitation (Eq. 4.26).

If the structure of the surroundings is well defined, as it may be for a chromophore in a protein (see, for example, the visual pigments shown in Fig. 4.25), the electric potential at each point in the chromophore can be estimated by summing the contributions from the charges and dipoles of the surrounding atoms:

$$V_i \approx \sum_{k \neq i \cdots} \left(\frac{Q_k}{|\boldsymbol{r}_{ik}|} + \frac{\boldsymbol{\mu}_k \cdot \boldsymbol{r}_{ik}}{|\boldsymbol{r}_{ik}|^3} \right). \tag{4.53}$$

Here Q_k is the charge on atom k of the surroundings, $\boldsymbol{r}_{ik}$ is the vector from atom k to atom i, and $\boldsymbol{\mu}_k$ is the electric dipole induced on atom k by the electric fields from all the charges and other induced dipoles in the system. The sums in this expression exclude any atoms whose interactions with atom i must be treated quantum mechanically. In addition to atom i itself, other atoms that are part of the chromophore or are connected to atom i by three or fewer bonds typically require such special treatment. The induced dipoles that are included in Eq. (4.53) can be calculated from the atomic charges and polarizabilities by an iterative procedure [84]. Shifts in the absorption spectrum thus can be used to measure the binding of prosthetic groups to proteins or to probe protein conformational changes in the region of a bound chromophore. Equations (4.52) and (4.53) involve significant approximations, however, because interactions with the surroundings ideally should be taken into account when the molecular orbitals and eigenvalues are obtained in the first place.

Figure 4.27 illustrates the calculated redistributions of charge that accompany excitation of the indole side chain of tryptophan. As discussed above, the $^1\mathrm{L_a}$ absorption band consists mainly of the configuration $\psi_2 \rightarrow \psi_3$ and a smaller contribution from $\psi_1 \rightarrow \psi_4$, where ψ_1 and ψ_2 are the second-highest and highest occupied molecular orbitals and ψ_3 and ψ_4 are the second-lowest and lowest unoccupied orbitals. Both these configurations result in transfer of electron density from the pyrrole ring to the benzyl portion of the indole sidechain. The energy of the $^1\mathrm{L_a}$ transition thus should be red-shifted by positively charged species near the benzene ring, and blue-shifted by negatively charged species in this region. The $^1\mathrm{L_b}$ absorption band consists largely of the configurations $\psi_2 \rightarrow \psi_4$ and $\psi_1 \rightarrow \psi_3$, the first of which results in a large shift of electron density to the benzyl ring. However, the $\psi_1 \rightarrow \psi_3$ transition moves electron density in the opposite direction, making the net change of dipole moment associated with the $^1\mathrm{L_b}$ band smaller than that associated with $^1\mathrm{L_a}$ (Fig. 4.27E, F). For a more quantitative analysis of these effects, contributions from higher-energy configurations also need to be considered [36, 37, 85]. The gradient operator may also be preferable to the dipole operator here, but this has not been studied extensively.

The situation takes on an additional dimension if the positions of the charged or polar groups near the chromophore fluctuate rapidly with time. One way to describe the effects of such fluctuations is to write the energies of the ground and excited electronic states (E_a and E_b) as harmonic functions of a generalized solvent coordinate (X):

$$E_a = E_a^o + (K/2)X^2, \tag{4.54}$$

and

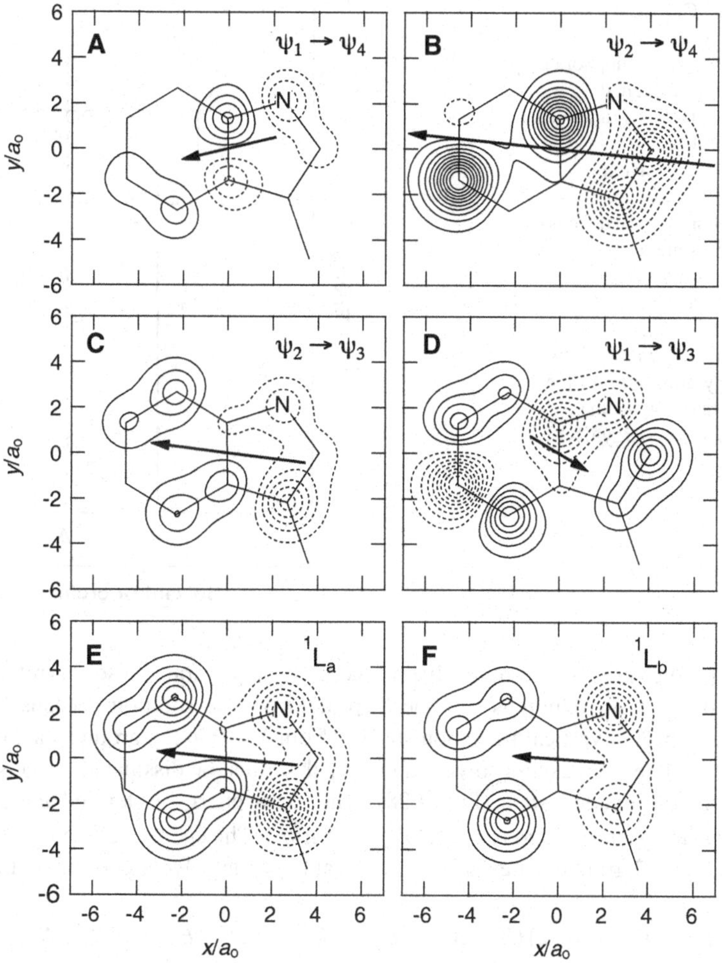

Fig. 4.27 Redistribution of charge upon excitation of 3-methylindole. (**A–D**) Contour plots show the changes in electron density (increases in negative charge) when an electron is excited from one of the two highest occupied molecular orbitals (ψ_1 or ψ_2) to one of the first two unoccupied orbitals (ψ_3 or ψ_4). The contour intervals are 0.01 $ea_o{}^3$. The planes of the drawings and the line types for positive and negative amplitudes are as in Fig. 4.11. (**E, F**) Similar plots for the combinations $0.841(\psi_2 \rightarrow \psi_3) + 0.116(\psi_1 \rightarrow \psi_4)$ and $0.536(\psi_1 \rightarrow \psi_3) + 0.402(\psi_2 \rightarrow \psi_4)$, which are approximately the contributions of these four orbitals in the ^1L_a and ^1L_b excitations, respectively. (Note that the coefficient for a given configuration here is the square of the corresponding coefficient for the transition dipole.) The contour intervals are 0.005 ea_o^3. The *arrows* indicate the changes in the permanent dipole ($\boldsymbol{\mu}_{bb} - \boldsymbol{\mu}_{aa}$) in units of eÅ/5a_o

$$E_b = E_b^o + (K/2)(X - \Delta)^2, \tag{4.55}$$

where E_a^o and E_b^o are the minimum energies of the two electronic states, Δ is the displacement of the energy minima along the solvent coordinate, and K is a force

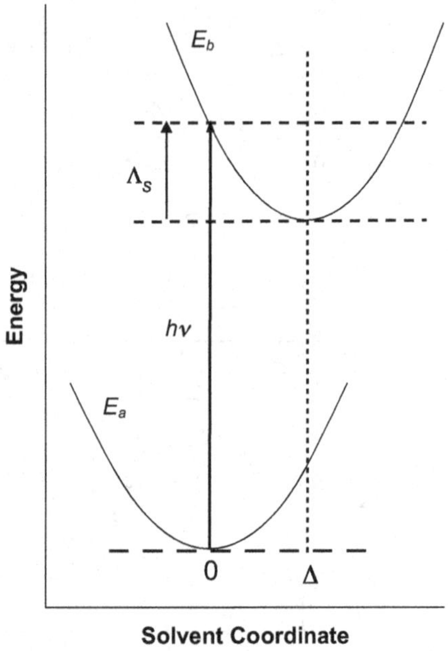

Fig. 4.28 Classical harmonic energy curves for a system with a chromophore in its ground and excited electronic states, as functions of a generalized, dimensionless solvent coordinate. If the chromophore is raised from the ground state to the excited electronic state with no change in nuclear coordinates, it has excess solvation energy relative to the energy minimum for the excited state. This is the reorganization energy (Λ_S)

constant. We use the term "solvent" here in a general sense to refer to a chromophore's surroundings in either a protein or free solution. Because of the displacement Δ, a vertical transition starting from $X = 0$ in the ground state creates an excited state with excess solvation energy that must be dissipated as the system relaxes in the excited state (Fig. 4.28). The extra energy is termed the *solvent reorganization energy*, and is given by $\Lambda_s = K\Delta^2/2$. The energy difference between the excited and ground states at any given value of the solvent coordinate thus is

$$E_b - E_a = E_b^o + (K/2)(X - \Delta)^2 - E_a^o - (K - 2)X^2 = E_o - KX\Delta + \Lambda_s, \quad (4.56)$$

where $E_o = E_b^o - E_a^o$.

If we equate potential energies approximately with free energies, we also can say that the relative probability of finding a particular value of X when the chromophore is in the ground state is $P(X) = \exp(-KX^2/2k_BT)$. Combining this expression with Eq. (4.56) and using the relationships $\Lambda_s = K\Delta^2/2$ and $h\nu = E_b - E_a$ gives an expression for the relative strength of absorption at energy $h\nu$:

$$\begin{aligned} P(h\nu) &= \exp\left\{-K(E_o + \Lambda - h\nu)^2/2(K\Delta)^2 k_BT\right\} \\ &= \exp\left\{-(E_o + \Lambda_s - h\nu)^2/4\Lambda_s k_BT\right\}. \end{aligned} \quad (4.57)$$

This Gaussian function of $h\nu$ peaks at an energy Λ_s above E_o and has a width at half-maximum amplitude (FWHM) of $2(\Lambda_s k_BT{\cdot}\ln 2)^{1/2}/\pi$, or $(2Kk_BT{\cdot}\ln 2)^{1/2}\Delta/\pi$.

Fluctuating interactions with the solvent thus broaden the vibronic absorption lines of the chromophore and shift them to higher energies relative to E_o. As discussed above, however, the mean energy of interaction can shift E_o either upward or downward depending on the chromophore and the solvent. We will discuss generalized solvent coordinates further in Chaps. 5 and 10.

Fluctuating electrostatic interactions can be treated microscopically by incorporating Eqs. (4.50–4.52) into molecular dynamics simulations (Box 6.1). The results can be used to construct potential energy surfaces similar to those of Eqs. (4.55a) and (4.55b), or can used in quantum calculations of the eigenvalues of the chromophore in the electric field from the solvent. Mercer et al. [86] were able to reproduce the width of the long-wavelength absorption band of bacteriochlorophyll in methanol well by this approach.

4.13 The Electronic Stark Effect

If an external electric field is applied across an absorbing sample, the absorption bands can be shifted to either higher or lower energies depending on the orientation of the chromophores with respect to the field. This is the *Stark* or *electrochromic effect*. The effect was discovered in 1913 by the physicist Johannes Stark, who found that electric fields on the order of 10^5 V/cm cause a splitting of the spectral lines of hydrogen into symmetrically placed components with different polarizations. The basic theoretical tools for extracting information on the dipole moment and polarizability of a molecule by Stark spectroscopy were worked out by Liptay [87–89]. They have been extended and applied to a variety of systems by the groups of Boxer [90–93], Nagae [94] and others. In one application, Premvardhan et al. found that excitation of photoactive yellow protein or its chromophore (a thioester of *p*-coumaric acid) in solution causes remarkably large changes in the dipole moment and the polarizability of the chromophore [95–97]. The change in dipole moment ($|\Delta\boldsymbol{\mu}| = 26$ debye) corresponds to moving an electric charge by 5.4 Å, and seems likely to contribute importantly to the structural changes that follow the excitation.

In the simplest situation for a molecular chromophore, the magnitude and direction of the shift depends on the dot product of the local electric field vector ($\mathbf{E}_{ext} = f\mathbf{E}_{app}$, where $\mathbf{E}_{app}$ is the applied field and f is the local-field correction factor) with the vector difference between the chromophore's permanent dipole moments in the excited and ground states ($\Delta\boldsymbol{\mu}$):

$$\Delta E = -\mathbf{E}_{ext} \cdot \Delta\boldsymbol{\mu} = -f\mathbf{E}_{app} \cdot \Delta\boldsymbol{\mu}. \tag{4.58}$$

The difference between the dipole moments in the two states can be related to the chromophore's molecular orbitals by the expression

$$\Delta\boldsymbol{\mu} = \boldsymbol{\mu}_{bb} - \boldsymbol{\mu}_{aa} \approx e\sum_{i}^{N} \boldsymbol{r}_i\left[\left(C_i^b\right)^2 - \left(C_i^a\right)^2\right], \tag{4.59}$$

where $\boldsymbol{r}_i$ is the position of atom i and C_i^a and C_i^b are the coefficients for this atom in the ground- and excited-state wavefunctions (Eqs. 4.22d–4.22e and 4.50–4.52). If the sample is isotropic, some molecules will be oriented so that the external field shifts their transition energies to higher values, while others will experience shifts to lower energies. The result will be a broadening of the overall absorption spectrum. If the system is anisotropic, on the other hand, the external field can shift the spectrum systematically to higher or lower frequencies.

Equation (4.59) assumes that the electric field does not cause the molecules to reorient, but simply shifts the energy difference between the ground and excited states without changing $\boldsymbol{\mu}_{aa}$ or $\boldsymbol{\mu}_{bb}$. The validity of this assumption depends on the molecule and the experimental apparatus. Although small polar molecules in solution can be oriented by external electric fields, this is less likely to occur for proteins, particularly if the direction of the field is modulated rapidly. It can be prevented by imobilizing the protein in a polyvinylalcohol film. But to the extent that the chromophore is polarizable, the field will create an additional *induced electric dipole* that depends on the strength of the field, and this dipole can change when the molecule is excited if the polarizability in the excited state differs from that in the ground state. In general, the molecular polarizability should be treated as a matrix, or more formally a second-rank tensor, because it depends on the orientation of the molecule relative to field and the induced dipole can have components that are not parallel to the field (Boxes 4.15 and 12.1); however, we'll assume here that the polarizability can be described adequately by a scalar quantity with dimensions of cm^{-3}. The induced dipole ($\boldsymbol{\mu}_{aa}^{ind}$ or $\boldsymbol{\mu}_{bb}^{ind}$ for the ground or excited state, respectively) then will be simply the product of the polarizability (α_{aa} or α_{bb}) and the total field, including both the external field ($\boldsymbol{E}_{ext}$) introduced by the applied field and the "internal" field from the molecule's surroundings ($\boldsymbol{E}_{int}$). Interactions of $\boldsymbol{E}_{ext}$ with the induced dipoles will change the transition energy by

$$\Delta E_{ind} = -\boldsymbol{E}_{ext}\cdot\left(\boldsymbol{\mu}_{bb}^{ind} - \boldsymbol{\mu}_{aa}^{ind}\right) = -\boldsymbol{E}_{ext}\cdot(\alpha_{bb} - \alpha_{aa})(\boldsymbol{E}_{ext} + \boldsymbol{E}_{\text{int}}) \tag{4.60a}$$

$$= -\Delta\alpha\left(\left|\boldsymbol{E}_{ext}\right|^2 + \boldsymbol{E}_{ext}\cdot\boldsymbol{E}_{\text{int}}\right), \tag{4.60b}$$

where $\Delta\alpha = \alpha_{bb} - \alpha_{aa}$.

The internal field $\boldsymbol{E}_{int}$ can have any orientation relative to the external field, and it usually has a considerably larger magnitude (typically on the order of 10^6 V/cm or more). But if the chromophore is bound to a highly structured system such as a protein, $\boldsymbol{E}_{int}$ will have approximately the same magnitude for all the molecules in a sample and its orientation will be relatively well fixed with respect to the individual molecular axes. We then can consider the factor $\Delta\alpha\boldsymbol{E}_{int}$ to be part of the dipole change $\Delta\boldsymbol{\mu}$ in Eq. (4.59), rather than including it separately in Eqs. (4.60a) and (4.60b). The additional contribution to the transition energy from dipoles induced by the external field $\boldsymbol{E}_{ext}$ then is just

$$\Delta E_{ind} = -\Delta\alpha|\boldsymbol{E}_{ext}|^2. \tag{4.61}$$

According to Eq. (4.61), interactions of the external field with induced dipoles will shift the transition energies for all the molecules of a sample in the same direction, depending on whether $\Delta\alpha$ is positive or negative. The change in the transition energy increases quadratically with the strength of the external field. The contribution to ΔE from induced dipoles is often termed a "quadratic" Stark effect to distinguish it from the "linear" contribution from $\Delta\boldsymbol{\mu}$, which depends linearly on $|\boldsymbol{E}_{ext}|$ as described by Eq. (4.58). As we will see shortly, however, the changes in the absorption spectrum attributable to $\Delta\alpha$ and $\Delta\boldsymbol{\mu}$ both depend quadratically on $|\boldsymbol{E}_{ext}|$.

Let's assume that the internal field $\boldsymbol{E}_{int}$ has a fixed magnitude and orientation with respect to the molecular axes, so that Eq. (4.61) is valid. If $\Delta\alpha$ and $\Delta\boldsymbol{\mu}$ both are non-zero, the total change in the transition frequency for an individual molecule then will be

$$\Delta\nu = -\left(\boldsymbol{E}_{ext}\cdot\Delta\boldsymbol{\mu} + |\boldsymbol{E}_{ext}|^2\Delta\alpha\right)/h. \tag{4.62}$$

The effect on the overall absorption spectrum of an isotropic sample can be evaluated by expanding the absorption spectrum as a Taylor's series in powers of $\Delta\nu$:

$$\begin{aligned}\varepsilon(\nu,\boldsymbol{E}) &= \varepsilon(\nu,0) + \frac{\partial\varepsilon(\nu,0)}{\partial\nu}\Delta\nu + \frac{1}{2}\frac{\partial^2\varepsilon(\nu,0)}{\partial\nu^2}|\Delta\nu|^2 + \cdots \\ &= \varepsilon(\nu,0) - \frac{\partial\varepsilon(\nu,0)}{\partial\nu}\left(\boldsymbol{E}_{ext}\cdot\Delta\boldsymbol{\mu} + |\boldsymbol{E}_{ext}|^2\Delta\alpha\right)h^{-1} \\ &+ \frac{1}{2}\frac{\partial^2\varepsilon(\nu,0)}{\partial\nu^2}\left(\boldsymbol{E}_{ext}\cdot\Delta\boldsymbol{\mu} + |\boldsymbol{E}_{ext}|^2\Delta\alpha\right)^2 h^{-2} + \cdots,\end{aligned} \tag{4.63}$$

where ε(ν,0) represents the absorption spectrum in the absence of the external field. We next need to average this expression over all orientations of the molecules with respect to the field. If the solution is isotropic, terms that depend on the first, third, or any odd power of $|\boldsymbol{E}_{ext}|$ average to zero while terms that depend on even powers of $|\boldsymbol{E}_{ext}|$ remain, and the average value of $(\boldsymbol{E}_{ext}\cdot\Delta\boldsymbol{\mu})^2$ becomes $(1/3)|\boldsymbol{E}_{ext}|^2|\Delta\boldsymbol{\mu}|^2$ (Box 4.6). The total effect of the external field on the molar extinction coefficient at frequency ν then will be:

$$\varepsilon(\nu,\boldsymbol{E}) - \varepsilon(\nu,0) = -\left(\frac{\partial\varepsilon(\nu,0)}{\partial\nu}\right)\frac{\Delta\alpha|\boldsymbol{E}_{ext}|^2}{h} + \frac{1}{2}\left(\frac{\partial^2\varepsilon(\nu,0)}{\partial\nu^2}\right)\frac{|\boldsymbol{E}_{ext}|^2|\Delta\boldsymbol{\mu}|^2}{3h^2} + \cdots. \tag{4.64}$$

This expression shows that the major contributions from $\Delta\alpha$ to the changes in the absorption spectrum depend on the first derivative of the spectrum with respect to ν, whereas the major contributions from $\Delta\boldsymbol{\mu}$ depend on the second derivative. As stated above, both depend on the square of $|\boldsymbol{E}_{ext}|$. Stark spectra often include contributions from both induced and permanent dipoles, which can be separated experimentally by fitting the measured difference spectrum to a sum of first- and second-derivative terms (Fig. 4.29). If the measuring light is polarized, the

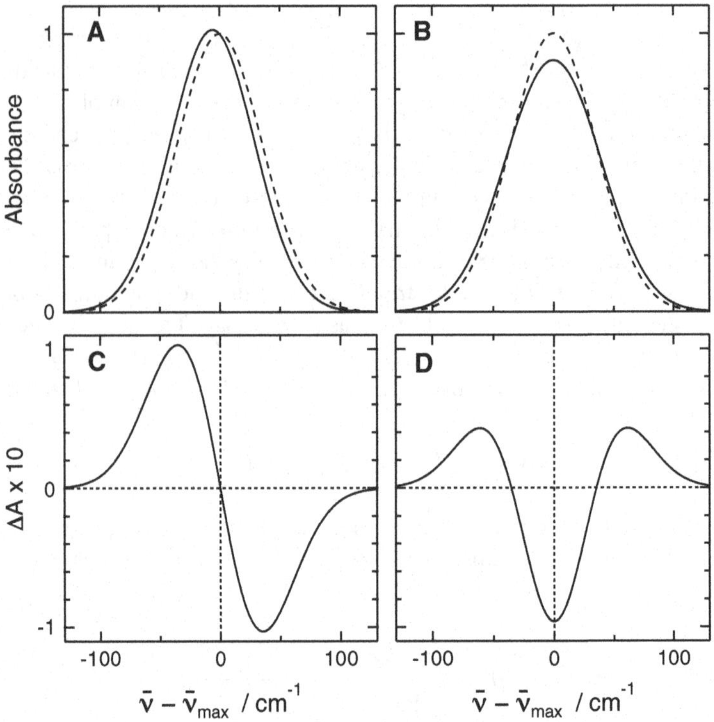

Fig. 4.29 Idealized absorption spectra of an isotropic system in the absence (*dashed curve*) and presence (*solid curve*) of an external electric field (**A**, **B**), and the changes in the spectra caused by the field (**C**, **D**). In (**A**) and (**C**), the chromophore is assumed to have the same dipole moment but a higher polarizability in the excited state than the ground state; the field shifts the spectrum to lower energies. In (**B**) and (**D**), the chromophore has the same polarizability but a higher dipole moment in the excited state than the ground state; the field broadens the spectrum and decreases the peak absorbance. Stark spectra often show a combination of these two effects

spectrum also depends on the angle between the polarization axis and the applied field, and this dependence can be used to determine the orientation of $\boldsymbol{\Delta\mu}$ relative to the transition dipole (Box 4.15).

Box 4.15 Electronic Stark Spectroscopy of Immobilized Molecules

In Liptay's treatment [87–89] as implemented by Boxer and coworkers [92], applying an external field $\boldsymbol{E}$ to a non-oriented but immobilized system changes the absorbance (A) at wavenumber $\bar{\nu}$ by

(continued)

Box 4.15 (continued)

$$\Delta A(\overline{\nu}) = |\boldsymbol{E}|^2 \left\{ A_\chi A(\overline{\nu}) + \frac{B_\chi}{15hc}\overline{\nu}\frac{\partial[A(\overline{\nu})/\overline{\nu}]}{\partial\overline{\nu}} + \frac{C_\chi}{30h^2c^2}\overline{\nu}\frac{\partial^2[A(\overline{\nu})/\overline{\nu}]}{\partial\overline{\nu}^2} \right\}. \tag{B4.15.1}$$

with

$$\begin{aligned} A_\chi = &\frac{1}{30|\boldsymbol{\mu}_{ba}|^2}\sum_{i,j}\left\{10(a_{ij})^2 + (3\cos^2\chi - 1)\left[3a_{ii}a_{jj} + (a_{ij})^2\right]\right\} \\ &+\frac{1}{15|\boldsymbol{\mu}_{ba}|^2}\sum_{i,j}\left\{10\mu_{ba(i)}b_{ijj} + (3\cos^2\chi - 1)\left[4\mu_{ba(i)}b_{ijj}\right]\right\}, \end{aligned} \tag{B4.15.2}$$

$$\begin{aligned} B_\chi = &\frac{5}{2}\mathrm{Tr}(\Delta\alpha) + (3\cos^2\chi - 1)\left[\frac{3}{2}\Delta\alpha_\mu - \frac{1}{2}\mathrm{Tr}(\Delta\alpha)\right] \\ &+\frac{1}{|\boldsymbol{\mu}_{ba}|^2}\sum_{i,j}\left\{10\mu_{ba(i)}a_{ij}\Delta\mu_j + (3\cos^2\chi - 1)\left[3\mu_{ba(i)}a_{jj}\Delta\mu_i + \mu_{ba(i)}a_{ij}\Delta\mu_j\right]\right\}, \end{aligned} \tag{B4.15.3}$$

and

$$C_\chi = |\boldsymbol{\mu}_{ba}|^2\left[5 + (3\cos^2\chi - 1)(3\cos^2\zeta - 1)\right]. \tag{B4.15.4}$$

These expressions include averaging over all orientations of the chromophore with respect to the field and the polarization of the measuring light. The factors a_{ij} and b_{ijj} in Eqs. (B4.15.2) and (B4.15.3) are elements of the *transition polarizability tensor* (**a**) and the *transition hyperpolarizability* (**b**), which describe the effects of the external field on the dipole strength of the absorption band. These effects probably are relatively minor in most cases, and are neglected in Eq. (4.64). The transition polarizability tensor **a** is a 3 × 3 matrix, whose nine elements are defined as $a_{ij} = \left(\partial\mu_{ba(i)}/\partial E_j\right)$, where $\mu_{ba(i)}$ is the i component of the transition dipole ($\boldsymbol{\mu}_{ba}$) and E_j is the j component of the field. The transition hyperpolarizability **b** is a a 3 × 3 × 3 cubic array, or third-rank tensor. Including both polarizability and hyperpolarizability, the change in the transition dipole ($\boldsymbol{\mu}_{ba}$) caused by the field is $\mathbf{a} \cdot \boldsymbol{E} + \boldsymbol{E} \cdot \mathbf{b} \cdot \boldsymbol{E}$. The change in the x component of $\boldsymbol{\mu}_{ba}$, for example, is $a_{xx}E_x + a_{xy}E_y + a_{xz}E_z + b_{xxx}|E_x|^2 + b_{yxy}|E_y|^2 + b_{zxz}|E_z|^2$. The hyperpolarizability terms are included in Eq. (B4.15.2) because, though small relative to $\boldsymbol{\mu}_{ba}$, they can dominate over the polarizability terms for strongly allowed

(continued)

Box 4.15 (continued)

transitions. This is because, as we discussed in Sect. 4.5, an electronic transition can be strongly allowed only if the initial and final states have different symmetries. The transition polarizability, by contrast, tends to be small for states with different symmetries because it depends on mixing of these states with other states ([92] and Box 12.1). If the initial and final states have different symmetries they generally cannot both mix well with a third state.

In Eqs. (B4.15.3) and (B4.15.4), $\Delta\boldsymbol{\mu}$ is the difference between the permanent dipole moments of the excited and ground states ($\boldsymbol{\mu}_{bb} - \boldsymbol{\mu}_{aa}$), and $\Delta\mu_x$, $\Delta\mu_y$ and $\Delta\mu_z$ are its components. $\Delta\boldsymbol{\alpha}$ is the difference between the polarizabilities of the excited and ground state, with the polarizabilities again described as second-rank tensors. To first order, the field changes $\Delta\boldsymbol{\mu}$ by $\Delta\boldsymbol{\alpha}{\cdot}\boldsymbol{E}_{ext}$. Tr($\Delta\boldsymbol{\alpha}$) is the sum of the three diagonal elements of $\Delta\boldsymbol{\alpha}$, and $\Delta\boldsymbol{\alpha}_{\mu}$ is the component of $\Delta\boldsymbol{\alpha}$ along the direction of $\boldsymbol{\mu}_{ba}\left(\boldsymbol{\mu}_{ba} \cdot \Delta\boldsymbol{\alpha} \cdot \boldsymbol{\mu}_{ba}/\left|\boldsymbol{\mu}_{ba}\right|^2\right)$. χ is the angle between $\boldsymbol{E}_{ext}$ and the polarization of the measuring light and ζ is the angle between $\boldsymbol{\mu}_{ba}$ and $\Delta\boldsymbol{\mu}$.

Stark effects usually are measured by modulating the external field at a frequency on the order of 1 kHz and using lock-in detection electronics to extract oscillations of a transmitted light beam at twice the modulation frequency. The three terms in the brackets on the right-hand side of Eq. (B4.15.1) can be separated experimentally by their dependence on, respectively, the 0th, 1st, and 2nd derivatives of $A/\bar{\nu}$ with respect to $\bar{\nu}$. $|\Delta\boldsymbol{\mu}_{ba}|^2$ and $|\zeta|$ then are obtained by measuring the dependence of C_{χ} on the experimental angle χ. The experiment thus gives the magnitude of $\Delta\boldsymbol{\mu}$ uniquely, but restricts the orientation of $\Delta\boldsymbol{\mu}$ only to a cone with half angle $\pm\zeta$ relative to $\Delta\boldsymbol{\mu}_{ba}$.

The factor B_{χ} depends on both $\Delta\boldsymbol{\alpha}$ and cross terms involving $\Delta\boldsymbol{\mu}$ and the transition polarizability (Eq. 4.15.3). Although the transition polarizability (**a**) is expected to be small for strongly allowed transitions, its products with $\Delta\boldsymbol{\mu}$ are not necessarily negligible relative to $\Delta\boldsymbol{\alpha}$. Conventional Stark measurements therefore do not yield unambiguous values for $\Delta\boldsymbol{\alpha}$. In some cases, it may be possible to obtain additional information by measuring the oscillations of the transmitted light beam at higher harmonics of the frequency at which the field is modulated and relating the signals to higher derivatives of the absorption spectrum [91, 92, 98, 99]. This technique is called *higher-order Stark spectroscopy*.

Equation (4.64) assumes that the absorption band responds homogeneously to the external field. This assumption can break down if the band represents several different transitions, particularly if these have nonparallel transition dipoles, but information about the individual components sometimes can be obtained by combining Stark spectroscopy with hole-burning [100]. An illustration is a study by

Gafert et al. [101] on mesoporphyrin-IX in horseradish peroxidase. The authors were able to evaluate the contribution of the internal field to $\Delta\mu$ and to relate varying local fields to different conformational states of the protein.

Pierce and Boxer [102] have described Stark effects on *N*-acetyl-L-tryptophanamide and the single tryptophan residue in the protein melittin. They separated effects on the ^1L_a and ^1L_b bands by taking advantage of the different fluorescence anisotropy of the two bands (Sect. 5.6). In agreement with Fig. 4.27, the ^1L_a band exhibited a relatively large $\Delta\mu$ of approximately $6/f$ Debye, where f is the unknown local-field correction factor. The $\Delta\mu$ for the ^1L_b band was much smaller.

Other interesting complexities can arise if the excited chromophore enters into a rapid photochemical reaction that is affected by the external field. This is the situation in photosynthetic bacterial reaction centers, where the excited bacteriochlorophyll dimer (P*) transfers an electron to a neighboring molecule (B) with a time constant on the order of 2 ps. The electron-transfer process generates an ion-pair state (P^+B^-) whose dipole moment is much larger than that of either P* or the ground state. An external electric field thus can shift the ion-pair state to substantially higher or lower energy depending on the orientation of the field relative to P and B, and this shift can alter the rate of electron transfer. The resulting changes in the absorption spectrum are not described well by a simple sum of first- and second-derivative terms (Eq. 4.64), but can be analyzed fruitfully by including higher-order terms [103–106].

Vibrational Stark spectroscopy is discussed in Sect. 6.4.

4.14 Metal-Ligand and Ligand-Metal Charge-Transfer Transitions and Rydberg Transitions

Complexes of transition metals with aromatic ligands often have strong absorption bands that reflect transfer of an electron from the metal to a ligand. These are termed *metal-ligand charge-transfer* (MLCT) bands. The complex of Ru(II) with three molecules of 2,2′-bipyridine provides an important example (Fig. 4.30). In water, tris(2,2′-bipyridine)ruthenium(II) chloride has a broad MLCT band at 452 nm with a peak extinction coefficient of $1.4 \times 10^4\ M^{-1}\ cm^{-1}$, in addition to a sharper absorption band of the bipyridine ligands at 285 nm [107]. The energy of the MLCT band in this and other similar complexes depends largely on the difference between the oxidation potential of the metal and the reduction potential of the ligand. The radical-pair created by the charge-transfer transition is born in a singlet state, but relaxes rapidly to a longer-lived triplet state. Resonance Raman studies and measurements of linear dichroism and emission anisotropy indicate that the electron that is transferred remains on an individual bipryridyl group on the time

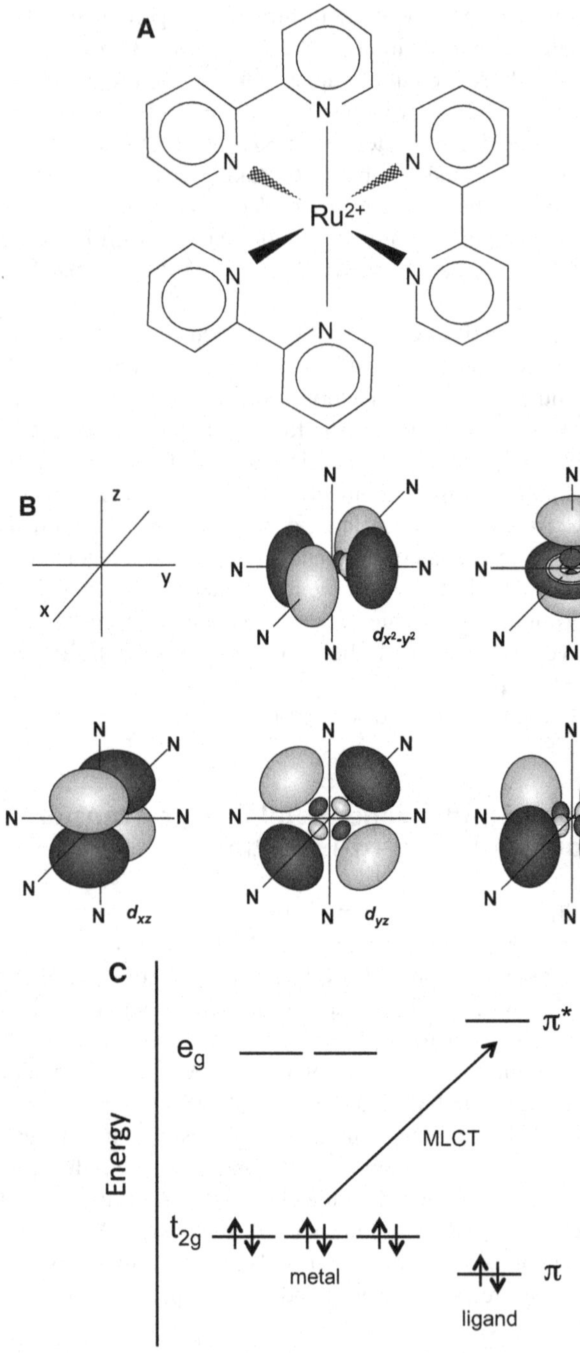

Fig. 4.30 (**A**) Structure of tris(2,2′-bipyridine)ruthenium(II). (**B**) A depiction of the boundary surfaces of the 4*d* orbitals of the Ru and the N atoms of its ligands. (**C**) The relative energies of the 4*d* orbitals (*metal*) and the HOMO (π) and LUMO (π*) of 2,2′-bipyridine (*ligand*). The Cartesian coordinate system used for naming the *d* orbitals is shown at the *upper left* in (**B**). In such

Fig. 4.30 (continued) octahedral complexes of transition metals, two of the d orbitals (d_{z^2} and $d_{x^2-y^2}$, collectively called e_g) have lobes that extend along either the z axis or both the x and y axes, and thus bring electrons close to electronegative atoms of the ligands (*upper row* in **B**). The other three orbitals (d_{xy}, d_{xz} and d_{yz}, collectively called t_{2g}) have nodes on the axes, keeping electrons farther from the ligands (*lower row*). Ru(II) in the trisbipyridyl complex has six t_{2g} electrons with paired spins (*small arrows* in **C**), leaving the e_g orbitals empty. Because the d orbitals are centrosymmetric and have gerade inversion symmetry, excitations from t_{2g} to e_g are forbidden by Laporte's rule although they can give rise to a weak absorption band in the near-IR (Chap. 9). Excitations from the HOMO to the LUMO of a ligand usually are allowed, and give an absorption band in the near UV. The metal and ligand orbitals overlap sufficiently so that excitations from t_{2g} to π^* also are allowed, and give an MLCT absorption band in the visible region of the spectrum

scale of molecular vibrations, hopping from one ligand to another with a time constant on the order of 50 ps [108–115]. As expected for excitations that involve substantial movements of electric charge, MLCT energies are strongly sensitive to solvatochromic and Stark effects [116–119].

The charge-transfer species created by MLCT excitations of Ru(II) complexes can pass an electron from the reduced ligand to a variety of secondary acceptors [107, 115, 120–122]. The oxidized Ru atom also can extract an electron from a secondary donor. These photochemical reactions are receiving intense study because of their potential applications in solar energy capture [115, 123–126].

Absorption bands reflecting charge transfer in the opposite direction (*ligand-metal charge-transfer*, or LMCT bands) are seen in some complexes containing metals in high oxidation states or ligands with electrons in relatively high-energy orbitals. The purple color of permanganate (MnO_4^-) ions, for example, results from an LMCT transition in which an electron moves from a filled p orbital of an oxygen atom to a d orbital of the central Mn(VII). Other types of charge-transfer transitions are discussed in Chap. 8.

Rydberg transitions are excitations to wavefunctions resembling the atomic orbitals of hydrogen. In the 1880s, Johannes Rydberg noted that the spectral lines of hydrogen and alkalai metals could be described empirically by the expression $1/\lambda = R(1/n_1^2 - 1/n_2^2)$ where R is a universal constant (now called the Rydberg constant), n_1 and n_2 are integers, and $n_1 < n_2$. The series with $n_1 = 2$ and $n_2 \geq 3$, for example, predicts the Balmer lines in the absorption spectrum of atomic hydrogen. A satisfactory rationalization had to await recognition that the energy of a spectral transition at wavelength λ is proportional to $1/\lambda$ and that reciprocal of the energy of an atomic orbital depends reciprocally on the square of the principle quantum number (n in Eq. (2.34)).

In molecules, a *Rydberg orbital* refers to an orbital that resembles an atomic orbital of an atomic fragment produced by dissociation, and a Rydberg transition represents excitation to such an orbital from a molecular orbital. Aldehydes and ketones have absorption bands in the vacuum UV that reflect Rydberg transitions from a nonbonding orbital of the carbonyl oxygen atom to a 3s atomic orbital [127–130].

4.15 Thermodynamics of Photoexcitation

Absorption of a photon with energy hv increases the internal energy (E_a) of the absorber by the same amount. Because the change in molecular volume usually is relatively small, the absorber's enthalpy ($H_a = E_a + PV$) also increases by approximately hv. However, the absorber's entropy (S_a) generally increases as well, making the increase in free energy ($\Delta G_a = \Delta H_a - T\Delta S_a$, the amount of useful work that the absorber could perform for each photon absorbed) less than hv [131]. Let's consider the entropy of an absorber immediately after an excitation, before vibrational relaxations have dissipated any of the photon's energy to the surroundings. Suppose the system includes N identical absorbing molecules, which are distributed among a set of m vibronic states with n_i molecules in state i. Assume that the system is illuminated continuously and uniformly, so that all the molecules have the same a priori probability of being in an excited state at any given time. As long as we exclude interactions with the surroundings, the absorbers constitute an isolated system with a fixed number of particles, or in the language of statistical mechanics, a microcanonical ensemble. The entropy of such an ensemble is given by Boltzmann's famous expression:

$$S_a = k_B \ln(\Omega), \tag{4.65}$$

where k_B is the Boltzmann constant (1.3806×10^{-23} J K^{-1}, or 8.6173×10^{-5} eV K^{-1}) and Ω, the multiplicity of the system, is the number of ways that the individual molecules could be assigned to the various states [132–136].

The multiplicity Ω in Eq. (4.65) is the product of two factors. Because the molecules could be distinguished by their positions, one factor is the number of ways of distributing N distinguishable particles among m groups so that n_i particles are in group m_i. This is $N!/\prod_{i=1}^{m} n_i$ (Fig. 4.31A). The other factor is $\prod_{i=1}^{m} \omega_i$, where ω_i is the degeneracy of state i (the number of substates with energy $E_i \pm \delta E$, where δE is an arbitrary, small range of energies). Combining the two factors, we have

$$S_a = k_B \ln\left[\left(\prod_{i=1}^{m} \omega_i\right)\left(N!/\prod_{i=1}^{m} n_i\right)\right] = k_B \sum_{i=1}^{m} \ln\omega_i + k_B \ln\left(N!/\prod_{i=1}^{m} n_i\right). \tag{4.66}$$

Now let n_g be the population of molecules in a ground electronic state with energy E_g, let n_e be the population of molecules in an excited state with energy $E_e = E_g + hv$, and call the degeneracies of these states ω_g and ω_e. If an excited molecule decays to the ground state, emitting a photon, the population of the excited state will decrease to $n_e - 1$ while that of the ground state increases to $n_g + 1$. This changes the entropy of the absorbers to

$$S_a^{'} = k_B \sum_{i=1}^{m} \ln\omega_i + k_B \ln(\omega_g/\omega_e) + k_B \ln\left(N!/\left(\frac{n_g+1}{n_e}\right)\prod_{i=1}^{m} n_i\right). \tag{4.67}$$

Excitation of one molecule therefore must change the absorbers' entropy by

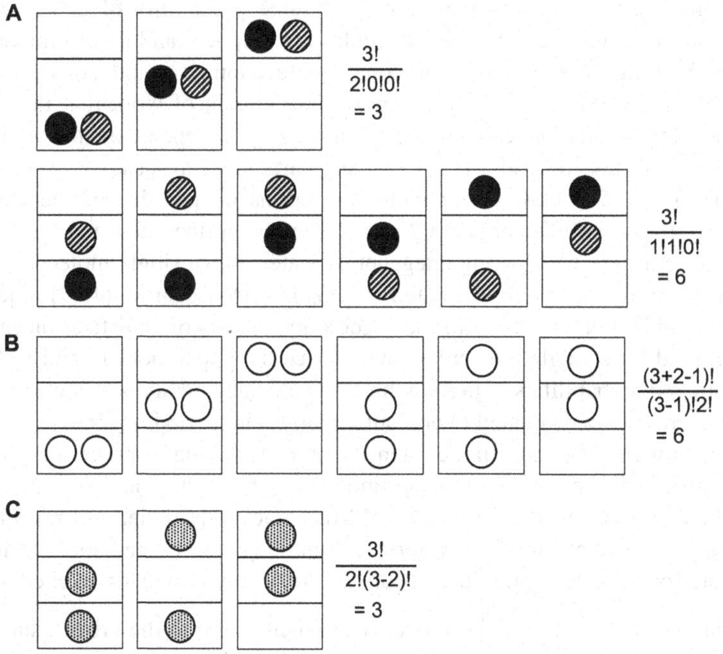

Fig. 4.31 The number of ways of distributing a set of objects among various states depends on whether the objects are distinguishable, and on how many objects can be in the same state. (**A**) The *filled and striped circles* represent two distinguishable molecules and the rectangles represent three states. The number of different ways of assigning N distinguishable objects to m states so that state m_i has n_i objects $\left(\sum_i n_i = N\right)$ is $N!/\prod_{i=1}^{m} n_i$. There are 3 ways of putting both molecules in the same state ($3!/2!0!0! = 3$), and 6 ways of putting them in different states ($3!/1!1!0! = 6$). (**B**) The *empty circles* represent two bosons (indistinguishable particles, any number of which can be in the same state). The number of ways of assigning N such particles to m states is $(m+N-1)!/(m-1)!\,N!$, or 6 in this example. (**C**) The *stippled circles* represent two fermions (indistinguishable particles, no more than one of which can be in the same state). The number of ways of assigning N fermions to m states is $m!/(N-\mathrm{m})!N!$, or 3 in this example

$$\Delta S_a = S_a - S_a' = \Delta S_a^0 + k_B \ln\left(\frac{n_g + 1}{n_e}\right), \tag{4.68}$$

with $\Delta S_a^0 = k_B \ln(\omega_e/\omega_g)$. In the usual situation that $n_g \gg 1$, Eq. (4.68) reduces to

$$\Delta S_a = \Delta S_a^0 + k_B \ln(n_g/n_e) = \Delta S_a^0 - k_B \ln(n_e/n_g). \tag{4.69}$$

The contribution from the term in n_e/n_g usually dominates strongly over that from ΔS_a^0.

Note again that the absorbers are not in thermal equilibrium with their surroundings at this stage of the analysis. The absorber ensemble is, however, at equilibrium in the sense that all the absorbers have the same probability of being excited, because we have assumed that the molecules are identical and the illumination is uniform. Moving an excitation from one site to another would not change the energy of the ensemble or affect the maximum amount of work that the system could do. The excitation entropy therefore does not depend on the ability of excitations to diffuse over all the sites on any particular time scale.

Equations (4.65)–(4.68) also put no restrictions on the size of the absorber ensemble. They apply to either a large ensemble or the idealized case of an individual molecule with two nondegenerate states. For a single molecule, $\Omega = 1$ whether or not the molecule is excited, so ΔS_a is zero. Equation (4.69) requires a larger ensemble, but according to the ergodic hypothesis of statistical mechanics, also will hold for a single molecule if we replace the populations n_e and n_g by the time-averaged probabilities of finding the molecule in the indicated states, provided that the averages are evaluated over a sufficiently long period of time.

To see how ΔS_a depends on the intensity of the radiation, suppose an ensemble of absorbers is exposed to continuous light with an irradiance of $I(\nu) = I_B(\nu) + I_r(\nu)$, where $I_B(\nu)$ is the diffuse blackbody radiation from the surroundings at the ambient temperature (T) and $I_r(\nu)$ is the radiation from any additional source. Molecules that absorb at frequency v will be excited with a first-order rate constant $k_e = \int I(\nu)\sigma(\nu)d\nu$ and will return to the ground state with rate constant $k_g = k_e + k_f + k_{nr}$, where $\sigma(\nu)$ is the absorption cross section (Sect. 4.3), k_f is the rate constant for fluorescence, and k_{nr} is the total rate constant for any non-radiative decay paths. The contribution of k_e to k_g represents stimulated emission. In a steady state, the ratio n_e/n_g will be

$$\frac{n_e}{n_g} = \frac{k_e}{k_g} = \frac{\int I(\nu)\sigma(\nu)d\nu}{\int I(\nu)\sigma(\nu)d\nu + k_f + k_{nr}}. \tag{4.70}$$

In the limit of very strong illumination, when the rates of excitation and stimulated emission are much greater than the rate of fluorescence and other decay processes $(k_e \gg k_f + k_{nr})$, n_e/n_g approaches 1. Equation (4.69) then gives $\Delta S_a \approx \Delta S_a^0$, which usually is much less than $h\nu/T$. In this limit, the free energy of excitation (ΔG_a) approaches $h\nu$. In the opposite limit of very weak radiation ($I \approx I_B$), when the absorbers approach thermal equilibrium with their surroundings, n_e/n_g goes to the ratio given by the Boltzmann distribution, $n_e/n_g = \exp(-h\nu/k_BT)$. Equation (4.69) then gives $\Delta S_a = \Delta S_a^0 + h\nu/T \approx h\nu/T$, and ΔG_a goes to zero. Figure 4.32 shows results for typical intermediate light intensities such that $I \gg I_B$ but $k_e \ll k_f + k_{nr}$. In this region, ΔS_a for absorption of visible light at room temperature is on the order of 25% of $h\nu/T$, depending largely on k_{nr}.

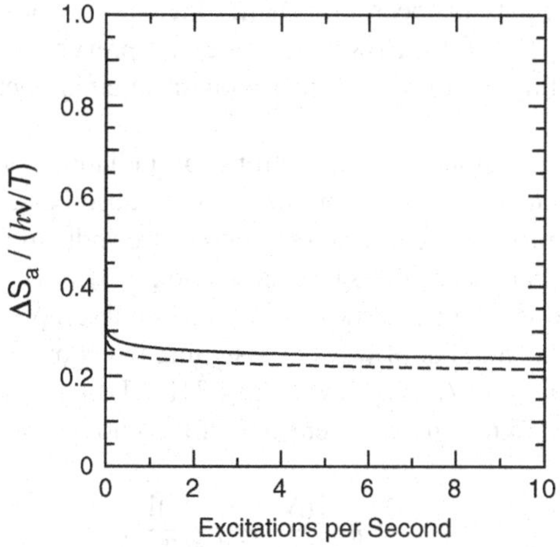

Fig. 4.32 Entropy change (ΔS_a) associated with excitation of an absorber as a function of the excitation rate constant $\left(k_e = \int I(\nu)\sigma(\nu)d\nu\right)$, with the decay rate constant k_g fixed at either 10^{10} (*dashed curve*) or 10^{11} s^{-1} (*solid curve*). The entropy is expressed relative to $h\nu/T$ for absorption at 500 nm ($h\nu = 2.48$ eV $= 3.98 \times 10^{-19}$ J) and $T = 300$ K. Possible contributions from degeneracy of the excited state (ΔS_a^0) are neglected. The values of k_g were chosen on the assumption that the excited absorber enters into a productive non-radiative reaction with a rate constant (k_{nr}) that is on the order of 100 times greater than typical values of the rate constant for fluorescence (k_f), as is necessary for efficient photochemistry. To put the abscissa scale in perspective, consider excitation in an absorption band at 500 nm with a mean extinction coefficient of 10^5 M^{-1} cm^{-1}, which is equivalent to a mean absorption cross section of 3.82×10^{-16} cm^2. Suppose the absorber is exposed to full sunlight, which has an irradiance at 500 nm of approximately 1.35 W m^{-2} nm^{-1} at the earth's surface, assuming a clear atomosphere and a solar zenith angle of 60°. If the absorption band has a width of 10 nm, the total irradiance is 1.35×10^{-3} J cm^{-2} s^{-1} or 3.40×10^{15} photons · cm^{-2} s^{-1}, giving $k_e = 1.30$ excitations per second, $\Delta S_a = 2.16 \times 10^{-3}$ eV K^{-1}, and $T\Delta S_a = 0.65$ eV

We now turn to the change in the entropy when a radiation field loses a photon (ΔSr). This term generally is negative, but smaller in magnitude than ΔS_a. The overall change in entropy therefore is greater than or equal to zero, in accord with the second law of thermodynamics. The simplest situation is the reversible loss of a photon from black-body radiation at temperature Tr (Sect. 3.2), for which the classical thermodynamic definition of the entropy change (the amount of heat transferred when a process occurs by a reversible path, divided by the temperature) gives

$$\Delta S_r = -h\nu/T_r. \tag{4.71}$$

The spectrum of solar radiation reaching the top of the earth's atmosphere is close to the black-body radiation curve for an emitter at 5,520 K. Absorption and scattering in the atmosphere decrease the apparent temperature to about 5,200 K at sea level. The overall entropy change $(\Delta S_a + \Delta S_r)$ associated with photoexcitation here is, therefore, on the order of $0.25h\nu/(300\ \text{K}) - h\nu/(5{,}200\ \text{K}) = 0.20h\nu/(300\ \text{K})$.

This entropy increase puts an upper limit on the efficiency of converting solar energy to useful work [131, 137–140]. However, it does not prevent an individual excited molecule from reacting to form a product with an internal energy of $h\nu$, or even somewhat higher.

More general expressions for the entropy of radiation can be obtained by considerations similar to those for an ensemble of absorbers, except that Bose-Einstein statistics must be used because photons are indistinguishbable and any number of photons can be in the same state (Box 2.6) [141–144]. Consider an ensemble of N photons with a narrow band of frequencies representing m different oscillation modes. The number of ways of assigning the photons to various oscillation modes is $(N + m - 1)!/(m - 1)!N!$ (Fig. 4.31B). Using this expression for the multiplicity Ω in Eq. (4.65) gives the entropy of the photons:

$$S_r = k_B \ln\left[\frac{(N+m-1)!}{(m-1)!N!}\right]. \tag{4.72}$$

For broad-band radiation, we would need the integral $\int S_r(\nu)d\nu$, in which N and m are functions of frequency.

Equation (4.72) can be put in a more convenient form in the common situation that $m \gg 1$, because we then can neglect the contributions of –1 relative to m in the arguments and can use Stirling's approximation to set $\ln(x!) \approx x\ln(x)$. With these approximations, some rearrangement of Eq. (4.72) leads to

$$S_r = k_B m\left[\left(1+\frac{N}{m}\right)\ln\left(1+\frac{N}{m}\right) - \left(\frac{N}{m}\right)\ln\left(\frac{N}{m}\right)\right]. \tag{4.73}$$

From Eq. (4.72), we find that loss of one photon changes the ensemble's entropy by

$$\Delta S_r = k_B \ln\left[\frac{N!}{(N-1)!}\right] - k_B \ln\left[\frac{(N+m-1)!}{(N+m-2)!}\right] = k_B \ln(N) - k_B \ln(N+m-1)$$
$$= -k_B \ln\left(1+\frac{m}{N}-\frac{1}{N}\right). \tag{4.74a}$$

This expression holds for any values of m and N, but simplifies to

$$\Delta S_r = -k_B \ln(1 + m/N) \tag{4.74b}$$

when $N \gg 1$. ΔSr goes to zero if $m = 1$.

For black-body radiation in a cube with volume V, the number of oscillation modes in frequency interval $d\nu$ is $m(\nu) = 8\pi n^3 \nu^2 V/c^3 d\nu$ (Eq. 3.39), and at temperature T_r the number of photons is $N(\nu) = (8\pi n^3 \nu^2 V/c^3 d\nu)/[\exp(h\nu/k_B T_r) - 1]$ (Eq. 3.41a). The ratio m/N for black-body radiation therefore is $[\exp(h\nu/k_B T_r) - 1]$. Inserting this expression for m/N in Eq. (4.74b) gives Eq. (4.71).

Following absorption of a photon, an excited molecule usually undergoes relaxations that distribute the vibrational reorganization energy among the molecule's family of vibrational modes and dissipate part of the excitation energy as heat to the surroundings. Intramolecular vibrational distributions increase the entropy of the absorber, whereas dissipation of heat reduces the entropy of the absorber but raises the entropy of the surroundings. In the harmonic approximation, the vibrational entropy of a molecule at thermal equilibrium can be written

$$S_V = k_B \sum_j \left\{ \frac{(h\upsilon_j/k_BT)\exp(-h\upsilon_j/k_BT)}{1-\exp(-h\upsilon_j/k_BT)} - \ln\left[1-\exp(-h\upsilon_j/k_BT)\right] \right\}, \quad (4.75)$$

where υ_j is the frequency of vibrational mode j [145]. Figure 4.33 shows the dependence of this function on υ_j for a single vibrational mode. Since the contribution to S_V rises steeply as υ_j goes to zero, dispersal of vibrational energy from a high-frequency vibrational mode of an absorber into many low-frequency modes of the surroundings results in an overall increase in entropy. There can, however, be an opposing entropy decrease due to ordering of the solvent if the absorber is more polar in the excited state than in the ground state. In some cases, the excited molecule itself also can relax into a more restricted geometry. 4-Methylbiphenyl and 4-methylbenzophenone, for example, have flexible, twisted geometries in the ground state, and more rigid, planar structures in the excited triplet state. Entropy changes associated with such structural changes have been measured from the temperature dependence of the equilibrium constant for transferring triplet excitations to or from other molecules with more constant geometries [146, 147].

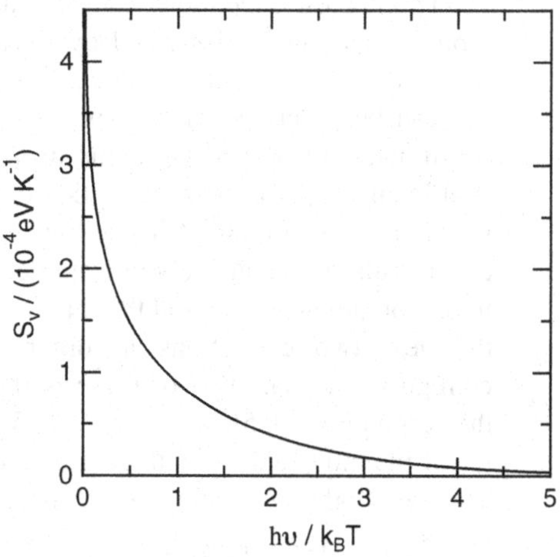

Fig. 4.33 Vibrational entropy of a system with a single harmonic mode with frequency υ, as calculated by Eq. (4.75)

Exercises

1. The highest occupied molecular orbital (HOMO) and the lowest unoccupied molecular orbital (LUMO) of ethylene can be represented as symmetric and antisymmetric combinations of atomic p_z orbitals (ψ_{p1} and ψ_{p2}) centered on the two carbon atoms: $\Psi_a = \frac{1}{\sqrt{2}}(\psi_{p1} + \psi_{p2})$ and $\Psi_b = \frac{1}{\sqrt{2}}(\psi_{p1} - \psi_{p2})$, respectively. Does the transition $\Psi_a \rightarrow \Psi_b$ change the permanent dipole moment of ethylene? Explain.
2. The graph below shows the absorption spectrum of a molecule in water (refractive index = 1.33). The spectrum is simplified for ease of integration. Calculate the dipole strength of the absorption band and the transition dipole moment. Specify the units of both quantities. Save your results for use in the exercises for Chaps. 5 and 6.

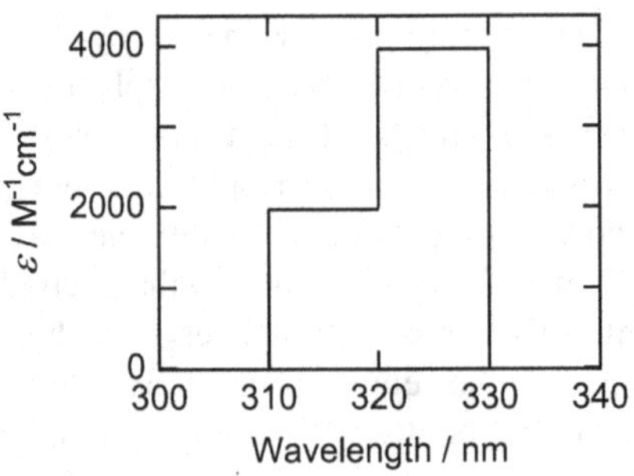

3. Describe the Born-Oppenheimer and Condon approximations, and explain why the Condon approximation is the more restrictive assumption.
4. The two highest filled molecular orbitals of *trans*-butadiene (ψ_1 and ψ_2) are π orbitals that can be described by linear combination of atomic p_z orbitals centered on the four carbon atoms. The first two unoccupied orbitals (ψ_3 and ψ_4) can be described in the same way. The table below gives the atomic coordinates of the carbon atoms in *trans*-butadiene (*x*, *y*, *z*, in Å) and the coefficients (C_i) for the contribution of the p_z orbital of atom *i* to each of the four molecular orbitals. (*a*) The lowest-energy excitation of *trans*-butadiene consists almost exclusively of a single configuration, $\psi_2 \rightarrow \psi_3$. Calculate the transition dipole vector and the dipole strength for this transition. (*b*) Each of the next two excitations in order of increasing energy includes two configurations. The first of these is $0.6574(\psi_2 \rightarrow \psi_4) - 0.7535(\psi_1 \rightarrow \psi_3)$; the second is $0.7535(\psi_2 \rightarrow \psi_4) + 0.6574(\psi_1 \rightarrow \psi_3)$. Show from the symmetry of the molecular orbitals that both of these transitions are forbidden. (*c*) The next higher excitation again consists almost exclusively of the configuration $\psi_1 \rightarrow \psi_4$. Explain qualitatively why the dipole strength for this transition will be considerably smaller than that for $\psi_2 \rightarrow \psi_3$. (You don't need to

calculate the dipole strength in order to see this qualitatively; just look at the symmetry of the orbitals.)

Atom	Coordinates			Coefficients			
	x	y	z	ψ_1	ψ_2	ψ_3	ψ_4
1	−1.731	−0.634	0.000	0.4214	−0.5798	−0.5677	0.4050
2	−0.390	−0.626	0.000	0.5677	−0.4050	0.4214	−0.5798
3	0.389	0.626	0.000	0.5679	0.4049	0.4215	0.5796
4	1.731	0.634	0.000	0.4215	0.5796	−0.5679	−0.4049

5. In aqueous solution, the reduced nicotinamide ring of NADH has an absorption band at 340 nm. The excitation could be described approximately as a movement of a nonbonding electron from the N atom of the pyridine ring to the amide O atom as shown in the valence-bond diagrams below. When NADH binds to liver alcohol dehydrogenase (ADH), the band shifts to 325 nm. Outline how you might calculate the effects of water and the protein on (*a*) the excitation energy and (*b*) the reorganization energy for this excitation, based on a crystal structure of the holoprotein. (*c*) What major approximations would limit the reliability of your calculation?

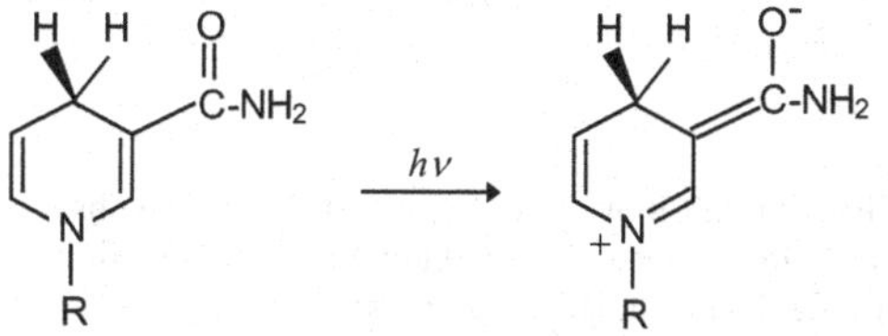

6. X-rays with an energy of about 7.1×10^6 eV are sufficiently energetic to dislodge an electron from the 1s shell of Fe. At higher energies, an electron leaves the atom, carrying the extra energy away as kinetic energy. Just below this edge, Fe has an absorption band associated with transitions from *1s* to the *3d* shell. The $1s \rightarrow 3d$ transition is very weak if the Fe atom binds six ligands in an octahedral arrangement, provided that the ligands on opposite sides of the Fe are similar. It becomes much stronger if Fe has four ligands in a tetrahedral arrangement, as happens in some metalloproteins such as rubredoxin. How could you explain these results? Hint: atomic *s* and *d* orbitals both have the same (even) inversion symmetry with respect to any axis passing through the center of the atom. For the purposes of this question, don't worry about the spin of a photon.
7. The C=C bond of ethylene has a mean length of 1.33 Å when a molecule is in the ground state. The bond length probably increases by about 0.05 Å in the first singlet $\pi - \pi^*$ excited state. (*a*) Assuming that the stretching of the bond is a harmonic vibrational mode with a frequency of 1,623 cm^{-1} in both the ground and the excited states, calculate the vibronic coupling strength

(Huang-Rhys factor) and the Debye-Waller factor for this mode. Making the same assumptions, calculate (*b*) the Franck-Condon factors for excitation from the lowest vibrational level ($m = 0$) of the ground electronic state to the first three vibrational levels ($m = 0$, 1 and 2) of the excited electronic state, and (*c*) the Franck-Condon factors for excitation from the $m = 1$ level of the ground state to the $m = 1$, 2 and 3 levels of the excited state.

8. When photosynthetic bacterial reaction centers are excited with light at 5 K, a bacteriochlorophyll dimer transfers an electron to a neighboring molecule with a time constant of about 1 ps. It has been suggested that the reaction is preceded by a transition of the excited dimer from its lowest $\pi - \pi^*$ state to an internal charge-transfer state. Observations relevant to this suggestion came from hole-burning studies, also performed at 5 K, which elicited a zero-phonon hole with a width of about 6 cm^{-1} in the dimer's long-wavelength absorption band. The width did not change significantly when the wavelength laser used for burning the hole was tuned over a broad region. Assuming that excitation in the long-wavelength band generates the $\pi - \pi^*$ state, what does the width of the zero-phonon hole imply with regard to the reaction mechanism?

References

1. Bakshiev, N.G., Girin, O.P., Libov, V.S.: Relation between the observed and true absorption spectra of molecules in a condensed medium III. Determination of the influence of an effective (internal) field according to the models of Lorentz and Onsager-Bötche. Opt. Spectrosc. **14**, 395–398 (1963)
2. Shipman, L.: Oscillator and dipole strengths for chlorophyll and related molecules. Photochem. Photobiol. **26**, 287–292 (1977)
3. Myers, A.B., Birge, R.R.: The effect of solvent environment on molecular electronic oscillator strengths. J. Chem. Phys. **73**, 5314–5321 (1980)
4. Alden, R.G., Johnson, E., Nagarajan, V., Parson, W.W.: Calculations of spectroscopic properties of the LH2 bacteriochlorophyll-protein antenna complex from Rhodopseudomonas sphaeroides. J. Phys. Chem. B **101**, 4667–4680 (1997)
5. Knox, R.S., Spring, B.Q.: Dipole strengths in the chlorophylls. Photochem. Photobiol. **77**, 497–501 (2003)
6. Cotton, F.A.: Chemical Applications of Group Theory, 3rd edn. Wiley, New York (1990)
7. Harris, D.C., Bertolucci, M.D.: Symmetry and Spectroscopy. Oxford Univ. Press, New York (1978) (reprinted by Dover, 1989)
8. Verméglio, A., Clayton, R.K.: *Orientation of chromophores in reaction centers of Rhodopseudomonas sphaeroides*. Evidence for two absorption bands of the dimeric primary electron donor. Biochim. Biophys. Acta **449**, 500–515 (1976)
9. Abdourakhmanov, I.A., Ganago, A.O., Erokhin, Y.E., Solov'ev, A.A., Chugunov, V.A.: Orientation and linear dichroism of the reaction centers from Rhodopseudomonas sphaeroides R-26. Biochim. Biophys. Acta **546**, 183–186 (1979)
10. Paillotin, G., Verméglio, A., Breton, J.: Orientation of reaction center and antenna chromophores in the photosynthetic membrane of Rhodopseudomonas viridis. Biochim. Biophys. Acta **545**, 249–264 (1979)

11. Rafferty, C.N., Clayton, R.K.: The orientations of reaction center transition moments in the chromatophore membrane of Rhodopseudomonas sphaeroides, based on new linear dichroism and photoselection measurements. Biochim. Biophys. Acta **546**, 189–206 (1979)
12. Breton, J.: Orientation of the chromophores in the reaction center of *Rhodopseudomonas viridis*. Comparison of low-temperature linear dichroism spectra with a model derived from X-ray crystallography. Biochim. Biophys. Acta **810**, 235–245 (1985)
13. Breton, J.: Low temperature linear dichroism study of the orientation of the pigments in reduced and oxidized reaction centers of Rps. viridis and Rb. sphaeroides. In: Breton, J., Verméglio, A. (eds.) The Photosynthetic Bacterial Reaction Center: Structure and Dynamics, pp. 59–69. Plenum Press, New York (1988)
14. Cone, R.A.: Rotational diffusion of rhodopsin in the visual receptor membrane. Nat. New Biol. **236**, 39–43 (1972)
15. Fotiadis, D., Liang, Y., Filipek, S., Saperstein, D.A., Engel, A., et al.: The G protein-coupled receptor rhodopsin in the native membrane. FEBS Lett. **564**, 281–288 (2004)
16. Junge, W., Lill, H., Engelbrecht, S.: ATP synthase: an electrochemical transducer with rotatory mechanics. Trends Biol. Sci. **22**, 420–423 (1997)
17. Sabbert, D., Engelbrecht, S., Junge, W.: Functional and idling rotatory motion within F_1-ATPase. Proc. Natl. Acad. Sci. **94**, 4401–4405 (1997)
18. Lim, M., Jackson, T.A., Anfinrud, P.A.: Binding of CO to myoglobin from a heme pocket docking site to form nearly linear Fe-C-O. Science **269**, 962–966 (1995)
19. Platt, J.R.: Molecular orbital predictions of organic spectra. J. Chem. Phys. **18**, 1168–1173 (1950)
20. Pariser, R., Parr, R.G.: A semi-empirical theory of the electronic spectra and electronic structure of complex unsaturated molecules. I. J. Chem. Phys. **21**, 466–471 (1953)
21. Pariser, R., Parr, R.G.: A semi-empirical theory of the electronic spectra and electronic structure of complex unsaturated molecules. II. J. Chem. Phys. **21**, 767–776 (1953)
22. Ito, H., I'Haya, Y.: The electronic structure of naphthalene. Theor. Chim. Acta **2**, 247–257 (1964)
23. Mataga, N., Kubota, T.: Molecular Interactions and Electronic Spectra. Dekker, New York (1970)
24. Pariser, R.: Theory of the electronic spectra and structure of the polyacenes and of alternant hydrocarbons. J. Chem. Phys. **24**, 250–268 (1956)
25. Warshel, A., Karplus, M.: Calculation of ground and excited state potential surfaces of conjugated molecules. I. Formulation and parametrization. J. Am. Chem. Soc. **94**, 5612–5625 (1972)
26. Gouterman, M.: Optical spectra and electronic structure of porphyrins and related rings. In: Dolphin, D. (ed.) The Porphyrins, pp. 1–165. Academic, New York (1978)
27. Gouterman, M.: Spectra of porphyrins. J. Mol. Spectrosc. **6**, 138–163 (1961)
28. McHugh, A.J., Gouterman, M., Weiss, C.J.P., XXIV: Energy, oscillator strength and Zeeman splitting calculations (SCMO-CI) for phthalocyanine, porphyrins, and related ring systems. Theor. Chim. Acta **24**, 346–370 (1972)
29. Parson, W.W., Warshel, A.: Spectroscopic properties of photosynthetic reaction centers. 2. Application of the theory to *Rhodopseudomonas viridis*. J. Am. Chem. Soc. **109**, 6152–6163 (1987)
30. Warshel, A., Parson, W.W.: Spectroscopic properties of photosynthetic reaction centers. 1. Theory. J. Am. Chem. Soc. **109**, 6143–6152 (1987)
31. Platt, J.R.: Classification of spectra of cata-condensed hydrocarbons. J. Chem. Phys. **17**, 484–495 (1959)
32. Weber, G.: Fluorescence-polarization spectrum and electronic energy transfer in tyrosine, tryptophan and related compounds. Biochem. J. **75**, 335–345 (1960)
33. Song, P.-S., Kurtin, W.E.: A spectroscopic study of the polarized luminescence of indoles. J. Am. Chem. Soc. **91**, 4892–4906 (1969)

34. Auer, H.E.: Far ultraviolet absorption and circular dichroism spectra of L-tryptophan and some derivatives. J. Am. Chem. Soc. **95**, 3003–3011 (1973)
35. Lami, H., Glasser, N.: Indole's solvatochromism revisited. J. Chem. Phys. **84**, 597–604 (1986)
36. Callis, P.R.: Molecular orbital theory of the ^1L_b and ^1L_a states of indole. J. Chem. Phys. **95**, 4230–4240 (1991)
37. Callis, P.R.: ^1L_a and ^1L_b transitions of tryptophan: applications of theory and experimental observations to fluorescence of proteins. Meth. Enzymol. **278**, 113–150 (1997)
38. McHugh, A.J., Gouterman, M.: *Oscillator strengths for electronic spectra of conjugated molecules from transition gradients* III. Polyacenes. Theor. Chim. Acta **13**, 249–258 (1969)
39. Chong, D.P.: Oscillator strengths for electronic spectra of conjugated molecules from transition gradients. I. Mol. Phys. **14**, 275–280 (1968)
40. Schlessinger, J., Warshel, A.: Calculations of CD and CPL spectra as a tool for evaluation of the conformational differences between ground and excited states of chiral molecules. Chem. Phys. Lett. **28**, 380–383 (1974)
41. Mulliken, R.S., Rieke, C.A., Orloff, D., Orloff, H.: Formulas and numerical tables for overlap integrals. J. Chem. Phys. **17**, 1248–1267 (1949)
42. Král, M.: Optical rotatory power of complex compounds. Matrix elements of operators Del and R x Del. Collect. Czech. Chem. Commun. **35**, 1939–1948 (1970)
43. Harada, N., Nakanishi, K.: Circular Dichroic Spectroscopy: Exciton Coupling in Organic Stereochemistry. University Science, Mill Valley, CA (1983)
44. Miller, J., Gerhauser, J.M., Matsen, F.A.: Quantum Chemistry Integrals and Tables. Univ. of Texas Press, Austin (1959)
45. Tavan, P., Schulten, K.: The low-lying electronic excitations in long polyenes: a PPP-MRD-CI study. J. Chem. Phys. **85**, 6602–6609 (1986)
46. Chadwick, R.R., Gerrity, D.P., Hudson, B.S.: Resonance Raman spectroscopy of butadiene: demonstration of a 2^1A_g state below the 1^1B_u V state. Chem. Phys. Lett. **115**, 24–28 (1985)
47. Koyama, Y., Rondonuwu, F.S., Fujii, R., Watanabe, Y.: Light-harvesting function of carotenoids in photosynthesis: the roles of the newly found 1^1B_u state. Biopolymers **74**, 2–18 (2004)
48. Birge, R.R.: 2-photon spectroscopy of protein-bound chromophores. Acc. Chem. Res. **19**, 138–146 (1986)
49. Koyama, Y., Kuki, M., Andersson, P.-O., Gilbro, T.: Singlet excited states and the light-harvesting function of carotenoids in bacterial photosynthesis. Photochem. Photobiol. **63**, 243–256 (1996)
50. Macpherson, A., Gilbro, T.: Solvent dependence of the ultrafast S_2-S_1 internal conversion rate of β-carotene. J. Phys. Chem. A **102**, 5049–5058 (1998)
51. Polivka, T., Herek, J.L., Zigmantas, D., Akerlund, H.E., Sundström, V.: Direct observation of the (forbidden) S_1 state in carotenoids. Proc. Natl. Acad. Sci. USA **96**, 4914–4917 (1999)
52. Born, M., Oppenheimer, R.: Zur Quantentheorie der Molekeln [On the quantum theory of molecules]. Ann. Phys. Lpz. **84**, 457–484 (1927)
53. Struve, W.S.: Fundamentals of Molecular Spectroscopy. Wiley Interscience, New York (1989)
54. Condon, E.U.: The Franck-Condon principle and related topics. Am. J. Phys. **15**, 365–374 (1947)
55. Manneback, C.: Computation of the intensities of vibrational spectra of electronic bands in diatomic molecules. Physica **17**, 1001–1010 (1951)
56. Lyle, P.A., Kolaczkowski, S.V., Small, G.J.: Photochemical hole-burned spectra of protonated and deuterated reaction centers of Rhodobacter sphaeroides. J. Phys. Chem. **97**, 6924–6933 (1993)
57. Zazubovich, V., Tibe, I., Small, G.J.: Bacteriochlorophyll a Franck-Condon factors for the S0 →S1(Qy) transition. J. Phys. Chem. B **105**, 12410–12417 (2001)

58. Sharp, T.E., Rosenstock, H.M.: Franck-Condon factors for polyatomic molecules. J. Chem. Phys. **41**, 3453–3463 (1964)
59. Sando, G.M., Spears, K.G.: Ab initio computation of the Duschinsky mixing of vibrations and nonlinear effects. J. Phys. Chem. A **104**, 5326–5333 (2001)
60. Sando, G.M., Spears, K.G., Hupp, J.T., Ruhoff, P.T.: Large electron transfer rate effects from the Duschinsky mixing of vibrations. J. Phys. Chem. A **105**, 5317–5325 (2001)
61. Jankowiak, R., Small, G.J.: Hole-burning spectroscopy and relaxation dynamics of amorphous solids at low temperatures. Science **237**, 618–625 (1987)
62. Volker, S.: Spectral hole-burning in crystalline and amorphous organic solids. Optical relaxation processes at low temperatures. In: Funfschilling, J. (ed.) Relaxation Processes in Molecular Excited States, pp. 113–242. Kluwer Academic Publ, Dordrecht (1989)
63. Friedrich, J.: Hole burning spectroscopy and physics of proteins. Meth. Enzymol. **246**, 226–259 (1995)
64. Reddy, N.R.S., Lyle, P.A., Small, G.J.: Applications of spectral hole burning spectroscopies to antenna and reaction center complexes. Photosynth. Res. **31**, 167–194 (1992)
65. Reddy, N.R.S., Picorel, R., Small, G.J.: B896 and B870 components of the Rhodobacter sphaeroides antenna: a hole burning study. J. Phys. Chem. **96**, 6458–6464 (1992)
66. Wu, H.-M., Reddy, N.R.S., Small, G.J.: Direct observation and hole burning of the lowest exciton level (B870) of the LH2 antenna complex of Rhodopseudomonas acidophila (strain 10050). J. Phys. Chem. **101**, 651–656 (1997)
67. Small, G.J.: On the validity of the standard model for primary charge separation in the bacterial reaction center. Chem. Phys. **197**, 239–257 (1995)
68. Johnson, E.T., Nagarajan, V., Zazubovich, V., Riley, K., Small, G.J., et al.: Effects of ionizable residues on the absorption spectrum and initial electron-transfer kinetics in the photosynthetic reaction center of Rhodopseudomonas sphaeroides. Biochemistry **42**, 13673–13683 (2003)
69. Palczewski, K., Kumasaka, T., Hori, T., Behnke, C.A., Motoshima, H., et al.: Crystal structure of rhodopsin: A G protein-coupled receptor. Science **289**, 739–745 (2000)
70. Stenkamp, R.E., Filipek, S., Driessen, C.A.G.G., Teller, D.C., Palczewski, K.: Crystal structure of rhodopsin: a template for cone visual pigments and other G protein-coupled receptors. Biochim. Biophys. Acta **1565**, 168–182 (2002)
71. Teller, D.C., Stenkamp, R.E., Palczewski, K.: Evolutionary analysis of rhodopsin and cone pigments: connecting the three-dimensional structure with spectral tuning and signal transfer. FEBS Lett. **2003**, 151–159 (2003)
72. Kochendoerfer, G.G., Kaminaka, S., Mathies, R.A.: Ultraviolet resonance Raman examination of the light-induced protein structural changes in rhodopsin activation. Biochemistry **36**, 13153–13159 (1997)
73. Kochendoerfer, G.G., Lin, S.W., Sakmar, T.P., Mathies, R.A.: How color visual pigments are tuned. Trends Biol. Sci. **24**, 300–305 (1999)
74. Ottolenghi, M., Sheves, M.: Synthetic retinals as probes for the binding site and photoreactions in rhodopsins. J. Membr. Biol. **112**, 193–212 (1989)
75. Aharoni, A., Ottolenghi, M., Sheves, M.: Retinal isomerization in bacteriorhodopsin is controlled by specific chromophore-protein interactions. A study with noncovalent artificial pigments. Biochemistry **40**, 13310–13319 (2001)
76. Nathans, J.: Determinants of visual pigment absorbance: identification of the retinylidene Schiff's base counterion in bovine rhodopsin. Biochemistry **29**, 9746–9752 (1990)
77. Asenjo, A.B., Rim, J., Oprian, D.D.: Molecular determinants of human red/green color discrimination. Neuron **12**, 1131–1138 (1994)
78. Ebrey, T.G., Takahashi, Y.: Photobiology of retinal proteins. In: Coohil, T.P., Velenzo, D.P. (eds.) Photobiology for the 21st Century, pp. 101–133. Valdenmar Publ, Overland Park, KS (2001)
79. Kamauchi, M., Ebrey, T.G.: Visual pigments as photoreceptors. In: Spudich, J., Briggs, W. (eds.) Handbook of Photosensory Receptors, pp. 43–76. Wiley-VCH, Weinheim (2005)

80. Deng, H., Callender, R.H.: A study of the Schiff base mode in bovine rhodopsin and bathorhodopsin. Biochemistry **26**, 7418–7426 (1987)
81. Sancar, A.: Structure and function of DNA photolyase and cryptochrome blue-light photoreceptors. Chem. Rev. **103**, 2203–2237 (2003)
82. Malhotra, K., Kim, S.T., Sancar, A.: Characterization of a medium wavelength type DNA photolyase: purification and properties of a photolyase from Bacillus firmus. Biochemistry **33**, 8712–8718 (1994)
83. Limantara, L., Sakamoto, S., Koyama, Y., Nagae, H.: Effects of nonpolar and polar solvents on the Q_x and Q_y energies of bacteriochlorophyll a and bacteriopheophytin a. Photochem. Photobiol. **65**, 330–337 (1997)
84. Lee, F.S., Chu, Z.T., Warshel, A.: Microscopic and semimicroscopic calculations of electrostatic energies in proteins by the POLARIS and ENZYMIX programs. J. Comp. Chem. **14**, 161–185 (1993)
85. Vivian, J.T., Callis, P.R.: Mechanisms of tryptophan fluorescence shifts in proteins. Biophys. J. **80**, 2093–2109 (2001)
86. Mercer, I.P., Gould, I.R., Klug, D.R.: A quantum mechanical/molecular mechanical approach to relaxation dynamics: calculation of the optical properties of solvated bacteriochlorophyll-a. J. Phys. Chem. B **103**, 7720–7727 (1999)
87. Liptay, W.: Dipole moments of molecules in excited states and the effect of external electric fields on the optical absorption of molecules in solution. In: Sinanoglu, O. (ed.) Modern Quantum Chemistry Part III: Action of Light and Organic Crystals. Academic, New York (1965)
88. Liptay, W.: Electrochromism and solvatochromism. Angew. Chem. Int. Ed. Engl. **8**, 177–188 (1969)
89. Liptay, W.: Dipole moments and polarizabilities of molecules in excited states. In: Lim, E.C. (ed.) Excited States, pp. 129–230. Academic, New York (1974)
90. Middendorf, T.R., Mazzola, L.T., Lao, K.Q., Steffen, M.A., Boxer, S.G.: Stark-effect (electroabsorption) spectroscopy of photosynthetic reaction centers at 1.5 K Evidence that the special pair has a large excited-state polarizability. Biochim. Biophys. **1143**, 223–234 (1993)
91. Lao, K., Moore, L.J., Zhou, H., Boxer, S.G.: Higher-order Stark spectroscopy: polarizability of photosynthetic pigments. J. Phys. Chem. **99**, 496–500 (1995)
92. Bublitz, G., Boxer, S.G.: Stark spectroscopy: applications in chemistry, biology and materials science. Annu. Rev. Phys. Chem. **48**, 213–242 (1997)
93. Boxer, S.G.: Stark realities. J. Phys. Chem. B **113**, 2972–2983 (2009)
94. Nagae, H. Theory of solvent effects on electronic absorption spectra of rodlike or disklike solute molecules: frequency shifts. J. Chem. Phys. **106** (1997)
95. Premvardhan, L.L., Buda, F., van der Horst, M.A., Lührs, D.C., Hellingwerf, K.J., et al.: Impact of photon absorption on the electronic properties of p-coumaric acid derivatives of the photoactive yellow protein chromophore. J. Phys. Chem. B **108**, 5138–5148 (2004)
96. Premvardhan, L.L., van der Horst, M.A., Hellingwerf, K.J., van Grondelle, R.: Stark spectroscopy on photoactive yellow protein, E46Q, and a nonisomerizing derivative, probes photo-induced charge motion. Biophys. J. **84**, 3226–3239 (2003)
97. Premvardhan, L.L., van der Horst, M.A., Hellingwerf, K.J., van Grondelle, R.: How light-induced charge transfer accelerates the receptor-state recovery of photoactive yellow protein from its signalling state. Biophys. J. **89**, L64–L66 (2005)
98. Bublitz, G.U., Laidlaw, W.M., Denning, R.G., Boxer, S.G.: Effective charge transfer distances in cyanide-bridged mixed-valence transform metal complexes. J. Am. Chem. Soc. **120**, 6068–6075 (1998)
99. Moore, L.J., Zhou, H.L., Boxer, S.G.: Excited-state electronic asymmetry of the special pair in photosynthetic reaction center mutants: absorption and Stark spectroscopy. Biochemistry **38**, 11949–11960 (1999)

100. Kador, L., Haarer, D., Personov, R.: Stark effect of polar and unpolar dye molecules in amorphous hosts, studied via persistent spectral hole burning. J. Chem. Phys. **86**, 213–242 (1987)
101. Gafert, J., Friedrich, J., Vanderkooi, J.M., Fidy, J.: *Structural changes and internal fields in proteins*. A hole-burning Stark effect study of horseradish peroxidase. J. Phys. Chem. **99**, 5223–5227 (1995)
102. Pierce, D.W., Boxer, S.G.: Stark effect spectroscopy of tryptophan. Biophys. J. **68**, 1583–1591 (1995)
103. Zhou, H., Boxer, S.G.: Probing excited-state electron transfer by resonance Stark spectroscopy. 1. Experimental results for photosynthetic reaction centers. J. Phys. Chem. B **102**, 9139–9147 (1998)
104. Zhou, H., Boxer, S.G.: Probing excited-state electron transfer by resonance Stark spectroscopy. 2. Theory and application. J. Phys. Chem. B **102**, 9148–9160 (1998)
105. Treynor, T.P., Andrews, S.S., Boxer, S.G.: Intervalence band Stark effect of the special pair radical cation in bacterial photosynthetic reaction centers. J. Phys. Chem. B **107**, 11230–11239 (2003)
106. Treynor, T.P., Boxer, S.G.: A theory of intervalence band stark effects. J. Phys. Chem. A **108**, 1764–1778 (2004)
107. Kalyanasundaram, K.: Photophysics, photochemistry and solar energy conversion with tris (bipyridyl)ruthenium(II) and its analogs. Coord. Chem. Rev. **46**, 159–244 (1982)
108. Dallinger, R.F., Woodruff, W.H.: Time-resolved resonance Raman study of the lowest (dπ*,3CT) excited state of tris(2,2–bipyridine)ruthenium(II). J. Am. Chem. Soc. **101**, 4391–4393 (1979)
109. Bradley, P.G., Kress, N., Hornberger, B.A., Dallinger, R.F., Woodruff, W.H.: Vibrational spectroscopy of the electronically excited state. 5. Time-resolved resonance Raman study of tris(bipyridine)ruthenium(II) and related complexes. Definitive evidence for the "localized" MLCT state. J. Am. Chem. Soc. **103**, 7441–7446 (1981)
110. Casper, J.V., Westmoreland, T.D., Allen, G.H., Bradley, P.G., Meyer, T.J., et al.: Molecular and electronic structure in the metal-to-ligand charge-transfer excited states of d6 transition-metal complexes in solution. J. Am. Chem. Soc. **106**, 3492–3500 (1984)
111. Smothers, W.K., Wrighton, M.S.: Raman spectroscopy of electronic excited organometallic complexes: a comparison of the metal to 2,2'-bipyridine charge-transfer state of fac-(2,2'-bipyridine)tricarbonylhalorhenium and tris(2,2'-bipyridine)ruthenium(II). J. Am. Chem. Soc. **105**, 1067–1069 (1983)
112. Felix, F., Ferguson, J., Güdel, J.U., Ludi, A.: The electronic spectrum of $Ru(bpy)_3^{2+}$. J. Am. Chem. Soc. **102**, 4096–4102 (1980)
113. De Armond, M.K., Myrick, M.L.: The life and times of $[Ru(bpy)_3]^{2+}$: localized orbitals and other strange occurrences. Acc. Chem. Res. **22**, 364–370 (1989)
114. Malone, R.A., Kelley, D.F.: Interligand electron transfer and transition state dynamics in Ru (II)trisbipyridine. J. Chem. Phys. **95**, 8970–8976 (1991)
115. Thompson, D.W., Ito, A., Meyer, T.J.: $[Ru(bpy)_3]^{2+*}$ and other remarkable metal-to ligand charge transfer (MLCT) excited states. Pure and Appl. Chem. **85**, 1257–1305 (2013)
116. Curtis, J.C., Sullivan, B.P., Meyer, T.J.: Hydrogen-bonding-induced solvatochromatism in the charge-transfer transitions of ruthenium(II) and ruthenium(III) complexes. Inorg. Chem. **22**, 224–236 (1983)
117. Riesen, H., Krausz, E.: Stark effects in the lowest-excited ^{3}MLCT states of $[Ru(bpy\text{-}d_8)_2]^{2+}$ in $[Zn(bpy)_3](ClO_4)_2$ (bpy = 2,2'-bipyridine). Chem. Phys. Lett. **260**, 130–135 (1996)
118. Coe, B.J., Helliwell, M., Peers, M.K., Raftery, J., Rusanova, D., et al.: Synthesis, structures, and optical properties of ruthenium(II) complexes of the Tris(1-pyrazolyl)methane ligand. Inorg. Chem. **53**, 3798–37811 (2014)
119. Oh, D.H., Boxer, S.G.: Stark effect spectra of $Ru(diimine)_3^{2+}$ complexes. J. Am. Chem. Soc. **111**, 1130–1132 (1989)

120. Sykora, J., Sima, J.: Development and basic terms of photochemistry of coordination compounds. Coord. Chem. Rev. **107**, 1–212 (1990)
121. Vogler, A., Kunkely, H.: Photoreactivity of metal-to-ligand charge transfer excited states. Coord. Chem. Rev. **177**, 81–96 (1998)
122. Vogler, A., Kunkely, H.: Photochemistry induced by metal-to-ligand charge transfer excitation. Coord. Chem. Rev. **208**, 321–329 (2000)
123. O'Regan, B., Grätzel, M.: A low-cost, high-efficiency solar cell based on dye-sensitized colloidal TiO_2 films. Nature **353**, 737–740 (1991)
124. Grätzel, M.: Photoelectrochemical cells. Nature **414**, 338–344 (2001)
125. Zhao, Y., Swierk, J.R., Megiatto Jr., J.D., Sherman, B., Youngblood, W.J., et al.: Improving the efficiency of water splitting in dye-sensitized solar cells by using a biomimetic electron transfer mediator. Proc. Natl. Acad. Sci. U.S.A. **109**, 15612–15616 (2012)
126. Alibabaei, L., Brennaman, M.K., Norris, M.R., Kalanyan, B., Song, W., et al.: Solar water splitting in a molecular photoelectrochemical cell. Proc. Natl. Acad. Sci. U.S.A. **110**, 20008–20013 (2013)
127. Robin, M.B.: Higher Excited States of Polyatomic Molecules, vol. III. Academic, New York (1985)
128. Shand, N.C., Ning, C.-L., Siggel, M.R.F., Walker, I.C., Pfab, J.: One- and two-photon spectroscopy of the $3s \leftarrow n$ Rydberg transition of propionaldehyde (propanal). J. Chem. Soc. Faraday Trans. **93**, 2883–2888 (1997)
129. Morisawa, Y., Ikehata, A., Higashi, N., Ozaki, Y.: Low-n Rydberg transitions of liquid ketones studied by attenuated total reflection far-ultraviolet spectroscopy. J. Phys. Chem. A **115**, 562–568 (2011)
130. Morisawa, Y., Yasunaga, M., Fukuda, R., Ehara, M., Ozaki, Y.: Electronic transitions in liquid amides studied by using attenuated total reflection far-ultraviolet spectroscopy and quantum chemical calculations. J. Chem. Phys. **139**, 154301 (2013)
131. Knox, R.S., Parson, W.W.: Entropy production and the second law in photosynthesis. Biochim. Biophys. Acta **1767**, 1189–1193 (2007)
132. Yourgrau, W., van der Merwe, A., Raw, G.: Treatise on Irreversible and Statistical Thermodynamics. Macmillan, New York (1966)
133. Kittel, C., Kroemer, H.: Thermal Physics. W.H. Freeman, San Francisco (1980)
134. Engel, T., Reid, P.: Thermodynamics, Statistical Thermodynamics, and Kinetics. Benjamin Cummings, San Francisco (2006)
135. Goldstein, S., Lebowitz, J.L.: On the (Boltzmann) entropy of non-equilibrium systems. Physica D **193**, 53–66 (2004)
136. Weinstein, M.A.: Thermodynamics of radiative emission processes. Phys. Rev. **119**, 499–501 (1960)
137. Knox, R.S.: Conversion of light into free energy. In: Gerischer, H., Katz, J.J. (eds.) Light Induced Charge Separation at Interfaces in Biology and Chemistry, pp. 45–59. Verlag Chemie, Weinheim (1979)
138. Duysens, L.N.M.: The path of light in photosynthesis. Brookhaven Symp. Biol. **11**, 18–25 (1958)
139. Parson, W.W.: Thermodynamics of the primary reactions of photosynthesis. Photochem. Photobiol. **28**, 389–393 (1978)
140. Ross, R.T.: Thermodynamic limitations on the conversion of radiant energy into work. J. Chem. Phys. **45**, 1–7 (1966)
141. Planck, M.: The Theory of Heat Radiation (Engl transl by M. Masius), 2nd edn. Dover, New York (1959)
142. Slater, J.C.: Introduction to Chemical Physics. McGraw-Hill, New York (1939)
143. Rosen, P.: Entropy of radiation. Phys. Rev. **96**, 555 (1954)
144. Ore, A.: Entropy of radiation. Phys. Rev. **98**, 887–888 (1955)
145. McQuarrie, D.A.: Statistical Mechanics. University Science, Sausalito, CA (2000)

146. Zhang, D., Closs, G.L., Chung, D.D., Norris, J.R.: Free energy and entropy changes in vertical and nonvertical triplet energy transfer processes between rigid and nonrigid molecules. A laser photolysis study. J. Am. Chem. Soc. **115**, 3670–3673 (1993)
147. Merkel, P.B., Dinnocenzo, J.P.: Thermodynamic energies of donor and acceptor triplet states. J. Photochem. Photobiol. A Chem. **193**, 110–121 (2008)
148. Connolly, J.S., Samuel, E.B., Franzen, A.F.: Effects of solvent on the fluorescence properties of bacteriochlorophyll a. Photochem. Photobiol. **36**, 565–574 (1982)
149. Warshel, A., Lappicirella, V.A.: Calculations of ground- and excited-state potential surfaces for conjugated heteroatomic molecules. J. Am. Chem. Soc. **103**, 4664–4673 (1981)
150. Slater, L.S., Callis, P.R.: Molecular orbital theory of the ^1L_a and ^1L_b states of indole. 2. An ab initio study. J. Phys. Chem. **99**, 4230–4240 (1995)
151. Stavenga, D.G., Smits, R.P., Hoenders, B.J.: Simple exponential functions describing the absorbance bands of visual pigment spectra. Vision Res. **33**, 1011–1017 (1993)

Fluorescence

5

5.1 The Einstein Coefficients

We have seen that light can excite molecules from their ground states to states with higher energies and can stimulate downward transitions from excited states to the ground state. But excited molecules also decay to the ground state even when the light intensity is zero. The extra energy of the excited molecule can be radiated as fluorescence, transferred to another molecule, or disippated to the surroundings as heat. In this chapter we consider fluorescence.

The rate constant for fluorescence can be related to the dipole strength for absorption by a line of reasoning that Einstein developed in the period 1914–1917 [1]. Consider a set of atoms with ground-state wavefunction Ψ_a and excited state wavefunction Ψ_b. Suppose that the atoms are enclosed in a box and are exposed only to the black-body radiation from the walls of the box. According to Eq. (4.8c), the rate at which the radiation causes upward transitions from Ψ_a to Ψ_b is

$$rate_{\uparrow} = \overline{|\boldsymbol{E}_o \cdot \boldsymbol{\mu}_{ba}|^2} \rho_\nu(\nu_{ba}) N_a / \hbar^2 \text{atoms cm}^{-3}\text{s}^{-1}, \tag{5.1}$$

where $\boldsymbol{\mu}_{ba}$ is the transition dipole for absorption, including nuclear factors, $\boldsymbol{E}_o$ is the amplitude of the oscillating electric field, $\rho_\nu(\nu_{ba})d\nu$ is the number of modes of oscillation of the field in a small interval of frequency $d\nu$ around the frequency of the optical transition (ν_{ba}), N_a is the number of molecules per cm^3 in state Ψ_a, and the bar over $|\boldsymbol{E}_o \cdot \boldsymbol{\mu}_{ba}|^2$ denotes averaging over all orientations of the molecules. Einstein cast this expression in terms of the energy density of the radiation, ρ, instead of $\boldsymbol{E}_o$ and ρ_ν. If the sample or the radiation field are isotropic (as the black-body radiation inside the box would be), so that the atoms are randomly oriented with respect to the field, the quantity $\overline{|\boldsymbol{E}_o \cdot \boldsymbol{\mu}_{ba}|^2} \rho_\nu(\nu_{ba})$ in Eq. (5.1) is $(2\pi f^2/3n^2)|\mu_{ba}|^2 \rho(\nu_{ba})$, where $\rho(\nu_{ba})$ is the energy density at ν_{ba} and f is the local-field correction factor (Eqs. 3.32, 3.37 and 4.13). The rate of upward transitions is, therefore,

W.W. Parson, *Modern Optical Spectroscopy*, DOI 10.1007/978-3-662-46777-0_5

$$rate_{\uparrow} = \left(2\pi f^2/3n^2\hbar^2\right)\left|\boldsymbol{\mu}_{ba}\right|^2\rho_{\nu}(\nu_{ba})N_a = \left(2\pi f^2/3n^2\hbar^2\right)D_{ba}\rho_{\nu}(\nu_{ba})N_a \tag{5.2a}$$

$$= B\rho(\nu_{ba})N_a. \tag{5.2b}$$

Equation (5.2b) simply defines a parameter B that is proportional to the dipole strength:

$$B \equiv \left(2\pi f^2/3n^2\hbar^2\right)D_{ba}. \tag{5.3}$$

B is called the *Einstein coefficient for absorption*. Because each transition from Ψ_a to Ψ_b removes an amount of energy $h\nu_{ba}$ from the radiation, the sample must absorb energy at the rate $B\rho(\nu_{ba})N_a h\nu_{ba}$. Equation (5.3) often is presented without the factors f^2 and n so that it refers to a sample in a vacuum.

Light also stimulates downward transitions from Ψ_b to Ψ_a, and from the symmetry of Eq. (4.7) the coefficient for this must be identical to the Einstein coefficient for upward transitions (B). If the excited atoms had no other way to decay, the rate of downward transitions would be

$$rate_{\downarrow} = B\rho(\nu_{ab})N_b, \tag{5.4}$$

where N_b is the number of atoms per cm^3 in the excited state. Let's restrict ourselves for now to an idealized atom that absorbs or emits at a single wavelength ($\nu_{ab} = \nu_{ba} = \nu$). The same value of ρ then will apply to upward and downward transitions: $\rho(\nu_{ba}) = \rho(\nu_{ab}) = \rho(\nu)$.

At equilibrium the rates of upward and downward transitions must be equal. But from Eqs. (5.2b) and (5.4), this would require that

$$N_b/N_a = 1, \tag{5.5}$$

no matter what the light intensity is. This can't be correct. At equilibrium, the number of atoms in the excited state is surely much smaller than the number in the ground state. There must be some other way for downward transitions to occur in addition to stimulated emission. Evidently, an excited atom also can emit light spontaneously (Fig. 5.1).

If spontaneous fluorescence occurs with a rate constant A (the *Einstein coefficient for fluorescence*), then the total rate of downward transitions will be

$$rate_{\downarrow} = [B\rho(\nu) + A]N_b. \tag{5.6}$$

The additional decay mechanism will decrease the ratio of N_b to N_a. At equilibrium, when the rates of upward and downward transitions are equal, we now find

$$N_b/N_a = B\rho(\nu)/[B\rho(\nu) + A]. \tag{5.7}$$

But we also know that, at thermal equilibrium:

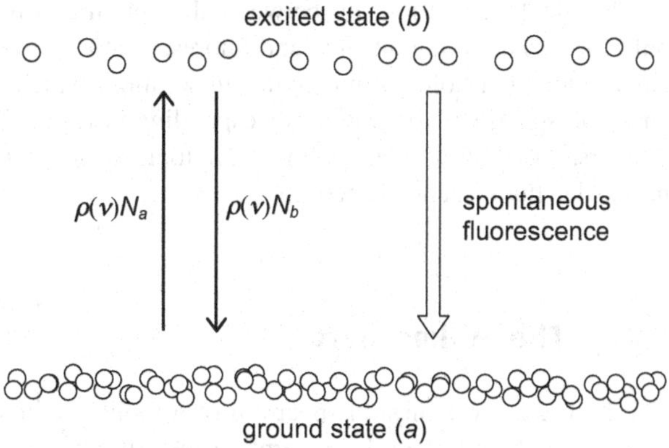

Fig. 5.1 The rate of absorption is proportional to $\rho(\nu)N_a$; the rate of stimulated emission is proportional to $\rho(\nu)N_b$. For the rates of upward and downward transitions to balance, downward transitions must occur by an additional mechanism that is independent of the light intensity. This is fluorescence

$$N_b/N_a = \exp[-(E_b - E_a)/k_BT] = \exp(-h\nu/k_BT), \tag{5.8}$$

where k_B is the Boltzmann constant and T the temperature. Combining Eqs. (5.7) and (5.8) reveals a relationship between A and B:

$$A/B = \rho(\nu)[1 - \exp(-h\nu/k_BT)]/\exp(-h\nu/k_BT). \tag{5.9}$$

Now let's use Planck's expression for the energy density of black-body radiation (Eq. 3.41a):

$$\rho(\nu) = \left(8\pi hn^3\nu^3/c^3\right)\exp(-h\nu/k_BT)/[1 - \exp(-h\nu/k_BT)]. \tag{5.10}$$

Substituting this expression for $\rho(\nu)$ in Eq. (5.9) gives

$$A = \left(8\pi hn^3\nu^3/c^3\right)B \tag{5.11}$$

$$= \left(8\pi hn^3\nu^3/c^3\right)\left(2\pi f^2/3n^2\hbar^2\right)D_{ba} = \left(32\pi^3 nf^2/3\hbar\lambda^3\right)D_{ba}. \tag{5.12}$$

Equation (5.12) says that the rate constant for spontaneous flourescence is proportional to the dipole strength for absorption. Strong absorbers are inherently strong emitters. But A also is inversely proportional to the cube of the fluorescence wavelength, so other things being equal, the fluorescence strength will increase as the absorption and emission move to shorter wavelengths. Although we have derived Eq. (5.12) for a system exposed to black-body radiation, the result should not depend on how the system actually is excited.

In his papers of 1914–1917, Einstein's actual line of reasoning was the reverse of the argument presented here. Einstein began with the assumption that an excited system can decay spontaneously as well as by stimulated emission. He also assumed that the relative populations of the ground and excited states follow the Boltzmann distribution (Eq. 5.8). With these assumptions, he obtained a simple derivation of the Planck black-body radiation law (Eqs. 3.41 and 5.10) and went on to show that absorption of light transfers momentum to the absorber.

The Einstein relationship between absorption and fluorescence is strictly valid only for a system that absorbs and fluoresces at a single frequency. This condition clearly does not hold for molecules in solution, which have broad absorption and emission spectra. Expressions corresponding to Eq. (5.12) can be derived for such systems, but before we do this let's look at some of the general features of molecular fluorescence spectra.

5.2 The Stokes Shift

The fluorescence emission spectrum of a molecule in solution usually peaks at a longer wavelength than the absorption spectrum because nuclear relaxations of the excited molecule and the solvent transfer some of the excitation energy to the surroundings before the molecule fluoresces. The red shift of the fluorescence is called the *Stokes shift* after George Stokes, a British mathematician and physicist who, in 1852, discovered that the mineral fluorspar emits visible fluorescence when it is illuminated with ultraviolet light. Stokes also described the red shift of the fluorescence of quinine, coined the term "fluorescence", and was the first to observe that a solution of hemoglobin changes from blue to red when the protein binds O_2. The Stokes shift reflects both intramolecular vibrational relaxations of the excited molecule and relaxations of the surrounding solvent. The contributions from intramolecular vibrations can be related to displacements of the vibrational potential energy curves between the ground and excited states. Figure 5.2 illustrates this relationship. Suppose that a particular bond has length b_g at the potential minimum in the ground state, and length b_e in the excited state. The *vibrational reorganization energy* (Λ_v) is the energy required to stretch or compress the bond by $b_e - b_g$. Classically, this energy is $(K/2)(b_e - b_g)^2$ where K is the vibrational force constant.

The quantum mechanical coupling strength (S) for a vibrational mode is defined as $\Delta^2/2$, where Δ is the dimensionless displacement of the potential surface in the excited state, $2\pi(m\upsilon/h)^{1/2}(b_e - b_g)$, and υ is the vibrational frequency (Eq. 4.42). We noted in Chap. 4 that when $|\Delta| > 1$, the Franck-Condon factors for absorption peak at an energy approximately $Sh\upsilon$ above the 0–0 transition energy (Fig. 4.21). The quantum mechanical reorganization energy for a strongly coupled vibrational mode thus is approximately $Sh\upsilon$, or $\Delta^2 h\upsilon/2$. If this is the only vibrational mode with a significant coupling strength, fluorescence emission will peak approximately $Sh\upsilon$ below the 0–0 transition energy, so the Stokes shift ($h\nu_{abs} - h\nu_{fl}$ in Fig. 5.2) will be roughly $2Sh\upsilon$, or $\Delta^2 h\upsilon$. If multiple vibrational modes are coupled to the transition, the vibrational Stokes shift is the sum of the individual contributions: $h\nu_{abs} - h\nu_{fl} \approx \sum_i |\Delta_i|^2 h\upsilon_i$ where Δ_i and υ_i are the displacement and frequency of mode i.

As we saw in Sect. 4.12, plots analogous to those in Fig. 5.2 also can be used to describe the dependence of the energies of the ground and excited states on a generalized solvent coordinate (Eqs. 4.57–4.59 and Fig. 4.28). Neglecting rotational energies, the overall reorganization energy is the sum of the vibrational and

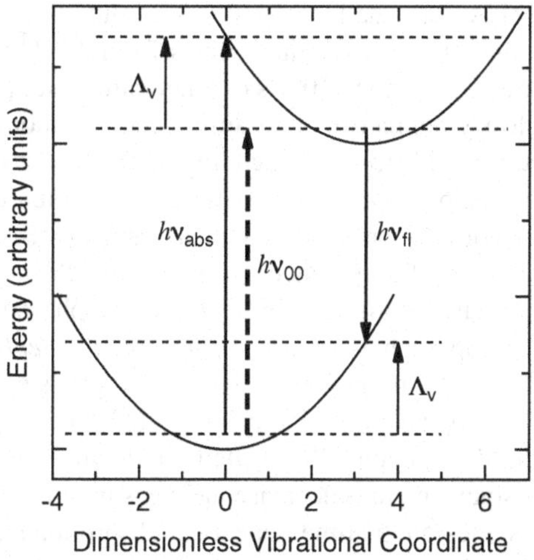

Fig. 5.2 Potential-energies and energy eigenvalues for a harmonic oscillator in ground and excited electronic states. The abscissa is the dimensionless vibrational coordinate $2\pi(m\upsilon/h)^{1/2}x$, where x is the Cartesian coordinate, m is the reduced mass of the vibrating atoms and υ is the classical vibration frequency. In this illustration, the vibrational frequencies are the same in the two states but the minimum is displaced by $\Delta = 3.25$ in the excited state. The vibrational reorganization energy (Λ_v) is indicated. (The reorganization energies of the ground and the excited state are the same if the vibrational frequency is the same in the two states.) The vibronic transitions labeled $h\nu_{abs}$ and $h\nu_{fl}$ have identical Franck-Condon factors and are displaced in energy above ($h\nu_{abs}$) and below ($h\nu_{fl}$) the 0–0 transition energy ($h\nu_{oo}$, *dashed arrow*) by Λ_v. The vibrational Stokes shift ($h\nu_{abs} - h\nu_{fl}$) is $2\Lambda_v$

solvent reorganization energies, and the total Stokes shift is the sum of the vibrational and solvent Stokes shifts.

The magnitude of the solvent Stokes shift usually increases with the polarity of the solvent, and decreases if the solvent is frozen. For example, the fluorescence emission spectra of indole in hexane, butanol and water peak near 300, 315 and 340 nm, respectively. The emission of tryptophan in proteins behaves similarly, usually peaking at longer wavelengths if the tryptophan is exposed to water and at shorter wavelengths if the tryptophan is located in a relatively nonpolar region of the protein. Shifts of tryptophan fluorescence to shorter or longer wavelengths thus can report on protein folding or unfolding. As discussed in Chap. 4, the 280-nm absorption band of the indole sidechain of tryptophan represents two excited states with similar energies. The ^1L_a state has the larger dipole strength and probably accounts for most of the fluorescence [2–4]. Because electron density moves from the pyrrole ring to the benzyl ring of the indole macrocycle when the ^1L_a state is created (Sects. 4.12 and 4.13 and Fig. 4.27), movements of nearby positively-charged species in this direction or movements of negatively charged species in the opposite direction will stabilize the excited state and shift the emission to the red.

Solvent Stokes shifts for small molecules in solution typically occur with multiphasic kinetics that stretch over time scales from 10^{-13} to 10^{-10} s. Similar multiphasic relaxations extending to 10^{-8} s or longer are seen in the fluorescence of proteins with a single tryptophan or with an exogenous fluorescent label that is sensitive to local electric fields [5–8]. Depending on the time range, changes in the emission spectrum on nanosecond or longer time scales often can be resolved by single-photon counting or frequency-domain measurements (Sect. 1.11), and faster components can be seen by fluorescence upconversion (Sect. 1.11), pump-probe measurements of stimulated emission (Fig. 11.7A), or photon-echo experiments (Sect. 11.4). Another approach is to use a *fluorescence-depletion* technique in which one measures the total flourescence when a sample is exposed to two short flashes of light separated by an adjustable delay [9–14]. The first flash prepares the excited state; the second can deplete this state by inducing stimulated emission or excitation to a higher state, and thus decrease the measured fluorescence. The effect of the second flash depends on the time-dependent absorption and emission spectra of the excited state, the frequencies and polarizations of the flashes, and the time between the two flashes. If the first flash excites the chromophore on the blue side of its absorption band, the fluorescence depletion caused by a flash at longer wavelengths will increase with time as relaxations move the excited chromophore's stimulated emission to the red. Applications of this technique in fluorescence microscopy are described in Sect. 5.10.

The bioluminescence of the Japanese firefly provides a striking illustration of a Stokes shift [15]. Bioluminescence occurs in a structurally diverse group of proteins called luciferases. Firefly luciferases bind a small molecule (luciferin), which reacts with O_2, ATP and Mg^{2+} to form the oxidized chromophore (oxyluciferin) in an excited electronic state. Emission occurs as the oxyluciferin decays to the ground state. The emission from the Japanese firefly normally is greenish yellow, peaking at 560 nm, but it shifts markedly to the red (605 nm) if a particular serine residue (S286) of luciferase is replaced by asparagine. The crystal structure of the S286N mutant protein bearing oxyluciferin is very similar to that of the wild-type protein. Crystal structures of the complexes with a model compound for an intermediate in the oxidation reaction, however, show significant differences. In the wild-type protein, an isoleucine side chain moves close to the chromophore, apparently providing a relatively rigid, hydrophobic environment that limits the size of the Stokes shift. In the mutant, the Ile side chain remains farther from the chromophore, leaving room for a larger relaxation before the excited chromophore emits.

5.3 The Mirror-Image Law

The fluorescence emission spectrum of a molecule usually is approximately a mirror image of the absorption spectrum, as illustrated in Fig. 5.3. Several factors contribute to this symmetry. First, if the Born-Oppenheimer approximation holds, and if the vibrational modes are harmonic and have the same frequencies in the ground and excited electronic states (all significant approximations), then the

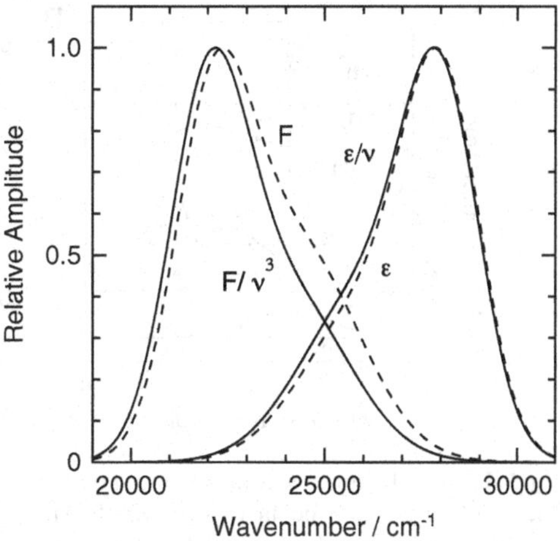

Fig. 5.3 When plotted against frequency (ν) and weighted by ν^{-3}, the normalized fluorescence emission spectrum usually is approximately a mirror image of the absorption spectrum weighted by ν^{-1}. The normalized spectra cross near the frequency of the 0–0 transition. The *solid curves* show fluorescence ($F\nu^{-3}$) and absorption ($\varepsilon\nu^{-1}$) spectra weighted in this way; the *dashed curves* are the unweighted fluorescence (F) and absorption (ε) spectra

energies of the allowed vibronic transitions in the absorption and emission spectra will be symmetrically located on opposite sides of the 0–0 transition energy, $h\nu_{oo}$. The solid vertical arrows in Fig. 5.2 illustrate such a pair of upward and downward transitions whose energies are, respectively, $h\nu_{oo} + 3h\upsilon$ and $h\nu_{oo} - 3h\upsilon$ where υ is the vibrational frequency. In general, for fluorescence at frequency $\nu = \nu_{oo} - \delta$, the corresponding absorption frequency is $\nu' = \nu_{oo} + \delta = 2\nu_{oo} - \nu$. Conversely, absorption at frequency ν' gives rise to fluorescence at $2\nu_{oo} - \nu'$.

Mirror-image symmetry also requires that the Franck-Condon factors be similar for corresponding transitions in the two directions, and this often is the case. Assuming again that the Born-Oppenheimer approximation holds and that the vibrational modes are harmonic and have fixed frequencies, the vibrational overlap integral for a downward transition from vibrational level m of the excited state to level n of the ground state is identical in magnitude to the overlap integral for an upward transition from level m of the ground state to level n of the excited state [16, 17]:

$$\left\langle \chi_{a(n)} \middle| \chi_{b(m)} \right\rangle = \pm \left\langle \chi_{b(n)} \middle| \chi_{a(m)} \right\rangle. \tag{5.13}$$

You can see this by examining the products of the harmonic-oscillator wavefunctions plotted in Fig. 5.4. Since $\chi_{b(2)}\chi_{a(0)}$ becomes superimposable on $\chi_{a(2)}\chi_{b(0)}$ if it is inverted with respect to the vibrational axis, the results of integrating the two products from $-\infty$ to ∞ over this axis must be identical. The product $\chi_{b(1)}\chi_{a(0)}$ becomes superimposable on $\chi_{a(1)}\chi_{b(0)}$ if it is inverted with respect to both the amplitude and vibrational axes, so their integrals over the vibrational coordinate must differ only in sign. The difference in sign disappears when the overlap integral is squared to form the Franck-Condon factor.

Corresponding upward and downward transitions with identical Franck-Condon factors thus occur at frequencies displaced equally on either side of ν_{oo}. But the

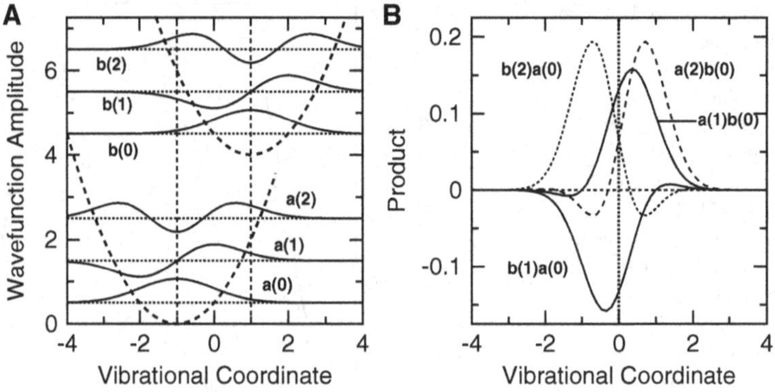

Fig. 5.4 (**A**) Vibrational wavefunctions for the ground and excited electronic states of a molecule. The *solid curves* are the first three wavefunctions for a harmonic oscillator with the same frequency in the two electronic states and a dimensionless displacement $\Delta = 2$ along the vibrational coordinate in the excited state ($\chi_{a(0)}, \chi_{a(1)}, \chi_{a(2)}, \chi_{b(0)}, \chi_{b(1)}$ and $\chi_{b(2)}$). The *bold, dotted curves* are the vibrational potential energies. The baseline for each wavefunction is shifted vertically by the vibrational energy; the energy of the excited electronic state relative to the ground state is arbitrary. (**B**) Products of the vibrational wavefunctions $\chi_{b(1)}\chi_{a(0)}, \chi_{a(1)}\chi_{b(0)}, \chi_{b(2)}\chi_{a(0)}$ and $\chi_{a(2)}\chi_{b(0)}$

emission strength at frequency ν also depends on $\nu^3 D_{ba}$ where D_{ba} is the electronic dipole strength (Eq. 5.12), whereas the molar absorption coefficient at frequency ν' depends on $\nu' D_{ba}$ (Eq. 4.15). Taking the factors of ν^3 and ν into account, the expected mirror-image relationship is

$$\frac{F(\nu_{oo} - \delta)}{[\nu_{oo} - \delta]^3} = \frac{F(\nu_{oo})}{\varepsilon(\nu_{oo})} \frac{\varepsilon(\nu_{oo} + \delta)}{[\nu_{oo} + \delta]} \tag{5.14}$$

for all δ. The factor $F(\nu_{oo})/\varepsilon(\nu_{oo})$ scales the emission and absorption spectra to the same amplitude at ν_{oo} and, if the mirror-image relationship holds, at their maxima. The spectra shown with solid lines in Fig. 5.3 are weighted by $\nu_{oo} + \delta$ or $[\nu_{oo} + \delta]^3$ as Eq. (5.14) prescribes.

In addition to requiring matching of the energies and Franck-Condon factors for corresponding upward and downward transitions, mirror-image symmetry requires the populations of the various vibrational sublevels from which downward transitions embark in the excited state to be similar to the populations of the sublevels where corresponding upward transitions originate in the ground state. This matching will flow from the similarity of the vibrational energies in the ground and excited states, provided that the vibrational sublevels of the excited state reach thermal equilibrium rapidly relative to the lifetime of the state. If the excited molecule decays before it equilibrates, the emission spectrum will depend on the excitation energy and typically will be shifted to higher energies than the mirror-image law predicts.

We have assumed that any coherence in the temporal parts of the individual vibrational wavefunctions is lost very rapidly relative to the lifetime of the excited state, so that we can treat the fluorescence from each vibrational level independently. This condition is assured if the vibrational levels have reached thermal equilibrium. We return to vibrational coherence in Chap. 11.

To examine how closely a molecule obeys the mirror-image relationship, the weighted absorption and emission spectra should be plotted on a an energy scale (frequency or wavenumber) as in Fig. 5.3, rather than on a wavelength scale. Fluorometers usually record signals that are proportional to $F(\lambda)\Delta\lambda$, which is the number of photons emitted per second in the wavelength interval $\Delta\lambda$ around wavelength λ. Because $d\nu/d\lambda = -c\lambda^{-2}$, switching an emission spectrum to the frequency scale requires the transformation $|F(\nu)\Delta\nu| = |cF(\lambda)\lambda^{-2}\Delta\lambda|$. Note that λ here is the wavelength of the emitted light passing through the detection monochromator (c/ν), not the wavelength in the solution ($c/n\nu$).

The mirror-image relationship can break down for a variety of reasons, including heterogeneity in the absorbing or emitting molecules, differences between the vibrational frequencies in the ground and excited states, and failure of the vibrational levels of the excited molecule to reacth thermal equilibrium.

5.4 The Strickler-Berg Equation and Other Relationships Between Absorption and Fluorescence

As we noted in Sect. 5.1, the Einstein relationship between absorption and fluorescence (Eq. 5.12) assumes that absorption and emission occur at a single frequency, which is not the case for molecules in solution. However, the overall rate of fluorescence by a molecule with broad absorption and emission bands can be related to the integrated absorption strength by expressions that were developed by Lewis and Kasha [18], Förster [19], Strickler and Berg [20], Birks and Dyson [16] and Ross [21].

Consider a set of molecules with ground and excited electronic states, a and b, in equilibrium with black-body radiation. Because the rates of absorption and emission at each frequency must balance, we can relate the rate constant for fluorescence at frequency ν ($k_{fl}(\nu)$) to the rate constant for excitation at this frequency ($k_{ex}(\nu)$) and the ratio of the populations of the two states (N_a and N_a):

$$k_{fl}(\nu) = k_{ex}(\nu)N_a/N_b. \tag{5.15}$$

Black-body radiation at moderate temperatures is so weak that stimulated emission is negligible compared to fluorescence. At thermal equilbrium, the population ratio is given by

$$\frac{N_a}{N_b} = \frac{Z_a}{Z_b}\exp\{h\nu_{oo}/k_BT\}, \tag{5.16}$$

where Z_a and Z_b are the vibrational partition functions of electronic states a and b, and $h\nu_{oo}$ is the energy difference between the lowest levels of the two state. The ratio Z_a/Z_b can differ from 1 if the energy of one or more vibrational modes are different in the two electronic states (Eqs. 4.48, B4.14.4a and B4.14.4b). A difference in the spin multiplicities (e.g., triplet versus singlet) also would cause Z_a/Z_b to differ from 1. We omitted the partition functions in deriving the Einstein relationship because we were considering a system with no vibrational or rotational sublevels, but a difference in spin multiplicity would have to be taken into account there as well.

In Chap. 4, we found an expression for the rate at which a material with a specified concentration (C) and molar absorption coefficient $\varepsilon(\nu)$ absorbs energy from a radiation field (Eq. 4.10). To obtain the rate constant for excitation at frequency ν in units of molecules per cm^3 per second ($k_{ex}(\nu)$), we need only divide that expression by the amount of energy absorbed on each upward transition ($h\nu$) and by the concentration of the absorber in molecules per cm^3 ($10^{-3}N_oC$, where N_A is Avogadro's number). This gives

$$k_{ex}(\nu) = \frac{10^3 \ln(10)}{N_A} \frac{\varepsilon(\nu)I(\nu)d\nu}{h\nu}, \tag{5.17}$$

where $I(\nu)$ is the irradiance in the interval $d\nu$ around frequency ν. Inserting the irradiance of black-body radiation (Eq. 3.41b), we have

$$k_{ex}(\nu)d\nu = \frac{8\pi 10^3 \ln(10)}{N_A} \left(\frac{n(\nu)\nu}{c}\right)^2 \frac{\exp(-h\nu/k_B T)}{1-\exp(-h\nu/k_B T)} \varepsilon(\nu)d\nu \tag{5.18a}$$

$$\approx \frac{8,000\pi \ln(10)}{N_A} \left(\frac{n(\nu)\nu}{c}\right)^2 \exp(-h\nu/k_B T)\varepsilon(\nu)d\nu. \tag{5.18b}$$

In Eq. (5.18b) we have assumed that $\exp[-h\nu/k_BT] \ll 1$, which is the case for absorption bands in the UV, visible or near-IR region at room temperature. (A 0–0 transition at 600 nm, for example, has $h\nu/k_BT = 24.4$ and $\exp[-h\nu/k_BT] = 2.6 \times 10^{-11}$ at 295 K.)

Combining Eqs. (5.15), (5.16) and (5.18b) gives the rate constant for fluorescence at frequency ν as a function of the absorption coefficient at the same frequency:

$$k_{fl}(\nu)d\nu = \frac{8,000\pi \ln(10)}{N_A} \frac{Z_a}{Z_b} \exp[h(\nu_{oo} - \nu)/k_B T] \left(\frac{n(\nu)\nu}{c}\right)^2 \varepsilon(\nu)d\nu. \tag{5.19}$$

To find the total rate constant for fluorescence (k_r) we could integrate Eq. (5.19) over the all the frequencies at which the molecule fluoresces. This is not very satisfactory, however, because most of the fluorescence occurs at frequencies below ν_{oo}, where $\varepsilon(\nu)$ is too small to measure acurately. A more practical procedure is to scale the measured fluorescence amplitude at each frequency relative to the fluorescence and ε at the 0–0 transition frequency, and then integrate over the fluorescence emission spectrum. The exponential factor then becomes 1, giving:

$$k_{fl}(\nu)d\nu = \frac{8,000\pi \ln(10)}{N_A}\frac{Z_a}{Z_b}\left(\frac{n(\nu_{oo})\nu_{oo}}{c}\right)^2 \frac{\varepsilon(\nu_{oo})}{F(\nu_{oo})}F(\nu)d\nu \tag{5.20}$$

and

$$k_r = \int k_{fl}(\nu)d\nu = \frac{8,000\pi \ln(10)}{N_A c^2}\frac{Z_a}{Z_b}\left(\frac{n(\nu_{oo})\nu_{oo}}{c}\right)^2 \frac{\varepsilon(\nu_{oo})}{F(\nu_{oo})}\int F(\nu)d\nu. \tag{5.21}$$

The rate constant k_r in Eq. (5.21) is a molecular analog of the Einstein *A* coefficient for spontaneous emission. As we reasoned in Sect. 5.1, this rate constant should not depend on how the excited electronic state is prepared, as long as the vibrational and rotational sublevels within the state reach thermal equilibrium among themselves.

Equation (5.21) provides a way to calculate the overall rate constant for fluorescence if we know the emission spectrum, the ratio of the vibrational partition functions, and the molar absorption coefficient at a reference wavelength. Apart from the problem of knowing Z_a/Z_b, its main shortcoming is the need to locate the 0–0 transition frequency accurately and to measure the fluorescence and ε here, where one or both of them can be weak. The fluorescence amplitude $F(\nu)$ can have an arbitrary magnitude but should have dimensions of photons s^{-1} cm^{-2} per unit frequency increment, not the dimensions of energy s^{-1} cm^{-2} per unit wavelength increment (fluorescence irradiance) that most fluorometers provide. The fact that the fluorescence amplitude can be scaled arbitrarily is significant because measuring the absolute intensity of fluorescence is technically difficult.

If the mirror-image law (Eq. 5.13) holds, an alternative approach is to recast Eq. (5.19) to give the rate constant for fluorescence at frequency $\nu_{oo} - \delta$ as a function of the absorption coefficient at frequency $\nu_{oo} + \delta$, and then to integrate over the absorption spectrum instead of emission. Letting the fluorescence frequency be $\nu' = 2\nu_{oo} - \nu$, this gives:

$$k_{fl}(\nu')d\nu = \frac{8,000\pi \ln(10)}{N_A}\frac{Z_a}{Z_b}\left(\frac{n}{c}\right)^2 \int \frac{\nu'^3\varepsilon(\nu')}{\nu}d\nu \tag{5.22}$$

and

$$k_r = \int k_{fl}(\nu')d\nu = \frac{8,000\pi \ln(10)n^2}{N_A c^2}\frac{Z_a}{Z_b}\left(\overline{\nu_f^{-3}}\right)^{-1}\int \frac{\varepsilon(\nu)}{\nu}d\nu, \tag{5.23}$$

where the factor $\left(\overline{\nu_f^{-3}}\right)^{-1}$ is the reciprocal of the average of ν^{-3} over the fluorescence emission spectrum,

$$\left(\overline{\nu_f^{-3}}\right)^{-1} = \int F(\nu)d\nu \Big/ \int \nu^{-3}F(\nu)d\nu. \tag{5.24}$$

As in Eq. (5.21), $F(\nu)$ in Eq. (5.24) has dimensions of photons $\cdot$ s^{-1} $\cdot$ cm^{-2} per unit frequency increment, but can be scaled arbitrarily. Note that the reciprocal of $\left\langle \nu_f^{-3} \right\rangle$ is not the same as the average of ν^3 ($\langle \nu_f^3 \rangle$), although the two quantities will converge if the emission spectrum is sufficiently sharp.

Except for the factor Z_a/Z_b, which Ross added later [21], Eq. (5.23) is the expression given by Strickler and Berg [20]. Strickler and Berg offered a somewhat different derivation that did not require mirror symmetry, but did assume that the ground and excited states have similar vibrational structure. The factor $\left\langle \nu_f^{-3} \right\rangle^{-1}$ has clearer physical significance in their derivation (Box 5.1).

Box 5.1 The ν^3 Factor in the Strickler-Berg Equation

In the derivation of Eq. (5.23) given in the text, the factor $\left(\overline{\nu_f^{-3}}\right)^{-1}$ is simply a mathematical consequence of the assumed mirror symmetry between the absorption and emission spectra [22]. The physical significance of this factor emerges more distinctly in the derivation given by Strickler and Berg [20], which does not require mirror symmetry.

Strickler and Berg expressed the total rate constant for fluorescence as a sum of rate constants for individual vibronic transitions. The contributions of the transitions from level n of the excited state to all vibrational levels of the ground state can be written as

$$k_{b,n\to a} = \sum_m k_{b,n\to a,m} \propto \sum_m (\nu_{b,n\to a,m})^3 \left|\langle \chi_{a,m} | \chi_{b,n} \rangle\right|^2 |\mu_{ba}|^2. \tag{B5.1.1}$$

Here $\langle \chi_{a,m} | \chi_{b,n} \rangle$ and $\nu_{b,n\to a,m}$ are the vibrational overlap integral and frequency for a transition from level n of the excited state to level m of the ground state and $|\mu_{ba}|^2$ is the electronic dipole strength (the square of $|\tilde{U}_{ba}|$ of Eq. (4.41)). Since the vibrational wavefunctions make up a complete set, Eq. (B5.1.1) can be divided by 1 in the form of $\sum_m \left|\langle \chi_{a,m} | \chi_{b,n} \rangle\right|^2$ to obtain

$$k_{b,n\to a} \propto \frac{\sum_m (\nu_{b,n\to a,m})^3 \left|\langle \chi_{a,m} | \chi_{b,n} \rangle\right|^2}{\sum_m \left|\langle \chi_{a,m} | \chi_{b,n} \rangle\right|^2} |\mu_{ba}|^2. \tag{B5.1.2}$$

Each of the terms in the numerator is proportional to the strength of one vibronic band in the emission spectrum, whereas each term in the

(continued)

Box 5.1 (continued)
demoninator is proportional to ν^{-3} times the strength of one vibronic band. Replacing the sums by integrals gives

$$k_{b,n\to a} \propto \frac{\int F(\nu)d\nu}{\int \nu^{-3}F(\nu)d\nu}|\boldsymbol{\mu}_{ba}|^2 = \left|\overline{\nu^{-3}}\right|^{-1}|\boldsymbol{\mu}_{ba}|^2. \tag{B5.1.3}$$

Inserting the numerical constants and summing over all the vibrational levels of the excited state gives Eq. (5.23).

The Strickler-Berg equation usually is written in terms of the wavenumber ($\bar{\nu} = 1/\lambda = \nu/c$) instead of the frequency:

$$k_r = \frac{8,000\pi \ln(10)cn^2}{N_A}\frac{Z_a}{Z_b}\left(\overline{\bar{\nu}_f^{-3}}\right)^{-1}\int\frac{\varepsilon(\bar{\nu})}{\bar{\nu}}d\bar{\nu}, \tag{5.25a}$$

$$= 2.880 \times 10^{-9}n^2\frac{Z_a}{Z_b}\left(\overline{\nu_f^{-3}}\right)^{-1}\int\frac{\varepsilon(\nu)}{\nu}d\nu, \tag{5.25b}$$

with

$$\left(\overline{\bar{\nu}_f^{-3}}\right)^{-1} = c^3\left(\overline{\nu_f^{-3}}\right)^{-1} = \int F(\bar{\nu})d\bar{\nu}\Big/\int \bar{\nu}^{-3}F(\bar{\nu})d\bar{\nu}. \tag{5.26}$$

The numerical factor in Eq. (5.25b) is for $\bar{\nu}$ in units of cm^{-1} and ε in $\text{M}^{-1}\ \text{cm}^{-1}$.

Equations (5.21) and (5.23–5.25b) give similar results for k_r when their underlying assumptions are valid. The Strickler-Berg equation generally appears to be somewhat more accurate than Eq. (5.21), having an error on the order of ±20% [17, 20–22]. The ratio of the partition functions that appears in both expressions (Z_a/Z_b) is rarely known independently, and usually is assumed to be unity. There are, however, cases in which the fluorescence rate constants obtained by the two methods agree reasonably well with each other, but are both significantly higher or lower than the experimental value. (We'll discuss experimental measurements of the fluorescence rate constant in Sect. 5.6.) Ross [21] suggested that these can be explained by departures of Z_a/Z_b from 1.

Figure 5.5 illustrates how an increase or decrease in vibrational frequency affects the Z_a/Z_b ratio. Note that changes in low-frequency modes are the most important in this context, because the vibrational partition function goes rapidly to 1 when the vibrational energy exceeds k_BT. At 295 K, k_BT is 205.2 cm^{-1}. Because the complete partition function is the product of the partition functions for all the different modes (Eqs. B4.14.4a and B4.14.4b), changes in the frequencies of two or

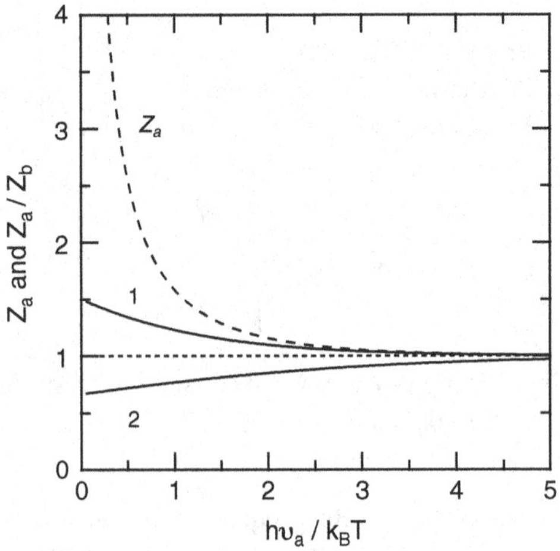

Fig. 5.5 Ratio of the vibrational partition functions (Z_a and Z_b) for the ground and excited states of a system with a single vibrational mode when the vibrational frequency in the excited state (υ_b) is 3/2 (*curve* 1) or 2/3 (*curve* 2) times the frequency in the ground state (υ_a). The abscissa is the vibrational energy in the ground state ($h\upsilon_a$) relative to k_BT. The *dashed curve* shows the ground-state partition function (Z_a)

more modes can either have a multiplicative effect or cancel, depending on whether they are in the same or opposite directions.

The mirror-image law and the fluorescence and absorption methods for obtaining k_r all assume that thermal equilibration of the excited molecule with the surroundings occurs rapidly relative to the lifetime of the excited state. If this assumption is valid, the ratio of fluorescence to absorbance at a given frequency should have a predictable dependence on temperature. Returning to Eq. (5.19) and collecting the terms that depend on ν, we see that

$$F(\nu)d\nu \quad \propto \quad \exp\{h[\nu_{oo} - \nu]/k_BT\}\left[\frac{n(\nu)\nu}{c}\right]^2 \varepsilon(\nu)d\nu, \tag{5.27}$$

so that

$$\ln(F(\nu)/\nu^2\varepsilon(\nu)) \quad \propto \quad -h\nu/k_BT. \tag{5.28}$$

A plot of $\ln[F(\nu)/\nu^2\varepsilon(\nu)]$ *vs* the frequency (ν) thus should be a straight line with slope $-h/k_BT$, where T is the temperature. This relationship was obtained independently by E.H. Kennard in 1918 and B.I. Stepanov in 1957, and is often called the *Kennard-Stepanov* expression. In many cases the slope of the plot turns out to be less than predicted, suggesting that the excited molecules emit before they reach thermal equilibrium with their surroundings. However, the "apparent temperature" derived from the slope is not very meaningful in this event because the relative populations of the vibrational levels of the excited state may not fit a Boltzmann distribution. The Kennard-Stepanov expression also can break down if the sample is inhomogeneous [23–25]. Plots of $\ln(F/\nu^2\varepsilon)$ *vs* ν sometimes show nonlinearities that may provide information on the dynamics of the excited state [25], but these features must be interpreted cautiously because they usually occur at frequencies where both F and ε are small.

We will return to the shapes of absorption and emission spectra in Chaps. 10 and 11.

5.5 Quantum Theory of Absorption and Emission

Although Einstein's theory accounts well for the relative amplitudes of absorption, fluorescence and stimulated emission, the notion that fluorescence occurs spontaneously is fundamentally inconsistent with the assertion we made in Chap. 2 that an isolated system is stable indefinitely in any one of its eigenstates. If the latter principle is correct, fluorescence must be caused by some perturbation we have neglected. The quantum theory of radiation provides a way out of this conundrum. As we discussed in Chap. 3, a radiation field has an eigenstate in which the number of photons is zero. Spontaneous fluorescence can be ascribed to perturbation of the excited molecule by the zero-point radiation field [26, 27]. Let's examine this rather unsettling idea.

We saw in Chap. 3 that electromagnetic radiation can be described by a vector potential $\boldsymbol{V}$ that is a periodic function of time and position, along with a scalar potential that can be made equal to zero by a judicious choice of the "guage" of the potentials. For linearly polarized radiation in a vacuum, the vector potential can be written

$$\boldsymbol{V} = \sum_j 2\pi^{1/2} c\ \hat{e}_j \{\exp[2\pi i(\nu_j t - \boldsymbol{k}_j \cdot \boldsymbol{r})] + \exp[-2\pi i(\nu_j t - \boldsymbol{k}_j \cdot \boldsymbol{r})]\} \quad (5.29a)$$

$$= \sum_j 2\pi^{1/2} c\ \hat{e}_j [q_j(t)\exp(-2\pi i \boldsymbol{k}_j \cdot \boldsymbol{r}) + q_j{}^*(t)\exp(2\pi i \boldsymbol{k}_j \cdot \boldsymbol{r})], \quad (5.29b)$$

where the unit vector $\hat{e}_j$ defines the polarization axis of oscillation mode j, and wavevector $\boldsymbol{k}_j$ has a magnitude of ν_j/c $(1/\lambda_j)$ and points in the direction of propagation of the oscillation of the mode (Eqs. 3.18 and 3.44). In Eq. (5.29b) we have replaced the time-dependent factor $\exp(2\pi i\nu_j t)$ and its complex conjugate by $q_j(t)$ and $q_j{}^*(t)$, respectively.

Now consider the interactions of a radiation field with an electron. In the absence of the radiation field, the Hamiltonian for the electron would be

$$\widetilde{\mathrm{H}} = \frac{1}{2m_e}\widetilde{\mathrm{p}}^2 + \widetilde{\mathrm{V}} = -\frac{\hbar^2}{2m_e}\widetilde{\nabla}^2 + \widetilde{\mathrm{V}}, \quad (5.30)$$

where $\widetilde{\mathrm{p}}$, $\widetilde{\nabla}$ and $\widetilde{\mathrm{V}}$ are the usual momentum, gradient and potential energy operators and m_e is the electron mass. In the presence of the radiation field, the Hamiltonian becomes

$$\widetilde{\mathrm{H}} = -\frac{1}{2m_e}\left(\widetilde{\mathrm{p}}^2 - \frac{e}{c}\boldsymbol{V}\right) + \widetilde{\mathrm{V}} \tag{5.31a}$$

$$= -\frac{\hbar^2}{2m_e}\widetilde{\nabla}^2 - \frac{\hbar e}{i2m_e c}\left(\widetilde{\nabla}\cdot\boldsymbol{V} + \boldsymbol{V}\cdot\widetilde{\nabla}\right) + \frac{e^2}{2m_e c^2}|\boldsymbol{V}|^2 + \widetilde{\mathrm{V}} \tag{5.31b}$$

$$= -\frac{\hbar^2}{2m_e}\widetilde{\nabla}^2 - \frac{\hbar e}{i m_e c}\boldsymbol{V}\cdot\widetilde{\nabla} + \frac{e^2}{2m_e c^2}|\boldsymbol{V}|^2 + \widetilde{\mathrm{V}}. \tag{5.31c}$$

Note that the vector potential combines vectorially with the momentum operator in Eq. (5.31a), rather than simply adding to the ordinary potential. The derivation of this formulation of the Hamiltonian is not straightforward, but the result can be justified by showing that it leads to the correct forces on a charged particle [26, 28, 29]. Combining $\widetilde{\nabla}\cdot\boldsymbol{V}$ and $\boldsymbol{V}\cdot\widetilde{\nabla}$ into a single term in Eq. (5.31c) is justified by the fact that $\widetilde{\nabla}$ and $\boldsymbol{V}$ commute, which follows from Eq. B3.1.16 [26].

By comparing Eq. (5.31c) with (5.30), expanding $\boldsymbol{V}$ as in Eqs. (5.29a) and (5.29b) and converting the functions $q_j(t)$ and $q_j^*(t)$ to operators, we see that the part of the Hamiltonian operator representing interactions of the electron with the field is

$$\widetilde{\mathrm{H}}' = -\frac{\hbar e}{i m_e c}\boldsymbol{V}\cdot\widetilde{\nabla} + \frac{e^2}{2m_e c^2}|\boldsymbol{V}|^2$$

$$= -\frac{e\hbar\pi^{1/2}}{i m_e}\sum_j\left(\hat{e}_j\cdot\widetilde{\nabla}\right)\left[\widetilde{\mathrm{q}}_j\exp(-2\pi i\boldsymbol{k}_j\cdot\boldsymbol{r}) + \widetilde{\mathrm{q}}_j{}^*\exp(2\pi i\boldsymbol{k}_j\cdot\boldsymbol{r})\right] \tag{5.32a}$$

$$+\left(\frac{2\pi e^2}{m_e c}\right)\sum_{j1}\sum_{j2}\left(\hat{e}_{j1}\cdot\hat{e}_{j2}\right)\left\{\widetilde{\mathrm{q}}_{j1}\widetilde{\mathrm{q}}_{j2}\exp\left[2\pi i\left(\boldsymbol{k}_{j1}+\boldsymbol{k}_{j2}\right)\cdot\boldsymbol{r}\right]\right.$$

$$+\widetilde{\mathrm{q}}_{j1}\widetilde{\mathrm{q}}_{j2}{}^*\exp\left[2\pi i\left(\boldsymbol{k}_{j1}-\boldsymbol{k}_{j2}\right)\cdot\boldsymbol{r}\right]$$

$$+\ \widetilde{\mathrm{q}}_{j1}\widetilde{\mathrm{q}}_{j2}{}^*\exp\left[2\pi i\left(\boldsymbol{k}_{j1}-\boldsymbol{k}_{j2}\right)\cdot\boldsymbol{r}\right]$$

$$\left.+\widetilde{\mathrm{q}}_{j1}{}^*\widetilde{\mathrm{q}}_{j2}{}^*\exp\left[-2\pi i\left(\boldsymbol{k}_{j1}+\boldsymbol{k}_{j2}\right)\cdot\boldsymbol{r}\right]\right\}. \tag{5.32b}$$

If the wavelengths of all the pertinent oscillation modes are long relative to the size of the chromophore holding the electron, so that $|2\pi i\boldsymbol{k}_j\cdot\boldsymbol{r}| \ll 1$, then $|\exp(\pm 2\pi i\boldsymbol{k}_j\cdot\boldsymbol{r})| \approx 1$ and Eq. (5.32b) simplifies to

$$\widetilde{\mathrm{H}}' \approx -\frac{e\hbar\pi^{1/2}}{im_e}\sum_j \left(\hat{e}_j\cdot\widetilde{\nabla}\right)\left\{\hat{\mathrm{q}}_j+\hat{\mathrm{q}}_j^*\right\} + \left(\frac{2\pi e^2}{m_e c}\right)\sum_{j1}\sum_{j2}\left(\hat{e}_{j1}\cdot\hat{e}_{j2}\right)\left\{\widetilde{\mathrm{q}}_{j1}\widetilde{\mathrm{q}}_{j2}+\widetilde{\mathrm{q}}_{j1}\widetilde{\mathrm{q}}_{j2}^*+\widetilde{\mathrm{q}}_{j1}^*\widetilde{\mathrm{q}}_{j2}+\widetilde{\mathrm{q}}_{j1}^*\widetilde{\mathrm{q}}_{j2}^*\right\}. \tag{5.33}$$

This description of the interaction energy is more general than the treatment we have used heretofore. The first line of Eq. (5.33) pertains to absorption or emission of a single photon, and the second line is for processes in which two photons are absorbed or emitted simultaneously. In the remainder of this chapter, we consider only one-photon processes. By setting the factor $\exp(\pm 2\pi i\boldsymbol{k}_j\cdot\boldsymbol{r})$ in Eq. (5.32b) equal to 1, we also continue to neglect interactions with the magnetic field of the radiation, as well as effects involving quadrupole and higher-order terms in the distribution of the charge. We return to the position-dependent factor $\exp(\pm 2\pi i\boldsymbol{k}_j\cdot\boldsymbol{r})$ in Chap. 9, and to two-photon processes in Chap. 12.

If the electron initially has wavefunction ψ_a and the radiation field wavefunction is ϑ_n, we can approximate the wavefunction for the system as a whole as the product

$$\Theta_{a,n} = \psi_a\vartheta_n. \tag{5.34}$$

We are interested in the rate of a transition to a new combined state $\Theta_{b,m}$, in which the molecule and the field have wavefunctions ψ_b and ϑ_m. Let's assume that the new radiation wavefunction ϑ_m differs from ϑ_n only in that oscillator j changes from $\chi_{j(n)}$ (level n_j) to $\chi_{j(m)}$ (level m_j), all the other oscillators being unchanged. The contribution of the field to the total energy then will change by $(m_j - n_j)h\nu_j$. We expect the transition to occur at a significant rate only if the total energy is approximately constant, so the change in the energy of the molecule must be approximately equal and opposite to the change in the energy of the field: $E_{b,m} - E_{a,n} \approx -(m_j - n_j)h\nu_j$.

Using the first line of Eq. (5.33), the interaction matrix element $\left\langle \psi_b\chi_{j(m)}\middle|\widetilde{\mathrm{H}}'\middle|\psi_a\chi_{j(n)}\right\rangle$ is

$$\left\langle \psi_b\chi_{j(m)}\middle|\widetilde{\mathrm{H}}'\middle|\psi_a\chi_{j(n)}\right\rangle = \frac{e\hbar\pi^{1/2}}{im_e}\left\langle \psi_b\chi_{j(m)}\middle|\left(\hat{e}_j\cdot\widetilde{\nabla}\right)\left\{\widetilde{\mathrm{q}}_j+\widetilde{\mathrm{q}}_j^*\right\}\middle|\psi_a\chi_{j(n)}\right\rangle. \tag{5.35}$$

The integral on the right side of this expression can be factored into the dot product of $\hat{e}_j$ with an integral involving $\widetilde{\nabla}$ and the electron wavefunctions, and a separate integral involving the radiation wavefunctions and $\widetilde{Q}_j$, where $\widetilde{Q}_j = \widetilde{\mathrm{q}}_j + \widetilde{q}_j^*$ as in Eq. (3.46):

$$\left\langle \psi_b\chi_{j(m)}\middle|\widetilde{\mathrm{H}}'\middle|\psi_a\chi_{j(n)}\right\rangle = -\frac{e\hbar\pi^{1/2}}{im_e}\left\langle \psi_b\middle|\hat{e}_j\cdot\widetilde{\nabla}\middle|\psi_a\right\rangle\left\langle \chi_{j(m)}\middle|\left\{\widetilde{\mathrm{q}}_j+\widetilde{\mathrm{q}}_j^*\right\}\middle|\chi_{j(n)}\right\rangle \tag{5.36a}$$

$$= -\frac{e\hbar\pi^{1/2}}{im_e}\hat{e}_j \cdot \left\langle \psi_b \middle| \tilde{\nabla} \middle| \psi_a \right\rangle \left\langle \chi_{j(m)} \middle| \tilde{Q}_j \middle| \chi_{j(n)} \right\rangle. \tag{5.36b}$$

In Chap. 4 we showed that, for exact wavefunctions, matrix elements of $\tilde{\nabla}$ are proportional to the corresponding matrix elements of the dipole operator $\tilde{\mu}$ (Sect. 4.8 and Box 4.10):

$$\left\langle \psi_b \middle| \tilde{\nabla} \middle| \psi_a \right\rangle = -\frac{2\pi m_e \nu_{ba}}{e\hbar}\boldsymbol{\mu}_{ba}. \tag{5.37}$$

In agreement with the semiclassical theory of absorption and emission, Eq. (5.36b) indicates that $\left\langle \psi_b \middle| \tilde{\nabla} \middle| \psi_a \right\rangle$ allows transitions between ψ_a and ψ_b only to the extent that it is parallel to the polarization of the radiation ($\hat{e}$).

What remains is to evaluate the factor $\left\langle \chi_{j(m)} \middle| \tilde{Q}_j \middle| \chi_{j(n)} \right\rangle$ in Eq. (5.36b). To do this, we need to write out eigenfunctions $\chi_{j(n)}$ and $\chi_{j(m)}$ of the radiation field explicitly. In Sect. 3.4 we showed that these are identical to the eigenfunctions of a harmonic oscillator with unit mass. For an oscillation mode with frequency ν,

$$\chi_{j(n)}(u_j) = N_{j(n)} H_n(u_j) \exp\left(-u_j^2/2\right), \tag{5.38}$$

where $n = 0, 1, \ldots,$ the dimensionless positional coordinate u_j is

$$u_j = Q_j\left(\hbar/2\pi\nu_j\right)^{-1/2}, \tag{5.39}$$

the $H_n(u_j)$ are the Hermite polynomials, and the normalization factor N_n is

$$N_{j(n)} = \left[\left(2\pi\nu_j/\hbar\right)^{1/2}/(2^n n!)\right]^{1/2} \tag{5.40}$$

(Eq. 2.30 and Box 2.5).

Simplifying the notation by dropping the subscript j and letting $\kappa = 2\pi\nu/\hbar$ so that $u = \kappa^{1/2} Q$, the matrix element we need is

$$\left\langle \chi_m \middle| \tilde{Q} \middle| \chi_n \right\rangle = \left\langle \chi_m \middle| Q\chi_n \right\rangle. \tag{5.41}$$

This integral can be evaluated by using the recursion formula for the Hermite polynomials (Box 2.5):

$$uH_n(u) = 1/2H_{n+1}(u) + nH_{n-1}(u), \tag{5.42}$$

According to this formula,

$$Q\chi_n = \kappa^{-1/2} u N_n \exp(-u^2/2) H_n \tag{5.43a}$$

$$= \kappa^{-1/2} N_n \exp(-u^2/2)\{(1/2)H_{n+1} + nH_{n-1}\} \tag{5.43b}$$

$$= \kappa^{-1/2} N_n \left\{ \frac{\chi_{n+1}}{2N_{n+1}} + n\frac{\chi_{n-1}}{N_{n-1}} \right\} = \{(n+1)/2\kappa\}^{1/2}\chi_{n+1} + \{n/2\kappa\}^{1/2}\chi_{n-1}. \tag{5.43c}$$

Incorporating this result in Eq. (5.41) gives

$$\left|\left\langle \chi_m \middle| \tilde{\mathrm{Q}} \middle| \chi_n \right\rangle\right| = \{(n+1)/2\kappa\}^{1/2}\langle \chi_m | \chi_{n+1}\rangle + \{n/2\kappa\}^{1/2}\langle \chi_m | \chi_{n-1}\rangle. \tag{5.44}$$

Because the harmonic-oscillator eigenfunctions are orthogonal and normalized, the integral $\langle \chi_m | \chi_{n+1}\rangle$ in Eq. (5.44) will be unity if $m = n+1$, and zero otherwise. $\langle \chi_m | \chi_{n-1}\rangle$, on the other hand, will be 1 if $m = n-1$, and zero otherwise. Equation (5.44) therefore can be rewritten as

$$\left|\left\langle \chi_m \middle| \tilde{\mathrm{Q}} \middle| \chi_n \right\rangle\right| = \{(n+1)/2\kappa\}^{1/2}\delta_{m,n+1} + \{n/2\kappa\}^{1/2}\delta_{m,n-1}, \tag{5.45}$$

where $\delta_{i,j}$ is the Kronecker delta function ($\delta_{i,j} = 1$ if and only if $i = j$).

The matrix element $\left\langle \chi_m \middle| \tilde{\mathrm{Q}}_j \middle| \chi_n \right\rangle$ thus consists of the sum of two terms, but only one of these can be non-zero for a given pair of wavefunctions. The first term on the right side of Eq. (5.45) applies if the number of photons increases from n to $n+1$. This term represents emission of light by the chromophore and is proportional to $[(n+1)/2\kappa]^{1/2}$. The second term, which applies if the number of photons decreases from n to $n-1$, represents absorption of light and is proportional to $(n/2\kappa)^{1/2}$. The absorption or emission strength depends on the square of the overall transition dipole, which according to Eq. (5.45) is proportional to $n/2\kappa$ for absorption or to $(n+1)/2\kappa$ for emission. The intensity of the radiation is proportional to n. Emission, therefore, includes both a component that depends on the light intensity (stimulated emission) and a component that is independent of the intensity (fluorescence).

An immediate corrolary of Eq. (5.45) is that $\left\langle \chi_n \middle| \tilde{\mathrm{Q}}_j \middle| \chi_n \right\rangle = 0$. Because the overall matrix element for a transition between two quantum states of a molecule (ψ_a and ψ_b) depends on the product $\left\langle \psi_b \middle| \tilde{\nabla} \middle| \psi_a \right\rangle \left\langle \chi_m \middle| \tilde{\mathrm{Q}}_j \middle| \chi_n \right\rangle$, light cannot cause such a transition unless one or more photons are absorbed or emitted ($m \pm n$).

The relationship described by Eq. (5.45) often is expressed in terms of photon *creation* and *annihilation* operators that convert wavefunction χ_n into χ_{n+1} or χ_{n-1}, respectively. The action of these operators on an oscillation mode is to increase or decrease the number of photons in the mode by one (Box 5.2).

Box 5.2 Creation and Annihilation Operators

We found in Eq. (5.43c) that the effect of the position operator $\widetilde{\mathrm{Q}}$ on χ_n is to generate a combination of χ_{n+1} and χ_{n-1}:

$$\widetilde{\mathrm{Q}}\chi_n = \{(n+1)/2\kappa\}^{1/2}\chi_{n+1} + \{n/2\kappa\}^{1/2}\chi_{n-1}, \tag{B5.2.1}$$

with $\kappa = 2\pi\nu/\hbar$. This transformation is the sum of the effects of $\widetilde{\mathrm{q}}^*$ and $\widetilde{\mathrm{q}}$, the two complex operators that together make up $\widetilde{Q}$ (Eqs. 3.46 and 5.36b).

To examine the action of $\widetilde{\mathrm{q}}^*$ or $\widetilde{\mathrm{q}}$ individually, we have to consider the effect of the radiation momentum operator $\widetilde{P}$ defined through Eq. (3.47), in addition to the effect $\widetilde{Q}$. By following an approach similar to that presented above for $\widetilde{Q}$ (Eqs. 5.41–5.44) and using the additional expression $P = (\partial/\partial Q)$, the action of $\widetilde{P}$ on χ_n is found to be:

$$\widetilde{\mathrm{P}}\chi_n = i\hbar\{\kappa(n+1)/2\}^{1/2}\chi_{n+1} - i\hbar\{\kappa n/2\}^{1/2}\chi_{n-1} \tag{B5.2.2}$$

[26]. The actions of $\widetilde{\mathrm{q}}^*$ and $\widetilde{\mathrm{q}}$ then can be obtained from Eqs. (3.46) and (3.47),

$$\widetilde{\mathrm{q}}^* = \widetilde{\mathrm{Q}} - \frac{i}{2\pi\nu}\widetilde{\mathrm{P}} = \widetilde{\mathrm{Q}} - \frac{i}{\kappa\hbar}\widetilde{\mathrm{P}}. \tag{B5.2.3}$$

$$\widetilde{\mathrm{q}}^*\chi_n = \{(n+1)/2\kappa\}^{1/2}\chi_{n+1}, \tag{B5.2.4}$$

$$\widetilde{\mathrm{q}} = \widetilde{\mathrm{Q}} + \frac{i}{2\pi\nu}\widetilde{\mathrm{P}} = \widetilde{\mathrm{Q}} + \frac{i}{\kappa\hbar}\widetilde{\mathrm{P}}, \tag{B5.2.5}$$

and

$$\widetilde{\mathrm{q}}\chi_n = \{n/2\kappa\}^{1/2}\chi_{n-1}. \tag{B5.2.6}$$

The operator $\widetilde{\mathrm{q}}^*$, which changes χ_n into $[(n+1)/2\kappa]^{1/2}\chi_{n+1}$, is called the *creation* or *raising operator* because it increases the number of photons in the radiation field by 1. The operator $\widetilde{\mathrm{q}}$ changes χ_n into $[n/2\kappa]^{1/2}\chi_{n-1}$ and is called the *annihilation* or *lowering operator*. Because the number of photons cannot be negative, $\widetilde{\mathrm{q}}$ acting on the zero-point wavefunction χ_0 is taken to give zero.

The creation and annihilation operators provide alternative forms for many quantum mechanical expressions, and they are used widely for phonons (vibrational quanta) as well as photons. For example, the Hamiltonian operator for an harmonic oscillator can be written

(continued)

Box 5.2 (continued)

$$\widetilde{\mathrm{H}} = (2\pi v)^2[\widetilde{q}\widetilde{q}^* + \widetilde{q}^*\widetilde{q}], \tag{B5.2.7}$$

which gives $\langle \chi_n | \widetilde{\mathrm{H}} | \chi_n \rangle = (n + 1/2)hv$ as you can see by evaluating the effects of performing the two operations in different orders:

$$\widetilde{\mathrm{q}}^*\widetilde{\mathrm{q}}\chi_n = \{n/2\kappa\}\chi_n = \frac{\hbar}{4\pi v} n\chi_n, \tag{B5.2.8}$$

and

$$\widetilde{\mathrm{q}}\widetilde{\mathrm{q}}^*\chi_n = \{(n+1)/2\kappa\}\chi_n = \frac{\hbar}{4\pi v}(n+1)\chi_n. \tag{B5.2.9}$$

Equations (B5.2.8) and (B5.2.9) also show that $\widetilde{q}^*$ and $\widetilde{q}$ do not commute with each other:

$$[\widetilde{\mathrm{q}}, \widetilde{\mathrm{q}}^*] = \widetilde{\mathrm{q}}\widetilde{\mathrm{q}}^* - \widetilde{\mathrm{q}}^*\widetilde{\mathrm{q}} = 1/2\kappa. \tag{B5.2.10}$$

5.6 Fluorescence Yields and Lifetimes

In Sect. 5.4, we derived expressions for the total rate constant for spontaneous fluorescence (k_r). The reciprocal of this rate constant is called the *radiative lifetime* (τ_r):

$$\tau_r = 1/A \equiv 1/k_r. \tag{5.46}$$

If the excited state decayed solely by fluorescence, its population would decrease exponentially with time and the time constant of the decay would be τ_r. However, other decay mechanisms compete with fluorescence, decreasing the lifetime of the excited state. The alternatives include formation of triplet states (*intersystem crossing*) with rate constant k_{isc}, nonradiative decay (*internal conversion*) to the ground state (k_{ic}), transfer of energy to other molecules (*resonance energy transfer*, k_{rt}), and *electron transfer* (k_{et}). These competing processes often are shown schematically in a diagram called a *Jablonski diagram* (Fig. 5.6).

Because rate constants for parallel processes add, the total rate constant for decay is

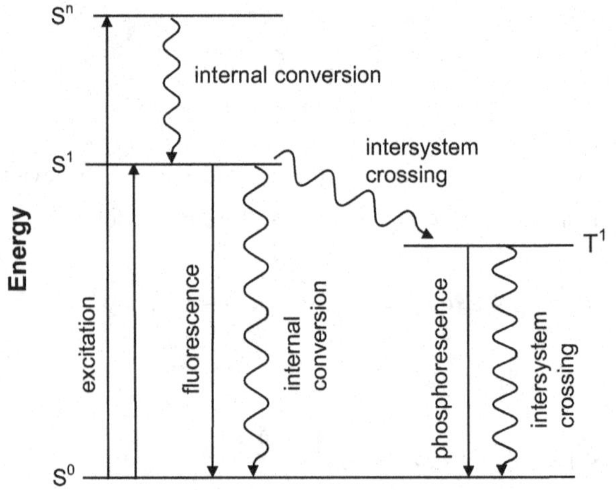

Fig. 5.6 A schematic representation of alternative pathways for formation and decay of excited states. Singlet states are labeled *S* and triplet states *T*; superscripts 0, 1, 2, ..., *n*, ... denote the ground state and excited states of increasing energy. Radiative processes (absorption, fluorescence, phosphorescence) are indicated with *solid arrows*; nonradiative processes (intersystem crossing, internal conversion, etc.), with *wavy arrows*. Internal conversion and intersystem crossing usually proceed via excited vibrational levels of the product state. Diagrams of this type were introduced by A. Jablonski in 1935 in a paper on the mechanism of phosphorescence [295]. The horizontal axis has no physical significance

$$k_{total} = k_r + k_{isc} + k_{ic} + k_{rt} + k_{et} + \cdots. \tag{5.47}$$

The actual lifetime of the excited state (the *fluorescence lifetime*,τ) thus is shorter than τ_r :

$$\tau = 1/k_{total} < 1/k_r. \tag{5.48}$$

In general, a sample will contain molecules that interact with their surroundings in a variety ways, for example because some of the fluorescing molecules are buried in the interior of a protein while others are exposed to the solvent. The fluorescence then decays with multiphasic kinetics that can be fit by a sum of exponential terms (Eq. 1.4). Fluorescence lifetimes can be measured by time-correlated photon counting, by fluorescence upconversion, or by modulating the amplitude of the excitation beam and measuring the modulation and phase shift of the fluorescence (Chap. 1). Pump-probe measurements of stimulated emission become the method of choice for sub-picosecond lifetimes (Chap. 11). For further information on these techniques and ways of analyzing the data see [30–34].

In most organic molecules, internal conversion from higher excited states to the lowest or "first" excited singlet state occurs much more rapidly than the decay from the lowest excited state to the ground state. The measured fluorescence thus occurs mainly from the lowest excited state even if the molecule is excited initially to a higher state (Fig. 5.6). This generalization is often called *Kasha's rule* after Michael Kasha, who first formalized it [35]. Kasha also pointed out that the rapid decay of higher excited states results in uncertainty broadening of their energies.

The *fluorescence yield* (ϕ) is the fraction of the excited molecules that decay by fluorescence (Sect. 1.11). For a homogeneous sample that emits exclusively from the first excited singlet state, this is simply the ratio of k_r to k_{total}:

$$\phi = \frac{photons\,emitted}{photons\,absorbed} = \frac{k_r}{k_{total}} = \frac{\tau}{\tau_r}. \tag{5.49}$$

The fluorescence yield from a homogeneous sample is, therefore, proportional to the fluorescence lifetime and can provide the same information. The fluorescence from a heterogeneous sample, however, can be dominated by the components with the longest lifetimes. Suppose, for example, that the fluorescing molecules in a sample all have the same radiative lifetime (τ_r), but are found in various environments with the fluorescence lifetime varying from site to site. The expression $\phi = \tau/\tau_r$ (Eq. 5.49) still holds in this situation, provided that the *mean fluorescence lifetime* is defined as

$$\tau = \int_0^\infty tF(t)dt \Bigg/ \int_0^\infty F(t)dt, \tag{5.50}$$

where $F(t)$ is the total fluorescence at time t. If the fluorescence decay kinetics in such a sample are described by a sum of exponentials, $F(t) = \sum A_i \exp(-t/\tau_i)$ where A_i is the amplitude of component i, then the integrated fluorescence from component i is $F_i = \int A_i \exp(-t/\tau_i)dt$, and the mean fluorescence lifetime is

$$\tau = \left(\sum_i \int_0^\infty tA_i\exp(-t/\tau_i)dt\right) \Bigg/ \left(\sum_i \int_0^\infty A_i\exp(-t/\tau_i)dt\right) = \left(\sum_i \tau_i^2 A_i\right) \Bigg/ \left(\sum_i \tau_i A_i\right). \tag{5.51}$$

Fluorescence can be quenched by a variety of agents, including O_2, I^- and acrylamide. Such quenching often involves electron transfer from the excited molecule to the quencher or vice versa, forming a charge-transfer complex that decays quickly by the return of an electron to the donor. The process depends on collisional contacts between the excited molecule and the quencher, and therefore usually is kinetically 1st-order in the concentration of the quencher ($[Q]$). The ratio of the fluorescence yields in the absence and presence of a quencher (ϕ_o/ϕ_q or F_o/F_q) is given by the *Stern-Volmer equation* [36]:

$$\begin{aligned}\phi_o/\phi_q &= F_o/F_q = (k_r/k_{total})/\{k_r/(k_{total}+k_q[Q])\} = (k_{total}+k_q[Q])/k_{total} \\ &= 1 + k_q[Q]/k_{total} = 1 + k_q\tau[Q],\end{aligned} \tag{5.52}$$

or $\phi_q = \phi_o/(1 + k_q\tau[Q])$. As in Eqs. (5.47–5.50), k_{total} here is the sum of the rate constants with which the excited state decays in the absence of the quencher and τ is the fluorescence lifetime in the absence of the quencher. The product $k_q\tau$ in Eq. (5.52) is called the *Stern-Volmer quenching constant* (K_Q):

$$K_Q \equiv k_q\tau. \tag{5.53}$$

According to Eq. (5.52), a plot of F_o/F_q versus $[Q]$ should give a straight line with a slope of K_Q and an intercept of 1, as shown by the solid line in Fig. 5.7B.

Measurements of K_Q can provide information on whether a fluorescent group in a protein is accessible to water-soluble quenchers in the solution or is sequestered in the interior of the protein. For example, tryptophan residues in different regions of a protein often exhibit very different quenching constants, particularly with ionic quenchers such as I^-. Reported values range from less than $5 \times 10^7\ M^{-1}\ s^{-1}$ for a deeply buried site in azurin to $4 \times 10^9\ M^{-1}\ s^{-1}$ for an exposed residue in a randomly coiled peptide [37]. Quenching of tryptophan fluorescence by phospholipids with attached nitroxide groups also can be used to study the depth of insertion of proteins in membranes [38, 39].

In some cases, quenching reflects formation of a complex between the fluorescent molecule and the quencher. If the formation and dissociation of the complex occur slowly relative to τ_r, the fluorescence decay kinetics may show two distinct components, a very short-lived fluorescence from molecules that are in complexes at the time of the excitation flash and a longer-lived fluorescence from free molecules. The decreased fluorescence of the complexes is termed *static quenching* to distinguish it from the *dynamic quenching* process assumed in Eq. (5.49). If the fluorescence yield from a complex is much lower than the yield from a free molecule, a plot of F_o/F_q versus $[Q]$ will still be linear, but the slope of the plot will be the association constant for the complex, rather than $k_q\tau$. A linear Stern-Volmer plot does not, therefore, imply that the quenching is dynamic. Static and dynamic quenching often can be distinguished by varying the viscosity of the solvent: dynamic quenching usually declines with increasing viscosity because it is limited by the rate of diffusion.

A mixture of static and dynamic quenching can result in a Stern-Volmer plot that curves upward as the concentration of the quencher is increased, as shown by the dotted curve in Fig. 5.7B. If we write $[C_T]/[C] = 1 + K_a[Q]$ where $[C]$ and $[C_T]$ are the concentrations of free and total chromophore and K_a is the association constant, and then let $F/F_q = (1 + k_q\tau[Q])[C_T]/[C]$ as in Eq. (5.52), we get a quadratic expression for the ratio of the fluorescence yields in the absence and presence of the quencher:

$$F_o/F_q = 1 + \left(K_q + K_a\right)[Q] + K_qK_a[Q]^2. \tag{5.54}$$

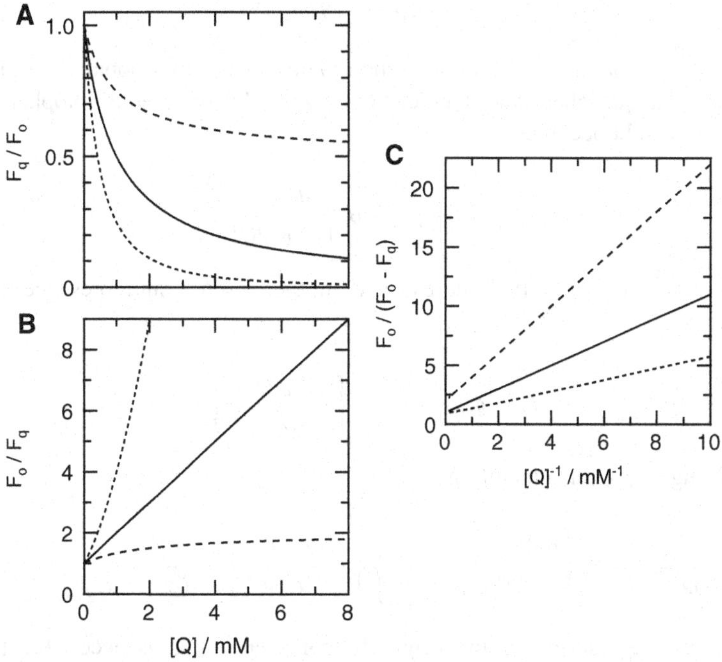

Fig. 5.7 Fluorescence measured in the presence (F_q) and absence (F_o) of a quencher, plotted as various functions of the concentration of the quencher ([Q]). The *solid curves* are for a homogeneous system in which all the fluorophores are accessible to the quencher, and the quenching is either purely dynamic or purely static. The Stern-Volmer quenching constant K_Q is taken to be 1 mM^{-1}. The *long dashes* are for an inhomogeneous system in which half the fluorophores are accessible to the quencher and half are not ($\zeta_a = 0.5$); the quenching again is either purely dynamic or purely static, K_Q is 1 mM^{-1}, and the rate constants for decay of the excited state in the absence of the quencher ($k_{total(a)}$ and $k_{total(i)}$) are assumed to be the same in the two pools. *Short dashes* are for a homogeneous system with an equal mixture of dynamic and static quenching; the K_Q for dynamic quenching and the association constant for static quenching (ζ_a) are both 1 mM^{-1}. In (**A**), F_q/F_o is plotted against [Q]; in (**B**) (a Stern-Volmer plot) F_o/F_q is plotted against [Q]; and in (**C**) (the modified Stern-Volmer plot) $F_o/(F_o - F_q)$ is plotted against 1/[Q]

In principle, both K_a and K_q can be obtained by fitting a plot of F_o/F_q versus [Q] to this expression.

Inhomogeneous systems can give Stern-Volmer plots that curve downward with increasing [Q], as shown by the dashed curve in Fig. 5.7B. In this situation, a *modified Stern-Volmer expression* is useful [40]. Consider a protein in which a fraction θ_a of the tryptophan residues is accessible to quenchers in the solvent and a fraction θ_i is inaccessible. In the absence of external quenchers, the overall fluorescence yield will be

$$\phi_o = \theta_i \phi_{o(i)} + \theta_a \phi_{o(a)}, \tag{5.55}$$

where $\phi_{o(i)}$ and $\phi_{o(a)}$ are the fluorescence yields in the two populations. In the presence of a quencher that interacts only with the exposed tryptophans, the fluorescence yield becomes

$$\phi_q = \theta_i \phi_{o(i)} + \theta_a \frac{\phi_{o(a)}}{1 + K_q[Q]}. \tag{5.56}$$

The difference between the fluorescence in the absence and presence of the quencher is

$$\phi_o - \phi_q = \theta_a \phi_{o(a)} \left\{ \frac{K_q[Q]}{1 + K_q[Q]} \right\}, \tag{5.57}$$

and dividing ϕ_o by this quantity gives

$$\frac{F_o}{\Delta F} = \frac{\phi_o}{\phi_o - \phi_q} = \left\{ \frac{\theta_i \phi_{o(i)} + \theta_a \phi_{o(a)}}{\theta_a \phi_{o(a)}} \right\} \left\{ 1 + \frac{1}{K_q[Q]} \right\} = \frac{1}{\zeta_a} + \frac{1}{\zeta_a K_q[Q]}, \tag{5.58}$$

where ζ_a is the fraction of the original fluorescence that is accessible to the quencher. A plot of $F_o/\Delta F$ versus $1/[Q]$ thus should give a straight line with an ordinate intercept of $1/\zeta_a$ and a slope of $1/\zeta_a K_q$, as shown by the dashed line in Fig. 5.7C.

The modified Stern-Volmer expression (Eq. 5.58) also gives a straight line in the case of a homogeneous system with either purely dynamic or purely static quenching (Fig. 5.7C, solid line). A mixture of dynamic and dynamic quenching gives a nonlinear plot, but the curvature may be apparent only at high quencher concentrations (Fig. 5.7C, dotted line). The slope and abscissa intercept of such a plot must, therefore, be interpreted cautiously and with comparisons to the curvature of an ordinary Stern-Volmer plot.

The sensitivity of fluoresence lifetimes and yields to interactions of chromophores with their surroundings makes fluorescence a versatile probe of macromolecular structure and dynamics. The fluorescence of tryptophan residues, which dominates the intrinsic fluorescence of most proteins following excitation in the region of 280 to 290 nm, is particularly well-suited for this, and is widely used to monitor protein folding and conformational changes [41–48]. The fluorescence yield varies from below 0.01 to about 0.35, depending on the local environment [42, 49–51]. The fluorescence of tryptophan 59 in cytochrome-*c*, for example, is quenched severely when the protein folds to a compact conformation that brings the tryptophan side chain close to the heme [52]. By contrast, the fluorescence yield of 3-methylindole in solution varies little in solvents ranging from hydrocarbons to water [53].

In addition to having a highly variable fluorescence yield, tryptophan often exhibits multiphasic fluorescence decay kinetics even in proteins that have only a single tryptophan residue. In general, the variable yield and multiphasic decay probably reflect variations in the rates of competing electron-transfer reactions in which an electron moves from the excited indole ring either to an amide group in the protein backbone or to the side chain of a neighboring residue. These reactions form charge-transfer (CT) states that decay quickly to the ground state. In addition to backbone amide groups, potential electron acceptors include cysteine thiol and disulfide groups, the amide groups of glutamine and asparagine side chains, protonated glutamic and aspartic acid carboxylic acid groups, and protonated histidine imidizolate rings [49, 51, 54–61]. The rate of electron transfer depends on the orientation and proximity of the indole ring relative to the electron acceptor, and often more importantly, on electrostatic interactions that modulate the energy of the CT state relative to the excited state [48, 50, 51, 58, 59, 62–69] (Box 5.3). A negatively charged aspartyl side chain, for example, could favor the electron-transfer reaction and thus decrease the fluorescence yield if it is located close to the indole ring, but have the opposite effect if it is near a potential electron acceptor [50]. Electron transfer to or from a tryptophan residue also can quench fluorescence of a dye attached to another residue in a protein, providing a probe of structural fluctuations that bring the two sites close together [70].

The electron-transfer reactions that quench tryptophan fluorescence can be prevented by modifying the indole side chain so as to lower the energy of the excited singlet state or raise the energy of the CT state. 7-Azatryptophan, in which the excited singlet state tends to be too low in energy to form a CT state, typically has a fluorescence lifetime of several hundred nanoseconds as compared to 1 to 6 ns for tryptophan [62, 65]. *Escherichia coli* incorporates added 7-azatryptophan into proteins with acceptable yields, and the substitution has relatively little effect on the activities of the enzymes that have been studied [71]. In proteins with a single 7-azatryptophan, fluorescence from the derivative is red-shifted by about 45 nm compared to that of the native protein. The fluorescence usually decays with a single time constant, although the emission spectrum remains highly sensitive to the polarity of the surroundings [65, 72]. Moving 7-azatryptophan from acetonitrile to water shifts the emission about 23 nm to the red, compared to the shift of 14 nm exhibited by tryptophan. 5-Fluorotryptophan, which has a higher ionization potential than tryptophan, also exhibits monoexponential fluorescence decay kinetics [69, 73]. To suppress its fluorescence in proteins, tryptophan can be replaced by 4-fluorotryptophan, which is essentially non fluorescent [74]. This can be useful if one is interested in the fluorescence of another component.

Lysine and Tyr side chains can quench tryptophan fluorescence by transferring a proton to the excited indole ring [59, 75]. The phenolate side chain of ionized tyrosine also can quench by resonance energy transfer and possibly by transferring an electron to the indole [51].

Fluorescence from tyrosine usually is strongly quenched both in proteins and in solution, although changes in this quenching have been used to monitor protein unfolding. Quenching of tyrosine fluorescence can reflect transfer of energy to tryptophan residues, proton transfer from the phenolic OH group of the excited tyrosine to other groups in the protein, or electron-transfer processes similar to those that occur with tryptophan [76–83]. The emission spectrum of tyrosine typically peaks near 300 nm and is less sensitive to solvent polarity than that of tryptophan, but shifts to the region of 335 nm in the tyrosinate anion [82]. Rapid quenching of fluorescence from a bound flavin has been ascribed to electron transfer to the flavin from a neighboring tyrosine residue in several proteins [84–89].

Box 5.3 Electron Transfer from Excited Molecules

Our understanding of the dynamics of electron-transfer reactions owes much to a general theory developed by R.A. Marcus [90–92]. In a classical picture, the free energies of the reactant and product states for electron transfer from donor D to acceptor A can be represented as parabolic functions of a distributed reaction coordinate that includes numerous solvent coordinates in addition to the intramolecular nuclear coordinates of D and A (Fig. 5.8A). The overall reorganization energy of the reaction (Λ) is the sum of contributions from both types of coordinates (Λ_s and Λ_v). If entropy changes in the reaction are negligible, the two parabolas will have the same curvature, making Λ the same for the reactant and product states. The reactant and product states here (DA and D^+A^-) are diabatic states, whose energies do not diagonalize the complete Hamiltonian of the system because they neglect electronic interactions between D and A (Sects. 2.3.6 and 8.1). Classically, electron transfer can occur only at the intersection of the two curves, where a transition from DA to D^+A^- requires no further change of geometry or energy. The activation free energy necessary to reach this point is $\Delta G^{\ddagger} = (\Delta G^{o} + \Lambda)^2/4\Lambda$, where ΔG^{o} is the free energy difference between the two minima (Figs. 4.28 and 5.8A and Eqs. 4.55 and 4.56). Transition-state theory then tells us that the probability of finding the reactants within a small energy interval (dE) around the transition point is

$$\rho\left(\Delta G^{\ddagger}\right)dE = (4\pi\Lambda k_B T)^{-1/2}\exp\left[-(\Delta G^{o} + \Lambda)^2/4\Lambda k_B T\right]dE. \qquad \text{(B5.3.1)}$$

The function $\rho(\Delta G^{\ddagger})$ is the *density of states* for which the reaction coordinate is the same for DA to D^+A^- and the energy change in the reaction (ΔE) is zero. Its pre-exponential factor, $(4\pi\Lambda k_B T)^{-1/2}$, normalizes the integrated probability of finding a given value of ΔE, which is a Gaussian function of ΔE (Eq. 4.56). The density of states has dimensions of reciprocal energy.

If the system has only a small chance of moving from the reactant to the product curve each time it passes through the transition point, the rate of electron transfer is simply

(continued)

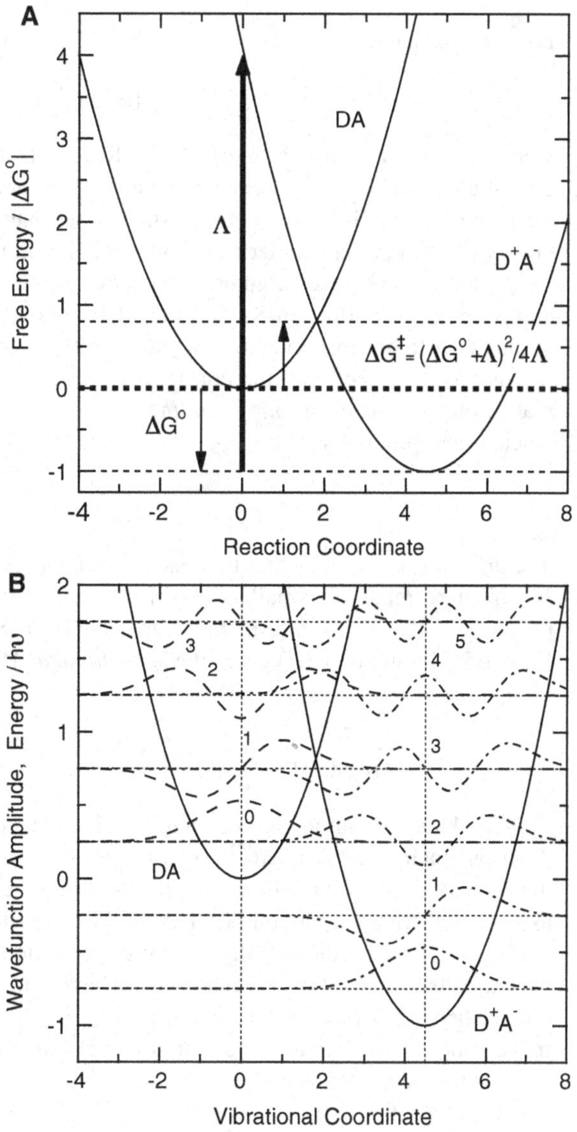

Fig. 5.8 (**A**) Classical parabolic free energy functions for diabatic reactant and product states of an electron-transfer reaction ($DA \rightarrow D^+A^-$) as functions of a dimensionless reaction coordinate. The standard free energy change (ΔG^0), reorganization energy (Λ), and activation free energy ($\Delta G^{\ddagger}$) are indicated (*vertical arrows*). The two functions are assumed to have the same curvature, which will be the case if there is no change in entropy in the reaction. (**B**) Potential energies (*solid curves*) and the first few vibrational wavefunctions (*dashed and dot-dashed curves*) and eigenvalues (*horizontal dotted lines*) for two electronic states (DA and D^+A^-), as functions of the dimensionless vibrational coordinate (u in Eq. 2.3.2) for a harmonic vibrational mode with the same frequency (υ) in the two states. The quantum number (n) of each wavefunction is indicated. The numbers on the ordinate and abscissa refer to DA. The potential minimum for D^+A^- is shifted down by $-h\upsilon$ and horizontally by $\Delta = 4.5$. The wavefunctions for both states are displaced vertically to align them with their eigenvalues as in Fig. 2.3A

Box 5.3 (continued)

$$k_{et} = k_o \rho(\Delta G^{\ddagger}), \tag{B5.3.2}$$

where k_o is a constant with units of s^{-1}. Reactions that meet this restriction are called *nonadiabatic* to indicate that a system reaching the transition point tends to remain in its initial diabatic state rather than following the adiabatic energy curve that diagonalizes the complete Hamiltonian.

Quantum mechanical arguments that we introduce in Sects. 2.5 and 4.2 and develop more fully in Sects. 7.2 and 10.4 concur that electron transfer requires ΔE to be close to zero. They also provide a quantitative relationship (*the golden rule*) between the constant k_o and the off-diagonal Hamiltonian matrix element or *electronic coupling factor* (H_{21}) that mixes the diabatic reactant and product states:

$$k_o = 2\pi |H_{21}|^2 / \hbar. \tag{B5.3.3}$$

The golden rule assumes that the reaction is nonadiabatic, which means that H_{21} must be relatively small. We will state this restriction somewhat more precisely below and return to it in detail in Sect. 10.4. Combining Eqs. (B5.3.2) and (B5.3.3) gives the *semiclassical Marcus equation*:

$$k_{et} = \frac{2\pi (H_{21})^2}{\hbar (4\pi \Lambda k_B T)^{1/2}} \exp\left[-(\Delta G^o + \Lambda)^2 / 4\Lambda k_B T\right]. \tag{B5.3.4}$$

The Marcus equation predicts that, for given values of H_{21} and Λ, $\log(k_{et})$ depends quadratically on $(\Delta G^0 + \Lambda)$. Because Λ is positive by definition, this means that the reaction will be slow if it is strongly exothermic ($\Delta G^0 \ll -\Lambda$), as well as if it is endothermic ($\Delta G^0 > 0$). The rate should be maximal when $\Delta G^0 = -\Lambda$. The prediction that making a reaction more exothermic can decrease its rate was unexpected when Marcus advanced it [91], and was not verified experimentally for many years [93–95]. The kinetics of electron transfer between reactants in solution can be complicated by formation of intermediate complexes and can be limited by the rate of diffusion of the reactants [96, 97].

To apply (B5.3.4) to electron transfer from excited tryptophan residues in proteins, Callis et al. incorporated calculations of H_{21}, ΔG^0 and Λ into hybrid classical-quantum mechanical molecular dynamics (MD) simulations [50, 68, 88, 98, 99]. (See Box 6.1 for an introduction to MD simulations, and Fig. 6.3 for an illustration of how calculated vibrational motions of a protein can affect the energy gap for electron transfer.) The calculated rate constants for electron transfer to backbone amide groups are in accord with the fluorescence quenching measured in many proteins, and suggest that variations in

(continued)

Box 5.3 (continued)
ΔG^0 usually are more important than differences in H_{21}. There are, however, proteins for which the calculated and measured fluorescence yields differ substantially. These cases could reflect the difficulty of calculating the energies accurately, or breakdowns of assumptions underlying the semiclassical Marcus equation.

The classical free energy curves in Fig. 5.8A neglect nuclear tunneling. Quantum mechanically, electron transfer can occur at points other than the intersection of the two potential energy surfaces, as long as the Franck-Condon factor for a pair of vibrational wavefunctions of the reactant and product electronic states is greater than zero. Figure 5.8B illustrates this for a harmonic vibrational mode that has the same frequency in the two states. The potential energy minimum for D^+A^- is shifted vertically so that its vibrational wavefunction with quantum number $n = 2\left(\chi_{D^+A^-(2)}\right)$ is isoenergetic with the DA wavefunction with $n = 0\left(\chi_{DA(0)}\right)$. The potential energy curves also are displaced along the vibrational coordinate so that they intersect approximately, but not exactly, at the energy of $\chi_{DA(1)}$ and $\chi_{D^+A^-(3)}$. Inspection of the figure shows that this pair of wavefunctions has a non-zero overlap integral, so that its Franck-Condon factor is non-zero. The Franck-Condon factors for $\chi_{DA(0)}$ with $\chi_{D^+A^-(2)}$, for $\chi_{DA(2)}$ with $\chi_{D^+A^-(4)}$, and for $\chi_{DA(3)}$ with $\chi_{D^+A^-(5)}$ also are non-zero. One consequence of tunneling between such wavefunctions is that the rate of electron transfer falls off less rapidly than Eq. (B5.3.4) predicts when $\Delta G^0 \ll -\Lambda$ [100]. The rate also does not necessarily go to zero at low temperatures.

Equations (B5.3.1) and (B5.3.4) also assume that the reacting system is in thermal equilibrium with its surroundings. This assumption can be problematic for a reactant that is created by photoexcitation. In addition, Eqs. (B5.3.3) and (B5.3.4) require electronic interactions of the donor and acceptor to be weak. The interaction matrix element H_{21} for electron transfer depends on the electronic orbital overlap of the reactants, which drops off rapidly as the intermolecular distance increases [101] but can be relatively strong for electron transfer from excited tryptophans to nearby residues and backbone amide groups in proteins [50, 51].

A more general approach is to write the electron-transfer rate constant as

$$k_{et} = \frac{2\pi}{\hbar}|H_{21}(t)|^2 \overline{\frac{\hbar\xi/2\pi}{|H_{21}(t)|^2 + \hbar\xi/2\pi}\rho(\Delta E(t))}. \qquad \text{(B5.3.5)}$$

[51]. Here $\Delta E(t)$ is the fluctuating energy difference between the diabatic reactant and product electronic states, as calculated during MD simulations of

(continued)

Box 5.3 (continued)
the system in the reactant state; $H_{21}(t)$ is the interaction matrix element at time t; $\rho(\Delta E(t))$ is a semi-empirical expression for the Franck-Condon-weighted density of product vibronic states that are degenerate with the reactant; ξ is the rate at which the product state moves out of resonance with the reactant by fluctuations of its energy or dissipation of energy to the surroundings; and the bar denotes an average over a suitable simulation period. The relaxation term ξ is assumed to be independent of time, although in principle it fluctuates along with H_{21} and ΔE. Equation (B5.3.5) reduces to the nonadiabatic limit (Eq. B5.3.3) when $\overline{|H_{21}(t)|^2} \ll \hbar\xi/2\pi$. In the opposite extreme, when $\overline{|H_{21}(t)|^2} \gg \hbar\xi/2\pi$, the rate constant goes to the adiabatic limit, $k_{et} = \overline{\xi\rho(\Delta E(t))}$, and no longer depends on H_{21}. Hybrid classical-quantum mechanical MD simulations using Eq. (B5.3.5) gave calculated rate constants that correlated well with fluorescence quenching in designed peptides containing a tryptophan in various surroundings [51]. The same procedure was used to treat changes in tryptophan fluorescence during unfolding of the headpiece of the protein villin [48].

Section 11.7 and Fig. 11.22 describe a density-matrix treatment that can be useful for electron transfer when information is available about the frequencies and displacements of the vibrational modes that are coupled to the reaction.

5.7 Fluorescent Probes and Tags

Fluorescent dyes have almost limitless applications as reporters for structural changes in proteins and other molecules, measuring intracellular components and tagging living cells for analysis or sorting. Figure 5.9 shows the structures of some commonly used dyes. Anilino-naphthalenesulfonate has been used extensively to probe protein folding because it binds nonspecifically to hydrophobic regions of proteins. Its fluorescence is strongly quenched in aqueous solution and usually increases markedly and shifts from green to blue when the dye binds to a protein [102, 103]. Similar shifts of emission with changes in the local environment occur with Prodan (6-propionyl-2-(dimethylamino)naphthalene) and related dyes [8, 104, 105]. Di-4-ANEPPS is one of a group of "voltage-sensitive" dyes that can be used to sense changes in membrane potentials in neurons and other electrically active cells. Changes in the membrane potential can alter the emission spectra or the fluorescence yields of these dyes by several mechanisms, including translation, reorientation or aggegregation of the dye, or electrochromic (Stark) effects [106–109]. Calcium Green 1 and related dyes are used to measure Ca^{2+} in tissues, cells and organelles. The fluorescence yield of Calcium Green 1 increases approximately tenfold when it binds Ca^{2+}. A great variety of other fluorescent molecules and

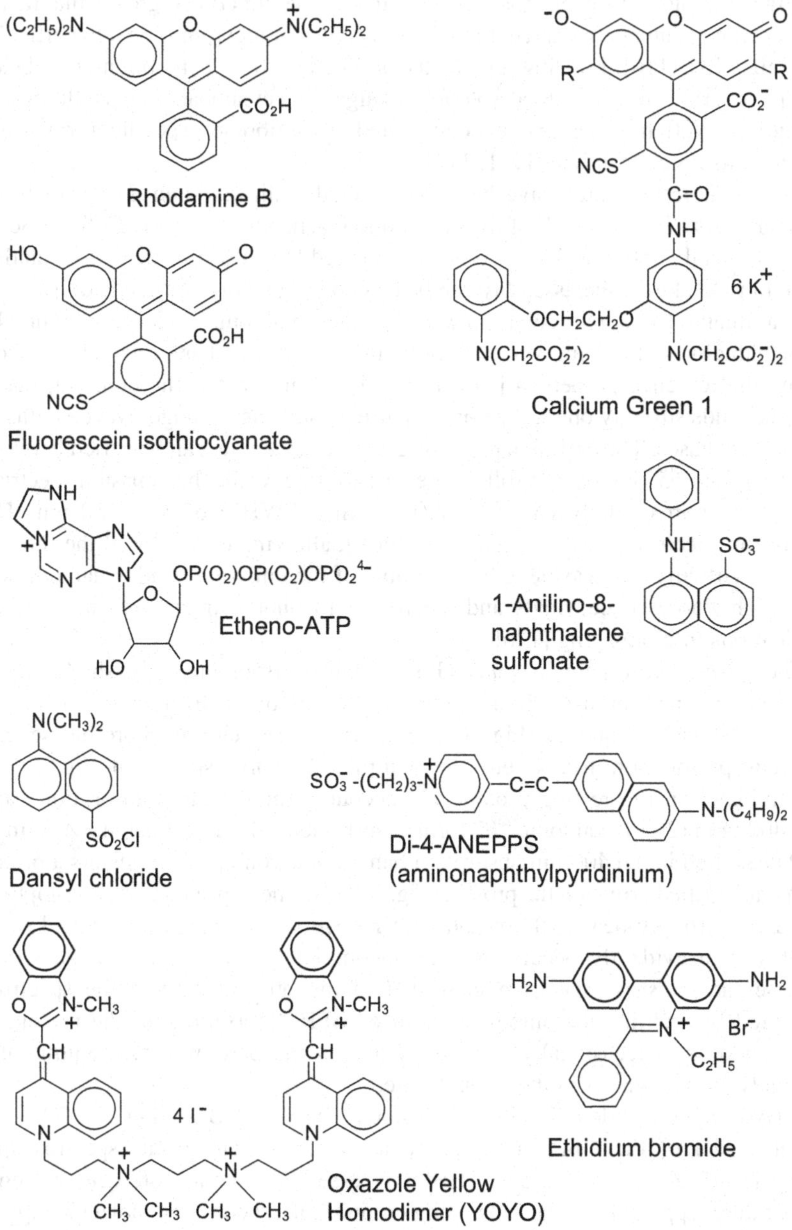

Fig. 5.9 Some common fluorescent dyes

substrate analogs have been synthesized with reactive functional groups that facilitate specific covalent attachment to macromolecules or incorporation into particular organelles [31, 110]. For flow cytometry and cell sorting, cells often are labeled with phycoerythrin, an antenna protein from algae and cyanobacteria that fluoresces strongly at 580 nm and can be conjugated to antibodies specific for various components on cell surfaces [111, 112].

Novel fluorescent tags have been constructed from *quantum dots*, which are "nanocrystals" or "clusters" of a semiconducting material such as CdS, CdSe or CdTe, typically 20 to 100 Å in diameter, embedded in a transparent, insulating medium [113, 114]. The assembly can be coated with a polymer that is derivatized with a ligand for a particular protein or other macromolecule. Confining the semiconductor's electronic wavefunctions in nanocrystals has remarkable effects on the fluorescence properties [115–119]. Most importantly, the emission maximum depends strongly on the size of the particle, shifting to longer wavelengths as the size increases. The emission peak for CdSe, for example, can be varied between about 525 and 655 nm by controlling the particle size, while the emission spectrum retains an approximately Gaussian shape with a FWHM of 30 to 50 nm. The absorption spectrum, by contrast, is very broad, allowing excitation of the fluorescence at virtually any wavelength to the blue of the emission band. Quantum dots have high fluorescence yields and are robust to photodamage, and have broad applications in bioimaging [120].

The *green fluorescent protein* (GFP) obtained from the jellyfish *Aequorea victoria* contains a built-in chromophore that forms by cyclization and oxidation of a Ser–Tyr–Gly sequence (Fig. 5.10) [121–125]. The chromophore has several different protonation forms, but the neutral "*A*" form shown in Fig. 5.10A predominates in the resting protein and accounts for the main absorption band near 400 nm [126]. An anionic "*B*" form absorbs near 475 nm. When the *A* form of GFP is excited in the 400-nm absorption band, the chromophore transfers a proton to an unidentified group on the protein (Fig. 5.10B). The deprotonated chromophore fluoresces strongly near 510 nm and then recovers a proton in the ground state. Fluorescence evidently occurs from an incompletely relaxed intermediate (I^*), because the emission spectrum is shifted to the blue relative to the spectrum generated by exciting the anionic *B* form of GFP. Relaxations of the protein in the excited state occasionally trap the system in the *B* form, which equilibrates relatively slowly with *A* in the ground state.

4-Hydroxybenzylidene-2,3-dimethyl-imidazolinone (HBDI, Fig. 5.10C) is a useful model of the GFP chromophore and shares many of its spectroscopic properties (Fig. 12.2). However, HBDI is essentially non-fluorescent at room temperature, apparently because twisting motions take the excited molecule rapidly to a configuration that favors internal conversion to the ground state [127–130]. The fluorescence yield increases about 1,000-fold at low temperatures, where the torsional distortions probably are blocked. The protein surrounding the chromophore must be relatively rigid to prevent similar distortions from occuring in GFP.

GFP enjoys wide use as a fluorescent marker because the gene encoding the protein can be fused with the gene for almost any other protein to provide a reporter

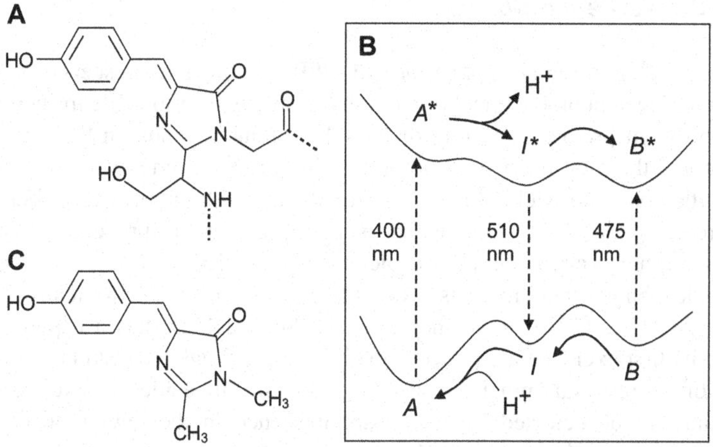

Fig. 5.10 (**A**) The chromophore of green fluorescent protein. This drawing shows one of several possible protonation states of the OH groups and the heterocyclic ring. *Dashed lines* indicate continuations of the protein main chain. (**B**) A schematic drawing of the potential energy surfaces of the main forms of the GFP chromophore. The neutral form (*A*) absorbs at 400 nm; the relaxed, anionic form (*B*), at 475 nm. Most of the green (510 nm) fluorescence comes from a deprotonated, but incompletely relaxed excited state (*I**). (**C**) 4-hydroxybenzylidene-2,3-dimethyl-imidazolinone (HBDI), a model for the GFP chromophore

for expression of the second gene [131–134]. Because the chromophore forms spontaneously after the fused protein is synthesized, the green fluorescence can be seen by microscopy without further additions. Variants of GFP that absorb and emit at shorter or longer wavelengths have been constructed [123, 133–141], and can be used in various combinations to determine the distance between the proteins to which they are fused (Chap. 7). Yellow variants (YFP's), for example, are obtained by combining the mutation S65G with T203F or T203Y along with substitutions at one to three other sites. Blue and cyan variants are obtained by substituting H, F or W for Y66, which forms the phenol group in GFP, in combination with additional mutations that increase the brightness and the stability of the protein. Red fluorescent proteins that are structurally homologous to GFP have been obtained from non-bioluminescent corals [142, 143], and have been optimized for brightness and stability by mutations [140, 141]. The "mCherry" variant, which absorbs maximally at 587 nm and emits at 610 nm, has been particularly useful. Another variant, PAmCherry, is non-fluorescent until it is activated by brief exposure to near-UV light. Sensors for cytosolic Ca^{2+} have been constructed by grafting calmodulin into a loop in GFP or a YFP [138, 144, 145]. There also is a remarkable variant of GFP (FP595) that becomes fluorescent in the red when it is illuminated with green light, and returns to a non-fluorescent state upon illumination with blue light [146–149]. An application of FP595 in ultra-high-resolution microscopy is discussed in Sect. 5.10.

5.8 Photobleaching

A limiting factor in many applications of GFP and other fluorescence probes is *photobleaching*, a photochemical process that converts the molecule irreversibly to a nonfluorescent product. The rate of photobleaching depends on the fluorescent molecule and the light intensity. Photobleaching of rhodamine dyes, for example, occurs with a quantum yield of only 10^{-7} to 10^{-6} at low light intensities, but with a much greater yield at high intensities, suggesting that the process is activated largely by higher excited singlet or triplet states [150, 151].

Photobleaching often involves oxidation by O_2 in an electronically excited singlet state [152, 153]. As we mentioned in Sect. 2.4, O_2 has a triplet ground state and its first excited state ($^1\Delta_g$) is a singlet state. Singlet O_2 can be formed by transfer of energy and spin from aromatic organic molecules in excited triplet states, and can be detected by its phosphorescence in the region of 1270 nm [154]. It reacts promiscuously at C=C double bonds to generate endoperoxides that rearrange into a variety of secondary photoproducts [155]. Photobleaching of some molecules can be minimized by removing O_2 from the solution or adding agents that quench triplet states [156–158].

Although photobleaching is a drawback in many uses of fluorescence, it can be turned to advantage for studying movements of biomolecules. In a technique termed *fluorescence recovery after photobleaching* (FRAP), fluorescent molecules in a small region of a sample are bleached by focusing a laser beam on the region for a short period of time (Fig. 5.11). The laser beam then is blocked, and fluorescence from the same region is measured as a function of time. Fluorescence reappears as unbleached molecules move into the focal region. A variation called *fluorescence loss in photobleaching* (FLIP) or *inverse FRAP* (iFRAP) is to monitor the fluorescence in a region that is not exposed to the bleaching beam. The fluorescence here will decrease with time if tagged molecules can move from this region to the

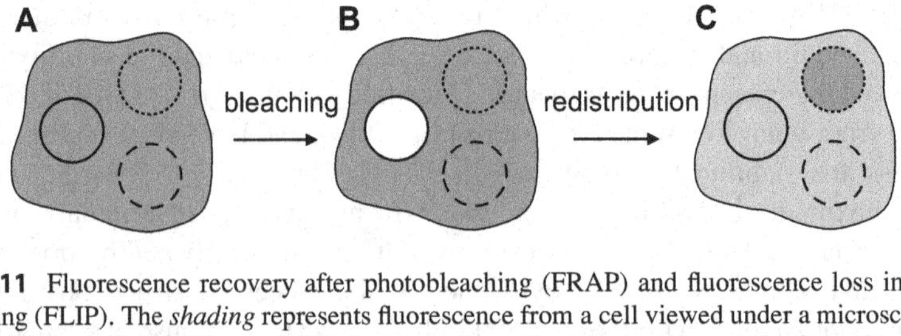

Fig. 5.11 Fluorescence recovery after photobleaching (FRAP) and fluorescence loss in photobleaching (FLIP). The *shading* represents fluorescence from a cell viewed under a microscope (**A**) before, (**B**) immediately after, and (**C**) at a longer time after a focused laser pulse is used to destroy the fluorophores in the region indicated by the *solid circle*. The fluorescence from the bleached region drops rapidly at the time of the bleaching and recovers with time as molecules migrate into this region from other parts of the cell. The *dashed circle* indicates a region that fluorescent molecules can leave to move to the bleached region; fluorescence here decreases gradually with time after the bleaching. The *dotted circle* indicates a region where the fluorescent molecules are confined or bound; the fluorescence here does not change

bleached region, but will remain constant if the tagged molecules cannot leave (Fig. 5.11).

FRAP was used initially to study the dynamics of lateral diffusion of lipids and proteins on cell surfaces [159–163]. It was combined with total internal reflection (Box 3.2 and Sect. 5.10) to study interactions of immunoglobulin fragments with planar bilayer membranes supported on glass surfaces [164] and binding of ligands to immobilized receptors [165]. Its applications expanded rapidly with the development of confocal microscopy (Sect. 5.10) and GFP tags [147, 166–168], and now include components of the nucleus [169, 170], mitochonrial matrix [171], endoplasmic reticulum [172], and Golgi apparatus [173].

5.9 Fluorescence Anisotropy

We now discuss the use of fluorescence to study rotational motions of molecules on a finer scale. Suppose we have a sample of randomly oriented molecules that we illuminate with linearly polarized light. Let the polarization be parallel to the laboratory's *z*-axis. The light will selectively excite molecules that have their transition dipole ($\boldsymbol{\mu}_{ba}$) oriented parallel to this same axis. However, molecules with off-axis orientations also will be excited with a probability that depends on $\cos^2\theta$, where θ is the angle from the *z*-axis (Eqs. 4.8a–4.8c).

The fraction of the molecules that have $\boldsymbol{\mu}_{ba}$ oriented with angle θ between θ and $\theta + d\theta$, and with ϕ between ϕ and $\phi + d\phi$, where ϕ is the angle of rotation in the *xy* plane, is proportional to the area element $\sin\theta\, d\theta\, d\phi$ on the surface of a sphere of unit radius (Box 4.6). The fraction of the excited molecules with this orientation, $W(\theta,\phi)d\theta d\phi$, is given by:

$$W(\theta,\phi)d\theta d\phi = \frac{\cos^2\theta \sin\theta d\theta d\phi}{\int_0^{2\pi} d\phi \int_0^{\pi} \cos^2\theta \sin\theta d\theta}. \tag{5.59}$$

The integral in the denominator, which simply counts all of the molecules that are excited, evaluates to $4\pi/3$, so

$$W(\theta,\phi)d\theta d\phi = (3/4\pi)cos^2\theta \sin\theta d\theta d\phi. \tag{5.60}$$

Now suppose that an excited molecule fluoresces. If absorption and emission involve the same electronic transition ($\Psi_a \leftrightarrow \Psi_b$) and the excited molecule does not change its orientation before it emits, the transition dipole for emission ($\boldsymbol{\mu}_{ba}$) will be parallel to the transition dipole for absorption ($\boldsymbol{\mu}_{ba}$). For each molecule in the sample, the probability that the emission is polarized along the *z*-axis (i.e., parallel to the excitation polarization) again depends on $\cos^2\theta$. Integrating over all possible orientations of the molecule gives:

$$\begin{aligned} F_{\parallel} &\propto \int_0^{2\pi} d\phi \int_0^{\pi} W(\theta,\phi) cos^2\theta d\theta \\ &= (3/4\pi) \int_0^{2\pi} d\phi \int_0^{\pi} \cos^4\theta \sin\theta d\theta = (3/4\pi)(4\pi/5) = 3/5. \end{aligned} \tag{5.61}$$

Similarly, the probability that a given molecule emits with polarization along the x-axis (i.e., perpendicular to the excitation) is proportional to $|\boldsymbol{\mu}_{ab} \cdot \hat{x}|^2$, which depends on $\sin^2\theta \cos^2\phi$. Integrating over all orientations again gives:

$$\begin{aligned} F_{\perp} &\propto \int_0^{2\pi} \cos^2\phi d\phi \int_0^{\pi} W(\theta,\phi) \sin^2\theta d\theta \\ &= (3/4\pi) \int_0^{2\pi} \cos^2\phi d\phi \int_0^{\pi} \cos^2\theta \sin^3\theta d\theta = 1/5. \end{aligned} \tag{5.62}$$

Comparing Eqs. (5.61) and (5.62) we see that the fluorescence polarized parallel to the excitation ($F_{\parallel}$) should be three times as strong as the fluorescence polarized perpendicular to the excitation ($F_{\perp}$). Two measures of this relative intensity have been used, the *fluorescence polarization* (P) and the *fluorescence anisotropy* (r):

$$P = \left(F_{\parallel} - F_{\perp}\right) / \left(F_{\parallel} + F_{\perp}\right) \tag{5.63}$$

and

$$r = \left(F_{\parallel} - F_{\perp}\right) / \left(F_{\parallel} + 2F_{\perp}\right). \tag{5.64}$$

In most cases, r is a more meaningful parameter than P. If a sample contains a mixture of components with different anisotropies, the observed anisotropy is simply the sum

$$r = \sum_i \Phi_i r_i, \tag{5.65}$$

where r_i is the anisotropy of component i and Φ_i is the fraction of the total fluorescence emitted by this component. Polarizations do not sum in this way. In addition, if the excited molecule can rotate, the time dependence of fluorescence anisotropy is simpler than the time dependence of the polarization. Although polarization was used in much of the early literature on fluorescence, most workers now use anisotropy.

The denominator in the definition of anisotropy ($F_{\parallel} + 2F_{\perp}$) is proportional to the total fluorescence, F_T, which includes components polarized along all three Cartesian axes: $F_T = F_x + F_y + F_z$. In our derivation of Eqs. (5.61) and (5.62), we excited

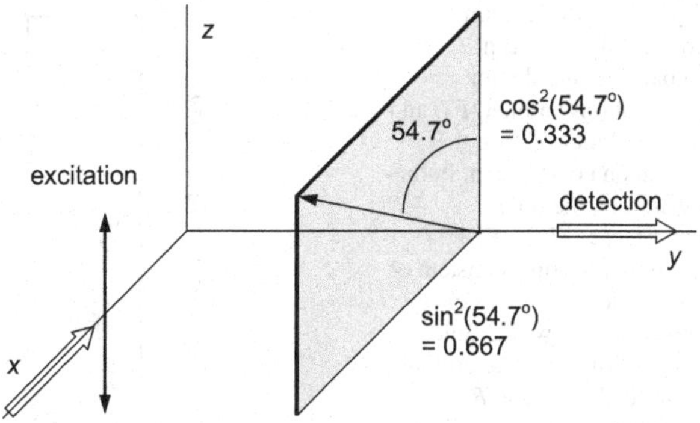

Fig. 5.12 If a sample is excited with light polarized parallel to the z axis, the total fluorescence is proportional to the fluorescence measured at right angles to the excitation through a polarizer at the "magic angle" of 54.7° with respect to z. This measurement weights z and x polarizations in the ratio 1:2

with z polarization and measured the fluorescence with polarizers parallel to the z and x axes, so $F_{\parallel} = F_z$ and $F_{\perp} = F_x$. Because the emission must be symmetrical in the xy plane, $F_y = F_x$. Thus $F_T = F_z + 2F_x = F_{\parallel} + 2F_{\perp}$. The total fluorescence also can be obtained by measuring the fluorescence through a polarizer set at the "magic angle" 54.7° from the z axis, as shown in Fig. 5.12. This is equivalent to combining z- and x-polarized measurements with weighting factors of $\cos^2(54.7°)$ and $\sin^2(54.7°)$, which have the appropriate ratio of 1:2. Fluorescence measured through a polarizer at the magic angle with respect to the excitation polarization is not affected by rotation of the emitting chromophore.

Inserting the values for $F_{\parallel}$ and $F_{\perp}$ from Eqs. (5.61) and (5.62) gives $r = 2/5$ for an immobile system. This is the maximum value of the anisotropy for a system with a single excited state. (In systems that have multiple excited states, the initial anisotropy could be as high as 1.0 at very early times, when an ensemble of excited molecules can include molecules that are in different eigenstates but have definite phase relationships between the time-dependent parts of their wavefunctions. We'll discuss such anomalously high anisotropies in Chap. 10. Interactions with the surroundings usually cause the wavefunctions to get out of phase within a few picoseconds so that r decays rapidly to 0.4 or lower.)

The observed anisotropy will be decreased if the emission and absorption transition dipoles are not parallel, or if the molecule rotates before it fluoresces. If the transition dipole for emission makes an angle ξ with respect to the transition dipole for absorption, $F_{\parallel}$, $F_{\perp}$ and r for an immobile molecule become (Box 10.5 and [174]):

$$F_{\parallel} = \left(1 + 2\cos^2\xi\right)/15, \tag{5.66}$$

$$F_{\perp} = \left(2 - \cos^2\xi\right)/15, \tag{5.67}$$

and

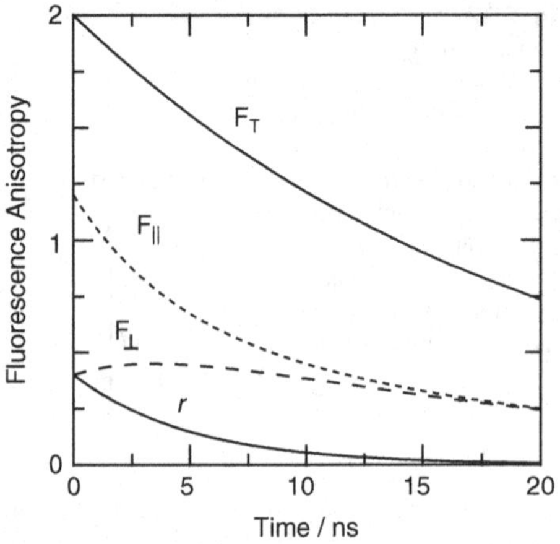

Fig. 5.13 Typical time courses of fluorescence signals measured with polarizations parallel ($F_{||}$) and perpendicular ($F_\perp$) to the excitation polarization. In this illustration the total fluorescence ($F_T = F_{||} + 2F_\perp$) decays with a time constant of 20 ns, and the fluorescence anisotropy (r) with a time constant of 5 ns. The ordinate scale for F_T, $F_{||}$ and $F_\perp$ is arbitrary

$$r' = \left(3\cos^2\xi - 1\right)/5. \tag{5.68}$$

These expressions come into play if the molecule relaxes to a different excited electronic state before it fluoresces. In bacteriochlorophyll, for example, the transition dipole for excitation to the second excited state (Q_x) is perpendicular to the transition dipole for the lowest state (Q_y). If a solution of bacteriochlorophyll in a viscous solvent such as glycerol is excited in the Q_x band at 575 nm, the emission from the Q_y band near 800 nm has an anisotropy of -0.2, which is the result given by Eq. (5.68) for $\xi = 90°$.

For an angle ξ of 45°, Eq. (5.68) gives $r' = 0.1$. An angle of 45° implies that the emission transition dipole, on average, has equal projections on the absorption transition dipole and on another axis that is orthogonal to the absorption dipole. An equivalent situation can arise if the excited molecule rapidly transfers its energy to an array of other molecules whose transition dipoles lie in a plane but take on all orientations within that plane. Some photosynthetic bacterial antenna systems contain such arrays of bacteriochlorophyll molecules. Following equilibration of the excitation over the array, fluorescence occurs with an anisotropy of 0.1 [175–177]. We'll return to the fluorescence anisotropy of such multimolecular systems in Chap. 10.

Now let's allow the excited molecule to rotate randomly so that the transition dipole explores all possible orientations. If the sample is excited with a single short flash at time $t = 0$, the initial fluorescence anisotropy immediately after the flash (r_o) will be 2/5 (or the value given by Eq. (5.68) if internal conversion to a different electronic state occurs very rapidly), but the anisotropy will decay to zero as the excited molecules rotate into new orientations (Fig. 5.13). By examining the decay kinetics of the anisotropy, we can learn how rapidly the molecule rotates and whether the rotational motions are isotropic or anisotropic.

If the fluorescing molecule is approximately spherical it will rotate more or less isotropically. The fluorescence anisotropy then will decay exponentially with a single time constant called the *rotational correlation time* (τ_c): $r = r_o\exp(-t/\tau_c)$. The rotational correlation time for a spherical molecule is given by the Debye-Stokes-Einstein relationship:

$$\tau_c = V\eta/k_BT, \tag{5.69}$$

where V is the hydrated volume, η the viscosity, and k_B the Boltzmann constant. Another "Stokes-Einstein expression" equates τ_c to $1/(6D)$, where D is the rotational diffusion coefficient. Thus, assuming that the emission and absorption transition dipoles are parallel, the time dependence of the fluorescence anisotropy for a spherical molecule is:

$$\begin{aligned} r &= \left[F_{\parallel}(t) - F_{\perp}(t)\right]/\left[F_{\parallel}(t) - 2F_{\perp}(t)\right] \\ &= r_o\exp(-t/\tau_c) = (2/5)\exp(-tk_BT/V\eta). \end{aligned} \tag{5.70}$$

(We'll discuss correlation functions more generally in Sect. 5.10 and in Chap. 10.)

If the molecule is more asymmetric its motions will be anisotropic and the fluorescence anisotropy usually will decay with time as a sum of exponential terms. The amplitudes and correlation times for these terms can provide information on the shape and flexibility of the molecule. Studying how the decay of the anisotropy depends on temperature and viscosity also can help to distinguish local motions of a fluorescent group from the slower tumbling of a macromolecule to which it is bound. For example, because of the greater restrictions on fluctuations of the structure, the anisotropy of fluorescence from tryptophan residues usually decays much more slowly in a folded protein than in an unfolded state. The anisotropy decay should speed up if a protein either unfolds or undergoes a conformational change that makes the overall structure more compact. Another example is the use of intercalated fluorescent dyes to explore bending motions of nucleic acids [178].

If a sample is excited with continuous light, so that the time dependence of the anisotropy is not resolved, an average value of the anisotropy will be measured. This is given by

$$r_{avg} = \left(\int_0^\infty r(t)F(t)dt\right) \Big/ \left(\int_0^\infty F(t)dt\right), \tag{5.71}$$

where $F(t)$ describes the time dependence of the total fluorescence probability for an excited molecule. For example, if $F(t) = F_o\exp(-t/\tau)$, and the molecule tumbles with a single rotational correlation time τ_c, then

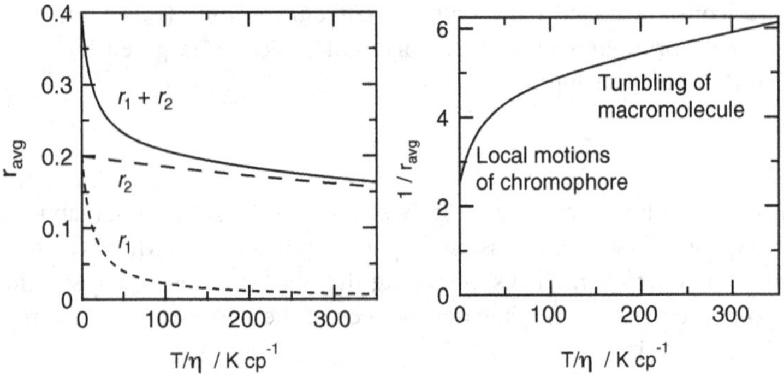

Fig. 5.14 Measurements of the average anisotropy (r_{avg}) of a chromophore bound to a macromolecule sometimes can distinguish local motions of the chromophore from tumbling of the entire macromolecule. Tumbling of the macromolecule freezes out at low temperature (T) or high viscosity (η). In this illustration, the overall anisotropy (r_{avg}) is the sum of two terms (r_1 and r_2) that make equal contributions to the initial anisotropy but have rotational correlation times differing by a factor of 100

$$r_{avg} = \frac{\int_0^\infty r_o \exp(-t/\tau_c) F_o \exp(-t/\tau) dt}{\int_0^\infty F_o \exp(-t/\tau) dt} = \frac{F_o r_o / (1/\tau + 1/\tau_c)}{F_o / (1/\tau)} = \frac{r_o}{1 + \tau/\tau_c}. \quad (5.72)$$

A plot of $1/r_{avg}$ vs. the fluorescence lifetime (τ) should be a straight line with a slope of $1/r_o\tau_c$ and an intercept of $1/r_o$:

$$1/r_{avg} = 1/r_o + (\tau/\tau_o\tau_c). \quad (5.73)$$

The initial anisotropy (r_o) and the rotational correlation time (τ_c) thus can be obtained by using a quencher such as O_2 to vary τ [179–181]. Note that a change in r_{avg} could reflect a change in either r_o, τ_c, or τ. For example, if the fluorescence lifetime is decreased by addition of a quencher, a measurement of r_{avg} will sample the anisotropy at earlier times after the molecules are excited, with the result that r_{avg} is closer to r_o.

In studies of small chromophores bound to macromolecules, plots of $1/r_{avg}$ vs T/η sometimes are biphasic as shown in the right-hand panel of Fig. 5.14. The anisotropy measured at high viscosity or low temperature (low T/η) is larger (i.e., $1/r_{avg}$ is smaller) than the value obtained by extrapolating measurements made at high T/η. This is because the two regions of the plot reflect different types of motions. At low T/η the macromolecule as a whole is essentially immobile but the bound chromophore may still have some freedom to move. The lower anisotropy seen at high T/η reflects the tumbling of the macromolecule as a whole. Time-resolved measurements would show that the anisotropy decay is multiphasic, with the faster component reflecting local motions of the chromophore and the slower component reflecting tumbling of the macromolecule.

If the chromophore can rotate such that the orientation of the emission dipole varies over an angle of ξ degrees, the fluorescence anisotropy will decay asymptotically from its initial value (r_o) to

$$r' = r_o \frac{3\overline{\cos^2\xi} - 1}{2}, \tag{5.74}$$

where $\overline{\cos^2\xi}$ is the average value of $\cos^2\xi$. The factor $\left(3\overline{\cos^2\xi} - 1\right)/2$, called the *order parameter*, is 1 for $\xi = 0$, and goes to zero for completely random motion when $\overline{\cos^2\xi} = 1/3$ (Box 4.6). Equation (5.74) is the same as Eq. (5.68) except that an average over all the accessible angles replaces a unique value of ξ. Box 10.5 discusses its origin in more detail.

In typical applications, fluorescence anisotropy has been used to study the effect of the amyloidogenic protein transthyretin on the fluidity of the plasma membrane of neuroblastoma cells [182] and changes in the mobility of various domains of a neurotoxin when the protein binds to the acetylcholine receptor [183].

5.10 Single-Molecule Fluorescence and High-Resolution Fluorescence Microscopy

Measurements of the fluorescence from a macroscopic sample provides information on average properties that may or may not resemble the properties of individual molecules. Biphasic fluorescence decay kinetics, for example, may reflect a heterogeneous population of molecules distributed between two different conformations or microenvironments. If the distribution is static on the nanosecond time scale, a measurement made on the macroscopic sample might allow us to deduce the relative populations of molecules in the two states, but generally would not say anything about how rapidly molecules change from one state to the other.

One way to dissect fluorescence from a heterogeneous sample is *fluorescence line narrowing*, in which the temperature is lowered to freeze out fluctuations of the solvent. The underlying idea is the same as in holeburning absorption spectroscopy (Sect. 4.11). A tunable dye laser with a narrow spectral bandwidth is used to excite the sample selectively and thus to pick out a subpopulation of molecules that absorb at a particular wavelength. The emission spectrum for the subpopulation can be much narrower than the spectrum for the ensemble as a whole, and it often shifts as the excitation wavelength is tuned over the inhomogeneous absorption band [184].

Techniques for detecting fluorescence have become sufficiently sensitive to measure the emission spectrum and fluorescence decay kinetics of individual molecules. There are a variety of ways that this can be done [185–188]. Small organic molecules can be frozen as dilute "guests" in a "host" solvent that forms a solid matrix at low temperatures, such as *p*-terphenyl. Molecules trapped in different environments can have sufficiently sharp absorption spectra so that individual molecules can be excited selectively [189–191]. A technique that can be applied to macromolecules is to flow a dilute solution through a capillary [192]. Molecules are excited as they flow past the focus of a pulsed laser. Fluorescence from individual molecules of the photosynthetic antenna protein phycoerythrin was studied in this way [193, 194]. However, this approach does not allow one to follow the fluorescence of the same molecule for an extended period of time.

In *near-field fluorescence spectroscopy and microscopy* [195–202], the excitation light is focused into a tapered optical fiber (Fig. 5.15A). The diameter of the spot of light at the tip of the fiber is determined by the size of the tip and can be as small as 15 nm, which is much less than the classical diffraction limit of about $\lambda/2$. If the fiber is translated laterally along a glass surface with a sparse coating of fluorescent molecules, the radiation at the tip will excite only the molecules within a small area at a given time. The fluorescence collected either through the same fiber or by a lens on the opposite side of the support can be analyzed until the molecule is destroyed by photobleaching [203].

Evanescent radiation created by total internal reflection (Box 3.2) also can be used to restrict excitation to a small region [204]. The excitation beam passes through a glass prism or lens to an interface with an aqueous medium as shown in Fig. 5.15B. The angle of incidence on the interface is made slightly greater than the critical angle set by the arcsin of the ratio of the refractive indices of the two media. If the aqueous phase is transparent, essentially all the radiation incident radiation is reflected back into the glass. However, the wave of evanescent radiation propagating along the interface penetrates into the water for a distance on the order of 500 nm and can excite fluorescence from molecules in this thin layer. In some applications, an objective lens with a large numerical aperture acts as the medium with the higher refractive index, and also serves to collect fluorescence emitted by the sample. Excitation schemes utilizing total internal reflection have been used to measure binding of fluorescent ligands to protein adsorbed on a glass surface [205–208], and to probe the orientation and motions of molecules in planar lipid bilayers deposited on such a surface [164, 165, 209].

An extension of the last technique is to coat the glass surface with gold or silver and tune the angle of incidence to the resonance angle for generating surface plasmons in the coating (Fig. 3.4E). As discussed in Box 3.2, the evanescent field associated with surface plasmons resembles that created by ordinary total internal reflection, but can be more than an order of magnitude stronger. Colloidal metal particles and lithographed surfaces patterned on a distance scale less than the wavelength of the light also can be used [210–213]. Although the fluorescence of the molecules close to the interface can be quenched by energy transfer and other interactions with the metal, some of these systems afford large increases in sensitivity [214].

In *confocal fluorescence microscopy* [215–219], an objective lens is used to form a magnified image of the light emitted from a small region (Fig. 5.15C). The image of light emitted from a circular aperture consists of concentric bright and dark rings around a central bright spot that is called the "Airy disk" after the English

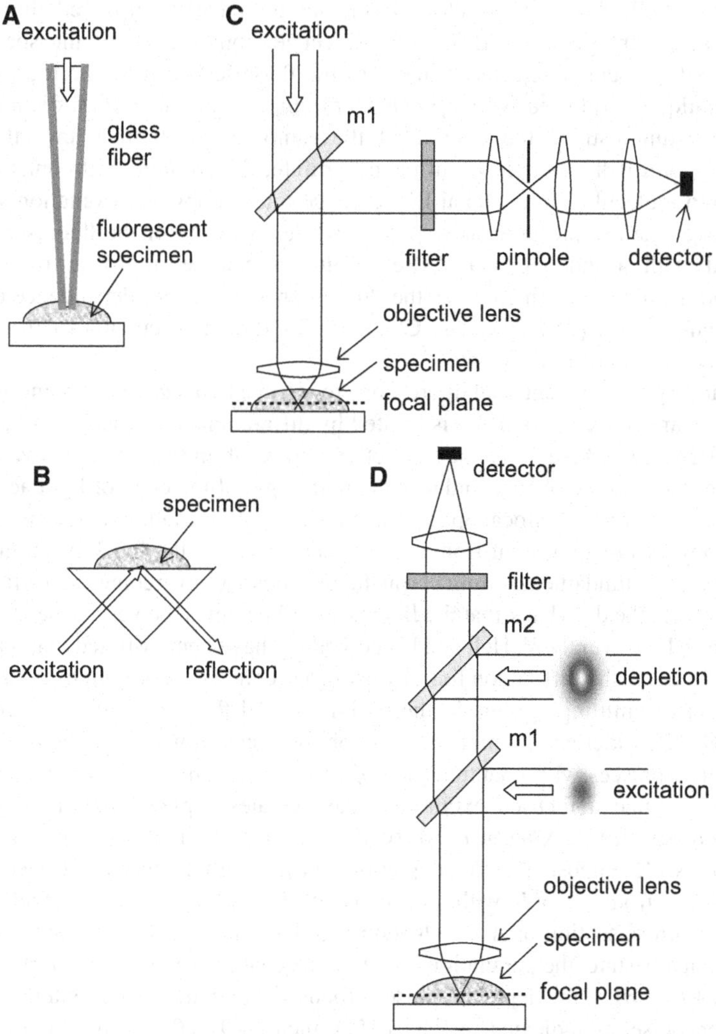

Fig. 5.15 Four ways of measuring fluorescence from small numbers of molecules. (**A**) Near-field fluorescence spectroscopy of individual molecules on a solid surface. The excitation light illuminates a small region close to the tip of a glass fiber. (**B**) Total internal reflection. Fluorescence is excited by the evanescent wave that penetrates only a short distance into the sample. (**C**) Confocal fluorescence spectroscopy and microscopy. An objective lens focuses the excitation light on the sample. Fluorescence is collected through the same lens, separated from the excitation light by a dichroic mirror (m1), refocused through a pinhole in the image plane, and transferred to a detector. The pinhole rejects florescence emitted by molecules above and below the focal plane in the sample (*dotted line*). (**D**) Fluorescence depletion microscopy. Excitation pulses from two synchronized lasers are focused through the same objective lens. One pulse (*excite*) raises fluorophores in the focal region to an excited state; the other (*deplete*) either returns the excited molecules to the ground state by stimulated emission, or converts the fluorophores to a form that does not absorb the excitation pulse. The two pulses are combined by dichroic mirrors (m1 and m2) and the depletion pulse is focused to a donut shape surrounding the excitation pulse

astronomer G.B. Airy (1801–1892). If the fluorescence is refocused through a pinhole in the image plane, diffuse fluorescence from regions of the specimen that are out of focus is rejected in favor of the fluorescence from the focal plain. This technique is referred to as *optical sectioning*. A pinhole with a diameter of 50 to 100 μm restricts the volume of the sampled region to a few attoliters (10^{-15} L), and if the sample is sufficiently dilute, this volume will contain only one or two molecules. In confocal fluorescence microscopy, the excitation spot is moved over the sample in a raster pattern while an image of the fluorescence is collected with a video camera. The resolution can be improved further by two-photon excitation, which gives the fluorescence a stronger dependence on the light intensity. See [220–225] and Chap. 12 for further discussion of these and related techniques of microscopy.

The ability of a conventional microscope to resolve two incoherent point sources of light separated by a distance d is limited by diffraction and generally is taken to be $d = \lambda/2n\sin\alpha$, where n is the refractive index of the medium between the objective lens and the objects and α is the half-angle of the cone of light accepted by the lens. Although confocal optics improve the spatial resolution of fluorescence microscopy by blocking light from objects that are out of the focal plane, they do not escape this fundamental limit. Near-field optics can overcome the diffraction limit, but with the drawback that the fluorescent object must be very close to the tip of the fiber. However, S.W. Hell and his coworkers have demonstrated that it also is possible to break the diffraction limit by using a second beam of light to deplete the population of emitting species in the region around the center of the excitation [226–231]. Consider an ensemble of fluorophores, each of which can be in either a form that fluoresces with quantum yield ϕ_F (A), or a form that is non-fluorescent (B). Assume that a focused excitation beam creates a population $A_E(\boldsymbol{r})$ of the fluorescent species peaking at $\boldsymbol{r} = 0$, so that the fluorescence from point $\boldsymbol{r} = 0$ is $\phi_F A_E(0)$. Now suppose that a "depletion" beam with intensity $I_D(\boldsymbol{r})$ converts fluorophores from A to B with rate constant $I_D(\mathbf{r})\sigma$, where σ is the effective absorption cross section of A. Let B return to A spontaneously with rate constant k_{BA}. In a steady state, the population of the fluorescent form at $\boldsymbol{r}$ will be $A(\boldsymbol{r}) = A_E(\boldsymbol{r}) k_{BA}/[k_{BA} + I_D(\mathbf{r})\sigma]$. If the depletion pulse is focused to a donut shape centered on the excitation pulse, as indicated in Fig. 5.15D, then $I_D(0) = 0$, and the fluorescence from $\boldsymbol{r} = 0$ is still $\phi_F A_E(0)$. But if $I_D(\boldsymbol{r})$ increases as $\boldsymbol{r}$ moves away from 0, so that $I_D(\boldsymbol{r})\sigma_D$ rapidly becomes much larger than k_{BA}, the fluorescence from regions away from $\boldsymbol{r} = 0$ is strongly attenuated.

As an illustration of the depletion technique, the solid curve in Fig. 5.16A shows an excitation function with a Gaussian dependence on position, $I_E(\boldsymbol{r}) = I_E^o \exp\left[-\left(|\boldsymbol{r}|/r_o\right)^2\right]$. The dotted curve shows the intensity of a depletion pulse with the shape $I_D(\boldsymbol{r}) = I_D^o \sin^2\left(\pi|\boldsymbol{r}|/4r_e\right)$. Panel B of Fig. 5.16 shows the functions $k_{BA}/[k_{BA} + I_D(\mathbf{r})\sigma]$ (dashed curve) and $I_E(\boldsymbol{r})k_{BA}/[k_{BA} + I_D(\boldsymbol{r})\sigma]$ (solid curve) for $I_D^o\sigma/k_{BA} = 10$, and panels C and D show the same for $I_D^o\sigma/k_{BA} = 100$ and 1,000. When $I_D^o\sigma \gg k_{BA}$, all the fluorescence comes from an arbitrarily small region around $\mathbf{r} = 0$, and the resolution has no theoretical limit. With depletion pulses of

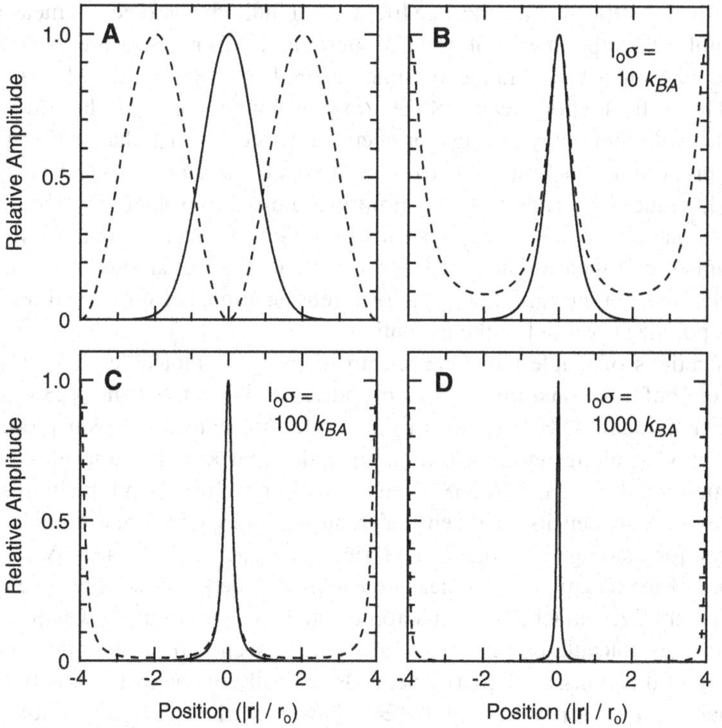

Fig. 5.16 Breaking the diffraction limit in fluorescence microscopy. (**A**) Relative intensities of a Gaussian excitation pulse $\left(I_E(\mathbf{r})/I_E^o = \exp\left[-(|\mathbf{r}|/r_o)^2\right]\right.$, *solid curve*) and a concentric, fluorescence-depleting pulse $\left(I_D(\mathbf{r})/I_D^o = sin^2(\pi|\mathbf{r}|/4r_o)\right)$, *dashed curve*), as functions of position relative to a focal point. I_o, r_o, and σ are arbitrary scale factors. (**B**), (**C**) and (**D**) The steady-state fraction of the fluorophores that are capable of fluorescing during the depleting pulse ($k_{BA}/[k_{BA}+I_D(\mathbf{r})\sigma]$, *dashed curves*), and the fluorescence amplitude ($I_E(\mathbf{r})k_{BA}/[k_{BA}+I_D(\mathbf{r})\sigma]$, *solid curves*), for $I_D^o\sigma/k_{BA} = 10$(**B**), 100 (**C**), and 1,000 (**C**)

intermediate intensities, the resolution limit is $d \approx \lambda/\left(2n\sin\alpha\sqrt{1+I_D/I_{D,sat}}\right)$, where $I_{D,sat}$ is the intensity that decreases the fluorescent population to $1/e$ [227].

One way to deplete the fluorescence, *stimulated emission depletion* (*STED*) microscopy, is to return the excited fluorophores to the ground state by stimulated emission as we discussed above in connection with temporal resolution of Stokes shifts (Sect. 5.2). Another approach is to use a variant of GFP that can be switched between fluorescent and non-flourescent states by light of different wavelengths (Sect. 5.7). This switching can be achieved with lower light intensities than are needed to induce stimulated emission [149, 229]. The essential requirements in either case are for the depletion pulse intensity to be zero at or near the focus of the excitation, and for the depletion process to approach saturation rapidly as the pulse intensity increases.

Emission spectra and fluorescence lifetimes of individual molecules measured in single-molecule experiments often vary substantially from molecule to molecule, and fluctuate on a broad range of time scales [186–188, 191, 192, 201, 232–234]. These fluctuations can reflect varying interactions of the fluorescing molecules with their surroundings, or chemical processes that change the spectroscopic properties, such as *cis-trans* isomerizations and transitions between oxidized, reduced or triplet states. Emitting molecules sometimes abruptly go dark, presumably as a result of irreversible photochemical reactions. In some cases, anisotropy measurements have shown that the fluctuations in the fluorescence yield are not due simply to rotational reorientation, because the fluorescence remains polarized parallel to the excitation.

Applications of single-molecule spectroscopy to biomolecules have included studies of conformational fluctuations of individual DNA molecules [233], hydrolysis of individual ATP molecules by a single molecule of myosin [204, 235], myosin moving along an actin rod [236], individual kinesin molecules moving along microtubules [235, 237, 238], interconversions of the bound flavin in cholesterol oxidase between its oxidized and reduced states [239], and elucidation of pathways for folding of proteins and ribozymes [234, 240–248]. A powerful extension of the technique is to measure resonance energy transfer between a pair of fluorescent dyes attached to a macromolecule. As we discuss in Chap. 7, excitation of one molecule can be followed by fluorescence from the other, and the efficiency of this transfer of energy depends critically on the distance between the donor and the acceptor. Willets et al. [249] have described a series of fluorophores that have the high fluorescence yields needed for single-molecule studies, but also are strongly sensitive to the local environment.

The 2014 Nobel Prize in Chemistry recognized the pioneering contributions of Eric Betzig [196–198, 202, 250], William Moerner [187, 189, 190, 235, 249] and Stefan Hell [226–230] to single-molecule fluorescence spectroscopy and high-resolution microscopy.

5.11 Fluorescence Correlation Spectroscopy

Fluctuations of the fluorescence as an individual molecule hops between different states or experiences varying interactions with the surroundings are averaged out in a conventional measurements of fluorescence from large populations of molecules. Between these extremes, fluorescence from a small number of molecules can fluctuate in a way that provides information on the dynamics of the individual molecules [215, 251–257]. Figure 5.17 illustrates some of the features of such fluctuations. In the model considered here, the system contains a specified number of molecules (N), each of which undergoes reversible transitions between a fluorescent ("*on*") a nonfluorescent ("*off*") state with first-order kinetics. If a molecule is *on* at a given time, the probability that it remains *on* after a short time interval Δt is $\exp(-k_{off}\Delta t)$; if the molecule is initially *off*, the probability of remaining *off* after time Δt is $\exp(-k_{on}\Delta t)$. The equilibrium constant is $K_{eq} = [on]_{eq}/[off]_{eq} = k_{on}/k_{off}$.

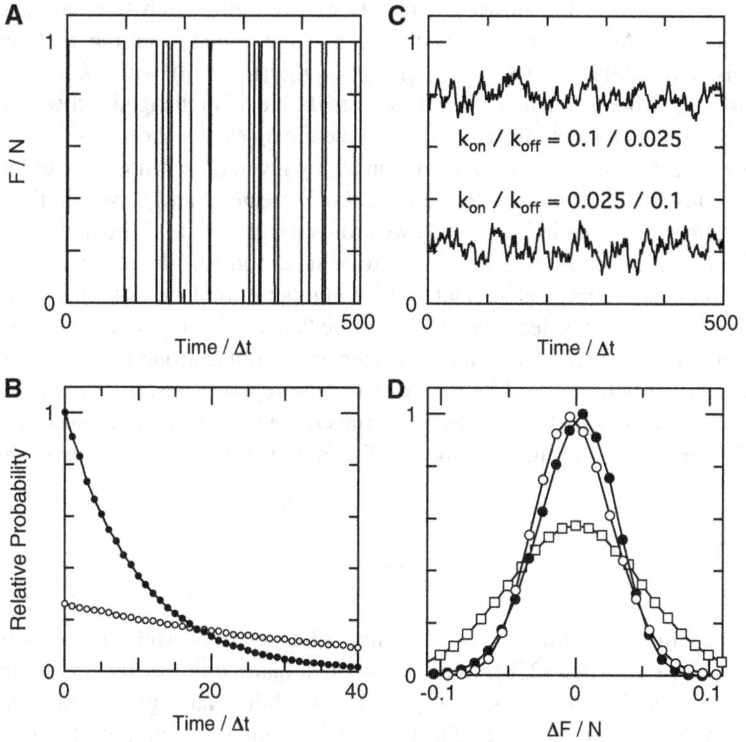

Fig. 5.17 Computer simulations of fluorescence fluctuations in a system with a specified number (N) of molecules, each of which can be in either a fluorescent (*on*) or nonfluorescent (*off*) state. On each time step (Δt), a molecule in the fluorescent state has a probability $\exp(-k_{off}\Delta t)$ of remaining in this state and a probability $1-\exp(-k_{off}\Delta t)$ of switching to the nonfluorescent state; a molecule in the nonfluorescent state remains in this state with probability $\exp(-k_{on}\Delta t)$ and switches to the fluorescent state with probability $1-\exp(-k_{on}\Delta t)$. The decision whether to change state is made by comparing $\exp(-k_{off}\Delta t)$ and $\exp(-k_{on}\Delta t)$ with a random number between 0 and 1. (**A**) Normalized fluorescence (F/N) as a function of time for a single molecule ($N=1$) with $k_{on}\Delta t=0.1$, and $k_{off}\Delta t=0.025$. (**B**) The probability that a molecule with $k_{on}\Delta t=0.1$ and $k_{off}\Delta t=0.025$ changes to a different state (*open circles, on→off; filled circles, off→on*) as a function of time after the molecule enters the first state. (**C**) Normalized fluorescence (F/N) as a function of time for 100 molecules with $k_{on}\Delta t=0.1$ and $k_{off}\Delta t=0.025$ (*upper trace*), or $k_{on}\Delta t=0.025$ and $k_{off}\Delta t=0.1$ (*lower trace*). (**D**) Relative probability of various deviations of the normalized fluorescence from the mean ($\Delta F/N$) for 100 molecules with $k_{on}\Delta t=0.01$ and $k_{off}\Delta t=0.1$ (*empty circles*), $k_{on}\Delta t$ and $k_{off}\Delta t$ both $=0.1$ (*empty squares*), or $k_{on}\Delta t=0.1$ and $k_{off}\Delta t=0.01$ (*filled circles*)

To simulate stochastic transitions between the two states, the algorithm used to generate the figure compares $\exp(-k_{on}\Delta t)$ or $\exp(-k_{off}\Delta t)$ to a random number between 0 and 1 after each time step. Figure 5.17A shows the calculated fluorescence from a single molecule with $k_{on}\Delta t=0.1$ and $k_{off}\Delta t=0.025$. The normalized fluorescence fluctuates randomly between 0 and 1, with the molecule remaining, on average, four times longer *on* than *off* (Fig. 5.17B).

Panel *C* of Fig. 5.17 shows the results of averaging such fluctuations over 100 molecules. As we would expect, the average signal over the time period shown is approximately 0.8 ($[on]/[on+off]=K_{eq}/[K_{eq}+1]$) when $k_{on}/k_{off}=4{:}1$, and drops to about 0.2 when the rate constants are interchanged. However, the signals fluctuate substantially around these mean values. As shown in Fig. 5.17D, the most probable deviation from the mean at any given time is not zero, unless the two rate constants happen to be the same. If k_{on} is greater than k_{off}, so that the mean is more than 0.5, the deviations are skewed to the positive side of the mean; if k_{on} is less than k_{off}, they are skewed to the negative side. Some of the asymmetry of the distribution function reflects the fact that the number of molecules in the *on* state at any given time cannot be less than zero or more than the total number of molecules.

In general, the deviations of the fluorescence amplitude around the mean follow a binomial distribution. The binomial distribution function describes the probability, $P(x,n,p)$, of having x "successes" in n trials if the probability of a success in any individual trial is p and the probability of a failure is $1-p$. This is given by the expression

$$P(x,n,p)=\frac{n!}{x!(n-x)!}p^x(1-p)^{n-x}, \tag{5.75}$$

in which the factor $n!/x!(n-x)!$ is the number of different possible combinations of n items taken x at a time [258]. In the situation at hand, n represents the number of molecules in the illuminated volume, p is the probability that a given molecule is in the fluorescent state at any particular time, and x is the total number of fluorescent molecules at that time. If we measure the fluorescence a large number of times, the average number of fluorescent molecules detected on each measurement will be np.

The mean ($\overline{x}$) of a binomial distribution is, indeed, equal to np:

$$\overline{x}=\sum_{x=0}^{n}x\frac{n!}{x!(n-x)!}p^x(1-p)^{n-x}=np, \tag{5.76}$$

as you can prove by defining $y=x+1$ and $m=n-1$ and noting that

$$\sum_{x=0}^{n}\frac{m!}{y!(m-y)!}p^y(1-p)^{m-y}=\sum_{x=0}^{n}P(y,m,p)=1 \tag{5.77}$$

[258]. The variance (σ^2, the average squared deviation from the mean) is given by

$$\sigma^2=\sum_{x=0}^{n}\left(x-\overline{x}^2\right)\frac{n!}{x!(n-x)!}p^x(1-p)^{n-x}=np(1-p). \tag{5.78}$$

There are several useful approximations for the binomial distribution that apply in certain limits of n and p (Box 5.4). However, neither of these limits is generally applicable to the situation we are considering here.

Box 5.4 Binomial, Poisson and Gaussian Distributions

The limit of a binomial distribution (Eq. 5.76) when the probability of success on an individual trial is very small ($p \ll 1$) is the *Poisson distribution*,

$$P_P(x, \bar{x}) = \frac{\bar{x}^x}{x!}\exp(-\bar{x}) = \frac{(np)^x}{x!}\exp(-np). \qquad \text{(B5.3.1)}$$

As in the underlying binomial distribution, $\bar{x}$ here is the mean value of x and is equal to np where n is the number of trials. The variance of a Poisson distribution also is np, as you can see by letting the factor $(1-p)$ in Eq. (5.78) go to 1. The standard deviation from the mean of a Poisson distribution (σ) thus is the square root of the mean.

The Poisson distribution is commonly used to analyze the statistics of photon-counting experiments. It also is useful for describing how a photophysical or photochemical process depends on the intensity of the excitation light. In a typical application, n represents the number of photons incident on a sample, p is the probability that a given incident photon results in a detectable process, and x is the total number of detectable events resulting from n photons. If absorbing more than one photon has the same result as absorbing a single photon, x follows a *cumulative one-hit Poisson distribution*, which expresses the probability that $x \neq 0$ as a continuous function of $\bar{x}$:

$$1 - P_P(0, \bar{x}) = 1 - \frac{\bar{x}^0}{0!}\exp(-\bar{x}) = 1 - \exp(-\bar{x}). \qquad \text{(B5.3.2)}$$

Binomial distributions have another important approximation in the limit that n is infinitely large and p also is large enough so that $np \gg 1$. This is the *Gaussian* or *normal distribution*:

$$P_G(x, \bar{x}, \sigma) = \frac{1}{\sigma\sqrt{2\pi}}\exp\left[-\frac{1}{2}\left(\frac{x - \bar{x}^2}{\sigma}\right)^2\right], \qquad \text{(B5.3.3)}$$

in which $\bar{x}$ again is the mean of the distribution and σ is the standard deviation from the mean. The Gaussian distribution, unlike the binomial and Poisson distributions, is a continuous function of x and is symmetrical around $\bar{x}$. See Eqs. (3.54), (3.56) and (4.59) for examples of Gaussian distributions and Fig.3.10 for an illustration.

Now consider the time dependence of the fluctuating fluorescence illustrated in Fig. 5.17C. Although the fluctuations are stochastic, they contain information on the dynamics of transitions between the on and off states in the model. This information can be extracted by calculating the *autocorrelation function* of the fluctuations, which is a measure of the correlation between the amplitude of the fluctuation at any given time with the amplitude at some later time. The autocorrelation function of the fluctuations of a time-dependent function $x(t)$ (sometimes called the *time-correlation function*) is defined as

$$C(t) \equiv \overline{\Delta x(t')\Delta x(t'+t)}, \tag{5.79}$$

where $\Delta x(t')$ is the deviation from the mean at time t' ($\Delta x(t') = x(t') - \overline{x}$) and the bar means an average over all times t'. At zero time, the autocorrelation function defined in this way gives the variance of the distribution:

$$C(0) \equiv \overline{\Delta x(t')\Delta x(t')} \equiv \sigma^2. \tag{5.80}$$

At long times, the autocorrelation function decays to zero, because the amplitudes of the fluctuations at two widely separated times are unrelated.

It often is useful to normalize an autocorrelation function relative to another parameter of the system. If the fluctuations follow a binomial distribution, dividing by the square of the mean and using Eqs. (5.76) and (5.78) gives

$$\frac{C(0)}{\overline{x}^2} = \frac{\overline{\Delta x^2}}{\overline{x}^2} = \frac{\sigma^2}{\overline{x}^2} = \frac{np[1-p]}{[np]^2} = \frac{1q}{np}. \tag{5.81}$$

With this normalization, $C(0)$ for the model described above is inversely proportional to the number of molecules in the illuminated volume ($n = N$). It also is inversely proportional to the ratio of the probabilities of finding a molecule in the fluorescent and nonfluorescent states ($q/p = k_{off}/k_{on} = 1/K_{eq}$).

Fluorescence correlation functions often are calculated by using the fluorescence signal itself rather than the deviations from the mean. After normalization relative to $\overline{x}^2$, the resulting function is the same as $C(t)/\overline{x}^2 + 1$:

$$\begin{aligned} G(t) &= \frac{\overline{x(t')x(t'+t)}}{\overline{x}^2} = \frac{\overline{[\overline{x}+\Delta x(t')][\overline{x}+\Delta x(t'+t)]}}{\overline{x}^2}, \\ &= \frac{\overline{x}^2 + \overline{x}\cdot\overline{\Delta x(t')} + \overline{x}\cdot\overline{\Delta x(t'+t)} + \overline{\Delta x(t')\Delta x(t'+t)}}{\overline{x}^2} = 1 + \frac{C(t)}{\overline{x}^2}. \end{aligned} \tag{5.82}$$

The terms $\Delta x(t')$ and $\Delta x(t'+t)$ average to zero.

Figure 5.18 shows autocorrelation functions calculated from several time courses similar to those in Fig. 5.17C but averaged over longer periods of time (2×10^5 time steps). The autocorrelation functions are normalized relative to $\overline{x}^2$ as in Eq. (5.81). As Eq. (5.81) predicts, the normalized values at zero time are inversely proportional to N (Fig. 5.18A) and K_{eq} (Fig. 5.18B). Fluorescence correlation spectroscopy thus provides a way to determine the number of fluorescent molecules in a small region of a sample, along with the equilibrium constant between states that have different fluorescence yields.

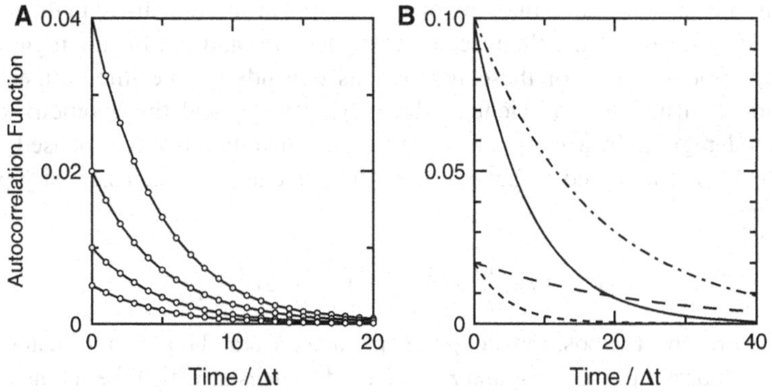

Fig. 5.18 (**A**) Normalized autocorrelation function of the fluorescence fluctuations for (from *top* to *bottom*) $N = 25$, 50, 100 or 200, all with $k_{on}\Delta t = k_{off}\Delta t = 0.1$. The autocorrelation functions are normalized relative to the square of the mean fluorescence amplitude. (**B**) Normalized autocorrelation function of the fluorescence fluctuations for $N = 50$, and $k_{on}\Delta t = 0.1$, $k_{off}\Delta t = 0.1$ (*short dashes*); $k_{on}\Delta t = 0.02$, $k_{off}\Delta t = 0.02$ (*long dashes*); $k_{on}\Delta t = 0.02$, $k_{off}\Delta t = 0.1$ (*solid curve*); or $k_{on}\Delta t = 0.01$, $k_{off}\Delta t = 0.05$ (*dot-dashed curve*). The results were averaged over 2×10^5 time steps

Inspection of Fig. 5.18 shows that the autocorrelation functions for this particular model decay exponentially with time, and that the rate constant for this decay is the sum of the rate constants for forward and backward transitions between the two states ($k_{on} + k_{off}$). The upper curve in Fig. 5.18B, for example, decays to $1/e$ (0.368) of its initial value in $16.67\Delta t$, which is the reciprocal of $(0.05 + 0.01)/\Delta t$. In classical kinetics, if a system with first-order reactions in the forward and backward directions is perturbed by an abrupt change in the concentration of one of the components, a change in temperature, or some other disturbance, it will relax to equilibrium with a rate constant given by the sum of the rate constants for the forward and backward reactions. The fact that the autocorrelation functions in Fig. 5.18 decay with the relaxation rate constant of the system is a general property of classical time-correlation functions [259–262]. One of the potential strengths of fluorescence correlation spectroscopy is that the relaxation dynamics can be obtained with the system at equilibrium; no perturbation is required.

Although the autocorrelation function provides the relaxation rate constant of the system, it does not immediately give the individual rate constants for the forward and backward reactions. These can be obtained by single-molecule studies as illustrated in Fig. 5.17. In some cases, they can be obtained by measuring the dependence of the relaxation rate constant on the concentration of one of the reactants. For example, the relaxation rate constant for a system with bimolecular reaction in one direction $\left(A + B \xrightarrow{k_1} C\right)$ and a unimolecular reaction in the reverse direction $\left(C \xrightarrow{k_{-1}} A + B\right)$ is $k_{+1}([A] + [B]) + k_{-1}$, which can be separated experimentally into parts that either depend on, or are independent of [A] and [B].

The fluorescence from a small region of a sample in solution also can fluctuate as a result of diffusion of the fluorescent molecules into and out of this region. The autocorrelation function of these fluctuations depends on the three-dimensional translational diffusion coefficient of the molecule (D) and the geometry of the illuminated region. In a commonly used model, excitation light focused by an objective lens is assumed to have a three-dimensional Gaussian intensity profile of the form

$$I = I_o \exp\left[-2\left(a^2 + b^2\right)/r^2 - 2c^2/l^2\right], \tag{5.83}$$

where c represents the position along the optical axis, a and b are coordinates in the focal plane normal to this axis, and r and l are distances at which the intensity falls to exp(-2) of its maximum value. The normalized autocorrelation function of fluorescence from molecules diffusing through such a region is given by [215]

$$\frac{C(t)}{\overline{x}^2} = \frac{1}{N}\left(\frac{1}{4Dt/r^2}\right)\left(\frac{1}{4Dt/l^2}\right)^{1/2}. \tag{5.84}$$

Fluorescence correlation spectroscopy thus provides a way to study processes that change the translational diffusion coefficient, such as binding of a small, fluorescent ligand to a macromolecule. However, the spatial dependence of the light intensity in the focal region can be more complex than Eq. (5.83) assumes and this can add spurious components to the autocorrelation function [263].

When diffusion and a reaction that affects the fluorescence yield occur on similar time scales, the fluorescence autocorrelation function consists of a product of factors of the forms given in Eqs. (5.81) and (5.84). Rather than trying to analyze the two components simultaneously, it may be best to slow diffusion by increasing the viscosity. Some enzymes have been shown to function well in agarose gels that restrict diffusion of the protein but allow the substrate and product to diffuse relatively freely [239]. However, diffusion may be necessary to replace chromophores that are bleached irreversibly by the excitation light.

Applications of fluorescence correlation spectroscopy have included studies of sparse molecules on cell surfaces [264, 265], diffusion of ligand-receptor complexes in cell membranes [266], conformational dynamics of DNA [150, 233], excited-state properties of flavins and flavoproteins [267], photodynamics of green- and red-fluorescent proteins [268, 269], protein unfolding pathways [270] and lipid-protein interactions [271].

An extension of fluorescence correlation spectroscopy is to measure the cross-correlation of fluorescence from two different fluorophores, such as a ligand and its receptors [272]. Strongly correlated fluctuations of the fluorescence from a small volume element indicate that the two species diffuse into and out of the element together, as would be expected for a complex. This approach has been used to study renaturation of complementary strands of nucleic acids [273], enzymatic fusion and cleavage kinetics [274], aggregation of prion protein [275], and DNA recombination [276] and repair [277].

In the related *photon counting histogram* technique, a histogram of the intensity of fluorescence from individual molecules is collected as molecules diffuse through the focal volume of a confocal microscope [278–280]. If the sample contains a mixture of molecules with different fluorescence properties, the histogram can reveal the relative amplitude of the fluorescence from a single molecule of each class, as well as the number of molecules in each class.

5.12 Intersystem Crossing, Phosphorescence, and Delayed Fluorescence

We saw in Chap. 4 that dipole interactions do not drive transitions between singlet and triplet states. But when a molecule is raised to an excited singlet state, relaxation into a triplet state often occurs with a rate constant (k_{isc} in Eq. (5.47)) on the order of 10^6 to 10^8 s^{-1}, and even more rapidly if the molecule contains a heavy atom such as Br. The excited molecule then can emit light (*phosphorescence*) as it decays from the triplet state back to the singlet ground state. This process usually is much slower than intersystem crossing from the excited singlet to the triplet state. For aromatic hydrocarbons, the radiative lifetime for phosphorescence typically is on the order of 30 s [281]. In the absence of quenchers such as O_2, radiationless intersystem crossing from the triplet state to the ground state also usually is slow relative to the rate of formation of the triplet state. Typical lifetimes of excited triplet states range from 10^{-5} s to more than 1 s.

As we discussed in Chap. 2, the angular momentum associated with an individual electron's spin must have a projection of either $\hbar/2$ (spin state α) or $-\hbar/2$ (spin state β) on the axis of a magnetic field. A molecule in an excited singlet state (spin wavefunction $2^{-1/2}\alpha(1)\beta(2) - 2^{-1/2}\alpha(2)\beta(1)$ with the numbers 1 and 2 denoting the two electrons) has no net angular momentum associated with electron spin, whereas a molecule in a triplet state with the spin wavefunction $\alpha(1)\alpha(2)$ or $\beta(1)\beta(2)$ has an angular momentum of $\pm\hbar$, respectively, in addition to any angular momentum associated with the orbital motions of the electrons. The third triplet spin wavefunction, $2^{-1/2}\alpha(1)\beta(2) + 2^{-1/2}\alpha(2)\beta(1)$, has no spin angular momentum projected along the field axis, but has angular momentum in the plane normal to this axis (Fig. 2.9). To conserve the total angular momentum of the system, the change in electron spin during a transition between singlet and triplet states must be coupled to a change in the angular momentum associated with another part of the system, such as orbital motions. It is the perturbation of the Hamiltonian by such *spin-orbit coupling* that drives intersystem crossing in most aromatic molecules.

The magnitude of spin-orbit coupling can be evaluated by considering the interaction of the magnetic dipole associated with an electron's spin with the intramolecular electric and magnetic fields, including the magnetic field created by the electron's own orbital motion [26, 281–286] . In some molecules

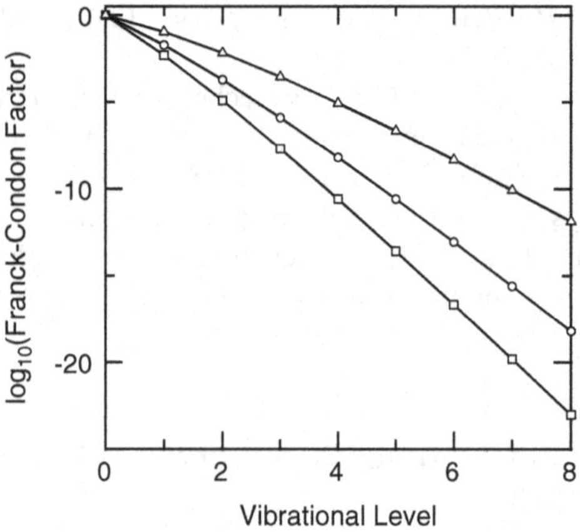

Fig. 5.19 Logarithm of the Franck-Condon factor $|\langle\chi_n|\chi_o\rangle|^2$ for transitions from the lowest vibrational level of one electronic state to level n of another state, when the transition is coupled to a single, harmonic vibrational mode with a displacement of 0.1 (*squares*), 0.2 (*circles*) or 0.5 (*triangles*). The vibration frequency (υ) is assumed to be the same in the two states. In a transition that is coupled to only one vibrational mode, energy conservation requires that $n \approx \Delta E_{oo}/h\upsilon$. More generally, ΔE_{oo} is partitioned among multiple vibrational modes of the molecule and the solvent

with π,π^* singlet and triplet states, the rate of intersystem crossing or phosphosphorescence depends on mixing with π,σ^* and σ,π^* states.

One reason that intersystem crossing from an excited singlet state ($^1\Psi_1$) to an excited triplet state ($^3\Psi_1$) usually occurs more rapidly than phosphorescence or radiationless decay of $^3\Psi_1$ to the ground state is that the energy gap between the zero-point vibrational levels of $^1\Psi_1$ and $^3\Psi_1$ usually is much smaller than the gap between $^3\Psi_1$ and the ground state. Within a related series of molecules, the rates of radiationless intramolecular transitions decrease approximately exponentially with the 0–0 energy difference between the initial and final states (ΔE_{oo}):

$$k = k_{oo}\exp(-\alpha\Delta E_{oo}), \tag{5.85}$$

with constants k_o and a that depend on the type of molecule. This is known as the *energy-gap law* [155, 281, 282]. The explanation is that if ΔE_{oo} is large, the product state must be formed in a highly excited vibrational level in order to conserve energy during the electronic transition. For a harmonic vibrational mode with a small displacement, the Franck-Condon factor, $|\langle\chi_n|\chi_o\rangle|^2$, for a transition from the lowest vibrational level of one state to level n of the other state falls off rapidly as n increases, as shown in Fig. 5.19. When the possible combinations of excitations in multiple vibrational modes are considered and some anharmonicity is introduced in the vibrations, the logarithm of the overall Franck-Condon factor becomes an approximately linear function of ΔE_{oo} [281, 287].

Triplet states also can form by back reactions of radical-pair states that are created by photochemical electron transfer. We discuss this process briefly in Box 10.2.

Energies of triplet states can be measured in several ways. The phosphorescence spectrum provides a direct measure of the energy difference between the triplet and ground states at the Franck-Condon maximum. Phosphorescence can be difficult to measure, however, because it is typically on the order of 10^6-times weaker than fluorescence. An alternative is to measure the *delayed fluorescence* that results from thermal excitation of the triplet state back to the excited singlet state. The ratio of the amplitudes of delayed and prompt fluorescence depends on the free energy difference between the zero-point vibrational levels these two states, and the temperature dependence of the ratio provides a measure of the enthalpy difference. Such measurements have been made for the bacteriochlorophyll-*a* dimer in photosynthetic bacterial reaction centers, where the phosphorescence spectrum puts the triplet state 0.42 eV below the excited singlet state [288]; the temperature dependence of the delayed fluorescence gives a similar energy gap of 0.40 eV [289]. For molecules in solution, another approach is to measure quenching of the triplet state by a series of molecules whose triplet energies are known. When the excited molecule collides with the quencher, energy and angular momentum can move from one molecule to the other by an exchange of electrons, provided that the acceptor's triplet state is similar to or somewhat lower than the energy of the donor. We will discuss the mechanism of this process in Chap. 7.

Delayed fluorescence also can report on the energies and dynamics of metastable states that are created photochemically by electron transfer or other processes. In photosynthetic reaction centers of purple bacteria or plant photosystem II, the amplitude of delayed fluorescence from an early ion-pair state decreases on picosecond and nanosecond time scales, while the population of the state remains essentially constant [290–293]. Both structural heterogeneity and relaxations of the protein around the ion-pair probably contribute to the complex time dependence of the delayed fluorescence.

Phosphorescence generally exhibits anisotropy similar to fluorescence anisotropy. In addition, as long as a molecule remains in an excited triplet state its ground-state absorption band will be bleached, giving an absorbance change that can have a definite anisotropy with respect to the excitation light. Triplet states thus can greatly extend the time scales over which linear dichroism and emission anisotropy report on molecular motions. An illustration is the use of eosin-5-maleimide to study rotational motions of proteins in erythrocyte plasma membranes [294]. Rotation of the anion transport protein occurs with a time constant between 20 and several hundred μs depending on the oligomeric state of the protein.

Exercises

The solid curve in the graph below is the absorption spectrum considered in Exercise 4.2. The dashed curve is the fluorescence emission spectrum of the same compound in water.

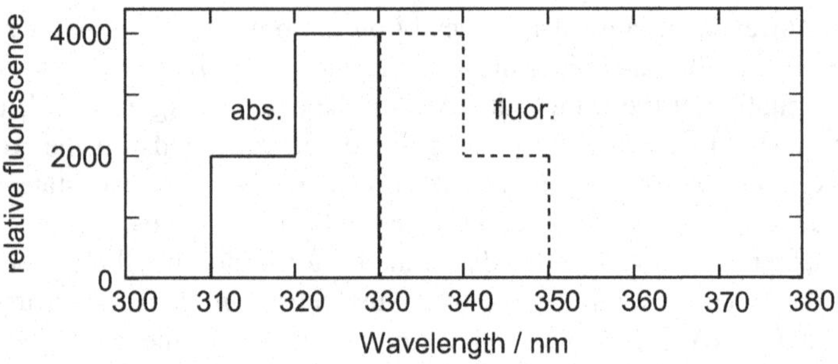

1. Using information from Exercise 4.2, calculate the radiative lifetime of the molecule.
2. In the absence of added quenchers, the fluorescence quantum yield is found to be 30%. Calculate the fluorescence lifetime under these conditions on the assumption that the fluorescence decay is monophasic.
3. 10 mM NaI causes the fluorescence yield and lifetime both to decrease to 10% of their original values. Calculate the Stern-Volmer quenching constant.
4. Calculate the bimolecular rate constant for the quenching by NaI.
5. How would your interpretation of the results differ if a quencher caused the fluorescence yield to decrease to 10% of its original value but had no effect on the measured fluorescence lifetime?
6. How would the amplitude of spontaneous fluorescence change if environmental effects shifted a chromophore's emission wavelength from 280 to 260 nm without altering the transition dipole or quenching by nonradiative processes?
7. Consider a quantized radiation field with frequency $\nu = 10^{15}\ \mathrm{s}^{-1}$. The Hamiltonian matrix element for interaction of the field with a molecule is proportional to $\left|\left\langle \chi_m \middle| \tilde{\mathrm{Q}} \middle| \chi_n \right\rangle\right|$, where $\tilde{\mathrm{Q}}$ is the position operator and χ_n and χ_m are eigenfunctions of the radiation field before and after the interaction, respectively. Quantum numbers n and m can be interpreted as the initial and final numbers of photons in the field. Evaluate $\left|\left\langle \chi_m \middle| \tilde{\mathrm{Q}} \middle| \chi_n \right\rangle\right|$ for fields with $n = 1{,}000{,}000$ and $m =$ (*a*) 999,999, (*b*) 1,000,000, and (*c*) 1,000,001.
8. A molecule is excited with polarized light and the fluorescence is measured at 90° to the excitation. In the absence of quenchers, the fluorescence measured through a polarizer perpendicular to the excitation polarizer is 80% as strong as that measured through a polarizer with parallel orientation. Calculate the fluorescence polarization and anisotropy.

9. Explain qualitatively how a quencher that decreases the measured fluorescence lifetime will affect the fluorescence anisotropy.
10. (*a*) Fluorescence of a homogeneous solution of molecules is measured through a confocal microscope and found to have a mean amplitude of 100 photons s^{-1} with a variance (σ^2) of 200 photons^2 s^{-2}. What is the mean number of fluorescent molecules in the illuminated region?
11. Explain how excitation with a second pulse of light can be used to increase the spatial resolution of fluorescence excited by the first pulse, so that the classical diffraction limit of resolution no longer holds.
12. Explain Kasha's rule.

References

1. Einstein, A.: Quantum theory of radiation. Physikal. Zeit. **18**: 121–128 (1917) (Eng. transl.: van der Waerden, B.L. (ed.). Sources of Quantum Mechanics, pp. 63–67. North Holland, Amsterdam (1967)
2. Callis, P.R.: Molecular orbital theory of the ^1L_b and ^1L_a states of indole. J. Chem. Phys. **95**, 4230–4240 (1991)
3. Callis, P.R.: ^1L_a and ^1L_b transitions of tryptophan: applications of theory and experimental observations to fluorescence of proteins. Meth. Enzymol. **278**, 113–150 (1997)
4. Vivian, J.T., Callis, P.R.: Mechanisms of tryptophan fluorescence shifts in proteins. Biophys. J. **80**, 2093–2109 (2001)
5. Gafni, A., DeToma, R.P., Manrow, R.E., Brand, L.: Nanosecond decay studies of a fluorescence probe bound to apomyoglobin. Biophys. J. **17**, 155–168 (1977)
6. Badea, M.G., Brand, L.: Time-resolved fluorescence measurements. Meth. Enzymol. **61**, 378–425 (1979)
7. Pierce, D.W., Boxer, S.G.: Stark effect spectroscopy of tryptophan. Biophys. J. **68**, 1583–1591 (1995)
8. Lakowicz, J.R.: On spectral relaxation in proteins. Photochem. Photobiol. **72**, 421–437 (2000)
9. Cote, M.J., Kauffman, J.F., Smith, P.G., McDonald, J.D.: Picosecond fluorescence depletion spectroscopy. 1. Theory and apparatus. J. Chem. Phys. **90**, 2865–2874 (1989)
10. Kauffman, J.F., Cote, M.J., Smith, P.G., McDonald, J.D.: Picosecond fluorescence depletion spectroscopy. 2. Intramolecular vibrational relaxation in the excited electronic state of fluorene. J. Chem. Phys. **90**, 2874–2891 (1989)
11. Kusba, J., Bogdanov, V., Gryczynski, I., Lakowicz, J.R.: Theory of light quenching. Effects on fluorescence polarization, intensity, and anisotropy decays. Biophys. J. **67**, 2024–2040 (1994)
12. Lakowicz, J.R., Gryczynski, I., Kusba, J., Bogdanov, V.: Light quenching of fluorescence. A new method to control the excited-state lifetime and orientation of fluorophores. Photochem. Photobiol. **60**, 546–562 (1994)
13. Zhong, Q.H., Wang, Z.H., Sun, Y., Zhu, Q.H., Kong, F.N.: Vibrational relaxation of dye molecules in solution studied by femtosecond time-resolved stimulated emission pumping fluorescence depletion. Chem. Phys. Lett. **248**, 277–282 (1996)
14. Nagarajan, V., Parson, W.: Femtosecond fluorescence depletion anisotropy: application to the B850 antenna complex of Rhodobacter sphaeroides. J. Phys. Chem. B **104**, 4010–4013 (2000)
15. Nakatsu, T., Ichiyama, S., Hiratake, J., Saldanha, A., Kobashi, N., et al.: Spectral difference in luciferase bioluminescence. Nature **440**, 372–376 (2006)

16. Birks, J.B., Dyson, D.J.: The relationship between absorption intensity and fluorescence lifetime of a molecule. Proc. Roy. Soc. Lond. Ser. A **275**, 135–148 (1963)
17. Birks, J.B.: Photophysics of Aromatic Molecules. Wiley-Interscience, New York (1970)
18. Lewis, G.N., Kasha, M.: Phosphorescence in fluid media and the reverse process of singlet-triplet absorption. J. Am. Chem. Soc. **67**, 994–1003 (1945)
19. Förster, T.: Fluoreszenz Organischer Verbindungen. Vandenhoeck & Ruprecht, Göttingen (1951)
20. Strickler, S.J., Berg, R.A.: Relationship between absorption intensity and fluorescence lifetime of a molecule. J. Chem. Phys. **37**, 814–822 (1962)
21. Ross, R.T.: Radiative lifetime and thermodynamic potential of excited states. Photochem. Photobiol. **21**, 401–406 (1975)
22. Seybold, P.G., Gouterman, M., Callis, J.B.: Calorimetric, photometric and lifetime determinations of fluorescence yields of fluorescein dyes. Photochem. Photobiol. **9**, 229–242 (1969)
23. van Metter, R.L., Knox, R.S.: Relation between absorption and emission spectra of molecules in solution. Chem. Phys. **12**, 333–340 (1976)
24. Becker, M., Nagarajan, V., Parson, W.W.: Properties of the excited singlet states of bacterio-chlorophyll a and bacteriopheophytin a in polar solvents. J. Am. Chem. Soc. **113**, 6840–6848 (1991)
25. Knox, R.S., Laible, P.D., Sawicki, D.A., Talbot, M.F.J.: Does excited chlorophyll a equilibrate in solution? J. Luminescence **72**, 580–581 (1997)
26. Hameka, H.: Advanced Quantum Chemistry. Addison-Wesley, Reading, MA (1965)
27. Sargent III, M., Scully, M.O., Lamb, W.E.J.: Laser Physics. Addison-Wesley, New York (1974)
28. Ditchburn, R.W.: *Light*. 3 ed. Academic, New York (1976)
29. Schatz, G.C., Ratner, M.A.: Quantum Mechanics in Chemistry, p. 325. Prentice-Hall, Englewood Cliffs, NJ (1993)
30. Lakowicz, J.R., Laczko, G., Cherek, H., Gratton, E., Limkeman, M.: Analysis of fluorescence decay kinetics from variable-frequency phase shift and modulation data. Biophys. J. **46**, 463–477 (1984)
31. Lakowicz, J.R.: Principles of Fluorescence Spectroscopy, 3rd edn. Springer, New York (2006)
32. Holzwarth, A.R.: Time-resolved fluorescence spectroscopy. Meth. Enzymol. **246**, 334–362 (1995)
33. Royer, C.A.: Fluorescence spectroscopy. Meth. Enzymol. **40**, 65–89 (1995)
34. Valeur, B.: Molecular Fluorescence. Wiley-VCH, Manheim (2002)
35. Kasha, M.: Characterization of electronic transitions in complex molecules. Faraday Discuss. Chem. Soc. **9**, 14–19 (1950)
36. Stern, O., Volmer, M.: The extinction period of fluorescence. Physikal. Zeit. **20**, 183–188 (1919)
37. Eftink, M.R., Ghiron, C.A.: Exposure of tryptophanyl residues in proteins. Quantitative determination by fluorescence quenching studies. Biochemistry **15**, 672–680 (1976)
38. Ren, J., Lew, S., Wang, Z., London, E.: Transmembrane orientation of hydrophobic α-helices is regulated both by the relationship of helix length to bilayer thickness and by the cholesterol concentration. Biochemistry **36**, 10213–10220 (1997)
39. Malenbaum, S.E., Collier, R.J., London, E.: Membrane topography of the T domain of diphtheria toxin probed with single tryptophan mutants. Biochemistry **37**, 17915–17922 (1998)
40. Lehrer, S.S.: Solute perturbation of protein fluorescence. The quenching of the tryptophanyl fluorescence of model compounds and of lysozyme by iodide ion. Biochemistry **10**, 3254–3263 (1971)
41. Beechem, J.M., Brand, L.: Time-resolved fluorescence of proteins. Annu. Rev. Biochem. **54**, 43–71 (1985)

42. Eftink, M.R.: Fluorescence techniques for studying protein structure. In: Schulter, C.H. (ed.) Methods in Biochemical Analysis, pp. 127–205. Wiley, New York (1991)
43. Millar, D.P.: Time-resolved fluorescence spectroscopy. Curr. Opin. Struct. Biol. **6**, 637–642 (1996)
44. Plaxco, K.W., Dobson, C.M.: Time-resolved biophysical methods in the study of protein folding. Curr. Opin. Struct. Biol. **6**, 630–636 (1996)
45. Royer, C.A.: Probing protein folding and conformational transitions with fluorescence. Chem. Rev. **106**, 1769–1784 (2006)
46. Kubelka, J., Eaton, W.A., Hofrichter, J.: Experimental tests of villin subdomain folding simulations. J. Mol. Biol. **329**, 625–630 (2003)
47. Kubelka, J., Henry, E.R., Cellmer, T., Hofrichter, J., Eaton, W.A.: Chemical, physical, and theoretical kinetics of an ultrafast folding protein. Proc. Natl. Acad. Sci. U.S.A. **105**, 18655–18662 (2008)
48. Parson, W.: Competition between tryptophan fluorescence and electron transfer during unfolding of the villin headpiece. Biochemistry **53**, 4503–4509 (2014)
49. Cowgill, R.W.: Fluorescence and the structure of proteins. I. Effects of substituents on the fluorescence of indole and phenol compounds. Arch. Biochem. Biophys. **100**, 36–44 (1963)
50. Callis, P.R., Liu, T.: Quantitative predictions of fluorescence quantum yields for tryptophan in proteins. J. Phys. Chem. B **108**, 4248–4259 (2004)
51. McMillan, A.W., Kier, B.L., Shu, I., Byrne, A., Andersen, N.H., et al.: Fluorescence of tryptophan in designed hairpin and Trp-cage miniproteins: measurements of fluorescence yields and calculations by quantum mechanical molecular dynamics simulations. J. Phys. Chem. B **117**, 1790–1809 (2013)
52. Shastry, M.C.R., Roder, H.: Evidence for barrier-limited protein folding kinetics on the microsecond time scale. Nat. Struct. Biol. **5**, 385–392 (1998)
53. Meech, S.R., Philips, D., Lee, A.G.: On the nature of the fluorescent state of methylated indole derivatives. Chem. Phys. **80**, 317–328 (1983)
54. Shinitsky, M., Goldman, R.: Fluorometric detection of histidine-trptophan complexes in peptides and proteins. Eur. J. Biochem. **3**, 139–144 (1967)
55. Steiner, R.F., Kirby, E.P.: The interaction of the ground and excited states of indole derivatives with electron scavengers. J. Phys. Chem. **73**, 4130–4135 (1969)
56. Ricci, R.W., Nesta, J.M.: Inter- and intramolecular quenching of indole fluorescence by carbonyl compounds. J. Phys. Chem. **80**, 974–980 (1976)
57. Loewenthal, R., Sancho, J., Fersht, A.R.: Fluorescence spectrum of barnase: contributions of three trptophan residues and a histidine-related pH dependence. Biochemistry **30**, 6775–7669 (1991)
58. Chen, Y., Barkley, M.D.: Toward understanding tryptophan fluorescence in proteins. Biochemistry **37**, 9976–9982 (1998)
59. Chen, Y., Liu, B., Yu, H.-T., Barkley, M.D.: The peptide bond quenches indole fluorescence. J. Am. Chem. Soc. **118**, 9271–9278 (1996)
60. DeBeuckeleer, K., Volckaert, G., Engelborghs, Y.: Time resolved fluorescence and phosphorescence properties of the individual tryptophan residues of barnase: evidence for protein-protein interactions. Proteins **36**, 42–53 (1999)
61. Qiu, W., Li, T., Zhang, L., Yang, Y., Kao, Y.-T., et al.: Ultrafast quenching of tryptophan fluorescence in proteins: Interresidue and intrahelical electron transfer. Chem. Phys. **350**, 154–164 (2008)
62. Petrich, J.W., Chang, M.C., McDonald, D.B., Fleming, G.R.: On the origin of non-exponential fluorescence decay in tryptophan and its derivatives. J. Am. Chem. Soc. **105**, 3824–3832 (1983)
63. Colucci, W.J., Tilstra, L., Sattler, M.C., Fronczek, F.R., Barkley, M.D.: Conformational studies of a constrained tryptophan derivative. Implications for the fluorescence quenching mechanism. J. Am. Chem. Soc. **112**, 9182–9190 (1990)

64. Arnold, S., Tong, L., Sulkes, M.: Fluorescence lifetimes of substituted indoles in solution and in free jets. Evidence for intramolecular charge-transfer quenching. J. Phys. Chem. **98**, 2325–2327 (1994)
65. Smirnov, A.V., English, D.S., Rich, R.L., Lane, J., Teyton, L., et al.: Photophysics and biological applications of 7-azaindole and its analogs. J. Phys. Chem. **101**, 2758–2769 (1997)
66. Sillen, A., Hennecke, J., Roethlisberger, D., Glockshuber, R., Engelborghs, Y.: Fluorescence quenching in the DsbA protein from Escherichia coli: complete picture of the excited-state energy pathway and evidence for the reshuffling dynamics of the microstates of tryptophan. Protein Sci. **37**, 253–263 (1999)
67. Adams, P.D., Chen, Y., Ma, K., Zagorski, M.G., Sönnichsen, F.D., et al.: Intramolecular quenching of tryptophan fluorescence by the peptide bond in cyclic hexapeptides. J. Am. Chem. Soc. **124**, 9278–9288 (2002)
68. Callis, P.R., Vivian, J.T.: Understanding the variable fluorescence quantum yield of tryptophan in proteins using QM-MM simulations. Quenching by charge transfer to the peptide backbone. Chem. Phys. Lett. **369**, 409–414 (2003)
69. Liu, T., Callis, P.R., Hesp, B.H., de Groot, M., Buma, W.J., et al.: Ionization potentials of fluoroindoles and the origin of nonexponential tryptophan fluorescence decay in proteins. J. Am. Chem. Soc. **127**, 4104–4113 (2005)
70. Doose, S., Neuweiler, H., Sauer, M.: Fluorescence quenching by photoinduced electron transfer: a reporter for conformational dynamics of macromolecules. Chemphyschem. **10**, 1389–1398 (2009)
71. Schlessinger, S.: The effect of amino acid analogues on alkaline phosphatase formation in Escherichia coli K-12. J. Biol. Chem. **243**, 3877–3883 (1968)
72. Ross, J.B.A., Szabo, A.G., Hogue, C.W.V.: Enhancement of protein spectra with tryptophan analogs: fluorescence spectroscopy of protein-protein and protein-nucleic interactions. Meth. Enzymol. **278**, 151–190 (1997)
73. Broos, J., Maddalena, F., Hesp, B.H.: In vivo synthesized proteins with monoexponential fluorescence decay kinetics. J. Am. Chem. Soc. **126**, 22–23 (2004)
74. Bronskill, P.M., Wong, J.T.: Suppression of fluorescence of tryptophan residues in proteins by replacement with 4-fluorotryptophan. Biochem. J. **249**, 305–308 (1988)
75. Yu, H.-T., Colucci, W.J., McLaughlin, M.L., Barkley, M.D.: Fluorescence quenching in indoles by excited-state proton transfer. J. Am. Chem. Soc. **114**, 8449–8454 (1992)
76. Feitelson, J.: On the mechanism of fluorescence quenching. Tyrosine and similar compounds. J. Phys. Chem. **68**, 391–397 (1964)
77. Cowgill, R.W.: Fluorescence and protein structure. X. Reappraisal of solvent and structural effects. Biochim. Biophys. Acta **133**, 6–18 (1967)
78. Tournon, J.E., Kuntz, E., El-Bayoumi, M.A.: Fluorescence quenching in phenylalanine and model compounds. Photochem. Photobiol. **16**, 425–433 (1972)
79. Laws, W.R., Ross, J.B.A., Wyssbrod, H.R., Beechem, J.M., Brand, L., et al.: Time-resolved fluorescence and ^{1}H NMR studies of tyrosine and tyrosine analogs: correlation of NMR-determined rotamer populations and fluorescence kinetics. Biochemistry **25**, 599–607 (1986)
80. Willis, K.J., Szabo, A.G.: Fluorescence decay kinetics of tyrosinate and tyrosine hydrogen-bonded complexes. J. Phys. Chem. **95**, 1585–1589 (1991)
81. Ross, J.B.A., Laws, W.R., Rousslang, K.W., Wyssbrod, H.R.: Tyrosine fluorescence and phosphorescence from proteins and peptides. In: Lakowicz, J.R. (ed.) Topics in Fluorescence Spectroscopy, pp. 1–63. Plenum, New York (1992)
82. Dietze, E.C., Wang, R.W., Lu, A.Y., Atkins, W.M.: Ligand effects on the fluorescence properties of tyrosine 9 in alpha 1-1 glutathione S-transferase. Biochemistry **35**, 6745–6753 (1996)
83. Mrozek, J., Rzeska, A., Guzow, K., Karolczak, J., Wiczk, W.: Influence of alkyl group on amide nitrogen atom on fluorescence quenching of tyrosine amide and N-acetyltyrosine amide. Biophys. Chem. **111**, 105–113 (2004)

84. van den Berg, P.A., van Hoek, A., Walentas, C.D., Perham, R.N., Visser, A.J.: Flavin fluorescence dynamics and photoinduced electron transfer in Escherichia coli glutathione reductase. Biophys. J. **74**, 2046–2058 (1998)
85. van den Berg, P.A.W., van Hoek, A., Visser, A.J.W.G.: Evidence for a novel mechanism of time-resolved flavin fluorescence depolarization in glutathione reductase. Biophys. J. **87**, 2577–2586 (2004)
86. Mataga, N., Chosrowjan, H., Shibata, Y., Tanaka, F., Nishina, Y., et al.: Dynamics and mechanisms of ultrafast fluorescence quenching reactions of flavin chromophores in protein nanospace. J. Phys. Chem. B **104**, 10667–10677 (2000)
87. Mataga, N., Chosrowjan, H., Taniguchi, S., Tanaka, F., Kido, N., et al.: Femtosecond fluorescence dynamics of flavoproteins: comparative studies on flavodoxin, its site-directed mutants, and riboflavin binding protein regarding ultrafast electron transfer in protein nanospaces. J. Phys. Chem. B **106**, 8917–8920 (2002)
88. Callis, P.R., Liu, T.Q.: Short range photoinduced electron transfer in proteins: QM-MM simulations of tryptophan and flavin fluorescence quenching in proteins. Chem. Phys. **326**, 230–239 (2006)
89. Merkley, E.D., Daggett, V., Parson, W.: A temperature-dependent conformational change of NADH oxidase from Thermus thermophilus HB8. Proteins: Struct. Funct. Bioinform. **80**, 546–555 (2011)
90. Marcus, R.A.: On the theory of oxidation-reduction reactions involving electron transfer I. J. Chem. Phys. **24**, 966–978 (1956)
91. Marcus, R.A.: Theory of oxidation-reduction reactions involving electron transfer. Part 4. A statistical-mechanical basis for treating contributions from the solvent, ligands and inert salt. Disc. Faraday Soc. **29**, 21–31 (1960)
92. Marcus, R.A.: Electron transfer reactions in chemistry. Theory and experiment. In: Bendall, D.S. (ed.) Protein Electron Transfer, pp. 249–272. BIOS Scientific Publishers, Oxford (1996)
93. Miller, J.R., Calcaterra, L.T., Closs, G.L.: Intramolecular long-distance electron transfer in radical anions. The effects of free energy and solvent on the reaction rates. J. Am. Chem. Soc. **106**, 3047–3049 (1984)
94. Gould, I.R., Ege, D., Mattes, S.L., Farid, S.: Return electron transfer within geminate radical pairs. Observation of the Marcus inverted region. J. Am. Chem. Soc. **109**, 3794–3796 (1987)
95. Mataga, N., Chosrowjan, H., Shibata, Y., Yoshida, N., Osuka, A., et al.: First unequivocal observation of the whole bell-shaped energy gap law in intramolecular charge separation from S_2 excited state of directly linked porphyrin-imide dyads and its solvent-polarity dependencies. J. Am. Chem. Soc. **123**, 12422–12423 (2001)
96. Rehm, D., Weller, A.: Kinetics of fluorescence quenching by electron and H-atom transfer. Isr. J. Chem. **8**, 259–271 (1970)
97. Farid, S., Dinnocenzo, J.P., Merkel, P.B., Young, R.H., Shukla, D., et al.: Reexamination of the Rehm-Weller data set reveals electron transfer quenching that follows a Sandros-Boltzmann dependence on the free energy. J. Am. Chem. Soc. **133**, 11580–11587 (2011)
98. Callis, P.R., Petrenko, A., Muino, P.L., Tusell, J.R.: Ab initio prediction of tryptophan fluorescence quenching by protein electric field enabled electron transfer. J. Phys. Chem. B **111**, 10335–10339 (2007)
99. Tusell, J.R., Callis, P.R.: Simulations of tryptophan fluorescence dynamics during folding of the villin headpiece. J. Phys. Chem. B **116**, 2586–2594 (2012)
100. Warshel, A., Chu, Z.-T., Parson, W.W.: Dispersed-polaron simulations of electron transfer in photosynthetic reaction centers. Science **246**, 112–116 (1989)
101. Moser, C.C., Dutton, P.L.: Engineering protein structure for electron transfer function in photosynthetic reaction centers. Biochim. Biophys. Acta **1101**, 171–176 (1992)
102. Weber, G., Daniel, E.: Cooperative effects in binding by bovine serum albumin. II. The binding of 1-anilino-8-naphthalenesulfonate. Polarization of the ligand fluorescence and quenching of protein fluorescence. Biochemistry **5**, 1900–1907 (1966)

103. Brand, L., Gohlke, J.R.: Fluorescence probes for structure. Annu. Rev. Biochem. **41**, 843–868 (1972)
104. Pierce, D.W., Boxer, S.G.: Dielectric relaxation in a protein matrix. J. Phys. Chem. **96**, 5560–5566 (1992)
105. Hiratsuka, T.: Prodan fluorescence reflects differences in nucleotide-induced conformational states in the myosin head and allows continuous visualization of the ATPase reactions. Biochemistry **37**, 7167–7176 (1998)
106. Waggoner, A.S., Grinvald, A.: Mechanisms of rapid optical changes of potential sensitive dyes. Annu. NY Acad. Sci. **303**, 217–241 (1977)
107. Loew, L.M., Cohen, L.B., Salzberg, B.M., Obaid, A.L., Bezanilla, F.: Charge-shift probes of membrane potential. Characterization of aminostyrylpyridinium dyes on the squid giant axon. Biophys. J. **47**, 71–77 (1985)
108. Fromherz, P., Dambacher, K.H., Ephardt, H., Lambacher, A., Mueller, C.O., et al.: Fluorescent dyes as probes of voltage transients in neuron membranes. Ber. Bunsen-Gesellsch. **95**, 1333–1345 (1991)
109. Baker, B.J., Kosmidis, E.K., Vucinic, D., Falk, C.X., Cohen, L.B., et al.: Imaging brain activity with voltage- and calcium-sensitive dyes. Cell. Mol. Neurobiol. **25**, 245–282 (2005)
110. Haugland, R.P.: Handbook of Fluorescent Probes and Research Chemicals, 6th edn. Molecular Probes Inc., Eugene, OR (1996)
111. Oi, V.T., Glazer, A.N., Stryer, L.: Fluorescent phycobiliprotein conjugates for analyses of cells and molecules. J. Cell Biol. **93**, 981–986 (1982)
112. Kronick, M.N., Grossman, P.D.: Immunoassay techniques with fluorescent phycobiliprotein conjugates. Clin. Chem. **29**, 1582–1586 (1983)
113. Alivisatos, A.P., Gu, W., Larabell, C.: Quantum dots as cellular probes. Annu. Rev. Biomed. Eng. **7**, 55–76 (2005)
114. Michalet, X., Pinaud, F.F., Bentolila, L.A., Tsay, J.M., Doose, S., et al.: Quantum dots for live cells, in vivo imaging, and diagnostics. Science **307**, 538–544 (2005)
115. Brus, L.E.: Electron-electron and electron-hole interactions in small semiconductor crystallites: the size dependence of the lowest excited electronic state. J. Chem. Phys. **80**, 4403–4409 (1984)
116. Brus, L.E.: Electronic wave functions in semiconductor clusters: experiment and theory. J. Phys. Chem. **90**, 2555–2560 (1986)
117. Nozik, A.J., Williams, F., Nenadovic, M.T., Rajh, T., Micic, O.I.: Size quantization in small semiconductor particles. J. Phys. Chem. **89**, 397–399 (1985)
118. Bawendi, M.G., Wilson, W.L., Rothberg, L., Carroll, P.J., Jedju, T.M., et al.: Electronic structure and photoexcited-carrier dynamics in nanometer-size CdSe clusters. Phys. Rev. Lett. **65**, 1623–1626 (1990)
119. Bruchez Jr., M., Moronne, M., Gin, P., Weiss, S., Alivisatos, A.P.: Semiconductor nanocrystals as fluorescent biological labels. Science **281**, 2013–2016 (1998)
120. Petryayeva, E., Algar, W.R., Medintz, I.L.: Quantum dots in bioanalysis: a review of applications across various platforms for fluorescence spectroscopy and imaging. Appl. Spectrosc. **67**, 215–252 (2013)
121. Shimomura, O., Johnson, F.H.: Intermolecular energy transfer in the bioluminescent system of Aequorea. Biochemistry **13**, 2656–2662 (1974)
122. Cody, C.W., Prasher, D.C., Westler, W.M., Prendergast, F.G., Ward, W.W.: Chemical structure of the hexapeptide chromophore of the Aequorea green-fluorescent protein. Biochemistry **9**, 1212–1218 (1979)
123. Ormö, M., Cubitt, A.B., Kallio, K., Gross, L.A., Tsien, R.Y., et al.: Crystal structure of the Aequorea victoria green fluorescent protein. Science **273**, 1392–1395 (1996)
124. Tsien, R.Y.: The green fluorescent protein. Annu. Rev. Biochem. **67**, 509–544 (1998)
125. Wachter, R.M.: The family of GFP-like proteins: structure, function, photophysics and biosensor applications. Introduction and perspective. Photochem. Photobiol. **82**, 339–344 (2006)

126. Chattoraj, M., King, B.A., Bublitz, G.U., Boxer, S.G.: Ultra-fast excited state dynamics in green fluorescent protein: multiple states and proton transfer. Proc. Natl. Acad. Sci. U.S.A. **93**, 8362–8367 (1996)
127. Weber, W., Helms, V., McCammon, J.A., Langhoff, P.W.: Shedding light on the dark and weakly fluorescent states of green fluorescent proteins. Proc. Natl. Acad. Sci. U.S.A. **96**, 6177–6182 (1999)
128. Webber, N.M., Litvinenko, K.L., Meech, S.R.: Radiationless relaxation in a synthetic analogue of the green fluorescent protein chromophore. J. Phys. Chem. B **105**, 8036–8039 (2001)
129. Mandal, D., Tahara, T., Meech, S.R.: Excited-state dynamics in the green fluorescent protein chromophore. J. Phys. Chem. B **108**, 1102–1108 (2004)
130. Martin, M.E., Negri, F., Olivucci, M.: Origin, nature, and fate of the fluorescent state of the green fluorescent protein chromophore at the CASPT2//CASSCF resolution. J. Am. Chem. Soc. **126**, 5452–5464 (2004)
131. Chalfie, M., Tu, Y., Euskirchen, G., Ward, W.W., Prasher, D.C.: Green fluorescent protein as a marker for gene expression. Science **263**, 802–805 (1994)
132. Miyawaki, A., Llopis, J., Heim, R., McCaffrey, J.M., Adams, J.A., et al.: Fluorescent indicators for Ca^{2+} based on green fluorescent proteins and calmodulin. Nature **388**, 882–887 (1997)
133. Zhang, J., Campbell, R.E., Ting, A.Y., Tsien, R.Y.: Creating new fluorescent probes for cell biology. Nat. Rev. Mol. Cell Biol. **3**, 906–918 (2002)
134. Nienhaus, G.U., Wiedenmann, J.: Structure, dynamics and optical properties of fluorescent proteins: perspectives for marker development. Chemphyschem. **10**, 1369–1379 (2009)
135. Heim, R., Tsien, R.Y.: Engineering green fluorescent protein for improved brightness, longer wavelengths and fluorescence resonance energy transfer. Curr. Biol. **6**, 178–182 (1996)
136. Wachter, R.M., King, B.A., Heim, R., Kallio, K., Tsien, R.Y., et al.: Crystal structure and photodynamic behavior of the blue emission variant Y66H/Y145F of green fluorescent protein. Biochemistry **36**, 9759–9765 (1997)
137. Miyawaki, A., Griesbeck, O., Heim, R., Tsien, R.Y.: Dynamic and quantitative Ca^{2+} measurements using improved cameleons. Proc. Natl. Acad. Sci. U.S.A. **96**, 2135–2140 (1999)
138. Griesbeck, O., Baird, G.S., Campbell, R.E., Zacharias, D.A., Tsien, R.Y.: Reducing the environmental sensitivity of yellow fluorescent protein. Mechanism and applications. J. Biol. Chem. **276**, 29188–29194 (2001)
139. Rizzo, M.A., Springer, G.H., Granada, B., Piston, D.W.: An improved cyan fluorescent protein variant usefule for FRET. Nat. Biotechnol. **22**, 445–449 (2004)
140. Shaner, N.C., Campbell, R.E., Steinbach, P.A., Giepmans, B.N.G., Palmer, A.E., et al.: *Improved monomeric red, orange and yellow fluorescent proteins derived from Discosoma sp. red fluorescent protein.* Nat. Biotechnol. **22**, 1567–1572 (2004)
141. Shaner, N.C., Steinbach, P.A., Tsien, R.Y.: A guide to choosing fluorescent proteins. Nat. Methods **2**, 905–909 (2005)
142. Matz, M.V., Fradkov, A.F., Labas, Y.A., Savitsky, A.P., Zaraisky, A.G., et al.: Fluorescent proteins from nonbioluminescent Anthozoa species. Nat. Biotechnol. **17**, 969–973 (1999)
143. Baird, G.S., Zacharias, D.A., Tsien, R.Y.: Biochemistry, mutagenesis, and oligomerization of DsRed, a red fluorescent protein from coral. Proc. Natl. Acad. Sci. U.S.A. **97**, 11984–11989 (2000)
144. Baird, G.S., Zacharias, D.A., Tsien, R.Y.: Circular permutation and receptor insertion within green fluorescent proteins. Proc. Natl. Acad. Sci. U.S.A. **96**, 11241–11246 (1999)
145. Nagai, T., Sawano, A., Park, E.S., Miyawaki, A.: Circularly permuted green fluorescent proteins engineered to sense Ca^{2+}. Proc. Natl. Acad. Sci. U.S.A. **98**, 3197–3202 (2001)
146. Lukyanov, K.A., Fradkov, A.F., Gurskaya, N.G., Matz, M.V., Labas, Y.A., et al.: Natural animal coloration can be determined by a nonfluorescent green fluorescent protein homolog. J. Biol. Chem. **275**, 25879–25882 (2000)

147. Lippincott-Schwartz, J., Altan-Bonnet, N., Patterson, G.H.: Photobleaching and photoactivation: following protein dynamics in living cells. Nat. Cell Biol. **5**, S7–S14 (2003)
148. Patterson, G.H., Lippincott-Schwartz, J.: Selective photolabeling of proteins using photoactivatable GFP. Methods **32**, 445–450 (2004)
149. Hess, S.T., Girirajan, T.P., Mason, M.D.: Ultra-high resolution imaging by fluorescence photoactivation localization microscopy. Biophys. J. **91**, 4258–4272 (2006)
150. Eggeling, C., Fries, J.R., Brand, L., Günther, R., Seidel, C.A.M.: Monitoring conformational dynamics of a single molecule by selective fluorescence spectroscopy. Proc. Natl. Acad. Sci. USA **95**, 1556–1561 (1998)
151. Eggeling, C., Volkmer, A., Seidel, C.A.: Molecular photobleaching kinetics of rhodamine 6G by one- and two-photon induced confocal fluorescence microscopy. Chemphyschem. **6**, 791–804 (2005)
152. Christ, T., Kulzer, F., Bordat, P., Basché, T.: Watching the photo-oxidation of a single aromatic hydrocarbon molecule. Angew. Chem. Int. Ed. **40**, 4192–4195 (2001)
153. Hoogenboom, J.P., van Dijk, E.M., Hernando, J., van Hulst, N.F., Garcia-Parajo, M.F.: Power-law-distributed dark states are the main pathway for photobleaching of single organic molecules. Phys. Rev. Lett. **95**, 097401 (2005)
154. Bilski, P., Chignell, C.F.: Optimization of a pulse laser spectrometer for the measurement of the kinetics of singlet oxygen O_2 ($\Delta^1{}_g$) decay in solution. J. Biochem. Biophys. Methods **33**, 73–80 (1996)
155. Turro, N.: Modern Molecular Photochemistry. Menlo Park CA, Benjamin/Cummings (1978)
156. Rasnik, I., McKinney, S.A., Ha, T.: Nonblinking and long-lasting single-molecule fluorescence imaging. Nat. Methods **3**, 891–893 (2006)
157. Vogelsang, J., Kasper, R., Steinhauer, C., Person, B., Heilemann, M., et al.: A reducing and oxidizing system minimizes photobleaching and blinking of fluorescent dyes. Angew. Chem. **47**, 5465–5469 (2008)
158. Campos, L.A., Liu, J., Wang, X., Ramanathan, R., English, D.S.: A photoprotection strategy for microsecond-resolution single-molecule fluorescence spectroscopy. Nat. Methods **8**, 143–146 (2011)
159. Axelrod, D., Koppel, D.E., Schlessinger, S., Elson, E., Webb, W.W.: Mobility measurement by analysis of fluorescence photobleaching recovery kinetics. Biophys. J. **16**, 1055–1069 (1976)
160. Jacobson, K., Derzko, Z., Wu, E.S., Hou, Y., Poste, G.: Measurement of the lateral mobility of cell surface components in single, living cells by fluorescence recovery after photobleaching. J. Supramol. Struct. **5**, 565–576 (1976)
161. Schlessinger, J., Koppel, D.E., Axelrod, D., Jacobson, K., Webb, W.W., et al.: Lateral transport on cell membranes: mobility of concanavalin A receptors on myoblasts. Proc. Natl. Acad. Sci. U.S.A. **73**, 2409–2413 (1976)
162. Wu, E.S., Jacobson, K., Szoka, F., Portis, J.A.: Lateral diffusion of a hydrophobic peptide, N-4-nitrobenz-2-oxa-1,3-diazole gramicidin S, in phospholipid multibilayers. Biochemistry **17**, 5543–5550 (1978)
163. Schindler, M., Osborn, M.J., Koppel, D.E.: Lateral diffusion of lipopolysaccharide in the outer membrane of Salmonella typhimurium. Nature **285**, 261–263 (1980)
164. Lagerholm, B.C., Starr, T.E., Volovyk, Z.N., Thompson, N.L.: Rebinding of IgE Fabs at haptenated planar membranes: measurement by total internal reflection with fluorescence photobleaching recovery. Biochemistry **39**, 2042–2051 (1999)
165. Thompson, N.L., Burghardt, T.P., Axelrod, D.: Measuring surface dynamics of biomolecules by total internal reflection fluorescence with photobleaching recovery or correlation spectroscopy. Biophys. J. **33**, 435–454 (1999)
166. Reits, E.A.J., Neefjes, J.J.: From fixed to FRAP: measuring protein mobility and activity in living cells. Nat. Cell Biol. **3**, E145–E147 (2001)
167. Klonis, N., Rug, M., Harper, I., Wickham, M., Cowman, A., et al.: Fluorescence photobleaching analysis for the study of cellular dynamics. Eur. Biophys. J. **31**, 36–51 (2002)

168. Houtsmuller, A.B.: Fluorescence recovery after photobleaching: application to nuclear proteins. Adv. Biochem. Eng. Biotechnol. **95**, 177–199 (2005)
169. Houtsmuller, A.B., Vermeulen, W.: Macromolecular dynamics in living cell nuclei revealed by fluorescence redistribution after photobleaching. Histochem. Cell Biol. **115**, 13–21 (2001)
170. Calapez, A., Pereira, H.M., Calado, A., Braga, J., Rino, J., et al.: The intranuclear mobility of messenger RNA binding proteins is ATP dependent and temperature sensitive. J. Cell Biol. **159**, 795–805 (2002)
171. Haggie, P.M., Verkman, A.S.: Diffusion of tricarboxylic acid cycle enzymes in the mitochondrial matrix in vivo. Evidence for restricted mobility of a multienzyme complex. J. Biol. Chem. **277**, 40782–40788 (2002)
172. Dayel, M.J., Hom, E.F., Verkman, A.S.: Diffusion of green fluorescent protein in the aqueous-phase lumen of endoplasmic reticulum. Biophys. J. **76**, 2843–2851 (1999)
173. Cole, N.B., Smith, C.L., Sciaky, N., Terasaki, M., Edidin, M., et al.: Diffusional mobility of Golgi proteins in membranes of living cells. Science **273**, 797–801 (1996)
174. van Amerongen, H., Struve, W.S.: Polarized optical spectroscopy of chromoproteins. Meth. Enzymol. **246**, 259–283 (1995)
175. Jimenez, R., Dikshit, S.N., Bradforth, S.E., Fleming, G.R.: Electronic excitation transfer in the LH2 complex of Rhodobacter sphaeroides. J. Phys. Chem. **100**, 6825–6834 (1996)
176. Pullerits, T., Chachisvilis, M., Sundström, V.: Exciton delocalization length in the B850 antenna of Rhodobacter sphaeroides. J. Phys. Chem. **100**, 10787–10792 (1996)
177. Nagarajan, V., Johnson, E., Williams, J.C., Parson, W.W.: Femtosecond pump-probe spectroscopy of the B850 antenna complex of Rhodobacter sphaeroides at room temperature. J. Phys. Chem. B **103**, 2297–2309 (1999)
178. Delrow, J.J., Heath, P.J.: Fujimoto, B.S. and Schurr, J.M. Effect of temperature on DNA secondary structure in the absence and presence of 0.5 M tetramethylammonium chloride. Biopolymers **45**, 503–515 (1998)
179. Lakowicz, J.R., Knutson, J.R.: Hindered depolarizing rotations of perylene in lipid bilayers. Detection by lifetime-resolved fluorescence anisotropy measurements. Biochemistry **19**, 905–911 (1980)
180. Lakowicz, J.R., Maliwal, B.P.: Oxygen quenching and fluorescence depolarization of tyrosine residues in proteins. J. Biol. Chem. **258**, 4794–4801 (1983)
181. Lakowicz, J.R., Maliwal, B.P., Cherek, H., Balter, A.: Rotational freedom of tryptophan residues in proteins and peptides. Biochemistry **22**, 1741–1752 (1983)
182. Hou, X., Richardson, S.J., Aguilar, M.I., Small, D.H.: Binding of amyloidogenic transthyretin to the plasma membrane alters membrane fluidity and induces neurotoxicity. Biochemistry **44**, 11618–11627 (2005)
183. Johnson, D.A.: C-terminus of a long alpha-neurotoxin is highly mobile when bound to the nicotinic acetylcholine receptor: a time-resolved fluorescence anisotropy approach. Biophys. Chem **116**, 213–218 (2005)
184. Fidy, J., Laberge, M., Kaposi, A.D., Vanderkooi, J.M.: Fluorescence line narrowing applied to the study of proteins. Biochim. Biophys. Acta **1386**, 331–351 (1998)
185. Nie, S., Zare, R.N.: Optical detection of single molecules. Annu. Rev. Biophys. Biomol. Struct. **26**, 567–596 (1997)
186. Xie, X.S., Trautman, J.K.: Optical studies of single molecules at room temperature. Annu. Rev. Phys. Chem. **49**, 441–480 (1998)
187. Moerner, W.E., Orrit, M.: Illuminating single molecules in condensed matter. Science **283**, 1670–1676 (1999)
188. Weiss, S.: Fluorescence spectroscopy of single biomolecules. Science **283**, 1676–1683 (1999)
189. Moerner, W.E., Kador, L.: Optical detection and spectroscopy of single molecule solids. Phys. Rev. Lett. **62**, 2535–2538 (1989)
190. Moerner, W.E., Basche, T.: Optical spectroscopy of individual dopant molecules in solids. Angew. Chem. **105**, 537–557 (1993)

191. Kulzer, F., Kettner, R., Kummer, S., Basché, T.: Single molecule spectroscopy: spontaneous and light-induced frequency jumps. Pure Appl. Chem. **69**, 743–748 (1997)
192. Goodwin, P.M., Ambrose, W.P., Keller, R.A.: Single-molecule detection in liquids by laser-induced fluorescence. Acc. Chem. Res. **29**, 607–613 (1996)
193. Nguyen, D.C., Keller, R.A., Jett, H., Martin, J.C.: Detection of single molecules of phycoerythrin in hydrodynamically focused flows by laser-induced fluorescence. Anal. Chem. **59**, 2158–2161 (1987)
194. Peck, K., Stryer, L., Glazer, A.N., Mathies, R.A.: Single-molecule fluorescence detection: autocorrelation criterion and experimental realization with phycoerythrin. Proc. Natl. Acad. Sci. U.S.A. **86**, 4087–4091 (1989)
195. Pohl, D.W., Denk, W., Lanz, M.: Optical stethoscopy: image recording with resolution l/20. Appl. Phys. Lett. **44**, 651–653 (1984)
196. Harootunian, A., Betzig, E., Isaacson, M., Lewis, A.: Super-resolution fluorescence near-field scanning optical microscopy. Appl. Phys. Lett. **49**, 674–676 (1986)
197. Betzig, E., Trautman, J.K.: Near-field optics: microscopy, spectroscopy, and surface modification beyond the diffraction limit. Science **257**, 189–195 (1992)
198. Betzig, E., Chichester, R.J., Lanni, F., Taylor, D.L.: Near-field fluorescence imaging of cytoskeletal actin. BioImaging **1**, 129–135 (1993)
199. Kopelman, R., Weihong, T., Birnbaum, D.: Subwavelength spectroscopy, exciton supertips and mesoscopic light-matter interactions. J. Lumin. **58**, 380–387 (1994)
200. Ha, T., Enderle, T., Ogletree, D.F., Chemla, D.S., Selvin, P.R., et al.: Probing the interaction between two single molecules: fluorescence resonance energy transfer between a single donor and a single acceptor. Proc. Natl. Acad. Sci. U.S.A. **93**, 6264–6268 (1996)
201. Meixner, A.J., Kneppe, H.: Scanning near-field optical microscopy in cell biology and microbiology. Cell. Mol. Biol. **44**, 673–688 (1998)
202. Betzig, E., Patterson, G.H., Sougrat, R., Lindwasser, O.W., Olenych, S., et al.: Imaging intracellular fluorescent proteins at nanometer resolution. Science **313**, 1642–1645 (2006)
203. Xie, X.S., Dunn, R.C.: Probing single molecule dynamics. Science **265**, 361–364 (1994)
204. Iwane, A.H., Funatsu, T., Harada, Y., Tokunaga, M., Ohara, O., et al.: Single molecular assay of individual ATP turnover by a myosin-GFP fusion protein expressed in vitro. FEBS Lett. **407**, 235–238 (1997)
205. Kalb, E., Engel, J., Tamm, L.K.: Binding of proteins to specific target sites in membranes measured by total internal reflection fluorescence microscopy. Biochemistry **29**, 1607–1613 (1990)
206. Poglitsch, C.L., Sumner, M.T., Thompson, N.L.: Binding of IgG to MoFc gamma RII purified and reconstituted into supported planar membranes as measured by total internal reflection fluorescence microscopy. Biochemistry **30**, 6662–6671 (1991)
207. Lieto, A.M., Cush, R.C., Thompson, N.L.: Ligand-receptor kinetics measured by total internal reflection with fluorescence correlation spectroscopy. Biophys. J. **85**, 3294–3302 (2003)
208. Lieto, A.M., Thompson, N.L.: Total internal reflection with fluorescence correlation spectroscopy: nonfluorescent competitors. Biophys. J. **87**, 1268–1278 (2004)
209. Sund, S.E., Swanson, J.A., Axelrod, D.: Cell membrane orientation visualized by polarized total internal reflection fluorescence. Biophys. J. **77**, 2266–2283 (1999)
210. Geddes, C.D., Parfenov, A., Gryczynski, I., Lakowicz, J.R.: Luminescent blinking of gold nanoparticles. Chem. Phys. Lett. **380**, 269–272 (2003)
211. Aslan, K., Lakowicz, J.R., Geddes, C.D.: Nanogold-plasmon-resonance-based glucose sensing. Anal. Biochem. **330**, 145–155 (2004)
212. Stefani, F.D., Vasilev, K., Boccio, N., Stoyanova, N., Kreiter, M.: Surface-plasmon-mediated single-molecule fluorescence through a thin metallic film. Phys. Rev. Lett. **94**, Art. 023005 (2005)
213. Wenger, J., Lenne, P.F., Popov, E., Rigneault, H., Dintinger, J., et al.: Single molecule fluorescence in rectangular nano-apertures. Opt. Express **13**, 7035–7044 (2005)

214. Lakowicz, J.R.: Radiative decay engineering: biophysical and biomedical applications. Anal. Biochem. **298**, 1–24 (2002)
215. Eigen, M., Rigler, R.: Sorting single molecules. Application to diagnostics and evolutionary biotechnology. Proc. Natl. Acad. Sci. USA **91**, 5740–5747 (1994)
216. Nie, S., Chiu, D.T., Zare, R.N.: Probing individual molecules with confocal fluorescence microscopy. Science **266**, 1018–1021 (1994)
217. Nie, S., Chiu, D.T., Zare, R.N.: Real-time detection of single molecules in solution by confocal fluorescence microscopy. Angew. Chem. **67**, 2849–2857 (1995)
218. Edman, L., Mets, U., Rigler, R.: Conformational transitions monitored for single molecules in solution. Proc. Natl. Acad. Sci. USA **93**, 6710–6715 (1996)
219. Macklin, J.J., Trautman, J.K., Harris, T.D., Brus, L.E.: Imaging and time-resolved spectroscopy of single molecules at an interface. Science **272**, 255–258 (1996)
220. Conn, P.M.: Confocal Microscopy. Methods in Enzymology, vol. 307. Academic, San Diego (1999)
221. Yuste, R., Konnerth, A.: Imaging in Neuroscience and Development: A Laboratory Manual. Cold Spring Harbor Laboratory Press, Cold Spring Harbor, N.Y. (2000)
222. Yuste, R.: Fluorescence microscopy today. Nat. Methods **2**, 902–904 (2005)
223. Lichtman, J.W., Conchello, J.-A.: Fluorescence microscopy. Nat. Methods **2**, 910–919 (2005)
224. Conchello, J.-A., Lichtman, J.W.: Optical sectioning microscopy. Nat. Methods **2**, 920–931 (2005)
225. Helmchen, F., Denk, W.: Deep tissue two-photon microscopy. Nat. Methods **2**, 932–940 (2005)
226. Hell, S.W., Wichmann, J.: Breaking the diffraction resolution by stimulated emission: stimulated emission depletion microscopy. Opt. Lett. **19**, 780–782 (1994)
227. Hell, S.W.: Toward fluorescence nanoscopy. Nat. Biotechnol. **21**, 1347–1355 (2003)
228. Hell, S.W., Jakobs, S., Kastrup, L.: Imaging and writing at the nanoscale with focused visible light through saturable optical transitions. Appl. Phys. A **77**, 859–860 (2003)
229. Hofmann, M., Eggeling, C., Jakobs, S., Hell, S.W.: Breaking the diffraction barrier in fluorescence microscopy at low light intensities by using reversibly photoswitchable proteins. Proc. Natl. Acad. Sci. U.S.A. **102**, 17565–17569 (2005)
230. Westphal, V., Hell, S.W.: Nanoscale resolution in the focal plane of an optical microscope. Phys Rev. Lett. **94**, 143903 (2005)
231. Hell, S.W.: Far-field optical nanoscopy. Science **316**, 1153–1158 (2007)
232. Lu, H.P., Xie, X.S.: Single-molecule spectral fluctuations at room temperature. Nature **385**, 143–146 (1997)
233. Wennmalm, S., Edman, L., Rigler, R.: Conformational fluctuations in single DNA molecules. Proc. Natl. Acad. Sci. USA **94**, 10641–10646 (1997)
234. Michalet, X., Weiss, S., Jäger, M.: Single-molecule fluorescence studies of protein folding and conformational dynamics. Chem. Rev. **106**, 1785–1813 (2006)
235. Peterman, E.J., Sosa, H., Moerner, W.E.: Single-molecule fluorescence spectroscopy and microscopy of biomolecular motors. Annu. Rev. Phys. Chem. **55**, 79–96 (2004)
236. Ohmachi, M., Komori, Y., Iwane, A.H., Fujii, F., Jin, T., et al.: Fluorescence microscopy for simultaneous observation of 3D orientation and movement and its application to quantum rod-tagged myosin V. Proc. Natl. Acad. Sci. U.S.A. **109**, 5294–5298 (2012)
237. Watanabe, T.M., Yanagida, T., Iwane, A.H.: Single molecular observation of self-regulated kinesin motility. Biochemistry **49**, 4654–4661 (2010)
238. Park, H., Toprak, E., Selvin, P.R.: Single-molecule fluorescence to study molecular motors. Q. Rev. Biophys. **40**, 87–111 (2007)
239. Lu, H.P., Xun, L., Xie, X.S.: Single-molecule enzymatic dynamics. Science **282**, 1877–1882 (1998)
240. Deniz, A.A., Laurence, T.A., Beligere, G.S., Dahan, M., Martin, A.B., et al.: Single-molecule protein folding: diffusion fluorescence resonance energy transfer studies of the denaturation of chymotrypsin inhibitor 2. Proc. Natl. Acad. Sci. U.S.A. **97**, 5179–5184 (2000)

241. Talaga, D.S., Lau, W.L., Roder, H., Tang, J., Jia, Y., et al.: Dynamics and folding of single two-stranded coiled-coil peptides studied by fluorescent energy transfer confocal microscopy. Proc. Natl. Acad. Sci. U.S.A. **97**, 13021–13026 (2000)
242. Zhuang, X., Bartley, L.E., Babcock, H.P., Russell, R., Ha, T., et al.: A single-molecule study of RNA catalysis and folding. Science **288**, 2048–2051 (2000)
243. Zhuang, X., Rief, M.: Single-molecule folding. Curr. Opin. Struct. Biol. **13**, 88–97 (2003)
244. Schuler, B., Lipman, E.Å., Eaton, W.A.: Probing the free-energy surface for protein folding with single-molecule fluorescence spectroscopy. Nature **419**, 743–747 (2002)
245. Chung, H.S., Cellmer, T., Louis, J.M., Eaton, W.A.: Measuring ultrafast protein folding rates from photon-by-photon analysis of single molecule fluorescence trajectories. Chem. Phys. **422**, 229–237 (2013)
246. Banerjee, P.R., Deniz, A.A.: Shedding light on protein folding landscapes by single-molecule fluorescence. Chem. Soc. Revs. **43**, 1172–1188 (2014)
247. Takei, Y., Iizuka, R., Ueno, T., Funatsu, T.: Single-molecule observation of protein folding in symmetric GroEL-$(GroES)_2$ complexes. J. Biol. Chem. **287**, 41118–41125 (2012)
248. Trexler, A.J., Rhoades, E.: Function and dysfunction of a-synuclein: probing conformational changes and aggregation by single molecule fluorescence. Mol. Neurobiol. **47**, 622–631 (2013)
249. Willets, K.A., Callis, P.R., Moerner, W.E.: Experimental and theoretical investigations of environmentally sensitive single-molecule fluorophores. J. Phys. Chem. B **108**, 10465–10473 (2004)
250. Betzig, E., Chichester, R.J.: Single molecules observed by near-field scanning optical microscopy. Science **262**, 1422–1425 (1993)
251. Magde, D., Elson, E., Webb, W.W.: Thermodynamic fluctuations in a reacting system - measurement by fluorescence correlation spectroscopy. Phys. Rev. Lett. **29**, 705–708 (1972)
252. Magde, D., Elson, E.L., Webb, W.W.: Fluorescence correlation spectroscopy. II. An experimental realization. Biopolymers **13**, 29–61 (1974)
253. Elson, E.L., Magde, D.: Fluorescence correlation spectroscopy. I. Conceptual basis and theory. Biopolymers **13**, 1–27 (1974)
254. Elson, E.: Fluorescence correlation spectroscopy: past, present, future. Biophys. J. **101**, 2855–2870 (2011)
255. Webb, W.W.: Fluorescence correlation spectroscopy: inception, biophysical experimentations and prospectus. Appl. Optics **40**, 3969–3983 (2001)
256. Fitzpatrick, J.A., Lillemeier, B.F.: Fluorescence correlation spectroscopy: linking molecular dynamics to biological function in vitro and in situ. Curr. Opin. Struct. Biol. **21**, 650–660 (2011)
257. Tian, Y., Martinez, M.M., Pappas, D.: Fluorescence correlation spectroscopy: a review of biochemical and microfluidic applications. Appl. Spectrosc. **65**, 115A–124A (2011)
258. Bevington, P.R., Robinson, D.K.: Data Reduction and Error Analysis for the Physical Sciences. McGraw-Hill, Boston (2003)
259. Kubo, R.: The fluctuation-dissipation theorem. Rept. Progr. Theor. Phys. **29**, 255–284 (1966)
260. Kubo, R., Toda, M., Hashitsume, N.: Statistical Physics II: Nonequilibrium Statistical Mechanics. Springer, Berlin (1985)
261. Parson, W.W., Warshel, A.: A density-matrix model of photosynthetic electron transfer with microscopically estimated vibrational relaxation times. Chem. Phys. **296**, 201–206 (2004)
262. Harp, G.D., Bern, B.J.: Time-correlation functions, memory functions, and molecular dynamics. Phys. Rev. A **2**, 975–996 (1970)
263. Hess, S.T., Webb, W.W.: Focal volume optics and experimental artifacts in confocal fluorescence correlation spectroscopy. Biophys. J. **83**, 2300–2317 (2002)
264. Maiti, S., Haupts, U., Webb, W.W.: Fluorescence correlation spectroscopy: diagnostics for sparse molecules. Proc. Natl. Acad. Sci. U.S.A. **94**, 11753–11757 (1997)
265. Jakobs, D., Sorkalla, T., Häberlein, H.: Ligands for fluorescence correlation spectroscopy on g protein-coupled receptors. Curr. Med. Chem. **19**, 4722–4730 (2012)

266. Widengren, J., Rigler, R.: Fluorescence correlation spectroscopy as a tool to investigate chemical reactions in solutions and on cell surfaces. Cell. Mol. Biol. **44**, 857–879 (1998)
267. van den Berg, P.A., Widengren, J., Hink, M.A., Rigler, R., Visser, A.J.: Fluorescence correlation spectroscopy of flavins and flavoenzymes: photochemical and photophysical aspects. Spectrochim. Acta A **57**, 2135–2144 (2001)
268. Haupts, U., Maiti, S., Schwille, P., Webb, W.W.: Dynamics of fluorescence fluctuations in green fluorescent protein observed by fluorescence correlation spectroscopy. Proc. Natl. Acad. Sci. U.S.A. **95**, 13573–13578 (1998)
269. Schenk, A., Ivanchenko, S., Röcker, C., Wiedenmann, J., Nienhaus, G.U.: Photodynamics of red fluorescent proteins studied by fluorescence correlation spectroscopy. Biophys. J. **86**, 384–394 (2004)
270. Chattopadhyay, K., Saffarian, S., Elson, E.L., Frieden, C.: Measuring unfolding of proteins in the presence of denaturant using fluorescence correlation spectroscopy. Biophys. J. **88**, 1413–1422 (2005)
271. Sanchez, S.A., Gratton, E.: Lipid-protein interactions revealed by two-photon microscopy and fluorescence correlation spectroscopy. Acc. Chem. Res. **38**, 469–477 (2005)
272. Felekyan, S., Sanabria, H., Kalinin, S., Kühnemuth, R., Seidel, C.A.: Analyzing Förster resonance energy transfer with fluctuation algorithms. Meth. Enzymol. **519**, 39–85 (2013)
273. Schwille, P., Meyer-Almes, F.J., Rigler, R.: Dual-color fluorescence cross-correlation spectroscopy for multicomponent diffusional analysis in solution. Biophys. J. **72**, 1878–1886 (1997)
274. Kettling, U., Koltermann, A., Schwille, P., Eigen, M.: Real-time enzyme kinetics of restriction endonuclease EcoR1 monitored by dual-color fluorescence cross-correlation spectroscopy. Proc. Natl. Acad. Sci. U.S.A. **95**, 1416–1420 (1998)
275. Bieschke, J., Giese, A., Schulz-Schaeffer, W., Zerr, I., Poser, S., et al.: Ultrasensitive detection of pathological prion protein aggregates by dual-color scanning for intensely fluorescent targets. Proc. Natl. Acad. Sci. U.S.A. **97**, 5468–5473 (2000)
276. Jahnz, M., Schwille, P.: An ultrasensitive site-specific DNA recombination assay based on dual-color fluorescence cross-correlation spectroscopy. Nucleic Acids Res. **33**, e60 (2005)
277. Collini, M., Caccia, M., Chirico, G., Barone, F., Dogliotti, E., et al.: Two-photon fluorescence cross-correlation spectroscopy as a potential tool for high-throughput screening of DNA repair activity. Nucleic Acids Res. **33**, e165 (2005)
278. Chen, Y., Müller, J.D., So, P.T.C., Gratton, E.: The photon counting histogram in fluorescence fluctuation spectroscopy. Biophys. J. **77**, 553–567 (1999)
279. Huang, B., Perroud, T.D., Zare, R.N.: Photon counting histogram: one-photon excitation. Chemphyschem. **5**, 1523–1531 (2004)
280. Perroud, T.D., Bokoch, M.P., Zare, R.N.: Cytochrome c conformations resolved by the photon counting histogram: watching the alkaline transition with single-molecule sensitivity. Proc. Natl. Acad. Sci. U.S.A. **102**, 17570–17575 (2005)
281. Siebrand, W.: Radiationless transitions in polyatomic molecules. I. Calculation of Franck-Condon factors. J. Chem. Phys. **46**, 440–447 (1967)
282. Siebrand, W.: Radiationless transitions in polyatomic molecules. II. Triplet-ground-state transitions in aromatic hydrocarbons. J. Chem. Phys. **47**, 2411–2422 (1967)
283. Henry, R.B., Siebrand, W.: Spin-orbit coupling in aromatic hydrocarbons. Analysis of nonradiative transitions between singlet and triplet states in benzene and naphthalene. J. Chem. Phys. **54**, 1072–1085 (1971)
284. Richards, W.G., Trivedi, H.P., Cooper, D.L.: Spin-orbit Coupling in Molecules. Clarendon, Oxford (1981)
285. McGlynn, S.P., Azumi, T., Kinoshita, M.: Molecular Spectroscopy of the Triplet State. Prentice Hall, Englewood Cliffs, NJ (1969)
286. Atkins, P.W.: Molecular Quantum Mechanics, 2nd edn. Oxford Univ. Press, Oxford (1983)
287. Shipman, L.: Oscillator and dipole strengths for chlorophyll and related molecules. Photochem. Photobiol. **26**, 287–292 (1977)

288. Takiff, L., Boxer, S.G.: Phosphorescence spectra of bacteriochlorophylls. J. Am. Chem. Soc. **110**, 4425–4426 (1988)
289. Shuvalov, V.A., Parson, W.W.: Energies and kinetics of radical pairs involving bacteriochlorophyll and bacteriopheophytin in bacterial reaction centers. Proc. Natl. Acad. Sci. U.S.A. **78**, 957–961 (1981)
290. Woodbury, N.W., Parson, W.W.: Nanosecond fluorescence from isolated reaction centers of Rhodopseudomonas sphaeroides. Biochim. Biophys. Acta **767**, 345–361 (1984)
291. Booth, P.J., Crystall, B., Ahmad, I., Barber, J., Porter, G., et al.: Observation of multiple radical pair states in photosystem 2 reaction centers. Biochemistry **30**, 7573–7586 (1991)
292. Ogrodnik, A., Keupp, W., Volk, M., Auermeier, G., Michel-Beyerle, M.E.: Inhomogeneity of radical pair energies in photosynthetic reaction centers revealed by differences in recombination diynamics of $P^{+}H_{A}^{-}$ when detected in delayed emission and absorption. J. Phys. Chem. **98**, 3432–3439 (1994)
293. Woodbury, N.W., Peloquin, J.M., Alden, R.G., Lin, X., Taguchi, A., Williams, J.C., et al.: Relationship between thermodynamics and mechanism during photoinduced charge separation in reaction centers from Rhodobacter sphaeroides. Biochemistry **33**, 8101–8112 (1994)
294. Che, A., Morrison, I.E., Pan, R., Cherry, R.J.: Restriction by ankyrin of band 3 rotational mobility in human erythrocyte membranes and reconstituted lipid vesicles. Biochemistry **36**, 9588–9595 (1997)
295. Jablonski, A.: Über den Mechanismus der Photolumineszenz von Farbstoffephosphoren. Z. Physik. **94**, 38–46 (1935)

Vibrational Absorption

6

6.1 Vibrational Normal Modes and Wavefunctions

Excitations of molecules to higher vibrational states typically occur in the mid-infrared region of the spectrum, between 200 and 5,000 cm^{-1} ($\lambda = 2.5$–50 μm). In this chapter we consider the main factors that determine the energies and strengths of vibrational excitations and describe several applications of infrared spectroscopy to macromolecules. Chapter 12 discusses Raman spectroscopy, in which vibrational transitions accompany the scattering of light at higher frequencies.

A molecule with N atoms has $3N$ degrees of motional freedom, of which 3 pertain to translation of the center of mass, 3 to rotation of the molecule as a whole, and the remaining $3N - 6$ to internal vibrations that leave the center of mass stationary. (A linear molecule has only 2 rotational degrees of freedom and $3N - 5$ vibrational modes.) In general, each vibrational mode of a complex molecule involves collective motions of many nuclei and cannot be described simply as the stretching or bending of an individual bond. However, analysis of the vibrations is simplified considerably if we assume that the vibrational potential energy is a harmonic function of the atomic coordinates. This means that the potential energy depends on quadratic terms such as x_{i}^2, $x_i x_j$, where x_i and x_j are displacements of any of the $3N$ Cartesian coordinates from their equilibrium values, but not on higher-order terms such as x_{i}^3, $x_{\mathrm{i}}^2 x_j$ or $x_i x_j x_k$. In this situation, it is possible to define a set of orthogonal *normal coordinates* made up of linear combinations of the individual atomic coordinates such that each vibrational mode, or *normal mode*, involves motion along a single normal coordinate. The vibrational potential energy of a molecule then can be written as

W.W. Parson, *Modern Optical Spectroscopy*, DOI 10.1007/978-3-662-46777-0_6

$$V = \frac{1}{2}\sum_i k_i \zeta_i^2, \tag{6.1}$$

where ζ_i is the normal coordinate for mode i and k_i is a force constant for this motion (Box 6.1).

As we discussed in Chap. 2, the solutions to the Schrödinger equation for a quadratic potential energy function of coordinate x are the harmonic-oscillator wavefunctions,

$$\chi_n = N_n H_n(u) exp\left(-u^2/2\right), \tag{6.2}$$

where the dimensionless coordinate u is $x/(\hbar/2\pi m_r \upsilon)^{1/2}$, m_r is the reduced mass of the system, υ is the classical vibrational frequency, $H_n(u_i)$ is a Hermite polynomial, and N_n is a normalization factor (Eq. 2.30 and Box 2.5). The frequency of a harmonic oscillator with force constant k_i is given by

$$\upsilon = (k/m_r)^{1/2}/2\pi. \tag{6.2}$$

Within the limits of the harmonic approximation, a vibrational wavefunction for a nonlinear molecule is simply a product of the wavefunctions of the $3N-6$ or $3N-5$ independent harmonic oscillators:

$$\mathrm{X}(x_1, x_2, \ldots) = \prod_{i=1}^{3N-6} \chi_{i(n)}(x_i). \tag{6.4}$$

And to the same approximation, the vibrational energy of a molecule is the sum of the energies of the individual normal modes:

$$E_{vib} = \sum_{i=1}^{3N-6} (n_i + 1/2) h\upsilon_i, \tag{6.5}$$

where n_i is the excitation level of mode i (Eq. 2.29).

Figure 6.1 illustrates the normal modes of linear and nonlinear triatomic molecules. Although the normal modes of larger molecules can be much more complicated and can involve movements of many atoms, low-frequency modes of proteins sometimes can be described as twisting or bending at hinges between domains, and such motions are often suggested to figure importantly in catalysis or other activities [1–4]. Higher-frequency modes also sometimes can be described fairly simply in terms of concerted motions of smaller groups of atoms.

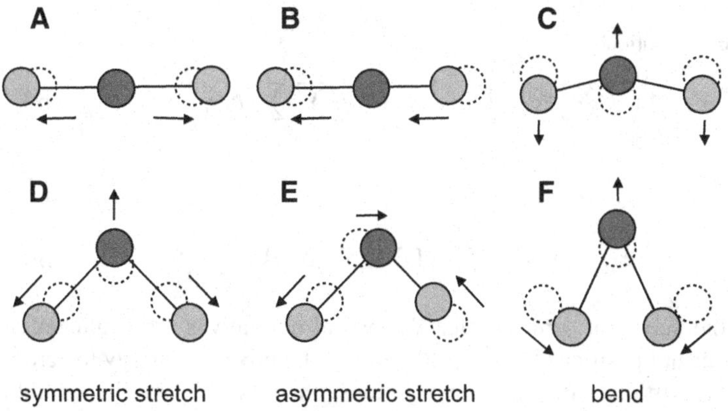

Fig. 6.1 The normal modes of linear (**A–C**) and nonlinear (**D–F**) triatomic molecules. The *open and filled circles* indicate positions of the atoms at two extremes of the motions; *arrows* show the movements in one direction between these extremes. The bending and symmetric stretching modes maintain reflection symmetry, whereas the asymmetric stretching mode destroys this symmetry. A linear triatomic molecule has an additional bending mode perpendicular to the plane of the drawing (not shown), but has only two rotational modes whereas a nonlinear triatomic has three. Both also have a single translational mode

Box 6.1 Normal Coordinates and Molecular-Dynamics Simulations

Consider a classical system with N particles and $3N$ Cartesian coordinates x_1, x_2, … x_{3N}. Let's define the coordinates of each atom relative to the equilibrium position of that atom, so that x_1, x_2 and x_3 are stated relative to the equilibrium position of atom 1; x_4, x_5 x_6, relative to the equilibrium position of atom 2; and so forth. The total kinetic energy of the system then is

$$T = (1/2)\sum_i m_i(\partial x_i/\partial t)^2, \quad \text{(B6.1.1)}$$

where m_i is the mass of the particle with coordinate x_i. We can simplify this expression by using "mass-weighted" coordinates, $\eta_i = (m_i)^{1/2}x_i$:

$$T = (1/2)\sum_i (\partial \eta_i/\partial t)^2. \quad \text{(B6.1.2)}$$

The potential energy of the system now can be expanded as a Taylor series in powers of the mass-weighted coordinates:

(continued)

Box 6.1 (continued)

$$V = V_0 + \sum_i (\partial V/\partial \eta_i)_0 \eta_i + (1/2) \sum_i \sum_j b_{ij} \eta_i \eta_j \ldots, \tag{B6.1.3a}$$

where

$$b_{ij} = \left(\partial^2 V/\partial \eta_i \partial \eta_j \right)_0, \tag{B6.1.3b}$$

and the subscript 0 means that the partial derivatives are evaluated at the equilibrium positions (η_i, $\eta_j = 0$). If we set the potential energy to zero when all the particles are at these positions, then $V_0 = 0$. At the equilibrium positions $\partial V/\partial \eta_i$ also must be zero for each coordinate, so neglecting cubic and higher-order terms, the potential energy is

$$V = (1/2) \sum_i \sum_j b_{ij} \eta_i \eta_j. \tag{B6.1.4}$$

Because the force acting along coordinate x_i is

$$F_i = -\partial V/\partial x_i = -m_i^{1/2} \partial V/\partial \eta_i = -m_i^{1/2} \sum_j b_{ij} \eta_j, \tag{B6.1.5}$$

and the acceleration on this coordinate is

$$\partial^2 x_i/\partial t^2 = m^{-1/2} \partial^2 \eta_i/\partial t^2, \tag{B6.1.6}$$

Newton's second law of motion can be written

$$\partial^2 \eta_i/\partial t^2 + \sum_j b_{ij} \eta_j = 0. \tag{B6.1.7}$$

We can solve Eq. (B6.1.7) immediately for η_i if the factors b_{ij} with $j \neq i$ are all zero, so that

$$\partial^2 \eta_i/\partial t^2 + b_{ij} \eta_j = 0. \tag{B6.1.8}$$

The solution is simply

$$\eta_i = \eta_i^o \sin\left(\sqrt{b_{ii}} t + \phi_i \right), \tag{B6.1.9}$$

where η_i^o is an arbitrary amplitude and ϕ_i is a phase shift.

(continued)

Box 6.1 (continued)

The matrix of second-order partial derivatives b_{ij} is referred to as the *Hessian* matrix of the system. Equation (B6.1.9) indicates that, if b_{ij} is zero for all $j \neq i$, then each of the mass-weighted coordinates will oscillate sinusoidally around its equilibrium value. The question now is whether it is always possible to convert Eq. (B6.1.7) into an expression with the form of (B6.1.8) by a linear transformation of variables, even if the b_{ij} with $j \neq i$ are not all zero. The answer is yes, as the following argument shows.

If the desired transformation *is* possible, the resulting normal coordinates (ζ_k with $k = 1, 2, \ldots 3N$) must allow us to write the kinetic and potential energies of the system as

$$T = (1/2)\sum_j \left(\partial_{\zeta_j}/\partial t\right)^2, \tag{B6.1.10}$$

and

$$V = (1/2)\sum_j \upsilon_j \zeta_j^2, \tag{B6.1.11}$$

where υ_j is a frequency that remains to be determined. The equation of motion for normal mode j then will be

$$\zeta_j = \zeta_j^o \sin\left(\sqrt{\upsilon_j} t + \phi_j\right). \tag{B6.1.12}$$

Suppose we start the system moving in a way that keeps all the amplitudes ζ_j^o zero except for one, ζ_k^o. The motions on normal coordinate k then will have the sinusoidal time dependence given by Eq. (B6.1.12). In principle, normal coordinate k could contain contributions from any or all of the individual mass-weighted coordinates (η_i). Conversely, since the η_i are linearly related to the normal coordinates, we can write

$$\eta_i = \sum_{k=1}^{3N} B_{ik}\zeta_k = \sum_{k=1}^{3N} B_{ik}\zeta_k^o \sin\left(\sqrt{\upsilon_k} t + \phi_k\right), \tag{B6.1.13}$$

where the coefficients B_{ik} also remain to be determined. Equation (B6.1.13) shows that any of the η_i for which B_{ik} is not zero must have the same dependence on time as normal coordinate k:

$$\eta_i = B_{ik}\zeta_k^o \sin\left(\sqrt{\upsilon_k} t + \delta_k\right). \tag{B6.1.14}$$

(continued)

Box 6.1 (continued)

All the nuclei thus will move with the same frequency and phase, but with varying amplitudes that depend on ζ_k^o and the coefficients B_{ik}.

Now consider the more general situation in which any of the normal-mode amplitudes (ζ_j^o) could be non-zero. Substituting Eq. (B6.1.14) in Eqs. (B6.1.10) and (B6.1.11) and using Newton's second law (Eq. B6.1.7) again gives

$$-\upsilon_j A_i + \sum_{i=1}^{3N} b_{ij} A_i = 0, \tag{B6.1.15}$$

where $A_i = B_{ij}\zeta_j^o$. This is a set of $3N$ simultaneous linear equations for the $3N$ unknown quantities A_i, and thus, once all the ζ_j^o are specified, for the coefficients B_{ij}. A trivial solution to these equations is that all the A_i are zero. As we show in Box 8.1, there will be one or more nontrivial solutions if, and only if, the determinant constructed from the b_{ij} and υ_j (the *secular determinant*) is zero:

$$\begin{vmatrix} b_{11}-\upsilon_j & b_{12} & \cdots & b_{13N} \\ b_{12} & b_{22}-\upsilon_j & \cdots & b_{23N} \\ \cdots & \cdots & \cdots & \cdots \\ b_{3N1} & b_{3N2} & \cdots & b_{3N3N}-\upsilon_j \end{vmatrix} = 0. \tag{B6.1.16}$$

Equation (B6.1.12) therefore must be a valid equation of motion for the system if frequency υ_j is one of the values that satisfy Eq. (B6.1.16). Once we obtain one of these roots, we can substitute it into Eq. (B6.1.15) to find all the A_i and hence the B_{ik}, subject to the initial conditions and the constraint

$$\sum_{i=1}^{3N} B_{ik}^2 = 1. \tag{B6.1.17}$$

This procedure gives $3N$ solutions to the equations of motion, one for each of the roots of Eq. (B6.1.16), although some of these solutions could be identical. Six of the solutions (or five for a linear molecule) describe translation or rotation of the molecule as a whole and have $\upsilon_j = 0$; the others give the normal modes we seek. The general solution for the motions of the system can be written as a sum of these particular solutions (Eq. B6.1.14), with amplitudes (ζ_j^o) and phases (δ_j) determined by the initial conditions. In practice, it is not necessary to work out the individual solutions one at a time in the manner outlined here; they can all be obtained directly by

(continued)

Box 6.1 (continued)
diagonalizing the Hessian matrix of partial derivatives (b_{ij}) corresponding to the determinant as described in Sect. 2.3.6.

A method for finding the normal modes of small molecules was devised by E.B. Wilson [5, 6]. The procedure involves constructing an $N \times N$ matrix of force constants (**F**) and a second matrix (**G**) whose elements depend on the molecular masses and the bond lengths and angles in the equilibrium geometry. Diagonalizing the product **FG** then gives the normal-mode vibrational frequencies and the coefficients for stretching or bending of various bonds in each of the normal modes. Information on the symmetry of the molecule can be used to facilitate setting up **G**, which usually is the most complex part of the problem [6–10].

This procedure bogs down for large molecules because the **F** and **G** matrices are based on internal molecular coordinates. It is much simpler to use Cartesian coordinates and to find the second derivatives of the potential energy with respect to the coordinates numerically. In this approach, which was pioneered by Lifson, Warshel and Levitt [11–13], the potential energy of a system is expressed as a sum of terms for bond stretching and bending, torsional twisting, van der Waals interactions, and electrostatic interactions (Fig. 6.2). The contribution of atom i to the potential energy thus can be written [14, 15]:

$$\begin{aligned} V_i(t) = {} & \sum_b^{\text{bonds}} \frac{k_b^l}{2}\left(l_b - l_b^0\right)^2 + \sum_a^{\text{angles}} \frac{k_a^\phi}{2}\left(\phi_b - \phi_a^o\right)^2 \\ & + \sum_b^{\text{torsions}} \frac{k_t^\theta}{2}\cos^2\left(n_t^o\theta_t - \theta_t^0\right) \\ & + \sum_{j \neq i}^{\text{atoms}} \left(A_i^0 r_{ij}^{-12} - B_i^0 r_{ij}^{-6} + q_i q_j r_{ij}^{-1} d_{ij}^{-1}\right) + \text{constant}. \end{aligned} \quad \text{(B6.1.18)}$$

The first sum in this expression runs over all the bonds to the atom, with l_b denoting the length of bond b and l_b^o the nominal length of a bond of this particular type. The second and third sums run over the bond angles (ϕ) and torsional (dihedral) angles (θ); the fourth, over the van der Waals and electrostatic interactions with atoms that are not bonded directly to atom i. In the final sum, q_i, q_j and r_{ij} are the partial charges and interatomic distance for atoms i and j, and d_{ij} is a dielectric screening function that increases with r_{ij}. The standard bond lengths and angles (l_b^o, ϕ_a^o, n_t^o and θ_t^o) are chosen to reflect crystallographic information, and the force constants (k_b^l, k_a^ϕ, and k_t^θ) are related to vibrational frequencies measured by IR spectroscopy. The atomic charges can be obtained from quantum calculations and measured

(continued)

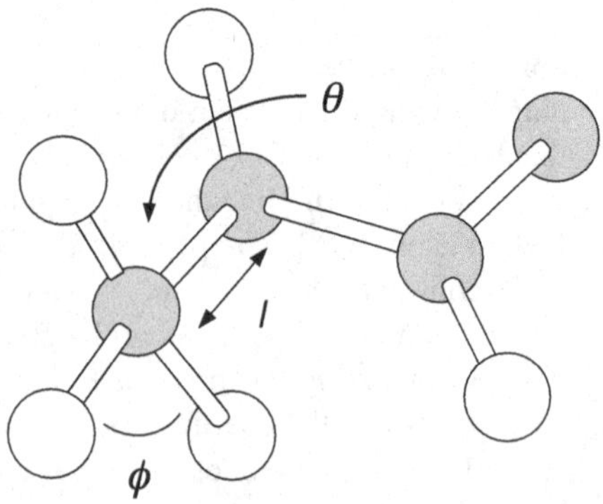

Fig. 6.2 The contributions of atom i to the classical potential energy of a system depend on the lengths (l) and angles (ϕ) of the bonds formed by the atom, the torsional rotations (θ) around bonds formed by neighboring atoms, and van der Waals and electrostatic interactions with other atoms

Box 6.1 (continued)

dipole moments, and the van der Waals parameters from molecular densities in crystals and liquids. Some treatments include additional terms that couple the bond-stretching energies to the bond-bending and torsional angles, or that treat dielectric screening by explicit induced dipoles rather than the screening function d_{ij}.

Using Eq. (B6.1.18), the partial derivatives (b_{ij}) that appear in Eqs. (B6.1.3–B6.1.7) can be computed by evaluating the changes in the potential energy that result from small changes in the atomic coordinates. Diagonalizing the Hessian matrix then gives the normal-mode frequencies and coordinates. Remember, however, that normal-mode analysis assumes that the system is at the potential-energy minimum. A macromolecule has a vast number of local energy minima, and is virtually never at the global minimum. For such a system, some of the energies obtained by diagonalizing the Hessian matrix will be negative, which means that the oscillation frequencies for those modes are imaginary numbers and the sine function in Eq. (B6.1.14) must be replaced by sinh. Useful information on collective properties of a system can still be obtained by considering only the modes with real frequencies [16].

Molecular-dynamics (MD) *simulations* [14, 17, 18] provide one way of taking a model of a complex structure to an energy minimum. In an MD simulation, the atoms in the starting model are assigned random initial velocities consistent with the total kinetic energy at a chosen temperature, and are allowed to move for a brief interval of time (typically 1 or 2 fs); the forces on each atom then are evaluated from the first derivatives of the potential-energy function (Eq. B6.1.18) with respect to the coordinates, and the resulting accelerations are calculated and used to update the velocities. The system is allowed to move for another interval, the forces are reevaluated, and this cycle is iterated many times with periodic corrections

(continued)

Box 6.1 (continued)
of the velocities if necessary to maintain a constant temperature. Runs at progressively decreasing temperatures sometimes are used to allow the system to explore a rugged energy surface on its way to (ideally) the region of the global minimum. Information on low-frequency normal modes can be incorporated in the forces to hasten evolution of the structure to a final form [19, 20]. The reliability of the results of course depends on the quality of the potential-energy function as well as the length of the simulation.

One use of MD simulations is to identify vibrational modes that are coupled to a process such as electron transfer. The fluctuating difference between the potential energies of the initial and final electronic states is evaluated at intervals during an MD trajectory. As discussed in Chaps. 10 and 11, a Fourier transform of the autocorrelation function of the fluctuations provides the frequencies of modes that are coupled to the transition [21, 22]. The amplitude of each peak in the transform is proportional to the displacement (Δ) of the potential curve for the product state along the vibrational coordinate for that mode. Figure 6.3A shows such a Fourier transform obtained from the calculated energy difference between the diabatic reactants and products of the initial electron-transfer step in photosynthetic bacterial reaction centers [22]. The cluster of modes with wavenumbers in the range of 300 to 400 cm^{-1} represent rotations of a tyrosine phenolic hydroxyl group that interacts electrostatically with the electron acceptor (Fig. 6.3B).

6.2 Vibrational Excitation

Excitation of a molecule to a higher vibrational state often can be described reasonably well as an elevation of an individual normal mode from quantum number n_i to a higher number, m_i. If the frequencies of the normal modes do not change significantly when the molecule is excited, the excitation energy is $(m_i - n_i)$ $h\upsilon_i$ where υ_i is the vibration frequency of the mode.

To investigate the selection rules and dipole strengths for vibrational excitations, we have to evaluate how the radiation field perturbs the vibrational energy and how this perturbation depends on the normal coordinate. Consider first a diatomic molecule so that we have only a single vibrational coordinate (x). The perturbation term in the Hamiltonian takes the form:

$$\widetilde{\mathrm{H}}'(x,t) \approx -\boldsymbol{E}(t) \cdot \boldsymbol{\mu}, \tag{6.6}$$

where $\boldsymbol{E}$ is the electric field and μ is the molecule's dipole, including contributions from both electronic and nuclear charges. With this expression for $\widetilde{\mathrm{H}}'$, the same line

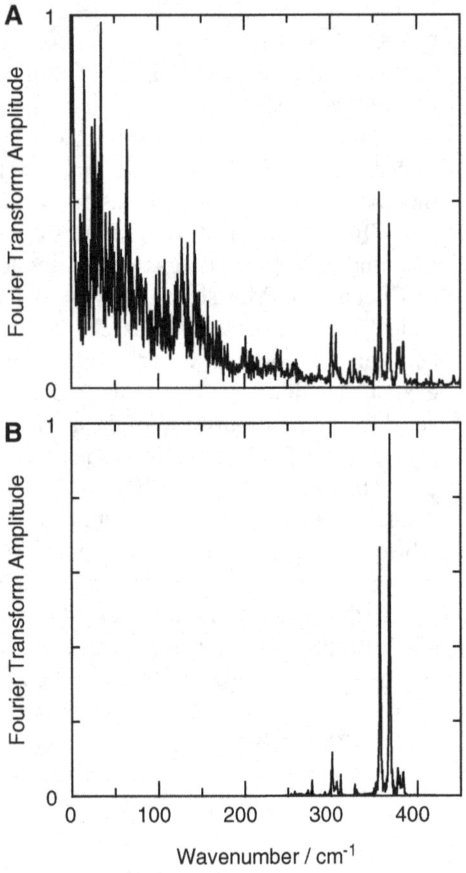

Fig. 6.3 Fourier transforms of autocorrelation functions of (**A**) the calculated energy difference between the reactants and products of the initial electron-transfer step in photosynthetic bacterial reaction centers, and (**B**) the C–C–O–H dihedral angle of the phenolic hydroxyl group of tyrosine M210. The amplitudes are scaled arbitrarily. Adapted from [22]

of reasoning that we applied to electronic transitions shows that the strength of an excitation from vibrational level n to level m depends on $|\boldsymbol{E}_o \cdot \boldsymbol{\mu}_{mn}|^2$, where $\boldsymbol{E}_o$ is the amplitude of the field and $\boldsymbol{\mu}_{mn}$ is the matrix element of the electric dipole operator,

$$\boldsymbol{\mu}_{mn} = \langle \chi_m | \tilde{\mu} | \chi_n \rangle. \tag{6.7}$$

The dipole moment of a molecule cannot be written rigorously as a simple analytical function of the normal coordinates. For one thing, because the individual atomic charges cannot be measured experimentally, any division of the molecule's total charge among the atoms depends on the choice of a model. In addition, changing the distance between two bonded atoms can alter the character of the bond, affect correlations among the electrons, and change the contribution of induced dipoles to the total dipole moment. For a polar bond betweeen atoms with different electronegativities, $|\boldsymbol{\mu}|$ generally increases with the length of the bond. For a diatomic molecule between more similar atoms, $|\boldsymbol{\mu}|$must go to zero at large distances when the free atoms dissociate, but can either increase or decrease

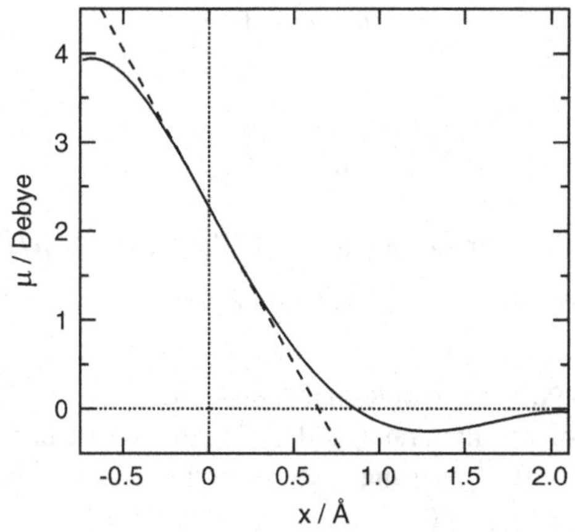

Fig. 6.4 Change in the calculated dipole ($\boldsymbol{\mu}$) of CS as the C–S bond is stretched or compressed. Adapted from [108]. The abscissa (x) is the change in the bond length from its equilibrium value of 1.539 Å, where the dipole moment is 2.69 D (C negative). The dipole reverses direction if x is increased above about 0.8 Å. The *dashed line* is the tangent to the curve at $x = 0$. See [108–110] for calculations on CO and other molecules

with smaller changes in distance (Fig. 6.4). We can, however, express $|\boldsymbol{\mu}|$ in a Taylor's series about the equilibrium value of the normal coordinate:

$$|\boldsymbol{\mu}(x)| = |\boldsymbol{\mu}(0)| + (\partial|\boldsymbol{\mu}|/\partial x)_0 x + \frac{1}{2}\left(\partial^2|\boldsymbol{\mu}|/\partial x^2\right)_0 x^2 + \cdots, \tag{6.8}$$

where x is the departure of the coordinate from its equilibrium value and the subscripts indicate that the derivatives are evaluated at the equilibrium position. The transition dipole can be written analogously:

$$\boldsymbol{\mu}_{mn} = \boldsymbol{\mu}(0)\langle\chi_m|\chi_n\rangle + (\partial\boldsymbol{\mu}/\partial x)_0\langle\chi_m|x|\chi_n\rangle + \frac{1}{2}\left(\partial^2\boldsymbol{\mu}/\partial x^2\right)_0\langle\chi_m|x^2|\chi_n\rangle + \cdots, \tag{6.9}$$

with the matrix elements also evaluated at $x = 0$.

The leading term in Eq. (6.9), $\boldsymbol{\mu}(0)\langle\chi_m|\chi_n\rangle$, is zero for $m \neq n$ because of the orthogonality of the eigenfunctions. The term $(\partial\boldsymbol{\mu}/\partial x)_0\langle\chi_m|x|\chi_n\rangle$ thus dominates the series, provided that $(\partial\boldsymbol{\mu}/\partial x)_0 \neq 0$. Examination of the eigenfunctions of a harmonic oscillator shows that the integral $\langle\chi_m|x|\chi_n\rangle$ in this term is non-zero only if $m = n \pm 1$ (Box 6.2). The formal selection rules for excitation of a harmonic vibration are, therefore:

$$(\partial\boldsymbol{\mu}/\partial x)_0 \neq 0, \tag{6.10}$$

$$m - n = \pm 1. \tag{6.11}$$

and

$$h\nu = h\upsilon, \tag{6.12}$$

where υ is the vibrational frequency at $x = 0$. The term $\frac{1}{2}\left(\partial^2 \boldsymbol{\mu}/\partial x^2\right)_0 \langle \chi_m | x^2 | \chi_n \rangle$ in Eq. (6.9) leads to a weak *overtone* transition at $2h\upsilon$.

Box 6.2 Selection Rules for Vibrational Transitions

The magnitude of the integral $\langle \chi_m | x | \chi_n \rangle$ in Eq. (6.9) can be evaluated for harmonic oscillators by using the recursion expression for Hermite polynomials:

$$uH_n(u) = (1/2)H_{n+1}(u) + nH_{n-1}(u). \tag{B6.2.1}$$

$$\begin{aligned} |\langle \chi_m | x | \chi_n \rangle| &= \int_{-\infty}^{\infty} \chi_m(u) u \kappa^{-1} \chi_n(u) dx = N_m N_n \kappa^{-1} \int_{-\infty}^{\infty} H_m(u) u H_n \exp(-u^2/2) dx \\ &= \frac{N_n \kappa^{-1/2}}{2N_{n+1}} \langle \chi_m | \chi_{n+1} \rangle + n \frac{N_n \kappa^{-1/2}}{N_{n-1}} \langle \chi_m | \chi_{n-1} \rangle \\ &= \left(\frac{n+1}{2\kappa} \right)^{1/2} \delta_{m,n+1} + \left(\frac{n}{2\kappa} \right)^{1/2} \delta_{m,n-1}. \end{aligned} \tag{B6.2.2}$$

(Eqs. 5.42–5.45). $\langle \chi_m | x | \chi_n \rangle$ thus is zero unless $m = n \pm 1$. The term $[(n+1)/2\kappa]^{1/2} \delta_{m,n+1}$ pertains to absorption (vibrational excitation), whereas $(n/2\kappa)^{1/2} \delta_{m,n-1}$ pertains to emission. Equation (B6.2.2) is identical to the expression we developed for emission and absorption of photons (Eq. 5.45). But note that here quantum numbers m and n refer to the excitation level of the molecular vibration, that is, the number of phonons in the mode, not to the density of photons in the incident radiation.

Similar arguments show that the integral $\langle \chi_m | x^2 | \chi_n \rangle$ in the second-order term in Eq. (6.9) is non-zero only for $m = n \pm 2$, and the integral $\langle \chi_m | x^3 | \chi_n \rangle$ in the third-order term is non-zero only for $m = n \pm 3$ [8]. The corresponding overtone absorption lines at $\nu = 2\upsilon$, 3υ and higher multiples of υ are weak for harmonic vibrational modes because $(\partial^2 \mu/\partial x^2)_0$, $(\partial^3 \mu/\partial x^3)_0$ and higher derivatives of μ usually are very small. However, such transitions can become increasingly allowed if the vibration is anharmonic. To illustrate this point, Fig. 6.5 shows the quantized wavefunctions and first-order transition dipoles ($\langle \chi_m | x | \chi_0 \rangle$) for an oscillator with the Morse potential that we considered in Fig. 2.1. Although the first few wavefunctions are qualitatively similar to those of a harmonic oscillator, those with higher quantum numbers are

(continued)

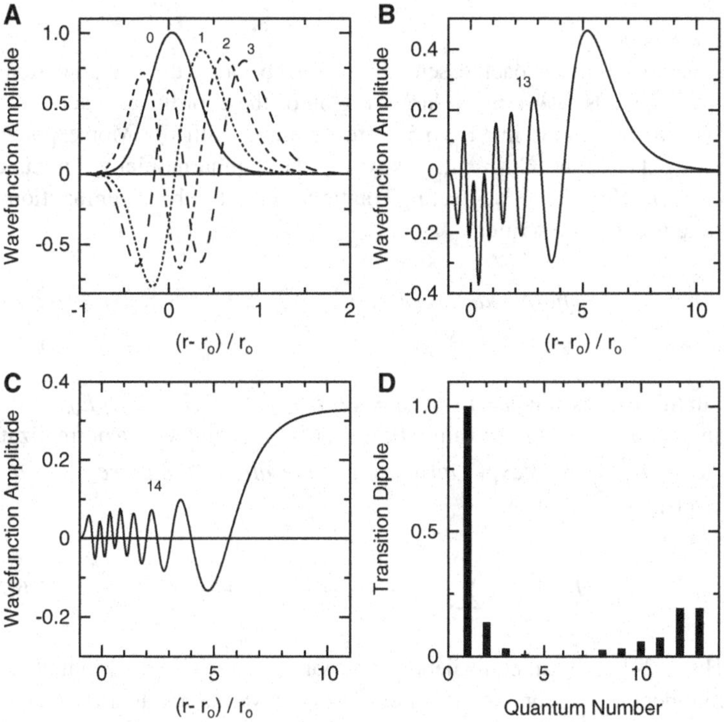

Fig. 6.5 Wavefunctions and transition dipole magnitudes for an anharmonic vibrational mode. (**A**) Relative amplitudes of wavefunctions 0–3 of an oscillator with the Morse potential illustrated in Fig. 2.1 (*curves 0, 1, 2* and *3*, respectively). Wavefunction 13 is shown in (**B**), and 14 in (**C**). The abscissa is the relative departure of the vibrational coordinate (r) from its equilibrium value (r_o). The curves are normalized to the same integrated probabilities (squares of the wavefunction amplitudes) in the range $0 < (r - r_o)/r_o \leq 11.5$, and are scaled relative to the peak of wavefunction 0. This normalization considers only part of wavefunction 14, which is at the dissociation energy and continues indefinitely off scale to the right. (**D**) The relative magnitudes of the transition dipoles ($\langle \chi_m | x | \chi_0 \rangle$) for excitation from the lowest level ($n = 0$) to each of the other levels below the dissociation limit. Most molecular vibrational potentials are more harmonic than the potential used for this illustration

Box 6.2 (continued)

weighted in favor of progressively larger values of the vibrational coordinate. As a result of this asymmetry, the matrix elements for the $0 \rightarrow 12$ and $0 \rightarrow 13$ transitions, which would be zero for a harmonic oscillator, are about 19% that for the $0 \rightarrow 1$ transition (Fig. 6.5D). The matrix element for the $0 \rightarrow 2$ transition (also zero for a harmonic oscillator) is about 14% that for $0 \rightarrow 1$. The matrix element for excitation to the highest quantized level ($m = 14$ in

(continued)

Box 6.2 (continued)
this illustration) drops back essentially to zero because the wavefunction for this level spreads over a much larger region of the coordinate space.

The wavefunctions in Fig. 6.5 were obtained with the Morse potential $V(r) = E_{diss}[1 - \exp(-2a(r - r_o)]^2$ with the anharmonicity parameter $a = 0.035/r_o$ (Fig. 2.1). Neglecting rotational effects, the eigenfunctions of this potential can be written

$$\psi_n(u) = N_n \exp[\eta exp(-au)][exp(-au)]^{(k-2n-1)/2} L_n^{(k-2n-1)}[2\eta exp(-au)]. \tag{B6.2.3}$$

[23]. In this expression, $u = r - r_o, \eta = (2\mu E_{diss})^{1/2}/a\hbar, k = 2(2\mu E_{diss})^{1/2}/a\hbar$, n is an integer with allowed values 0, 1, ... $(k-1)/2$, and N_n is a normalization constant. $L_n^{(k-2n-1)}[2\eta \exp(-au)]$ is a *generalized Laguerre polynomial* defined by

$$L_n^{(\alpha)}[x] = \sum_{m=0}^{n} \frac{(n+\alpha)!}{(n-m)!(m+\alpha)!} \frac{(-1)^m}{m!} x^m. \tag{B6.2.4}$$

Ter Haar [24] gives an equivalent expression in terms of a polynomial called the confluent hypergeometric function. The eigenvalues are, to a close approximation, $E_n = h\upsilon_o[(n + 1/2) - (h\upsilon_o/4E_{diss})(n + 1/2)^2]$ with $\upsilon_o = a(2E_{diss}/M)^{1/2}$, where M is the reduced mass of the system. Sage [25, 26] and Spirko et al. [27] give analytical expressions for $\langle\chi_m|x|\chi_n\rangle$ and matrix elements with higher powers of x.

We have assumed that the normal modes of the chromophore do not change significantly when the molecule is excited. For a more general treatment that covers breakdowns of this assumption (*Duschinsky effects*), the vibrational modes of the excited molecule can be written as linear combinations of the modes in the ground state [28–30].

The main conclusion that emerges from this analysis is that excitation of a harmonic vibrational mode requires $(\partial\mu/\partial x)_0$ to be non-zero: the transition is allowed only if the vibration perturbs the equilibrium geometry in a way that changes the molecule's dipole moment. A homonuclear diatomic molecule such as O_2 retains a dipole moment of zero for small changes in the bond length, and so cannot undergo pure vibrational excitations. In addition, because the vector $\langle\chi_m|x|\chi_n\rangle$ is directed along the bond axis in a diatomic molecule, a vibrational absorption band exhibits linear dichroism with respect to this axis. The extinction coefficient of the major IR absorption band of an oriented CO molecule thus will depend on $\cos^2\theta$, where θ is the angle between the electric field of the light and the C–O axis.

If the selection rule $m = n + 1$ is satisfied and $(\partial\boldsymbol{\mu}/\partial x)_0$ is non-zero, the dipole strength of the corresponding IR absorption band depends on $|\langle\chi_m|x|\chi_n\rangle|^2$, which for an harmonic oscillator is proportional to $(n + 1)$ (Box 6.2). The extinction coefficient for $(n = 1, m = 2)$ is, therefore, approximately twice that for $(n = 0, m = 1)$. However, only the $0 \rightarrow 1$ absorption band is seen in most cases because the population of the zero-point vibrational level $(n = 0)$ usually is much greater than the populations of higher levels. Excitation from level 1 to level 2 can be measured by two-dimensional IR spectroscopy as disussed in Chapt. 11. Such absorption typically occurs at a somewhat lower frequency than the $0 \rightarrow 1$ band because the vibrational potential surface is not strictly harmonic. For the stretching of CO or NO bound to the heme of myoglobin, the energy gap between levels 1 and 2 is about 1.5% smaller than that between 0 and 1. (The Morse potential used in Figs. 2.1 and 6.5 has a considerably greater anharmonicity of about 7.5%.) Excited-state absorption also is seen transiently in some systems that are created in excited vibrational states. Photodissociation of CO-myoglobin, for example, generates loosely bound CO with an IR absorption band at 2,085 cm^{-1}, and a weak satelite band at 2,059 cm^{-1} [31, 32]. The amplitude of the satelite band decays with a time constant of about 600 ps as the C=O vibration equilibrates thermally with other vibrational modes of the protein and solvent.

A Taylor's series expansion of the transition dipole for a polyatomic molecule leads to a an expression similar to Eq. (6.9):

$$\boldsymbol{\mu}_{mm} = \sum_{i=1}^{3N-6} (\partial\boldsymbol{\mu}/\partial x_i)_0 \langle\chi_m|x_i|\chi_n\rangle + \frac{1}{2}\sum_{i=1}^{3N-6}\sum_{j=1}^{3N-6}\left(\partial^2\boldsymbol{\mu}/\partial x_i\partial x_j\right)_0 \left\langle\chi_{m(i)}\chi_{m(j)}|x_i x_j|\chi_{n(i)}\chi_{n(j)}\right\rangle + \cdots, \tag{6.13}$$

where $\boldsymbol{\mu}$ refers to the total dipole of the molecule and the sums run over all the normal modes. Again, the dominant absorption bands represent the terms $(\partial\boldsymbol{\mu}/\partial x_i)_0\langle\chi_{m(i)}|x_i|\chi_{n(i)}\rangle$, which are non-zero if $m_i - n_i = \pm 1$ and $(\partial\boldsymbol{\mu}/\partial x_i)_0 \neq 0$; the higher-order terms lead to weaker overtone and combination excitations. The requirement for $(\partial\boldsymbol{\mu}/\partial x_i)_0$ to be non-zero means that infra-red transitions are forbidden for vibrations that are totally symmetric with respect to the molecular structure. In CO_2 or H_2O, for example, the antisymmetric stretching mode changes the dipole moment of the molecule and gives an allowed IR transition but the symmetric stretching mode does not (Fig. 6.1). We'll return to this point in Sect. 12.4 when we discuss selection rules for Raman scattering. The antisymmetric stretching mode of water has a molar extinction coefficient on the order of 100 to 100 M^{-1} cm^{-1}, and occurs frequencies between 2,700 and 3,700 cm^{-1}, shifting to higher energy as the strength of hydrogen bonding increases.

Table 6.1 IR absorption bands of peptides

Vibrational mode	Energy (cm^{-1})[a]	Dichroism[b]
N–H stretch		
α-helix	3,290–3,300	‖
β-sheet	3,280–3,300	⊥
Amide I (C=O stretch)		
α-helix	1,650–1,660	‖
β-sheet	1,620–1,640	⊥
Amide II (in-plane N–H bond)		
α-helix	1,540–1,550	⊥
β-sheet	1,520–1,525	‖

[a] In hydrogen-bonded peptide groups
[b] Polarization with respect to the peptide chain axis

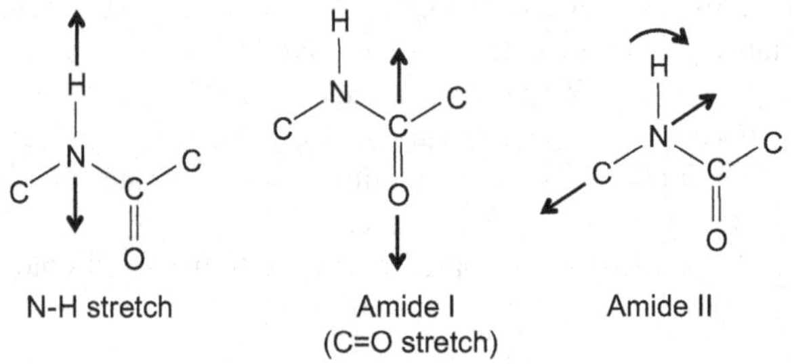

Fig. 6.6 The main vibrational modes of a peptide bond

6.3 Infrared Spectroscopy of Proteins

IR spectroscopy is widely used for structural analysis because many functional groups have characteristic vibrational frequencies. Vibrations of the peptide bond give rise to three major infrared (IR) absorption bands (Table 6.1, Fig. 6.6):

1. The N–H bond-stretching mode of hydrogen-bonded amide groups occurs in the region of 3,280 to 3,300 cm^{-1}. This absorption band is sometimes called the "amide A" band. In α-helical structures, The absorption is polarized parallel to the N-H bond, which is parallel to the helix axis in α-helical structures and perpendicular to the polypeptide chain in β-sheets. The band shifts to lower frequency as the strength of the hydrogen bonding increases [33].
2. The C=O stretching mode, with contributions from in-phase bending of the N-H bond and stretching of the C–N bond, occurs in the region of 1,650 to 1,660 cm^{-1} in α-helical structures and between 1,620 and 1,640 cm^{-1} in β-sheets [34, 35]. This band is often called the "amide I" band. Its polarization in α and β secondary structures is the same as that of the N–H stretching mode.
3. The "amide II" band occurs at lower energies than the amide I band, between 1,540 and 1,550 cm^{-1} in α-helices and between 1,520 and 1,525 cm^{-1} in β-sheets. This vibration involves in-plane bending of the N–H bond, coupled

to C–N and C_α–C stretching and C=O in-plane bending. It is polarized approximately along the C–N bond (nearly perpendicular to the axis of an α-helix and nearly parallel to the polypeptide chain in a β-sheet). In non-hydrogen bonded peptides, the N–H stretching and amide I mode shift to higher energies by 20 to 100 cm^{-1}, and the amide II mode shifts to a slightly lower energy.

IR spectra of polypeptides and proteins are affected by coupling of the vibrations of neighboring peptide groups [33, 36–39]. This coupling is analogous to exciton interactions of electronic transitions (Chap. 8), and it results in a similar splitting of the absorption bands. However, the situation is complicated by coupling of the transitions through both covalent bonds and hydrogen bonds as well as through space. In spite of this complexity, α-helices still exhibit a strong band polarized parallel to the helix axis corresponding to the amide I mode, and a perpendicularly polarized band at lower energy corresponding to the amide II mode. β-Sheet structures have diagnostic amide I bands that involve coupled motions of four peptide groups. Although amide-I absorption bands of regions of a protein with different secondary structure usually are not well resolved, the overall band sometimes can be dissected by fitting to a sum of gaussians [40, 41]. The IR absorption spectra of amino acid sidechains have been described by Venyaminov et al. [42].

IR spectroscopy has been used extensively to study the conformations of polypeptides and proteins in solution [40, 43–48]. In a seminal study [49, 50], Naik and Krimm determined the secondary structure of the ionophore gramicidin A in a variety of environments. Time-resolved measurements of changes in the amide IR bands have been used to study the dynamics of refolding of myoglobin following denaturation by a temperature jump [51–53]. However, the amide absorption bands are broad and their components usually are not well resolved. This problem can be addressed by labeling the amide groups of selected residues with ^{13}C or $^{13}C{=}^{18}O$ to make their vibration frequencies distinguishable from those of the unlabeled residues [54–58]. Two-dimensional spectroscopic techniques discussed in Chap. 11 provide another way to address the problem.

Infrared spectroscopic studies of macromolecules became increasingly powerful with the development of Fourier transform techniques [44, 47, 48, 59–67]. (See Chap. 1 for a description of an FTIR spectrometer.) FTIR measurements can be used to probe changes in the bonding or interactions of individual amino acid side chains in proteins. Bacteriorhodopsin provides an illustration. When bacteriorhodopsin is illuminated, its protonated retinylidine Schiff base chromophore isomerizes and then transfers a proton to a group in the protein. FTIR measurements showed the formation of an absorption band at 1,763 cm^{-1} in addition to a set of absorption changes attributable to the chromophore [63, 68]. In bacteriorhodopsin that was enriched in [4-^{13}C]-aspartic acid the band appeared at 1,720 cm^{-1}, and an additional shift to 1,712 cm^{-1} was obtained when the sovent was replaced by D_2O. These observations indicated that the band reflected C=O stretching of a protonated aspartic acid, leading to identification of a particular aspartic acid residue as the H^+ acceptor for deprotonation of the chromophore.

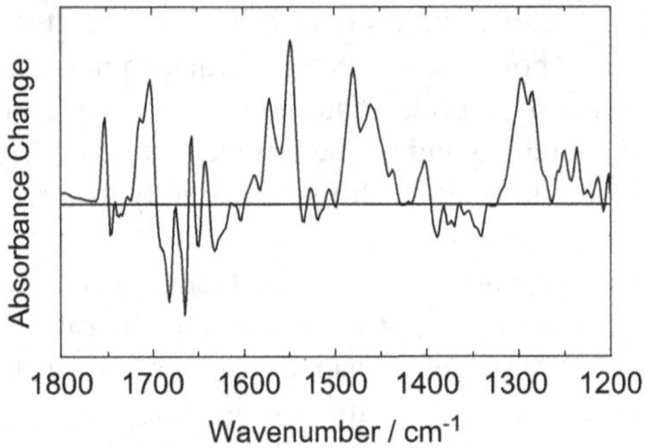

Fig. 6.7 FTIR difference spectrum (light-minus-dark) of the absorbance changes associated with electron transfer from the special pair of bacteriochlorophylls (P) to a quinone (Q_A) in photosynthetic reaction centers of *Rhodobacter sphaeroides*. The negative absorption changes result mainly from loss of absorption bands of P; the positive changes, from the absorption bands of the oxidized dimer (P^+). These measurements were made with a thin film of reaction centers at 100 K. The amplitudes are scaled arbitrarily. Adapted from [101]

In photosynthetic reaction centers, FTIR measurements combined with site-directed mutagenesis and isotopic substitutions have been used to identify residues that interact with the electron carriers, or that bind a proton when one of the carriers is reduced [61, 64, 69–74]. Figure 6.7 shows a spectrum of the absorbance changes that result from illumination of bacterial reaction centers at low temperatures. The absorption increases at 1,703 and 1,713 cm^{-1} are assigned to the C=O stretching mode of the 13^1-keto groups of the two bacteriochlorophylls that make up the photochemical electron donor (P), when the dimer is in its oxidized form (P^+). The absorption decreases at 1,682 and 1,692 cm^{-1} mark the positions of the same vibrations when the dimer is in its neutral form. The bands of the two bacteriochlorophylls appear at somewhat different frequencies as a result of differences in hydrogen bonding and local electrical fields (Sect. 6.4).

As we mentioned in Chap. 4, the linear dichroism of IR absorption bands was used to determine the orientation of CO bound to the heme of myoglobin in solution and the change in orientation that occurs upon photodissociation [75]. A 35-ps pulse of polarized, 527-nm light was used to dissociate bound CO, and a 0.2-ps pulse of IR light was used to probe the C–O stretching mode. An absorption band attributable to this mode occurs near 1,900 cm^{-1} when CO is bound to the heme and near 2,100 cm^{-1} when the CO is released from the heme but remains in a pocket of the protein. The polarization of the flash-induced absorbance changes showed that the C–O bond of the bound molecule is nearly normal to the plane of the heme, whereas the released CO is oriented approximately parallel to the plane. These observations necessitated a rethinking of ideas about how myoglobin and hemoglobin discriminate against binding of CO in favor of O_2. Shifts in the IR absorption band also revealed that the photodissociated CO can bind to the protein in several different ways [76, 77].

Other applications of polarized IR spectroscopy in molecular biophysics have included studies of the orientations of tryptophan side chains in a filamentous virus [78] and studies of folding of integral membrane proteins [48, 79].

Chiral molecules exhibit vibrational circular dichroism (VCD) analogous to electronic CD, and the underlying theory is essentially the same (Chap. 9). Measurements of VCD, though still much less common than measurements of electronic CD, have been stimulated by improvements in instrumentation and procedures for predicting the VCD spectra of small molecules [80–83], and have been correlated with secondary structural elements in proteins [46, 84–86]. Stretching modes of heme-ligand bonds in some mutant hemoglobins and myoglobins exhibit an unusually strong VCD that is sensitive to interactions of the ligand with the protein [87].

6.4 Vibrational Stark Effects

Because the vibrations that underlie IR absorption spectra must affect the electric dipole of a molecule, we would expect the frequencies of these modes to be sensitive to local electric fields, and this is indeed the case. Shifts in vibration frequencies caused by external electric fields can be measured in essentially the same manner as electronic Stark shifts, by recording oscillations of the IR transmission in the presence of oscillating fields. The *Stark tuning rate* is defined as $\delta_\upsilon = \partial\overline{\upsilon}/\partial E_\upsilon$, where $\overline{\upsilon}$ is the wavenumber of the mode and E_υ is the projection of the field ($\boldsymbol{E}$) on the normal coordinate [88, 89]. To a first approximation, δ_υ is given by $-\hat{u}\cdot(\boldsymbol{\Delta\mu}+\boldsymbol{E}\cdot\Delta\alpha)/hc$, where $\hat{u}$ is a unit vector parallel to the normal coordinate, $\boldsymbol{\Delta\mu}$ is the difference between the molecule's dipole moments in the excited and ground states, and $\Delta\alpha$ is the difference between the polarizability tensors in the two states (Sect. 4.13, Box 4.15 and Box 12.1). However, anharmonicity and geometrical distortions caused by the field also can contribute to vibrational Stark effects [90, 91].

Typical Stark tuning rates for carbonyl stretching modes are on the order of $(1/f)$ $cm^{-1}/(MV\ cm^{-1})$ where f is the local-field correction factor [90, 92]. The C=O stretching mode of CO bound to the heme Fe of myoglobin has a relatively large Stark tuning rate about $2.4/f\ cm^{-1}/(MV\ cm^{-1})$ [92]. The frequency of this mode varies between 1,937 and 1,984 cm^{-1} in mutant myoglobins, cytochromes and other heme proteins, and differences in the local field probably account for much of this variation [92–98]. Measurements of the Stark effect with CO bound to a Ni surface showed that the Stark tuning rate for the CO vibration is positive when the field makes the potential more positive at the C atom relative to the O [88]. Boxer and coworkers have measured Stark tuning rates for a the stretching modes of C–F, C–D, and a variety of other chemical bonds that can be introduced at specific sites in proteins as potential reporters for local electric fields [90, 99, 100]. Mutations of photosynthetic bacterial reaction centers shift the C=O vibration frequencies of the

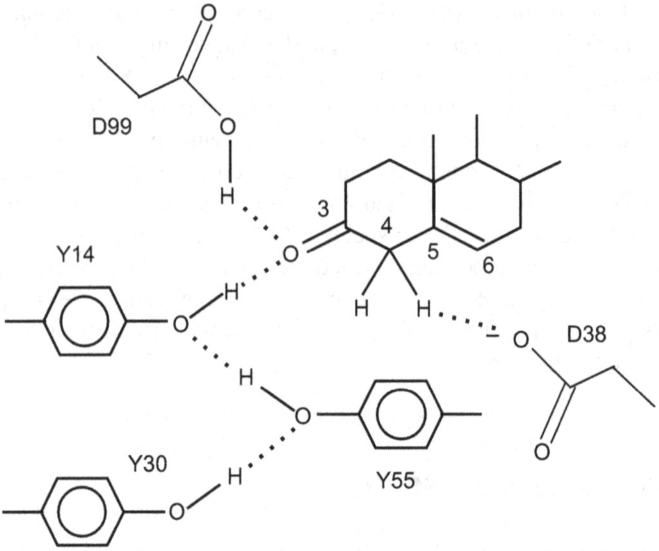

Fig. 6.8 Residues in the active site of Δ^5-3-ketosteroid isomerase from *Escherichia coli* [111]. Aspartic acid (*D*) residues 38 and 99, tyrosines (*Y*) 14, 30 and 55, and atoms 3–6 of a bound steroid substrate are labeled. (Add 2 to the residue numbers to get those in the widely studied enzyme from *Pseudomonas putida*.) *Dotted lines* indicate likely hydrogen bonds. The enzyme catalyzes rearrangement of the C5–C6 double bond to C4–C5 by facilitating enolization of the keto group and transfer of a proton from C4 to C6, probably via D38. The shifted C=O stretching frequencies of bound 19-nortestosterone point to a strong local electric field that favors the enolization [106]

13^1-keto groups of the two bacteriochlorophylls of the special pair, and the shifts correlate well with calculated changes in the local fields [101].

The bacterial enzyme Δ^5-3-ketosteroid isomerase shifts the C=O stretching mode of 19-nortestosterone (a 3-ketosteroid substrate analog) dramatically from 1,638 to 1,612 cm^{-1} [102, 103]. This shift does not occur in the enzymatically inactive Y14F mutant, and is diminished in other mutants with reduced activities. Figure 6.8 shows some of the residues surrounding the bound substrate in the enzyme's crystal structure. Measurements and MD simulations of the C=O stretching frequency of 19-nortestosterone in a variety of solvents indicate that Y14 and other groups in the active site of the isomerase create a strong electric field of approximately 144 $MVcm^{-1}$ parallel to the C=O axis, and that this field accounts for most of the decrease in the activation energy for the catalytic reaction [104–107]. Although Y14 forms a hydrogen bond to the keto O, the phenolic C-O stretching mode of Y14 itself is not shifted greatly, probably because of balancing effects of H-bonds from Y30 and Y55.

Exercises

1. In the molecular vibrations diagrammed below, ↔ denotes bonds that are stretched at a particular time, and ↔ denotes bonds that are compressed at this time. Which of these vibrations can result in infrared absorption?

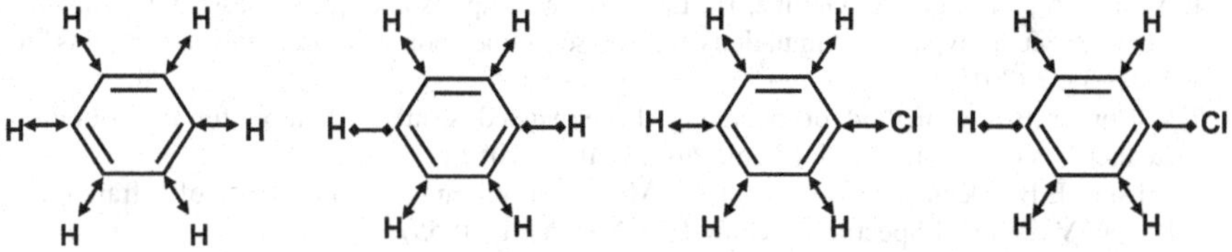

2. Consider the excitations (0 → 1), (0 → 2) and (1 → 2) of a weakly anharmonic oscillator, where the two numbers for each excitation denote the initial and final vibrational quantum numbers. (a) If the (0 → 1) excitation has a relative dipole strength of 1.0, what is the expected relative dipole strength of the (1 → 2) excitation? (b) Suppose the (0 → 1) absorption has an energy of 200 cm^{-1}. Estimate the ratio of the observed strengths of the (1 → 2) and (0 → 1) absorption bands at 295 K. (c) Would the excitation energy of the (1 → 2) band be the same as, smaller than, or greater than that of the (0 → 1) band?
3. The figure below shows hypothetical IR absorption (—) and linear dichroism (- - -) spectra of membrane vesicles containing a protein that transports a sugar and Na^+ ions across a cellular membrane. The vesicles are flattened by deposition and partial drying on a glass slide, and linear dichroism is measured with respect to an axis normal to the slide. In the presence of the sugar (*S*), the linear dichroism in the region of 1,650 cm^{-1} becomes more positive, and that around 1,560 cm^{-1} becomes more negative. Suggest an interpretation of the spectra and the effect of the sugar.

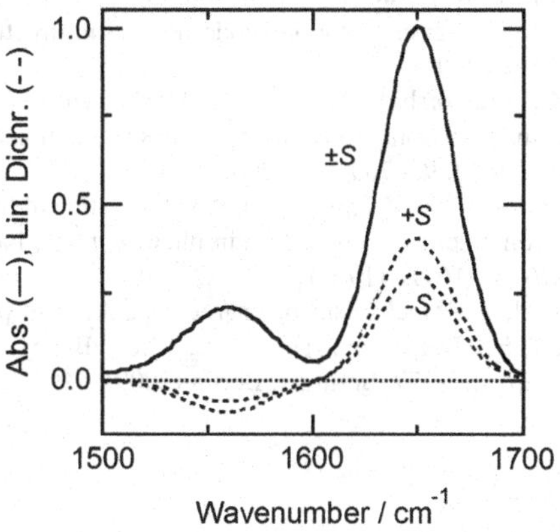

References

1. Levitt, M., Sander, C., Stern, P.S.: Protein normal-mode dynamics: trypsin inhibitor, crambin, ribonuclease and lysozyme. J. Mol. Biol. **181**, 423–447 (1985)
2. Bruccoleri, R.E., Karplus, M., McCammon, J.A.: The hinge-bending mode of a lysozyme-inhibitor complex. Biopolymers **25**, 1767–1802 (1986)
3. Ma, J.: Usefulness and limitations of normal mode analysis in modeling dynamics of biomolecular complexes. Structure **13**, 373–380 (2005)
4. Ahmed, A., Villinger, S., Gohlke, H.: Large-scale comparison of protein essential dynamics from molecular dynamics simulations and coarse-grained normal mode analysis. Proteins **78**, 3341–3352 (2010)
5. Wilson, E.B.: A method of obtaining the expanded secular equation for the vibration frequencies of a molecule. J. Chem. Phys. **7**, 1047–1052 (1939)
6. Wilson, E.B., Decius, J.C., Cross, P.C.: Molecular Vibrations. The Theory of Infrared and Raman Vibrational Spectra. McGraw-Hill, New York (1955)
7. Painter, P.C., Coleman, M.M., Koenig, J.L.: The Theory of Vibrational Spectroscopy and Its Application to Polymeric Materials. Wiley Interscience, New York (1982)
8. Struve, W.S.: Fundamentals of Molecular Spectroscopy. Wiley Interscience, New York (1989)
9. Cotton, F.A.: Chemical Applications of Group Theory, 3rd edn. Wiley, New York (1990)
10. McHale, J.L.: Molecular Spectroscopy. Upper Saddle River, NJ, Prentice Hall (1999)
11. Lifson, S., Warshel, A.: Consistent force field for calculations of conformations, vibrational spectra, and enthalpies of cycloalkane and n-alkane molecules. J. Chem. Phys. **49**, 5116–5129 (1968)
12. Warshel, A., Levitt, M., Lifson, S.: Consistent force field for calculation of vibrational spectra and conformations of some amides and lactam rings. J. Mol. Spectrosc. **33**, 84–99 (1970)
13. Warshel, A., Lifson, S.: Consistent force field calculations. II. Crystal structures, sublimation energies, molecular and lattice vibrations, molecular conformations, and enthalpies of alkanes. J. Chem. Phys. **53**, 582–594 (1970)
14. Warshel, A.: Computer Modeling of Chemical Reactions in Enzymes and Solutions. Wiley, New York (1991)
15. Rapaport, D.C.: The Art of Molecular Dynamics Simulation. Cambridge Univ. Press, Cambridge (1997)
16. Buchner, M., Ladanyi, B.M., Stratt, R.M.: The short-time dynamics of molecular liquids. Instantaneous-normal-mode theory. J. Chem. Phys. **97**, 8522–8535 (1992)
17. Hansson, T., Oostenbrink, C., van Gunsteren, W.: Molecular dynamics simulations. Curr. Opin. Struct. Biol. **12**, 190–196 (2002)
18. Adcock, S.A., McCammon, J.A.: Molecular dynamics: survey of methods for simulating the activity of proteins. Chem. Rev. **106**, 1589–1615 (2006)
19. Bahar, I., Rader, A.J.: Course-grained normal mode analysis in structural biology. Curr. Opin. Struct. Biol. **15**, 586–592 (2005)
20. Isin, B., Schulten, K., Tajkhorshid, E., Bahar, I.: Mechanism of signal propagation upon retinal isomerization: insights from molecular dynamics simulations of rhodopsin restrained by normal modes. Biophys. J. **95**, 789–803 (2008)
21. Alden, R.G., Parson, W.W., Chu, Z.T., Warshel, A.: Orientation of the OH dipole of tyrosine (M)210 and its effect on electrostatic energies in photosynthetic bacterial reaction centers. J. Phys. Chem. **100**, 16761–16770 (1996)
22. Parson, W.W., Warshel, A.: Mechanism of charge separation in purple bacterial reaction centers. In: Hunter, C.N., Daldal, F., Thurnauer, M.C., Beatty, J.T. (eds.) The Purple Phototropic Bacteria, pp. 355–377. Springer, Berlin (2009)

23. Morse, P.M.: Diatomic molecules according to the wave mechanics. II. Vibrational levels. Phys. Rev. **34**, 57–64 (1929)
24. ter Haar, D.: The vibrational levels of an anharmonic oscillator. Phys. Rev. **70**, 222–223 (1946)
25. Sage, M.L.: Morse oscillator transition probabilities for molecular bond modes. Chem. Phys. **35**, 375–380 (1978)
26. Sage, M.L., Williams, J.A.I.: Energetics, wave functions, and spectroscopy of coupled anharmonic oscillators. J. Chem. Phys. **78**, 1348–1358 (1983)
27. Spirko, V., Jensen, P., Bunker, P.R., Cejhan, A.: The development of a new Morse-oscillator based rotation vibration Hamiltonian for H^{3+}. J. Mol. Spectrosc. **112**, 183–202 (1985)
28. Sharp, T.E., Rosenstock, H.M.: Franck-Condon factors for polyatomic molecules. J. Chem. Phys. **41**, 3453–3463 (1964)
29. Sando, G.M., Spears, K.G.: Ab initio computation of the Duschinsky mixing of vibrations and nonlinear effects. J. Phys. Chem. A **104**, 5326–5333 (2001)
30. Sando, G.M., Spears, K.G., Hupp, J.T., Ruhoff, P.T.: Large electron transfer rate effects from the Duschinsky mixing of vibrations. J. Phys. Chem. A **105**, 5317–5325 (2001)
31. Sagnella, D.E., Straub, J.E.: A study of vibrational relaxation of B-state carbon monoxide in the heme pocket of photolyzed carboxymyoglobin. Biophys. J. **77**, 70–84 (1999)
32. Sagnella, D.E., Straub, J.E., Jackson, T.A., Lim, M., Anfinrud, P.A.: Vibrational population relaxation of carbon monoxide in the heme pocket of photolyzed carbonmonoxy myoglobin: comparison of time-resolved mid-IR absorbance experiments and molecular dynamics simulations. Proc. Natl. Acad. Sci. U.S.A. **96**, 14324–14329 (1999)
33. Krimm, S., Bandekar, J.: Vibrational spectroscopy and conformation of peptides, polypeptides, and proteins. Adv. Prot. Chem. **38**, 181–364 (1986)
34. Chirgadze, Y.N., Nevskaya, N.A.: Infrared spectra and resonance interaction of amide-I vibration of the antiparallel-chain pleated sheet. Biopolymers **15**, 607–625 (1976)
35. Nevskaya, N.A., Chirgadze, Y.N.: Infrared spectra and resonance interactions of amide-I and II vibration of alpha-helix. Biopolymers **15**, 637–648 (1976)
36. Miyazawa, T., Shimanouchi, T., Mizushima, J.: Normal vibrations of N-methylacetamide. J. Chem. Phys. **29**, 611–616 (1958)
37. Miyazawa, T.: Perturbation treatment of the characteristic vibrations of polypeptide chains in various configurations. J. Chem. Phys. **32**, 1647–1652 (1960)
38. Brauner, J.W., Dugan, C., Mendelsohn, R.: ^{13}C labeling of hydrophobic peptides. Origin of the anomalous intensity distribution in the infrared amide I spectral region of b-sheet structures. J. Am. Chem. Soc. **122**, 677–683 (2000)
39. Brauner, J.W., Flach, C.R., Mendelsohn, R.: A quantitative reconstruction of the amide I contour in the IR spectra of globular proteins: from structure to spectrum. J. Am. Chem. Soc. **127**, 100–109 (2005)
40. Byler, D.M., Susi, H.: Examination of the secondary structure of proteins by deconvolved FTIR spectra. Biopolymers **25**, 469–487 (1986)
41. Susi, H., Byler, D.M.: Resolution-enhanced Fourier transform infrared spectroscopy of enzymes. Meth. Enzymol. **25**, 469–487 (1986)
42. Venyaminov, S.Y., Yu, S., Kalnin, N.N.: Quantitative IR spectrophotometry of peptide compounds in water (H_2O) solutions. I. Spectral parameters of amino acid residue absorption bands. Biopolymers **30**, 1243–1257 (1990)
43. Surewicz, W.K., Mantsch, H.H., Chapman, D.: Determination of protein secondary structure by Fourier-transform infrared spectroscopy: a critical assessment. Biochemistry **32**, 389–394 (1993)
44. Siebert, F.: Infrared spectroscopy applied to biochemical and biological problems. Methods Enzymol. **246**, 501–526 (1995)
45. Jackson, M., Mantsch, H.: The use and misuse of FTIR spectroscopy in the determination of protein structure. Crit. Rev. Biochem. Mol. Biol. **30**, 95–120 (1995)

46. Baumruk, V., Pancoska, P., Keiderling, T.A.: Predictions of secondary structure using statistical analyses of electronic and vibrational circular dichroism and Fourier transform infrared spectra of proteins in H_2O. J. Mol. Biol. **259**, 774–791 (1996)
47. Kötting, C., Gerwert, K.: Proteins in action monitored by time-resolved FTIR spectroscopy. Chemphyschem. **6**, 881–888 (2005)
48. Haris, P.I.: Probing protein-protein interaction in biomembranes using Fourier transform infrared spectroscopy. Biochim. Biophys. Acta **1828**, 2265–2271 (2013)
49. Naik, V.M., Krimm, S.: Vibrational analysis of peptides, polypeptides, and proteins. 33. Vibrational analysis of the structure of gramicidin A. 1. Normal mode analysis. Biophys. J. **46**, 1131–1145 (1986)
50. Naik, V.M., Krimm, S.: Vibrational analysis of peptides, polypeptides, and proteins. 34. Vibrational analysis of the structure of gramicidin A. 2. Vibrational spectra. Biophys. J. **49**, 1147–1154 (1986)
51. Gilmanshin, R., Williams, S., Callender, R.H., Woodruff, W.H., Dyer, R.B.: Fast events in protein folding: relaxation dynamics of the I form of apomyoglobin. Biochemistry **36**, 15006–15012 (1997)
52. Gilmanshin, R., Callender, R.H., Dyer, R.B.: The core of apomyoglobin E-form folds at the diffusion limit. Nature Struct. Biol. **5**, 363–365 (1998)
53. Callender, R.H., Dyer, R.B., Gilmanshin, R., Woodruff, W.H.: Fast events in protein folding: the time evolution of primary processes. Ann. Rev. Phys. Chem. **49**, 173–202 (1998)
54. Brewer, S.H., Song, B.B., Raleigh, D.P., Dyer, R.B.: Residue specific resolution of protein folding dynamics using isotope-edited infrared temperature jump spectroscopy. Biochemistry **46**, 3279–3285 (2007)
55. Nagarajan, S., Taskent-Sezgin, H., Parul, D., Carrico, I., Raleigh, D.P., et al.: Differential ordering of the protein backbone and side chains during protein folding revealed by site-specific recombinant infrared probes. J. Am. Chem. Soc. **133**, 20335–20340 (2007)
56. Hauser, K., Krejtschi, C., Huang, R., Wu, L., Keiderling, T.A.: Site-specific relaxation kinetics of a tryptophan zipper hairpin peptide using temperature-jump IR spectroscopy and isotopic labeling. J. Am. Chem. Soc. **130**, 2984–2992 (2008)
57. Ihalainen, J.A., Paoli, B., Muff, S., Backus, E.H.G., Bredenbeck, J., et al.: α-Helix folding in the presence of structural constraints. Proc. Natl. Acad. Sci. U.S.A. **105**, 9588–9593 (2008)
58. Kubelka, G.S., Kubelka, J.: Site-specific thermodynamic stability and unfolding of a de novo designed protein structural motif mapped by ^{13}C isotopically edited IR spectroscopy. J. Am. Chem. Soc. **136**, 6037–6048 (2014)
59. Griffiths, P.R., deHaseth, J.A.: Fourier Transform Infrared Spectrometry. Wiley, New York (1986)
60. Braiman, M.S., Rothschild, K.J.: Fourier transform infrared techniques for probing membrane protein structure. Annu. Rev. Biophys. Biophys. Chem. **17**, 541–570 (1988)
61. Mäntele, W.: Infrared vibrational spectroscopy of the photosynthetic reaction center. In: Deisenhofer, J., Norris, J.R. (eds.) The Photosynthetic Reaction Center, pp. 240–284. Academic, San Diego (1993)
62. Slayton, R.M., Anfinrud, P.A.: Time-resolved mid-infrared spectroscopy: methods and biological applications. Curr. Opin. Struct. Biol. **7**, 717–721 (1997)
63. Gerwert, K.: Molecular reaction mechanisms of proteins monitored by time-resolved FTIR-spectroscopy. Biol. Chem. **380**, 931–935 (1999)
64. Berthomieu, C., Hienerwadel, R.: Fourier transform (FTIR) spectroscopy. Photosynth. Res. **101**, 157–170 (2009)
65. Nienhaus, K., Nienhaus, G.U.: Ligand dynamics in heme proteins observed by Fourier transform infrared-temperature derivative spectroscopy. Biochim. Biophys. Acta **1814**, 1030–1041 (2011)
66. Li, J.J., Yip,. C.M.: Super-resolved FT-IR spectroscopy: strategies, challenges, and opportunities for membrane biophysics. Biochim. Biophys. Acta **1828**, 2272–2282 (2013)

67. Lewis, R.N., McElhaney, R.N.: Membrane lipid phase transitions and phase organization studied by Fourier transform infrared spectroscopy. Biochim. Biophys. Acta **1828**, 2347–2358 (2013)
68. Engelhard, M., Gerwert, K., Hess, B., Kreutz, W., Siebert, F.: Light-driven protonation changes of internal aspartic acids of bacteriorhodopsin - an investigation by static and time-resolved infrared difference spectroscopy using [4-^{13}C] Aspartic acid labeled purple membrane. Biochemistry **24**, 400–407 (1985)
69. Mäntele, W., Wollenweber, A., Nabedryk, E., Breton, J.: Infrared spectroelectrochemistry of bacteriochlorophylls and bacteriopheophytins. Implications for the binding of the pigments in the reaction center from photosynthetic bacteria. Proc. Natl. Acad. Sci. U.S.A. **85**, 8468–8472 (1988)
70. Leonhard, M., Mantele, W.: Fourier-transform infrared spectroscopy and electrochemistry of the primary electron donor in Rhodobacter sphaeroides and Rhodopseudomonas viridis reaction centers. Vibrational modes of the pigments in situ and evidence for protein and water modes affected by P^+ formation. Biochemistry **32**, 4532–4538 (1993)
71. Breton, J., Nabedryk, E., Allen, J.P., Williams, J.C.: Electrostatic influence of Q_A reduction on the IR vibrational mode of the 10a-ester C=O of H_A demonstrated by mutations at residues Glu L104 and Trp L100 in reaction centers from Rhodobacter sphaeroides. Biochemistry **36**, 4515–4525 (1997)
72. Breton, J., Nabedryk, E., Leibl, W.: FTIR study of the primary electron donor of photosystem I (P700) revealing delocalization of the charge in P700$^+$ and localization of the triplet character in ^{3}P700. Biochemistry **38**, 11585–11592 (1999)
73. Breton, J.: Fourier transform infrared spectroscopy of primary electron donors in type I photosynthetic reaction centers. Biochim. Biophys. Acta **1507**, 180–193 (2001)
74. Noguchi, T., Fukami, Y., Oh-Oka, H., Inoue, Y.: Fourier transform infrared study on the primary donor P798 of Heliobacterium modesticaldum: Cysteine S-H coupled to P798 and molecular interactions of carbonyl groups. Biochemistry **36**, 12329–12336 (1997)
75. Lim, M., Jackson, T.A., Anfinrud, P.A.: Binding of CO to myoglobin from a heme pocket docking site to form nearly linear Fe-C-O. Science **269**, 962–966 (1995)
76. Lim, M.H., Jackson, T.A., Anfinrud, P.A.: Modulating carbon monoxide binding affinity and kinetics in myoglobin: the roles of the distal histidine and the heme pocket docking site. J. Biol. Inorg. Chem. **2**, 531–536 (1997)
77. Lehle, H., Kriegl, J.M., Nienhaus, K., Deng, P.C., Fengler, S., et al.: Probing electric fields in protein cavities by using the vibrational Stark effect of carbon monoxide. Biophys. J. **88**, 1978–1990 (2005)
78. Tsuboi, M., Overman, S.A., Thomas Jr., G.J.: Orientation of tryptophan-26 in coat protein subunits of the filamentous virus Ff by polarized Raman microspectroscopy (1996). Biochemistry **35**, 10403–10410 (1996)
79. Hunt, J.F., Earnest, T.N., Bousche, O., Kalghatgi, K., Reilly, K., et al.: A biophysical study of integral membrane protein folding. Biochemistry **36**, 15156–15176 (1997)
80. Stephens, P.J.: Theory of vibrational circular dichroism. J. Phys. Chem. **89**, 748–752 (1985)
81. Buckingham, A.D., Fowler, P.W., Galwas, P.A.: Velocity-dependent property surfaces and the theory of vibrational circular dichroism. Chem. Phys. **112**, 1–14 (1987)
82. Amos, R.D., Handy, N.C., Drake, A.F., Palmieri, P.: The vibrational circular dichroism of dimethylcyclopropane in the C-H stretching region. J. Chem. Phys. **89**, 7287–7297 (1988)
83. Stephens, P.J., Devlin, F.J.: Determination of the structure of chiral molecules using ab initio vibrational circular dichroism spectroscopy. Chirality **12**, 172–179 (2000)
84. Keiderling, T.A.: Protein and peptide secondary structure and conformational determination with vibrational circular dichroism. Curr. Opin. Chem. Biol. **6**, 682–688 (2002)
85. Pancoska, P., Wang, L., Keiderling, T.A.: Frequency analysis of infrared absorption and vibrational circular dichroism of proteins in D_2O solution. Protein Sci. **2**, 411–419 (1993)
86. Matsuo, K., Hiramatsu, H., Gekko, K., Namatame, H., Taniguchi, M., et al.: Characterization of intermolecular structure of β_2-microglobulin core fragments in amyloid fibrils by vacuum-

ultraviolet circular dichroism spectroscopy and circular dichroism theory. J. Phys. Chem. B **118**, 2785–2795 (2014)

87. Bormett, R.W., Asher, S.A., Larkin, P.J., Gustafson, W.G., Ragunathan, N., et al.: Selective examination of heme protein azide ligand-distal globin interactions by vibrational circular dichroism. J. Am. Chem. Soc. **114**, 6864–6867 (1992)
88. Lambert, D.K.: Vibrational Stark effect of carbon monoxide on nickel(100), and carbon monoxide in the aqueous double layer: experiment, theory, and models. J. Chem. Phys. **89**, 3847–3860 (1988)
89. Boxer, S.G.: Stark realities. J. Phys. Chem. B **113**, 2972–2983 (2009)
90. Park, E.S., Boxer, S.G.: Origins of the sensitivity of molecular vibrations to electric fields: carbonyl and nitrosyl stretches in model compounds and proteins. J. Phys. Chem. B **106**, 5800–5806 (2002)
91. Brewer, S.H., Franzen, S.: A quantitative theory and computational approach for the vibrational Stark effect. J. Chem. Phys. **119**, 851–858 (2003)
92. Park, E.S., Andrews, S.S., Hu, R.B., Boxer, S.G.: Vibrational stark spectroscopy in proteins: A probe and calibration for electrostatic fields. J. Phys. Chem. B **103**, 9813–9817 (1999)
93. Park, K.D., Guo, K., Adebodun, F., Chiu, M.L., Sligar, S.G., et al.: Distal and proximal ligand interactions in heme proteins: correlations between C-O and Fe-C vibrational frequencies, oxygen-17 and carbon-13 nuclear magnetic resonance chemical shifts, and oxygen-17 nuclear quadrupole coupling constants in $C^{17}O$- and ^{13}CO-labeled species. Biochemistry **30**, 2333–2347 (1991)
94. Jewsbury, P., Kitagawa, T.: The distal residue-CO interaction in carbonmonoxy myoglobins: a molecular dynamics study of two distal histidine tautomers. Biophys. J. **67**, 2236–2250 (1994)
95. Li, T., Quillin, M.L., Phillips, G.N.J., Olson, J.S.: Structural determinants of the stretching frequency of CO bound to myoglobin. Biochemistry **33**, 1433–1446 (1994)
96. Ray, G.B., Li, X.-Y., Ibers, J.A., Sessler, J.L., Spiro, T.G.: How far can proteins bend the FeCO unit? Distal polar and steric effects in heme proteins and models. J. Am. Chem. Soc. **116**, 162–176 (1994)
97. Laberge, M., Vanderkooi, J.M., Sharp, K.A.: Effect of a protein electric field on the CO stretch frequency. Finite difference Poisson-Boltzmann calculations on carbonmonoxycytochromes c. J. Phys. Chem. **100**, 10793–10801 (1996)
98. Phillips, G.N.J., Teodoro, M.L., Li, T., Smith, B., Olson, J.S.: Bound CO is a molecular probe of electrostatic potential in the distal pocket of myoglobin. J. Phys. Chem. B **103**, 8817–8829 (1999)
99. Chattopadhyay, A., Boxer, S.G.: Vibrational Stark-effect spectroscopy. J. Am. Chem. Soc. **117**, 1449–1450 (1995)
100. Suydam, I.T., Boxer, S.G.: Vibrational Stark effects calibrate the sensitivity of vibrational probes for electric fields in proteins. Biochemistry **42**, 12050–12055 (2003)
101. Johnson, E.T., Müh, F., Nabedryk, E., Williams, J.C., Allen, J.P., et al.: Electronic and vibronic coupling of the special pair of bacteriochlorophylls in photosynthetic reaction centers from wild-type and mutant strains of Rhodobacter sphaeroides. J. Phys. Chem. B **106**, 11859–11869 (2002)
102. Austin, J.C., Kuliopulos, A., Mildvan, A.S., Spiro, T.G.: Substrate polarization by residues in Δ^5-3-ketosteroid isomerase probed by site-directed mutagenesis and UV resonance Raman spectroscopy. Protein Sci. **1**, 259–270 (1992)
103. Austin, J.C., Zhao, Q., Jordan, T., Talalay, P., Mildvan, A.S., et al.: Ultraviolet resonance Raman spectroscopy of Δ^5-3-ketosteroid isomerase revisited: substrate polarization by active-site residues. Biochemistry **34**, 4441–4447 (1995)
104. Fried, S.D., Bagchi, S., Boxer, S.G.: Measuring electrostatic fields in both hydrogen-bonding and non-hydrogen-bonding environments using carbonyl vibrational probes. J. Am. Chem. Soc. **135**, 11181–11192 (2013)

105. Fried, S.D., Wang, L.-P., Boxer, S.G., Ren, P., Pande, V.S.: Calculations of electric fields in liquid solutions. J. Phys. Chem. B **117**, 16236–16248 (2013)
106. Fried, S.D., Bagchi, S., Boxer, S.G.: Extreme electric fields power catalysis in the active site of ketosteroid isomerase. Science **346**, 1510–1514 (2014)
107. Hildebrandt, P.: More than fine tuning: local electric fields accelerate an enzymatic reaction. Science **346**, 1456–1457 (2014)
108. Harrison, J.F.: Relationship between the charge distribution and dipole moment functions of CO and the related molecules CS SiO and SiS. J. Phys. Chem. A **110**, 10848–10857 (2006)
109. Harrison, J.F.: A Hirschfeld-I interpretation of the charge distribution, dipole and quadrupole moments of the halogenated acetylenes FCCH, ClCCH BrCCH and ICCH. J. Chem. Phys. **133**, 214103 (2010)
110. Cheam, T.C., Krimm, S.: Infrared intensities of amide modes of N-methylacetamide and poly (glycine I) from ab initio calculations of dipole moment derivatives of N-methylacetamide. J. Chem. Phys. **82**, 1631–1641 (1985)
111. Kim, S.W., Cha, S.-S., Cho, H.-S., Kim, J.-S., Ha, N.-C., et al.: High-resolution crystal structures of Δ^5-3-ketosteroid isomerase with and without a reaction intermediate analogue. Biochemistry **36**, 14030–14036 (1997)

Resonance Energy Transfer

7

7.1 Introduction

One way that an excited molecule can return to the ground state is to transfer the excitation energy to another molecule. This process, *resonance energy transfer*, plays a particularly important role in photosynthetic organisms. Extended arrays of pigment-protein complexes in the membranes of plants and photosynthetic bacteria absorb sunlight and transfer energy to the reaction centers, where the energy is trapped in electron-transfer reactions [1, 2]. In other organisms, photolyases, which use the energy of blue light to repair ultraviolet damage in DNA, contain a pterin or deazaflavin that transfers energy efficiently to a flavin radical in the active site [3]. A similar antenna is found in cryptochromes, which appear to play a role in circadian rhythms [4]. Because the rate of resonance energy transfer depends on the distance between the energy donor and acceptor, the process also is used experimentally to probe intermolecular distances in biophysical systems [5]. Typical applications are to measure the distance between two proteins in a multienzyme complex or between ligands bound at two sites on a protein, or to examine the rate at which components from two membrane vesicles mingle in a fused vesicle. An inquiry into the mechanism of resonance energy transfer also provides a springboard for discussing other time-dependent processes such as electron transfer.

Consider two identical molecules for which the wavefunctions of the ground states are $\phi_{1a}\chi_{1a}$ and $\phi_{2a}\chi_{2a}$, where ϕ and χ represent electronic and nuclear wavefunctions, respectively, and subscripts 1 and 2 denote the molecules. Suppose, first, that the two molecules do not interact. The Hamiltonian for the dimer then is just the sum of the Hamiltonians for the individual molecules:

$$\tilde{\mathrm{H}} = \tilde{\mathrm{H}}_1 + \tilde{\mathrm{H}}_2, \tag{7.1}$$

where $\tilde{\mathrm{H}}_1$ operates only on molecule 1 and $\tilde{\mathrm{H}}_2$ operates only on molecule 2. The Schrödinger equation for the ground state of the combined system is satisfied by writing the wavefunction as a simple product of the molecular wavefunctions:

W.W. Parson, *Modern Optical Spectroscopy*, DOI 10.1007/978-3-662-46777-0_7

$$\Psi_A = \phi_{1a}\chi_{1a}\phi_{2a}\chi_{2a}. \tag{7.2}$$

If each of the individual molecules has energy E_a in the ground state, the energy of the dimer's ground state is simply $2E_a$:

$$\begin{aligned}
&\langle \phi_{1a}\chi_{1a}\phi_{2a}\chi_{2a}|\tilde{\mathrm{H}}_1 + \tilde{\mathrm{H}}_2|\phi_{1a}\chi_{1a}\phi_{2a}\chi_{2a}\rangle \\
&\quad = \langle \phi_{1a}\chi_{1a}|\tilde{\mathrm{H}}_1|\phi_{1a}\chi_{1a}\rangle\langle \phi_{2a}\chi_{2a}|\phi_{2a}\chi_{2a}\rangle + \langle \phi_{2a}\chi_{2a}|\tilde{\mathrm{H}}_2|\phi_{2a}\chi_{2a}\rangle\langle \phi_{1a}\chi_{1a}|\phi_{1a}\chi_{1a}\rangle \\
&\quad = E_a + E_a = 2E_a.
\end{aligned} \tag{7.3}$$

If either molecule can be raised to an excited state $\phi_b\chi_b$ with energy E_b, there are two possible excited states of the dimer:

$$\psi_1 = \phi_{1b}\chi_{1b}\phi_{2a}\chi_{2a} \quad \text{(molecule 1 excited)}, \tag{7.4a}$$

or

$$\psi_1 = \phi_{1a}\chi_{1a}\phi_{2b}\chi_{2b} \quad \text{(molecule 2 excited)}. \tag{7.4b}$$

As long as the two molecules do not interact, both ψ_1 and ψ_2 are eigenfunctions of the total Hamiltonian, and both states will have the same energy, $E_a + E_b$. In addition,

$$\langle \psi_1|\tilde{\mathrm{H}}_1 + \tilde{\mathrm{H}}_2|\psi_2\rangle = \langle \psi_2|\tilde{\mathrm{H}}_1 + \tilde{\mathrm{H}}_2|\psi_1\rangle = 0. \tag{7.5}$$

This means that states ψ_1 and ψ_2 are stationary states: the excitation has no tendency to hop from one molecule to the other.

The wavefunction for the excited dimer also could be written as a linear combination of ψ_1 and ψ_2:

$$\Psi_B = C_1\psi_1 + C_2\psi_2 = C_1\phi_{1b}\chi_{1b}\phi_{2a}\chi_{2a} + C_2\phi_{1a}\chi_{1a}\phi_{2b}\chi_{2b}, \tag{7.6}$$

with $|C_1|^2 + |C_2|^2 = 1$. In this representation, $|C_1|^2$ is the probability that molecule 1 is excited and $|C_2|^2$ the probability that molecule 2 is excited. The energy of the excited state comes out the same ($E_a + E_b$) if we use this representation and is independent of the values of C_1 and C_2 as long as long as the sum of their squares is 1. The coefficients can have any magnitude between -1 and $+1$, and they could be complex numbers. But if we excite molecule 1 so that $|C_1|^2 = 1$, $|C_2|$ will remain zero indefinitely and *vice versa*.

Now let's allow the two molecules to interact. This adds a new term to the Hamiltonian:

$$\tilde{\mathrm{H}} = \tilde{\mathrm{H}}_1 + \tilde{\mathrm{H}}_2 + \tilde{\mathrm{H}}_{21}. \quad (7.7)$$

Because of the perturbation represented by $\tilde{\mathrm{H}}_{21}$, ψ_1 and ψ_2 are no longer stationary states. The interaction term allows the system to change between states ψ_1 and ψ_2, so that the excitation energy moves back and forth between the two molecules. This is resonance energy transfer. The interactions also could change the total energy of the system, but we'll put off a discussion of this point for now and focus on the rate of transfer of energy between the molecules when the interactions are too weak to affect the energy significantly.

7.2 The Förster Theory

Suppose we know that the excitation is on molecule 1 at zero time. How fast will it move to molecule 2? Let's start by describing the system by Eq. (7.6) with $C_1 = 1$ and $C_2 = 0$. Then, using Eq. (2.61), we can write

$$\partial C_2/\partial t = -(i/\hbar)H_{21}\exp[i(E_2 - E_1)t/\hbar], \quad (7.8)$$

where E_1 and E_2 are the energies of ψ_1 and ψ_2, and H_{21} is the interaction matrix element, $\langle\psi_2|\tilde{\mathrm{H}}_{21}|\psi_1\rangle$. To find the probability that the excitation appears on molecule 2 after a short interval of time (τ), we can obtain $C_2(\tau)$ by integrating Eq. (7.8) from $t = 0$ to τ, and then evaluate $C_2^*(\tau)C_2(\tau)$.

If H_{21} is independent of time, and we restrict ourselves to intervals that are short enough so that $|C_2|^2 \ll 1$, Eq. (7.8) can be integrated immediately:

$$C_2(\tau) = H_{21}(1 - \exp[i(E_2 - E_1)t/\hbar])/(E_2 - E_1). \quad (7.9)$$

As we found when we considered the absorption of light, Eq. (7.9) implies that $C_2^*(\tau)C_2(\tau)$ can have a significant magnitude only if $|E_2 - E_1|$ is close to zero; this is the resonance condition. In the case of absorption, one of the energies of the total system includes the photon energy $h\nu$ in addition to the energy of an unexcited molecule; in resonance energy transfer we simply have two different ways of placing an excitation in a system containing two molecules. The requirement for matching the energies of the two states is illustrated in Fig. 7.1.

If we measure the rate of resonance energy transfer in a population of many donor-acceptor pairs the energy difference $E_2 - E_1$ will vary from pair to pair because the molecules will be in many different vibrational states. In addition, the energy distributions will be broadened by relaxations of the excited states (Sects. 2.6 and 10.7) and by inhomogeneous interactions with the surroundings. The measured rate will depend on an integral over the distribution of energies. For each value of E_1 in a particular donor-acceptor pair, we first need to integrate over all possible values of E_2. This integral can be evaluated as outlined in Box 4.6, and the results are similar: the amount of energy transfer occurring in time τ is

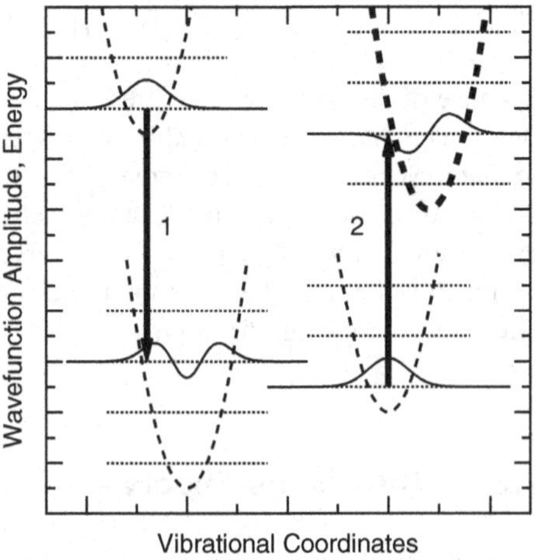

Fig. 7.1 Resonance energy transfer requires coupled downward and upward vibronic transitions. The energy lost by molecule 1 must match the energy gained by molecule 2 so that energy is conserved in the overall process, and the Franck-Condon factors must be nonzero for both transitions

$$\int_{-\infty}^{\infty} C_2^*(\tau, E_{21}) C_2(\tau, E_{21}) \rho_{s2}(E_2) dE_2 = \frac{2\pi\tau}{\hbar} |H_{21}|^2 \rho_{s2}(E_1), \tag{7.10}$$

where $E_{21} = E_2 - E_1$ and $\rho_{s2}(E)$ is the *density of final states* defined so that $\rho_{s2}(E)dE$ is the number of final states with energies in the small interval between E and $E + dE$. We are interested specifically in $\rho_{s2}(E_1)$, the density of states around the energy of the initial state, E_1. Equation (7.10) is *Fermi's golden rule*. The expression we derived in Chap. 4 for the rate of absorption of light (Eq. 4.8c) is the result of the golden rule for the case that $|H_{21}|^2 = |\mathbf{E}_o \cdot \boldsymbol{\mu}|^2$ and the density of radiation states per unit energy is h^{-1} times the density of oscillation modes per unit frequency ($\rho_\nu(\nu)$). We also used the golden rule in discussing electron-transfer reactions in Chap. 5 (Eq. B.5.3.3), and in Sect. 10.4 we'll derive it by a different approach and reexamine the limits under which it holds.

To obtain the overall rate constant, we must integrate the expression in Eq. (7.10) over the distribution of the initial energies, $\rho_{s1}(E_1)$, and then divide by τ:

$$k_{rt} = \frac{2\pi}{\hbar} \int_{-\infty}^{\infty} |H_{21}|^2 \rho_{s2}(E_1) \rho_{s1}(E_1) dE_1. \tag{7.11}$$

The interaction matrix element H_{21} in the integrand must consider the initial and final nuclear states of the energy donor and acceptor in addition to the electronic wavefunctions. However, to the extent that the Born-Oppenheimer approximation is valid, the nuclei will not move significantly during the instant when the excitation energy jumps between the molecules. H_{21} thus can be approximated as a product of a purely electronic interaction matrix element ($H_{21(el)}$) and two nuclear overlap integrals [*cf.* Eq. (4.42)]:

$$H_{21} = \langle \phi_{1a}\phi_{2b}|\tilde{\mathrm{H}}_{21}|\phi_{1b}\phi_{2a}\rangle\langle\chi_{1a}|\chi_{1b}\rangle\langle\chi_{2b}|\chi_{2a}\rangle \\ = H_{21(el)}\langle\chi_{1a}|\chi_{1b}\rangle\langle\chi_{2b}|\chi_{2a}\rangle. \tag{7.12}$$

In general, the nuclear wavefunctions χ_b and χ_a could represent any of many different vibrational states of the system. We have to weight the contribution from each of these nuclear states by the appropriate Boltzmann factor. Taking the Boltzmann factors into account gives the following expression for the rate constant:

$$k_{rt} = \frac{2\pi}{\hbar}|H_{21(el)}|^2\left\{\sum_n\sum_m \frac{\exp(-E_{n(1)}/k_BT)}{Z_1}|\langle\chi_{1m}|\chi_{1n}\rangle|^2 \right. \\ \left. \times\sum_u\sum_w \frac{\exp(-E_{u(2)}/k_BT)}{Z_2}|\langle\chi_{2w}|\chi_{2u}\rangle|^2\right\}\delta(\Delta E_1 - \Delta E_2). \tag{7.13}$$

Here $|\langle\chi_{1m}|\chi_{1n}\rangle|^2$ is the Frank-Condon factor for vibrational levels n and m in the excited and ground states, respectively, of molecule 1; $E_{n(1)}$ is the energy of vibrational level n relative to the zero-point level of molecule 1's excited state; and Z_1 is the vibrational partition function for this molecule's excited state (Eqs. B4.14.4*a*, *b*). Similarly, $|\langle\chi_{2w}|\chi_{2u}\rangle|^2$ is the Franck-Condon factor for vibrational levels u and w of molecule 2's ground and excited states; $E_{u(2)}$ is the energy of vibrational level u relative to the zero-point level of molecule 2's ground state; and Z_2 is the vibrational partition function for this molecule's ground state. The first two sums in Eq. (7.13) run over all of molecule 1's vibrational states, and the second two sums run over all the vibrational states of molecule 2. Finally, the Dirac delta function, $\delta(\Delta E_1 - \Delta E_2)$, is 1 if the energy of molecule 1's downward vibronic transition (ΔE_1) is the same as the energy of molecule 2's upward transtion (ΔE_2), and zero otherwise. The delta function insures that energy is conserved in the overall process. We have assumed that the different donor-acceptor pairs in the sample act independently, and that the Franck-Condon factors and the electronic term H_{21} are the same for all the donor-acceptor pairs.

Although the double sums in Eq. (7.13) look forbidding, we will see below that they are related to the absorption and emission spectra of the energy donor and acceptor, so in many cases we don't need to evaluate them term by term.

Now consider the electronic interaction energy $H_{21(el)}$. Let's assume that there is no orbital overlap between the two molecules, so that electrons can be assigned unambiguously to one molecule or the other, we do not have to consider intermolecular exchange of electrons, and the motions of the electrons in one molecule are not correlated with those in the other. The dominant electronic interactions then are simply Coulombic. If we have a reasonably good description of the molecular orbitals, we can estimate the magnitude of the interactions as follows:

$$H_{21(el)} = \langle \phi_{1a}\phi_{2b} | \tilde{\mathrm{H}}_{21} | \phi_{1b}\phi_{2a} \rangle \approx (f^2/n^2) \int \int \phi_{1a}^{*}\phi_{2b}^{*} \frac{e^2}{r_{21}} \phi_{1b}\phi_{2a} d\boldsymbol{r}_1 d\boldsymbol{r}_2, \quad (7.14)$$

where r_{21} is the distance between electron 1 (on molecule 1) and electron 2 (on molecule 2), and the integration parameters $\boldsymbol{r}_1$ and $\boldsymbol{r}_2$ are the coordinates of the two electrons. The factor (f^2/n^2) represents (approximately) the local-field effect and dielectric screening in a medium with refractive index n (Box 7.1). Evaluating Eq. (7.14) is straightforward if we write the molecular orbitals as linear combinations of atomic p_z orbitals as in Eqs. (4.19–4.22). The expression for $H_{21(el)}$ then becomes a sum of "transition monopole" terms:

$$H_{21(el)} \approx (2f^2/n^2) \sum_s \sum_t C_s^{1a} C_s^{1b} C_t^{2a} C_t^{2b} (e^2/r_{st}), \quad (7.15)$$

where r_{st} is the distance from atom s of molecule 1 to atom t of molecule 2 and the C's are the coefficients for the p_z orbitals on these atoms in molecular orbitals ϕ_{1a}, ϕ_{1b}, ϕ_{2a} and ϕ_{2b}.

A useful approximate expression for $H_{21(el)}$ often can be obtained by breaking the intermolecular electrostatic interactions into monopole-monopole, monopole-dipole and dipole-dipole terms. If the molecules have no net charges, and are sufficiently far apart relative to the molecular dimensions, the main contributions to $H_{21(el)}$ usually come from dipole-dipole interactions. $\tilde{\mathrm{H}}_{21}$ then can be replaced by the operator for the energy of interaction of electric "point" dipoles located at the centers of the two chromophores. The dipole-dipole operator is:

$$\tilde{\mathrm{V}}_{21} = (f^2/n^2) \left\{ (\tilde{\mu}_1 \cdot \tilde{\mu}_2) |\boldsymbol{R}_{21}|^{-3} - 3(\tilde{\mu}_1 \cdot \boldsymbol{R}_{21})(\tilde{\mu}_2 \cdot \boldsymbol{R}_{21}) |\boldsymbol{R}_{21}|^{-5} \right\}, \quad (7.16)$$

where $\tilde{\mu}_1$ and $\tilde{\mu}_2$ are the dipole operators for electrons on the two molecules and $\boldsymbol{R}_{21}$ is the vector from the center of molecule 1 to the center of molecule 2 (Fig. 7.2 and Box 7.1).

Box 7.1 Dipole-Dipole Interactions

The classical energy of interaction between two dipoles $\boldsymbol{\mu}_1$ and $\boldsymbol{\mu}_2$ can be found by evaluating the potential energy of $\boldsymbol{\mu}_2$ in the electric field $\boldsymbol{E}_1$ generated by $\boldsymbol{\mu}_1$: the interaction energy is $-\boldsymbol{E}_1 \cdot \boldsymbol{\mu}_2$. The field at a point with coordinates (x,y,z) is $\mathrm{E}_1(x, y, z) = -\tilde{\nabla}\,[V_1(x, y, z)]$, where $V_1(x,y,z)$ is the electric potential (a scalar) and $\tilde{\nabla} V$, the gradient of the potential, is the vector $(\partial V/\partial x, \partial V/\partial y, \partial V/\partial z)$ (Eqs. 2.4 and 3.9).

To find $V_1(x,y,z)$, let's represent $\boldsymbol{\mu}_1$ by a positive charge q and an equal negative charge $-q$ separated by a fixed distance. Put the origin of the coordinate system midway between the two charges. For the general case

(continued)

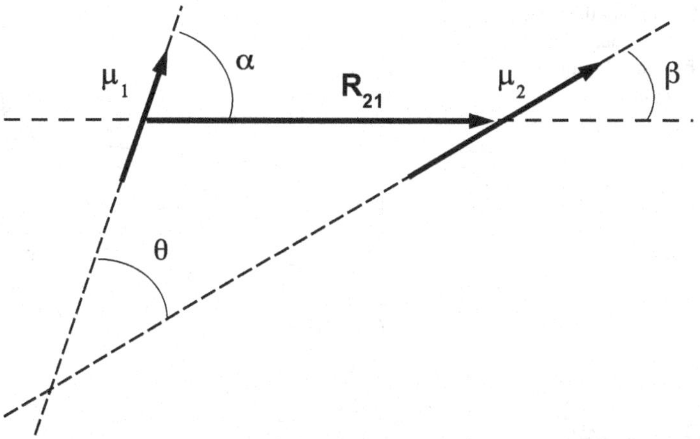

Fig. 7.2 The energy of interaction of two dipoles depends on the magnitudes of the dipoles ($|\boldsymbol{\mu}_1|$ and $|\boldsymbol{\mu}_2|$), the distance between the two centers of charge ($|\boldsymbol{R}_{21}|$), the angle between the dipoles (θ), and the angles with respect to $\boldsymbol{R}_{21}$ (α and β)

Box 7.1 (continued)
of an arbitrary distribution of charges in a vacuum, the electric potential at a point $\boldsymbol{R} = (x, y, z)$ is

$$V(x, y, z) = \sum_i q_i/|\boldsymbol{r}_i|$$
$$= \sum_i q_i \Big/ \left\{(x - x_i)^2 + (y - y_i)^2 + (z - z_i)^2 + \right\}^{1/2}, \quad \text{(B7.1.1)}$$

in which q_i and (x_i, y_i, z_i) are the charge and location of charge i, and $\boldsymbol{r}_i$ is the vector from charge i to the point (x,y,z) (Fig. 7.3).

The contribution from charge q_i to $V(\boldsymbol{R})$ can be expanded in a Taylor series in $|\boldsymbol{r}_i|^{-1}$ by the procedure described in Box 4.1 for finding the energy of a set of charges in an external field:

$$V = \sum_i q_i|\boldsymbol{R}|^{-1} + \sum_i q_i\left\{x_i\partial\left(|\boldsymbol{r}_i|^{-1}\right)/\partial x_i + y_i\partial\left(|\boldsymbol{r}_i|^{-1}\right)/\partial y_i + z_i\partial\left(|\boldsymbol{r}_i|^{-1}\right)/\partial z_i\right\}$$
$$+\frac{1}{2}\sum_i q_i\left\{x_ix_i\partial^2\left(|\boldsymbol{r}_i|^{-1}\right)/\partial x_i^2 + x_iy_i\partial^2\left(|\boldsymbol{r}_i|^{-1}\right)/\partial x_i\partial y_i + x_iz_i\partial^2\left(|\boldsymbol{r}_i|^{-1}\right)/\partial x_i\partial z_i\right.$$
$$+x_iy_i\partial^2\left(|\boldsymbol{r}_i|^{-1}\right)/\partial y_i\partial x_i + y_iy_i\partial^2\left(|\boldsymbol{r}_i|^{-1}\right)/\partial y_i^2 + y_iz_i\partial^2\left(|\boldsymbol{r}_i|^{-1}\right)/\partial y_i\partial z_i$$
$$\left.+x_iz_i\partial^2\left(|\boldsymbol{r}_i|^{-1}\right)/\partial z_i\partial x_i + y_iz_i\partial^2\left(|\boldsymbol{r}_i|^{-1}\right)/\partial z_i\partial y_i + z_iz_i\partial^2\left(|\boldsymbol{r}_i|^{-1}\right)/\partial z_i^2\right\} + \ldots.$$
$$\text{(B7.1.2)}$$

As in the problem we considered in Chap. 4, such a multipole expansion is most useful if the charges are not too far from the origin. The first sum gives

(continued)

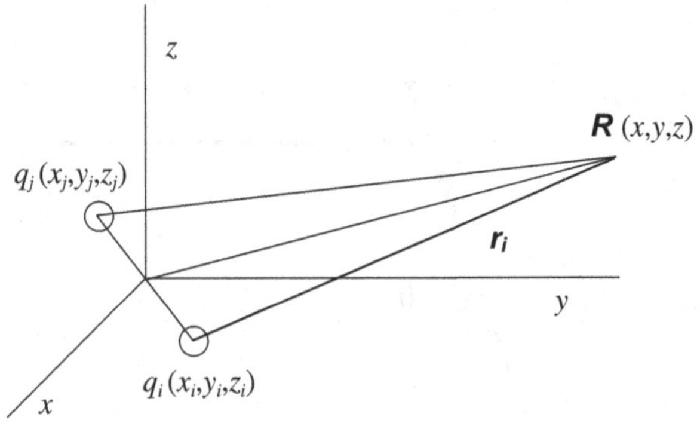

Fig. 7.3 Evaluating the field at $\boldsymbol{R}$ from a set of charges near the origin by summing the fields from the individual charges

Box 7.1 (continued)

the potential if all the charges are located exactly at the origin; this is the monopole term. The second sum contains the dot product of $\tilde{\nabla}\left(|\boldsymbol{r}_i|^{-1}\right)$ with the dipole moment of the charge distribution ($\sum_i q_i\boldsymbol{r}_i$), while the third sum involves the nine components of the quadrupole distribution. If the system contains equal numbers of positive and negative charges, the first sum vanishes and the dipolar interaction usually is the leading term. The quadrupolar and higher terms are zero if the charge distribution consists of only two charges, but can be important in larger systems if both the monopole and dipole terms are zero or small. (Figure 4.1B shows a charge distribution with a quadrupole moment but no dipole moment. The linear CO_2 molecule also has a quadrupole moment but no dipole moment.)

The derivatives of $|\boldsymbol{r}_i|^{-1}$ with respect to x_i, y_i and z_i in the dipole term can be rewritten as derivatives with respect to the coordinates of the point of interest, and then can be simplified if the charges are sufficiently close to the origin so that $\boldsymbol{r}_i \approx \boldsymbol{R}$:

$$\partial\left(|\boldsymbol{r}_i|^{-1}\right)\Big/\partial x_i = -\partial\left(|\boldsymbol{r}_i|^{-1}\right)\Big/\partial x \approx -\partial\left(|\boldsymbol{R}|^{-1}\right)\Big/\partial x = x\Big/|\boldsymbol{R}|^3. \quad \text{(B7.1.3)}$$

The potential at $\boldsymbol{R}$ from a dipole $\boldsymbol{\mu}_1$ centered at the origin thus is:

$$V(\boldsymbol{R}) = \frac{1}{|\boldsymbol{R}|^3}\sum_i [q_i x_i x + q_i y_i y + q_i z_i z] = \frac{\boldsymbol{\mu}_1 \cdot \boldsymbol{R}}{|\boldsymbol{R}|^3}, \quad \text{(B7.1.4)}$$

and the field at $\boldsymbol{R}$ is:

(continued)

Box 7.1 (continued)

$$E_1(\boldsymbol{R}) = -\nabla\left(|\boldsymbol{R}|^{-3}\boldsymbol{\mu}_1 \cdot \boldsymbol{R}\right) \tag{B7.1.5a}$$

$$\begin{aligned} &= -\boldsymbol{\mu}_1|\boldsymbol{R}|^{-3} - (\boldsymbol{\mu}_1 \cdot \boldsymbol{R})\nabla\left(|\boldsymbol{R}|^{-3}\right) \\ &= -\boldsymbol{\mu}_1|\boldsymbol{R}|^{-3} + 3(\boldsymbol{\mu}_1 \cdot \boldsymbol{R})\boldsymbol{R}|\boldsymbol{R}|^{-5}. \end{aligned} \tag{B7.1.5b}$$

Finally, the energy of interaction between dipole $\boldsymbol{\mu}_2$ and the field from $\boldsymbol{\mu}_1$ is:

$$V = (\boldsymbol{\mu}_1 \cdot \boldsymbol{\mu}_2)|\boldsymbol{R}|^{-3} - 3(\boldsymbol{\mu}_1 \cdot \boldsymbol{R})(\boldsymbol{\mu}_2 \cdot \boldsymbol{R})|\boldsymbol{R}|^{-5}. \tag{B7.1.6}$$

Corrections to this expression are needed if the dipoles are embedded in a dielectric medium. For dipoles that fluctuate in position or orientation very rapidly relative to the time scale of nuclear motions, the field in a homogeneous medium with refractive index n is reduced by a factor of $1/n^2$. (The field from a dipole that fluctuates slowly would be screened by the low-frequency dielectric constant of the medium rather than the high-frequency dielectric constant, n^2.) However, to be consistent with our treatment of molecular transition dipoles in Chaps. 4 and 5, we should view the interacting molecular dipoles as residing in small cavities that correspond roughly to the molecular volumes. Moving dipole 1 into a spherical cavity increases the effective field in the surrounding medium by a factor of approximately $3n^2/(2n^2+1)$ [6]. This is the cavity-field correction factor f_c (Eq. 3.35). Moving dipole 2 into its cavity increases the local field acting on the dipole by this same factor. The overall correction to the interaction energy thus is approximately f_c^2/n^2, or f_L^2/n^2 if we use the Lorentz correction factor f_L (Eq. 3.36) instead of the cavity-field factor. A more realistic treatment would require a microscopic analysis of the molecular shapes and of the surrounding material. This is not necessary for our present purposes because the dielectric correction factors cancel out of the main results derived below (Eqs. 7.24 and 7.27).

If we approximate $\tilde{\mathrm{H}}_{21}$ by $\tilde{\mathrm{V}}_{21}$, the electronic interaction matrix element $H_{21(el)}$ becomes:

$$\begin{aligned} H_{21(el)} &= \frac{f^2}{n^2}\left\{\frac{\langle\phi_{1b}|\boldsymbol{\mu}_1|\phi_{1a}\rangle\cdot\langle\phi_{2b}|\boldsymbol{\mu}_2|\phi_{2a}\rangle}{|\boldsymbol{R}_{21}|^3} - 3\frac{(\langle\phi_{1b}|\boldsymbol{\mu}_1|\phi_{1a}\rangle\cdot\boldsymbol{R}_{21})(\langle\phi_{2b}|\boldsymbol{\mu}_2|\phi_{2a}\rangle\cdot\boldsymbol{R}_{21})}{|\boldsymbol{R}_{21}|^5}\right\} \\ &= (f^2/n^2)\left\{\sqrt{D_{ba(1)}}\sqrt{D_{ba(2)}}(\cos\theta - 3\cos\alpha\cos\beta)|\boldsymbol{R}_{21}|^{-3}\right\} \\ &= (f^2/n^2)\sqrt{D_{ba(1)}}\sqrt{D_{ba(2)}}\kappa|\boldsymbol{R}_{21}|^{-3}. \end{aligned} \tag{7.17}$$

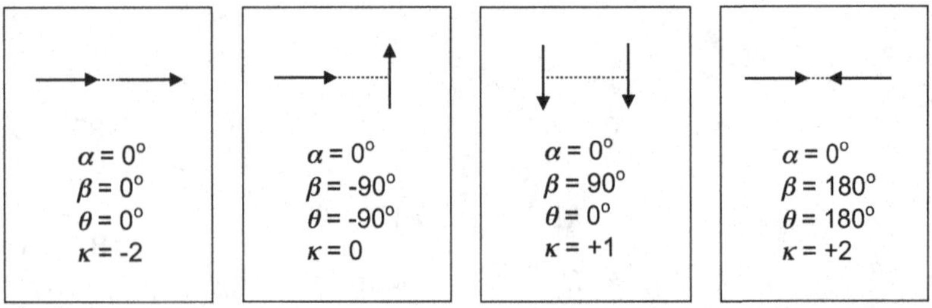

Fig. 7.4 The orientation factor κ in the dipole-dipole interaction energy can vary from -2 to $+2$. The rate of resonance energy transfer is proportional to κ^2

Here $D_{ba(1)}$ and $D_{ba(2)}$ are the dipole strengths for the transitions of the two monomers; θ is the angle between the two transition dipoles; α and β are the angles that the transition dipoles make with $\boldsymbol{R}_{21}$; and $\kappa = (\cos\theta - 3\cos\alpha\,\cos\beta)$ (Fig. 7.2). This is the *point-dipole approximation.*

Because the orientation factor κ in Eq. (7.17) $(\cos\theta - 3\cos\alpha\,\cos\beta)$ can vary from -2 to $+2$ depending on the angles θ, α and β, $H_{21(el)}$ can be either positive, negative or zero. Figure 7.4 illustrates the limiting cases. The sign of κ is of no consequence for the rate of energy transfer because the rate depends on $|H_{21(el)}|^2$, which is proportional to κ^2. If the molecules tumble isotropically on a time scale that is fast relative to the lifetime of the excited donor, κ^2 has an average value of 2/3.

If the two molecules are identical, with $D_{ba(1)} = D_{ba(2)} = D_{ba}$, the factor $(D_{ba(1)}D_{ba(2)})^{1/2}$ in Eq. (7.17) is simply D_{ba}. If D_{ba} is given in units of debye^2 and $|\boldsymbol{R}_{21}|$ in Å, then

$$H_{21(el)} \approx 5.03 \times 10^3 \left(f^2/n^2\right) D_{ba}\, \kappa |\boldsymbol{R}_{21}|^{-3} \tag{7.18}$$

in units of cm^{-1}. [1 cm^{-1} is 1.99×10^{-16} erg, 1.24×10^{-4} eV, or 2.86 cal/mole. (1 debye $\times\, 10^{-18}$ esu cm/debye)^2(1 Å $\times\, 10^{-8}$ cm/Å)$^{-3} = 10^{-12}$ esu^2 cm $= 10^{-12}$ erg $\times\, 5.03 \times 10^{15}$ cm^{-1}/erg $= 5.03 \times 10^3$ cm^{-1}.]

Figure 7.5 shows the values of $H_{21(el)}$ calculated for a pair of *trans*-butadiene molecules by the point-dipole (Eq. 7.18) and transition-monopole (Eq. 7.15) expressions. In panel A, the orientations of both molecules are held fixed while the center-to-center distance is varied; in B the second molecule is rotated at a fixed distance. As a rule of thumb, the point-dipole approximation is reasonably satisfactory if the intermolecular distance is more than 4 or 5 times the length of the chromophores, although the relative error can still be substantial in some situations.

Equation (7.13) says that the rate of transitions from state Ψ_1 to state Ψ_2 is proportional to $|H_{21(el)}|^2$, which according to Eqs. (7.17) and (7.18) decreases with the sixth power of $|\boldsymbol{R}_{21}|$. To focus on the effect of varying the intermolecular distance, it is useful to write the rate constant for resonance energy transfer (k_{rt}) in the form

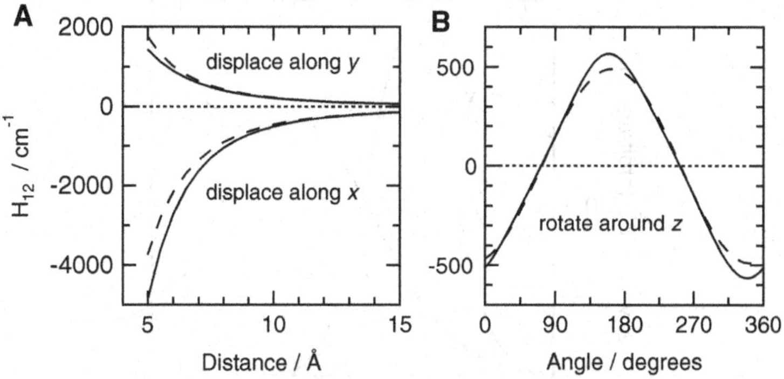

Fig. 7.5 Electronic interaction matrix elements for two *trans*-butadiene molecules as calculated by the transition-monopole (Eq. (7.15), *solid curves*) or point-dipole (Eq. 7.18, *dashed curves*) expressions. Molecule 1 is fixed in position in the *xy* plane, centered at the origin, with its long axis parallel to the *x* axis. The transition dipole for its lowest-energy excitation lies in the *xy* plane at an angle of 169° from the positive *x* axis. f^2/n^2 is taken to be 1. In (**A**), molecule 2 is centered at various points along either *x* or *y*, in the same orientation as molecule 1; the abscissa gives the center-to-center distance. In **B**, molecule 2 is centered at (10 Å, 0, 0) and is rotated in the *xy* plane; the abscissa indicates the angle (θ) between the two transition dipoles

$$k_{rt} = \tau^{-1}(|\mathbf{R}_{21}|/R_o)^{-6}, \tag{7.19}$$

where τ is the fluorescence lifetime of the energy donor (molecule 1) in the absence of the acceptor (molecule 2). R_o is the center-to-center distance at which k_{rt} is equal to the overall rate constant for the decay of the excited state by all other mechanisms including fluorescence ($1/\tau$), so that 50% of the decay involves energy transfer. R_o is called the *Förster radius* after Th. Förster, who first showed how the value of R_o for a given donor-acceptor pair can be calculated from the spectroscopic properties of the individual molecules [7, 8]. Förster's theory can be developed as follows.

Equations (7.13) and (7.17) indicate that, for widely separated molecules with a given intermolecular distance and orientation, the rate of resonance energy transfer is proportional to the product of the dipole strengths ($D_{ba(1)}$ and $D_{ba(2)}$). The rate also depends on the thermally-weighted Franck-Condon factors for the pairs of downward and upward vibronic transitions that satisfy the resonance condition. These facts suggest that we can relate the rate to the absorption spectrum of the acceptor and the emission spectrum of the donor (Fig. 7.6).

Consider the emission spectrum of the energy donor (molecule 1). In the absence of energy transfer or other decay mechanisms, the rate constant for fluorescence at frequency v would be:

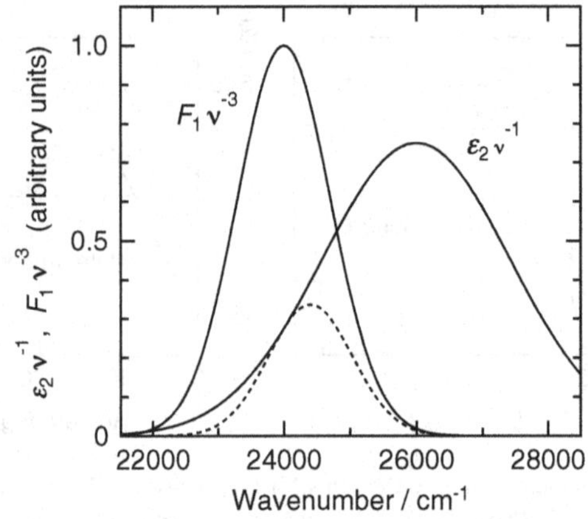

Fig. 7.6 In the Förster theory, the rate of energy transfer from molecule 1 to molecule 2 is proportional to the overlap integral, $\int \varepsilon_2(v)F_1(v)v^{-4}dv$. Contributions to the integral come only from the spectral region where the weighted emission spectrum of molecule 1 (F_1v^{-3}) overlaps the absorption spectrum of molecule 2 (ε_2v^{-1}). The product $\varepsilon_2(v)F_1(v)v^{-4}$ (*dotted curve*) can be viewed as a spectrum of resonance energy transfer

$$F_1(v)dv = \left[32\pi^3nf^2v^3/3\hbar c^3\right]D_{ba(1)}\mathrm{X}_1(E)\rho_1(v)dv, \tag{7.20}$$

where $\rho_1(v)$ is the density of excited donor states around frequency v (states per unit frequency) and $X_1(E)$ is a thermally-weighted Franck-Condon factor for emission with energy $E = hv$ (Eq. 5.12). The number of excited donor states in energy interval ΔE corresponding to frequency interval Δv is $\rho_1(E)\Delta E = \rho_1(v)\Delta v$, where $\rho_1(E) = \rho_1(v)/h$ and $\Delta E = h\Delta v$. The total rate constant for fluorescence integrated over the emission spectrum, $\int F_1(v)dv$, is the reciprocal of the radiative time constant, τ_r (Eq. 5.39). Because $\tau_r = \tau/\Phi$, where τ and Φ are the fluorescence lifetime and yield in the absence of energy transfer (Eq. 5.42), Eq. (7.20) implies that:

$$D_{ba(1)}\mathrm{X}(E)\rho_1(E)dE = \left(\frac{3\hbar c^3\Phi}{32\pi^3nf^2\tau}\right)\frac{F_1(v)v^{-3}}{\int F_1(v)dv}. \tag{7.21}$$

Now consider the absorption spectrum of the acceptor (molecule 2). From Eqs. (4.15–4.16a) and (4.37–4.40), we can write:

$$D_{ba(2)}\mathrm{X}_2(E)\rho_2(E) = D_{ba(2)}\mathrm{X}_2(E)\frac{\rho_2(v)}{h} = \frac{1}{h}\left(\frac{3000\ln 10\,nhc}{8\pi^3f^2N_A}\right)\frac{\varepsilon_2(v)}{v}, \tag{7.22}$$

where $\rho_2(v)$ is the density of acceptor states on a frequency scale, $X_2(E)$ is the thermally-weighted Franck-Condon factor for excitations with energy $E = hv$, and N_A is Avogadro's number.

Förster's expression for the rate of resonance energy transfer is obtained by combining Eqs. (7.11), (7.17), (7.21), and (7.22):

$$\begin{aligned} k_{rt} &= \frac{2\pi}{\hbar}\left\{\frac{f^4}{n^4}D_{ba(1)}D_{ba(2)}\kappa^2|\boldsymbol{R}_{21}|^{-6}\right\}\int \mathrm{X}_1(E)\rho_1(v)\mathrm{X}_2(E)\rho_2(v)dE \\ &= \frac{2\pi}{\hbar}\frac{f^4}{n^4}\kappa^2|\boldsymbol{R}_{21}|^{-6}\left(\frac{3000\ln 10\, nhc}{8\pi^3 f^2 N_A h}\right)\left(\frac{3\hbar c^3\Phi}{32\pi^3 n f^2 \tau}\right)\frac{\int F_1(v)\varepsilon_2(v)v^{-4}dv}{\int F_1(v)dv}. \end{aligned} \tag{7.23}$$

Collecting the constants gives

$$k_{rt} = \left(\frac{9000\ln 10\kappa^2 c^4\Phi}{128\pi^5 n^4 N_A \tau}\right)|\boldsymbol{R}_{21}|^{-6}J, \tag{7.24}$$

where J is an *overlap integral* of the absorption and fluorescence spectra with the contribution in each frequency interval weighted by v^{-4}:

$$J = \frac{\int F_1(v)\varepsilon_2(v)v^{-4}dv}{\int F_1(v)dv}. \tag{7.25}$$

As noted above, the donor's fluorescence lifetime τ and yield Φ in Eq. (7.24) are the values measured in the absence of energy transfer. The fluorescence amplitude $F_1(v)dv$ in Eq. (7.25) can be in any convenient units because the fluorescence is normalized by the integral in the denominator. The acceptor's molar extinction coefficient ε_2 has its usual units of $\mathrm{M}^{-1}\mathrm{cm}^{-1}$. If $|\boldsymbol{R}_{21}|$ is given in units of cm and v is in s^{-1}, then the overlap integral J has units of $\mathrm{M}^{-1}\mathrm{cm}^{-1}\ \mathrm{s}^4$ and k_{rt} is in s^{-1}.

The overlap integral often is defined for absorption and fluorescence spectra on a wavenumber scale ($\overline{v} = v/c = 1/\lambda$) with $\overline{v}$ in units of cm^{-1}:

$$\overline{J} = \frac{\int F_1(\overline{v})\varepsilon_2(\overline{v})\overline{v}^{-4}d\overline{v}}{\int F_1(\overline{v})d\overline{v}} = \frac{\int F_1(\lambda)\varepsilon_2(\lambda)\lambda^2 d\lambda}{\int F_1(\lambda)\lambda^{-2}d\lambda}. \tag{7.26}$$

In this expression, $\overline{J}$ is Jc^{-4} and has units of $\mathrm{M}^{-1}\ \mathrm{cm}^3$. If this definition is used and $|\boldsymbol{R}_{21}|$ is given in Å, then Eq. (7.24) becomes:

$$k_{rt} = \left(\frac{9000\ln 10\kappa^2\Phi}{128\pi^5 n^4 N_A \tau}\right)\left(\frac{|\boldsymbol{R}_{21}|}{1\times 10^8}\right)^{-6}\overline{J} = \left(\frac{8.78\times 10^{23}\kappa^2\Phi}{n^4\tau}\right)|\boldsymbol{R}_{21}|^{-6}\overline{J}s^{-1}. \tag{7.27}$$

Finally, combining Eqs. (7.27) and (7.19) gives a value for the Förster radius:

$$R_o = 9.80 \times 10^3 \left(\kappa^2 \Phi n^{-4} \bar{J}\right)^{1/6} \text{Å}. \tag{7.28}$$

The overlap integral $\bar{J}$ sometimes is written as $\int F_1(\lambda)\varepsilon_2(\lambda)\lambda^4 d\lambda / \int F_1(\lambda)d\lambda$. This is not strictly correct. Because $d\bar{\nu} = -\lambda^{-2}d\lambda$, the correct expression for $\bar{J}$ on a wavelength scale is $\int F_1(\lambda)\varepsilon_2(\lambda)\lambda^2 d\lambda / \int F_1(\lambda)\lambda^{-2}d\lambda$ (Eq. 7.26). However, the error may be negligible if the two spectra overlap only over a narrow region.

Although the Förster radius is evaluated from the overlap of the absorption and emission spectra of the two molecules, note that resonance energy transfer does *not* involve emission and reabsorption of light. It occurs by a resonance between two states of the entire system. Because the energy donor does not fluoresce in the process, the common jargon "fluorescence resonance energy transfer" (FRET) is somewhat misleading. The same acronym, however, can be used more accurately for "Förster resonance energy transfer."

The dependence of the rate on the sixth power of the intermolecular distance makes resonance energy transfer particularly useful for exploring the locations of binding sites for ligands on macromolecules, as well as conformational changes that follow ligand binding [5, 9–12]. By chosing appropriate donor-acceptor pairs, rates of energy transfer can be measured over distances ranging from 10 to more than 100 Å. Yet, the rate for a given donor-acceptor pair can be sufficiently sensitive to distance to afford a resolution on the order of 1 Å. In favorable cases, the rate of energy transfer can be obtained by simply measuring the quenching of fluorescence from the donor. From Eq. (5.45), the ratio of the fluorescence yields in the absence (Φ) and presence (Φ_q) of the acceptor is:

$$\frac{\Phi}{\Phi_q} = \frac{k_r/k_{tot}}{k_r/(k_{tot} + k_{rt})} = 1 + \frac{k_{rt}}{k_{tot}} = 1 + k_{rt}\tau, \tag{7.29}$$

where k_r is the radiative rate constant and k_{tot} is the sum of the rate constants for all decay processes other than energy transfer ($k_{tot} = \tau^{-1}$). Once the product $k_{rt}\tau$ and the Förster radius are known, Eq. (7.19) gives the distance between the donor and acceptor:

$$|\mathbf{R}_{21}| = R_o(k_{rt}\tau)^{-1/6} = R_o\left(\Phi/\Phi_q - 1\right)^{-1/6} \tag{7.30}$$

However, it is important to show that the quenching of the fluorescence reflects energy transfer rather than other processes such as electron transfer. This can be done by exciting the sample at a wavelength where only the energy donor absorbs significantly and measuring fluorescence at a wavelength where only the acceptor emits.

Stryer and Haugland [13] verified the dependence of the energy-transfer rate on the inverse sixth power of the distance by using oligomers of dansyl-

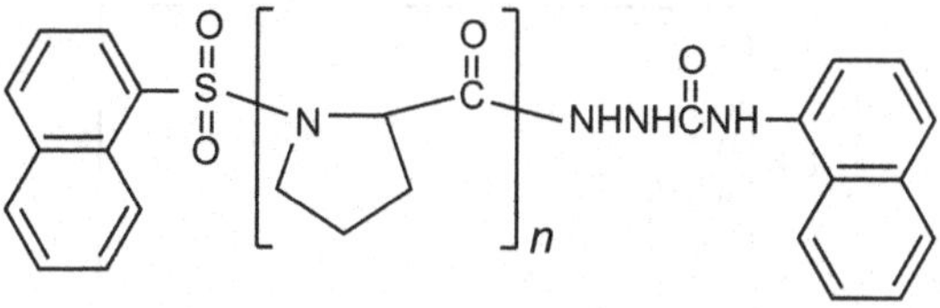

Fig. 7.7 Dansyl-(L-prolyl)$_n$-α-naphthylsemicarbazide. The polyprolyl chain provides a rigid spacer between the dansyl and naphthyl groups (*left* and *right*, respectively), each of which can rotate relatively freely

polyproline-α-naphthylsemicarbazide (Fig. 7.7) to position the energy donor and acceptor (naphthylene and a dansyl derivative, respectively) at distances ranging from 12 to 46 Å. The rate was found to vary as $R^{-5.9 \pm 0.3}$. Haugland et al. [14] demonstrated that the rate also was proportional to J over a 40-fold range of J.

There are several possible complications in using the rate of resonance energy transfer to determine intermolecular distances. First, the measured rate depends on the average value of κ^2, which usually is not known accurately. Fortunately, a fairly limited rotational mobility of the chromophores is sufficient to make the average value of κ^2 approach 2/3, so that the uncertainties in calculated distances become relatively small [9, 15]. The motional freedom of the chromophores can be assessed from measurements of fluorescence anisotropy and linear dichroism as discussed in Chaps. 4 and 5.

If the sample has a distribution of donor-acceptor distances, the rate of energy transfer will be weighted in favor of structures in which the distance is smaller than average, making the fluorescence decay kinetics nonexponential. Figure 7.8 illustrates this effect for a Gaussian distribution of distances. If the distribution is symmetric, Eq. (7.30) still gives a reasonably accurate estimate of the mean donor-acceptor distance. The error in the case shown in Fig. 7.8 would be only about 1%, even though the distribution is quite broad.

If the signal/noise ratio in the data is sufficiently high, a distribution of donor-acceptor distances ($P(R)$) can be extracted by fitting the fluoresence decay kinetics to a function of the form

$$F(t) = \int_0^\infty P(R) e^{-(k_{tot} + k_{rt}(R))t} dR = \int_0^\infty P(R) e^{-k_{tot}\left(1 + (R/R_o)^{-6}\right)t} dR. \qquad (7.31)$$

Haas et al. [16] used this approach to investigate the distribution of end-to-end lengths in polyglutamyl peptides that were labeled with naphthalene at one end and a dansyl group at the other. Both the mean and the width of the distribution were found to depend on the solvent. In similar studies, other workers have probed the conformational heterogeneity of oligonucleotides labeled at opposite ends with fluorescein and rhodamine [17], of dansyl-labeled Zn-finger peptides with and

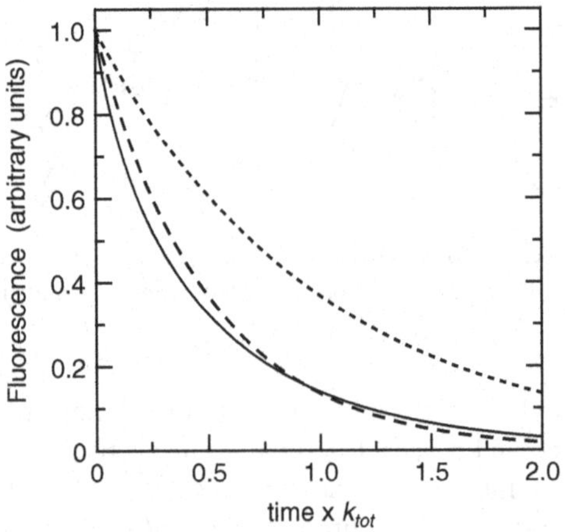

Fig. 7.8 Calculated decay kinetics of fluorescence from an ensemble of energy donors with a Gaussian distribution of donor-acceptor distances. In this illustration, the distribution is centered at R_o and has a width (FWHM) of either $0.5R_o$ (*solid line*) or zero (*long dashes*). The time course of the fluorescence was calculated by Eq. (7.31). Also shown is the exponential fluorescence decay in the absence of energy transfer (*short dashes*). A FWHM of $0.5R_o$ corresponds to a standard deviation of $0.212R_o$

without bound Zn^{2+} [18], and of calmodulin in the presence and absence of Ca^{2+} [19]. Haas et al. [20, 21] also have treated the situation that the intermolecular distance fluctuates during the lifetime of the donor's excited state. Rhoades et al. [22, 23] have examined the structural heterogeneity of several proteins by measuring the distribution of energy-transfer rates between dyes bound to different sites in single protein molecules, which they trapped in lipid vesicles. Jang et al. [24] have presented a generalized version of the Förster theory for systems that contain multiple types of chromophores and heterogeneous donor-acceptor distances, such as the mixtures of chlorophylls and carotenoids found in photosynthetic antenna complexes.

Several additional points concerning the Förster theory need mentioning. First, using the spectral overlap integral J as in Eqs. (7.23), (7.25), and (7.26) is strictly correct only for a homogeneous system such as a single donor-acceptor pair. As we discussed in Chaps. 4 and 5, absorption and emission spectra usually represent averages over large populations of molecules in various environments. Whether or not the entire inhomogeneously-broadened spectra should be included in the overlap integral depends on how rapidly the environments of the individual molecules in the ensemble fluctuate relative to the lifetime of the donor's excited state. If the environmental shifts of the individual spectra fluctuate rapidly on this time scale, then (assuming that the fluctuations in the donor and acceptor are not correlated) the entire spectra of the ensemble are properly included in J. If these shifts are frozen during the lifetime of the excited state, then the overlap integral will vary among the donor-acceptor pairs in the ensemble and the individual values of J should be weighted to reflect their contributions to the experimental measurement. However, such weighting probably would have relatively small effects on the calculated donor-acceptor distance in most cases.

The refractive index n in Eqs. (7.20) and (7.21) pertains to the bulk medium in which the absorption and emission spectra of the chromophores are measured. This factor cancels out when Eqs. (7.20) and (7.21) are combined in Eq. (7.23). The factor of n^{-4} that remains in Eqs. (7.23), (7.24), (7.27), and (7.28) enters through $|H_{21(el)}|^2$ and reflects the attenuation of high-frequency electrostatic interactions between the energy donor and acceptor. It therefore pertains to the protein or other material that surrounds the interacting molecules, and not to the solvent used to measure the individual spectra. The appropriate value of the refractive index usually is not known accurately, but for proteins probably is comparable to that of *N*-methylacetamide (1.43 at 589.3 nm). Using the refractive index of water (1.33) would increase the calculated rate constant for energy transfer $[(1.43/1.33)^4 \approx 1.34]$, but the resulting error in R_o or $|\mathbf{R}_{21}|$ would be relatively small $[(1.43/1.33)^{4/6} \approx 1.05]$.

Finally, it is important to remember that the Förster theory assumes that the donor and acceptor molecules interact only very weakly. This means that the interactions must not change the absorption or emission spectrum significantly. It also means that, once the excitation energy moves from the donor to the acceptor, the energy must have little probability of returning to the donor. In most cases, vibrational and rotational relaxations of the acceptor and the solvent rapidly dissipate some of the excitation energy rapidly as heat, so that the overlap integral J for energy transfer back to the original donor is small; the excited acceptor instead either fluoresces or decays by some other path. We'll discuss this point in more detail in Chap. 7.

Resonance energy transfer is used widely to detect the formation of complexes between proteins. Variants of green fluorescent protein (GFP) make particularly useful energy donors and acceptors because genes for GFPs with different absorption and emission spectra can be fused with the genes for the proteins of interest (Sects. 5.7 and 5.10). The built-in chromophore makes it unnecessary to label the proteins with exogenous fluorescent dyes. Muller et al. [25] have used this approach to determine the three-dimensional arrangement of five different proteins in the yeast spindle-pole body. They used fluorescence microscopy to measure energy transfer between GFP fused at either the N-terminal or C-terminal end of one protein and and a cyan-fluorescent variant (CFP) fused at either end of another protein. Their paper describes a procedure for correcting the measurements for "spillover" fluorescence from the donor that contaminates signals from the acceptor.

Sensors for Ca^{2+} called "cameleons" have been constructed by fusing CFP with the Ca^{2+}-binding protein calmodulin, a calmodulin-binding peptide, and a green- or yellow-fluorescing variety of GFP (YFP) [26, 27]. Binding of Ca^{2+} to the calmodulin causes a conformational change that increases resonance energy transfer between the CFP and YFP's, changing the color of the fluorescence. To examine the distribution of Ca^{2+}-conducting glutamate receptors in neuromuscular junctions, a cameleon was fused with a protein that localizes in the postsynaptic terminal [28].

A variation on the resonance energy transfer technique is to use a bioluminescence system as the energy donor [29, 30]. The gene for a luciferase, a protein that catalyses a bioluminescence reaction, is fused with the gene for one of the proteins of interest, while a GFP gene is fused with that for the other protein. If formation of a complex brings the luciferase and the GFP sufficiently close together, energy transfer will shift the emission to the longer wavelength characteristic of the GFP. In another variation, a chelated lanthanide ion serves as the donor [31–35]. Lanthanide excited states have lifetimes on the order of milliseconds and typically have very low emission anisotropy, making energy transfer insensitive to the orientation of the donor.

7.3 Using Energy Transfer to Study Fast Processes in Single Protein Molecules

Section 5.10 describes several techniques for measuring fluorescence from individual molecules. Measurements of resonance energy transfer in single molecules lend themselves well to studying the dynamics of protein folding. The energy donor and acceptor can be attached to sites that are far apart in the unfolded protein but come together during folding. In a typical experiment, the temperature or the concentration of a denaturant might be adjusted so that the equilibrium constant for folding is on the order of 1. Fluctuations in the efficiency of energy transfer in an individual molecule then would signal transitions between folded and unfolded conformations. The time constant for transitions in one direction or the other can be obtained from a histogram of the length of time the efficiency remains more or less constant before it changes to a value characteristic of the other conformation [22, 23, 36–48].

The signal/noise ratio in such measurements depends on the mean number of photons that are detected before a transition, and so decreases as the transitions occur more frequently. Raising the light intensity can increase the signal, but tends to cause more rapid photodestruction of the fluorophores, making it necessary to replace the sample at frequent intervals (Sect. 5.8).

When the time between transitions is insufficient for construction of a satisfactory histogram, the most promising approach appears to be to record the wavelengths and timing of individual photons. Gopich and Szabo have developed a *maximum likelihood* method for relating a kinetic model to such photon-by-photon records [49–51]. Consider a molecule with two states (N and U for "native" and "unfolded"), which have different apparent efficiencies of resonance energy transfer (ξ_n and ξ_u). We can use a column vector $\mathbf{p}(t) = \begin{pmatrix} N(t) \\ U(t) \end{pmatrix}$ to express the probability of finding the molecule in a given state at time t. If transitions from N to U occur with microscopic rate constant k_u, and transitions from U to N with rate constant k_n, then $dN/dt = -k_uN + k_nU$ and $dU/dt = k_uN - k_nU$, or more compactly, $d\mathbf{p}/dt = \mathbf{K}\cdot\mathbf{p}$, with $\mathbf{K} = \begin{bmatrix} -k_u & k_n \\ k_u & -k_n \end{bmatrix}$. This equation has the formal solution $\mathbf{p}(t) = e^{\mathbf{K}t}\cdot\mathbf{p}(0)$, in which the matrix exponential function is a 2×2 matrix defined by the power series $e^{\mathbf{K}t} = \mathbf{I} + \mathbf{K}t + \frac{1}{2!}(\mathbf{K}t)^2 + \cdots$, and $\mathbf{I}$ is the identity matrix $\begin{bmatrix} 1 & 0 \\ 0 & 1 \end{bmatrix}$. Solutions to the rate equation also can be expressed in terms of the eigenvalues of $\mathbf{K}$. These are the macroscopic rate constants with which the system approaches equilibrium. For the simple two-state system we are considering, the eigenvalues are 0 and $k_u + k_n$, and the equilibrium values of N and U are, respectively, $k_n/(k_n + k_u)$ and $k_u/(k_n + k_u)$. The apparent efficiency of energy transfer (ξ_n or ξ_u) when the molecule is in a particular state is given by the ratio $n_A/(n_A + n_D)$, where n_A and n_D are the rates of detecting photons emitted by the energy acceptor (A) and donor (D) when the donor is excited continuously.

Suppose the molecule is at equilibrium, and call vector $\mathbf{p}$ for this situation $\mathbf{p}_{\text{eq}}$. If we construct a diagonal matrix $\mathbf{Z}$ with the apparent efficiencies of energy transfer on the diagonal, $\mathbf{Z} = \begin{bmatrix} \xi_n & 0 \\ 0 & \xi_u \end{bmatrix}$, the probability that the first photon we detect will come from the energy acceptor (A) is $\mathbf{I}\cdot\mathbf{Z}\cdot\mathbf{p}_{\text{eq}}$, while the probability that the photon will come from D is $\mathbf{I}\cdot(\mathbf{I}\text{-}\mathbf{Z})\cdot\mathbf{p}_{\text{eq}}$. The multiplication by $\mathbf{I}$ on the left in these expressions sums over the two conformational states. If we detect another photon immediately after the first one, these probabilities would not have changed significantly. With time, however, the molecule could change from N to U or *vice versa*. The kinetic equation $\mathbf{p}(t) = e^{\mathbf{K}t}\cdot\mathbf{p}(0)$ describes these dynamics. If a time interval τ_2 elapses between detection of the first and second photons, the probability that the second photon comes from A is either $\mathbf{I}\cdot\mathbf{Z}\cdot e^{\mathbf{K}\tau_2}\cdot\mathbf{Z}\cdot\mathbf{p}_{eq}$ (if the first photon also came from A) or $\mathbf{I}\cdot\mathbf{Z}\cdot e^{\mathbf{K}\tau_2}\cdot(\mathbf{I}-\mathbf{Z})\cdot\mathbf{p}_{eq}$ (if the first one came from D). Similarly, the probability that the second photon comes from D is either $\mathbf{I}\cdot(\mathbf{I}-\mathbf{Z})\cdot e^{\mathbf{K}\tau_2}\cdot\mathbf{Z}\cdot\mathbf{p}_{eq}$ or $\mathbf{I}\cdot(\mathbf{I}-\mathbf{Z})\cdot e^{\mathbf{K}\tau_2}\cdot(\mathbf{I}-\mathbf{Z})\cdot\mathbf{p}_{eq}$, depending again on the source of the first photon. By continuing this reasoning, you can find the probability of observing the particular sequence of photons shown in Fig. 7.9. If sequences of photons are recorded from several individual molecules, the overall probability is the product of the probabilities of the individual sequences. To find the best values of k_n, k_u, ξ_n and ξ_u, one adjusts these parameters to maximize the overall probability of the observations.

Chung et al. have used this technique to measure rate constants for folding and unfolding of several proteins [51–54]. They also explore how quickly individual transitions between U and N states can occur, as measured by the minimum time between detection of photons with wavelengths indicative of different states. The transition time is much less than the macroscopic time constant for folding and unfolding.

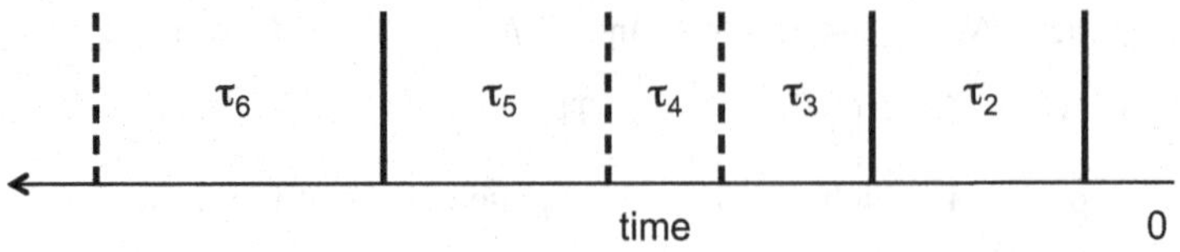

Fig. 7.9 A possible sequence of the wavelengths and timing of six photons detected in measurements of fluorescence from the energy donor (D) or acceptor (A) in a single molecule with two conformational states. *Solid vertical lines* reflect photons with wavelengths characteristic of D; *dashed lines*, photons with wavelengths characteristic of A. Time runs from right to left. The time intervals between detecting the photons are τ_2, τ_3, etc. The probability of observing this particular sequence of photons is $\mathbf{I}\cdot\mathbf{Z}\cdot e^{\mathbf{K}\tau_6}\cdot(\mathbf{I}-\mathbf{Z})\cdot e^{\mathbf{K}\tau_5}\cdot\mathbf{Z}\cdot e^{\mathbf{K}\tau_4}\cdot\mathbf{Z}\cdot e^{\mathbf{K}\tau_3}\cdot(\mathbf{I}-\mathbf{Z})\cdot e^{\mathbf{K}\tau_2}\cdot(\mathbf{I}-\mathbf{Z})\cdot\mathbf{p}_{eq}$, where matrices $\mathbf{I}$, $\mathbf{Z}$, $\mathbf{K}$ and $\mathbf{p}_{eq}$ are explained in the text.

7.4 Exchange Coupling

Förster's theory describes the rate of energy transfer between chromophores that are relatively far enough apart. In addition to using the point-dipole approximation, the theory assumes that the intermolecular interactions have no significant effect on the absorption or fluorescence spectra of the molecules. In the next chapter we'll discuss what happens when these assumptions to break down. However, it is pertinent to mention one other limitation of the Förster treatment here. The treatment considers only the spatial parts of the molecular wavefunctions; the spin wavefunctions are implicitly assumed to be constant. The theory thus is applicable to a transition in which a molecule in an excited singlet state returns to the singlet ground state and another molecule is elevated to an excited singlet state. It also could apply to excited triplet states but only if the ground states also are triplet states, which is not usually the case. It would not allow a transition in which a molecule in an excited triplet state decays to a singlet ground state, raising another molecule from a singlet to a triplet state. As predicted by the theory, this last type of energy transfer does not occur at significant rates between widely separated molecules. However, it does occur between molecules that are in close contact, and an additional mechanism clearly is needed to account for this fact. Such a mechanism was proposed by D.L. Dexter [55].

To see how triplet-triplet energy transfer can occur, we need to expand Eq. (7.14) to include spin wavefunctions. Let's refine the notation used there so that ϕ_{1a} and ϕ_{1b} now explicitly denote spatial wavefunctions of molecule 1, ϕ_{2a} and ϕ_{2b} denote spatial wavefunctions of molecule 2, and σ_{1a}, σ_{1b}, σ_{2a} and σ_{2b} denote corresponding spin wavefunctions. As we discussed in Chap. 4, the overall wavefunctions must be antisymmetric for an exchange of labels between any two electrons:

$$\Psi_1 = 2^{-1/2}\{\phi_{1b}(1)\sigma_{1b}(1)\phi_{2a}(2)\sigma_{2a}(2) - \phi_{1b}(2)\sigma_{1b}(2)\phi_{2a}(1)\sigma_{2a}(1)\}, \quad (7.32a)$$

and

$$\Psi_2 = 2^{-1/2}\{\phi_{1a}(1)\sigma_{1a}(1)\phi_{2b}(2)\sigma_{2b}(2) - \phi_{1a}(2)\sigma_{1a}(2)\phi_{2b}(1)\sigma_{2b}(1)\}, \quad (7.32b)$$

where (1) and (2) denote the coordinates of electrons 1 and 2. With these wavefunctions, the electronic interaction matrix element becomes

$$H_{21(el)} = H_{21}^{Coulomb} + H_{21}^{exchange}, \quad (7.33)$$

$$\begin{aligned} H_{21}^{Coulomb} &= \left\langle \phi_{1a}(1)\phi_{2b}(2)\middle|\widetilde{\mathrm{H}}_{21}\middle|\phi_{1b}(1)\phi_{2a}(2)\right\rangle \langle\sigma_{1a}(1)|\sigma_{1b}(1)\rangle\langle\sigma_{2b}(2)|\sigma_{2a}(2)\rangle \\ &= \left\langle \phi_{1a}(1)\phi_{2b}(2)\middle|\frac{e^2}{r_{21}}\middle|\phi_{1b}(1)\phi_{2a}(2)\right\rangle \langle\sigma_{1a}(1)|\sigma_{1b}(1)\rangle\langle\sigma_{2b}(2)|\sigma_{2a}(2)\rangle, \end{aligned} \quad (7.34a)$$

and

$$\begin{aligned} H_{21}^{exchange} &= -\left\langle \phi_{1a}(2)\phi_{2b}(1)\middle|\widetilde{\mathrm{H}}_{21}\middle|\phi_{1b}(1)\phi_{2a}(2)\right\rangle \langle\sigma_{2b}(1)|\sigma_{1b}(1)\rangle\langle\sigma_{1a}(2)|\sigma_{2a}(2)\rangle \\ &= -\left\langle \phi_{1a}(2)\phi_{2b}(1)\middle|\frac{e^2}{r_{21}}\middle|\phi_{1b}(1)\phi_{2a}(2)\right\rangle \langle\sigma_{2b}(1)|\sigma_{1b}(1)\rangle\langle\sigma_{1a}(2)|\sigma_{2a}(2)\rangle. \end{aligned} \quad (7.34b)$$

The term $H_{21}^{Coulomb}$, which is the one that the Förster theory considers, is called the *Coulomb* or *direct* interaction. It pertains to a process in which one electron moves from ϕ_{1b} to ϕ_{1a} while the other electron moves from ϕ_{2a} to ϕ_{2b}. Each electron thus remains on its original molecule (Fig. 7.10A). The product of spin integrals in this term will be non-zero only if $\sigma_{1a} = \sigma_{1b}$ and $\sigma_{2a} = \sigma_{2b}$; there must be no change of spin on either molecule. The term $H_{21}^{exchange}$ is called the *exchange*

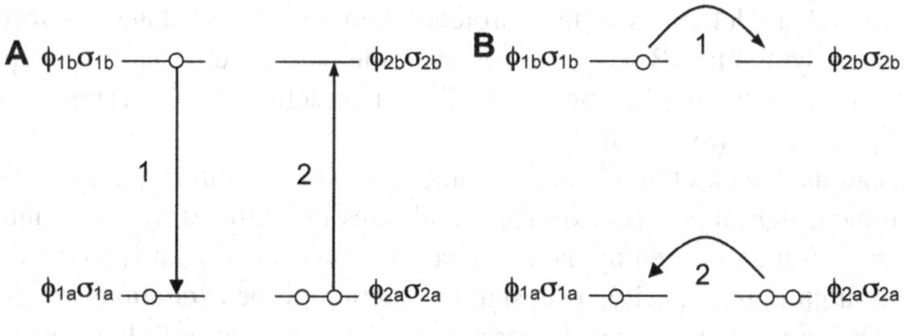

Fig. 7.10 The Coulomb contribution to the interaction matrix element $H_{21(el)}$ reflects a two-electron transition in which both electrons (*circles*) remain on their original molecules (*lines*) (**A**). This process requires conservation of spin on each molecule. The exchange contribution to H_{21} represents a transition in which electrons are interchanged between the molecules (**B**). The electron spins associated with the two molecules also are interchanged

interaction. Here one electron moves from ϕ_{1b} to ϕ_{2b} (from molecule 1 to molecule 2) while the other electron moves from ϕ_{2a} to ϕ_{1a} (from molecule 2 to molecule 1) as shown in Fig. 7.10B. In this case, the product of spin integrals will be non zero if $\sigma_{1b}=\sigma_{2b}$ and $\sigma_{2a}=\sigma_{1a}$. The electron spins of the two molecules are interchanged. This condition is consistent with either singlet or triplet energy transfer because σ_{1a} does not necessarily have to be the same as σ_{1b} and σ_{2a} does not have to be the same as σ_{2b}.

Equation (7.34b) indicates that $H_{21}^{exchange}$ can be appreciable only if there is a region of space where both ϕ_{1a} and ϕ_{2a} are significantly different from zero, and the same for ϕ_{1b} and ϕ_{2b}. This means that $H_{21}^{exchange}$ requires significant overlap of the two molecular orbitals. The magnitude of $H_{21}^{exchange}$ depends on the details of the molecular orbitals but generally is taken to fall off approximately as $\exp(-R_{edge}/L)$, where R_{edge} is the edge-to-edge intermolecular distance and L is on the order of 1 A. This is a much stronger dependence on distance than the R^{-6} dependence of $H_{21}^{Coulomb}$. $H_{21}^{Coulomb}$ therefore usually dominates heavily over $H_{21}^{exchange}$ as long as the edge-to-edge distance is more than 4 or 5 Å. However, in addition to accounting for triplet transfer, the exchange term can contribute to singlet energy transfer when the Förster mechanism is ineffective because the donor or acceptor has a very low dipole strength.

In addition to $H_{21}^{Coulomb}$ and $H_{21}^{exchange}$, H_{12} can include other higher-order terms that reflect mixing of the excited singlet states ψ_1 and ψ_2 with triplet states [56]. Again, these terms usually are significant only when $H_{21}^{Coulomb}$ is small.

7.5 Energy Transfer to and from Carotenoids in Photosynthesis

Excited carotenoid molecules can transfer energy rapidly to nearby chlorophyll molecules in photosynthetic antenna complexes [57]. It has been suggested that this process involves exchange coupling, because the radiative transitions from the lowest excited singlet states of the carotenoids to the ground states are forbidden by molecular symmetry (Box 4.12). However, the rate could possibly be explained by considering the full expression for the direct interaction (Eq. 7.14) instead of just dipole-dipole interactions [58].

Plants adjust the carotenoids in their antenna complexes in response to changes in the ambient light intensity and other conditions [57, 59]. At low light intensity, the energy of light absorbed by the antenna is transferred efficiently to the reaction centers; at high light intensity, a substantial fraction of the excitations are diverted to heat. The "non-photochemical quenching" that occurs at high light intensity is important for preventing destructive side reactions that occur if the reaction centers are oversaturated with excitations. The process involves, in part, enzymatic conversion of the carotenoid zeaxanthin to antheraxanthin and then violaxanthin by

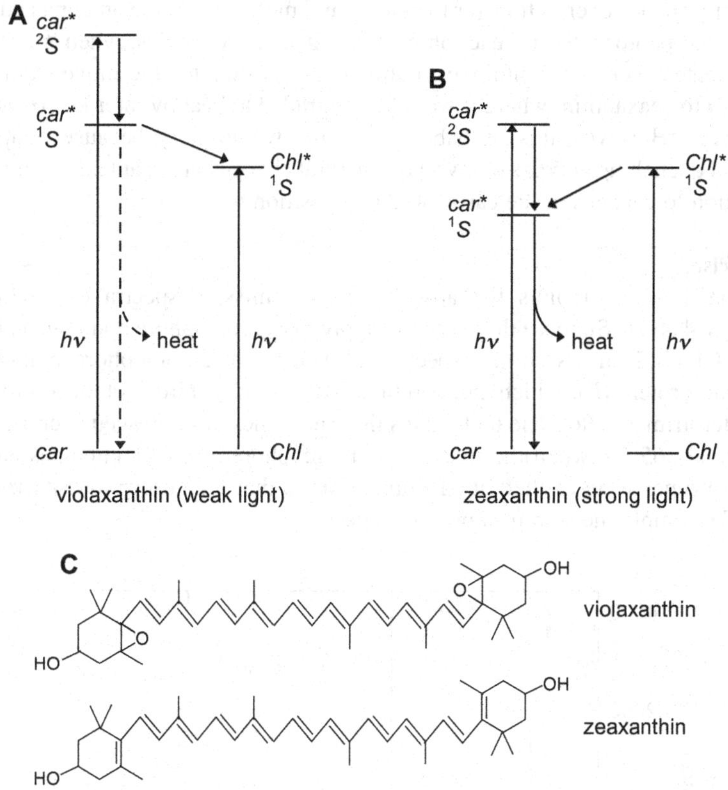

Fig. 7.11 The relative energies of excited states of chlorophyll-*a* (*Chl*) and the carotenoids (*car*) violaxanthin (**A**) and zeaxanthin (**B**). The structures of violoxanthin and zeaxanthin are shown in (**C**). Plants convert violaxanthin to zeaxanthin when they grow in strong light, and convert zeaxanthin to violaxanthin in weak light. The transition dipole for excitation of either carotenoid from the ground state to the second excited singlet state (^2S) is much larger than the transition dipole for the lowest excited state (^1S), but the excited molecule relaxes rapidly from ^2S to ^1S [61]. A possible mechanism of nonphotochemical quenching in chloroplasts is the transfer of energy from excited chlorophyll (Chl*) to zeaxanthin, followed by decay of zeaxanthin from ^1S to the ground state by internal conversion. The ^1S state of violaxanthin may be too high in energy to quench Chl*. Instead, zeaxanthin would transfer energy to chlorophyll

succesive epoxidation reactions in strong light, and the regeneration of violaxanthin by de-epoxidations in weak light (Fig. 7.11).

Epoxidation decreases the number of conjugated bonds in the carotenoid, which raises the energy of the lowest excited state. Violaxanthin has 9 conjugated double bonds; antheraxanthin, 10; and zeaxanthin, 11. The lowest excited singlet state of violaxanthin thus lies above that of zeaxanthin by about 300 cm^{-1} [60–62]. Although the absolute energies are subject to some uncertainty, the excitation energy of violaxanthin is close to that of chlorophyll-*a* in its lowest excited singlet state. At low light intensities, energy absorbed by violaxanthin could flow by resonance energy transfer to chlorophyll molecules in the antenna and from there to the photosynthetic reaction centers. At high intensities, when zeaxanthin predominates over violaxanthin, excitation energy would tend to move from chlorophyll-*a* to zeaxanthin, where it would be degraded to heat by internal conversion (Fig. 7.11). However, this probably is not the whole story because nonphotochemical quenching appears to involve structural changes in the antenna complexes in addition to changes in the carotenoid composition [63].

Exercises

Graph 1 below replots the absorption and emission spectra considered in Chaps. 4 and 5. Suppose these spectra apply to a chromophore bound to a protein (*A*). Graph 2 shows similar spectra for a different chromophore bound to a second protein (*B*), which can combine with *A* and a third protein (*C*) to form a heterotrimer, *ABC*. The table gives the amplitudes of fluorescence emission at 335 and 365 nm when the individual chromophore-protein complexes and the heterotrimer were excited at 315 nm. Assume that the absorption spectrum of *ABC* is simply the sum of those of *A* and *B*.

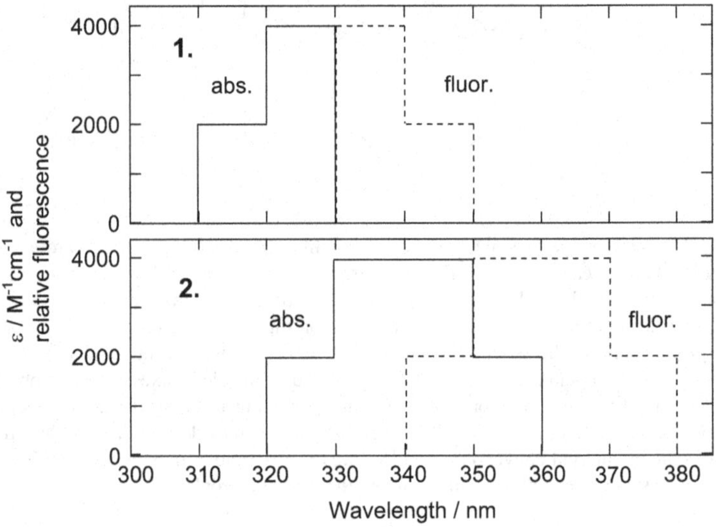

System	Fluorescence amplitude (arbitrary units)	
	335 nm	365 nm
A	4,000	10
B	<10	10
ABC	1,000	2,000

1. Calculate the emission-absorption overlap integral $(\overline{J})$ for A and B. (Be sure to give the units.)
2. Using your results from Exercise 4.2 and the fluorescence yield given in Exercise 5.2, calculate the Förster radius (R_o) for resonance energy transfer from A to B.
3. Estimate the distance between the chromophores bound to A and B in the ABC complex, assuming that the change in the fluorescence yield of A relative to that for monomeric A is due solely to resonance energy transfer. Assume also that the transition dipoles of both chromophores rotate rapidly and isotropically on the timescale of fluorescence.
4. Estimate the distance between the chromophores as in question 3 but assuming that the transition dipoles are fixed in position along the intermolecular axis (⟷⋯⋯⟷).
5. Does the fluorescence of the ABC complex at 365 nm verify the assumption that the change in the fluorescence yield of A results solely from resonance energy transfer? Why or why not?
6. If each of the fluorescence amplitudes in the table has an uncertainty of ±10%, what would be the uncertainty in the distance calculated in Exercise 3?
7. In the diagram below, the circles represent positive or negative unit charges located as indicated by the grid; distances are given in Å. Calculate the vacuum interaction energy between the two electric dipoles indicated by the dashed lines, using (*a*) the point-dipole approximation and (*b*) point charges. Specify the sign and units of your answers.

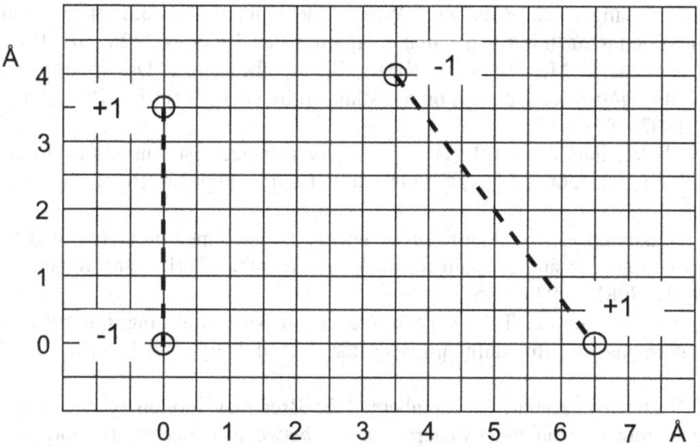

8. Under what circumstances is resonance energy transfer between two molecules likely to occur more rapidly by exchange-coupling than by dipole-dipole coupling?

References

1. van Amerongen, H., Valkunas, L., van Grondelle, R.: Photosynthetic Excitons. World Scientific, Singapore (2000)
2. Green, B.R., Parson, W.W. (eds.): Light-Harvesting Antennas in Photosynthesis. Advances in Photosynthesis and Respiration, Govindjee, ed., vol. 13. Dordrecht: Kluwer Academic (2003)
3. Sancar, A.: Structure and function of DNA photolyase and cryptochrome blue-light photoreceptors. Chem. Rev. **103**, 2203–2237 (2003)
4. Saxena, V.P., Wang, H., Kavakli, H., Sancar, A., Zhong, D.: Ultrafast dynamics of resonance energy transfer in cryptochrome. J. Am. Chem. Soc. **127**, 7984–7985 (2005)
5. van der Meer, B.W., Coker, G.I., Chen, S.-Y.S.: Resonance Energy Transfer: Theory and Data. VCH, New York (1994)
6. Böttcher, C.J.F.: Theory of Electric Polarization, vol. 1, 2nd edn. Elsevier, Amsterdam (1973)
7. Förster, T. *Zwischenmoleckulare energiewanderung und fluoreszenz.* Ann. Phys. **2**, 55–75 (1948) (for an English translation with some corrections, see *Biological Physics*, Mielczarek, E.V., Greenbaum, E. and Knox, R.S., eds., New York: Am. Inst. Phys., pp. 148–160 (1993))
8. Förster, T.: Delocalized excitation and excitation transfer. In: Sinanoglu, O. (ed.) Modern Quantum Chemistry, Pt. III., pp. 93–137. Academic, New York (1965)
9. Stryer, L.: Fluorescence energy transfer as a spectroscopic ruler. Annu. Rev. Biochem. **47**, 819–846 (1978)
10. Selvin, P.R.: Fluorescence resonance energy transfer. Methods Enzymol. **246**, 300–334 (1995)
11. Ward, R.J., Milligan, G.: Structural and biophysical characterization of G protein-coupled receptor ligand binding using resonance energy transfer and fluorescent labelling techniques. Biochim. Biophys. Acta **1838**, 3–14 (2014)
12. Sridharan, R., Zuber, J., Connelly, S.M., Mathew, E., Dumont, M.E.: Fluorescent approaches for understanding interactions of ligands with G protein coupled receptors. Biochim. Biophys. Acta **1838**, 15–33 (2014)
13. Stryer, L., Haugland, R.P.: Energy transfer: a spectroscopic ruler. Proc. Natl. Acad. Sci. USA **58**, 719–726 (1967)
14. Haugland, R.P., Yguerabide, J., Stryer, L.: Dependence of the kinetics of singlet-singlet energy transfer on spectral overlap. Proc. Natl. Acad. Sci. USA **63**, 23–30 (1969)
15. Dale, R.E., Eisinger, J., Blumberg, W.E.: Orientational freedom of molecular probes. The orientation factor in intramolecular energy transfer. Biophys. J. **26**, 161–194 (1979)
16. Haas, E., Wilcheck, M., Katchalski-Katzir, E., Steinberg, I.Z.: Distribution of end-to-end distances of oligopeptides in solution as estimated by energy transfer. Proc. Natl. Acad. Sci. USA **72**, 1807–1811 (1975)
17. Parkhurst, K.M., Parkhurst, L.J.: Donor-acceptor distance distributions in a double-labeled fluorescent oligonucleotide both as a single strand and in duplexes. Biochemistry **34**, 293–300 (1995)
18. Eis, P.S., Lakowicz, J.R.: Time-resolved energy transfer measurements of donor-acceptor distance distributions and intramolecular flexibility of a CCHH zinc finger peptide. Biochemistry **32**, 7981–7993 (1995)
19. Sun, H., Yin, D., Squire, T.C.: Calcium-dependent structural coupling between opposing globular domains of calmodulin involves the central helix. Biochemistry **38**, 7981–7993 (1999)
20. Haas, E., Katchalski-Katzir, E., Steinberg, I.Z.: Brownian motion of ends of oligopeptide chains in solution as estimated by energy transfer between chain ends. Biopolymers **17**, 11–31 (1978)
21. Beechem, J.M., Haas, E.: Simultaneous determination of intramolecular distance distributions and conformational dynamics by global analysis of energy transfer measurements. Biophys. J. **55**, 1225–1236 (1989)
22. Rhoades, E., Gussakovsky, E., Haran, G.: Watching proteins fold one molecule at a time. Proc. Natl. Acad. Sci. USA **100**, 3197–3202 (2003)

23. Rhoades, E., Cohen, M., Schuler, B., Haran, G.: Two-state folding observed in individual protein molecules. J. Am. Chem. Soc. **126**, 14686–14687 (2004)
24. Jang, S.J., Newton, M.D., Silbey, R.J.: Multichromophoric Förster resonance energy transfer. Phys. Rev. Lett. **92**, Art. No. 218301 (2004)
25. Muller, E.G., Snydsman, B.E., Novik, I., Hailey, D.W., Gestaut, D.R., et al.: The organization of the core proteins of the yeast spindle pole body. Mol. Biol. Cell **16**, 3341–3352 (2005)
26. Miyawaki, A., Llopis, J., Heim, R., McCaffrey, J.M., Adams, J.A., et al.: Fluorescent indicators for Ca^{2+} based on green fluorescent proteins and calmodulin. Nature **388**, 882–887 (1997)
27. Miyawaki, A., Griesbeck, O., Heim, R., Tsien, R.Y.: Dynamic and quantitative Ca^{2+} measurements using improved cameleons. Proc. Natl. Acad. Sci. USA **96**, 2135–2140 (1999)
28. Guerrero, G., Rieff, D.F., Agarwal, G., Ball, R.W., Borst, A., et al.: Heterogeneity in synaptic transmission along a Drosophila larval motor axon. Nat. Neurosci. **8**, 1188–1196 (2005)
29. Xu, Y., Piston, D.W., Johnson, C.H.: A bioluminescence resonance energy transfer (BRET) system: application to interacting circadian clock proteins. Proc. Natl. Acad. Sci. USA **96**, 151–156 (1999)
30. Angers, S., Salahpour, A., Joly, E., Hilairet, S., Chelsky, D., et al.: Detection of β_2-adrenergic receptor dimerization in living cells using bioluminescence resonance energy transfer (BRET). Proc. Natl. Acad. Sci. USA **97**, 3684–3689 (2000)
31. Selvin, P.R., Rana, T.M., Hearst, J.E.: Luminescence resonance energy transfer. J. Am. Chem. Soc. **116**, 6029–6030 (1994)
32. Heyduk, T., Heyduk, E.: Luminescence energy transfer with lanthanide chelates: interpretation of sensitized acceptor decay amplitudes. Anal. Biochem. **289**, 60–67 (2001)
33. Selvin, P.R.: Principles and biophysical applications of lanthanide-based probes. Annu. Rev. Biophys. Biomol. Struct. **31**, 275–302 (2002)
34. Reifenberger, J.G., Snyder, G.E., Baym, G., Selvin, P.R.: Emission polarization of europium and terbium chelates. J. Phys. Chem. B **107**, 12862–12873 (2003)
35. Posson, D.J., Ge, P., Miller, C., Benzanilla, F., Selvin, P.R.: Small vertical movement of a K^+ channel voltage sensor measured with luminescence energy transfer. Nature **436**, 848–851 (2005)
36. Talaga, D.S., Lau, W.L., Roder, H., Tang, J., Jia, Y., et al.: Dynamics and folding of single two-stranded coiled-coil peptides studied by fluorescent energy transfer confocal microscopy. Proc. Natl. Acad. Sci. USA **97**, 13021–13026 (2000)
37. Schuler, B., Lipman, E.Å., Eaton, W.A.: Probing the free-energy surface for protein folding with single-molecule fluorescence spectroscopy. Nature **419**, 743–747 (2002)
38. Kuzmenkina, E.V., Heyes, C.D., Nienhaus, G.U.: Single-molecule Förster resonance energy transfer study of protein dynamics under denaturing conditions. Proc. Natl. Acad. Sci. USA **102**, 15471–15476 (2005)
39. Huang, F., Sato, S., Sharpe, T.D., Ying, L., Fersht, A.R.: Distinguishing between cooperative and unimodal downhill protein folding. Proc. Natl. Acad. Sci. USA **104**, 123–127 (2007)
40. Merchant, K.A., Best, R.B., Louis, J.M., Gopich, I.V., Eaton, W.A.: Characterizing the unfolded states of proteins using single-molecule FRET spectroscopy and molecular simulations. Proc. Natl. Acad. Sci. USA **104**, 1528–1533 (2007)
41. Nettels, D., Gopich, I.V., Hoffmann, A., Schuler, B.: Ultrafast dynamics of protein collapse from single-molecule photon statistics. Proc. Natl. Acad. Sci. USA **104**, 2655–2660 (2007)
42. Kinoshita, M., Kamagata, K., Maeda, A., Goto, Y., Komatsuzaki, T., et al.: Development of a technique for the investigation of folding dynamics of single proteins for extended time periods. Proc. Natl. Acad. Sci. USA **104**, 10453–10458 (2007)
43. Huang, F., Ying, L., Fersht, A.R.: Direct observation of barrier-limited folding of BBL by single-molecule fluorescence resonance energy transfer. Proc. Natl. Acad. Sci. USA **106**, 16239–16244 (2009)

44. Yang, L.-L., Kao, M.W.-P., Chen, H.-L., Lim, T.-S., Fann, W., et al.: Observation of protein folding/unfolding dynamics of ubiquitin trapped in agarose gel by single-molecule FRET. Eur. Biophys. J. **41**, 189–198 (2012)
45. Rieger, R., Nienhaus, G.U.: A combined single-molecule FRET and tryptophan fluorescence study of RNase H folding under acidic conditions. Chem. Phys. **396**, 3–9 (2012)
46. Schuler, B., Hofmann, H.: Single-molecule spectroscopy of protein folding dynamics – expanding scope and timescales. Curr. Opin. Struct. Biol. **23**, 36–47 (2013)
47. Mashaghi, A., Kramer, G., Lamb, D.C., Mayer, M.P., Tans, S.J.: Chaperone action at the single-molecule level. Chem. Rev. **114**, 660–676 (2014)
48. Brucale, M., Schuler, B., Samori, B.: Single-molecule studies of intrinsically disordered proteins. Chem. Rev. **114**, 3281–3317 (2014)
49. Gopich, I.V., Szabo, A.: Single-molecule FRET with diffusion and conformational dynamics. J. Phys. Chem. B **111**, 12925–12932 (2007)
50. Gopich, I.V., Szabo, A.: Decoding the pattern of photon colors in single-molecule FRET. J. Phys. Chem. B **113**, 10965–10973 (2009)
51. Chung, H.S., Gopich, I.V.: Fast single-molecule FRET spectroscopy: theory and experiment. Phys. Chem. Chem. Phys. **16**, 18644–18657 (2014)
52. Chung, H.S., Louis, J.M., Eaton, W.A.: Experimental determination of upper bound for transition path times in protein folding from single-molecule photon-by-photon trajectories. Proc. Natl. Acad. Sci. USA **106**, 11837–11844 (2009)
53. Chung, H.S., McHale, K., Louis, J.M., Eaton, W.A.: Single-molecule fluorescence experiments determine protein folding transition path times. Science **335**, 981–984 (2012)
54. Chung, H.S., Cellmer, T., Louis, J.M., Eaton, W.A.: Measuring ultrafast protein folding rates from photon-by-photon analysis of single molecule fluorescence trajectories. Chem. Phys. **422**, 229–237 (2013)
55. Dexter, D.L.: A theory of sensitized luminescence in solids. J. Chem. Phys. **21**, 836–850 (1953)
56. Struve, W.S.: Theory of electronic energy transfer. In: Blankenship, R.E., Madigan, M.T., Bauer, C.E. (eds.) Anoxygenic photosynthetic bacteria, pp. 297–313. Kluwer Academic, Dordrecht (1995)
57. Young, A.J., Phillip, D., Ruban, A.V., Horton, P., Frank, H.A.: The xanthophyll cycle and carotenoid-mediated dissipation of excess excitation energy in photosynthesis. Pure Appl. Chem. **69**, 2125–2130 (1997)
58. Krueger, B.P., Scholes, G.D., Jimenez, R., Fleming, G.R.: Electronic excitation transfer from carotenoid to bacteriochlorophyll in the purple bacterium Rhodopseudomonas acidophila. J. Phys. Chem. B **102**, 2284–2292 (1998)
59. Frank, H.A., Cua, A., Chynwat, V., Young, A., Gosztola, D., et al.: Photophysics of the carotenoids associated with the xanthophyll cycle in photosynthesis. Photosynth. Res. **41**, 389–395 (1994)
60. Martinsson, P., Oksanen, J.A., Hilgendorff, M., Hynninen, P.H., Sundström, V., et al.: Dynamics of ground and excited state chlorophyll a molecules in pyridine solution probed by femtosecond transient absorption spectroscopy. Chem. Phys. Lett. **309**, 386–394 (1999)
61. Polivka, T., Herek, J.L., Zigmantas, D., Akerlund, H.E., Sundström, V.: Direct observation of the (forbidden) S_1 state in carotenoids. Proc. Natl. Acad. Sci. USA **96**, 4914–4917 (1999)
62. Frank, H.A., Bautista, J.A., Josue, J.S., Young, A.J.: Mechanism of nonphotochemical quenching in green plants: Energies of the lowest excited singlet states of violaxanthin and zeaxanthin. Biochemistry **39**, 2831–2837 (2000)
63. Ruban, A.V., Phillip, D., Young, A.J., Horton, P.: Excited-state energy level does not determine the differential effect of violaxanthin and zeaxanthin on chlorophyll fluorescence quenching in the isolated light-harvesting complex of photosystem II. Photochem. Photobiol. **68**, 829–834 (1998)

Exciton Interactions

8

8.1 Stationary States of Systems with Interacting Molecules

The Förster theory we considered in the last chapter applies to molecules that are far enough apart so that intermolecular interactions are very weak. Jumping of excitations from one molecule to the other is slow relative to the vibrational relaxation and dephasing that determine the homogeneous widths of the absorption bands, and it has little effect on the absorption spectra of the molecules. If the energy donor and acceptor are distinguishable we could examine the overall absorption or stimulated-emission spectrum of the system and, at least in principle, determine which molecule is excited at any given time. But suppose we move the molecules together so that the time required for energy to hop from one to the other becomes shorter and shorter. At some point, it will be impossible to say which molecule is excited. In this situation, we might expect that resonance between multiple excited states could cause the absorption spectrum of an oligomer to differ from the spectra of the individual molecules, and indeed this turns out to be the case.

The term *exciton* means an excitation that is delocalized over more than one molecule, or that moves rapidly from molecule to molecule. *Exciton interactions*, the intermolecular interactions that cause the excitation to spread over several molecules, are physically just the same as the weak interactions that result in stochastic jumping of excitations by resonance energy transfer; they are just stronger because the molecules are closer together or have larger transition dipoles. As a result, the absorption, fluorescence, and circular dichroism of the system can be significantly different from those of the individual molecules. But we are not yet in the region where overlap of the molecular orbitals allows new bonds to form and the definition of the molecules themselves becomes blurred.

Our discussion will concern mainly what are called *Frenkel excitons*, in which an electron that has been excited to a normally empty molecular orbital remains associated with a vacancy or "hole" in a normally filled orbital as they migrate from

W.W. Parson, *Modern Optical Spectroscopy*, DOI 10.1007/978-3-662-46777-0_8

one molecule to another. In *Wannier excitons*, the electron and hole can be on separate molecules, although they usually are not far apart.

As long as the two molecules are not so close together as to become covalently bonded, it still makes sense to describe the wavefunctions of a dimer with combinations of the individual molecular wavefunctions, as we did in Eqs. (7.2), (7.4), and (7.6):

$$\Psi_A = \phi_{1a}\chi_{1a}\phi_{2a}\chi_{2a} \quad (ground\ \ state) \tag{8.1}$$

and

$$\begin{aligned}\Psi_B &= C_1\psi_1 + C_2\psi_2 \\ &= C_1\phi_{1b}\chi_{1b}\phi_{2a}\chi_{2a} + C_2\phi_{1a}\chi_{1a}\phi_{2b}\chi_{2b} \quad (excited\ \ state).\end{aligned} \tag{8.2}$$

As before, the basis states ψ_1 and ψ_2 for Ψ_B represent states in which the excitation is localized on molecule 1 or 2, respectively. These are not stationary states because excitations hop back and forth between the two molecules. Setting one of the coefficients C_1 or C_2 to 1 and the other to zero cannot, therefore, describe the system well on time scales comparable to, or longer than the oscillation period. But we will see that it is possible to find values of the coefficients that make Ψ_B into a stationary state. As we might expect, the values of C_1 and C_2 that are required depend strongly on the interaction term in the Hamiltonian, $\widetilde{\mathrm{H}}'$, which is the term that mixes the two basis states (Sects. 2.3.6 and 7.2).

Let's consider how the energies of the ground and excited states described by Eqs. (8.1) and (8.2) depend on $\widetilde{\mathrm{H}}'$. The electronic energy of the ground state is:

$$\begin{aligned}E_A &= \left\langle \phi_{1a}\phi_{2a} \middle| \widetilde{\mathrm{H}}_1 + \widetilde{\mathrm{H}}_2 + \widetilde{\mathrm{H}}' \middle| \phi_{1a}\phi_{2a} \right\rangle \\ &= E_{1a} + E_{2a} + \left\langle \phi_{1a}\phi_{2a} \middle| \widetilde{\mathrm{H}}' \middle| \phi_{1a}\phi_{2a} \right\rangle,\end{aligned} \tag{8.3}$$

where E_{1a} and E_{2a} are the energies of the individual molecules in their ground states. Similarly, the electronic energy of the excited state is:

$$\begin{aligned}E_B &= \left\langle C_1\psi_1 + C_2\psi_2 \middle| \widetilde{\mathrm{H}}_1 + \widetilde{\mathrm{H}}_2 + \widetilde{\mathrm{H}}' \middle| C_1\psi_1 + C_2\psi_2 \right\rangle \\ &= |C_1|^2 E_1 + |C_2|^2 E_2 + \left\langle C_1\psi_1 + C_2\psi_2 \middle| \widetilde{\mathrm{H}}' \middle| C_1\psi_1 + C_2\psi_2 \right\rangle,\end{aligned} \tag{8.4}$$

where $E_1 = E_{1b} + E_{2a}$ and $E_2 = E_{1a} + E_{2b}$. $\left\langle \psi_1 \middle| \widetilde{\mathrm{H}}_2 \middle| \psi_1 \right\rangle$ and $\left\langle \psi_2 \middle| \widetilde{\mathrm{H}}_1 \middle| \psi_2 \right\rangle$ are zero because $\widetilde{\mathrm{H}}_2$ operates only on ψ_2 and $\widetilde{\mathrm{H}}_1$ acts only on ψ_1; $\left\langle \psi_1 \middle| \widetilde{\mathrm{H}}_2 \middle| \psi_2 \right\rangle$ and $\left\langle \psi_2 \middle| \widetilde{\mathrm{H}}_1 \middle| \psi_1 \right\rangle$ evaluate to $E_2\langle\psi_1|\psi_2\rangle$ and $E_1\langle\psi_2|\psi_1\rangle$, which also are zero if ψ_1 and ψ_2 are orthogonal. In the absence of interactions, the ground-state energy would be simply $E_{1a}+E_{2a}$, and the two excited states would have energies E_1 and E_2.

If molecules 1 and 2 are uncharged and have only small permanent dipole moments, the term $\left\langle \phi_{1a}\phi_{2a}\middle|\widetilde{\mathrm{H}}'\middle|\phi_{1a}\phi_{2a}\right\rangle$ in Eq. (8.3) will be relatively small and the energy of the ground state will still be approximately $E_{1a}+E_{2a}$. The effect on the excited state often is more significant. To evaluate E_B and to see how it is related to the coefficients C_1 and C_2, we can follow essentially the same approach that we used for time-dependent perturbations, with the simplification that here we're interested only in stationary states. First, write the time-independent Schrödinger equation for the excited dimer:

$$\left(\widetilde{\mathrm{H}}_1+\widetilde{\mathrm{H}}_2+\widetilde{\mathrm{H}}'\right)(C_1\psi_1+C_2\psi_2)=E_B(C_1\psi_1+C_2\psi_2). \tag{8.5a}$$

Now multiply both sides of this equation by ψ_1*, integrate over all space, and drop the terms that are zero ($\left\langle\psi_1\middle|\widetilde{\mathrm{H}}_2\middle|\psi_1\right\rangle$, $\left\langle\psi_1\middle|\widetilde{\mathrm{H}}_2\middle|\psi_2\right\rangle$ and $\left\langle\psi_1\middle|\widetilde{\mathrm{H}}_1\middle|\psi_2\right\rangle$). On the left side this gives

$$C_1\left[\left\langle\psi_1\middle|\widetilde{\mathrm{H}}_1\middle|\psi_1\right\rangle+\left\langle\psi_1\middle|\widetilde{\mathrm{H}}'\middle|\psi_1\right\rangle\right]+C_2\left\langle\psi_1\middle|\widetilde{\mathrm{H}}'\middle|\psi_2\right\rangle, \tag{8.5b}$$

and on the right side,

$$E_B[C_1\langle\psi_1|\psi_1\rangle+C_2\langle\psi_1|\psi_2\rangle]=E_BC_1. \tag{8.5c}$$

Equating the quantities obtained from the two sides, we have

$$C_1\left[\left\langle\psi_1\middle|\widetilde{\mathrm{H}}_1\middle|\psi_1\right\rangle+\left\langle\psi_1\middle|\widetilde{\mathrm{H}}'\middle|\psi_1\right\rangle-E_B\right]+C_2\left\langle\psi_1\middle|\widetilde{\mathrm{H}}'\middle|\psi_2\right\rangle=0. \tag{8.6a}$$

Similarly, multiplying both sides of Eq. (8.5a) by ψ_2* and integrating leads to

$$C_1\left\langle\psi_2\middle|\widetilde{\mathrm{H}}'\middle|\psi_1\right\rangle+C_2\left[\left\langle\psi_2\middle|\widetilde{\mathrm{H}}_2\middle|\psi_2\right\rangle+\left\langle\psi_2\middle|\widetilde{\mathrm{H}}'\middle|\psi_2\right\rangle-E_B\right]=0. \tag{8.6b}$$

We now have two simultaneous equations relating E_B to C_1 and C_2 (Eqs. 8.6a and 8.6b), which we can rewrite in the following compact form:

$$C_1(H_{11}-E_B)+C_2H_{12}=0 \tag{8.7a}$$

and

$$C_1H_{21}+C_2(H_{22}-E_B)=0, \tag{8.7b}$$

where

$$H_{11} \equiv \left\langle \psi_1 \middle| \widetilde{\mathrm{H}}_1 + \widetilde{\mathrm{H}}' \middle| \psi_1 \right\rangle = E_1 + \left\langle \psi_1 \middle| \widetilde{\mathrm{H}}' \middle| \psi_1 \right\rangle, \tag{8.8a}$$

$$H_{22} \equiv \left\langle \psi_2 \middle| \widetilde{\mathrm{H}}_2 + \widetilde{\mathrm{H}}' \middle| \psi_2 \right\rangle = E_2 + \left\langle \psi_2 \middle| \widetilde{\mathrm{H}}' \middle| \psi_2 \right\rangle, \tag{8.8b}$$

and

$$H_{21} \equiv \left\langle \psi_2 \middle| \widetilde{\mathrm{H}}' \middle| \psi_1 \right\rangle, \quad H_{12} \equiv \left\langle \psi_1 \middle| \widetilde{\mathrm{H}}' \middle| \psi_2 \right\rangle. \tag{8.8c}$$

Because the Hamiltonian operator is Hermitian (Box 2.1), $H_{12} = H_{21}{}^*$. In the cases of interest here, these matrix elements generally are real numbers and are identical. In many cases, H_{11} and H_{22} are approximately equal to E_1 and E_2. The additional term $\left\langle \psi_1 \middle| \widetilde{\mathrm{H}}' \middle| \psi_1 \right\rangle$ in H_{11}, for example, represents the effect that molecule 2 has on the energy of the excited molecule 1 when molecule 2 remains in the ground state; if the molecules are not charged, this often will be a relatively small effect and will be comparable to the term $\left\langle \phi_{1a}\phi_{2a} \middle| \widetilde{\mathrm{H}}' \middle| \phi_{1a}\phi_{2a} \right\rangle$ in the energy of the ground state (Eq. 8.3). The small difference between H_{11} and E_1 thus will make little contribution to the excitation energy, $E_B - E_A$. However, our derivation does not require that this effect be negligible.

A trivial solution to Eqs. (8.7a and 8.7b) is $C_1 = C_2 = 0$. There can be another, nonzero solution only if the determinant formed from the Hamiltonian matrix elements and energies in the equations is equal to zero:

$$\begin{vmatrix} (H_{11} - E_B) & H_{21} \\ H_{21} & (H_{22} - E_B) \end{vmatrix} = 0 \tag{8.9}$$

(Box 8.1). This is called the *secular* determinant. Expanding the determinant gives a quadratic equation for E_B in terms of H_{21}, H_{11} and H_{22}. The quadratic equation has two possible solutions for E_B, which we'll call E_{B+} and E_{B-}. After we find these solutions, we can plug either E_{B+} or E_{B-} back into Eqs. (8.8a and 8.8b) to find the corresponding values of C_1 and C_2.

Box 8.1 Why Must the Secular Determinant be Zero?

The term "secular" is used in classical mechanics for a state or motion that persists for a long period of time. In quantum mechanics, the set of simultaneous equations that describe a long-lasting state (Eqs. (8.7a and 8.7b) in the situation we are considering) are called the secular equations of the system.

A set of simultaneous linear equations can be solved conveniently by Cramer's rule, which involves finding the quotient of two determinants. Given the equations

(continued)

Box 8.1 (continued)

$$\begin{aligned} a_1x + b_1y + c_1z + \cdots &= m_1 \\ a_2x + b_2y + c_2z + \cdots &= m_2 \\ a_3x + b_3y + c_3z + \cdots &= m_3 \\ &\cdots, \end{aligned} \tag{B8.1.1}$$

the solutions for $x, y, z, \ldots$ are

$$x = \begin{vmatrix} m_1 & b_1 & c_1 & \cdots \\ m_2 & b_2 & c_2 & \cdots \\ m_3 & b_3 & c_3 & \cdots \\ \vdots & \vdots & \vdots & \ddots \end{vmatrix} \Bigg/ \begin{vmatrix} a_1 & b_1 & c_1 & \cdots \\ a_2 & b_2 & c_2 & \cdots \\ a_3 & b_3 & c_3 & \cdots \\ \vdots & \vdots & \vdots & \ddots \end{vmatrix}, \tag{B8.1.2a}$$

$$y = \begin{vmatrix} a_1 & m_1 & c_1 & \cdots \\ a_2 & m_2 & c_2 & \cdots \\ a_3 & m_3 & c_3 & \cdots \\ \vdots & \vdots & \vdots & \ddots \end{vmatrix} \Bigg/ \begin{vmatrix} a_1 & b_1 & c_1 & \cdots \\ a_2 & b_2 & c_2 & \cdots \\ a_3 & b_3 & c_3 & \cdots \\ \vdots & \vdots & \vdots & \ddots \end{vmatrix}, \quad \text{etc.} \tag{B8.1.2b}$$

Applying Cramer's rule to Eqs. (8.7a and 8.7b) yields the following solution for C_1 and C_2:

$$C_1 = \begin{vmatrix} 0 & H_{21} \\ 0 & (H_{22} - E_B) \end{vmatrix} \Bigg/ \begin{vmatrix} (H_{11} - E_B) & H_{21} \\ H_{21} & (H_{22} - E_B) \end{vmatrix}, \tag{B8.1.3a}$$

and

$$C_2 = \begin{vmatrix} (H_{11} - E_B) & 0 \\ H_{21} & 0 \end{vmatrix} \Bigg/ \begin{vmatrix} (H_{11} - E_B) & H_{21} \\ H_{21} & (H_{22} - E_B) \end{vmatrix}. \tag{B8.1.3b}$$

The determinants in the numerators evaluate to 0 because each of them has a column of zeros. It follows that there can be a non-zero solution for C_1 and C_2 only if the determinant in the denominator also is zero.

A more powerful procedure is to use matrix algebra to solve Eqs. (8.7a and 8.7b) simultaneously for the two sets of E_B, C_1 and C_2. Mathematically, the problem is to diagonalize the matrix **H** of H_{ij} terms (Sect. 2.3.6). It is the off-diagonal terms H_{12} and H_{21} in the matrix that cause the system to oscillate between the nonstationary basis states ψ_1 and ψ_2. Diagonalization converts **H** to a matrix that characterizes a stationary state because all the off-diagonal terms are zero. In the process, it transforms the basis states from ψ_1 and ψ_2 into Ψ_{B+} and Ψ_{B-}.

The solutions to Eqs. (8.7a and 8.7b) depend on how the magnitude of the interaction term H_{21} compares to the difference between H_{11} and H_{22}. The general solutions can be written:

$$\Psi_{B+} = \sqrt{\frac{(1+s)}{2}}\,\psi_1 + \sqrt{\frac{(1-s)}{2}}\,\psi_2, \quad E_{B+} = E_0 + \frac{1}{2}\sqrt{\delta^2 + 4(H_{12})^2}, \quad (8.10a)$$

$$\Psi_{B-} = \sqrt{\frac{(1-s)}{2}}\,\psi_1 - \sqrt{\frac{(1+s)}{2}}\,\psi_2, \quad E_{B-} = E_0 - \frac{1}{2}\sqrt{\delta^2 + 4(H_{12})^2}, \quad (8.10b)$$

in which

$$s = \delta/\sqrt{\delta^2 + 4(H_{12})^2}, \quad (8.10c)$$

where E_0 is the average of the energies of the two basis states, and δ is the difference between these energies: $\delta = (H_{11} - H_{22})$ and $E_0 = (H_{11} + H_{22})/2$.

Figures 8.1 and 8.2 show how the coefficients and energies of the eigenstates depend on the ratio of $|\delta|$ to $|H_{21}|$. If the two basis states are widely separated in energy ($|\delta| \gg |H_{21}|$), so that $s \approx 1$, then the coefficients C_1 and C_2 for Ψ_{B+} approach 1.0 and zero, respectively (Fig. 8.1). At the same time, C_1 and C_2 for Ψ_{B-} (not shown in the figure) go to zero and -1.0, respectively. Ψ_{B+} and Ψ_{B-} thus approach ψ_1 and $-\psi_2$:

$$\Psi_{B+} \rightarrow \psi_1, \quad E_{B+} \rightarrow H_{11}, \quad (8.11a)$$

and

$$\Psi_{B-} \rightarrow -\psi_2, \quad E_{B-} \rightarrow H_{22}. \quad (8.11b)$$

In this region, C_2 for Ψ_{B+} is given approximately by $-H_{21}/(H_{22} - H_{11})$, and C_1 for Ψ_{B-} by $-H_{12}/(H_{11} - H_{22})$ (Box 12.1). On the other hand, if the basis states are located close together ($|\delta| \ll |H_{21}|$), so that $s \approx 0$, then the two solutions become

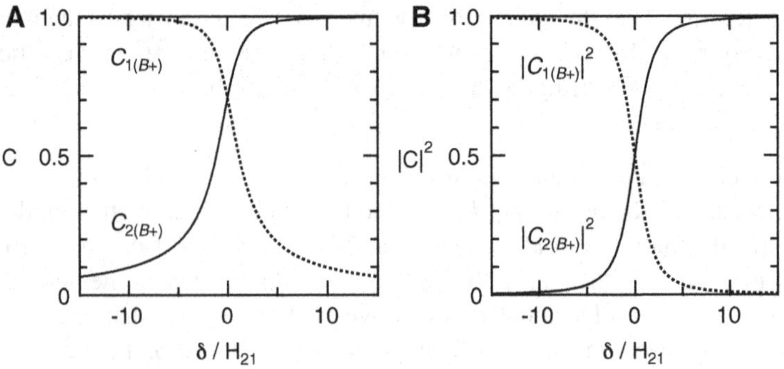

Fig. 8.1 The coefficients ($C_{1(B+)}$ and $C_{2(B+)}$, *left*) and their squares ($|C_{1(B+)}|^2$ and $|C_{2(B+)}|^2$, *right*) for one of the excited states of a dimer, as a function of the energy difference between the two basis states ψ_1 and ψ_2 ($\delta = H_{11} - H_{22}$). The interaction energy H_{21} is held constant as δ is varied. When $|\delta| \gg |H_{21}|$, one of the coefficients for each eigenstate goes to ± 1.0 and the other goes to zero. When $|\delta| \ll |H_{21}|$, $|C_1|^2 = |C_2|^2 = 1/2$

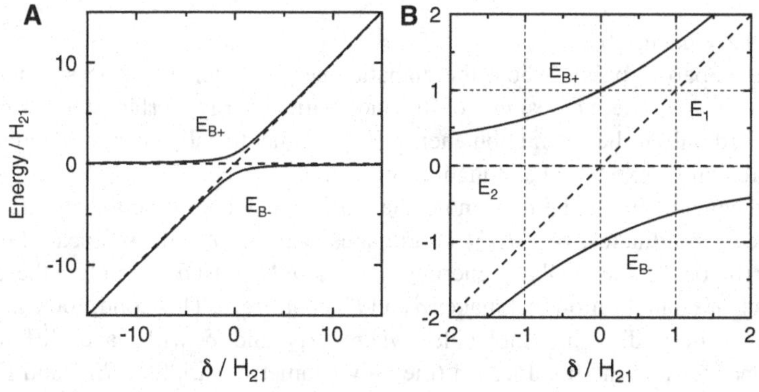

Fig. 8.2 Energies of the excited states of a dimer ($E_{B\pm}$, *solid curves*) as functions of the energy difference (δ) between the basis states ψ_1 and ψ_2. The energy scales are in units of the interaction energy H_{21}, which is held constant as in Fig. 8.1. The energy of ψ_2 (E_2) is fixed at zero while that of ψ_1 (E_1) is varied (*dotted lines*). The plot on the *right* shows an expanded view of the region around $\delta \approx 0$, where E_{B+} and E_{B-} are separated by $2H_{21}$. Away from this region, E_{B+} and E_{B-} approach E_1 and E_2

$$\Psi_{B+} = 2^{-1/2}[\psi_1 + \psi_2], \quad E_{B+} = E_0 + H_{21}, \tag{8.12a}$$

and

$$\Psi_{B-} = 2^{-1/2}[\psi_1 - \psi_2], \quad E_{B-} = E_0 - H_{21}. \tag{8.12b}$$

In this case, the two excited states (Ψ_{B+} and Ψ_{B-}) are symmetric and antisymmetric combinations of ψ_1 and ψ_2.

When $|\delta| \gg |H_{21}|$ the energies of the eigenstates approach H_{11} and H_{22} (Fig. 8.2). But in the opposite limit, when $|\delta| \ll |H_{21}|$, the energies of the eigenstates are not simply the average of H_{11} and H_{22}, and the two energies do not become identical. Instead, the energies split apart to lie above and below E_0 by $\pm H_{21}$ (Eqs. 8.12a and 8.12b). This *resonance splitting* of the energy is a purely quantum mechanical effect that results from mixing of the two basis states and distribution of the excitation over the two molecules (Box 8.2).

Box 8.2 Real and Avoided Crossings of Energy Surfaces
The basis states ψ_1 and ψ_2 are diabatic states, which means that they do not diagonalize the Hamiltonian of the system. Ψ_{B+} and Ψ_{B-}, which *do* diagonalize $\mathbf{H}$, are adiabatic. Equations (8.10a–8.10b) for the adiabatic energies apply not only to the particular situation we have discussed, in which an excitation is distributed over two molecules, but to any two-state system. However, the energy diagram in Fig. 8.2, in which the adiabatic energies

(continued)

Box 8.2 (continued)
remain apart at points where the diabatic energies intersect, is less general. Such an "avoided crossing" of the adiabatic energy surfaces is strictly required only if the interaction energy H_{12} and the diabatic energy difference δ both can be expressed as functions of a single variable.

In general, H_{12} and δ for a molecule with N nuclei are functions of (3N-6) nuclear coordinates, or (3N-5) coordinates if the molecule is linear. From Eqs. (8.10a, b), the adiabatic energies E_{B+} and E_{B-} must differ unless there is a nuclear configuration that makes δ and H_{12} both zero. This condition cannot be met in a diatomic molecule, where H_{12} and δ will have different dependences on the bond length (the only geometric variable). E_{B+} and E_{B-} therefore must have an avoided crossing in a diatomic molecule. However, a larger system with M geometric variables will have an (M-2)-dimensional hyperline along which both δ and H_{12} for a given pair of diabatic states are zero and the adiabatic energy surfaces have real, rather than avoided crossings. Such crossings are called *conical intersections* because if the energy surfaces are plotted as functions of the two remaining geometric variables they resemble a double cone. A system with two geometric variables, for example, will have a single point at which both δ and H_{12} are zero and the adiabatic energies are the same. But this conical intersection is not necessarily in an accessible region of the configuration space; depending on the molecule, it could involve a geometry with a very high energy. For further discussion of conical intersections and their possible roles in photochemical processes such as the isomerization of rhodopsin, see [1–7].

The eigenfunctions Ψ_{B+} and Ψ_{B-} given by Eqs. (8.10a, b) are normalized and orthogonal:

$$\langle \Psi_{B+} | \Psi_{B+} \rangle = \langle \Psi_{B-} | \Psi_{B-} \rangle = 1 \tag{8.13a}$$

and

$$\langle \Psi_{B+} | \Psi_{B-} \rangle = \langle \Psi_{B+} | \Psi_{B-} \rangle = 0. \tag{8.13b}$$

In addition

$$\left\langle \Psi_{B+} \middle| \widetilde{\mathrm{H}}' \middle| \Psi_{B-} \right\rangle = \left\langle \Psi_{B-} \middle| \widetilde{\mathrm{H}}' \middle| \Psi_{B+} \right\rangle = 0 \tag{8.13c}$$

(Box 8.3). This means that the two excited states Ψ_{B+} and Ψ_{B-} obtained by diagonalizing the Hamiltonian matrix are stationary states. $\widetilde{\mathrm{H}}'$ causes oscillations between the two basis states ψ_1 and ψ_2, but not between Ψ_{B+} and Ψ_{B-}. This is because we included $\widetilde{\mathrm{H}}'$ in the Hamiltonian that we used to find Ψ_{B+} and Ψ_{B-}.

It is instructive to compare these results with the results we obtained in earlier chapters where we used time-dependent perturbation theory. In the present chapter we have found two, and only two states that diagonalize the 2×2 Hamiltonian matrix, **H**. In Chap. 6, we found an infinite number of states that satisfied the time-dependent Schrödinger equation for the same Hamiltonian. Although we focused mainly on the region where $C_1(t) \approx 1$ and $C_2(t) \approx 0$, the coefficients were continuous functions of time and thus could have any values. The only restriction was that $|C_1(t)|^2 + |C_2(t)|^2 = 1$. The difference is that here we have found the two stationary states of the system; the continuum of states obtained by time-dependent perturbation theory are nonstationary states.

Box 8.3 Exciton States Are Stationary in the Absence of Further Perturbations

To see that the states described by Eqs. (8.10a, b) are stationary in the absence of other purturbations, let the coefficients C_1 and C_2 for state Ψ_{B+} be $C_{1(B+)}$ and $C_{2(B+)}$, and the coefficients for Ψ_{B-} be $C_{1(B-)}$ and $C_{2(B-)}$. Using Eq. (8.5a) to replace $\left(\widetilde{\mathrm{H}}_1 + \widetilde{\mathrm{H}}_2 + \widetilde{\mathrm{H}}'\right)\left(C_{1(B-)}\psi_1 + C_{2(B-)}\psi_2\right)$ by $E_{B-}\left(C_{1(B-)}\psi_1 + C_{2(B-)}\psi_2\right)$ then gives:

$$\left\langle \Psi_{B+}\middle|\widetilde{\mathrm{H}}\middle|\Psi_{B-}\right\rangle = \left\langle C_{1(B+)}\psi_1 + C_{2(B+)}\psi_2\middle|\widetilde{\mathrm{H}}_1 + \widetilde{\mathrm{H}}_2 + \widetilde{\mathrm{H}}'\middle|C_{1(B-)}\psi_1 + C_{2(B-)}\psi_2\right\rangle \tag{B8.4.1}$$

$$= E_{B-}\left\langle C_{1(B+)}\psi_1 + C_{2(B+)}\psi_2\middle|C_{1(B-)}\psi_1 + C_{2(B-)}\psi_2\right\rangle. \tag{B8.4.2}$$

If the basis wavefunctions ψ_1 and ψ_2 are orthogonal and normalized, this expression reduces to

$$\left\langle \Psi_{B+}\middle|\widetilde{\mathrm{H}}\middle|\Psi_{B-}\right\rangle = E_{B-}\left[C_{1(B+)}C_{1(B-)} + C_{2(B+)}C_{2(B-)}\right], \tag{B8.4.3}$$

which evaluates to zero when we insert the expressions for the coefficients given in Eqs. (8.10a, b).

8.2 Effects of Exciton Interactions on the Absorption Spectra of Oligomers

What effects do the intermolecular interactions have on the absorption spectrum of the dimer? Note first that, for each absorption band of the individual molecules, the dimer has two absorption bands representing the transitions $\Psi_A \rightarrow \Psi_{B+}$ and $\Psi_A \rightarrow \Psi_{B-}$. If $|\delta| \ll |H_{21}|$, the two absorption bands are separated in energy by $2H_{21}$ (Eqs. 8.12a and 8.12b). The former transition has an energy of $(H_{11} + H_{22})/$

$2 + H_{21}$, and the latter has an energy of $(H_{11} + H_{22})/2 - H_{21}$, so which energy is higher depends on the sign of H_{21}.

Once the coefficients C_1 and C_2 for the excited state are known, the transition dipoles and dipole strengths of the two exciton bands can be related straightforwardly to the spectroscopic properties of the monomers:

$$\boldsymbol{\mu}_{BA\pm} = \langle \Psi_{B\pm} | \tilde{\mu}_1 + \tilde{\mu}_2 | \Psi_A \rangle$$
$$= \left\langle C_{1(B\pm)} \phi_{1b} \phi_{2a} + C_{2(B\pm)} \phi_{1a} \phi_{2b} | \tilde{\mu}_1 + \tilde{\mu}_2 | \phi_{1a} \phi_{2a} \right\rangle \tag{8.14a}$$

$$= C_{1(B\pm)} \langle \phi_{1b} | \tilde{\mu}_1 | \phi_{1a} \rangle + C_{2(B\pm)} \langle \phi_{2b} | \tilde{\mu}_2 | \phi_{2a} \rangle = C_{1(B\pm)} \boldsymbol{\mu}_1 + C_{2(B\pm)} \boldsymbol{\mu}_2 \tag{8.14b}$$

and

$$D_{BA\pm} = \left(C_{1(B\pm)} \boldsymbol{\mu}_{ba(1)} + C_{2(B\pm)} \boldsymbol{\mu}_{ba(2)} \right) \cdot \left(C_{1(B\pm)} \boldsymbol{\mu}_{ba(1)} + C_{2(B\pm)} \boldsymbol{\mu}_{ba(2)} \right) \tag{8.15a}$$

$$= \left(C_{1(B\pm)} \right)^2 D_{ba(1)} + \left(C_{2(B\pm)} \right)^2 D_{ba(2)} + 2 C_{1(B\pm)} C_{2(B\pm)} \boldsymbol{\mu}_{ba(1)} \cdot \boldsymbol{\mu}_{ba(2)}, \tag{8.15b}$$

where $\boldsymbol{\mu}_{ba(1)}$ and $\boldsymbol{\mu}_{ba(2)}$ are the transition dipole vectors of the two individual molecules, and $D_{ba(1)}$ and $D_{ba(2)}$ are the individual dipole strengths. Because the coefficients for Ψ_{B+} ($C_{1(B+)}$ and $C_{2(B+)}$) generally are different from the coefficients for Ψ_{B-} ($C_{1(B-)}$ and $C_{2(B-)}$), the transition dipoles and dipole strengths of the two absorption bands will differ.

For a dimer of two identical molecules ($\delta = 0$ and $D_{ba(1)} = D_{ba(2)} = D_{ba}$) Eqs. (8.14a), (8.14b), (8.15a), and (8.15b) become:

$$\boldsymbol{\mu}_{BA+} = 2^{-1/2} \left(\boldsymbol{\mu}_{ba(1)} + \boldsymbol{\mu}_{ba(2)} \right), \tag{8.16a}$$

$$\boldsymbol{\mu}_{BA-} = 2^{-1/2} \left(\boldsymbol{\mu}_{ba(1)} - \boldsymbol{\mu}_{ba(2)} \right), \tag{8.16b}$$

$$D_{BA+} = (1/2)[D_{ba} + D_{ba} + 2D_{ba}(1 + \cos\theta)] = D_{ba}(1 + \cos\theta), \tag{8.16c}$$

and

$$D_{BA-} = (1/2)[D_{ba} + D_{ba} + 2D_{ba}(1 - \cos\theta)] = D_{ba}(1 - \cos\theta). \tag{8.16d}$$

These expressions show that the transition dipoles of the dimer's two bands are proportional to the vector sum and difference of the transition dipoles of the monomers (Fig. 8.3). D_{BA+} and D_{BA-} thus can range from 0 to $2D_{ba}$ depending on the angle θ between the transition dipoles of the monomers; however, the sum of D_{BA+} and D_{BA-} is always simply $2D_{ba}$.

Figure 8.4 illustrates the application of these results to dimers of identical molecules in several different arrangements. The point-dipole approximation (Eq. 7.17) was used to calculate the signs and relative magnitudes of the interaction

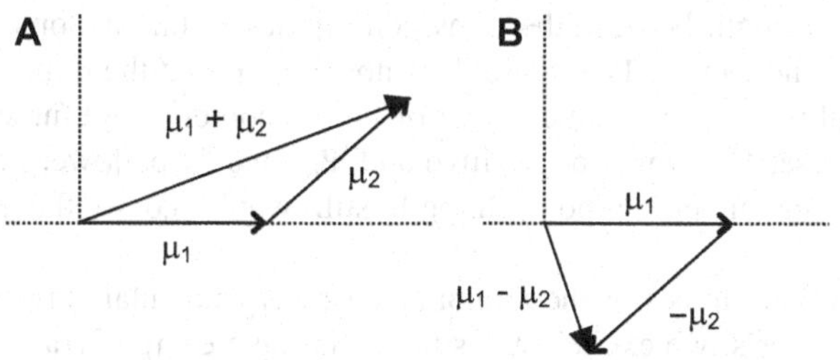

Fig. 8.3 The transition dipoles for a dimer are the vector sum and difference of the transition dipoles of the two monomeric chromophores. The sum and difference of any two vectors are always perpendicular to each other

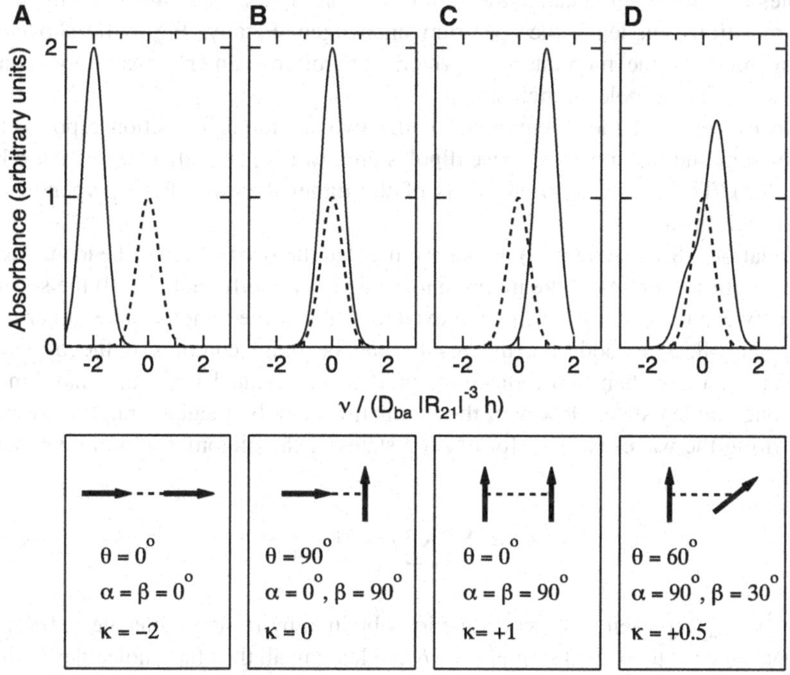

Fig. 8.4 Calculated absorption spectra of homodimers with four different geometries. The orientations of the transition dipoles and the geometrical factor κ (Fig. (7.17)) are indicated in the *box below* each spectrum. The absorption spectrum of the monomer is shown with a *dotted line* in each panel and the spectrum of the dimer with a *solid line*. From left to right, the relative dipole strengths of the Ψ_{B-} and Ψ_{B+} transitions are 0:2, 1:1, 0:2 and 0.5:1.5. (Ψ_{B-} is the high-energy transition in **A** and the low-energy transition in **C** and **D**.) The sum of the dipole strengths is always twice the dipole strength of the monomer. The exciton bands have been given Gaussian shapes with an arbitrary width. See Fig. 9.7A for another illustration

matrix element (H_{21}) for the different geometries. If the transition dipoles of the monomers are parallel and are aligned along the intermolecular axis (Fig. 8.4A) H_{21} is negative. Ψ_{B+} then lies lower in energy than Ψ_{B-} and all of the dimer's dipole strength is associated with excitation to Ψ_{B+}. The higher-energy excitation to Ψ_{B-}

has no dipole strength because the transition dipoles of the monomers enter with opposite signs and cancel. The figure does not show it, but the dimer's absorption spectrum would be just the same if one of the monomers were turned around by 180°. In that case, H_{21} would be positive and Ψ_{B-} would be lower in energy than Ψ_{B+}, but all the dimer's dipole strength still would go to the lower-energy transition.

If the transition dipoles of the monomers are perpendicular (Fig. 8.4B), H_{21} is zero and the dimer's two excited states have the same energy. Transitions to both states are allowed but are indistinguishable. If the monomer transition dipoles are parallel to each other but both perpendicular to the intermolecular axis (Fig. 8.4C), H_{21} is positive. In this case, Ψ_{B+} is the higher-energy state and all of the dimer's dipole strength goes to the higher-energy excitation. Again, rotating one of the dipoles by 180° would change the sign of H_{21} and interchange the assignments of the transitions but leave the spectrum unchanged. Finally, Fig. 8.4D shows an arrangement of the monomers in which both of the dimer's absorption bands have significant dipole strengths.

No matter what the orientations of the two monomer transition dipoles, the vector sum and difference of these dipoles are always perpendicular to each other (Fig. 8.3). The two absorption bands of the dimer therefore have perpendicular linear dichroism.

Equations (8.14) and (8.15) do not require that the two molecules be identical. If the excitation energies of the individual molecules are different ($\delta \neq 0$) the separation between the energies of the two excited states of the dimer will be larger than $2H_{21}$ (Fig. 8.2). In addition, the results can be more complex if the oligomer consists of more than two molecules, or if the individual molecules have more than one excited state. However, these situations can be treated straightforwardly by writing the wavefunctions for excited states of the oligomer in a more general way:

$$\Psi_B = \sum_m \sum_b C^B_{b(m)} \, {}^1\psi_{ba(m)}. \tag{8.17}$$

Here ${}^1\psi_{ba(m)}$ represents the wavefunction obtained by raising molecule m from its ground state (a) to excited singlet state b and leaving all the other molecules in their ground states. Equation (8.17) allows each of the individual molecules in the oligomer to have any number of excited states ($b = 1, 2, 3, \ldots$). Ψ_B includes contributions from all these states, although some of the contributions may be negligible depending on the geometry of the oligomer and the energies of the transitions in the monomers. An oligomer comprised of m monomeric subunits, each with b excited singlet states, has $m \times b$ excited singlet states, which are termed *exciton* states.

The coefficients $C^B_{b(m)}$ in Eq. (8.17) can be found by solving a set of simultaneous equations analogous to Eqs. (8.7a and 8.7b). Again, this is done by diagonalizing the interaction matrix **H**, in which the diagonal terms are the excitation energies of the monomers and the off-diagonal terms are the interaction matrix elements:

$$\mathbf{H} = \begin{bmatrix} H_{11} & H_{12} & \cdots & H_{1n} \\ H_{21} & H_{22} & \cdots & H_{2n} \\ & \vdots & & \\ H_{n1} & H_{n1} & \cdots & H_{nn} \end{bmatrix}. \tag{8.18}$$

The transition dipole vector for the oligomer's absorption band associated with exciton wavefunction Ψ_B is obtained by summing over the monomer transitions:

$$\boldsymbol{\mu}_{BA} = \sum_m \sum_b C^B_{b(m)} \boldsymbol{\mu}_{ba(m)}. \tag{8.19}$$

The dipole strength then is $D_{BA} = |\boldsymbol{\mu}_{BA}|^2$. If the monomeric subunits have several different excited states, the oscillator strengths of these bands are redistributed among all the exciton bands of the oligomer. This can result in *hyperchromism* or *hypochromism*, which means an excess or deficiency of the oligomer's absorbance in a particular region of the spectrum compared to the absorbance of the monomeric constituents [8, 9]. For an oligomer made up of m monomers, D_{BA} for any one of the exciton bands can be greater (or less) than m times the dipole strength of the corresponding absorption band of the monomers (mD_{BA}), provided that some other exciton band is decreased (or increased) in strength correspondingly. The sum of the dipole strengths of all the bands is constant (Box 8.4).

Box 8.4 The Sum Rule for Exciton Dipole Strengths

The sum of the dipole strengths of the exciton bands of an oligomer is the same as the sum of the dipole strengths for the individual molecules. This is easy to see for a dimer of identical molecules (Eqs. 8.16c and 8.16d). To show that the statement can be generalized to larger oligomers, let's rewrite the coefficients $C^B_{b(m)}$ as a matrix **C** with elements C_{ik}, where i indicates an exciton state and j indicates one of the monomer transitions. The C_{ik} are the eigenvectors of the interaction matrix **H**. With this notation, the sum of the exciton dipole strengths is

$$\sum_i D_i = \sum_i \sum_j \sum_k C_{ij} C_{ik} \left(\boldsymbol{\mu}_j \cdot \boldsymbol{\mu}_k\right)$$

$$= \sum_j \sum_k \left(\boldsymbol{\mu}_j \cdot \boldsymbol{\mu}_k\right) \sum_i C^T_{ji} C_{ik} = \sum_j \sum_k \left(\boldsymbol{\mu}_j \cdot \boldsymbol{\mu}_k\right) G_{jk}, \tag{B8.5.1}$$

where $\mathbf{G} = \mathbf{C}^T \cdot \mathbf{C}$ and $\mathbf{C}^T$ is the transpose of **C** ($C^T_{ji} = C_{ij}$) and we have used the definition of matrix multiplication (Appendix A.2).

(continued)

Box 8.4 (continued)

Because **H** is both real and symmetric, we now can make use of the fact that the matrix of eigenvectors of any real, symmetric matrix is orthogonal, which means that $\mathbf{C}^T = \mathbf{C}^{-1}$ (Appendix A2). Therefore, $\mathbf{G} = \mathbf{C}^{-1} \cdot \mathbf{C} = \mathbf{1}$, where **1** is a matrix with all the diagonal terms equal to 1 and all the off-diagonal terms zero. The double sum in Eq. (B8.5.1) thus reduces to

$$\sum_j (\boldsymbol{\mu}_j \cdot \boldsymbol{\mu}_j) = \sum_j |\boldsymbol{\mu}_j|^2, \tag{B8.5.2}$$

which is the sum of the monomer dipole strengths.

The sum rule for exciton absorption bands sometimes is stated in terms of oscillator strengths rather than dipole strengths. This is strictly correct only if the intermolecular interactions are very weak, so that the exciton transitions have essentially the same energies as the corresponding monomer transitions.

Figure 8.5 illustrates the transfer of dipole strength between absorption bands in different regions of the spectrum for dimers with two different structures. The side-by-side geometry (Fig. 8.5A) is qualitatively similar to that of the bases in double-stranded DNA. With this alignment, most of the dipole strength in each region of the spectrum goes to the exciton band that lies above the monomer transition in energy (Fig. 8.4C). In addition, mixing of the transitions in different regions of the spectrum results in transfer of dipole strength from absorption bands in the lower-energy region to bands at higher energies. The high-energy bands of the nucleotide bases occur too far into the UV to be measured conveniently, whereas the low-energy bands occur in the more accessible region around 260 nm. The loss of dipole strength in the 260-nm region is therefore seen experimentally as hypochromism [8, 9]. Similar hypochromism is seen in the region of 200 nm when peptides such as polylysine adopt a helical conformation.

Photosynthetic systems provide examples of chlorophyll or bacteriochlorophyll oligomers with alignments similar to that shown in Fig. 8.5B. In this geometry, most of the dipole strength in each region of the absorption spectrum goes to the exciton band that lies below the monomer transition (Fig. 8.4A). Mixing of the transitions in different regions now transfers dipole strength from the higher-energy region around 400 nm to lower-energy bands in the region of 700 to 1,000 nm [10, 11]. The low-energy region of the spectrum thus has hyperchromism.

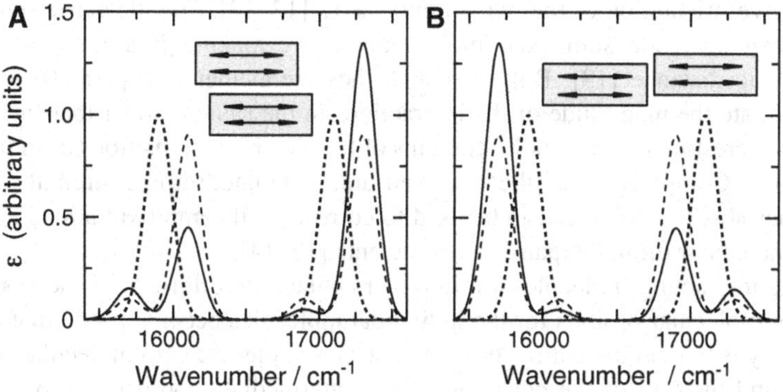

Fig. 8.5 Hyper- and hypochromism in dimers with two geometries. Each monomeric unit is assumed to have two absorption bands with transition dipoles oriented along the long axis of the molecule (*double arrows* in *shaded boxes*). In (**A**) the two monomers are aligned side-by-side; in (**B**), one is displaced along an axis parallel to the transition dipoles. The monomer absorption spectra are assumed to peak at 15,900 and 17,100 cm^{-1} and are shown (multiplied by 2) with *short dashes*. The spectra drawn with *long dashes* consider exciton interactions only between monomer transitions in the same region of the spectrum. The spectra with *solid lines* include interactions of all of the transitions. In the side-by-side dimer (**A**), the spectrum in the high-energy region is strengthened at the expense of the low-energy region; in the in-line structure (**B**), the spectrum is strengthened at low energies

8.3 Transition-Monopole Treatments of Interaction Matrix Elements and Mixing with Charge-Transfer Transitions

For purposes of illustration in Figs. 8.4 and 8.5, we have used the point-dipole approximation to evaluate the interaction matrix element H_{12} (Eqs. 7.16 and 7.17). This is a good approximation as long as the molecules are far apart relative to the molecular dimensions. For molecules that are closer together, a *transition-monopole* treatment based on Eqs. (7.14) and (7.15) is, in principle, more accurate. To illustrate the transition-monopole treatment, suppose that the relevant transitions of the monomers both involve only a single configuration, excitation from the monomer's HOMO to the LUMO, and that the molecular orbitals can be written as linear combinations of atomic p_z orbitals as in Eq. (2.35). Inserting the antisymmetrized singlet wavefunctions for two electrons on each molecule (Eqs. 2.48, 2.54, 4.20, and 4.21) leads straightforwardly to the expression:

$$H_{12} \approx 2\left(f^2/n^2\right)\sum_s \sum_t C_s^{HOMO} C_s^{LUMO} C_t^{HOMO} C_t^{LUMO}\left(e^2/r_{st}\right), \tag{8.20}$$

where r_{st} is the distance between atom s on molecule 1 and atom t on molecule 2, and f^2/n^2 is an approximate correction for the dielectric effects of a homogeneous medium. Equation (8.20) can be refined by replacing the simple $1/r_{st}$ by a parametrized semiempirical function of r_{st} and by incorporating a dependence on the relative orientation of the two atomic z axes [12, 13]. The dielectric screening factor can be made somewhat more realistic by making it a function of the interatomic distance [13]. But even with these refinements, Eq. (8.20) tends to overestimate the magnitude of the interactions in the same manner that Eq. (4.26) usually overestimates the dipole strengths of the electronic transitions of individual molecules. Comparisons of the observed and calculated dipole strengths of the monomer absorption bands can be used to correct for the approximations inherent in the molecular orbital expansion coefficients [12–14].

If the interacting molecules are in direct contact, descriptions of the system in terms of molecular orbitals for the individual molecules become problematic. One possibility is to treat the entire complex as a supermolecule with molecular orbitals that extend over all of the constituents. An intermediate approach is to augment Eq. (8.17) with charge-transfer transitions in which an electron moves from one molecule to the other [12, 15, 16].

If we consider only the HOMO and LUMO of the two molecules, a dimer will have two charge-transfer (CT) states in addition to the two exciton states that we have considered above. An electron can move from the HOMO of molecule 1 to the LUMO of molecule 2, as shown schematically in Fig. 8.6, or from the HOMO of 2 to the LUMO of 1. The two CT states can lie either above or below the corresponding exciton states of the dimer, depending mainly on the electrostatic interactions of the species with each other and with the surroundng medium.

We discussed CT transitions of transition-metal complexes in Chap. 4. Charge-transfer transitions between more widely separated molecules usually have very little intrinsic dipole strength because, to a first approximation, the initial and final molecular orbitals have no atoms in common; at the level of Eq. (4.26), the transition dipole is zero. However, CT states can mix with exciton states to give hybrid eigenstates that are shifted up or down in energy and have varying amounts of CT character.

The strength of the mixing of CT and intramolecular transitions depends on the extent of orbital overlap between the interacting molecules [12]. The coefficients

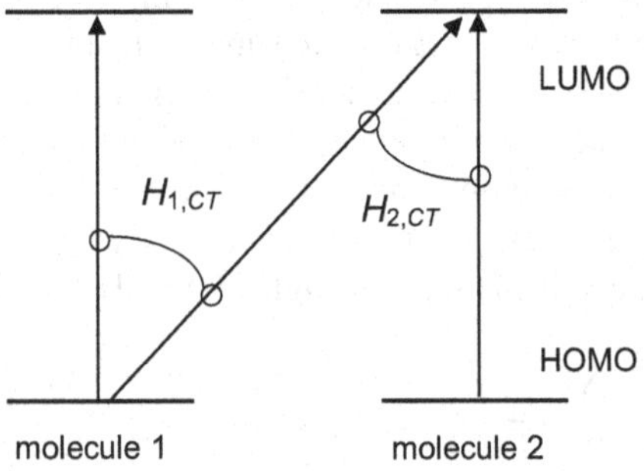

Fig. 8.6 Mixing of a charge-transfer transition with intramolecular transitions. The *diagonal arrow* represents a CT transition in which an electron moves from the HOMO of molecule 1 to the LUMO of molecule 2. This transition is mixed with the two intramolecular transitions by off-diagonal terms in the interaction matrix ($H_{1,CT}$ and $H_{2,CT}$)

and energies for the four excited states of a dimer are obtained by diagonalizing a 4×4 interaction matrix in which the CT transition energies appear on the diagonal along with the energies of the two intramolecular transitions. The off-diagonal interaction matrix element that mixes the intramolecular excitation of molecule 1 with a CT transition in which an electron moves from the HOMO of molecule 1 to the LUMO of molecule 2 takes the form:

$$H_{1,CT} \approx \sum_s \sum_t C_s^{LUMO} C_t^{LUMO} \beta_{st}. \tag{8.21}$$

Here β_{st} is a semiempirical atomic resonance integral that depends on the orbital overlap of atom s of molecule 1 with atom t of molecule 2 [12, 13]. The corresponding matrix element for mixing the same CT transition with the intramolecular excitation of molecule 2 is:

$$H_{2,CT} \approx -\sum_s \sum_t C_s^{HOMO} C_t^{HOMO} \beta_{st}. \tag{8.22}$$

In the first case (Eq. 8.21), where the intra- and intermolecular transitions both remove an electron from the same orbital (the HOMO of molecule 1), the matrix element depends on the overlap of the two orbitals to which the electron can be promoted. In the second (Eq. 8.22), where both transitions promote an electron to the same orbital (the LUMO of molecule 2), the matrix element depends on the overlap of the two orbitals where the electron could originate. The off-diagonal matrix element for two CT transitions in opposite directions is zero.

Semiempirical expressions have been developed for the atomic resonance integrals as functions of interatomic distance and orientation [12, 13]. Because the resonance integrals drop off very rapidly with distance at separations greater than about 3.5 Å, the off-diagonal interaction matrix elements involving CT transitions usually are much smaller than the terms that mix purely intramolecular transitions. Absorption bands that are predominantly CT in character typically are very broad because their energies depend strongly on electrostatic interactions of the charged species with each other and with the surroundings. The final absorption spectrum thus includes broad, weak bands representing transitions that are mostly CT in character and stronger exciton-type bands with smaller admixtures of CT character. However, each exciton-type band is shifted up or down in energy, depending on the location of the CT transition with which it mixes most strongly. An exciton-type transition interacting with a CT transition that lies above it in energy will be shifted down in energy, while one interacting with a CT transition lying at lower energies will be shifted up.

Charge-transfer transitions play an important role in the long-wavelength absorption band of photosynthetic bacterial reaction centers. These pigment-protein complexes contain four molecules of bacteriochlorophyll (BChl) and two bacteriopheophytins. Two of the BChls are located close together, with their macrocyclic planes approximately parallel and about 3.8 Å apart. The center-to-center distance is about 6 Å. When the reaction center is excited by light or by resonance energy transfer from an antenna complex, it is this "special pair" of BChls that undergoes photooxidation. Whereas the long-wavelength (Q_y) absorption band of monomeric BChl solution is in the region of 770 nm, the corresponding band of the special pair of BChls in the reaction center occurs at 865 nm (Fig. 4.12A). The shift of the band to longer wavelengths probably results in part from exciton interactions and interactions with the protein, but also from mixing of the exciton transitions with CT transitions [12, 15–20]. This interpretation has been tested experimentally by site-directed mutations that move the CT states of the dimer upward or downward in energy by changing the hydrogen bonding of the BChls [18–20]. In agreement with the idea that the lowest excited state has substantial CT character, Stark measurements have shown that the excitation is coupled to a relatively large change in dipole moment [18–21].

8.4 Exciton Absorption Band Shapes and Dynamic Localization of Excitations

The expressions we have derived for the absorption spectrum of an oligomer apply as well to stimulated emission, and they can be extended to spontaneous fluorescence by using the Einstein relationships (Chap. 5). However, we have not yet considered relaxations that follow excitation of an oligomer to an exciton state. As we discussed in Chaps. 4 and 5, excitation of an individual molecule changes the molecule's electrostatic interactions with the surroundings. The energy of the excited state usually decreases with time as the solvent relaxes in response to the new distribution of charge. To the extent that an exciton is distributed over several molecules, the changes in interactions with the solvent will be smaller than they would be for a localized excitation. But relaxations still should occur, and in some cases they can cause the excitation to become increasingly localized with time.

Consider first the case that the exciton interaction energy between the two molecules of a homodimer is large relative to the solvent reorganization energy of either molecule alone. Figure 8.7A illustrates how the energies of the excited states might depend on a nuclear coordinate of the solvent. The coordinate could represent the rotational orientation of a particular nearby water molecule, for example, or a composite of many independent degrees of freedom. For any value of the solvent coordinate, the energy of the lower exciton state (E_{B-}) in Fig. 8.7A has a single minimum that lies below the minima for the localized excited states. If the dimer is raised from its ground state to Ψ_{B-} by a short flash when the solvent coordinate is somewhat displaced from 0, the ensuing relaxations of the solvent will stabilize Ψ_{B-} but will not cause the excitation to localize on an individual molecule. Note also in Fig. 8.7A that the curvature of E_{B-} is less than the curvatures of H_{11} and H_{22}. This difference in curvature reduces the Stokes shift for the dimer relative to a monomer. By contrast, if $|H_{12}|$ is small compared to the reorganization energy of the monomer, E_{B-} will have two minima that coincide closely with the energy minima for the two localized states (Fig. 8.7B). In this case, relaxations will move

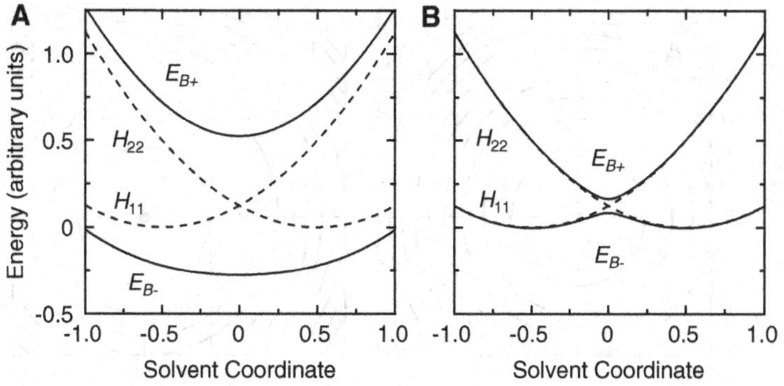

Fig. 8.7 Dependence of the excited-state energy surfaces on a generalized solvent coordinate in cases of strong (**A**) and weak (**B**) exciton interactions. H_{11} and H_{22} (*dashed lines*) are the excited-state energies of individual molecules, which are assumed to be identical. E_{B+} and E_{B-} (*solid lines*) are the energies of the two exciton states as given by Eqs. (8.9b–8.9d). (Exciton state Ψ_{B+} is assumed arbitrarily to have a higher energy than Ψ_{B-}.) The ground-state energy is off scale at the *bottom*. In the units of energy used for the ordinate scale, H_{12} is 0.4 in (**A**) and 0.04 in (**B**)

the system out of the region of resonance so that the eigenstates become more localized with time.

Exciton absorption bands typically are sharper than the corresponding absorption bands of the monomeric chromophores [22, 23]. Figure 8.8 shows the origin of this effect. In panel *A*, the energy of the ground state of a homodimer is represented by a contour plot as a function of two orthogonal coordinates (x and y) that could be either solvent coordinates or intramolecular vibrational coordinates. The energy minimum is at the origin. The energy surface of the lowest excited state one of the monomeric units of the dimer is displaced along x as shown in Fig. 8.8B, and that of the other is displaced by the same amount (Δ) along y (Fig. 8.8C). If the two molecules do not interact, the energy difference between the ground state and the excited state therefore depends only on x if molecule 1 is excited and only on y if molecule 2 is excited. Panel D shows the contour plot for the energy of the dimer's lowest exciton state when the molecules interact strongly enough so that the excited state has a single minimum, as in Fig. 8.7A. The energy difference between the ground and excited state still can be described in terms of a single coordinate, but this is now the linear combination $2^{-1/2}(x+y)$ indicated by the solid diagonal line in Fig. 8.8D. The minimum is equidistant from the minima of H_{11} and H_{22}, so its displacement from the origin along this coordinate is $2^{-1/2}\Delta$. The displacement of the minimum along the perpendicular coordinate indicated by the dashed line in Fig. 8.8D, $2^{-1/2}(x-y)$, is zero. Because the width of an absorption band that is strongly coupled to a nuclear coordinate is proportional to the displacement of the minimum along the coordinate (Eq. 4.57), the dimer's absorption band will be narrower than the monomer band by approximately a factor of $\sqrt{2}$.

At room temperature, the narrowing of exciton bands sets in when the interaction energy H_{21} is on the order of the reorganization energy of the monomer absorption

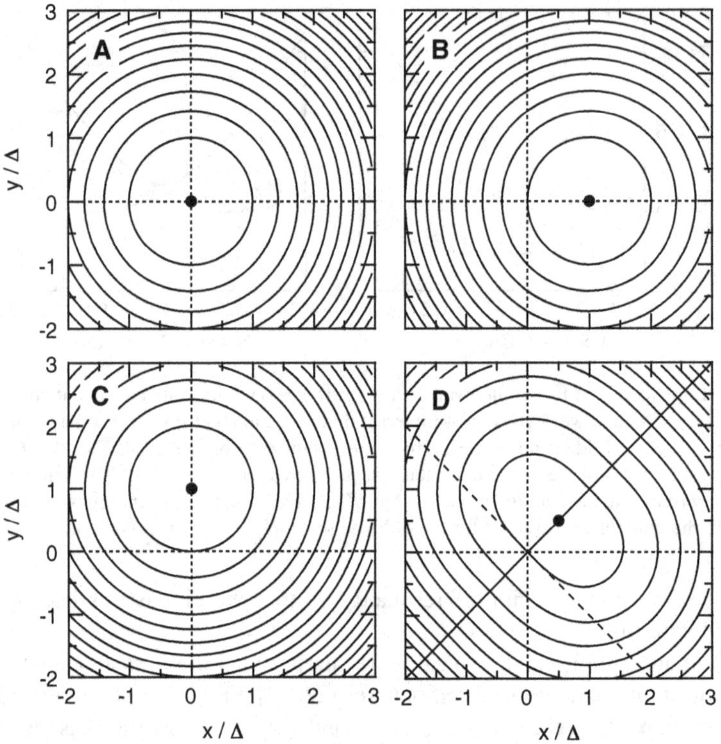

Fig. 8.8 Contour plots of energy surfaces of the ground state (**A**) and lowest exciton state (**D**) of a homodimer and of the lowest excited states of the monomeric units (**B** and **C**), as functions of two orthogonal coordinates (x and y). The energy minimum (*filled circle*) of the excited state of one of the monomers is displaced from the ground-state minimum by Δ along x (**B**), and that of the other monomer is displaced similarly along y (**C**). The energy surfaces for the monomers are assumed to increase quadratically with distance from the minimum. The excited-state energy of the dimer was calculated by Eqs. (8.10a, b) with H_{12} equal to twice the reorganization energy of the monomer. In this model, the energy minimum in the dimer's excited state is displaced by $2^{-1/2}\Delta$ along the diagonal indicated by the *solid line* in **D**. The displacement along the orthogonal coordinate (*dashed line*) is zero

band (Λ_m). (In Fig. 8.8, $H_{21} = 2\Lambda_m$.) Weaker interactions give an energy surface with a double minimum as shown in Fig. 8.7B, and the exciton band width approaches the width of the monomer band.

This analysis of the shapes of exciton absorption bands neglects mixing of different vibrational levels of the chromophores. A more general approach is to write the excited states of the dimer as in Eq. (8.17), using Born-Oppenheimer vibronic states of the individual chromophores as the basis. Each off-diagonal element in the interaction matrix **H** (Eq. 8.18) then consists of a product of an electronic interaction energy and the vibrational overlap integral for a particular pair of basis states. A treatment along these lines was used for a chlorophyll dimer with two effective vibrational modes [24]. Vibrational states of the surroundings could be included in the same manner, if the interaction energies and overlap integrals for these were known. Treatments using spectral density functions (Chap. 10) provide more practical ways of incorporating these states, and have been used successfully for a protein complex with four molecules of chlorophyll [25–27].

Because charge-transfer (CT) transitions generally involve substantial changes in dipole moment, CT absorption bands are broadened by both strong vibronic coupling and inhomogeneous interactions of the chromophores with the surroundings. Absorption bands that reflect mixed exciton and CT transitions are, therefore, typically much broader than the bands of the monomeric chromophores. Friesner and coworkers [15, 28, 29], Zhou and Boxer [18–20] and Renger [16] have described methods for treating the shapes of such mixed bands. Renger's treatment appears to account well for the temperature dependence of the long-wavelength absorption band of photosynthetic bacterial reaction centers.

8.5 Exciton States in Photosynthetic Antenna Complexes

The "LH2" or "B800-850" antenna complexes of purple photosynthetic bacteria illustrate many of the points discussed in the previous sections of this chapter. These pigment-protein complexes absorb light and transfer the excitation energy to the photosynthetic reaction centers that carry out the initial electron-transfer reactions of photosynthesis (Sects. 4.7 and 4.11). The LH2 complex from *Rhodopseudomonas acidophila* contains nine copies each of two small proteins arranged in a cylindrical structure with C_9 rotational symmetry, along with 27 bacteriochlorophyll (BChl) molecules and 18 carotenoids [30]. *Rhodospirillum molischianum* has a similar complex containing eight copies of each protein, 24 BChls, eight carotenoids and C_8 symmetry [31]. The BChl's in each complex form two rings: an inner ring of 16 or 18 "B850" BChl's with center-to-center distances of about 9 Å and an outer ring, offset toward one side of the membrane, containing eight or nine "B800" BChl's with center-to-center distances of about 21 Å. The Q_y transition dipoles of the B850 BChls lie approximately in the plane of the membrane, while those of the B800 BChls are approximately normal to the membrane. The complexes have a strong absorption band at 850 nm that is assigned to the B850 BChls, and a band at 800 attributed mainly to the B800 BChls (Fig. 8.9). Their spectroscopic properties have been studied extensively by both theoretical and experimental approaches [32, 33].

Figure 8.10 shows calculated contributions of the Q_y excitations of the 18 individual B850 molecules to the overall transition dipoles for forming the first three excited states of the *Rps. acidophila* complex [13]. The arrows are the weighted vectors $C^{\mathrm{B}}_{\mathrm{b(m)}}\boldsymbol{\mu}_{ba(m)}$ that appear in Eq. (8.19), which together with minor additional contributions from the higher-energy Q_x, B_x, B_y and charge-transfer (CT) transitions form the total transition dipole for each excited state ($\boldsymbol{\mu}_{BA}$). (See Sect. 4.7 for an explanation of the Q_x, B_x and B_y states of BChl and related molecules.) The

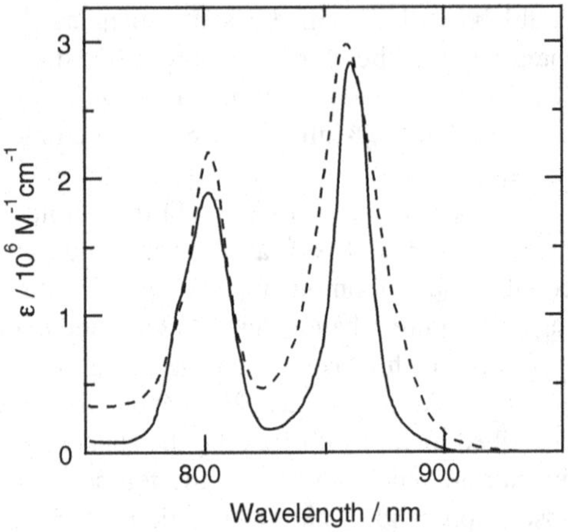

Fig. 8.9 Observed (*dashed curve*) and calculated (*solid curve*) absorption spectra of the LH2 antenna complex of *Rhodopseudomonas acidophila* [13]. See the text for a description of the calculations

calculations used for Fig. 8.10 assume that the Q_y transition energies of all the individual BChls are the same. In the lowest excited state (Fig. 8.10A), the contributions from the different BChls all have approximately the same magnitude and the weighted transition dipoles point in a consistent direction around the ring. The total electric transition dipole for forming this state (the result of a complete cycle around the ring) is zero, so (neglecting interactions with the magnetic field of light) excitation into the first excited state is formally forbidden. In the second and third excited states (Fig. 8.10B, C), monomer transition dipoles with large coefficients collect on opposite sides of the ring, with the contributions from the two sides adding vectorially to give large total transition dipoles. The energies of these two states are the same, and their transition dipoles are orthogonal. Excitations to the second or third states have large transition dipoles with perpendicular orientations, accounting for the strong absorption band in the region of 850 nm [13]. Above these states are an additional 15 Q_y exciton states, including two that resemble the second and third states except that the transition dipoles on opposite sides of the ring oppose each other rather than cooperating. The calculated dipole strengths of all these states are much smaller than those of the second and third states. The nine B800 BChls contribute a similar pair of strong bands in the 800-nm region, along with seven weaker bands. Bands that come largely from Q_x, B_x, B_y or CT transitions are predicted for higher energies.

The model just described neglects perturbations of the symmetry of the complex by heterogeneity in the energies of the individual pigments. Although the model considers coupling to a set of vibrational modes with experimentally measured frequencies and coupling factors [13], the predicted width of the absorption bands is much sharper than observed. Figure 8.9 shows the result of repeating the calculations many times with random variations in the energies of the individual transitions, on the assumption that each of the monomer transitions has a Gaussian distribution of energies. The width of the energy distribution at half maximum amplitude was taken to be 100 cm^{-1}. Introducing such disorder in the monomer

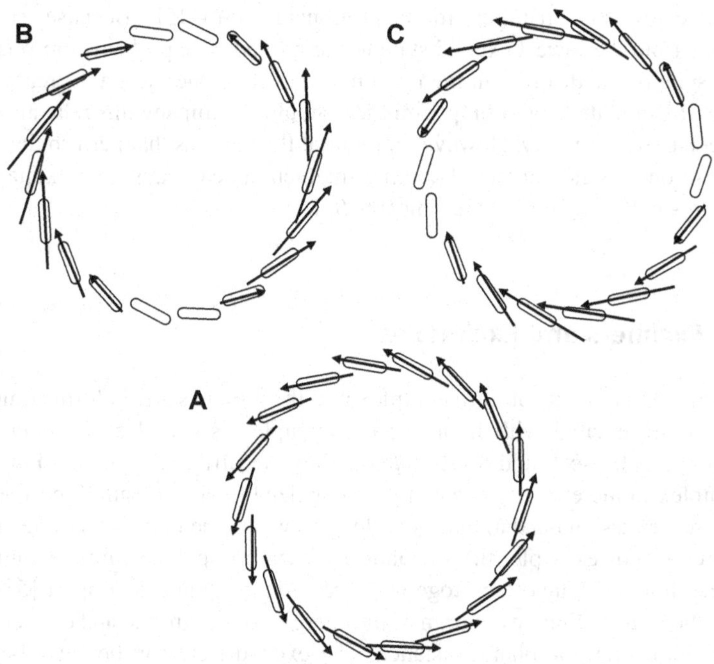

Fig. 8.10 Calculated coefficients for monomer Q_y excitations in the first three excited singlet states of a photosynthetic bacterial antenna complex (**A**–**C**, respectively) [13]. The complex contains a ring of 18 BChls, represented here by *rounded rectangles*, and a larger, concentric ring of nine BChls that interact relatively weakly with the ring of 18 (not shown). Alternating BChls in the ring shown have slightly different conformations. The *arrows* represent the products of the coefficient for the Q_y excitation of each molecule (m) in the excited state ($C^{\mathrm{B}}_{\mathrm{b(m)}}$) and the molecule's Q_y electric transition dipole ($\boldsymbol{\mu}_{ba(m)}$). In the lowest excited state (**A**) the weighted vectors sum to zero; in the next two states (**B** and **C**) the magnitudes of $C^{\mathrm{B}}_{\mathrm{b(m)}}\ \boldsymbol{\mu}_{ba(m)}$ peak on opposite sides of the ring and combine to give strong total transition dipoles that are orthogonal for the two states

transition energies causes the two strong transitions of the complex to split apart in energy and lose some of their dipole strength, while the previously weak transitions become stronger. The agreement with the observed spectrum now is reasonably satisfactory, although the predicted 850-nm band remains somewhat too narrow while the 800-nm band is too broad. Disorder in the interaction energies would have a qualitatively similar effect.

Aartsma and coworkers tested the exciton model by measuring fluorescence excitation spectra of individual LH2 complexes at low temperatures [34–42]. In agreement with the picture presented above, the excitation spectrum of an individual complex typically included two dominant components with orthogonal polarizations, and the transition energies varied from complex to complex.

Following the absorption of light, an ensemble of excited LH2 complexes probably relaxes rapidly to a thermally equilibrated mixture of states in which the lowest excited state features more prominantly [43–45]. Because radiative transitions from the lowest excited state to the ground state probably are relatively weak in spite of the disorder in the monomer transition energies, a decrease in the intensity of stimulated and prompt emission should accompany this relaxation, and this is seen experimentally. However, structural fluctuations that perturb the monomer transition energies or the electronic interactions can cause the excitation to localize in smaller regions of the ring [46–52].

8.6 Excimers and Exciplexes

Some molecules that do not form complexes in the ground state do form complexes when they are excited with light. Such a complex is called an *excimer* if the molecules are identical, and an *exciplex* if they are different [53, 54]. Formation of a complex in the excited state can be recognized by a concentration-dependent fluorescence emission band at long wavelengths where the individual molecules do not fluoresce. For example, the aromatic hydrocarbon pyrene forms excimers in which the aromatic rings stack together face-to-face about 3.5 Å apart [53]. Figure 8.11 shows a qualitative diagram of the energies of the ground and excited states as functions of the interplanar distance. The excited state can be described as a combination of an antisymmetric exciton state ($2^{-1/2}\Psi_1 - 2^{-1/2}\Psi_2$, where Ψ_1 and Ψ_2 are the lowest excited singlet states of the individual molecules) and an antisymmetric "charge resonance" state ($2^{-1/2}\Psi_{\pm} - 2^{-1/2}\Psi_{\mp}$, where $\Psi_{\pm}$ and $\Psi_{\mp}$

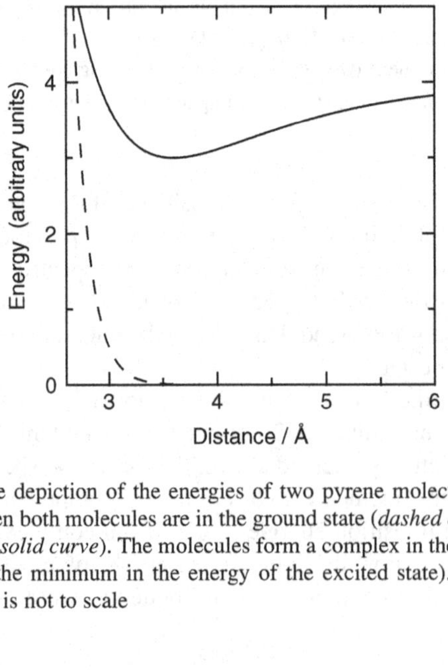

Fig. 8.11 A qualitative depiction of the energies of two pyrene molecules as functions of the interplanar distance when both molecules are in the ground state (*dashed curve*) or when one is in an excited singlet state (*solid curve*). The molecules form a complex in the excited state but not in the ground state (note the minimum in the energy of the excited state). The energy separation between the two curves is not to scale

are charge-transfer states in which an electron has moved from the HOMO of molecule 1 to the LUMO of molecule 2 or *vice versa*) [55].

Pyrene excimers fluoresce with a broad, unstructured emission band peaking between 450 and 500 nm, whereas the monomer has a highly structured emission spectrum with peaks near 385 and 400 nm. If the concentration of a solution of pyrene is increased, the excimer emission grows in strength relative to the monomer emission. Changing the concentration has little effect on the absorption spectrum because the molecules do not form a complex in the ground state. When a solution of 1 mM pyrene in cyclohexane is excited with light, the emission spectrum changes with time as excited and ground-state molecules that encounter each other by diffusion form excimers [56]. At 1 ns after the excitation, the fluorescence comes mainly from pyrene monomers; by 100 ns it comes mainly from excimers.

Measurements of excimer or exciplex fluorescence can be used to determine whether a pair of macromolecules or two regions of a macromolecule are able to come in close contact during the lifetime of the excited state. Derivatives of pyrene that can be attached to various functional groups in proteins, lipids or polysaccharides lend themselves well to such studies [57–61]. In one application, the *N*-terminus of the EcoR1 restriction endonuclease was labeled with *N*-(1-pyrenyl)iodoacetamide [62]. A broad excimer emission band at 480 nm indicated that the *N*-termini of two molecules come into close proximity when the protein dimerizes. The *N*-termini are essential for enzymatic activity but are too disordered to be seen in a crystal structure of the protein.

Exercises

Consider a molecule that has an absorption band at 700 nm with a dipole strength of 25 debye^2. Suppose two such molecules form a co-planar complex with a center-to-center distance of 8 Å as sketched below. The arrows represent intramolecular vectors that are aligned with the transition dipoles.

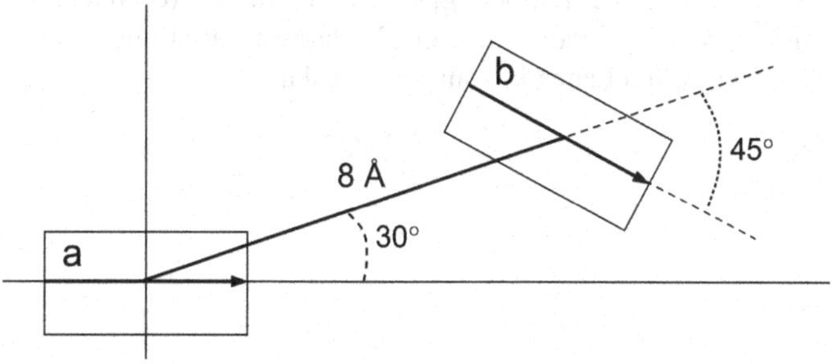

1. Calculate the dipole-dipole interaction energy (H_{ab}) in the point-dipole approximation for a medium with index of refraction $n = 1.2$. Specify the sign and units of your answer. Save your result for use in Chap. 10.
2. What are the excitation energies for the dimers two excited states?
3. Calculate the dipole strengths of the two exciton absorption bands of the dimer. (Assume that each monomeric molecule has only one excited state.)
4. (a) Sketch the predicted absorption spectrum of the dimer, indicating the wavelengths and relative intensities of the two bands. (b) What would happen to the absorption spectrum if one of the molecules is rotated 180° about an axis normal to the molecular plane, so that the transition dipole points in the opposite direction?
5. Sketch the predicted absorption spectrum of the dimer in the situation that interactions with the surroundings shift one of the monomer absorption bands to 690 nm and the other to 710 nm.
6. Consider a protein in which the amino group of the N-terminal tryptophan (Trp) residue is acetylated. Table E.1 below gives the atomic expansion coefficients [the $C_{\mathrm{n}}^{\mathrm{k}}$ of Eq. (2.36)] for a π molecular orbital description of the Trp residue and its two peptide links. Atoms C5 and O6 are the acetyl carboxyl C and O atoms; N7 is the Trp amino nitrogen; C14–C28 and N17 are in the indole ring; and C29, O30 and N31 represent the Trp carboxyl group and the amino N of the next residue. Wavefunctions $\psi_6 - \psi_9$ are the four highest occupied π molecular orbitals, and $\psi_{10} - \psi_{14}$ the five lowest unoccupied orbitals. Table E.2 gives the configuration-interaction coefficients [the $A_{\mathrm{j,k}}^{\mathrm{a,b}}$ of Eq. (4.26)] for the first three singlet excitations of the system. (a) Qualitatively, how would you describe wavefunctions ψ_6, ψ_9 and ψ_{11}? (b) What types of processes do excitations 1 and 3 represent? (c) Verify your answer to (b) by calculating the changes of charge in the –CONH– and indole groups. (d) What is the dipole strength of excitation 3 (within the accuracy of Eqs. (4.22a–e) and the numbers given in the tables)? (e) Which of the three excitations probably would undergo the largest variations in energy as the protein structure fluctuates in solution? Explain.

Table E.1 Atomic coefficients

Atom	ψ_6	ψ_7	ψ_8	ψ_9	ψ_{10}	ψ_{11}	ψ_{12}	ψ_{13}	ψ_{14}
C5	0.184					0.820			
O6	0.726					−0.485			
N7	−0.663					−0.304			
C14		0.481	−0.025	−0.508			−0.288	−0.008	0.518
C15		0.195	−0.380	−0.383			0.479	−0.175	−0.466
N17		−0.304	−0.290	0.361			−0.197	0.246	0.214
C19		−0.181	0.485	0.123			−0.004	−0.513	−0.178
C20		−0.415	0.234	−0.360			0.462	0.168	0.378
C22		−0.314	−0.308	−0.302			−0.400	0.303	−0.392
C24		0.068	−0.479	0.205			−0.085	−0.588	0.223
C26		0.373	−0.124	0.426			0.451	0.378	0.057
C28		0.437	0.377	0.077			−0.256	0.192	−0.295
C29					0.839				
O30					−0.521				
N31					−0.155				

Table E.2 CI coefficients

Excitation	Configurations and coefficients
1	$0.761(\psi_8 \rightarrow \psi_{12}) + 0.498(\psi_9 \rightarrow \psi_{13}) - 0.298(\psi_9 \rightarrow \psi_{12})$ $+0.166(\psi_7 \rightarrow \psi_{13}) + 0.142(\psi_8 \rightarrow \psi_{13}) + 0.106(\psi_9 \rightarrow \psi_{14}) + \ldots$
2	$0.924(\psi_9 \rightarrow \psi_{12}) + 0.241(\psi_9 \rightarrow \psi_{13}) + 0.235(\psi_8 \rightarrow \psi_{12})$ $- 0.129(\psi_8 \rightarrow \psi_{13})$ - $0.057(\psi_7 \rightarrow \psi_{12}) + 0.055(\psi_8 \rightarrow \psi_{14}) + \ldots$
3	$0.979(\psi_9 \rightarrow \psi_{10}) + 0.132(\psi_8 \rightarrow \psi_{10}) - 0.118(\psi_7 \rightarrow \psi_{10})$ $+ 0.088(\psi_5 \rightarrow \psi_{10}) + \ldots$

References

1. Salem, L.: Electrons in Chemical Reactions: First Principles. Wiley-Interscience, New York (1982)
2. Bonacic-Koutecky, V., Kouteckey, J., Michl, J.: Neutral and charged biradicals, zwitterions, funnels in S_1, and proton translocation. Their role in photochemistry, photophysics, and vision. Angew. Chem. Int. Ed. Engl. **26**, 170–189 (1987)
3. Klessinger, M.: Conical intersections and the mechanism of singlet photoreactions. Angew. Chem. Int. Ed. Engl. **34**, 549–551 (1995)
4. Garavelli, M., Vreven, T., Celani, P., Bernardi, F., Robb, M.A., et al.: Photoisomerization path for a realistic retinal chromophore model: the nonatetraeniminium cation. J. Am. Chem. Soc. **120**, 1285–1288 (1998)
5. Toniolo, A., Granucci, G., Martínez, T.J.: Conical intersections in solution: a QM/MM study using floating occupation semiempirical configuration interaction wave functions. J. Phys. Chem. A **107**, 3822–3830 (2003)
6. Toniolo, A., Olsen, S., Manohar, L., Martínez, T.J.: Conical intersection dynamics in solution: the chromophore of green fluorescent protein. Farad. Disc. **127**, 149–163 (2004)
7. Martin, M.E., Negri, F., Olivucci, M.: Origin, nature, and fate of the fluorescent state of the green fluorescent protein chromophore at the CASPT2//CASSCF resolution. J. Am. Chem. Soc. **126**, 5452–5464 (2004)

8. Tinoco, I., Jr.: Hypochromism in polynucleotides. J. Am. Chem. Soc. **82**, 4785–4790 (Erratum J. Am. Chem. Soc. 4784, 5047 (1961)) (1961)
9. Tinoco Jr., I.: Theoretical aspects of optical activity part two: polymers. Adv. Chem. Phys. **4**, 113–160 (1962)
10. Scherz, A., Parson, W.: Oligomers of bacteriochlorophyll and bacteriopheophytin with spectroscopic properties resembling those found in photosynthetic bacteria. Biochim. Biophys. Acta **766**, 653–665 (1984)
11. Scherz, A., Parson, W.: Exciton interactions of dimers of bacteriochlorophyll and related molecules. Biochim. Biophys. Acta **766**, 666–678 (1984)
12. Warshel, A., Parson, W.W.: Spectroscopic properties of photosynthetic reaction centers. 1. Theory. J. Am. Chem. Soc. **109**, 6143–6152 (1987)
13. Alden, R.G., Johnson, E., Nagarajan, V., Parson, W.W.: Calculations of spectroscopic properties of the LH2 bacteriochlorophyll-protein antenna complex from Rhodopseudomonas sphaeroides. J. Phys. Chem. B **101**, 4667–4680 (1997)
14. Murrell, J.N., Tanaka, J.: The theory of the electronic spectra of aromatic hydrocarbon dimers. Mol. Phys. **7**, 363–380 (1964)
15. Lathrop, E.J.P., Friesner, R.A.: Simulation of optical spectra from the reaction center of Rhodobacter sphaeroides. Effects of an internal charge-separated state of the special pair. J. Phys. Chem. **98**, 3050–3055 (1994)
16. Renger, T.: Theory of optical spectra involving charge transfer states: dynamic localization predicts a temperature-dependent optical band shift. Phys. Rev. Lett. **93**, Art. 188101 (2004)
17. Parson, W.W., Warshel, A.: Spectroscopic properties of photosynthetic reaction centers. 2. Application of the theory to Rhodopseudomonas viridis. J. Am. Chem. Soc. **109**, 6152–6163 (1987)
18. Zhou, H., Boxer, S.G.: Charge resonance effects on electronic absorption line shapes: application to the heterodimer absorption of bacterial photosynthetic reaction centers. J. Phys. Chem. B **101**, 5759–5766 (1997)
19. Zhou, H., Boxer, S.G.: Probing excited-state electron transfer by resonance Stark spectroscopy. 1. Experimental results for photosynthetic reaction centers. J. Phys. Chem. B **102**, 9139–9147 (1998)
20. Zhou, H., Boxer, S.G.: Probing excited-state electron transfer by resonance Stark spectroscopy. 2. Theory and application. J. Phys. Chem. B **102**, 9148–9160 (1998)
21. Lösche, M., Feher, G., Okamura, M.Y.: The Stark effect in reaction centers from Rhodobacter sphaeroides R-26 and Rhodopseudomonas viridis. Proc. Natl. Acad. Sci. USA **84**, 7537–7541 (1987)
22. Simpson, W.T., Peterson, D.L.: Coupling strength for resonance force transfer of electronic energy in van der Waals solids. J. Chem. Phys. **26**, 588–593 (1957)
23. Förster, T.: Delocalized excitation and excitation transfer. In: Sinanoglu, O. (ed.) Modern Quantum Chemistry, Pt. III, pp. 93–137. Academic, New York (1965)
24. Renger, T., May, V.: Multiple exciton effects in molecular aggregates: application to a photosynthetic antenna complex. Phys. Rev. Lett. **78**, 3406–3409 (1997)
25. Renger, T., Marcus, R.A.: On the relation of protein dynamics and exciton relaxation in pigment-protein complexes: an estimation of the spectral density and a theory for the calculation of optical spectra. J. Chem. Phys. **116**, 9997–10019 (2002)
26. Renger, T., Trostmann, I., Theiss, C., Madjet, M.E., Richter, M., et al.: Refinement of a structural model of a pigment-protein complex by accurate optical line shape theory and experiments. J. Phys. Chem. B **111**, 10487–10501 (2007)
27. Dinh, T.-C., Renger, T.: Towards an exact theory of linear absorbance and circular dichroism of pigment-protein complexes: importance of non-secular contributions. J. Chem. Phys. **142**, 034104 (2015)
28. Friesner, R.A.: Green functions and optical line shapes of a general 2-level system in the strong electronic coupling limit. J. Chem. Phys. **76**, 2129–2135 (1982)

29. Lagos, R.E., Friesner, R.A.: Calculation of optical line shapes for generalized multilevel systems. J. Chem. Phys. **81**, 5899–5905 (1984)
30. McDermott, G., Prince, S.M., Freer, A.A., Hawthornthwaite-Lawless, A.M., Papiz, M.Z., et al.: Crystal structure of an integral membrane light-harvesting complex from photosynthetic bacteria. Nature **374**, 517–521 (1995)
31. Koepke, J., Hu, X.C., Muenke, C., Schulten, K., Michel, H.: The crystal structure of the light-harvesting complex II (B800-850) from Rhodospirillum molischianum. Structure **4**, 581–597 (1996)
32. Pearlstein, R.M.: Theoretical interpretation of antenna spectra. In: Scheer, H. (ed.) Chlorophylls, pp. 1047–1078. CRC Press, Boca Raton, FL (1991)
33. van Amerongen, H., Valkunas, L., van Grondelle, R.: Photosynthetic Excitons. World Scientific, Singapore (2000)
34. van Oijen, A.M., Ketelaars, M., Kohler, J., Aartsma, T.J., Schmidt, J.: Unraveling the electronic structure of individual photosynthetic pigment-protein complexes. Science **285**, 400–402 (1999)
35. van Oijen, A.M., Ketelaars, M., Kohler, J., Aartsma, T.J., Schmidt, J.: Spectroscopy of individual LH2 complexes of Rhodopseudomonas acidophila: localized excitations in the B800 band. Chem. Phys. **247**, 53–60 (1999)
36. van Oijen, A.M., Ketelaars, M., Kohler, J., Aartsma, T.J., Schmidt, J.: Spectroscopy of individual light-harvesting 2 complexes of Rhodopseudomonas acidophila: diagonal disorder, intercomplex heterogeneity, spectral diffusion, and energy transfer in the B800 band. Biophys. J. **78**, 1570–1577 (2000)
37. Ketelaars, M., van Oijen, A.M., Matsushita, M., Kohler, J., Schmidt, J., et al.: Spectroscopy on the B850 band of individual light-harvesting 2 complexes of Rhodopseudomonas acidophila I. Experiments and Monte Carlo simulations. Biophys. J. **80**, 1591–1603 (2001)
38. Kohler, J., van Oijen, A.M., Ketelaars, M., Hofmann, C., Matsushita, M., et al.: Optical spectroscopy of individual photosynthetic pigment protein complexes. Int. J. Mod. Phys. B **15**, 3633–3636 (2001)
39. Matsushita, M., Ketelaars, M., van Oijen, A.M., Kohler, J., Aartsma, T.J., et al.: Spectroscopy on the B850 band of individual light-harvesting 2 complexes of Rhodopseudomonas acidophila II. Exciton states of an elliptically deformed ring aggregate. Biophys. J. **80**, 1604–1614 (2001)
40. Hofmann, C., Ketelaars, M., Matsushita, M., Michel, H., Aartsma, T.J., et al.: Single-molecule study of the electronic couplings in a circular array of molecules: light-harvesting-2 complex of Rhodospirillum molischianum. Phys. Rev. Lett. **90**, 013004 (2003)
41. Hofmann, C., Aartsma, T.J., Kohler, J.: Energetic disorder and the B850-exciton states of individual light-harvesting 2 complexes from Rhodopseudomonas acidophila. Chem. Phys. Lett. **395**, 373–378 (2004)
42. Ketelaars, M., Segura, J.M., Oellerich, S., de Ruijter, W.P.F., Magis, G., et al.: Probing the electronic structure and conformational flexibility of individual light-harvesting 3 complexes by optical single-molecule spectroscopy. J. Phys. Chem. B **110**, 18710–18717 (2006)
43. Nagarajan, V., Alden, R.G., Williams, J.C., Parson, W.W.: Ultrafast exciton relaxation in the B850 antenna complex of Rhodobacter sphaeroides. Proc. Natl. Acad. Sci. USA **93**, 13774–13779 (1996)
44. Nagarajan, V., Johnson, E., Williams, J.C., Parson, W.W.: Femtosecond pump-probe spectroscopy of the B850 antenna complex of Rhodobacter sphaeroides at room temperature. J. Phys. Chem. B **103**, 2297–2309 (1999)
45. Wu, H.-M., Reddy, N.R.S., Small, G.J.: Direct observation and hole burning of the lowest exciton level (B870) of the LH2 antenna complex of Rhodopseudomonas acidophila (strain 10050). J. Phys. Chem. **101**, 651–656 (1997)
46. Jimenez, R., Dikshit, S.N., Bradforth, S.E., Fleming, G.R.: Electronic excitation transfer in the LH2 complex of Rhodobacter sphaeroides. J. Phys. Chem. **100**, 6825–6834 (1996)

47. Kumble, R., Palese, S., Visschers, R.W., Dutton, P.L., Hochstrasser, R.M.: Ultrafast dynamics within the B820 subunit from the core (LH-1) antenna complex of Rs. rubrum. Chem. Phys. Lett. **261**, 396–404 (1996)
48. Kühn, O., Sundström, V.: Pump–probe spectroscopy of dissipative energy transfer dynamics in photosynthetic antenna complexes: a density matrix approach. J. Chem. Phys. **107**, 4154–4164 (1997)
49. Kühn, O., Sundstrom, V., Pullerits, T.: Fluorescence depolarization dynamics in the B850 complex of purple bacteria. Chem. Phys. **275**, 15–30 (2002)
50. Meier, T., Chernyak, V., Mukamel, S.: Multiple exciton coherence sizes in photosynthetic antenna complexes viewed by pump-probe spectroscopy. J. Phys. Chem. B **101**, 7332–7342 (1997)
51. Monshouwer, R., Abrahamsson, M., van Mourik, F., van Grondelle, R.: Superradiance and exciton delocalization in bacterial photosynthetic light-harvesting systems. J. Phys. Chem. B **101**, 7241–7248 (1997)
52. Yang, M., Agarwal, R., Fleming, G.R.: The mechanism of energy transfer in the antenna of photosynthetic purple bacteria. J. Photochem. Photobiol. A Chem. **142**, 107–119 (2001)
53. Förster, T.: Excimers. Angew. Chem. Int. Ed. Engl. **8**, 333–343 (1969)
54. Gordon, M., Ware, W.R. (eds.): The Exciplex. Academic, New York (1975)
55. McGlynn, S.P., Armstrong, A.T., Azumi, T.: Interaction of molecular exciton, charge resonance states, and excimer luminescence. In: Sinanoglu, O. (ed.) Modern Quantum Chemistry Part III: Action of Light and Organic Crystals, pp. 203–228. Academic, New York (1965)
56. Yoshihara, K., Kasuya, T., Inoue, A., Nagakura, S.: Time-resolved spectra of pyrene excimer and pyrene-dimethylaniline exciplex. Chem. Phys. Lett. **9**, 469–471 (1971)
57. Betcher-Lange, S.L., Lehrer, S.S.: Pyrene excimer fluorescence in rabbit skeletal alphaalpha-tropomyosin labeled with N-(1-pyrene)maleimide. A probe of sulfhydryl proximity and local chain separation. J. Biol. Chem. **253**, 3757–3760 (1978)
58. Pal, R., Barenholz, Y., Wagner, R.R.: Pyrene phospholipid as a biological fluorescent probe for studying fusion of virus membrane with liposomes. Biochemistry **27**, 30–36 (1988)
59. Stegmann, T., Schoen, P., Bron, R., Wey, J., Bartoldus, I., et al.: Evaluation of viral membrane fusion assays. Comparison of the octadecylrhodamine dequenching assay with the pyrene excimer assay. Biochemistry **32**, 11330–11337 (1993)
60. Jung, K., Jung, H., Kaback, H.R.: Dynamics of lactose permease of Escherichia coli determined by site-directed fluorescence labeling. Biochemistry **33**, 3980–3985 (1994)
61. Sahoo, D., Narayanaswami, V., Kay, C.M., Ryan, R.O.: Pyrene excimer fluorescence: a spatially sensitive probe to monitor lipid-induced helical rearrangement of apolipophorin III. Biochemistry **39**, 6594–6601 (2000)
62. Liu, W., Chen, Y., Watrob, H., Bartlett, S.G., Jen-Jacobson, L., et al.: N-termini of EcoRI restriction endonuclease dimer are in close proximity on the protein surface. Biochemistry **37**, 15457–15465 (1998)

Circular Dichroism

9

9.1 Magnetic Transition Dipoles and n-π^* Transitions

In our analysis of how an electromagnetic radiation field interacts with electrons, we have, to this point, considered only the oscillating electric field, $\boldsymbol{E}(t)$. We set aside possible effects of the magnetic field, $\boldsymbol{B}(t)$, on the grounds that they usually are much smaller than the effects of the electric field. With this assumption we found that the strength of the absorption band for a transition between two states with wavefunctions ψ_a and ψ_b depends on the square of the dot product of $\boldsymbol{E}_o$ with the electric dipole matrix element, $\boldsymbol{\mu}_{ba}$. There are, however, cases in which the symmetry of the wavefunctions makes $\boldsymbol{\mu}_{ba}$ zero, and yet the transition still has a measurable dipole strength. The absorption in these cases sometimes results from quadrupole, octupole, or other small terms that we have neglected in using the dipole operator, but in other cases it can be traced to interactions with the magnetic field. In addition, coupled interactions involving both $\boldsymbol{E}(t)$ and $\boldsymbol{B}(t)$ can cause the dipole strength of a transition to be different for left- and right-circularly polarized light. This is *circular dichroism.*

To treat effects of the magnetic field of radiation by the same semiclassical approach that we used for the electric field, we can expand Eq. (4.2) so that the time-dependent perturbation term in the Hamiltonian $\left(\widetilde{\mathrm{H}}'\right)$ includes a term proportional to $\boldsymbol{B}(t)$:

$$\widetilde{\mathrm{H}}'(t) = -\boldsymbol{E}(t) \cdot \widetilde{\mu} - \boldsymbol{B}(t) \cdot \widetilde{\mathrm{m}}. \tag{9.1}$$

Here $\widetilde{\mathrm{m}}$ is the *magnetic dipole operator*, which is given by

$$\widetilde{\mathrm{m}} = \frac{e}{2m_e c}(\boldsymbol{r} \times \widetilde{\mathrm{p}} + g_e \boldsymbol{S}), \tag{9.2}$$

W.W. Parson, *Modern Optical Spectroscopy*, DOI 10.1007/978-3-662-46777-0_9

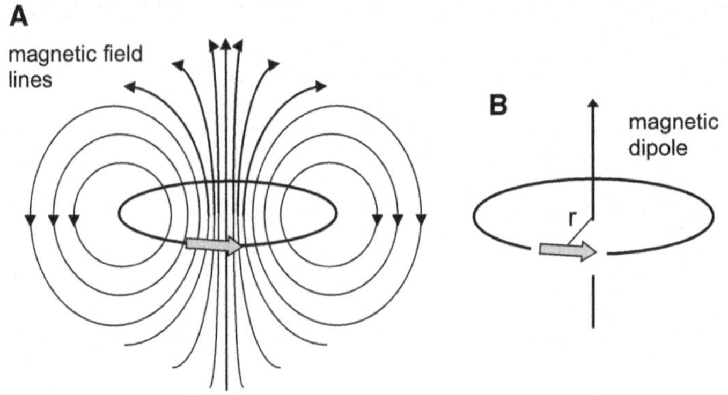

Fig. 9.1 (A) An electric current (*shaded arrow*) flowing through a wire creates a circular magnetic field around the wire. If the wire forms a ring, the magnetic field lines have the same shape as those around a uniformly magnetized disk with the same area. (**B**) At large distances, the magnetic field lines are equivalent to those created by a magnetic dipole whose magnitude is the product of the current and the area of the ring. Classically, the angular momentum associated with circular motion of a particle is given by $\boldsymbol{r} \times \boldsymbol{p}$, where $\boldsymbol{r}$ is the vector from the center of the *circle* to a point on the trajectory and $\boldsymbol{p}$ is the particle's linear momentum at that point

where e and m_e are the electron charge and mass, c is the speed of light, $\boldsymbol{r}$ is the position of the electron, $\tilde{\mathrm{p}}$ is the linear momentum operator $\left(-i\hbar\tilde{\nabla}\right)$, g_e is the electron g-factor (2.00232), $\boldsymbol{S}$ is the angular momentum associated with the electron spin, and "×" denotes a vector cross product (Appendix A.2). The factor $-e/2m_e$ (a positive number, since we use the convention that e is negative) is called the electronic *gyromagnetic ratio*. The cross product $\boldsymbol{r} \times \tilde{\mathrm{p}}$ in Eq. (9.2) is the orbital *angular momentum operator*, and the product $(e/2m_e c)\boldsymbol{r} \times \tilde{\mathrm{p}}$ corresponds to the classical expression for the magnetic moment generated by a current in a coil of wire with radius $|\boldsymbol{r}|$ (Fig. 9.1). The term $(e/2m_e c)g_e\boldsymbol{S}$ usually does not contribute significantly to molecular absorption or CD spectra, and we will neglect it.

The term for magnetic interactions in Eq. (9.1) gives rise to a magnetic transition dipole matrix element ($\boldsymbol{m}_{ba}$) for a transition of a system between different states. A magnetic transition dipole is a vector analogous to an electric transition dipole but with the magnetic-dipole operator $\tilde{\mathrm{m}}$ replacing $\tilde{\mu}$:

$$\boldsymbol{m}_{ba} = \langle\psi_b|\tilde{\mathrm{m}}|\psi_a\rangle. \tag{9.3}$$

If $\boldsymbol{m}_{ba}$ is non-zero, perturbations of the molecule by the magnetic field of light can cause transitions between the two states. The strength of such a transition is proportional to $|\boldsymbol{m}_{ba}|^2|\boldsymbol{B}|^2\cos^2\theta$, where θ is the angle between the magnetic field and $\boldsymbol{m}_{ba}$. This is analogous to Eq. (4.8c) for transitions driven by the electric field of light.

Magnetic dipoles and magnetic transition dipoles commonly are stated in units of $(-e\hbar/2me)$, which is called the *Bohr magneton* (μ_B). One Bohr magneton is the magnetic dipole that would result from the classical circular motion of a 1s electron in the Bohr model of a hydrogen atom. It also is the fundamental unit of the magnetic moment associated with an electron's quantized orbital angular momentum. The component of the angular momentum in a specified direction is constrained to integer multiples of $\hbar$, which restricts the corresponding component of the orbital magnetic moment to multiples of $-e\hbar/2me$. One μ_B is 9.274×10^{-21} emu·cm or 9.274×10^{-21} erg·G^{-1} in cgs units (9.274×10^{-24} J·T^{-1} in SI units). For comparison, the debye unit for electric dipoles (D) is about two orders of magnitude larger (10^{-18} esu·cm). The molar extinction coefficient for absorption driven by an electric transition dipole of 1 D is, therefore, on the order of 10^4 times that for the absorption driven by a magnetic transition dipole of 1 μ_B.

The magnetic transition dipole $\boldsymbol{m}_{ba}$ for a molecular excitation to an excited singlet state can be evaluated by writing the molecular orbitals as linear combinations of atomic orbitals. For a system constructed with atomic $2p$ orbitals, $\boldsymbol{m}_{ba}$ takes a form similar to Eq. (4.28):

$$\boldsymbol{m}_{ba} = -i\frac{\sqrt{2}\hbar e}{2m_e c}\sum_s \sum_t C_s^b C_t^a \left\langle p_s \middle| \boldsymbol{r} \times \tilde{\nabla} \middle| p_t \right\rangle. \tag{9.4}$$

As in Eqs. (4.22e) and (4.28), p_t here denotes a $2p$ orbital centered on atom t, C_t^a is the expansion coefficient for this atomic orbital in molecular wavefunction Ψ_a, and $\boldsymbol{r}$ is the position vector in a coordinate system with its origin at the center of the rotation. To examine the ways that the expression in Eq. (9.4) can give a non-zero magnetic transition dipole, it is helpful to write the position vector $\boldsymbol{r}$ in the integral $\left\langle p_s \middle| \boldsymbol{r} \times \tilde{\nabla} \middle| p_t \right\rangle$ as the sum of two vectors, $\boldsymbol{R}_t$ and $\boldsymbol{r}_t$, where $\boldsymbol{R}_t$ is the location of atom t and $\boldsymbol{r}_t$ describes the position of an electron relative to the center of atom t. The constant factor $\boldsymbol{R}_t$ then can be factored out of its term in the matrix element, giving

$$\boldsymbol{m}_{ba} = -i\frac{\sqrt{2}\hbar e}{2m_e c}\sum_s \sum_t C_s^b C_t^a \left(\left\langle p_s \middle| \boldsymbol{r}_t \times \tilde{\nabla} \middle| p_t \right\rangle + \left\langle p_s \middle| \boldsymbol{R}_t \times \tilde{\nabla} \middle| p_t \right\rangle \right) \tag{9.5a}$$

$$= -i\frac{\sqrt{2}\hbar e}{2m_e c}\sum_s \sum_t C_s^b C_t^a \left(\left\langle p_s \middle| \boldsymbol{r}_t \times \tilde{\nabla} \middle| p_t \right\rangle + \boldsymbol{R}_t \times \left\langle p_s \middle| \tilde{\nabla} \middle| p_t \right\rangle \right). \tag{9.5b}$$

The first term in the parentheses on the right side of Eq. (9.5b) pertains mainly to pairs of atomic orbitals centered on the same atom; the second term relates to the relative positions of the atoms that make up the chromophore. Writing out the x, y and z components of $\tilde{\nabla}$ and the vector cross product in the first term gives

$$\left\langle p_s \middle| \boldsymbol{r}_t \times \tilde{\nabla} \middle| p_t \right\rangle = \left\langle p_s \middle| y\frac{\partial}{\partial z} - z\frac{\partial}{\partial y} \middle| p_t \right\rangle \hat{x} - \left\langle p_s \middle| x\frac{\partial}{\partial z} - z\frac{\partial}{\partial x} \middle| p_t \right\rangle \hat{y}$$
$$+ \left\langle p_s \middle| x\frac{\partial}{\partial y} - y\frac{\partial}{\partial x} \middle| p_t \right\rangle \hat{z}, \tag{9.6}$$

where $\hat{x}$, $\hat{y}$ and $\hat{z}$ are unit vectors in a coordinate system centered on atom t. Consider the integral $\langle p_s|y\partial/\partial z - z\partial/\partial y|p_t\rangle$, which contributes the x component of $\left\langle p_s \middle| \boldsymbol{r}_t \times \tilde{\nabla} \middle| p_t \right\rangle$. Figure 9.2 illustrates the functions that enter into this integral when p_s and p_t are, respectively, $2p_y$ and $2p_z$ orbitals of the same atom. Differentiating $2p_z$ with respect to z and then multiplying by y gives the function $y\partial(2p_z)/\partial z$, as shown in panels B and D of the figure. Panels C and E show the functions generated by differentiating with respect to y and then multiplying by z. In both cases, the result is an even function of z and an odd function of y. Multiplying these functions by $2p_y$ generates the completely symmetric functions illustrated in panels G and H. Further, because the two final functions have opposite signs, they combine constructively if we subtract one from the other as prescribed in Eq. (9.6). In fact, the combination $y\partial(2p_z)/\partial z - z\partial(2p_z)/\partial y$ has exactly the same spatial distribution as $2p_y$ [1, 2]. The x component of $\left\langle 2p_{y(t)} \middle| \boldsymbol{r}_t \times \tilde{\nabla} \middle| 2p_{z(t)} \right\rangle$ therefore is proportional to $\langle 2p_{y(t)}|2p_{y(t)}\rangle$, which is 1. The magnetic dipole matrix element $\boldsymbol{m}_{ba}$ thus has a non-zero x component. A similar analysis shows that the y and z components of $\left\langle 2p_{y(t)} \middle| \boldsymbol{r}_t \times \tilde{\nabla} \middle| 2p_{z(t)} \right\rangle$, and thus the y and z components of $\boldsymbol{m}_{ba}$, are zero for this pair of orbitals, so $\boldsymbol{m}_{ba}$ is oriented on the x axis.

The results are qualitatively the same if p_s is a $2p_y$ orbital of a different atom, except that the x component of $\left\langle 2p_{y(s)} \middle| \boldsymbol{r}_t \times \tilde{\nabla} \middle| 2p_{z(t)} \right\rangle$ is smaller for $s \neq t$ because the overlap of the two wavefunctions falls off with the interatomic distance. The overlap integral can be evaluated by procedures similar to those described in Box 4.11 [2]. By studying Fig. 9.2 you also can see that all three components of $\left\langle 2p_{z(s)} \middle| \boldsymbol{r}_t \times \tilde{\nabla} \middle| 2p_{z(t)} \right\rangle$ (*i.e.*, the corresponding matrix element for two atomic $2p_z$ orbitals with parallel z-axes) are zero.

An example of a transition that involves $2p_y$ and $2p_z$ orbitals of the same atom is an n-π* excitation of a carbonyl group, in which an electron is excited from a nonbonding (n) orbital of the oxygen atom to a π* molecular orbital constructed of oxygen and carbon $2p$ orbitals (Fig. 9.3). In the absence of structural distortions, the electric transition dipole ($\boldsymbol{\mu}_{ba}$) for such an excitation is zero by symmetry. The magnetic transition dipole ($\boldsymbol{m}_{ba}$) for an n-π* excitation is non-zero and is directed along the C=O bond, which means that the transition can be driven by an oscillating magnetic field with this orientation. However, n-π* transitions typically are much weaker than π-π* transitions of common chromophores. Saturated ketones have n-π* transitions with molar extinction coefficients of 20–30 M^{-1} cm^{-1} in the region of 280 nm. The n-π* transition of acetamide in water, which occurs at 214 nm, has an extinction coefficient of 60 M^{-1} cm^{-1}. n-π* transitions of the

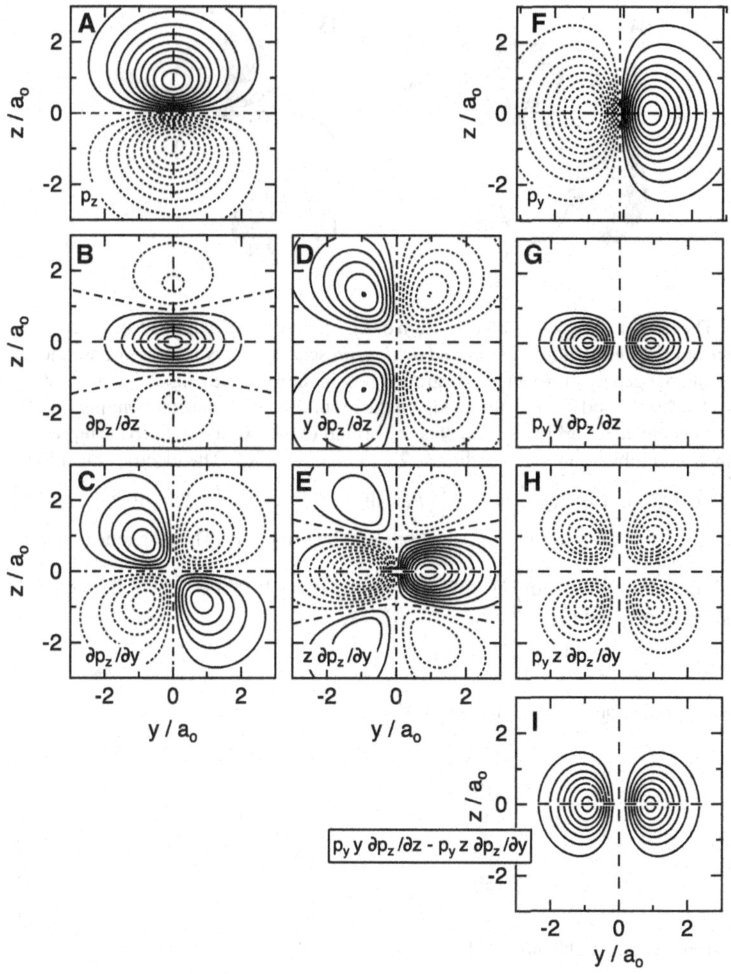

Fig. 9.2 Construction of the x component of the magnetic transition matrix element for $2p$ atomic orbitals. (**A**, **F**) Contour plots in the yz plane of the amplitude of atomic $2p_z$ and $2p_y$ orbitals (p_z and p_y); (**B**, **C**) the derivatives $\partial p_z/\partial z$ and $\partial p_z/\partial y$, respectively; (**D**, **E**) the products $y \cdot \partial p_z/\partial z$ and $z \cdot \partial p_z/\partial y$. Note that the last two products are odd functions of y. (**G**, **H**) The functions $p_y \cdot y \cdot \partial p_z/\partial z$ and $p_y \cdot z \cdot \partial p_z/\partial y$, respectively; (**I**) $p_y \cdot y \cdot \partial p_z/\partial z - p_y \cdot z \cdot \partial p_z/\partial y$. *Solid lines* denote positive amplitudes; *dotted lines*, negative. The contour intervals are arbitrary. See Fig. 4.17C, D for similar plots of $\partial \psi_a/\partial y$ and $\partial \psi_a/\partial z$ for the first $\pi-\pi^*$ excitation of ethylene

peptide bonds contribute to the absorption by proteins in the region of 200–220 nm (Table 9.1), but probably owe most of their dipole strengths to excitonic mixing with π-π^* transitions rather than to their magnetic transition dipoles directly [3–6].

Returning to Eq. (9.5b), we still need to consider the term $\boldsymbol{R}_t \times \left\langle p_s \middle| \tilde{\nabla} \middle| p_t \right\rangle$. If the atomic orbitals that participate in the molecular wavefunctions are dominated by a single atom, as is the case for the n-π^* transition just discussed, the atomic

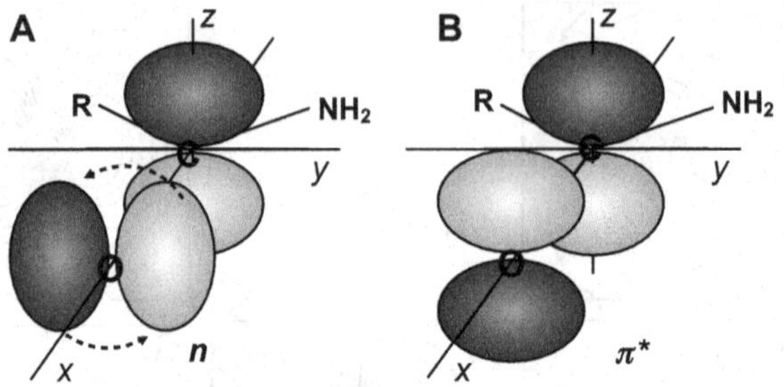

Fig. 9.3 The lowest-energy $n \rightarrow \pi^*$ transition of a carbonyl group requires rotational displacement of an electronic wavefunction. The *shaded ovals* represent boundary surfaces of atomic $2p$ orbitals of carbon and oxygen atoms of an amide group lying in the xy plane (Fig. 2.5). The C atom is at the origin and the C=O bond is along the x axis. In the ground state (**A**) the nonbonding (n) electrons of the oxygen occupy a $2p_y$ orbital ($p_{y(O)}$). The π^* orbital of the excited state (**B**) is represented by an antisymmetric combination of $2p_z$ orbitals, $2^{-1/2}(p_{z(C)} - p_{z(O)})$. The electric transition dipole is $2^{-1/2}\left\langle \left(p_{z(C)} - p_{z(O)}\right)|\tilde{\mu}|p_{y(O)}\right\rangle = 2^{-1/2}\left\langle p_{z(C)}|\tilde{\mu}|p_{y(O)}\right\rangle - 2^{-1/2}\left\langle -p_{z(O)}|\tilde{\mu}|p_{y(O)}\right\rangle$. The integrals $\left\langle p_{z(C)}|\tilde{\mu}|p_{y(O)}\right\rangle$ and $\left\langle -p_{z(O)}|\tilde{\mu}|p_{y(O)}\right\rangle$ are both zero by symmetry. The magnetic transition dipole, $2^{-1/2}\left\langle \left(p_{z(C)} - p_{z(O)}\right)|\tilde{m}|p_{y(O)}\right\rangle$, is nonzero because $p_{y(O)}$ is perpendicular to $p_{z(C)}$ and $p_{z(O)}$, as explained in the text and Fig. 9.2

Table 9.1 Amide transitions in the near UV

Assignment	λ (nm)	$\lvert\mu\rvert$ (D)	$\lvert\boldsymbol{m}\rvert$ (μ_B)
n-π^*	210	0	0.6
π-π^*	190	3.1	0
n-π^*	165	1.4	0.2
n-σ^*	150	1.8	0.8
π-π^*	125	1.7	0

Adapted from Woody and Tinoco [6].

center will be the center of the rotation, which makes $\boldsymbol{R}_t$ zero. The term $\boldsymbol{R}_t \times \left\langle p_s\middle|\tilde{\nabla}\middle|p_t\right\rangle$ then disappears. But if the molecular orbitals involve multiple atoms, Eq. (9.5b) calls for a sum of $C_s^b C_t^a \boldsymbol{R}_t \times \left\langle p_s\middle|\tilde{\nabla}\middle|p_t\right\rangle$ over all the pairs of atoms, and the result depends on the geometry of the chromophore. The factor $\left\langle p_s\middle|\tilde{\nabla}\middle|p_t\right\rangle$ is the matrix element of the gradient operator for $2p$ atomic orbitals on atoms s and t. We discussed this matrix element in Chap. 4, where we showed that it is oriented along the line between the two atoms (Eqs. 4.27–4.29; Boxes 4.10 and 4.11; Figs. 4.17–4.19). Its contribution to the overall magnetic transition matrix element ($\boldsymbol{m}_{ba}$) depends the coefficients for atoms s and t in the initial and final molecular orbitals (C_t^a and C_s^b) and on the position of atom t ($\boldsymbol{R}_t$). The sum over all pairs of atoms in Eq. (9.5b) usually is treated by an expression with the form of Eq. (4.29). If the atomic p_z orbitals are all close to parallel, so that the first term in the braces in Eq. (9.5b) is effectively zero for all the atoms, we then have

$$\boldsymbol{m}_{ba} = -i\frac{\sqrt{2}\hbar e}{2m_e c}\sum_{s>t}\sum_{t} 2\left(C_s^a C_t^b - C_s^b C_t^a\right)\boldsymbol{R}_{mid(s,t)} \times \left\langle p_s\middle|\tilde{\nabla}\middle|p_t\right\rangle, \qquad (9.7)$$

where $\boldsymbol{R}_{mid(s,t)}$ is the vector from the origin of the coordinate system to a point midway between atoms s and t.

The dependence of Eq. (9.7) on the positions of the atoms is potentially problematic because it means that we might get spurious results simply by shifting the origin of the coordinate system. But this non-physical sensitivity of the calculated $\boldsymbol{m}_{ba}$ to the choice of the coordinate system occurs only if excitation from ψ_a to ψ_b has a non-zero electric transition dipole. You can see this by considering what happens if we shift the coordinate system by adding any arbitrary vector $\boldsymbol{R}_{arb}$ to $\boldsymbol{R}_t$ for all the atoms in the chromophore. The calculated value of $\boldsymbol{m}_{ba}$ becomes

$$\boldsymbol{m}'_{ba} = \boldsymbol{m}^0_{ba} - i\frac{\sqrt{2}\hbar e}{2m_e c}\boldsymbol{R}_{arb} \times \left[\sum_{s>t}\sum_{t} 2\left(C_s^a C_t^b - C_s^b C_t^a\right)\left\langle p_s\middle|\tilde{\nabla}\middle|p_t\right\rangle\right], \qquad (9.8)$$

where $\boldsymbol{m}^0_{ba}$ is result obtained before we change the coordinate system. We showed in Sect. 4.8 that the factor in square brackets is proportional to the electric transition dipole, $\boldsymbol{\mu}_{ba}$. So shifting the coordinate system will have no effect on $\boldsymbol{m}_{ba}$ if $|\boldsymbol{\mu}_{ba}|$ is zero. This argument applies to either exact or inexact wavefunctions as long as we obtain $\boldsymbol{\mu}_{ba}$ with the gradient operator $\left(\tilde{\nabla}\right)$ rather than $\tilde{\mu}$.

Values of $\boldsymbol{m}_{ba}$ calculated by Eq. (9.7) do depend on the choice of the coordinate system if the electric transition dipole is not zero. But this is basically no different from the formula for angular momentum in classical physics ($\boldsymbol{r} \times \boldsymbol{p}$), which assumes that the motion is circular and that we know the center of the rotation: the result is physically meaningless if the motion is linear or if we use the wrong center. As we will see later in this chapter, we usually are less concerned with $\boldsymbol{m}_{ba}$ itself than with the dot product of $\boldsymbol{m}_{ba}$ and $\boldsymbol{\mu}_{ba}$. This product is independent of the choice of the coordinate system as long as $\boldsymbol{m}_{ba}$ and $\boldsymbol{\mu}_{ba}$ are calculated consistently by using $\tilde{\nabla}$ in the same coordinate system.

As an illustration of these points, consider *trans*-butadiene. The pertinent molecular orbitals are shown in Fig. 4.19, along with vector diagrams of the weighted atomic matrix elements $\left(C_s^b C_t^a \left\langle p_s\middle|\tilde{\nabla}\middle|p_t\right\rangle\right)$ that combine to make the electric transition dipoles for the first four excitations. Figure 9.4 reproduces the vector diagrams for excitations from the HOMO (ψ_2) to the two lowest unoccupied molecular orbitals (ψ_3 and ψ_4). The figure also shows the position vectors $\boldsymbol{R}_{mid(s,t)}$ that are needed to calculate $\boldsymbol{m}_{ba}$ by Eq. (9.7). For the excitation $\psi_2 \rightarrow \psi_3$, combining the weighted $\left\langle p_s\middle|\tilde{\nabla}\middle|p_t\right\rangle$ vectors for atom pairs $(s,t) = (2,1)$, $(3,2)$ and $(4,3)$ gives an

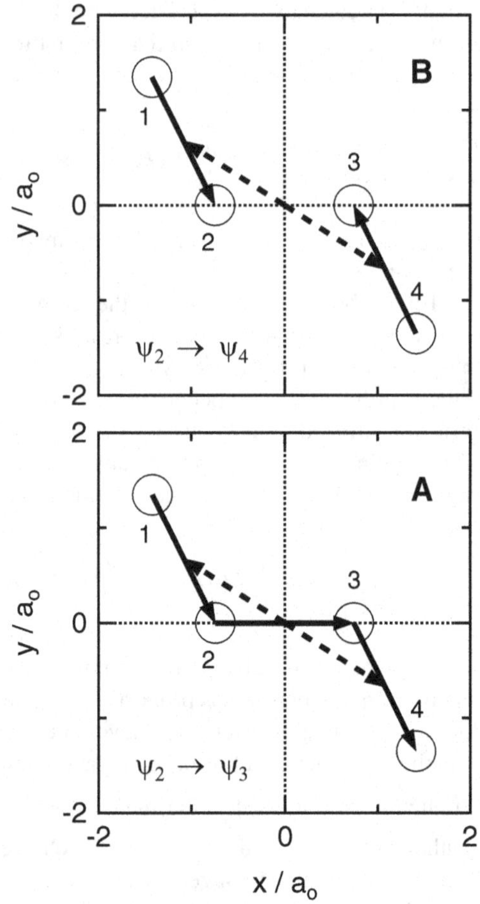

Fig. 9.4 Vector diagrams of contributions to the transition-gradient matrix elements for the first two excitations of *trans*-butadiene. The initial and final molecular wavefunctions in the four-orbital model used for Fig. 4.19 are indicated. The atoms are labeled 1–4 by the *empty circles* indicating the positions. The *solid arrows* indicate the directions and relative magnitudes of $C_s^b C_t^a \langle p_s | \tilde{\nabla} | p_t \rangle$ for pairs of bonded atoms. C_s^b and C_t^a are the coefficients for atomic $2p_z$ orbitals of atoms *s* and *t* in the final and initial wavefunction, respectively; $\langle p_s | \tilde{\nabla} | p_t \rangle$ is the matrix element of the gradient operator for the two atomic orbitals. The vectors ($\boldsymbol{R}_{mid(s,t)}$) from the origin of the coordinate system to points midway between the pairs of bonded atoms are indicated with *dashed arrows*. The transition gradient matrix element for each excitation is given by the vector sum of the *solid arrows*. (Including contributions from the pairs of nonbonded atoms would not change the results.) The magnetic transition dipole ($\boldsymbol{m}_{ba}$) is obtained by summing the cross products $\boldsymbol{R}_{mid(s,t)} \times C_s^b C_t^a \langle p_s | \tilde{\nabla} | p_t \rangle$

electric transition dipole that points along the principle axis of the molecule (Figs. 4.19E and 9.4A). The magnetic transition dipole calculated by Eq. (9.7) is zero for this excitation if we center the coordinate system at the molecule's center of symmetry as shown in the figure. The contribution from atom pair (3,2) $\left(\boldsymbol{R}_{mid(3,2)}C_3^bC_2^a\left\langle p_3\middle|\tilde{\nabla}\middle|p_2\right\rangle\right)$ is zero because the position vector $\boldsymbol{R}_{mid(3,2)}$ is zero in this coordinate system; the contributions from atom pairs (2,1) and (4,3) sum to zero because $C_2^bC_1^a\left\langle p_2\middle|\tilde{\nabla}\middle|p_1\right\rangle = C_4^bC_3^a\left\langle p_4\middle|\tilde{\nabla}\middle|p_3\right\rangle$ while $\boldsymbol{R}_{mid(2,1)} = -\boldsymbol{R}_{mid(4,3)}$. The lack of a magnetic transition dipole for this excitation is not surprising, because the electron motions depicted by the vectors do not constitute a rotation around the molecular center of symmetry, or indeed around any point at a finite distance. Equation (9.7) would, however, give a non-zero result with no physical significance if we shifted the coordinate system off the center of symmetry.

The situation is reversed for $\psi_2 \rightarrow \psi_4$. Here, the electric transition dipole is zero because $C_3^bC_2^a\left\langle p_3\middle|\tilde{\nabla}\middle|p_2\right\rangle = 0$ and $C_2^bC_1^a\left\langle p_2\middle|\tilde{\nabla}\middle|p_1\right\rangle = -C_4^bC_3^a\left\langle p_4\middle|\tilde{\nabla}\middle|p_3\right\rangle$ (Figs. 4.19F and 9.4B), whereas the magnetic transition dipole is non-zero. Inspection of Fig. 9.4B shows that, although the electron motions represented by the vectors do not complete a circle, they occur in opposite directions on either side of the center of symmetry. The opposite signs of $C_2^bC_1^a\left\langle p_2\middle|\tilde{\nabla}\middle|p_1\right\rangle$ and $C_4^bC_3^a\left\langle p_4\middle|\tilde{\nabla}\middle|p_3\right\rangle$ rectify the inversion of $\boldsymbol{R}_{mid}$, making the sum of $\boldsymbol{R}_{mid(s,t)} \times C_s^b C_t^a\left\langle p_s\middle|\tilde{\nabla}\middle|p_t\right\rangle$ over the atoms non-zero. As we would expect, shifting the origin of the coordinate system does not affect the calculated value of $\boldsymbol{m}_{ba}$ for this excitation.

Transitions involving magnetic transition dipoles also can be treated by the quantum-mechanical approach we used in Chap. 5 to account for fluorescence. Remarkably, although the final results are the same and the problem of chosing the coordinate system remains, the quantum theory does not require distinguishing between the magnetic and electric fields of the radiation. Instead, as explained in Box 9.1, "magnetic dipole-allowed" transitions are related to the change in the amplitude of the vector potential with position across the chromophore. This approach provides a different perspective on the question of why magnetic-dipole transitions typically are much weaker than electric-dipole transitions. If the wavelength of the radiation is large compared to the size of the chromophore, as is usually the case in electronic spectroscopy, the variation in the vector potential across the chromophore will be relatively insignificant.

Box 9.1 Quantum Theory of Magnetic-Dipole and Electric-Quadrupole Transitions

To see how the quantum theory accounts for interactions of a chromophore with the magnetic field of light, we return to the expressions we used in

(continued)

Box 9.1 (continued)
Chap. 5 for the energy of interaction of an electron with a linearly polarized radiation field (Eqs. 5.32a and 5.32b):

$$\widetilde{\mathrm{H}}' = -\frac{\hbar e}{i m_e c} \boldsymbol{V} \cdot \widetilde{\nabla} + \frac{e^2}{2 m_e c^2} |\boldsymbol{V}|^2 \tag{B9.1.1a}$$

$$= -\frac{e\hbar\pi^{1/2}}{i m_e} \sum_j \left(\hat{e}_j \cdot \widetilde{\nabla}\right) \left[\widetilde{\mathrm{q}}_j \exp(2\pi i \boldsymbol{k}_j \cdot \boldsymbol{r}) + \widetilde{\mathrm{q}}_j^* \exp(-2\pi i \boldsymbol{k}_j \cdot \boldsymbol{r})\right] + \cdots. \tag{B9.1.1b}$$

Here $\boldsymbol{V}$ is the vector potential of the radiation field; $\hat{e}_j$ and $\boldsymbol{k}_j$ are, respectively, the polarization axis and wavevector of radiation mode j; and m_e is the electron mass. The elipsis in Eq. (B9.1.1b) represents two-photon processes that we continue to defer to Chap. 12.

In Sect. 5.5, our next step was to assume that the wavelength of the radiation was long enough so that the factors $\exp(\pm 2\pi i \boldsymbol{k}_j \cdot \boldsymbol{r})$ could be set equal to 1 (Eq. 5.33). Here we include the second-order term in a power-series expansion of the exponential,

$$\exp\left(\pm 2\pi i \boldsymbol{k}_j \cdot \boldsymbol{r}\right) = 1 \pm 2\pi i \boldsymbol{k}_j \cdot \boldsymbol{r} + \frac{1}{2}\left(2\pi i \boldsymbol{k}_j \cdot \boldsymbol{r}\right)^2 + \cdots. \tag{B9.1.2}$$

For 300-nm light and a chromophore with dimensions in the range of 1–3 nm, the product $\boldsymbol{k}_j \cdot \boldsymbol{r}$ is on the order of 0.01, which is small enough to neglect in most cases. But suppose that the electric transition dipole $\boldsymbol{\mu}_{ba}$ for a transition of an electron between wavefunctions ψ_a and ψ_b is zero by symmetry. The first term on the right in Eq. (B9.1.2) then does not contribute to the overall matrix element for the transition, so we must consider the second term $(\pm 2\pi i \boldsymbol{k}_j \cdot \boldsymbol{r})$. Using this term with Eq. (B9.1.1b) gives

$$\left\langle \psi_b \chi_{j(m)} \middle| \widetilde{\mathrm{H}}' \middle| \psi_a \chi_{j(n)} \right\rangle = -\frac{2\pi e\hbar\pi^{1/2}}{m_e} \left\langle \psi_b \middle| (\boldsymbol{k}_j \cdot \boldsymbol{r}) \left(\hat{e}_j \cdot \widetilde{\nabla}\right) \middle| \psi_a \right\rangle \left\langle \chi_{j(m)} \middle| \widetilde{\mathrm{Q}} \middle| \chi_{j(n)} \right\rangle, \tag{B9.1.3}$$

where $\widetilde{\mathrm{Q}}$ is the photon position operator and $\chi_{j(n)}$ denotes the photon wavefunction for the nth excitation level of mode j. We analyzed the factor $\left\langle \chi_{j(m)} \middle| \widetilde{\mathrm{Q}}_j \middle| \chi_{j(n)} \right\rangle$ in Sect. 5.5, and found that it includes separate terms for absorption $(m = n - 1)$ and emission $(m = n + 1)$. Our interest now is the factor $\left\langle \psi_b \middle| (\boldsymbol{k}_j \cdot \boldsymbol{r}) \left(\hat{e}_j \cdot \widetilde{\nabla}\right) \middle| \psi_a \right\rangle$.

(continued)

Box 9.1 (continued)

Consider a radiation mode that propagates along the y axis with wavelength λ_{ba} and is polarized in the z direction. The dot product $\boldsymbol{k}_j \cdot \boldsymbol{r}$ then reduces to the y component of $\boldsymbol{r}$; $\hat{e}_j \cdot \tilde{\nabla}$ reduces to $\partial/\partial z$; and the matrix element for excitation of ψ_a to ψ_b becomes

$$-\frac{2e\hbar\pi^{3/2}}{m_e}\left\langle\psi_b\middle|(\boldsymbol{k}_j\cdot\boldsymbol{r})\left(\hat{e}_j\cdot\tilde{\nabla}\right)\middle|\psi_a\right\rangle$$

$$= -\frac{2e\hbar\pi^{3/2}}{m_e}\frac{1}{\lambda_{ba}}\left\langle\psi_b\middle|(\hat{y}\cdot\boldsymbol{r})\left(\hat{z}\cdot\tilde{\nabla}\right)\middle|\psi_a\right\rangle \quad \text{(B9.1.4a)}$$

$$= -\frac{2e\hbar\pi^{3/2}}{m_e\lambda_{ba}}\left\langle\psi_b\middle|y\frac{\partial}{\partial z}\middle|\psi_a\right\rangle. \quad \text{(B9.1.4b)}$$

The integral on the right side of Eq. (B9.1.4b) can be manipulated to give:

$$-\frac{2e\hbar\pi^{3/2}}{m_e\lambda_{ba}}\left\langle\psi_b\middle|y\frac{\partial}{\partial z}\middle|\psi_a\right\rangle$$

$$= -\frac{2e\hbar\pi^{3/2}}{m_e\lambda_{ba}}\frac{1}{2}\left[\left(\left\langle\psi_b\middle|y\frac{\partial}{\partial z}\middle|\psi_a\right\rangle - \left\langle\psi_b\middle|z\frac{\partial}{\partial y}\middle|\psi_a\right\rangle\right) + \left(\left\langle\psi_b\middle|y\frac{\partial}{\partial z}\middle|\psi_a\right\rangle\right.\right.$$
$$\left.\left.+\left\langle\psi_b\middle|z\frac{\partial}{\partial y}\middle|\psi_a\right\rangle\right)\right]. \quad \text{(B9.1.5)}$$

The quantity in the first set of parentheses on the right side of this expression is, except for multiplicative constants, the x component of the magnetic transition dipole:

$$-\frac{2e\hbar\pi^{3/2}}{m_e\lambda_{ba}}\frac{1}{2}\left(\left\langle\psi_b\middle|y\frac{\partial}{\partial z}\middle|\psi_a\right\rangle - \left\langle\psi_b\middle|z\frac{\partial}{\partial y}\middle|\psi_a\right\rangle\right) = -\frac{e\hbar\pi^{3/2}}{m_e\lambda_{ba}}\left\langle\psi_b\middle|\boldsymbol{r}\times\tilde{\nabla}\middle|\psi_a\right\rangle$$

$$= -\frac{e\hbar\pi^{3/2}}{m_e\lambda_{ba}}\left(\frac{2m_e c}{-i\hbar e}\right)(\boldsymbol{m}_{ba})_x = \frac{2\pi^{3/2}c}{i\lambda_{ba}}(\boldsymbol{m}_{ba})_x = -i2\pi^{3/2}\nu_{ba}(\boldsymbol{m}_{ba})_x. \quad \text{(B9.1.6)}$$

Now look at the quantity in the second set of square brackets on the right side of Eq. (B9.1.5). By using the relationship between the matrix elements of the gradient operator and the commutator of the Hamiltonian and dipole operators (Box 4.10), we can relate this term to the matrix element of the product yz [7, 8]. This gives

(continued)

Box 9.1 (continued)

$$-\frac{2e\hbar\pi^{3/2}}{m_e\lambda_{ba}}\frac{1}{2}\left[\left\langle\psi_b\middle|y\frac{\partial}{\partial z}\middle|\psi_a\right\rangle+\left\langle\psi_b\middle|z\frac{\partial}{\partial y}\middle|\psi_a\right\rangle\right]=\frac{2e\hbar\pi^{3/2}}{m_e\lambda_{ba}}\frac{1}{2}\left(\frac{2\pi m_e\nu_{ba}}{\hbar}\right)\langle\psi_b|yz|\psi_a\rangle$$
$$=\frac{2\pi^{5/2}\nu_{ba}}{\lambda_{ba}}e\langle\psi_b|yz|\psi_a\rangle. \tag{B9.1.7}$$

Further, the factor $e\langle\psi_b|yz|\psi_a\rangle$ in Eq. (B9.1.7) is recognizable as the yz element of the electric quadrupole interaction matrix (Eqs. 4.5 and B4.2.4).

A similar analysis for radiation propagating along the x or z axis instead of y, or polarized along y or x instead of z, gives corresponding results with the y or z component of $\boldsymbol{m}_{ba}$ replacing the x component in Eq. (B9.1.6), and/or with different components of the quadrupole matrix replacing the yz element in Eq. (B9.1.7) [7, 8]. By including the second-order term in the dependence of the vector potential $\boldsymbol{V}$ on position we thus obtain a transition matrix element that commonly is ascribed to a magnetic transition dipole (Eq. B9.1.6), along with a matrix element representing the quadrupolar distribution of the wavefunction (Eq. B9.1.7). Including the third-order term in the distance dependence would add a matrix element for octupolar interactions. Since the magnetic-dipole and electric-quadrupole matrix elements both arise from the same term in the distance dependence, they should have comparable magnitudes and generally should be much smaller than electric-dipole matrix elements.

As pointed out above, the quantum theory of absorption differs from the semiclassical theory in that it does not distinguish explicitly between the electric and magnetic fields of electromagnetic radiation. Although the vector potential was obtained originally as a solution to Maxwell's equations, which generalize a large body of experimental observations on electric and magnetic effects, the distinction between electric and magnetic interactions no longer seems as clear as it did in classical physics.

9.2 The Origin of Circular Dichroism

The difference between the dipole strengths for left- and right-circularly polarized light is characterized experimentally by the *rotational strength* $\mathfrak{R}$ of an absorption band:

$$\mathfrak{R} = \frac{3000\ln(10)hc}{32\pi^3 N_A}\left(\frac{n}{f^2}\right)\int \frac{\Delta\varepsilon(\nu)}{\nu}d\nu \tag{9.9a}$$

$$\approx 0.248\left(\frac{n}{f^2}\right)\int \frac{\Delta\varepsilon(\nu)}{\nu}d\nu \;(debye\,Bohr\,magneton)/\left(M^{-1}cm^{-1}\right), \tag{9.9b}$$

where N_A is Avogadro's number, $\Delta\varepsilon$ is the difference between the molar extinction coefficients for left- and right-circularly polarized light ($\varepsilon_l - \varepsilon_r$) in units of M^{-1} cm^{-1}, ν is the frequency, n is the refractive index and f is the local-field correction. Rotational strengths commonly are expressed in units of debye·Bohr magnetons (9.274×10^{-39} $esu^2 \cdot cm^2$ or 9.274×10^{-39} $erg \cdot cm^3$). Unlike the dipole strength, $\mathfrak{R}$ can be either positive or negative.

For historical reasons, circular dichroism often is described in terms of the *molar ellipticity*, $[\theta]_M$, in the arcane units of $100 \times$ degree·M^{-1} cm^{-1} (degree·decimol^{-1}·cm^2). The angular units reflect the fact that plane-polarized light passing through an optically active sample emerges with elliptical polarization (Box 9.2 and Fig. 9.5). Ellipticity is defined as the arc tangent of the ratio d_{min}/d_{max}, where d_{min} and d_{max} are the short and long axes of the ellipse. The relationship between $[\theta]_M$ and $\varepsilon_l - \varepsilon_r$ is

$$[\theta]_M = \frac{100\ln(10)180°\Delta\varepsilon}{4\pi} = 3298\Delta\varepsilon. \tag{9.10}$$

Early CD spectrometers actually measured ellipticity, but most modern instruments measure $\Delta\varepsilon$ more directly and sensitively by switching a light beam rapidly between right- and left-circular polarization (Chap. 1). In studies of proteins or polynucleotides, the molar ellipticity usually is divided by the number of amino acid residues or nucleotide bases in the macromolecule to obtain the *mean residue ellipticity*.

Circular dichroism (CD) is a very small effect, typically amounting to a difference of only about 1 part in 10^3 or 10^4 between the extinction coefficients for light with left- or right-circular polarization. But in spite of its small magnitude, CD proves to be a very sensitive probe of molecular structure.

Box 9.2 Ellipticity and Optical Rotation

As we discussed in Chap. 3, the electric and magnetic fields linearly polarized light can be viewed as superpositions of fields from right- and left-circularly polarized light (Figs. 3.8, 3.9 and 9.5A). Consider a beam of linearly polarized light after it has passed through 1 cm of a 1 M solution of an optically active material. If the molar absorption coefficient for left-circular polarization exceeds that for right-circular polarization by $\Delta\varepsilon$, the ratio of the intensities of the transmitted light with left- and right-circular polarization

(continued)

Box 9.2 (continued)
will be $I_l/I_r = \exp[-\ln(10)\Delta\varepsilon]$. Since the electric field amplitude is proportional to the square root of the intensity, the ratio of the fields will be

$$|\boldsymbol{E}_l|/|\boldsymbol{E}_r| = \exp[-\ln(10)\Delta\varepsilon/2]. \tag{B9.2.1}$$

The resultant field will oscillate in amplitude from $|\boldsymbol{E}_{max}| = |\boldsymbol{E}_r| + |\boldsymbol{E}_l|$ when the two fields are aligned in parallel (the times labeled 0 and 4 in Fig. 9.5B) to $|\boldsymbol{E}_{min}| = |\boldsymbol{E}_r| - |\boldsymbol{E}_l|$ when they are antiparallel (times 2 and 6). The minor and major half-axes of the elipse swept out by the resultant field (Fig. 9.5B) are $|\boldsymbol{E}_{min}|$ and $|\boldsymbol{E}_{max}|$, which have the ratio

$$\frac{|\boldsymbol{E}_{min}|}{|\boldsymbol{E}_{max}|} = \frac{1 - \exp[-\ln(10)\Delta\varepsilon/2]}{1 + \exp[-\ln(10)\Delta\varepsilon/2]} \approx \frac{\ln(10)\Delta\varepsilon/2}{2} = \ln(10)\Delta\varepsilon/4 \tag{B9.2.2}$$

when $\Delta\varepsilon << 1$. Eqation (9.10) is obtained by making the approximation $\arctan(\phi) \approx \phi$, which holds when $\phi << 1$, multiplying by $180°/\pi$ to convert from radians to degrees, and multiplying by 100 to give the conventional units.

Optically active materials also have different indices of refraction for right- and left-circularly polarized light. As a result, the ellipse shown in Fig. 9.5B, C will be rotated slightly relative to the orientation of the original linearly polarized beam. This is the phenomenon of *optical rotation*, and its dependence on wavelength is *optical rotatory dispersion* (ORD). CD and ORD spectra are related by general expressions called the Kramers-Kronig transforms [9, 10]. The contribution of a particular absorption band to the ORD of a sample has opposite signs on either side of the band and extends to wavelengths far from the band. We will not discuss ORD in further detail because it provides little additional information or experimental advantage and because overlapping contributions from distant absorption bands complicate its interpretation.

L. Rosenfeld [11] showed that the rotational strength of a transition depends on the dot product of the magnetic transtion dipole $\boldsymbol{m}_{ba}$ and the electric transition dipole $\boldsymbol{\mu}_{ba}$:

$$\mathfrak{R}_{ba} = -\mathrm{Im}(\langle\Psi_b|\tilde{\mathrm{m}}|\Psi_a\rangle \cdot \langle\Psi_b|\tilde{\mu}|\Psi_a\rangle) = -\mathrm{Im}(\boldsymbol{m}_{ba} \cdot \boldsymbol{\mu}_{ba}), \tag{9.11}$$

where, as before, Im(. . .) means the imaginary part of the quantity in parentheses. Because $\boldsymbol{m}_{ba}\cdot\boldsymbol{\mu}_{ba}$ is imaginary, $\mathfrak{R}_{ba}$ is real.

Before we consider the derivation of Eq. (9.11), let's look at what the equation says: the rotational strength depends on the extent to which the magnetic and electric transition dipoles cooperate or oppose each other. More precisely, $\mathfrak{R}_{ba}$ can be nonzero only if $\boldsymbol{m}_{ba}$ and $\boldsymbol{\mu}_{ba}$ have components that are parallel or antiparallel.

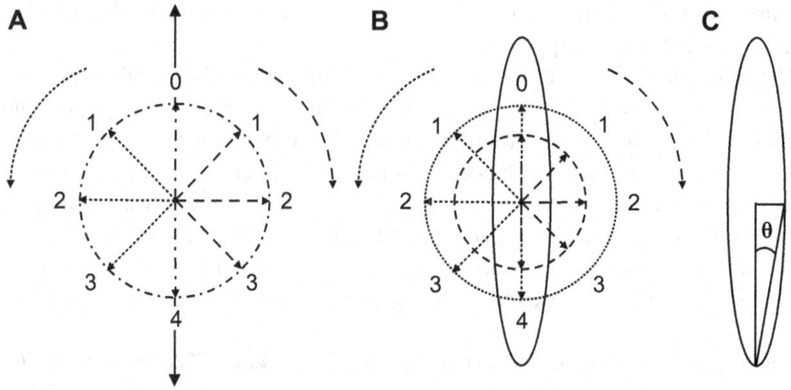

Fig. 9.5 A linearly-polarized beam of light passing through an optically active material becomes elliptically polarized. As in Figs. 3.8 and 3.9, the *straight arrows* in these diagrams represent the electric fields of light propagating away from the observer; the numbers indicate equal increasing intervals of time. (**A**) The electric field of linearly polarized light (*double-headed solid arrow*) can be viewed as the resultant of equal fields from left- (*dotted arrows*) and right- (*dashed arrows*) circularly polarized light. The *circles* indicate the bounds of the rotating fields and the *curved arrows* indicate the directions of rotation. (**B**) After passage through an optically active material, light with one of the two circular polarizations (here, the right-circular polarization) is attenuated relative to that with the other. The resultant of the two fields now sweeps out an *ellipse*. (**C**) The ellipticity (θ) is defined as the arctangent of the ratio of the minor to to the major half-axes of the *ellipse*. This figure greatly exaggerates the major axis of the *ellipse* relative to the minor. It neglects rotation of the *ellipse* (optical rotation) caused by the difference between the refractive indices for left- and right-circularly polarized light

This immediately implies that planar molecules cannot show CD. If a molecule is planar, $\boldsymbol{\mu}_{ba}$ must lie in this plane but $\boldsymbol{m}_{ba}$ will be normal to the plane (Figs. 9.1 and 9.2). The electric and magnetic transition dipoles hence will be perpendicular to each other. In a helical molecule, on the other hand, $\boldsymbol{m}_{ba}$ and $\boldsymbol{\mu}_{ba}$ both could have components along the axis of the helix. Such a molecule can have a nonzero rotational strength, and we might expect that the sign of $\mathfrak{R}_{ba}$ will depend on whether the helix has a left- or right-handed twist.

More generally, Eq. (9.11) implies that an isotropic solution of an absorbing molecule can exhibit CD only if the molecule is chiral, which means that it is distinguishable from its mirror image (enantiomer). If a molecule is transformed into its mirror image, $\boldsymbol{\mu}_{ba}$ changes sign but $\boldsymbol{m}_{ba}$ does not, so $\mathfrak{R}_{ba}$ becomes $-\mathfrak{R}_{ba}$. (Remember that $\boldsymbol{m}_{ba}$ depends on a sum of cross products of the position vector ($\boldsymbol{r}$) with matrix elements of the gradient operator for each pair of atoms $(\boldsymbol{\nabla}_{st})$; converting a molecule to its mirror image changes the sign of both $\boldsymbol{r}$ and $\boldsymbol{\nabla}_{st}$, leaving the sign of the cross product the same.) If the original molecule and the mirror image are indistinguishable, $\mathfrak{R}_{ba}$ must be equal to $-\mathfrak{R}_{ba}$, which can be true only if $\mathfrak{R}_{ba}$ is zero. In particular, chirality requires that the molecule not have a plane of symmetry or a center of inversion, because a molecule with either of these symmetry elements can be superimposed on its mirror image. Modern quantum chemical methods have made it possible to use comparisons of calculated and measured CD to determine the absolute stereochemistry of moderately sized organic molecules [12–16].

The principle that only chiral molecules can be optically active does not necessarily hold in anisotropic systems. Molecules that would be indistinguishable in solution can be distinguished by their fixed orientations in some crystal forms. A crystal of an achiral material thus can display optical activity when is illuminated along a particular crystallographic axis. Such optical activity was first seen in crystalline $AgGaS_2$ and $CdGa_2S_4$ [17, 18], and has been demonstrated in other achiral materials including pentaerythritol ($C(CH_2OH)_4$) [19]. However, it is difficult to measure in the presence of the strong linear dichroism that is typical of crystals.

Equation (9.11) also informs us that a molecule can have CD only if $\langle \Psi_b | \widetilde{\mu} | \Psi_a \rangle$ is nonzero. Further, the CD spectrum of a single molecule will have approximately the same shape as the absorption spectrum, though a much smaller absolute amplitude. CD spectra differ in this regard from optical rotatory dispersion (ORD) spectra, which reflect differences between the indices of refraction for right- and left-circularly polarized light (Box 9.2).

As in previous chapters, the wavefunction for the initial state is written on the right in the matrix elements in Eq. (9.11), and the wavefunction for the final state on the left. For the electric dipole operator the order is immaterial because $\langle \Psi_b | \widetilde{\mu} | \Psi_a \rangle = \langle \Psi_a | \widetilde{\mu} | \Psi_b \rangle$, but this is not the case for the magnetic dipole operator. Here, interchanging the orbitals changes the sign of the integral. Thus, because $\langle \Psi_b | \widetilde{m} | \Psi_a \rangle = -\langle \Psi_a | \widetilde{m} | \Psi_b \rangle$, the Rosenfeld equation (Eq. 9.11) also can be written:

$$\mathfrak{R}_{ba} = \mathrm{Im}(\langle \Psi_a | \widetilde{m} | \Psi_b \rangle \cdot \langle \Psi_b | \widetilde{\mu} | \Psi_a \rangle) = \mathrm{Im}(\boldsymbol{m}_{ab} \cdot \boldsymbol{\mu}_{ba}). \tag{9.12}$$

To derive the Rosenfeld equation, we must consider the linked time dependence of the electric and magnetic fields of circularly polarized light. For light propagating along the z axis, the electric field can be written:

$$\boldsymbol{E}_{\pm}(t) = 2I_c[\cos(2\pi\nu t)\hat{x} \pm \sin(2\pi\nu t)\hat{y}] \tag{9.13a}$$

$$= I_c\{[\exp(2\pi i\nu t) + \exp(-2\pi i\nu t)]\hat{x} \mp i[\exp(2\pi i\nu t) - \exp(-2\pi i\nu t)]\hat{y}\}, \tag{9.13b}$$

where I_c is a scalar magnitude that depends on the light intensity and the local-field correction, and $\hat{x}$ and $\hat{y}$ are unit vectors in the x and y directions. $\boldsymbol{E}_+$ describes left-circular polarization; $\boldsymbol{E}_-$, right-circular polarization (Fig. 3.9). Similarly,

$$\boldsymbol{B}_{\pm}(t) = 2B_c[-\cos(2\pi\nu t)\hat{y} \pm \sin(2\pi\nu t)\hat{x}] \tag{9.14a}$$

$$= B_c\{-[exp(2\pi i\nu t) + exp(-2\pi i\nu t)]\hat{y} \mp i[exp(2\pi i\nu t) - exp(-2\pi i\nu t)]\hat{x}\}, \tag{9.14b}$$

where B_c is a scalar magnitude of the magnetic field. The electric and magnetic fields are perpendicular to each other, and rotate together around the z axis at frequency ν.

Now consider an isotropic solution of a molecule with electronic states a and b. If the molecule is in state a at zero time ($C_a(0) = 1$ and $C_b(0) = 0$), illumination with circularly polarized light will cause C_b to grow with time according to Eq. (2.61), with $\widetilde{\mathrm{H}}'$ given by Eq. (9.1) and $\boldsymbol{E}(t)$ and $\boldsymbol{B}(t)$ given by Eqs. (9.13a,b) and (9.14a,b). By proceding as we did to derive Eq. (4.7), we can obtain an expression for $C_b(\tau)$ that has separate terms for $E_b > E_a$ (absorption) and $E_b < E_a$ (stimulated emission). The result for absorption in time τ is as follows:

$$C_b^{\pm}(\tau) = \left(\frac{\exp[i(E_b - E_a - h\nu)\tau/\hbar]}{E_b - E_a - h\nu}\right)\left(I_c\mu_{ba}^x\hat{x} \pm iI_c\mu_{ba}^y\hat{y} - B_c m_{ba}^y\hat{y} \pm iB_c m_{ba}^x\hat{x}\right). \tag{9.15}$$

Here μ_{ba}^x and μ_{ba}^y are the x and y components of $\boldsymbol{\mu}_{ba}$, and m_{ba}^x and m_{ba}^y are the x and y components of $\boldsymbol{m}_{ba}$. We now need to form the complex conjugate of $C_b^{\pm}(\tau)$, remembering that the complex conjugate of $\boldsymbol{m}_{ba}$ is $-\boldsymbol{m}_{ba}$, and then integrate the product $C_b^{\pm}{}^*(\tau)C_b^{\pm}(\tau)$ over a range of frequencies as in Eqs. (4.8a)–(4.8c) (Box 4.4). This gives

$$\begin{aligned}
&\int_0^{\infty} C_b^{\pm *}(\tau)C_b^{\pm}(\tau)\rho_\nu(\nu)d\nu \\
&= \left[\rho_\nu(\nu)\tau/\hbar^2\right]\left[I_c\mu_{ba}^x\hat{x} \mp iI_c\mu_{ba}^y\hat{y} + B_c m_{ba}^y\hat{y} \pm iB_c\mu_{ba}^x\hat{x}\right]\left[I_c\mu_{ba}^x\hat{x} \pm iI_c\mu_{ba}^y\hat{y} - B_c m_{ba}^y\hat{y} \pm iB_c\mu_{ba}^x\hat{x}\right] \\
&= \left[\rho_\nu(\nu)\tau/\hbar^2\right]\left\{I_c^2\left(\left|\mu_{ba}^x\right|^2 + \left|\mu_{ba}^y\right|^2\right) + B_c^2\left(\left|m_{ba}^x\right|^2 + \left|m_{ba}^y\right|^2\right) \mp 2I_cB_c\mathrm{Im}\left[\mu_{ba}^x m_{ba}^x + \mu_{ba}^y m_{ba}^y\right]\right\},
\end{aligned} \tag{9.16}$$

which, upon averaging over all orientations of the molecular axes, becomes:

$$\overline{\int_0^{\infty} C_b^{\pm}{}^*(\tau)C_b^{\pm}(\tau)\rho_\nu(\nu)d\nu} = \frac{2\rho_\nu(\nu)\tau}{3\hbar^2}\left[I_c^2|\boldsymbol{\mu}_{ba}|^2 + B_c{}^2|\boldsymbol{m}_{ba}|^2 \mp 2I_cB_c\mathrm{Im}(\boldsymbol{\mu}_{ba}\cdot\boldsymbol{m}_{ba})\right]. \tag{9.17}$$

The terms containing $|\boldsymbol{\mu}_{ba}|^2$ and $|\boldsymbol{m}_{ba}|^2$ in Eq. (9.17) represent the electric and magnetic dipole strengths. These terms have positive signs whether the light is left- or right-circularly polarized. The term containing the dot product $\boldsymbol{\mu}_{ba}\cdot\boldsymbol{m}_{ba}$, however, enters Eq. (9.17) with a positive sign for right-circular polarization but with a negative sign for left-circular polarization. The difference between the rates of absorption of left- and right-circularly polarized light is therefore proportional to $-\left[8\rho_\nu(\nu)\tau/3\hbar^2\right)]I_cB_c\mathrm{Im}(\boldsymbol{\mu}_{ba}\cdot\boldsymbol{m}_{ba})$. Evaluating the factors $\rho_\nu(\nu)$, I_c and B_c just as we did to obtain Eqs. (4.16a) and (4.16b) gives Eqs. (9.9a), (9.9b) and the Rosenfeld equation (Eq. 9.11).

9.3 Circular Dichroism of Dimers and Higher Oligomers

Dimers and larger oligomers often exhibit relatively strong CD even when the individual molecules do not [13, 20–25]. To see how this CD arises, let's describe the excited state of an oligomer by a linear combination of the excited states of the individual molecules as in Eqs. (8.17) and (8.19):

$$\Psi_B = \sum_n \sum_b C^B_{b(n)}\,{}^1\psi_{ba(n)}. \tag{9.18}$$

We then can write the oligomer's electric and magnetic transition dipoles as

$$\boldsymbol{\mu}_{BA} = \sum_n \sum_b C^B_{b(n)} \boldsymbol{\mu}_{ba(n)} \tag{9.19}$$

and

$$\boldsymbol{m}_{BA} = \sum_n \sum_b C^B_{b(n)} \boldsymbol{m}_{ba(n)}, \tag{9.20}$$

where $\boldsymbol{\mu}_{ba(n)}$ and $\boldsymbol{m}_{ba(n)}$ are the electric and magnetic transition dipoles for exciting subunit n of the oligomer from its ground state (a) to excited state b. The Rosenfeld equation (Eq. 9.11) then gives the rotational strengths of the oligomer's exciton bands:

$$\mathfrak{R}_{BA} = -\mathrm{Im}(\boldsymbol{m}_{BA} \cdot \boldsymbol{\mu}_{BA}). \tag{9.21}$$

Following the procedure we used above in Eqs. (9.5a)–(9.7), we can evaluate the magnetic transition dipoles of the subunits by breaking the position vector for electron i in subunit n ($\boldsymbol{r}_{i(n)}$) into two parts:

$$\boldsymbol{r}_{i(n)} = \boldsymbol{R}_n + \boldsymbol{r}'_i. \tag{9.22}$$

Here $\boldsymbol{R}_n$ is a vector from the origin of the coordinate system to the center of subunit n, and $\boldsymbol{r}'_i$ is a vector from the center of this subunit to the position of the electron (Fig. 9.6). With this decomposition of $\boldsymbol{r}_{i(n)}$, $\boldsymbol{m}_{ba(n)}$ becomes the sum of two terms:

$$\boldsymbol{m}_{ba(n)} = (e/2mc)\left\langle \psi_{b(n)} \middle| \boldsymbol{r}_{i(n)} \times \tilde{\mathrm{p}} \middle| \psi_{a(n)} \right\rangle \tag{9.23a}$$

$$= (e/2mc)\left\langle \psi_{b(n)} \middle| \boldsymbol{R}_n \times \tilde{\mathrm{p}} \middle| \psi_{a(n)} \right\rangle + (e/2mc)\left\langle \psi_{b(n)} \middle| \boldsymbol{r}'_i \times \tilde{\mathrm{p}} \middle| \psi_{a(n)} \right\rangle \tag{9.23b}$$

$$= (e/2mc)\boldsymbol{R}_n \times \left\langle \psi_{b(n)} \middle| \tilde{\mathrm{p}} \middle| \psi_{a(n)} \right\rangle + \boldsymbol{m}'_{ba(n)}. \tag{9.23c}$$

From Eqs. (9.20) and (9.23c), the oligomer's magnetic transtion dipole is

Fig. 9.6 The position vector $\boldsymbol{r}_{i(n)}$ for electron i in subunit n of an oligomer is the sum of a vector from the origin of the coordinate system to the center of the subunit ($\boldsymbol{R}_n$) and a vector from the subunit's center to the location of electron i ($\boldsymbol{r}_i'$)

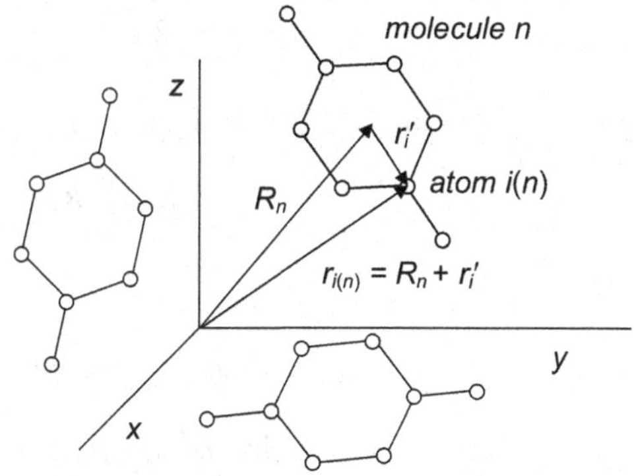

$$\boldsymbol{m}_{BA} = \sum_n \sum_b C^B_{b(n)} \left[\boldsymbol{m}'_{ba(n)} + (e/2mc)\boldsymbol{R}_n \times \left\langle \psi_{b(n)} | \tilde{\mathrm{p}} | \psi_{a(n)} \right\rangle \right]. \tag{9.24}$$

The term $\boldsymbol{m}'_{ba(n)}$ in this expression is an intrinsic property of subunit n and is independent of where this molecule is located in the oligomer. The term $(e/2mc)$ $\boldsymbol{R}_n \times \left\langle \psi_{b(n)} | \tilde{\mathrm{p}} | \psi_{a(n)} \right\rangle$ depends on the position and orientation of monomer n in the oligomer.

The integral $\left\langle \psi_{b(n)} | \tilde{\mathrm{p}} | \psi_{a(n)} \right\rangle$ in Eq. (9.24) can be evaluated directly from the molecular orbitals $\psi_{b(n)}$ and $\psi_{a(n)}$ if these are written as linear combinations of atomic orbitals (Box 4.11; Eqs. 9.5a–9.7). Alternatively, it can be obtained from the electric transition dipole by using the expression

$$\left\langle \psi_{b(n)} | \tilde{\mathrm{p}} | \psi_{a(n)} \right\rangle = \left(-2\pi imc/e\lambda_{ba(n)}\right) \left\langle \psi_{b(n)} | \tilde{\mu} | \psi_{a(n)} \right\rangle, \tag{9.25}$$

where $\lambda_{ba(n)}$ is the wavelength of the monomer's absorption band. This expression is useful if the orbitals are not known accurately, because the magnitude of the electric transition dipole can be obtained experimentally by measuring the dipole strength. Equation (9.25) follows directly from the relationship

$$\left\langle \psi_b \middle| \tilde{\nabla} \middle| \psi_a \right\rangle = -\left[(E_b - E_a)m/\hbar^2 e\right] \langle \psi_b | \tilde{\mu} | \psi_a \rangle, \tag{9.26}$$

which we discussed in Chap. 4 (Eq. 4.27; Box 4.10).

Inserting Eq. (9.25) in Eq. (9.24) gives:

$$\boldsymbol{m}_{BA} = \sum_n \sum_b C^B_{b(n)} \left[\boldsymbol{m}'_{ba(n)} - \left(i\pi/\lambda_{ba(n)}\right) \boldsymbol{R}_n \times \boldsymbol{\mu}_{ba(n)} \right]. \tag{9.27}$$

For a dimer of identical molecules with one excited state, Eqs. (9.18), (9.19), and (9.27) yield the following results for the two exciton bands:

$$\boldsymbol{\mu}_{BA\pm} = 2^{-1/2}\left(\boldsymbol{\mu}_{ba(1)} \pm \boldsymbol{\mu}_{ba(2)}\right), \tag{9.28}$$

$$\begin{aligned} \boldsymbol{m}_{BA\pm} &= 2^{-1/2}\left(\boldsymbol{m}'_{ba(1)} \pm \boldsymbol{m}'_{ba(2)}\right) \\ &\quad - \left(i\pi/\sqrt{2}\lambda_{ba}\right)\left[\boldsymbol{R}_1 \times \boldsymbol{\mu}_{ba(1)} \pm \boldsymbol{R}_2 \times \boldsymbol{\mu}_{ba(2)}\right], \end{aligned} \tag{9.29}$$

and

$$\begin{aligned} \mathfrak{R}_{BA\pm} &= -\mathrm{Im}(\boldsymbol{m}_{BA\pm} \cdot \boldsymbol{\mu}_{BA\pm}) \\ &= -\frac{1}{2}\mathrm{Im}\left(\boldsymbol{m}'_{ba(1)} \cdot \boldsymbol{\mu}_{ba(1)} + \boldsymbol{m}'_{ba(2)} \cdot \boldsymbol{\mu}_{ba(2)}\right) \\ &\quad \mp \frac{1}{2}\mathrm{Im}\left(\boldsymbol{m}'_{ba(1)} \cdot \boldsymbol{\mu}_{ba(2)} + \boldsymbol{m}'_{ba(2)} \cdot \boldsymbol{\mu}_{ba(1)}\right) \\ &\quad \pm (\pi/2\lambda_{ba})(\boldsymbol{R}_2 - \boldsymbol{R}_1) \cdot \left(\boldsymbol{\mu}_{ba(2)} \times \boldsymbol{\mu}_{ba(1)}\right). \end{aligned} \tag{9.30}$$

Here the $\pm$ signs refer to symmetric and antisymmetric combinations of the excited states of the two monomers, as in Eqs. (8.11) and (8.16): the positive signs are for Ψ_{B+}, and the negative signs for Ψ_{B-}. In deriving Eq. (9.30) we have used the facts that interchanging the order of any two vectors in a triple product $\boldsymbol{c}\cdot\boldsymbol{a} \times \boldsymbol{b}$ changes the sign of the product, and that the product is zero if any two of the vectors are parallel (Appendix A.1).

Equation (9.30) reveals that there are three contributions to a dimer's rotational strength, $\mathfrak{R}_{mon}$, $\mathfrak{R}_{e-m}$ and $\mathfrak{R}_{ex}$:

$$\mathfrak{R}_{mon} = -\frac{1}{2}\mathrm{Im}\left(\boldsymbol{m}'_{ba(1)} \cdot \boldsymbol{\mu}_{ba(1)} + \boldsymbol{m}'_{ba(2)} \cdot \boldsymbol{\mu}_{ba(2)}\right), \tag{9.31a}$$

$$\mathfrak{R}_{e-m} = \mp\frac{1}{2}\mathrm{Im}\left(\boldsymbol{m}'_{ba(1)} \cdot \boldsymbol{\mu}_{ba(2)} + \boldsymbol{m}'_{ba(2)} \cdot \boldsymbol{\mu}_{ba(1)}\right), \tag{9.31b}$$

$$\mathfrak{R}_{ex} = \pm(\pi/2\lambda_{ba})(\boldsymbol{R}_2 - \boldsymbol{R}_1) \cdot \left(\boldsymbol{\mu}_{ba(2)} \times \boldsymbol{\mu}_{ba(1)}\right). \tag{9.31c}$$

$\mathfrak{R}_{mon}$ is simply the sum of the intrinsic rotational strengths of the individual molecules, with the contribution of each subunit weighted by the square of the corresponding coefficient in $\Psi_{B\pm}$. This term, sometimes called the "*one-electron*" contribution, is independent of how the two molecules are arranged with respect to each other in the dimer, assuming that the intermolecular interactions do not affect the magnetic and electronic transition dipoles of the individual molecules. However, it could reflect perturbations of $\boldsymbol{\mu}_{ba}$ or $\boldsymbol{m}'_{ba}$ by the electrostatic environment in the dimer.

The second term, $\mathfrak{R}_{e-m}$, reflects coupling between the electric transition dipole of one molecule and the magnetic transition dipole of the other molecule. This is called *electric-magnetic coupling*.

Finally, $\mathfrak{R}_{ex}$, the *exciton* or *coupled-oscillator* term, depends on the two electric transition dipoles and on the geometry of the dimer. We can write this term more succinctly as

$$\mathfrak{R}_{ex} = \pm(\pi/2\lambda_{ba})\boldsymbol{R}_{21} \cdot \boldsymbol{\mu}_{ba(2)} \times \boldsymbol{\mu}_{ba(1)} \tag{9.32a}$$

$$= \pm(\pi/2\lambda_{ba})|\boldsymbol{R}_{21}|D_{ba} \sin\theta \cos\phi, \tag{9.32b}$$

where $\boldsymbol{R}_{21}$ is the vector from the center of molecule 1 to the center of molecule 2, D_{ba} is the dipole strength of the monomeric absorption band, θ is the angle between $\boldsymbol{\mu}_{ba(1)}$ and $\boldsymbol{\mu}_{ba(2)}$, and ϕ is the angle between $\boldsymbol{R}_{21}$ and the cross product of $\boldsymbol{\mu}_{ba(1)}$ and $\boldsymbol{\mu}_{ba(2)}$. If $|\boldsymbol{R}_{21}|$ and λ_{ba} are given in the same units (*e.g.*, Å) and the transition dipoles are in debyes,

$$\mathfrak{R}_{ex} \approx \pm(171/\lambda_{ba})\boldsymbol{R}_{21} \cdot \boldsymbol{\mu}_{ba(2)} \times \boldsymbol{\mu}_{ba(1)}(D\,\mu_B)D^{-2} \tag{9.33}$$

For electronically allowed transitions, $\mathfrak{R}_{ex}$ often dominates over $\mathfrak{R}_{mon}$ and $\mathfrak{R}_{e-m}$ because the dipole strength D_{ba} usually is much larger than either $\text{Im}[\boldsymbol{m}'_{ba(1)}\cdot\boldsymbol{\mu}_{ba(1)} + \boldsymbol{m}'_{ba(2)}\cdot\boldsymbol{\mu}_{ba(2)}]$ or $\text{Im}[\boldsymbol{m}'_{ba(1)}\cdot\boldsymbol{\mu}_{ba(2)} + \boldsymbol{m}'_{ba(2)}\cdot\boldsymbol{\mu}_{ba(1)}]$. $\mathfrak{R}_{mon}$ will be zero if the individual molecules are not optically active, whereas $\mathfrak{R}_{ex}$ does not require that either molecule exhibit optical activity. But as Eq. (9.32b) indicates, $\mathfrak{R}_{ex}$ is extremely sensitive to the geometry of the dimer. It will be zero if any two of the three vectors ($\boldsymbol{\mu}_{ba(1)}$, $\boldsymbol{\mu}_{ba(2)}$, and $\boldsymbol{R}_{21}$) are parallel or if all three vectors lie in the same plane.

Equations (9.30), (9.32a), and (9.32b) show that the contributions of $\mathfrak{R}_{ex}$ to the dimer's two exciton bands are equal in magnitude but opposite in sign (Fig. 9.7B). This is in contrast to the dipole strengths of the two exciton bands, which are always positive but can differ substantially in magnitude (Eqs. 8.16c and 8.16d; Figs. 8.4 and 9.7A).

$\mathfrak{R}_{e-m}$, like $\mathfrak{R}_{ex}$, contributes rotational strengths with equal magnitudes and opposite signs to the two exciton bands, whereas the contributions from $\mathfrak{R}_{mon}$ have the same sign. A CD spectrum in which the positive and negative rotational strengths of corresponding bands balance is called *conservative*. A *nonconservative* CD spectrum can reflect either significant contributions from $\mathfrak{R}_{mon}$ or mixing with other excited states at higher or lower energies. If the total rotational strength is integrated from $\nu = 0$ to ∞, so that excitations from the ground state to the complete set of excited states is included, the integral must be zero [26]. We saw an analogous sum rule for absorption spectra in Chap. 8: the sum of the dipole strengths for excitation of one electron from the ground state to all possible excited states of a dimer is the same as the sum of the dipole strengths for the two monomers.

According to Eqs. (9.32a), (9.32b), and (9.33), $|\mathfrak{R}_{ex}|$ increases linearly and indefinitely with $|\boldsymbol{R}_{21}|$. This seems counter to intuition. How can the rotational strength arising from the interactions of two molecules be large if the molecules are far apart? The observed strengths of the CD bands, however, do not behave this way. They go to zero at large values of $|\boldsymbol{R}_{21}|$ because the two bands have opposite

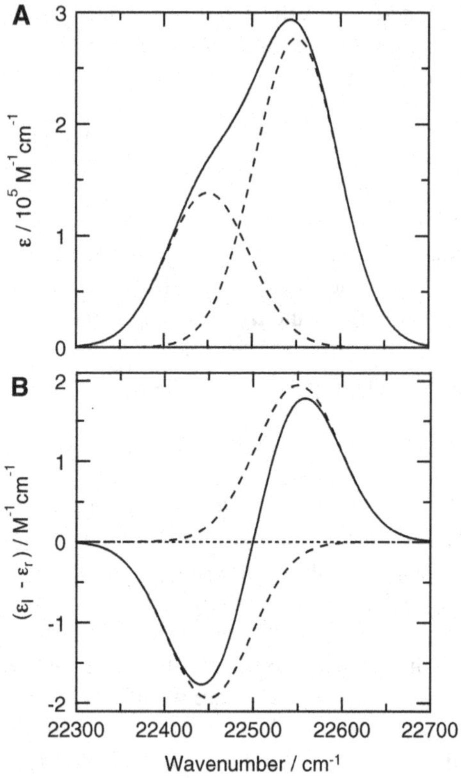

Fig. 9.7 $\mathfrak{R}_{ex}$ contributes opposite rotational strengths to the exciton bands of a dimer. The *dashed lines* in (**A**) show the exciton absorption bands of a dimer; the *solid curve* is the total absorption spectrum. In (**B**), the *dashed lines* are the circular dichroism (CD) of the two bands and the *solid curve* is the total CD spectrum. The spectra are for a homodimer with $D_{ba(1)} = D_{ba(2)} = 10\ \text{D}^2$, $|\boldsymbol{R}_{21}| = 7\ \text{Å}$, $\theta = 71°$, $\alpha = \beta = 90°$ (Fig. 7.2) and $\lambda_{ba} = 4{,}444\ \text{Å}$. This geometry makes H_{21} positive ($H_{21} = 50\ \text{cm}^{-1}$ in the point-dipole approximation) and gives the Ψ_{B+} exciton band the higher transition energy, the larger dipole strength, and a positive rotational strength. For purposes of illustration, the exciton bands were assigned Gaussian shapes with arbitrary widths

signs and the separation between the bands ($2H_{21}$) decreases as $|\boldsymbol{R}_{21}|^{-3}$ (Eq. 7.17; Fig. 7.5). When the molecules are far apart, the opposite rotational strengths of the two overlapping bands cancel. Figure 9.8 illustrates this point. In the example shown, the net CD increases as $|\boldsymbol{R}_{21}|$ is raised from 7 to 10 Å (Fig. 9.8A), but then decreases at larger distances (Fig. 9.8B).

Examination of Fig. 9.8B shows that when the two exciton bands overlap the positions of the positive and negative CD peaks are relatively insensitive to $|\boldsymbol{R}_{21}|$. This points out a general problem in interpreting experimental CD spectra: complexes with different structures can give similar spectra. CD spectra also can be complicated by mixing with higher excited states such as states in which more than one of the monomeric units are excited. For these reasons and because of the uncertainties in calculations of exciton-interaction matrix elements, attempts to deduce the structure of an oligomer primarily on the basis of a CD spectrum must be viewed critically. However, measurements of both the CD and absorption spectra in the region of several absorption bands may provide enough information to rule out some of the possible structures with good confidence.

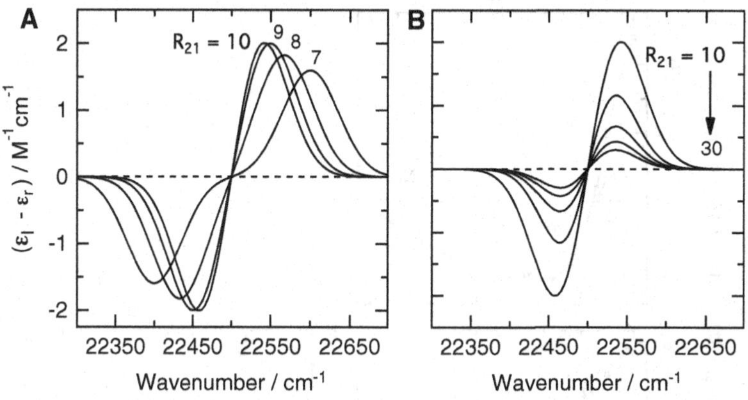

Fig. 9.8 Calculated exciton CD spectra of homodimers with $|\boldsymbol{R}_{21}| = 7$, 8, 9 or 10 Å (**A**), and 10, 15, 20, 25 or 30 Å (**B**). Except for the variable separation $|\boldsymbol{R}_{21}|$, the dimer geometries are as in Fig. 9.7; the dipole strengths are 20 D^2. The bands are given Gaussian shapes with arbitrary widths. For $|\boldsymbol{R}_{21}| = 7$ Å, 10 Å and 30 Å, H_{21} (in the point-dipole approximation) is 100 cm^{-1}, 34 cm^{-1} and 1.3 cm^{-1}, respectively, while $\mathfrak{R}_{ex} = 5.0$, 7.1 and 21.4 Dμ_B

9.4 Circular Dichroism of Proteins and Nucleic Acids

The CD of proteins between 180 and 230 nm arises primarily from coupled transitions of multiple peptide groups. Because the coupling depends strongly on the relative positions and orientations of these groups, CD provides a sensitive probe of protein secondary structure [13, 25, 27–34]. Common applications include measurements of protein folding or unfolding and of conformational changes caused by ligand binding [30–32, 35, 36].

Circular dichroism spectra representative of α-helical, β-sheet and unordered ("random coil") conformations have been obtained by studying polypeptides that adopt different structures depending on the pH, ionic strength and temperature, and also by decomposing CD spectra of sets of proteins with known crystal structures [27, 28]. Figure 9.9B illustrates the first of these approaches; Fig. 9.9C, the second. Right-handed α-helical elements in proteins typically have a CD band with positive rotational strength at 190 nm and bands with negative rotational strengths near 205 and 222 nm, whereas β-sheets have a positive band near 195 and a single negative band at 215 nm. Random coils have a very different spectrum with negative CD to the blue of 210 nm and positive CD to the red. Poly(proline)-I structure (a right-handed helix with *cis* peptide bonds that is stable only in relatively nonpolar solvents) gives a positive CD peak at 215 nm and a weaker negative peak near 200 nm; poly(proline)-II (a left-handed helix with *trans* peptide bonds that is the more stable structure in water) gives a broader negative band near 205 nm [37–39].

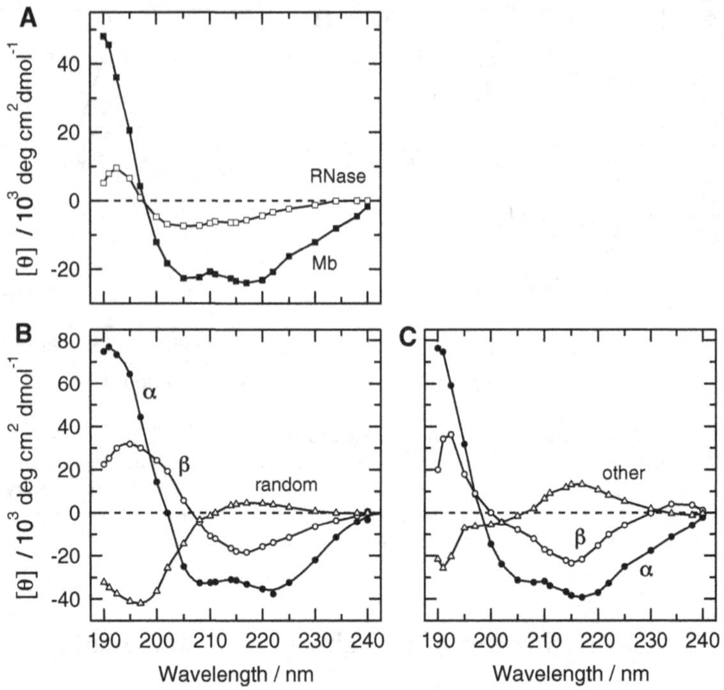

Fig. 9.9 (**A**) CD spectra of myoglobin (*filled symbols*) and ribonuclease A (*open symbols*). In crystal structures, myoglobin is comprised of approximately 71% α-helix, no β-sheet, and 29% other secondary structure such as turns. Ribonuclease A has approximately 12% α-helix, 36% β-sheet, and 52% "other." (**B**) CD spectra of polylysine under conditions that cause the polypeptide to adopt α-helix (*filled circles*), β-sheet (*open circles*), or random coil (*triangles*) structure. The α-helix was obtained at pH 11.1, 22 °C; β-sheet, at the same pH but after heating to 52 °C and recooling to 22 °C; and random coil, at pH 5.7. (At low pH, electrostatic interactions of the positively charged Lys sidechains prevent the polypeptide from packing into a compact structure.) (**C**) CD basis spectra for α-helix (*filled circles*), β-sheet (*open circles*) and other (*triangles*) secondary structure, obtained by fitting the observed spectra of myglobin, ribonuclease A and lysozyme. (Lysozyme contains intermediate amounts of α-helix and β-sheet compared to myoglobin and ribonuclease A.) [(**A**) Adapted from [64] and [28]. (**B**) Adapted from [27]. (**C**) Adapted from [28])

Except for peptides with many aromatic amino acids, the CD spectra of α-helical polypeptides comprised of more than about 10 amino acid residues are relatively independent of the exact amino acid composition [31]. A minimum of 2–3 turns of helix (7–11 residues) appears to be sufficient to generate a typical spectrum [40]. The CD of β-sheet structure is more variable, probably mainly as a result of different amounts of twisting [40]. β-turns also have highly variable spectra.

The contributions that various secondary structural elements make to a protein of unknown structure can be estimated by fitting the observed CD spectrum with a sum of basis spectra like those shown in Fig. 9.9B, C [27–29, 41–44]. A web-server providing a variety of algorithms for this analysis is available [33, 45].

Attempts to predict the CD spectra of proteins from first principles are complicated by the fact that each peptide group has five, poorly resolved π-π^* or n-π^* transitions between 125 and 210 nm (Table 9.1). The strongest π-π^* transition occurs near 190 nm and has a dipole strength of about 9 debye^2. The lowest-energy transition is the n-π^* transition commencing from a non-bonding $2p$ oxygen orbital. As discussed above, this transition has essentially no electric dipole strength, but is weakly allowed by its magnetic transition dipole; it can gain dipole strength by exciton interactions with the π-π^* transitions of neighboring residues. It is seen in UV absorption spectra of α-helical proteins as a very weak shoulder in the region of 230 nm [46, 47].

In simple models of an α-helix, the π-π^* transitions of the peptide groups are predicted to generate three main exciton bands: a band with negative CD in the region of 205 nm, and two bands with positive CD at shorter wavelengths [6, 21, 48–51]. The 205-nm band should be polarized parallel to the helix axis, while both of higher-energy bands should be perpendicular to this axis. Mixing with higher-energy π-π^* and n-π^* transitions is expected to split one of the perpendicularly polarized bands into a complex band with positive CD on the red side and negative CD on the blue side. The low-energy n-π^* transitions of the peptide groups are predicted to gain dipole strength by exciton interactions with the π-π^* transitions, giving rise to an absorption band with negative CD near 220 nm. Figure 9.10A shows the calculated contributions of the individual bands to the CD spectrum of α-helical poly-L-alanine, and Fig. 9.10B shows a comparison of the total theoretical spectrum with the measured CD spectrum [49]. The agreement of the predicted and observed spectra is remarkably good.

In β-sheets, exciton interactions of the π-π^* transitions are predicted to generate a band with positive rotational strength in the region of 200 nm, along with a band with negative rotational strength at shorter wavelengths. The n-π^* transitions again are expected to gain dipole strength by mixing with the π-π^* transitions, giving rise to a band with negative CD near 215 nm. However, theoretical calculations of the spectra of proteins that are rich in β-sheets generally are less satisfactory than those for proteins that are predominantly α-helical [6, 40, 48–50, 52], possibly because the calculated spectrum for a β-sheet is very sensitive to twisting of the sheet.

Exciton interactions of tryptophan sidechains can give rise to pairs of CD bands in the region of 220 nm, along with weaker bands in the near UV [53, 54]. The bands around 220 nm stem from the indole ^1B_b transition. Coupling of peptide transitions with the transitions of tryptophan and tyrosine sidechains also can contribute significantly to the CD spectra of proteins [44, 55, 56].

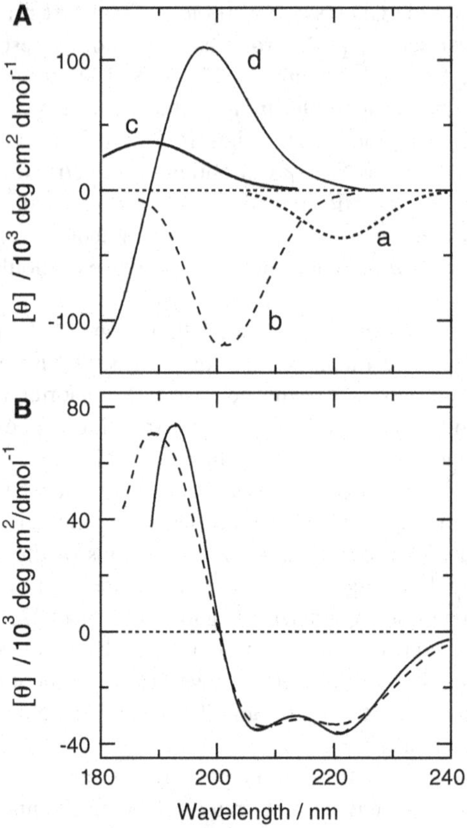

Fig. 9.10 (**A**) Calculated contributions to the CD spectrum of α-helical poly-L-alanine from the lowest-energy *n*-π* transition (*a*), the π-π* transition polarized parallel to the helix axis (*b*), and the π-π* transitions with perpendicular polarization (*c* and *d*). (**B**) The total calculated spectrum (*dashed curve*) and the measured CD spectrum (*solid curve*). The widths of the calculated bands were adjusted to optimize the agreement of the calculated and measured spectra. Adapted from [49]

Circular dichroism spectra of DNA and RNA also depend strongly on the molecular structure. As shown in Fig. 9.11, B-form DNA typically has a CD band with positive rotational strength near 185 nm and negative bands in the regions of 200 and 250 nm, while Z-form DNA has positive bands at 180 and 260 nm and negative bands at 195 and 290 nm.

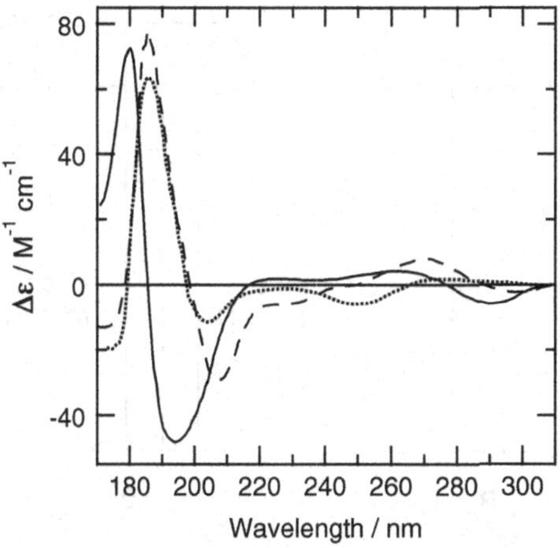

Fig. 9.11 Circular dichroism spectra of double-stranded poly[d(G-C)] in three different helical forms [65]. *Dotted curve*, B form (10 mM phosphate); *dashed curve*, A form (0.67 mM phosphate, 80% trifluoroethanol); *solid curve*, Z form (10 mM phosphate, 2 M $NaClO_4$). The ordinate scale is the mean difference between the molar extinction coefficients for left- and right-circularly polarized light ($\varepsilon_l - \varepsilon_r$), with molarity referring to the concentration of nuclotide base pairs

9.5 Magnetic Circular Dichroism

Although only chiral molecules have intrinsic circular dichroism, almost any substance can be made to exhibit CD by placing it in a magnetic field. *Magnetic circular dichroism* (MCD) is the difference between the extinction coefficients measured with left- and right-circularly polarized light when a magnetic field is imposed along the axis of the measuring beam. MCD is a manifestation of the *Zeeman effect*, which is a shift of transition energy by a magnetic field. Like ordinary CD, MCD is a very small effect. However, it has been particularly useful for probing the binding sites of Fe and other metals in metalloproteins.

The physical origins of MCD have been discussed in detail by Stephens [57, 58]. Atomic $2s \to 2p$ transitions will illustrate the principle. An imposed magnetic field changes the relative energies of the three p orbitals because these orbitals have different angular momenta, and thus different magnetic moments, along the axis of the field. The magnitude of the orbital angular momentum is given by

$$|\boldsymbol{L}| = \sqrt{l(l+1)}\ \hbar, \tag{9.34}$$

where l is the azimuthal quantum number (Sect. 2.3.4), and the component of the angular momentum parallel to the field takes the values $L_z = -\hbar$, 0, or $+\hbar$ for $l = 1$ and m_l (the magnetic quantum number) $= -1$, 0, or 1, respectively. A magnetic field

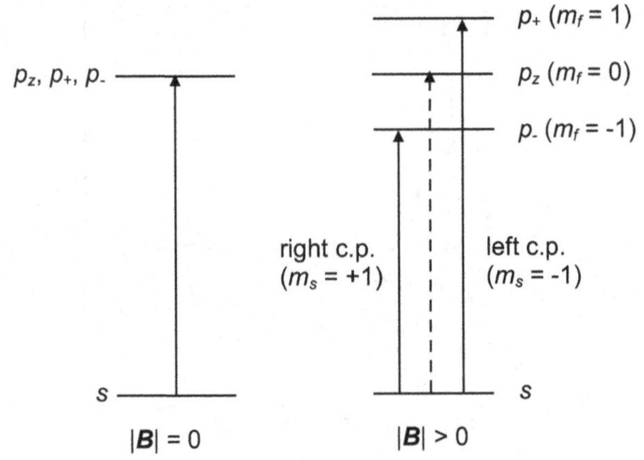

Fig. 9.12 *Left*: In the absence of a magnetic field ($|\boldsymbol{B}| = 0$), the three atomic $2p$ orbitals are degenerate. *Right*: A magnetic field along the z axis ($|\boldsymbol{B}| > 0$) splits the p orbitals into different energies but has no effect on s orbitals. To conserve angular momentum, a $2s \rightarrow 2p_-$ excitation with light propagating along the z axis requires right-circularly polarized (c.p.) light; a $2s \rightarrow 2p_+$ excitation requires left-circularly polarized light. The $2s \rightarrow 2p_z$ excitation is z-polarized and is forbidden for light propagating in this direction

$\boldsymbol{B}$ along the z axis raises the p_+ orbital in energy by $(-e/2m)|\boldsymbol{B}|\hbar$ and lowers p_- by the same amount (Fig. 9.12). At a field strength of 1 T, the separation between p_+ and p_- is about 0.5 cm^{-1}. The $2s$ orbital, with $l = 0$, has no orbital angular momentum and is unaffected by magnetic fields.

To conserve angular momentum when an electron is excited, the change in the orbital angular momentum must match the angular momentum contributed by the photon. (We assume again that no change in electron spin occurs during the excitation.) Right-circularly polarized light propagating along the z axis has an angular momentum of $+\hbar$ on this axis, whereas left-circularly polarized light has an angular momentum of $-\hbar$. This means that right-circularly polarized light will drive excitations from $2s$ to $2p_-$ orbitals, while left-circularly polarized light drives excitations to $2p_+$ (Fig. 9.12). In the absence of a magnetic field the $2s \rightarrow 2p$ absorption band has no net CD because the transitions to $2p_+$ and $2p_-$ are degenerate. In the presence of a field, transitions to $2p_+$ give a net positive rotational strength on the high-energy side of the band, whereas transitions to $2p_-$ give a net negative rotational strength on the low-energy side of the band. The resulting MCD spectrum (Fig. 9.13) resembles the first derivative of the absorption spectrum.

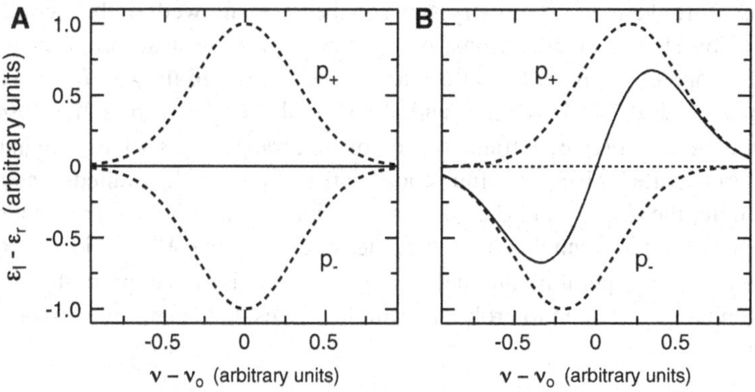

Fig. 9.13 (**A**) In the absence of a magnetic field, the $2s \rightarrow 2p_+$ and $2s \rightarrow 2p_-$ transitions contribute opposite rotational strengths at the same energies (*dashed lines*), so the net CD (*solid line*) is zero. (**B**) In the presence of a field, the two transitions occur at slightly different frequencies (*dashed lines*), giving a net positive MCD signal on the high-energy side of the absorption band and a negative signal on the low-energy side (*solid line*)

The effects of a magnetic field are somewhat different if the field splits the ground state rather than the excited state. The ground-state sublevel that moves to lower energies then is populated preferentially, depending on the temperature. The unequal populations of the different sublevels give rise to a net MCD signal that has the same shape as the overall absorption spectrum, rather than the derivative shape shown in Fig. 9.13B. The magnitude of the MCD increases with decreasing temperature, but levels off at high fields or very low temperatures when essentially all the systems are in the lowest sublevel.

Magnetic fields also can modify the mixing of the ground and excited states with other states at higher or lower energy [57, 58]. However, this effect probably makes a relatively minor contribution to the MCD spectrum in most cases.

MCD has been particularly informative in studies of ferrous iron-proteins [59–62]. Interactions with the surrounding ligands modify the relative energy levels of the five *d* orbitals of the Fe in a way that depends sensitively on the geometry of the complex. As we noted in Chap. 4, the two e_g orbitals (d_{z^2} and $d_{x^2-y^2}$) of transition metals in octahedral complexes have lobes that come close to electronegative atoms of the ligands, while the t_{2g} orbitals (d_{xz}, d_{yz} and d_{xy}) are farther away. Electrostatic repulsion therefore pushes d_{z^2} and $d_{x^2-y^2}$ up in energy relative to d_{xz}, d_{yz} and d_{xy} (Fig. 4.30). The splitting between the two groups is on the order of 10,000 cm^{-1}, depending on the electronegativity and positions of the ligands. Transitions from t_{2g}

to e_g are forbidden by symmetry but can become allowed if the symmetry is disrupted by structural distortions, giving rise to a weak absorption band in the near IR. If one of the ligands on the z axis moves closer to the Fe, d_{z^2}, d_{xz}, and d_{yz} shift up in energy while $d_{x^2-y^2}$ and d_{xy} shift down, modifying the absorption spectrum. Square planar, tetrahedral, or other arrangements of the ligands give still different patterns [63]. In a four-coordinate complex with tetrahedral geometry, for example, the d_{xz}, d_{yz} and d_{xy} orbitals have lobes that approach the ligands, and have higher energies than d_{z^2} and $d_{x^2-y^2}$. Measurements of the MCD in the IR bands as a function of temperature and field strength can be used to explore the geometry of metal binding sites and to probe structural changes that occur on interaction with substrates.

Exercises

1. Calculate the rotational strength ($\mathfrak{R}$) of the transition represented by the CD spectrum of Fig. E.1.
2. What are the exciton rotational strengths ($\mathfrak{R}_{ex}$) of the two exciton bands of the dimer discussed in Exercise 8.1? Explain your answer.
3. The dimer sketched in Fig. E.2 is similar to that of Exercise 8.1, except that molecules a and b lie in two parallel planes ($x_a y_a$ and $x_b y_b$) that are separated by 4 Å along the z axis. (Axes x_a and x_b project toward the viewer.) Again, the monomeric molecule absorbs at 700 nm with a dipole strength of 25 D^2. (*a*) Calculate the contributions of $\mathfrak{R}_{ex}$ to the rotational strengths of the two exciton bands of this dimer. (*b*) Sketch the predicted absorption and CD spectra of the dimer on the assumption that $\mathfrak{R}_{ex}$ dominates the rotational strengths. Indicate the wavelengths and relative intensities of the two bands in each spectrum. (*c*) What would happen to the CD spectrum if you rotated one molecule by 180° about an axis normal to the molecular plane?
4. Molecules a and b in the dimer shown in Fig. E.3 are the same as those of Fig. E.2, and also lie in planes separated by 4 Å in the z direction. Axes x_a and x_b again project out toward the viewer. (*a*) What is the stereochemical relation between this structure and that of Fig. E.2? (*b*) Calculate the contributions of $\mathfrak{R}_{ex}$ to the rotational strengths of the two exciton bands of this dimer. (*c*) Sketch the predicted absorption and CD spectra, assuming again that $\mathfrak{R}_{ex}$ dominates the rotational strengths.

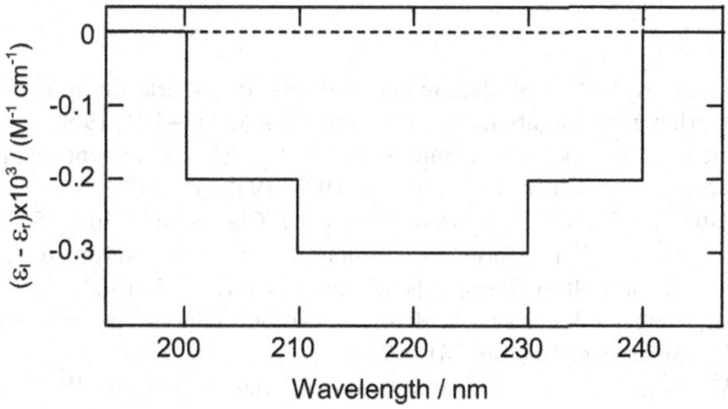

Fig. E.1

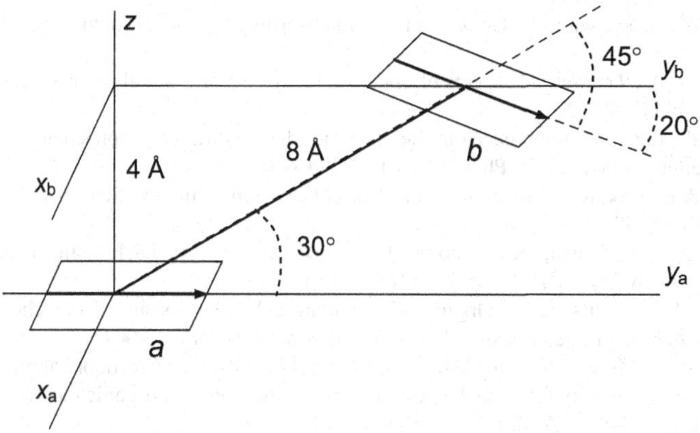

Fig. E.2

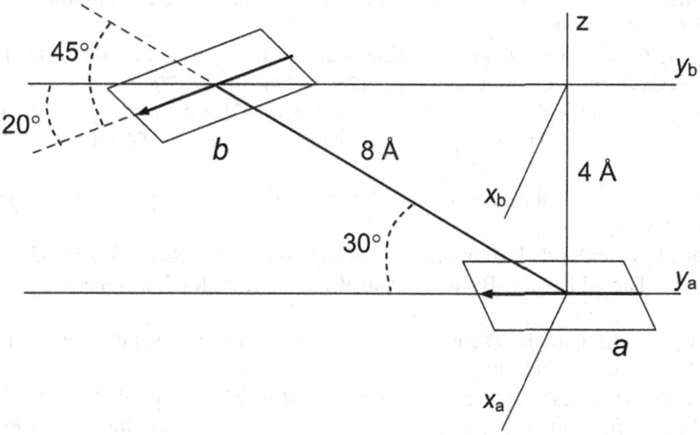

Fig. E.3

References

1. Hansen, A.E.: On evaluation of electric and magnetic dipole transition moments in the zero differential overlap approximation. Theor. Chim. Acta **6**, 341–349 (1966)
2. Král, M.: Optical rotatory power of complex compounds. Matrix elements of operators Del and R x Del. Collect. Czech. Chem. Commun. **35**, 1939–1948 (1970)
3. Ham, J.S., Platt, J.R.: Far U.V. spectra of peptides. J. Chem. Phys. **20**, 335–336 (1952)
4. Barnes, E.E., Simpson, W.T.: Correlations among electronic transitions for carbonyl and for carboxyl in the vacuum ultraviolet. J. Chem. Phys. **39**, 670–675 (1963)
5. Callomon, J.H., Innes, K.K.: Magnetic dipole transition in the electronic spectrum of formaldehyde. J. Mol. Spectrosc. **10**, 166–181 (1963)
6. Woody, R.W., Tinoco Jr., I.: Optical rotation of oriented helices. III. Calculation of the rotatory dispersion and circular dichroism of the alpha and 3_{10}-helix. J. Chem. Phys. **46**, 4927–4945 (1967)
7. Hameka, H.: Advanced Quantum Chemistry. Addison-Wesley, Reading, MA (1965)
8. Schatz, G.C., Ratner, M.A.: Quantum Mechanics in Chemistry, p. 325. Prentice-Hall, Englewood Cliffs, NJ (1993)
9. Moffitt, W., Moscowitz, A.: Optical activity in absorbing media. J. Chem. Phys. **30**, 648–660 (1959)
10. Moscowitz, A.: Theoretical aspects of optical activity part one: small molecules. Adv. Chem. Phys. **4**, 67–112 (1962)
11. Rosenfeld, L.Z.: Quantenmechanische Theorie der natürlichen optischen Aktivität von Flüssigkeiten und Gasen. Z. Phys. **52**, 161–174 (1928)
12. Hansen, A.E., Bak, K.L.: Ab-initio calculations of electronic circular dichroism. Enantiomer **4**, 1024–2430 (1999)
13. Berova, N., Nakanishi, K., Woody, R.W. (eds.): Circular Dichroism. Principles and Applications. Wiley-VCH, New York, NY (2000)
14. Lightner, D.A., Gurst, J.E.: Organic Conformational Analysis and Stereochemistry from Circular Dichroism Spectroscopy. Wiley-VCH, New York, NY (2000)
15. Autschbach, J., Ziegler, T., van Gisbergen, S.J.A., Baerends, E.J.: Chirooptical properties from time-dependent density functional theory. I. Circular dichroism of organic molecules. J. Chem. Phys. **116**, 6930–6940 (2002)
16. Diedrich, C., Grimme, S.: Systematic investigation of modern quantum chemical methods to predict electronic circular dichroism spectra. J. Phys. Chem. A **107**, 2524–2539 (2003)
17. Hobden, M.V.: Optical activity in a nonenantiomorphous crystal silver gallium sulfide. Nature **216**, 678 (1967)
18. Hobden, M.V.: Optical activity in a non-enantiomorphous crystal cadmium gallium sulfide. Nature **220**, 781 (1968)
19. Claborn, K., Cedres, J.H., Isborn, C., Zozulya, A., Weckert, E., et al.: Optical rotation of achiral pentaerythritol. J. Am. Chem. Soc. **128**, 14746–14747 (2006)
20. Kirkwood, J.G.: On the theory of optical rotatory power. J. Chem. Phys. **5**, 479–491 (1937)
21. Moffitt, W.: Optical rotatory dispersion of helical polymers. J. Chem. Phys. **25**, 467–478 (1956)
22. Tinoco Jr., I.: Theoretical aspects of optical activity part two: polymers. Adv. Chem. Phys. **4**, 113–160 (1962)
23. Schellman, J.: Circular dichroism and optical rotation. Chem. Rev. **75**, 323–331 (1975)
24. Charney, E.: The Molecular Basis of Optical Activity. Wiley-Interscience, New York, NY (1979)
25. Fasman, G.D. (ed.): Circular Dichroism and the Conformational Analysis of Macromolecules. Plenum, New York, NY (1996)
26. Condon, E.U.: Theories of optical rotatory power. Rev. Mod. Phys. **9**, 432–457 (1937)
27. Greenfield, N., Fasman, G.D.: Computed circular dichroism spectra for the evaluation of protein conformation. Biochemistry **8**, 4108–4116 (1969)

28. Saxena, V.P., Wetlaufer, D.B.: A new basis for interpreting the circular dichroic spectra of proteins. Proc. Natl. Acad. Sci. U. S. A. **68**, 969–972 (1971)
29. Johnson Jr., W.C.: Analysis of circular dichroism spectra. Methods Enzymol. **210**, 426–447 (1992)
30. Ramsay, G.D., Eftink, M.R.: Analysis of multidimensional spectroscopic data to monitor unfolding of proteins. Methods Enzymol. **240**, 615–645 (1994)
31. Woody, R.W.: Circular dichroism. Methods Enzymol. **246**, 34–71 (1995)
32. Plaxco, K.W., Dobson, C.M.: Time-resolved biophysical methods in the study of protein folding. Curr. Opin. Struct. Biol. **6**, 630–636 (1996)
33. Whitmore, L., Wallace, B.A.: Protein secondary structure analyses from circular dichroism. Biopolymers **89**, 392–400 (2008)
34. Wallace, B.A., Janes, R.W.: Modern Techniques for Circular Dichroism and Synchrotron Radiation Circular Dichroism Spectroscopy. IOS, Amsterdam (2009)
35. Pan, T., Sosnick, T.R.: Intermediates and kinetic traps in the folding of a large ribozyme revealed by circular dichroism and UV absorbance spectroscopies and catalytic activity. Nat. Struct. Biol. **4**, 931–938 (1997)
36. Settimo, L., Donnini, S., Juffer, A.H., Woody, R.W., Marin, O.: Conformational changes upon calcium binding and phosphorylation in a synthetic fragment of calmodulin. Biopolymers **88**, 373–385 (2007)
37. Bovey, F.A., Hood, F.P.: Circular dichroism spectrum of poly-L-proline. Biopolymers **5**, 325–326 (1967)
38. Woody, R.W.: Circular dichroism of unordered polypeptides. Adv. Biophys. Chem. **2**, 37–79 (1992)
39. Woody, R.W.: Circular dichroism spectrum of peptides in the polyPro II conformation. J. Am. Chem. Soc. **131**, 8234–8245 (2009)
40. Manning, M.C., Illangasekare, M., Woody, R.W.: Circular dichroism studies of distorted α-helices, twisted β-sheets, and β-turns. Biophys. Chem. **31**, 77–86 (1988)
41. Provencher, S.W., Glockner, J.: Estimation of globular protein secondary structure from circular dichroism. Biochemistry **20**, 33–37 (1981)
42. van Stokkum, I.H., Spoelder, H.J., Bloemendal, M., van Grondelle, R., Groen, F.C.: Estimation of protein secondary structure and error analysis from circular dichroism spectra. Anal. Biochem. **191**, 110–118 (1990)
43. Andrade, M.A., Chacon, P., Merelo, J.J., Moran, F.: Evaluation of secondary structure of proteins from UV circular dichroism spectra using an unsupervised learning neural network. Protein Eng. **6**, 383–390 (1993)
44. Sreerama, N., Manning, M.C., Powers, M.E., Zhang, J.-X., Goldenberg, D.P., et al.: Tyrosine, phenylalanine, and disulfide contributions to the circular dichroism of proteins: circular dichroism spectra of wild-type and mutant bovine pancreatic trypsin inhibitor. Biochemistry **38**, 10814–10822 (1999)
45. Whitmore, L., Wallace, B.A.: DICHROWEB, an online server for protein secondary structure analyses from circular dichroism spectroscopic data. Nucleic Acids Res. **32 (Web Server issue)**, W668–W673 (2004)
46. Gratzer, W.B., Holzwarth, G.M., Doty, P.: Polarization of the ultraviolet absorption bands in a-helical polypeptides. Proc. Natl. Acad. Sci. U. S. A. **47**, 1785–1791 (1961)
47. Rosenheck, K., Doty, P.: The far ultraviolet absorption spectra of polypeptide and protein solutions and their dependence on conformation. Proc. Natl. Acad. Sci. U. S. A. **47**, 1775–1785 (1961)
48. Tinoco Jr., I., Woody, R.W., Bradley, D.F.: Absorption and rotation of light by helical polymers: the effect of chain length. J. Chem. Phys. **38**, 1317–1325 (1963)
49. Woody, R.W.: Improved calculation of the np* rotational strength in polypeptides. J. Chem. Phys. **49**, 4797–4806 (1968)
50. Sreerama, N., Woody, R.W.: Computation and analysis of protein circular dichroism spectra. Methods Enzymol. **383**, 318–351 (2004)

51. Hirst, J.D., Colella, K., Gilbert, A.T.B.: Electronic circular dichroism of proteins from first-principles calculations. J. Phys. Chem. B **107**, 11813–11819 (2003)
52. Hirst, J.D.: Improving protein circular dichroism calculations in the far-ultraviolet through reparametrizing the amide chromophore. J. Chem. Phys. **109**, 782–788 (1998)
53. Grishina, I.B., Woody, R.W.: Contributions of tryptophan side chains to the circular dichroism of globular proteins: exciton couplets and coupled oscillators. Faraday Discuss. **99**, 245–267 (1994)
54. Cochran, A.G., Skelton, N.J., Starovasnik, M.A.: Tryptophan zippers: stable, monomeric beta-hairpins. Proc. Natl. Acad. Sci. U. S. A. **98**, 5578–5583 (2001)
55. Ohmae, E., Matsuo, K., Gekko, K.: Vacuum-ultraviolet circular dichroism of Escherichia coli dihydrofolate reductase: insight into the contribution of tryptophan residues. Chem. Phys. Lett. **572**, 111–114 (2013)
56. Matsuo, K., Hiramatsu, H., Gekko, K., Namatame, H., Taniguchi, M., et al.: Characterization of intermolecular structure of β_2-microglobulin core fragments in amyloid fibrils by vacuum-ultraviolet circular dichroism spectroscopy and circular dichroism theory. J. Phys. Chem. B **118**, 2785–2795 (2014)
57. Stephens, P.J.: Theory of magnetic circular dichroism. J. Chem. Phys. **52**, 3489–3516 (1970)
58. Stephens, P.J.: Magnetic circular dichroism. Ann. Rev. Phys. Chem. **25**, 201–232 (1974)
59. Thomson, A.J., Cheesman, M.R., George, S.J.: Variable-temperature magnetic circular dichroism. Methods Enzymol. **226**, 199–232 (1993)
60. Solomon, E.I., Pavel, E.G., Loeb, K.E., Campochiaro, C.: Magnetic circular dichroism spectroscopy as a probe of the geometric and electronic structure of nonheme ferrous enzymes. Coord. Chem. Rev. **144**, 369–460 (1995)
61. Kirk, M.L., Peariso, K.: Recent applications of MCD spectroscopy to metalloenzymes. Curr. Opin. Chem. Biol. **7**, 220–227 (2003)
62. McMaster, J., Oganesyan, V.S.: Magnetic circular dichroism spectroscopy as a probe of the structures of the metal sites in metalloproteins. Curr. Opin. Struct. Biol. **20**, 615–622 (2010)
63. Companion, A.L., Komarynsky, M.A.: Crystal field splitting diagrams. J. Chem. Ed. **41**, 257–262 (1964)
64. Quadrifoglio, F., Urry, D.M.: Ultraviolet rotatory properties of peptides in solution. I. Helical poly-L-alanine. J. Am. Chem. Soc. **90**, 2755–2760 (1968)
65. Riazance, J.H., Baase, W.A., Johnson Jr., W.C., Hall, K., Cruz, P., et al.: Evidence for Z-form RNA by vacuum UV circular dichroism. Nucleic Acids Res. **13**, 4983–4989 (1985)

Coherence and Dephasing

10

10.1 Oscillations Between Quantum States of an Isolated System

The time-dependent perturbation theory that we have used to treat resonance energy transfer and absorption of light assumes that we know that a system is in a given state (state 1), so that the coefficient associated with this state (C_1) is 1, while the coefficient for finding the system in a different state (C_2) is zero. The resulting expression for the rate of transitions to state 2 (Eq. 2.61 or 7.8) neglects the possibility of a return to state 1. It can continue to hold at later times only if the transition to state 2 is followed by a relaxation that takes the two states out of resonance. Without such relaxations, the system would oscillate between the two states as described by the coupled equations

$$\partial C_1/\partial t = -(i/\hbar)\left\langle \psi_1|\widetilde{\mathrm{H}}'|\psi_2\right\rangle \exp[i(E_1 - E_2)t/\hbar]C_2 \\ - (i/\hbar)\left\langle \psi_1|\widetilde{\mathrm{H}}'|\psi_1\right\rangle C_1 \tag{10.1a}$$

and

$$\partial C_2/\partial t = -(i/\hbar)\left\langle \psi_2|\widetilde{\mathrm{H}}'|\psi_1\right\rangle \exp[i(E_2 - E_1)t/\hbar]C_1 \\ - (i/\hbar)\left\langle \psi_2|\widetilde{\mathrm{H}}'|\psi_2\right\rangle C_2. \tag{10.1b}$$

Equations (10.1a, 10.1b) just restate Eq. (2.59) for a two-state system with spatial wavefunctions ψ_1 and ψ_2 that are mixed by perturbation operator $\widetilde{\mathrm{H}}'$. E_1 and E_2 are the energies of the states in the absence of the perturbation. The factor $\left\langle \psi_1|\widetilde{\mathrm{H}}'|\psi_1\right\rangle$ in Eq. (10.1a) represents any shift in the energy of state 1 caused by the perturbation, and $\left\langle \psi_2|\widetilde{\mathrm{H}}'|\psi_2\right\rangle$ in Eq. (10.1b) is the corresponding shift in the energy

W.W. Parson, *Modern Optical Spectroscopy*, DOI 10.1007/978-3-662-46777-0_10

of state 2. In resonance energy transfer, for example, interactions between the two molecules could shift the excitation energy of the donor molecule, even though the acceptor remains in its ground state. Such shifts in the excitation energies are analogous to the effects of varying the solvent, and usually are relatively small.

Equations (10.1a and 10.1b) simplify if $\left\langle\psi_1\middle|\widetilde{\mathrm{H}}'\middle|\psi_1\right\rangle$ and $\left\langle\psi_2\middle|\widetilde{\mathrm{H}}'\middle|\psi_2\right\rangle$ are zero, because then the rate of change of C_1 depends only on C_2 and *vice versa*. If these matrix elements are not zero, we can achieve the same simplification by defining adjusted basis-state energies H_{11} and H_{22} that include $\left\langle\psi_1\middle|\widetilde{\mathrm{H}}'\middle|\psi_1\right\rangle$ and $\left\langle\psi_2\middle|\widetilde{\mathrm{H}}'\middle|\psi_2\right\rangle$:

$$H_{11} = \left\langle\psi_1\middle|\widetilde{\mathrm{H}}_1 + \widetilde{\mathrm{H}}'\middle|\psi_1\right\rangle = E_1 + \left\langle\psi_1\middle|\widetilde{\mathrm{H}}'\middle|\psi_1\right\rangle \tag{10.2a}$$

and

$$H_{22} = \left\langle\psi_2\middle|\widetilde{\mathrm{H}}_2 + \widetilde{\mathrm{H}}'\middle|\psi_2\right\rangle = E_2 + \left\langle\psi_2\middle|\widetilde{\mathrm{H}}'\middle|\psi_2\right\rangle. \tag{10.2b}$$

The differential equations for C_1 and C_2 then are

$$\partial C_1/\partial t = -(i/\hbar)H_{12}\exp(iE_{12}t/\hbar)C_2 \tag{10.3a}$$

and

$$\partial C_2/\partial t = -(i/\hbar)H_{21}\exp(iE_{21}t/\hbar)C_1, \tag{10.3b}$$

where $H_{21} = \left\langle\psi_2\middle|\widetilde{\mathrm{H}}'\middle|\psi_1\right\rangle$ as before and $E_{21} = H_{22} - H_{11}$.

Equations (10.3a and 10.3b) can be solved by differentiating each of them again with respect to time; straightforward substitutions then yield separate differential equations for C_1 and C_2 (see, e.g., [1]). Assuming again that the system is in state 1 at time zero, so that $C_1(0) = 1$ and $C_2(0) = 0$, the solutions are

$$C_1(t) = \{\cos(\Omega t/\hbar) - i(E_{12}/2\Omega)\sin(\Omega t/\hbar)\}\exp(iE_{21}t/2\hbar) \tag{10.4a}$$

and

$$C_2(t) = -i(H_{21}/\Omega)\sin(\Omega t/\hbar)\exp(-iE_{21}t/2\hbar), \tag{10.4b}$$

with

$$\Omega = (1/2)\left[(E_{21})^2 + 4(H_{21})^2\right]^{1/2}. \tag{10.4c}$$

From Eq. (10.4b), the probability of finding the system in state 2 at time t is

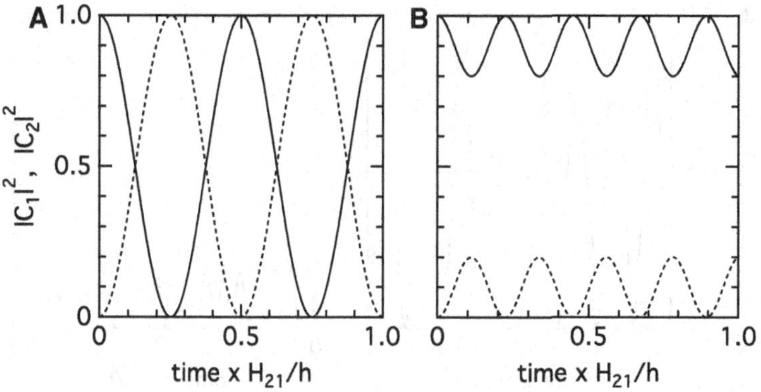

Fig. 10.1 Oscillations of $|C_1(t)|^2$ (*solid curves*) and $|C_2(t)|^2$ (*dotted curves*), the probabilities that the excitation is found on molecule 1 or 2, respectively, as functions of time after excitation of molecule 1. The plots in (**A**) were obtained with $E_{21} = 0$, which reduces Eqn. (10.5) to (10.6); those in (**B**) were obtained by Eq. (10.7) with $E_{21} = -4|H_{21}|$

$$\begin{aligned} |C_2(t)|^2 &= (H_{21}/\Omega)^2 \sin^2(\Omega t/\hbar) \\ &= (1/2)(H_{21}/\Omega)^2[1 - \cos(2\Omega t/\hbar)]. \end{aligned} \tag{10.5}$$

This oscillates with a period of $2\pi\hbar/2\Omega$, or $h/2\Omega$, with the mean probability being $(H_{21}/\Omega)^2/2$.

Figure 10.1A shows the predicted oscillations of C_1 and C_2 when the adjusted energies of the two basis states are the same ($E_{21} = 0$). In this case, $\Omega = H_{21}$, and Eq. (10.5) simplifies to

$$|C_2(t)|^2 = (1/2)[1 - \cos(2H_{21}t/\hbar)], \tag{10.6}$$

which oscillates between 0 and 1 at a frequency of $2H_{21}/h$. A remarkable feature of this result is that the system is guaranteed to be in state 2 at certain times (e.g., at $t = h/4H_{21}$), even if H_{21} is very small. This is a consequence of exact resonance and does not hold for any value of E_{21} other than 0.

In the opposite limit, when $|E_{21}|^2 \gg 4|H_{21}|^2$, we have $\Omega = E_{12}/2$, and Eq. (10.5) becomes

$$|C_2(t)|^2 = 2(H_{21}/E_{21})^2[1 - \cos(E_{21}t/\hbar)]. \tag{10.7}$$

Now the oscillation frequency increases linearly with $|E_{21}|$, while the amplitude of the oscillations and the average probability of finding the excitation on molecule 2 both decrease. If $|E_{21}| \gg |H_{21}|$, the system has very little probability of moving to state 2. Figure 10.1B illustrates the results for the case $E_{12} = -4|H_{21}|$.

Oscillations of the type described by Eqs. (10.4a–10.5) usually occur for, at most, only a short period on the picosecond or sub-picosecond time scale. There are several reasons for this. First, these expressions pertain to a single system with sharply defined energies. In most experimental measurements of resonance energy

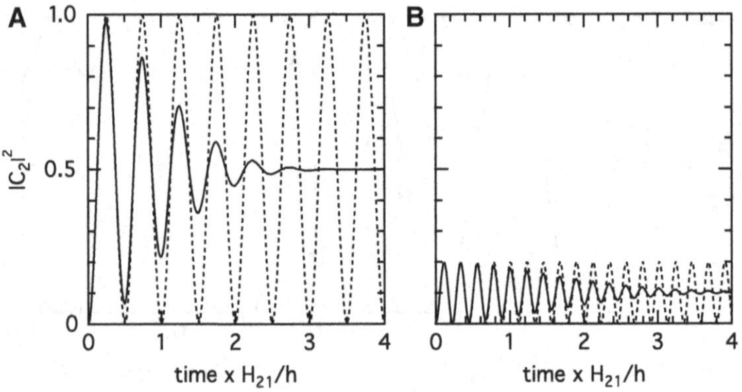

Fig. 10.2 Oscillations of $|C_2(t)|^2$ as functions of time in an ensemble of donor-acceptor pairs. The *dotted curves* show Eq. (10.5) for a single donor-acceptor pair with $E_{21} = 0$ (**A**) or $-4H_{21}$ (**B**); these are the same as the dotted curves in Fig. 10.1. The *solid curves* were obtained by averaging Eq. (10.5) over uncorrelated Gaussian distributions of E_{21} and H_{21}. The mean values of H_{21} and E_{21} in the distributions ($\overline{H_{21}}$ and $\overline{E_{21}}$) were the same as the corresponding values for the dotted curves. Both distributions had widths (*FWHM*) of $0.2\overline{H_{21}}$ (standard deviation $\sigma = \text{FWHM}/(8\ln 2)^2 = 0.084\overline{H_{21}}$)

transfer, we record the average behavior of a very large number of molecules with a range of energies and various values of H_{21}. The individual donor-acceptor pairs in such an inhomogeneous ensemble may oscillate as prescribed by Eqs. (10.4a–10.5), but at various frequencies. Even if they all begin with precise timing, the oscillations soon get out of phase. As a result, the probability of finding the excitation on molecule 2 undergoes damped oscillations converging on $\overline{(H_{21}/\Omega)^2}/2$, where the bar denotes the average value for the ensemble. Figure 10.2 illustrates this effect for ensembles in which E_{21} and H_{21} have Gaussian distributions around the values used in Fig. 10.1. The oscillations of an ensemble are said to be *coherent* if the individual microscopic systems that comprise the ensemble oscillate in phase, and *incoherent* or *stochastic* if the phases of the microscopic systems are random.

A second problem is that Eqs. (10.4a–10.5) apply only to an isolated system. If the system can transfer energy to or from the surroundings, it will relax toward an equilibrium in which the relative probabilities of being in various states are given by the Boltzmann equation. Equations (10.4a–10.5) provide no mechanism for this relaxation: the system apparently will continue to oscillate forever. In Fig. 10.2B, for example, the probability of finding the system in state 2 after a long time is only about 10%, even though this state has the lower energy. We would expect to find the system mostly in state 2 at equilibrium, unless the temperature is greater than E_{12}/k_B.

As we discussed in Chaps. 4 and 5, relaxations of the product state also will broaden the distribution of energies of the state. This homogeneous broadening applies to an ensemble of molecules as well as to an individual molecule or a single donor–acceptor pair. It will damp the measured oscillations of $|C_2|^2$ in a manner that is qualitatively similar to the damping shown in Fig. 10.2 for an inhomogeneous distribution of E_{12}. However, we will see in Chap. 11 that the detailed time course of the damping will be somewhat different.

Förster's theory of resonance energy transfer depends implicitly on thermal equilibration to trap the excitation on the acceptor. But the Förster theory is silent concerning the oscillations predicted by Eqs. (10.4a–10.5), and it does not address how rapidly the system equilibrates with the surroundings; it simply assumes that equilibration occurs rapidly compared to the rate at which the excitation can return to the donor.

Clearly, we need a more complete theory that bridges the gap between the Förster theory and the coherent oscillations described by Eqs. (10.4a–10.5). If the oscillations of the individual systems in an ensemble start out in phase, we would like to understand how this coherence decays, and how this decay affects the observed properties of the ensemble. To address such questions, we must deal with large ensembles of molecules that interact stochastically with their surroundings. We will develop some basic tools for this in the next three sections.

10.2 The Density Matrix

Consider, again, an individual system with wavefunction $\Psi(t) = C_1(t)\psi_1 exp[-i(H_{11}t/\hbar + \zeta_1)] + C_2(t)\psi_2 exp[-i(H_{22}t/\hbar + \zeta_2)]$, where ψ_1 and ψ_2 are spatial wavefunctions for two basis states with energies H_{11} and H_{22}, and ζ_n is a phase shift that depends on when state n is created. Let's combine the time-dependent coefficients C_1 and C_2 with the corresponding time-dependent exponential factors by defining a new set of coefficients c_1 and c_2:

$$c_n(t) = C_n(t)\exp[-i(H_{nn}t/\hbar + \zeta_n)]. \tag{10.8}$$

The wavefunction for the system then can be written simply as

$$\Psi(t) = c_1(t)\psi_1 + c_2(t)\psi_2. \tag{10.9}$$

In principle, we can find the expectation value of any dynamic property A of the system by evaluating the expression

$$\begin{aligned}\langle A(t)\rangle &= \left\langle \Psi(t)\middle|\tilde{\mathrm{A}}\middle|\Psi(t)\right\rangle = \left\langle c_1(t)\psi_1 + c_2(t)\psi_2\middle|\tilde{\mathrm{A}}\middle|c_1(t)\psi_1 + c_2(t)\psi_2\right\rangle \\ &= c_1^*(t)\,c_1(t)\left\langle\psi_1\middle|\tilde{\mathrm{A}}\middle|\psi_1\right\rangle + c_1^*(t)\,c_2(t)\left\langle\psi_1\middle|\tilde{\mathrm{A}}\middle|\psi_2\right\rangle + c_2^*(t)c_1(t)\left\langle\psi_2\middle|\tilde{\mathrm{A}}\middle|\psi_1\right\rangle \\ &\quad + c_2^*(t)\,c_2(t)\left\langle\psi_2\middle|\tilde{\mathrm{A}}\middle|\psi_2\right\rangle \\ &= c_2^*(t)c_1(t)A_{11} + c_1^*(t)c_2(t)A_{12} + c_2^*(t)c_1(t)A_{21} + c_2^*(t)c_2(t)A_{22},\end{aligned} \tag{10.10}$$

where $\tilde{A}$ is the corresponding operator. We used this approach in Box 4.5 to find the electric dipole of a superposition state. More generally, for any system that can be described by a linear combination of basis states, the expectation value of A is

$$\langle A(t)\rangle = \sum_m \sum_n c_m^*(t) c_n(t) \left\langle \psi_m \middle| \tilde{\mathrm{A}} \middle| \psi_n \right\rangle = \sum_m \sum_n c_m^*(t) c_n(t) A_{mn}, \tag{10.11}$$

where ψ_m and ψ_n again represent purely spatial wavefunctions. All the observable time dependence of the system thus resides in the array of products of the coefficients, $c_{\mathrm{m}}^* c_n$.

Equation (10.11) can be written more succinctly if we define a matrix $\boldsymbol{\rho}$ whose terms are

$$\rho_{nm}(t) = c_n(t) c_m^*(t). \tag{10.12}$$

The double sum over n and m in Eq. (10.11) then can be related to the product of matrices $\boldsymbol{\rho}$ and $\mathbf{A}$. The definition of a matrix product is given in Appendix A2: for $\mathbf{W} = \boldsymbol{\rho}\mathbf{A}$, $W_{nm} = \sum_k \rho_{nk} A_{km}$. With this definition and Eqs. (10.11) and (10.12), we have

$$\begin{aligned} \langle A(t)\rangle &= \sum_n \sum_m c_n(t) c_m^*(t) A_{mn} = \sum_n \sum_m \rho_{nm}(t) A_{mn} \\ &= \sum_n (\rho A)_{nn} = Tr(\boldsymbol{\rho}\mathbf{A}), \end{aligned} \tag{10.13}$$

where $Tr(\boldsymbol{\rho}\mathbf{A})$ means the sum of the diagonal elements of the matrix product $\boldsymbol{\rho}\mathbf{A}$ (Appendix A2). $\boldsymbol{\rho}$ is called the *density matrix* of the system.

The definition of the density matrix given in Eq. (10.12), in which in c_n and c_{m}^* include the factors $\exp(-iH_{nn}t/\hbar)$ and $\exp(-iH_{mm}t/\hbar)$ along with the coefficients $C_n(t)$ and $C_{\mathrm{m}}^*(t)$, uses what is called the "Schrödinger representation." An alternative formulation called the "interaction representation" is $\rho_{nm} = C_n(t) C_m^*(t)$. In the interaction representation, the factors $\exp(-iH_{nn}t/\hbar)$ and $\exp(-iH_{mm}t/\hbar)$ must be introduced separately in order to obtain the complete time-dependence of the system. Both representations are widely used, and the choice is mostly a matter of personal preference. We will use the Schrödinger representation.

Equation (10.13) is a remarkably general expression. The density matrix $\boldsymbol{\rho}$ could refer to any system that can be described with a linear combination of basis wavefunctions. Further, $\mathbf{A}$ could represent the matrix of expectation values of the operator for any dynamic property. The only requirement, but an important one to note, is that the matrix elements of $\mathbf{A}$ and $\boldsymbol{\rho}$ must be expressed in terms of the same set of basis wavefunctions.

With a slight modification, Eq. (10.13) provides a powerful way to deal with an ensemble of many systems. The expectation value of observable A for an ensemble is

$$\langle A(t)\rangle = Tr\left(\overline{\boldsymbol{\rho}(t)\mathbf{A}}\right), \tag{10.14}$$

where $\overline{\boldsymbol{\rho}(t)\mathbf{A}}$ means an average of the product $\boldsymbol{\rho}\mathbf{A}$ for all the systems in the ensemble at time t. Equation (10.14) follows from (10.13) because the trace obeys the distributive law of arithmetic. For more formal justifications see [2–6].

We have used the diagonal elements of the density matrix implicitly in previous chapters to represent the probabilities of finding a system in various basis states $\left(\rho_{nn} = c_n(t)c_n^*(t) = C_n(t)C_n^*(t)\right)$. When the density matrix is averaged over the ensemble, the diagonal elements can be interpreted as the relative populations of the corresponding states. The time-dependent factors from the basis wavefunctions $[exp(-iH_{nn}t/\hbar)]$ cancel out in the diagonal elements of $\boldsymbol{\rho}$ because each of the exponential factors is multiplied by its complex conjugate. The diagonal elements hence are always real numbers with positive signs, and all their time dependence comes from the coefficients $|C_n(t)|^2$. The off-diagonal elements, however, can be complex numbers, either positive or negative in sign. They generally oscillate at a frequency that increases with the energy difference between states n and m. They consist of products of the form

$$\rho_{nm} = C_n(t)C_m^*(t)\exp(-iE_{nm}t/\hbar)\exp(-i\zeta_{nm}), \tag{10.15}$$

with $E_{nm} = E_n - E_m$, $\zeta_{nm} = \zeta_n - \zeta_m$, and $n \neq m$. The off-diagonal elements also depend on any interaction matrix elements that couple the two states. We will see that when the density matrix is averaged over the ensemble, the magnitudes of the off-diagonal elements provide information on the coherence of the ensemble.

As an illustration, consider an ensemble of N molecules, each of which has a ground state (state 1) and an excited electronic state (2). Suppose all the molecules initially are in state 1; any given molecule (j) thus has coefficients $C_{1(j)} = 1$ and $C_{2(j)} = 0$. If we now expose the ensemble to a short pulse of light, any given molecule (k) might be promoted to state 2 so that $C_{2(k)} = 1$ and $C_{1(k)} = 0$, while another molecule (j) remains in state 1. The ensemble-average of ρ_{22} at time t after the excitation pulse then is

$$\overline{\rho_{22}}(t) = N^{-1}\sum_{k=1}^{N} c_{2(k)}(t)c_{2(k)}{}^*(t) = |\overline{c_2}(t)|^2 = \left|\overline{C_2}(t)\right|^2, \tag{10.16}$$

which is just the average probability of finding a molecule in state 2 at time t. The ensemble-average of ρ_{12} at time t is, similarly,

$$\overline{\rho_{12}}(t) = N^{-2}\sum_{j=1}^{N}\sum_{k=1}^{N} C_{1(j)}(t)C_{2(k)}{}^*(t)\ \exp\left(-iE_{1(j)2(k)}t/\hbar\right), \tag{10.17}$$

where $E_{1(j)2(k)}$ is the difference between the energies of molecules j and k when molecule j is in state 1 and k is in state 2 $(E_{1(j)} - E_{2(k)})$.

If the energy difference between states 1 and 2 is the same for all the molecules ($E_{1(j)2(k)} = E_{12}$), Eq. (10.17) simplifies to

$$\overline{\rho_{12}}(t) = \overline{C_{1(j)}}(t)\ \overline{C^*_{2(k)}}(t)\ \exp(-iE_{12}t/\hbar). \tag{10.18}$$

The factor $\exp(-iE_{12}t/\hbar)$ in this expression contains real and imaginary parts, both of which oscillate with period $|h/E_{12}|$. The factor $\overline{C_1}(t)\overline{C_2^*}(t)$ also can oscillate if the interaction matrix element H_{12} is non-zero, and can evolve with time as the excited molecules decay from state 2 back to state 1.

On the other hand, if $E_{1(j)2(k)}$ varies from molecule to molecule, the oscillation frequencies in Eq. (10.17) will vary, and after a sufficiently long time $\rho_{12}(t)$ will average to zero. The off-diagonal elements of $\overline{\boldsymbol{\rho}}$ thus reflect the *phase coherence* of the ensemble as a whole, that is, the extent to which the members of the ensemble remain in phase. Using a broad excitation flash would spread out the times at which molecules are excited to state 2, which would impart random phase shifts to the oscillations of ρ_{12} for the individual molecules and cause $\overline{\rho_{12}}$ to average to zero more rapidly. To anticipate the discussion below, we might expect that fluctuating interactions with the surroundings also will cause the off-diagonal elements of $\overline{\boldsymbol{\rho}}$ to decay to zero with time. At this point, however, we are still considering systems that are isolated from their surroundings.

A differential equation describing the time dependence of both the diagonal and the off-diagonal elements of $\boldsymbol{\rho}$ can be obtained as follows. Let the wavefunction for the system again be

$$\Psi(t) = \sum_n c_n(t)\psi_n(\boldsymbol{r}), \tag{10.19}$$

where the ψ_n are spatial wavefunctions and the c_n include all the time dependence. Applying the time-dependent Schrödinger equation as we did to derive Eqs. (2.58), (10.1a), and (10.1b) (still neglecting interactions with the surroundings) yields

$$\frac{\partial c_m}{\partial t} = -(i/\hbar)\sum_k H_{mk}c_k(t), \tag{10.20}$$

where the index k runs over all the basis states including state m. From the definition of ρ_{nm} (Eq. 10.12), using Eq. (10.20) and the product rule for differentiation gives

$$\frac{\partial \rho_{nm}}{\partial t} = \frac{\partial\left(c_n c_m^*\right)}{\partial t} = c_n\frac{\partial c_m^*}{\partial t} + c_m^*\frac{\partial c_n}{\partial t} \tag{10.21a}$$

$$= (i/\hbar)\left(c_n\sum_k H_{mk}^* c_k^* - c_m^*\sum_k H_{nk}c_k\right)$$

$$= (i/\hbar)\sum_k \left(H_{km}\rho_{nk} - H_{nk}\rho_{km}\right) \tag{10.21b}$$

$$= (i/\hbar)\sum_k \{\rho_{nk}H_{km} - H_{nk}\rho_{km}\}. \tag{10.21c}$$

In Eq. (10.21b), we have taken advantage of the facts that $\rho_{mn} = \rho^*_{nm}$ (by definition) and $H^*_{mk} = H_{km}$ (because the Hamiltonian operator is Hermitian).

Equation (10.21c) can be rewritten in terms of products of the matrices $\boldsymbol{\rho}$ and $\mathbf{H}$ by using the expression for matrix multiplication again and noting that $\boldsymbol{\rho}\mathbf{H} - \mathbf{H}\boldsymbol{\rho}$ is the commutator $[\boldsymbol{\rho},\mathbf{H}]$. Using the notation $[\mathbf{A}]_{nm}$ for element A_{nm} of matrix $\mathbf{A}$,

$$\frac{\partial \rho_{nm}}{\partial t} = (i/\hbar)\{[\boldsymbol{\rho}\mathbf{H}]_{nm} - [\mathbf{H}\boldsymbol{\rho}]_{nm}\} \equiv (i/\hbar)[\boldsymbol{\rho},\mathbf{H}]_{nm}. \tag{10.22}$$

The commutator of two matrices is defined analogously to the commutator of two operators (Box 2.2).

Finally, we can cast Eq. (10.22) symbolically in terms of the matrices themselves rather than particular matrix elements:

$$\frac{\partial \boldsymbol{\rho}}{\partial t} = (i/\hbar)[\boldsymbol{\rho},\mathbf{H}], \tag{10.23}$$

or for the average density matrix in an ensemble,

$$\frac{\partial \overline{\boldsymbol{\rho}}}{\partial t} = (i/\hbar)[,\overline{\boldsymbol{\rho}},\mathbf{H}]. \tag{10.24}$$

Equation (10.24) is called the *von Neumann equation* after the mathematician John von Neumann, who originated the concept of the density matrix. It also is known as the *Liouville equation* because of its parallel to Liouville's classical statistical mechanical theorem on the density of dynamic variables in phase space.

The time dependence of the nine elements of $\boldsymbol{\rho}$ for an isolated system with three basis states is written out in Box 10.1.

Box 10.1 Time Dependence of the Density Matrix for an Isolated Three-State System

The elements of the density matrix for an isolated system with three basis states evolve with time as described by Eq. (10.21c):

$$\begin{aligned}\partial\rho_{11}/\partial t &= (i/\hbar)(\rho_{11}H_{11} + \rho_{12}H_{21} + \rho_{13}H_{31} - \rho_{11}H_{11} - \rho_{21}H_{12} - \rho_{31}H_{13})\\ &= (i/\hbar)(\rho_{12}H_{21} - \rho_{21}H_{12} + \rho_{13}H_{31} - \rho_{31}H_{13}).\end{aligned} \tag{B10.1.1}$$

(continued)

Box 10.1 (continued)

$$\begin{aligned}\partial\rho_{22}/\partial t &= (i/\hbar)(\rho_{21}H_{12} + \rho_{22}H_{22} + \rho_{23}H_{32} - \rho_{12}H_{21} - \rho_{22}H_{22} - \rho_{32}H_{23}) \\ &= (i/\hbar)(\rho_{21}H_{12} - \rho_{12}H_{21} + \rho_{23}H_{32} - \rho_{32}H_{23}).\end{aligned} \quad \text{(B10.1.2)}$$

$$\begin{aligned}\partial\rho_{33}/\partial t &= (i/\hbar)(\rho_{31}H_{13} + \rho_{32}H_{23} + \rho_{33}H_{33} - \rho_{13}H_{31} - \rho_{23}H_{32} - \rho_{33}H_{33}) \\ &= (i/\hbar)(\rho_{31}H_{13} - \rho_{13}H_{31} + \rho_{32}H_{23} - \rho_{23}H_{32}).\end{aligned} \quad \text{(B10.1.3)}$$

$$\begin{aligned}\partial\rho_{12}/\partial t &= (i/\hbar)(\rho_{11}H_{12} + \rho_{12}H_{22} + \rho_{13}H_{32} - \rho_{12}H_{11} - \rho_{22}H_{12} - \rho_{32}H_{13}) \\ &= (i/\hbar)[(\rho_{11} - \rho_{22})H_{12} + \rho_{12}(H_{22} - H_{11}) + \rho_{13}H_{32} - \rho_{32}H_{13}].\end{aligned} \quad \text{(B10.1.4)}$$

$$\begin{aligned}\partial\rho_{21}/\partial t &= (i/\hbar)(\rho_{21}H_{11} + \rho_{22}H_{21} + \rho_{23}H_{31} - \rho_{11}H_{21} - \rho_{21}H_{22} - \rho_{31}H_{23}) \\ &= (i/\hbar)[(\rho_{22} - \rho_{11})H_{21} + \rho_{21}(H_{11} - H_{22}) + \rho_{23}H_{31} - \rho_{31}H_{23}].\end{aligned} \quad \text{(B10.1.5)}$$

$$\begin{aligned}\partial\rho_{13}/\partial t &= (i/\hbar)(\rho_{11}H_{13} + \rho_{12}H_{23} + \rho_{13}H_{33} - \rho_{13}H_{11} - \rho_{23}H_{12} - \rho_{33}H_{13}) \\ &= (i/\hbar)[(\rho_{11} - \rho_{33})H_{13} + \rho_{13}(H_{33} - H_{11}) + \rho_{12}H_{23} - \rho_{23}H_{12}].\end{aligned} \quad \text{(B10.1.6)}$$

$$\begin{aligned}\partial\rho_{31}/\partial t &= (i/\hbar)(\rho_{31}H_{11} + \rho_{32}H_{21} + \rho_{33}H_{31} - \rho_{11}H_{31} - \rho_{21}H_{32} - \rho_{31}H_{33}) \\ &= (i/\hbar)[(\rho_{33} - \rho_{11})H_{31} + \rho_{31}(H_{11} - H_{33}) + \rho_{32}H_{21} - \rho_{21}H_{32}].\end{aligned} \quad \text{(B10.1.7)}$$

$$\begin{aligned}\partial\rho_{23}/\partial t &= (i/\hbar)(\rho_{21}H_{13} + \rho_{22}H_{23} + \rho_{23}H_{33} - \rho_{13}H_{21} - \rho_{23}H_{22} - \rho_{33}H_{23}) \\ &= (i/\hbar)[(\rho_{22} - \rho_{33})H_{23} + \rho_{23}(H_{33} - H_{22}) + \rho_{21}H_{13} - \rho_{13}H_{21}].\end{aligned} \quad \text{(B10.1.8)}$$

$$\begin{aligned}\partial\rho_{32}/\partial t &= (i/\hbar)(\rho_{31}H_{12} + \rho_{32}H_{22} + \rho_{33}H_{32} - \rho_{12}H_{31} - \rho_{22}H_{32} - \rho_{32}H_{33}) \\ &= (i/\hbar)[(\rho_{33} - \rho_{22})H_{32} + \rho_{32}(H_{22} - H_{33}) + \rho_{31}H_{12} - \rho_{12}H_{31}].\end{aligned} \quad \text{(B10.1.9)}$$

The same expressions hold for the corresponding elements of $\overline{\boldsymbol{\rho}}$ for an ensemble of systems. Note that, because $\rho_{nm} = \rho^*_{mn}$, $\partial\rho_{nm}/\partial t = \partial\rho^*_{mn}/\partial t = (\partial\rho_{mn}/\partial t)^*$.

Inspection of Eqs. (B10.1.1–B10.1.3) shows that any change of the population of one of the states (i.e., a change in one of the diagonal elements of $\boldsymbol{\rho}$) is equal and opposite to the sum of the changes of population in the other states, as it must be to maintain mass balance in an isolated system:

$$\partial\overline{\rho_{mm}}/\partial t = -\sum_{n \neq m} \partial\overline{\rho_{nn}}/\partial t. \tag{10.25}$$

These equations also show that the rate of change of a diagonal element of $\boldsymbol{\rho}$ depends on the off-diagonal elements, but does not depend directly on the other diagonal elements. So, if the ensemble starts in state 1 (i.e., with $\rho_{11}(0) = 1$ and all the other elements of $\boldsymbol{\rho}$ zero), population cannot appear immediately in state 2 or 3 (ρ_{22} or ρ_{33}); there first must be a buildup of one or more off-diagonal terms such as ρ_{12} or ρ_{21}. This is consistent with the $\sin^2(t)$ dependence of $|C_2|^2$ in Eqs. (10.3a and 10.3b) for a two-state system (Fig. 10.1). Finally, Eqs. (B10.1.4–B10.1.9) indicate that if H_{12}, H_{13}, and H_{23} are zero, so that ψ_1, ψ_2 and ψ_3 are stationary states, then the rate of change of an off-diagonal element ρ_{nm} depends only on the product of itself and the energy difference between states n and m. This means that once all the off-diagonal elements of $\boldsymbol{\rho}$ have gone to zero they must remain there. Coherence, once lost, cannot be recovered without external perturbations.

10.3 The Stochastic Liouville Equation

Now let our ensemble of quantum systems interact with the surroundings. Transfer of energy to or from the surroundings should cause the ensemble to approach thermal equilibrium. In principle, we could use Eq. (10.23) to describe this process, provided that the density matrix includes the states of the surroundings and the Hamiltonian matrix includes a term for each interaction. But this usually would require an astronomically large matrix. Specifying all the possible quantum states of a large number of solvent molecules would be virtually impossible. It is more practical to use the explicit elements of $\overline{\boldsymbol{\rho}}$ and $\mathbf{H}$ only for an ensemble of individual quantum systems, as we have done heretofore, and to introduce interactions with the surroundings in a statistical way. A density matrix that is restricted in this way is called a *reduced* density matrix.

What becomes of the various elements of the averaged reduced density matrix $\overline{\boldsymbol{\rho}}$ as the ensemble approaches equilibrium? Consider the diagonal elements, which represent the populations of the basis states of the quantum systems. The equilibrium populations $\left(\overline{\rho_{nn}^o}\right)$ should depend on the Boltzmann factors for these states:

$$\overline{\rho_{nn}^o} = Z^{-1}\exp(-E_n/k_B T), \tag{10.26}$$

where Z is the partition function. Classical kinetic theory suggests that $\overline{\rho_{nn}}$ will evolve toward $\overline{\rho_{nn}^o}$ at a rate that depends on a set of rate constants for interconversions between state n and all the other states. If all the reaction steps follow first-order kinetics, we can write:

$$\{\partial\overline{\rho_{nn}}/\partial t\}_{stochastic} = \sum_{m\neq n}(k_{nm}\overline{\rho_{mm}} - k_{mn}\overline{\rho_{nn}}). \tag{10.27}$$

Here k_{mn} and k_{nm} are the microscopic classical rate constants for conversion of state n to state m and *vice versa*. The subscript "stochastic" indicates that we are considering relaxations that depend on random fluctuations of the surroundings, not the oscillatory, quantum-mechanical phenomena described by Eq. (10.23). The ensemble will relax to a Boltzmann distribution of populations if the ratio k_{mn}/k_{nm} is given by $\exp(-E_{nm}/k_BT)$. According to Eq. (10.27), relaxations of the diagonal elements toward thermal equilibrium do not depend on the off-diagonal elements of $\overline{\boldsymbol{\rho}}$, which is in accord with classical treatments of kinetic processes simply in terms of populations.

For a two-state system, any changes of the populations of the two states must always be equal and opposite, which means that $\overline{\rho_{11}} - \overline{\rho_{11}^o}$ and $\overline{\rho_{22}} - \overline{\rho_{22}^o}$ must both decay to zero with the same time constant, T_1:

$$\left[\partial\left(\overline{\rho_{nn}} - \overline{\rho_{nn}^o}\right)/\partial t\right]_{stochastic} = -\frac{1}{T_1}\left(\overline{\rho_{nn}} - \overline{\rho_{nn}^o}\right). \tag{10.28}$$

The relaxation time constant for a two-state system is the reciprocal of the sum of the forward and backward rate constants k_{12} and k_{21} ($1/T_1 = k_{12} + k_{21}$). In NMR and EPR spectroscopy, T_1 is called the *longitudinal relaxation time* or the *spin-lattice relaxation time*.

Assuming that the basis states used to define $\overline{\boldsymbol{\rho}}$ are stationary, the off-diagonal elements of $\overline{\boldsymbol{\rho}}$ must go to zero at equilibrium. There are several reasons for this. First, stochastic fluctuations of the diagonal elements will cause an ensemble to lose coherence. This is because stochastic kinetic processes modify the coefficients (c_k) of the individual systems at unpredictable times, imparting random phase shifts, ζ_k. We'll show in Sect. 10.5 that a relaxation of $\overline{\rho_{11}}$ and $\overline{\rho_{22}}$ occurring with rate constant $1/T_1$ causes the off-diagonal elements $\overline{\rho_{12}}$ and $\overline{\rho_{21}}$ to decay to zero with a rate constant of $1/(2T_1)$.

Inhomogeneity in the energies of the individual systems also causes the off-diagonal elements of $\overline{\boldsymbol{\rho}}$ to decay to zero. As we pointed out above, the oscillation frequency for an off-diagonal element ρ_{nm} in an individual system depends on the energy difference between states n and m in that system (Eqs. 10.15 and 10.18), and the oscillations will get out of phase if E_{nm} varies from system to system in the ensemble. This mechanism of loss of coherence is called *pure dephasing*. In a two-state system, pure dephasing can be characterized by a first-order time constant that is generally called "T_2*." The off-diagonal matrix elements then decay to zero with a composite time constant (T_2) that depends on both T_2* and $2T_1$:

$$\{\partial\overline{\rho_{nm}}/\partial t\}_{stochastic} = -\frac{1}{T_2}\overline{\rho_{nm}}, \tag{10.29a}$$

where

$$1/T_2 \approx 1/T_2^* + 1/2T_1. \tag{10.29b}$$

In magnetic resonance spectroscopy, T_2 is called the *transverse relaxation time*.

Equations (10.25), (10.27), (10.29a), and (10.29b) suggest that the dynamics of both the diagonal and off-diagonal elements of a reduced density matrix can be described by a general expression of the form

$$\partial \overline{\rho_{nm}}/\partial t = (i/\hbar)[, \overline{\boldsymbol{\rho}}, \mathbf{H}]_{nm} + \sum_{j,k} R_{nm.jk}\overline{\rho_{jk}}, \tag{10.30}$$

where the bars again denote averages over the ensemble, the commutator describes the quantum mechanical processes that underlie the von Neumann equation (Eq. 10.24), and the $R_{nm,jk}$ are a set of rate constants for stochastic processes. More abstractly, we can write

$$\partial \overline{\boldsymbol{\rho}}/\partial t = (i/\hbar)[\overline{\boldsymbol{\rho}}, \mathbf{H}] + \mathbf{R}\overline{\boldsymbol{\rho}}. \tag{10.31}$$

Equation (10.31) is called the *stochastic Liouville equation* and **R** is the *relaxation matrix*.

10.4 Effects of Stochastic Relaxations on the Dynamics of Quantum Transitions

We now are ready to examine how stochastic relaxations affect the dynamics of transitions between two quantum states. We will use the stochastic Liouville equation with phenomenological first-order relaxation time constants $1/T_1$ and $1/T_2$ for the diagonal and off-diagonal density matrix elements, respectively, although we'll see later that the off-diagonal elements generally have a more complicated time dependence in systems with more than two states.

Figure 10.3 shows the calculated dynamics of transitions between two states that are separated by an energy gap E_{12} and coupled by an interaction matrix element (H_{12}) of $E_{12}/4$. Stochastic transitions between the two quantum states are assumed to be negligible ($T_1 = 10^4\, h/H_{21}$), so the ensemble loses coherence only by pure dephasing. The curves shown in the figure were generated by using a Runge-Kutta procedure to integrate Eq. (10.30) numerically, starting with $\overline{\rho_{11}} = 1.0$ and $\overline{\rho_{12}} = \overline{\rho_{21}} = \overline{\rho_{22}} = 0$ at time 0. (In a Runge-Kutta integration of a set of differential equations, the derivatives with respect to the independent variable are evaluated at the initial value of the variable and after a series of small increments in the variable. The results are used to approximate the dependent variables after the last increment, and the process is repeated until the integral converges or reaches a limit.) Panel *A* shows the results when the time constant for pure dephasing (T_2^*) is much longer than $|h/H_{12}|$. An ensemble starting in state 1 then undergoes the sustained oscillations we saw in Fig. 10.1B. The real parts of the off-diagonal elements of $\overline{\boldsymbol{\rho}}$

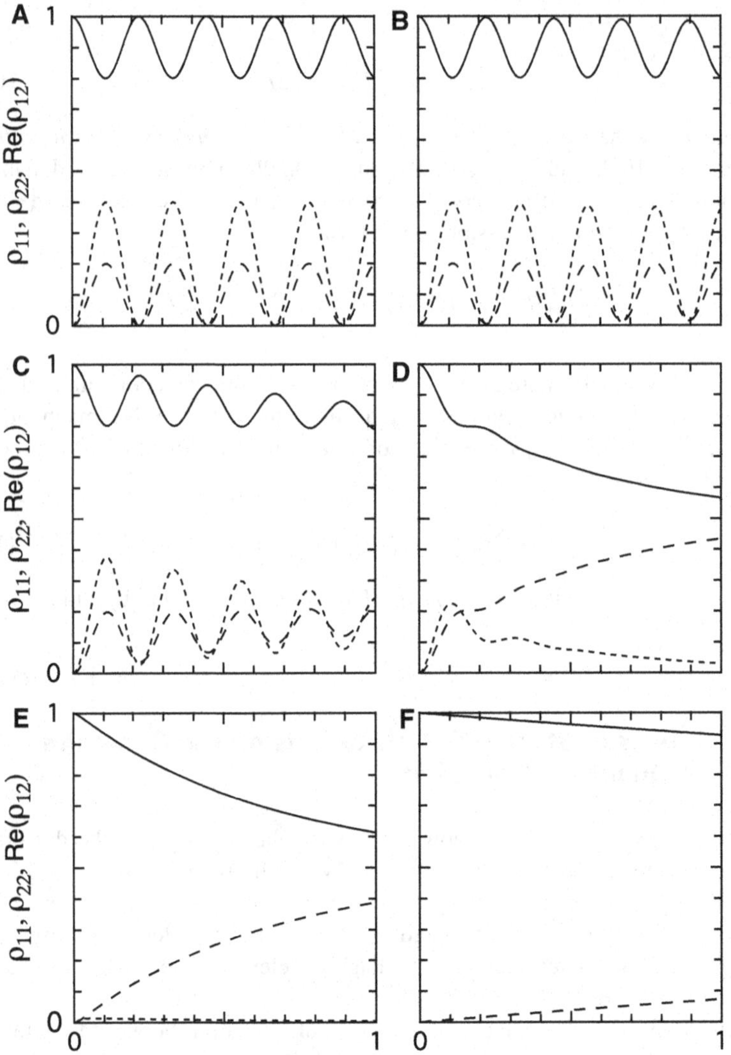

Fig. 10.3 Elements of $\overline{\rho}$ for an ensemble of systems with two states that are coupled quantum mechanically by an interaction matrix element (H_{21}) of $E_{12}/4$, where E_{12} is the energy gap between the states ($H_{11} - H_{22}$). The *solid curves* show ρ_{11} (the population of state 1); the *long dashes*, ρ_{22} (the population of state 2); and the *short dashes*, the real part of ρ_{12} or ρ_{21} [$(\rho_{12}+\rho_{21})/2$]. The time constant for pure dephasing (T_2^*) was 10^3 h/H_{21} (**A**), 10 h/H_{21} (**B**), h/H_{21} (**C**), 0.1 h/H_{21} (**D**), 0.01 h/H_{21} (**E**), or 0.001 h/H_{21} (**F**). The time constant for stochastic transitions between the two states (T_1) was 10^4 h/H_{21} in all cases. If $E_{12} = 100$ cm^{-1} and $H_{21} = 25$ cm^{-1}, the oscillation period ($|h/2\Omega|$ in Eq. (10.6)) is 0.3 ps [(3.33 × 10^{-11} cm^{-1} s)/(111.8 cm^{-1})] and the full time scale shown is 1.33 ps

(*dotted curve*) oscillate in phase with $\overline{\rho_{22}}$. In the opposite limit, when pure dephasing occurs very rapidly $\left(T_2^* \ll |h/H_{12}|\right)$, $\overline{\rho_{12}}$ and $\overline{\rho_{21}}$ are held close to zero, the oscillations of $\overline{\rho_{11}}$ and $\overline{\rho_{22}}$ are severely damped, and interconversions of states 1 and 2 are suppressed (Fig. 10.3F). The last effect can be understood by recalling that the rates of change of $\overline{\rho_{11}}$ and $\overline{\rho_{22}}$ depend on the off-diagonal elements $\overline{\rho_{12}}$ and $\overline{\rho_{21}}$ (Eqs. B10.1.1 and B10.1.2). If the off-diagonal elements of $\overline{\boldsymbol{\rho}}$ are quenched rapidly, changes in the diagonal elements must slow down. Figure 10.3B–E shows results for intermediate situations in which the oscillations are damped less severely. Under these conditions, the population of state 1 decreases asymptotically with time, while the population of state 2 increases.

Note that the ensembles considered in Fig. 10.3 all evolve toward an equal mixture of states 1 and 2, without regard to the temperature or the value of E_{12}. This is because the stochastic decay of $\overline{\rho_{12}}$ and $\overline{\rho_{21}}$ caused by pure dephasing is equally likely to trap the quantum system in either state. Thermal equilibrium would be attained only at times exceeding T_1, which in Fig. 10.3 is taken to be very long.

Now consider two states that are coupled by the same interaction matrix element (H_{21}) but have equal energies ($E_{12} = 0$), and suppose that state 2 can relax stochastically to a third state of much lower energy. The ensemble starts in state 1, and we are interested in how rapidly population appears in state 3. This is a simple model for resonance energy transfer, which can be pulled along by thermal equilibration of the products with the surroundings. Figure 10.4 shows the time dependence of the populations calculated with various values of the time constant for stochastic relaxation of state 2 to state 3 (T_1). To focus on the effects of the relaxation, the time constant for pure dephasing (T_2*) is assumed to be very long. We also assume, as in Fig. 10.3, that interconversion of states 1 and 2 is driven only by the quantum mechanical coupling factor H_{21}, so that T_1 applies only to interconversion of 2 and 3 ($R_{11,22}$ and $R_{22,11} = 0$); $|H_{13}|$ also is taken to be zero.

If the thermal equilibration of states 2 and 3 is very slow, the oscillations between states 1 and 2 continue indefinitely (Fig. 10.4A). As the time constant for conversion of state 2 to state 3 is decreased, the oscillations are damped and state 3 is formed more rapidly (Fig. 10.4B–D). But when T_1 becomes much less than $|h/H_{21}|$, the rate of formation of state 3 decreases again (Fig. 10.4E,F)! This quantum mechanical effect is completely contrary to what one would expect from a classical kinetic model of a two-step process, where increasing the rate constant for conversion of the intermediate state to the final product can only speed up the overall reaction (Box 10.2). In the stochastic Liouville equation, the slowing of the overall process results from very rapid quenching of the off-diagonal terms of $\overline{\boldsymbol{\rho}}$ by the stochastic decay of state 2. This is essentially the same as the slowing of equilibration of two quantum states when T_2^* is much less than $|h/H_{12}|$, which we saw in Fig. 10.3.

In Fig. 10.5, the relaxation rates of the ensembles considered in Figs. 10.3 and 10.4 are plotted as functions of $1/T_2^*$ and $1/T_1$, respectively. The left-hand limb of each curve shows the customary speeding up of the relaxation as $1/T_2^*$ or $1/T_1$ increases; the right-hand limb shows the slowing when $1/T_2^*$ or $1/T_1$ becomes much larger than H_{21}/h.

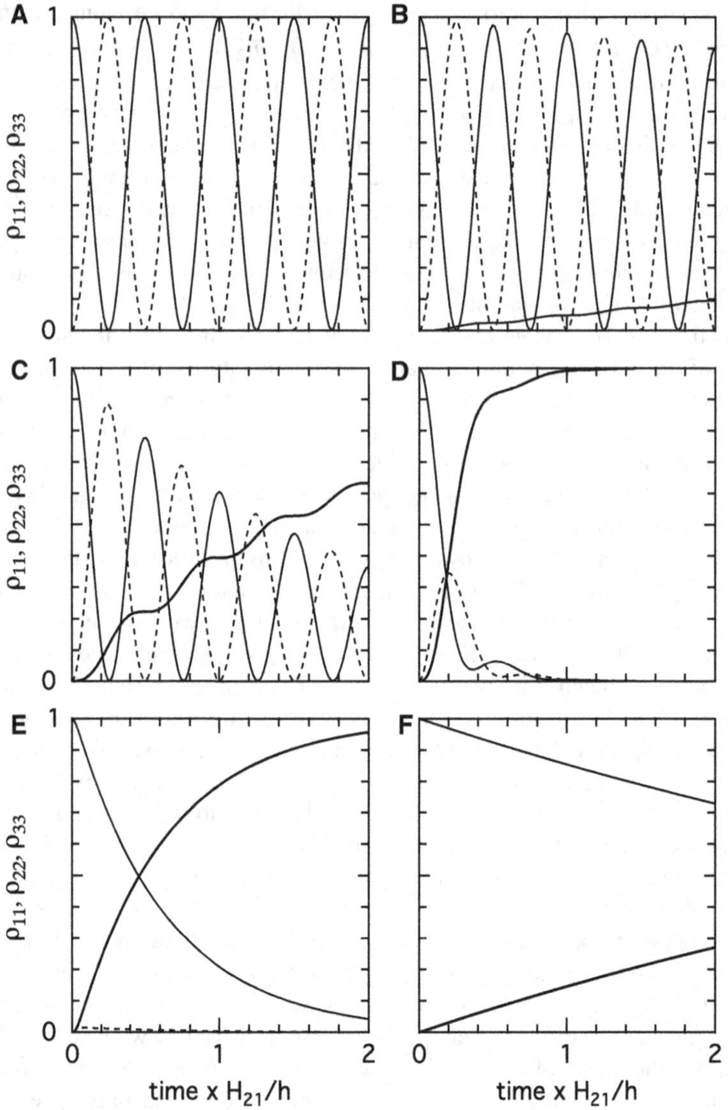

Fig. 10.4 Populations of states 1, 2 and 3 in an ensemble of systems in which states 1 (*thin, solid curve*) and 2 (*dashed curve*) are isoenergetic ($E_{21} = 0$) and are coupled quantum mechanically by an interaction matrix element H_{21}, and state 3 (*thick, solid curve*) lies below 1 and 2 by $40 \times |H_{21}|$. State 3 is formed stochastically from state 2 with a time constant T_1 of ∞ (**A**), $10\,h/H_{21}$ (**B**), h/H_{21} (**C**), $0.1\,h/H_{21}$ (**D**), $0.01\,h/H_{21}$ (**E**), or $0.001\,h/H_{21}$ (**F**). The temperature is assumed to be much less than less than E_{23}/k_B, making the decay of state 2 to state 3 effectively irreversible. The time constant for pure dephasing (T_2^*) was $750\,h/H_{21}$. If $H_{21} = 25$ cm^{-1} and $E_{23} = 1{,}000$ cm^{-1}, the oscillation period ($|h/2H_{21}|$ in Eq. (10.7)) is 0.67 ps [(3.33×10^{-11} cm^{-1} s)/(50 cm^{-1})], $T_2^* = 10^3$ ps, and the full time scale shown is 2.66 ps

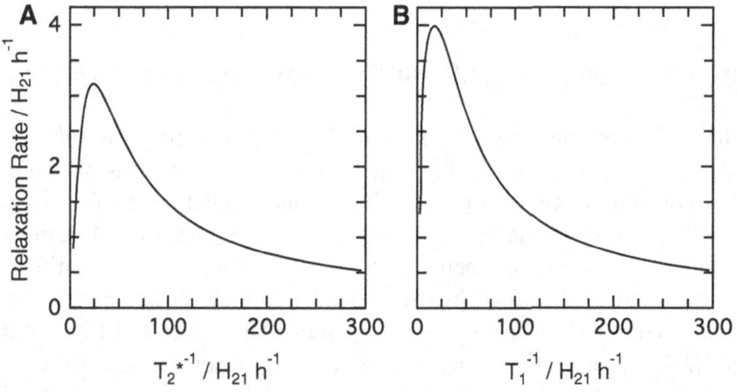

Fig. 10.5 (**A**) Relaxation rate for the ensemble described in Fig. 10.3 as a function of $1/T_2^*$. The quantity plotted is the reciprocal of the time required for the population of state 1 to fall from 1 to 0.816 $(1-0.5/e)$. (**B**) Relaxation rate for the ensemble described in Fig. 10.4 as a function of $1/T_1$. The quantity plotted is the reciprocal of the time for the population of state 3 to rise from 0 to 0.632 $(1-1/e)$

Box 10.2 The "Watched-Pot" or "Quantum Zeno" Paradox

As Figs. 10.3 and 10.5 illustrate, the stochastic Liouville equation predicts that the overall rate of a two-step process will decrease if the rate of stochastic equilibration of states 2 and 3 becomes very large. Although its mathematical origin in the stochastic Liouville equation is straightforward, there are several ways of rationalizing this strange effect. In one view, the stochastic process tests the state of each system in the ensemble at intervals of approximately T_1. All systems that are indentified as being in state 2 are converted to state 3. After each such sorting, the clock for the coherent oscillations between states 1 and 2 is reset to zero for all the systems that were found to be in state 1. Because the probability of being in state 2 at very early times increases with $\sin^2 t$, the overall rate of depletion of state 1 is approximately proportional to $(1/T_1)\sin^2(T_1)$, which becomes approximately equal to T_1 as T_1 approaches 0. This reasoning recalls the adage that the more closely you watch a pot of water on a stove, the longer the water takes to boil. It also calls to mind Zeno's classic paradox in which a tortoise argued that he could outrun Achilles by dividing time and distance into sufficiently small intervals.

Another way to view the situation is that a very rapid decay of state 2 broadens the homogeneous distribution of energies associated with this state. As Fig. 10.1 illustrates, the amplitude of oscillations between states 1 and 2 falls off rapidly as the two states move out of resonance. According to Fermi's golden rule (Eq. 7.10) the average rate depends on the density of

(continued)

Box 10.2 (continued)
states in the region of $E_{12}=0$, which decreases as the energy distribution broadens.

The three-state model considered in Fig. 10.4 is pertinent to the radical-pair states that are formed in the initial electron-transfer steps of photosynthesis. Transfer of a electron from the excited singlet state of an electron donor (D^*) to a nearby acceptor (A) generates a radical pair (D^+A^-), in which the spins of the unpaired electrons on the two radicals remain antiparallel. The radical pair thus is born in a singlet state overall. However, because the unpaired electrons are on separate molecules, the singlet and triplet radical pairs ($^1[D^+A^-]$ and $^3[D^+A^-]$) are similar in energy. Interactions of the electron spins with nuclear spins of the two molecules can cause the radical pair to oscillate back and forth between the singlet and triplet states. If secondary electron-transfer reactions are blocked, the radical pair eventually decays by a back reaction in which an electron moves from A^- to D^+. Back reactions that occur from $^3[D^+A^-]$ place the original electron donor in a triplet excited state (^3D); back reactions occurring from $^1[D^+A^-]$ regenerate the (singlet) ground state. From this analysis, we would expect the yield of ^3D at first to increase with an increase in the rate constant for the step $^3[D^+A^-] \rightarrow {}^3D$, and then to decrease again as the rate constant becomes very large. See [7–17] for further discussion of the quantum Zeno effect and the "anti-Zeno" effect (acceleration of a decay process by frequent measurements) in this and other systems. The effect was first observed experimentally for the escape of cold Na atoms trapped in a potential well [18].

The Marcus equation for nonadiabatic electron-transfer reactions (Eq. B5.3.4), and the Förster theory that we discussed in Chap. 7 apply only to systems with weak intermolecular interactions, which we now can define more precisely as meaning that $|H_{21}|/h \ll 1/T_2$. It is instructive to derive a steady-state approximation to the stochastic Liouville equation for a two-state reaction in this limit. From Eqs. (B10.1.15), (10.29a), (10.29b), and (10.30), we have

$$\partial\overline{\rho_{12}}/\partial t = (i/\hbar)\{(\overline{\rho_{11}} - \overline{\rho_{22}})H_{12} + \overline{\rho_{12}}(H_{22} - H_{11})\} - \overline{\rho_{12}}/T_2. \tag{10.32}$$

If $|H_{21}|/h \ll 1/T_2$, then $|\partial\overline{\rho_{12}}/\partial t|$ will be close to zero in the steady state, so from Eq. (10.32),

$$\overline{\rho_{12}} \approx \frac{(i/\hbar)(\overline{\rho_{22}} - \overline{\rho_{11}})H_{12}}{(i/\hbar)(H_{22} - H_{11}) - 1/T_2} = \frac{(\overline{\rho_{22}} - \overline{\rho_{11}})H_{12}}{(H_{22} - H_{11}) + i\hbar/T_2} \tag{10.33a}$$

and

$$\overline{\rho_{21}} = \overline{\rho_{12}}^* = \frac{(\overline{\rho_{22}} - \overline{\rho_{11}})H_{21}}{(H_{22} - H_{11}) - i\hbar/T_2}. \tag{10.33b}$$

Assuming that states 1 and 2 are coupled only by H_{21}, we also have

$$\partial\overline{\rho_{22}}/\partial t = (i/\hbar)(\overline{\rho_{21}}H_{12} - \overline{\rho_{12}}H_{21}), \tag{10.34}$$

from Eq. (B10.1.2). Inserting the steady-state values of $\overline{\rho_{12}}$ and $\overline{\rho_{21}}$ (Eqs. 10.33a and 10.33b) into Eq. (10.34) gives

$$\partial\overline{\rho_{22}}/\partial t = (\overline{\rho_{11}} - \overline{\rho_{22}})\,|H_{21}|^2\left(\frac{2\hbar/T_2}{(E_{12})^2 + (\hbar/T_2)^2}\right). \tag{10.35}$$

The steady-state rate of population of state 2 thus has a Lorentzian dependence on the energy gap E_{12}. As we discussed in Chap. 2, the Lorentzian function can be equated to the homogeneous distribution of E_{12} when the mean value of E_{12} is zero and state 2 has a lifetime of $T_2/2$. Note that, according to Eq. (10.29b), $T_2/2 = T_1$ when pure dephasing is negligible. If we identify the time constant T_2 in Eq. (10.35) with $2T$ in Eq. (2.71), and identify the energy difference E_{12} with $(E - E_a)$, then the factor in the second set of parentheses in Eq. (10.35) must be $2\pi/\hbar$ times the distribution function Re[$G(E)$] in Eq. (2.71).

Finally, if we define $\rho_s(0)dE_{12}$ as the number of states for which the energy gap falls within a small interval (dE_{12}) around zero, we have

$$\partial\overline{\rho_{22}}/\partial t = (\overline{\rho_{11}} - \overline{\rho_{22}})\,|H_{21}|^2\frac{2\pi}{\hbar}\rho_s(0). \tag{10.36}$$

This result is formally identical to Fermi's "golden rule" (Eqs. B5.3.3 and 7.10).

Equation (10.36) has the form of a classical first-order kinetic expression. In the limit of weak coupling where the golden rule applies, any quantum mechanical oscillations are damped so strongly that we can describe the kinetics simply in terms of stochastic transitions of the system between the reactant and product states. The net rate in the forward or backward direction then is proportional to the population difference between the two states $(\overline{\rho}_{11} - \overline{\rho}_{22})$. Persistent oscillations of the populations evidently occur only in the opposite limit, when $|H_{21}|/h$ is comparable to, or larger than $1/T_2$.

10.5 A Density-Matrix Treatment of Absorption of Weak, Continuous Light

In the last section, we used the stochastic Liouville equation to find the steady-state rate of transitions between two weakly-coupled quantum states, on the assumption that coherence decayed rapidly relative to the rate of the transitions. The resulting expression (Eq. 10.36) reproduces Fermi's golden rule. We can use the same approach to find the rate of absorption of weak, continuous light. Consider an ensemble of systems with two electronic states (a and b) that are stationary in the absence of external perturbations. In the presence of electromagnetic radiation with angular frequency ω ($\omega = 2\pi\nu$) and electric field amplitude $\boldsymbol{E}_o$, the electric-dipole matrix element that mixes the two states is

$$V_{ab} = -\boldsymbol{\mu}_{ab} \cdot \boldsymbol{E}(t) = -\boldsymbol{\mu}_{ab} \cdot \left[\boldsymbol{E}_o(t)\ \exp(i\omega t) + \boldsymbol{E}_o^*(t)\ \exp(-i\omega t)\right], \qquad (10.37)$$

where $\boldsymbol{\mu}_{ab}$ is the transition dipole $\langle\psi_a|\tilde{\mu}|\psi_b\rangle$. We have written the radiation field in a general way that will be useful in Chap. 11, although we usually will assume that its amplitude is real ($\boldsymbol{E}_o^* = \boldsymbol{E}_o$) and has only a weak dependence on time compared to the complex exponential factors. Let's also assume for now that all the molecules in the ensemble have the same transition dipole and orientation with respect to $\boldsymbol{E}_o$ so that V_{ab} is the same for all the molecules. The differential equation for $\overline{\rho_{ab}}$ (Eq. 10.32) then takes the form

$$\begin{aligned} \partial\overline{\rho_{ab}}/\partial t &= (i/\hbar)\left\{(\overline{\rho_{bb}} - \overline{\rho_{aa}})\boldsymbol{\mu}_{ab} \cdot \left[\boldsymbol{E}_o \exp(i\omega t) + \boldsymbol{E}_o^* \exp(-i\omega t)\right] + \overline{\rho_{ab}} E_{ba}\right\} - \overline{\rho_{ab}}/T_2 \\ &= (i/\hbar)(\overline{\rho_{bb}} - \overline{\rho_{aa}})\boldsymbol{\mu}_{ab} \cdot \left[\boldsymbol{E}_o \exp(i\omega t) + \boldsymbol{E}_o^* \exp(-i\omega t)\right] + (i\omega_{ba} - 1/T_2)\overline{\rho_{ab}}, \end{aligned} \qquad (10.38)$$

where E_{ba} is the energy difference between the two states ($H_{bb} - H_{aa}$) and we have defined $\omega_{ba} = E_{ba}/\hbar$ to simplify the notation. The populations $\overline{\rho_{bb}}$ and $\overline{\rho_{aa}}$ are written without an explicit time dependence here, on the assumption that at low light intensities they change only slowly relative to $\overline{\rho_{ab}}$.

As discussed in Sect. 4.2, one of the two exponential terms in the square brackets will dominate, depending on whether E_{ba} is positive or negative. If we retain the term $\boldsymbol{E}_o\exp(i\omega t)$, which dominates when $H_{bb} > H_{aa}$, then $V_{ab} = -\boldsymbol{\mu}_{ab} \cdot \boldsymbol{E}_o\exp(i\omega t)$, and Eq. (10.38) reduces to

$$\partial\overline{\rho_{ab}}/\partial t = (i/\hbar)(\overline{\rho_{bb}} - \overline{\rho_{aa}})\boldsymbol{\mu}_{ab} \cdot \boldsymbol{E}_o \exp(i\omega t) + (i\omega_{ba} - 1/T_2)\overline{\rho_{ab}}. \qquad (10.39)$$

The coherence $\overline{\rho_{ab}}$ described by Eq. (10.39) is a function of both frequency and time. Because the ensemble is driven by an oscillating electric field, we would expect the solution to this equation to include a factor that oscillates at the same frequency as the field. Let's try writing

$$\overline{\rho_{ab}}(\omega, t) = \overline{\overline{\rho_{ab}}}\exp(i\omega t), \qquad (10.40)$$

where $\overline{\overline{\rho_{ab}}}$ is an average of $\overline{\rho_{ab}}$ over an oscillation period of the field ($1/\omega$). $\overline{\overline{\rho_{ab}}}$ might change rapidly at short times after the light is turned on, but presumably becomes constant at long times ($t > T_2$ and $\gg 1/\omega$). With this substitution, Eq. (10.39) gives

$$
\begin{aligned}
\partial\overline{\rho_{ab}}/\partial t &= \exp(i\omega t)\partial\overline{\overline{\rho_{ab}}}/\partial t + i\omega\cdot\exp(i\omega t)\overline{\overline{\rho_{ab}}} \\
&= (i/\hbar)(\overline{\rho_{bb}} - \overline{\rho_{aa}})\boldsymbol{\mu}_{ab}\cdot\boldsymbol{E}_o\exp(i\omega t) \\
&\quad + (i\omega_{ba} - 1/T_2)\overline{\overline{\rho_{ab}}}\exp(i\omega t).
\end{aligned} \tag{10.41}
$$

Equation (10.41) can be solved immediately for $\overline{\overline{\rho_{ab}}}$ in a steady state, when $\partial\overline{\overline{\rho_{ab}}}/\partial t = 0$:

$$
\overline{\overline{\rho_{ab}}} = \frac{(i/\hbar)(\overline{\rho_{bb}} - \overline{\rho_{aa}})\boldsymbol{\mu}_{ab}\cdot\boldsymbol{E}_o}{i(\omega - \omega_{ba}) + 1/T_2}. \tag{10.42a}
$$

Similarly, setting $\overline{\rho_{ba}} = \overline{\overline{\rho_{ba}}}\exp(-i\omega t)$ gives

$$
\overline{\overline{\rho_{ba}}} = \frac{(i/\hbar)(\overline{\rho_{bb}} - \overline{\rho_{aa}})\boldsymbol{\mu}_{ba}\cdot\boldsymbol{E}_o^*}{i(\omega - \omega_{ba}) - 1/T_2}. \tag{10.42b}
$$

Here we retained only the $\exp(-i\omega t)$ component of $\boldsymbol{E}$, making $V_{ba} = -\mu_{ba}\cdot E_o^*\exp(-i\omega t)$. This is consistent with the hermiticity of the Hamiltonian operator: if $V_{ab} = -\boldsymbol{\mu}_{ab}\cdot\boldsymbol{E}_o\exp(i\omega t)]$, then $V_{ba} = V_{ab}^* = -\boldsymbol{\mu}_{ba}\cdot\boldsymbol{E}_o^*\exp(-i\omega t)$.

To find the steady-state rate of excitation of molecules from state *a* to state *b* in the absence of stochastic decay processes, we now can use Eq. (10.21c):

$$
\begin{aligned}
\partial\overline{\rho_{bb}}/\partial t &= (i/\hbar)(\overline{\rho_{ba}}V_{ab} - \overline{\rho_{ab}}V_{ba}) \\
&\approx (i/\hbar)\left\{\left[\exp(-i\omega t)\overline{\overline{\rho_{ba}}}\right]\left[-\boldsymbol{\mu}_{ab}\cdot\boldsymbol{E}_o^*\exp(i\omega t)\right]\right. \\
&\quad \left. - \left[\exp(i\omega t)\overline{\overline{\rho_{ab}}}\right]\left[-\boldsymbol{\mu}_{ba}\cdot\boldsymbol{E}_o\exp(-i\omega t)\right]\right\} \\
&= (\overline{\rho_{bb}} - \overline{\rho_{aa}})\frac{|\boldsymbol{\mu}_{ba}\cdot\boldsymbol{E}_o|^2}{\hbar^2}\left(\frac{1}{i(\omega_{ba}+\omega) - 1/T_2} - \frac{1}{i(\omega_{ba}-\omega) + 1/T_2}\right) \\
&= (\overline{\rho_{aa}} - \overline{\rho_{bb}})\frac{|\boldsymbol{\mu}_{ba}\cdot\boldsymbol{E}_o|^2}{\hbar^2}\left(\frac{2/T_2}{(\omega_{ba}-\omega)^2 + (1/T_2)^2}\right).
\end{aligned} \tag{10.43}
$$

Equation (10.43) is just the same as Eq. (10.35) for the particular case of absorption of light. It predicts again that the absorption spectrum will be a Lorentzian function of frequency. The width of the Lorentzian corresponds to the homogeneous distribution of transition energies when the effective lifetime of the excited state is $T_2/2$, as we discussed in Sect. 10.4. However, this result depends on our assumption that dephasing of the ensemble is described adequately by an exponential decay with single time constant, T_2. As we will discuss in Sect. 10.7, the absorption band shape depends on a Fourier transform of the dephasing dynamics, which usually is more complex than we have assumed here. Note also that Eq. (10.43) pertains only to times greater than T_2. We'll consider shorter times in the following chapter.

The total steady-state rate of excitation also can be written as

$$\partial \overline{\rho_{bb}}/\partial t = -\partial \overline{\rho_{aa}}/\partial t = (2\pi/\hbar)(\overline{\rho_{aa}} - \overline{\rho_{bb}})|\boldsymbol{\mu}_{ba} \cdot \boldsymbol{E}_o|^2 \rho_s(E_{ba}), \tag{10.44}$$

where $\rho_s(E_{ba})dE$ is the number of states for which the excitation energy is within dE of E_{ba}. This again is equivalent to Fermi's golden rule (Eqs. B5.3.3, 7.10 and 10.36). As in Eq. (10.36), the density of states ρ_s here has units of reciprocal energy (e.g., states per cm^{-1}). Equation (4.8) has an additional factor of h in the denominator because ρ_v, the corresponding density of oscillation modes in units of reciprocal frequency (modes per Hz^{-1}), is $h\rho_s$.

If the molecules in the ensemble have different orientations relative to the excitation, then Eqs. (10.43) and (10.44) require additional averaging over the orientations. For an isotropic sample, the average value of $|\boldsymbol{\mu}_{21} \cdot \boldsymbol{E}_o|^2$ is $(1/3)|\boldsymbol{\mu}_{21}|^2|\boldsymbol{E}_o|^2$ (Eq. 4.11; Boxes 4.6 and 10.5).

Equations (10.43) and (10.44) also describe the rate of stimulated emission. The corresponding rates for the situation that $H_{bb} < H_{aa}$ are obtained in the same manner by retaining the term $\exp(-i\omega t)$ instead of $\exp(i\omega t)$ in V_{ab} and *vice versa* for V_{ba}.

10.6 The Relaxation Matrix

Our discussion so far has focused on systems with only two quantum states, and our treatment of the time constants T_1 and T_2* has been entirely phenomenological. We now discuss the elements of the relaxation matrix **R** in a more general way and consider how they depend on the strengths and dynamics of interactions with the surroundings.

Following work by Pound, Bloch and others, Alfred Redfield [19] explored descriptions of relaxations of the density matrix by the stochastic Liouville equation (Eqs. 10.24 and 10.28–10.31). Redfield's treatment showed clearly how the main elements of the relaxation matrix **R** depend on the frequencies and strengths of fluctuating interactions with the surroundings. His basic approach was to write the Hamiltonian matrix for the system as $\mathbf{H}(t) = \mathbf{H_0} + \mathbf{V}(t)$, where $\mathbf{H_0}$ is independent of time and $\mathbf{V}(t)$ represents interactions with fluctuating electric or magnetic fields from the surroundings. Redfield then used the von Neumann-Liouville equation (Eq. 10.24) to find the effects of $\mathbf{V}(t)$ on $\overline{\boldsymbol{\rho}}$ for the system. In the following outline of the derivation, we use the Schrödinger representation of the density matrix for simplicity and consistency with Sects. 10.2 and 10.3. Redfield [19] and Slichter [20], who provided an excellent introduction to the theory, used the interaction representation and then returned to the Schrödinger picture for the final expressions.

If we know the averaged reduced density matrix for the ensemble at zero time ($\overline{\boldsymbol{\rho}}(0)$), we can obtain an estimate of $\overline{\boldsymbol{\rho}}$ for a later time (t) by integrating the von Neumann-Liouville equation (Eq. 10.24), using $\overline{\boldsymbol{\rho}}(0)$ in the commutator:

$$\overline{\boldsymbol{\rho}}(t) = \overline{\boldsymbol{\rho}}(0) + \frac{i}{\hbar}\int_0^t \left[\overline{\boldsymbol{\rho}(0), \mathbf{H}(t_1)}\right] dt_1. \tag{10.45}$$

For a better estimate, we could use Eq. (10.45) to find $\overline{\boldsymbol{\rho}}$ at an intermediate time (t_2) and then use $\overline{\boldsymbol{\rho}}(t_2)$ instead of $\overline{\boldsymbol{\rho}}(0)$ in the commutator:

$$\overline{\boldsymbol{\rho}}(t) = \frac{i}{\hbar}\int_0^t \left[\overline{\left(\overline{\boldsymbol{\rho}}(0) + \frac{i}{\hbar}\int_0^{t_2} [\boldsymbol{\rho}(0), \mathbf{H}(t_1)]\, dt_1\right), \mathbf{H}(t_2)}\right] dt_2. \tag{10.46}$$

Differentiating this expression gives $\partial\overline{\boldsymbol{\rho}}/\partial t$ at time t:

$$\frac{\overline{\partial \boldsymbol{\rho}(t)}}{\partial t} = \frac{i}{\hbar}\left[\overline{\boldsymbol{\rho}(0), \mathbf{H}(t)}\right] + \left(\frac{i}{\hbar}\right)^2 \int_0^t \left[\overline{[\boldsymbol{\rho}(0), \mathbf{H}(t_1)], \mathbf{H}(t)}\right] dt_1, \tag{10.47}$$

where, as indicated above, $\mathbf{H}(t) = \mathbf{H_0} + \mathbf{V}(t)$.

According to the ergodic hypothesis of statistical mechanics, the ensemble averages indicated by bars in Eq. (10.47) are equivalent to averages for a single system over a long period of time. If the fluctuating interactions with the surroundings ($V(t)$) vary randomly between positive and negative values, the matrix elements of **V** will average to zero:

$$\overline{V_{nm}(t)} = \overline{\left\langle \psi_n \middle| \widetilde{\mathrm{V}}(t) \middle| \psi_m \right\rangle} = 0. \tag{10.48}$$

(Any constant interactions with the surroundings could be included in $\mathbf{H_0}$ and used simply to redefine the basis states.) $\left[\overline{\boldsymbol{\rho}(0), \mathbf{V}(t)}\right]$ thus is zero and does not contribute to $\partial\overline{\boldsymbol{\rho}}(t)/\partial t$. The average value of $(V_{nm}(t))^2$ or $V_{nm}(t)V_{jk}(t)$, however, generally is not zero, and these factors contribute to $\partial\, \overline{\boldsymbol{\rho}}\, (t)/\partial t$ through the integral on the right side of Eq. (10.47).

The integrand in Eq. (10.47) includes a sum of terms of the form $\overline{\boldsymbol{\rho}}(0)\overline{V_{nm}(t_1)V_{jk}(t_2)}$, each of which involves the product of the density matrix at zero time with a *correlation function* or *memory function*, $M_{nm,\,jk}(t_1, t_2) = \overline{V_{nm}(t_1)V_{jk}(t_2)}$. $M_{nm,jk}(t_1,t_2)$ is the ensemble average of the product of V_{nm} at time t_1 with V_{jk} at time t_2. If V_{nm} and V_{jk} vary randomly, their average product should not depend on the particular times t_1 and t_2, but only on the difference, $t = t_2 - t_1$. The correlation function $M_{nm,jk}$ therefore can be written as a function of this single variable:

$$M_{nm,\,jk}(t) = \overline{V_{nm}(t_1)V_{jk}(t_1 + t)}, \tag{10.49}$$

where the bar implies an average over time t_1 as well as an average over the ensemble. For $jk = nm$, $M_{nm,jk}(t)$ is the same as the autocorrelation function of

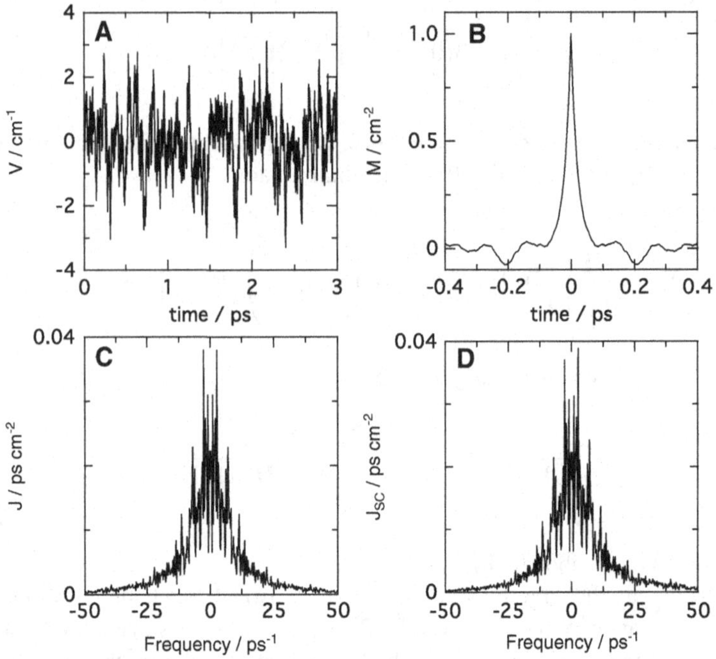

Fig. 10.6 (**A**) A fluctuating interaction energy, $V(t)$, with a mean value $\left(\overline{V}\right)$ of zero and a mean square $\left(M(0) = \overline{V^2}\right)$ of 1.0 cm^{-2} (root mean square 1.0 cm^{-1}). (**B**) The autocorrelation function, $M(t)$, of $V(t)$. (**C**) The spectral density function, $J(\nu)$, of $V(t)$ as defined by Eq. (10.52). (Some authors use "spectral density function" to refer the product $\nu J(\nu)$ or $\omega J(\omega)$.) (**D**) A semiclassical spectral density function $J_{SC}(\nu) = J(\nu)/[1 + \exp(-h\nu/k_BT)]$ for $T = 295$ K ($k_BT = 205$ cm^{-1}). Note the slight asymmetry of $J_{SC}(\nu)$ relative to $J(\nu)$

V_{nm} (Eq. 5.79). Panel *B* of Fig. 10.6 shows the autocorrelation function of the fluctuating quantity shown in panel *A*.

The initial value of a correlation function is the mean product of the fluctuating quantities, $M_{nm,\,jk}(0) = \overline{V_{nm}V_{jk}}$, which for an autocorrelation function is simply the mean-square amplitude of the fluctuations, $M_{nm,nm}(0) = \overline{|V_{nm}|^2}$. At long times, $M_{nm,jk}$ approaches the product of the two means, $M_{nm,\,jk}(\infty) = \overline{V_{nm}}\;\overline{V_{jk}}$, which is zero (Eq. 10.48). In the simplest model, $M_{nm,jk}(t)$ decays exponentially to zero with a single time constant τ_c called the *correlation time*: $M_{nm,jk}(t) = M_{nm,jk}(0)\exp(-t/\tau_c)$. The correlation time for more complex decays can be defined by the expression

$$\tau_c = \frac{1}{M_{nm,\,jk}(0)}\int_0^\infty M_{nm,\,jk}(t)dt, \qquad (10.50)$$

which reduces to the exponential time constant if the decay is monoexponential. Redfield's analysis does not make any assumptions about the detailed time course of the decay, but it does assume that the correlation functions decay rapidly relative to the phenomenological relaxation time constants T_1 and T_2^* that we introduced in Eqs. (10.28), (10.29a), and (10.29b) (i.e., that $\tau_c \ll T_1, T_2^*$).

The correlation function $M_{nm,jk}(t)$ describes the fluctuations of $V_{nm}V_{jk}$ as a function of time: slow fluctuations have long correlation times. The same fluctuations also can be described as a function of frequency (ν) or angular frequency ($\omega = 2\pi\nu$) by taking a Fourier transform of $M_{nm,jk}(t)$. According to the *Wiener-Khinchin theorem* of statistical mechanics, the Fourier transform of the autocorrelation function of a fluctuating quantity is the *power spectrum* of the fluctuations, where the power at a given frequency is the mean square amplitude of the fluctuations at that frequency. To convert $M_{nm,jk}(t)$ to a suitable function of frequency, Redfield [19] defined the *spectral density function*, $J_{nm,jk}(\omega)$, as

$$J_{nm,\,jk}(\omega) = \int_0^{\infty} M_{nm,\,jk}(t)\exp(i\omega t)dt. \tag{10.51}$$

In this formulation, the integral on the right includes only positive frequencies. But if the correlation function is a real and even function of time $[M_{nm,jk}(-t) = M_{nm,jk}(t)]$, then $J_{nm,jk}(\omega)$ also is real and is just half the integral from $t = -\infty$ to ∞,

$$J_{nm,\,jk}(\omega) = \frac{1}{2}\int_{-\infty}^{\infty} M_{nm,\,jk}(t)\exp(i\omega t)dt. \tag{10.52}$$

This is, except for a normalization factor, the Fourier transform of the correlation function (Appendix A3). The Fourier transform in Eq. (10.51) is sometimes called a "half" Fourier transform.

Figure 10.6C shows the spectral density function obtained from the correlation function shown in Fig. 10.6B. If the correlation function $M_{nm,jk}$ decays exponentially with a single time constant τ_c, $J_{nm,jk}(\omega)$ is a Lorentzian peaking at $\omega = 0$ (Appendix A3 and Eq. (2.70)). The Lorentzian has a peak amplitude of $\overline{V_{nm}V_{jk}}$ and a width at half-maximal amplitude of $2/\tau_c$.

With these definitions, the elements of the relaxation matrix $\mathbf{R}$ can be written [19–21]

$$R_{nm,\,jk} = \frac{1}{2\hbar^2}\left[J_{nj,mk}(\omega_{mk}) + J_{nj,mk}(\omega_{nj})\right] - \delta_{mk}\sum_i J_{ij,in}(\omega_{ij})$$
$$- \delta_{nj}\sum_i J_{im,ik}(\omega_{ik}), \tag{10.53}$$

where $\omega_{jk} = (H_{jj} - H_{kk})/\hbar = E_{jk}/\hbar$, δ_{jk} is the Kronecker delta function (1 if $j = k$, 0 if $j \neq k$), and $J_{nm,jk}(\omega_{jk})$ means the value of $J_{nm,jk}$ at frequency ω_{jk}. The relaxation matrix described by Eqs. (10.51) and (10.53) cannot be entirely correct, however, because it does not take the ensemble to a Boltzmann equilibrium at long times. At equilibrium, the diagonal elements of $\overline{\boldsymbol{\rho}}$ (the populations of the various basis states) should depend on the relative energies of the states, so that $\overline{\rho}_{nn}/\overline{\rho}_{mm} = \exp(-\hbar\omega_{nm}/k_BT)$. This requires the rate constants for forward and backward transitions between two states to have the relationship $R_{nn,mm}/R_{mm,nn} = \exp(\hbar\omega_{nm}/k_BT)$. Instead, Eq. (10.53) gives $R_{nn,mm} = R_{mm,nn}$, which makes all the diagonal elements of $\overline{\boldsymbol{\rho}}$ eventually become equal, in conflict with the principle of detailed balance. The problem is that Eq. (10.53) considers only the density matrix of the *system*. It neglects transfer of energy to or from the surroundings, which must occur in order for the system to reach thermal equilibrium. We encountered the same problem in Fig. 10.3, where we found that pure dephasing alone does not take an ensemble of two-state systems to a Boltzmann equilibrium.

To satisfy detailed balance, Redfield [19] multiplied the spectral density function by the factor $\exp(-\hbar\omega_{nm}/2k_BT)$. More recent authors [22–24] have used "semiclassical" spectral density functions of the form

$$J^{SC}_{nm,\,jk}\left(\omega_{\alpha\beta}\right) = J_{nm,\,jk}\left(\omega_{\alpha\beta}\right)\left[1 + \exp\left(\hbar\omega_{\alpha\beta}/k_BT\right)\right]^{-1}, \tag{10.54}$$

where $J_{nm,jk}$ is the classical spectral density function defined by Eq. (10.51). Figure 10.6D shows such a modified spectral density function. These semiclassical spectral density functions incorporate the quantum mechanical zero-point energy, and they meet the requirement for detailed balance, $J^{SC}_{nn,mm}(\omega_{nm}) = J^{SC}_{mm,nn}\exp(\hbar\omega_{mn}/k_BT)$, as you can see by the following algebraic manipulations:

$$\begin{aligned} J^{SC}_{nn,mm}(\omega_{nm})\exp(\hbar\omega_{nm}/k_BT) &= J_{nn,mm}(\omega_{nm})\frac{\exp(\hbar\omega_{nm}/k_BT)}{1+\exp(\hbar\omega_{nm}/k_BT)} \\ &= J_{nn,mm}(\omega_{nm})[1+\exp(-\hbar\omega_{nm}/k_BT)]^{-1} \\ &= J_{mm,nn}(\omega_{mn})[1+\exp(\hbar\omega_{mn}/k_BT)]^{-1} = J^{SC}_{mm,nn}(\omega_{mn}). \end{aligned} \tag{10.55}$$

The semiclassical spectral density function $J^{SC}(\omega)$ can be written as a sum of two components that are, respectively, even and odd functions of ω. The fact that the odd component is not zero implies that the correlation function obtained by an inverse Fourier transform of $J^{SC}(\omega)$ is complex (Appendix A3). The imaginary component of the correlation function can be viewed as conveying the directionality of time that is missing in classical physics. We'll return to this point in Sect. 10.7.

Replacing the classical spectral density functions by their semiclassical analogs in Eq. (10.53) gives the following expression for the relaxation matrix:

$$R_{nm,jk} = \frac{1}{2\hbar^2}\left(\frac{J_{nj,mk}(\omega_{mk})}{1+\exp(\hbar\omega_{mk}/k_BT)} + \frac{J_{nj,mk}(\omega_{nj})}{1+\exp(\hbar\omega_{nj}/k_BT)}\right)$$
$$- \delta_{mk}\sum_i \frac{J_{ij,in}(\omega_{ij})}{1+\exp(\hbar\omega_{ij}/k_BT)} - \delta_{nj}\sum_i \frac{J_{im,ik}(\omega_{ik})}{1+\exp(\hbar\omega_{ik}/k_BT)}. \tag{10.56}$$

The elements of **R** for a two-state system are given in Box 10.3.

Box 10.3 The Relaxation Matrix for a Two-State System

Equation (10.56) gives the following results for a two-state system. We have written out all the terms in the sums only for $R_{11,11}$, $R_{11,22}$, and $R_{12,12}$; the other matrix elements are obtained from these easily by symmetry.

$$R_{11,11} = \frac{1}{2\hbar^2}\left(J_{11,11}(0) + J_{11,11}(0) - J_{11,11}(0) - \frac{J_{21,21}(\omega_{21})}{1+\exp(\hbar\omega_{21}/k_BT)}\right.$$
$$\left. - J_{11,11}(0) - \frac{J_{21,21}(\omega_{21})}{1+\exp(\hbar\omega_{21}/k_BT)}\right)$$
$$= -\frac{1}{\hbar^2}\left(\frac{J_{21,21}(\omega_{21})}{1+\exp(\hbar\omega_{21}/k_BT)}\right), \tag{B10.3.1a}$$

$$R_{22,22} = -\frac{1}{\hbar^2}\left(\frac{J_{12,12}(\omega_{12})}{1+\exp(\hbar\omega_{12}/k_BT)}\right), \tag{B10.3.1b}$$

$$R_{11,22} = -\frac{1}{2\hbar^2}\left(\frac{J_{12,12}(\omega_{12})}{1+\exp(\hbar\omega_{12}/k_BT)} + \frac{J_{12,12}(\omega_{12})}{1+\exp(\hbar\omega_{12}/k_BT)}\right)$$
$$= -\frac{1}{\hbar^2}\left(\frac{J_{12,12}(\omega_{12})}{1+\exp(\hbar\omega_{12}/k_BT)}\right), \tag{B10.3.2a}$$

$$R_{22,11} = -\frac{1}{\hbar^2}\left(\frac{J_{21,21}(\omega_{21})}{1+\exp(\hbar\omega_{21}/k_BT)}\right), \tag{B10.3.2b}$$

$$R_{12,12} = \frac{1}{2\hbar^2}\left(J_{11,22}(0) + J_{11,22}(0) - J_{11,11}(0) - \frac{J_{21,21}(\omega_{21})}{1+\exp(\hbar\omega_{21}/k_BT)}\right.$$
$$\left. - \frac{J_{12,12}(\omega_{12})}{1+\exp(\hbar\omega_{12}/k_BT)} - J_{22,22}(0)\right)$$
$$= \frac{-1}{2\hbar^2}[J_{11,11}(0) - 2J_{11,22}(0) + J_{22,22}(0)] + \frac{1}{2}[R_{22,11} + R_{11,22}], \tag{B10.3.3a}$$

(continued)

Box 10.3 (continued)

$$R_{21,21} = \frac{-1}{2\hbar^2}[J_{22,22}(0) - 2J_{22,11}(0) + J_{11,11}(0)] + \frac{1}{2}[R_{11,22} + R_{22,11}], \quad \text{(B10.3.3b)}$$

$$R_{12,21} = \frac{1}{2\hbar^2}[J_{12,21}(0) + J_{12,21}(0)] = \frac{1}{\hbar^2}J_{12,21}(0), \quad \text{(B10.3.4a)}$$

and

$$R_{12,21} = \frac{1}{\hbar^2}J_{21,12}(0). \quad \text{(B10.3.4b)}$$

If the interaction matrix elements V_{12} and V_{21} are real, $J_{21,21}(\omega) = J_{12,12}(\omega)$ and $J_{21,12}(\omega) = J_{12,21}(\omega)$.

Inspection of Eqs. (B10.3.1a), (B10.3.1b), (B10.3.2a) and (B10.3.2b) shows that the rate constant for conversion of state 1 to state 2 is $k_{21} = -R_{11,11} = \hbar^{-2}J_{21,21}(\omega_{21})[1 + \exp(\hbar\omega_{21}/k_BT)]^{-1} = \hbar^{-2}J^{SC}_{21,21}(\omega_{21})$. The rate constant thus depends on the spectral density of electric or magnetic fields from the surroundings that fluctuate at angular frequency ω_{21}, which is the frequency corresponding to the energy difference between the two states $(E_{21} = \hbar\omega_{21})$. The amplitude of $J^{SC}_{21,21}(\omega_{21})$ in an ensemble of systems is proportional to the mean-squared amplitude of the fluctuating interaction matrix element, $\overline{|V_{12}|^2}$. The parallel between these results and the expressions we derived for the rate of absorption or emission of light should be evident.

The rate constant for the reverse reaction is, similarly, $k_{12} = -R_{22,22} = \hbar^{-2}J_{12,12}(\omega_{12})[1 + \exp(\hbar\omega_{12}/k_BT)]^{-1} = \hbar^{-2}J^{SC}_{12,12}(\omega_{12})$. From Eq. (10.56), the ratio of the rate constants for the forward and backward reactions is $k_{21}/k_{12} = \exp(\hbar\omega_{12}) = exp[(H_{11} - H_{22})/k_BT]$, as required for detailed balance. The equilibration time constant T_1 (Eq. 10.29b) is given by $1/T_1 = k_{21} + k_{12} = \hbar^{-2}\left[J^{SC}_{21,21}(\omega_{21}) + J^{SC}_{12,12}(\omega_{12})\right]$.

Turning to $R_{12,12}$, which pertains to the first-order decay of $\overline{\rho_{12}}$ in a two-state system, each of the terms in Eq. (B10.3.3a) that are evaluated at $\omega = 0$ [$J_{11,11}(0)$, $J_{22,22}(0)$ and $J_{11,22}(0)$] consists of a product of a correlation time (τ_c) and the amplitude of one of the correlation functions at zero time (Eq. 10.51). More generally,

$$J_{nm,\,jk}(\omega = 0) = \int_0^\infty M_{nm,\,jk}(t)dt = \tau_c M_{nm,\,jk}(t = 0). \quad (10.57)$$

Because $M_{nm,\,jk}(0) = \overline{V_{nm}V_{jk}}$, the quantity $[J_{11,11}(0) - 2J_{11,22}(0) + J_{22,22}(0)]$ in Eq. (B10.3.3a) is the product of τ_c and the mean square of the fluctuating component of the energy gap:

$$[J_{11,11}(0) - 2J_{11,22}(0) + J_{22,22}(0)] = \tau_c\left(\overline{V_{11}V_{11}} - 2\overline{V_{11}V_{22}} + \overline{V_{22}V_{22}}\right)$$
$$= \tau_c\overline{(V_{11} - V_{22})^2}. \qquad (10.58)$$

We can identify this part of $-R_{12,12}$ with the rate constant for pure dephasing ($1/T_2$*), which is the first term on the right-hand side of Eq. (10.29b):

$$\frac{1}{T_2^*} = \frac{\tau_c}{2\hbar^2}\overline{(V_{11} - V_{22})^2}. \qquad (10.59)$$

Because the mean value of $V_{11} - V_{22}$ is zero, $\overline{(V_{11} - V_{22})^2}$ is simply the variance (the square of the standard deviation) of the energy gap.

The expression for $R_{12,12}$ (Eq. B10.3.3a) also contains $(1/2)[R_{11,22} + R_{22,11}]$. This is one-half the sum of the rate constants for interconversions of the two states, or $1/2\mathrm{T}_1$, which is the second term on the right-hand side of Eq. (10.29b). If the system had additional basis states, $R_{12,12}$ would include further terms of the form (1/2) $[R_{11,33} + R_{11,44} + \ldots + R_{22,33} + R_{22,44} + \ldots]$, or in general,

$$\frac{1}{T_2} = \frac{1}{T_2*} - \frac{1}{2}\sum_{j \neq n,m}\left(R_{nn,\,jj} + R_{mm,\,jj}\right). \qquad (10.60)$$

According to the stochastic Liouville equation (Eq. 10.31), the rate of change of a given element of $\overline{\boldsymbol{\rho}}$ $(\partial\overline{\rho_{nm}}/\partial t)$ depends on the sum of the product of $R_{nm,\,jk}\overline{\rho_{jk}}$ for all the other elements. This can make evaluation of the dynamics arduous for systems with many states, because the size of the density matrix increases quadratically with the number of states. For an off-diagonal element ($\overline{\rho_{nm}}$ with $n \neq m$), however, the effects of another off-diagonal matrix element $\left(\overline{\rho_{jk}},\, j \neq k\right)$ usually are greatest if $\omega_{jk} \approx \omega_{nm}$. The two matrix elements then oscillate more or less in synchrony, so that $R_{nm,\,jk}\overline{\rho_{jk}}$ consistently either increases or decreases $\partial\overline{\rho_{nm}}/\partial t$ when $\overline{\rho_{nm}}$ is positive and has the opposite effect when $\overline{\rho_{nm}}$ is negative. The effects of a matrix element that does not meet this resonance condition tend to average to zero. Retaining only the terms of **R** for which $\omega_{jk} \approx \omega_{nm}$ is called the *rotating-wave approximation.*

For systems with three or more states, Redfield [19] also derived terms of the relaxation matrix for transfer of coherence between two pairs of states. These terms allow one off-diagonal density matrix element (ρ_{jk}) to increase at the expense of another (ρ_{nm}). Again, these terms are most important if $\omega_{jk} \approx \omega_{nm}$.

We have found that fluctuations of the surroundings at angular frequency ω_{21} can cause transitions between states that differ in energy by $\hbar\omega_{21}$. Suppose that the pertinent correlation function for these fluctuations $M_{21,21}(t)$ is an exponential

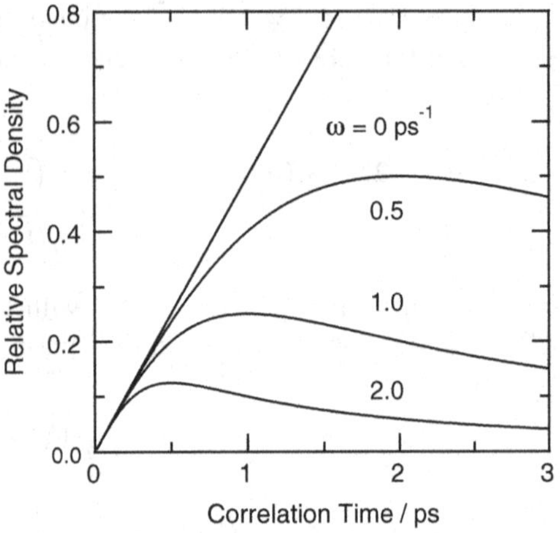

Fig. 10.7 Spectral densities ($J_{nm,jk}$) at several frequencies ($\omega = 0$, 0.5, 1 and $2 \times 10^{12}\ s^{-1}$) as functions of the correlation time (τ_c) of the corresponding correlation function ($M_{nm,jk}$). $M_{nm,jk}$ is assumed to decay exponentially with a time constant of τ_c, making $J_{nm,jk}$ a Lorentzian function of ω

function of time, so that the classical spectral density function $J_{21,21}(\omega)$ is Lorentzian. Decreasing the correlation time τ_c (i.e., speeding up the fluctuations) will broaden the spectral density function, which increases its amplitude at high frequencies but decreases the amplitude at low frequencies (Fig. 2.12). Figure 10.7 illustrates how the classical spectral densities at several frequencies vary with τ_c. In general, the spectral density at frequency ω_{21} peaks when $\tau_c = 1/\omega_{21}$. Speeding up the fluctuations thus increases the rate constant for transitions between the two states as long as $\tau_c > 1/\omega_{21}$, but decreases the rate constant if $\tau_c = 1/\omega_{21}$. At low frequencies ($\omega << 1/\tau_c$), $J(\omega)$ is proportional to τ_c and independent of ω.

According to Eqs. (B10.3.3a, B10.3.3b, and 10.57–10.59), the rate of pure dephasing ($1/T_2^*$) depends on the spectral densities at $\omega = 0$, which increase linearly with τ_c (Fig. 10.7). Perhaps counter to intuition, slowing the fluctuations increases the rate of pure dephasing. One way to view this result is that during a period when energy difference between states 1 and 2 for a particular system deviates from the mean value of this difference in the ensemble, the oscillations of ρ_{21} for this system build up a phase difference relative to the average oscillations of ρ_{21}. The longer the deviation of the energy gap persists, the greater the accumulated deviation of the phase. Pure dephasing thus results from heterogeneity of the oscillation frequencies during the periods when the energies of the individual systems do not change greatly. In the limit of extremely rapid fluctuations, the oscillations remain in phase because all the systems experience the same average energy difference (E_o). In magnetic resonance, this effect is termed *motional narrowing*.

In addition to assuming that the fluctuations of the surroundings occur rapidly relative to T_1 and T_2^*, Redfield's theory applies only to times that are long compared to τ_c. Box 10.4 shows that the kinetics of dephasing are different in the opposite limit of very slow fluctuations: the off-diagonal density matrix elements then decay with a Gaussian dependence on time. In the next section we'll discuss more general relaxation functions that bridge these two limits.

Box 10.4 Dephasing by Static Inhomogeneity

Consider the decay of coherence between two states, j and k, whose energies (H_{jj} and H_{kk}) vary from system to system in an ensemble but are essentially constant for any given system. If the factors C_j and C_k^* also are constant, the average of ρ_{jk} for the ensemble can be written

$$\overline{\rho_{jk}}(t) = \overline{\rho_{jk}}(0) \int_{-\infty}^{\infty} G(E_{jk}) \exp(-iE_{jk}t/\hbar) dE_{jk}, \tag{B10.4.1}$$

where $\overline{\rho_{jk}}(0)$ is the mean value of ρ_{jk} at zero time, $E_{jk} = H_{jj} - H_{kk}$ for a given system, and $G(E_{jk})$ is the normalized distribution of E_{jk} for the ensemble.

Suppose $G(E_{jk})$ is a Gaussian with mean E_o and standard deviation σ. If we define $x = t/\hbar$, $\alpha = E_{jk} - E_o$, and $f(\alpha) = (1/\sigma)\exp(-\alpha^2/2\sigma^2)$, then

$$\begin{aligned}\frac{\overline{\rho}_{jk}(t)}{\overline{\rho}_{jk}(0)} &= \frac{1}{\sqrt{2\pi}\sigma} \int_{-\infty}^{\infty} \left\{ \exp\left[-(E_{jk} - E_o)^2/2\sigma^2\right] \exp(-iE_{jk}t/\hbar) \right\} d(E_{jk} - E_o) \\ &= \frac{\exp(-iE_o t/\hbar)}{\sqrt{2\pi}\sigma} \int_{-\infty}^{\infty} \left\{ \exp(-\alpha^2/2\sigma^2) \exp(-i\alpha t/\hbar) \right\} d\alpha \\ &= \frac{\exp(-iE_o t/\hbar)}{\sqrt{2\pi}} \int_{-\infty}^{\infty} f(\alpha) \exp(-i\alpha x)\, d\alpha = \exp(-iE_o t/\hbar)\ F(x),\end{aligned} \tag{B10.4.2}$$

where $F(x)$ is the Fourier transform of $f(\alpha)$. The Fourier transform of this function is $\exp(-\sigma^2 x^2/2)$, another Gaussian (Appendix A3), so

$$\overline{\rho_{jk}}(t)/\overline{\rho_{jk}}(0) = \exp(-iE_o t/\hbar)\ \exp\left(-\sigma^2 t^2/2\hbar^2\right). \tag{B10.4.3}$$

If the ensemble is homogeneous ($\sigma = 0$), then Eq. (B10.4.3) gives $\overline{\rho_{jk}}(t)/\overline{\rho_{jk}}(0) = \exp(-iE_o t/\hbar)$, which oscillates at frequency E_o/h with a constant amplitude of 1. Inhomogeneity in the energies ($\sigma > 0$) causes the magnitude of the oscillations to decay to zero as $\exp\left(-\sigma^2 t^2/2\hbar^2\right)$, which gives a Gaussian dependence on time. $\left|\overline{\rho_{jk}}(t)\right|$ falls to half its initial value in a time $\tau_{1/2} = (2\ln 2)^{1/2}\hbar/\sigma = 1.177\hbar/\sigma$. A standard deviation ($\sigma$) of 1 cm^{-1} gives $\tau_{1/2} = 1.177 \times 5.31 \times 10^{-12}$ (cm^{-1} s)/(cm^{-1}) = 6.2 ps.

Figure 10.8 shows plots of $\overline{\rho_{jk}}(t)/\overline{\rho_{jk}}(0)$ for Gaussian distributions of E_{jk} with several values of σ. Gaussian decays of coherence also can be seen in Fig. 10.2, but the models considered there include static heterogeneity in the interaction matrix element that couples the two quantum states (H_{21}) in addition to the energy difference between the states.

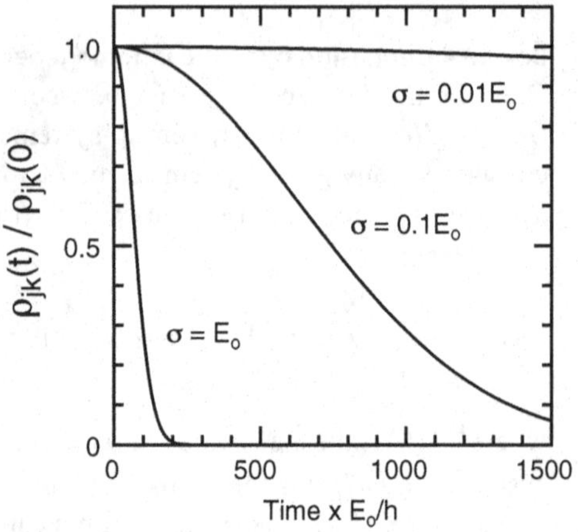

Fig. 10.8 The time dependence of $\overline{\rho_{jk}}(t)$ in ensembles in which the energy difference between states j and k (E_{jk}) has a Gaussian distribution with a mean of E_o and a standard deviation (σ) of $0.01E_o$, $0.1E_o$ or E_o. The oscillation period of $\overline{\rho_{jk}}$ is h/E_o

10.7 More General Relaxation Functions and Spectral Lineshapes

We have seen that if the interactions of a quantum system with its surroundings fluctuate rapidly, the decay of coherence in an ensemble of such systems can be described by a relaxation matrix of microscopic first-order rate constants. But if the energies of the basis states vary statically from system to system, coherence decays with a Gaussian dependence on time (Box 10.4; Figs. 10.2 and 10.8). We now seek a more general expression for pure dephasing to connect the domains in which the fluctuations of the surroundings are either very slow or very fast.

To start, consider an off-diagonal density matrix element ρ_{nm} of an individual system when the energies of states n and m (H_{nn} and H_{mm}) vary with time (Eq. 10.15). Assume that n and m are stationary states so that the factors C_n and C^*_{m} are independent of time. The initial value of ρ_{nm} then is $\rho_{nm}(0) = C_n C^*_m$. ρ_{nm} now begins to oscillate with an angular frequency $\omega_{nm}(0) = (H_{nn} - H_{mm})/\hbar$, which we take to be constant for a short interval of time Δt. By the end of this interval, ρ_{nm} has become

$$\rho_{nm}(t = \Delta t) = \rho_{nm}(0)\exp\{-i\,[\omega_{nm}(0)\Delta t]\}. \tag{10.61}$$

Suppose that after time Δt, the oscillation frequency changes to some new value, $\omega_{nm}(1)$, which then persists for the same increment of time. At the end of this second interval, ρ_{nm} is

$$\rho_{nm}(t = 2\Delta t) = \rho_{nm}(0)\exp\{-i\,[\omega_{nm}(0)\Delta t + \omega_{nm}(1)\Delta t]\}, \tag{10.62}$$

and in general,

$$\rho_{nm}(t) = \rho_{nm}(0)\exp\left(-i\int_0^t \omega_{nm}(\tau)d\tau\right). \tag{10.63}$$

In an ensemble of such systems, the detailed time dependence of ω_{nm} will vary randomly from one system to the next. Let the average frequency for the ensemble be $\overline{\omega_{nm}}$, and call the variable part for our particular system w_{nm}, so that $\omega_{nm}(\tau) = \overline{\omega_{nm}} + w_{nm}(\tau)$. Then, for an individual system,

$$\rho_{nm}(t)\rho^*_{nm}(0) = \rho_{nm}(0)\rho^*_{nm}(0)\exp\left(-i\int_0^t [\overline{\omega_{nm}} + w_{nm}(\tau)]d\tau\right) \tag{10.64a}$$

$$= |\rho_{nm}(0)|^2\exp\left(-i\overline{\omega_{nm}}t - i\int_0^t w_{nm}(\tau)d\tau\right). \tag{10.64b}$$

Assuming that $|\rho_{nm}(0)|^2$ is not correlated with the fluctuations of ω_{nm}, we can obtain a correlation function for the ensemble-averaged density matrix element by averaging the integral in Eq. (10.64b):

$$\frac{\overline{\rho_{nm}(t)\,\rho^*_{nm}(0)}}{\overline{|\rho_{nm}(0)|^2}} = \exp[-i\overline{\omega_{nm}}t - g_{nm}(t)] = \exp(-i\overline{\omega_{nm}}t)\phi_{nm}(t), \tag{10.65}$$

where

$$g_{nm}(t) = i\overline{\int_0^t w_{nm}(\tau)d\tau}. \tag{10.66a}$$

and

$$\phi_{nm}(t) = \exp[-g_{nm}(t)], \tag{10.66b}$$

The dephasing of the ensemble thus is contained in the *relaxation function* $\phi_{nm}(t) = \exp(-g_{nm}(t))$, with $g_{nm}(t)$ (sometimes called the *lineshape* or *line-broadening function*) defined in Eq. (10.66a).

Ryogo Kubo [2, 3] showed that if w_{nm} has a Gaussian distribution about zero, $g_{nm}(t)$ is given by

$$
\begin{aligned}
g_{nm}(t) &= \int_0^t d\tau_1 \int_0^{\tau_1} \overline{w_{nm}(\tau_1) w_{nm}(\tau_2)}\, d\tau_2 \\
&= \sigma_{nm}^2 \int_0^t d\tau_1 \int_0^{\tau_1} M_{nm}(\tau_2)\, d\tau_2.
\end{aligned}
\tag{10.67}
$$

Here σ_{nm}^2 is the mean square of the frequency fluctuations ($\sigma_{nm}^2 = \left|\overline{w_{nm}^2}\right|$ in units of radian^2 s^{-2}) and $M_{nm}(t)$ is a normalized autocorrelation function of the fluctuations:

$$
M_{nm}(t) = \frac{1}{\sigma_{nm}^2} \overline{w_{nm}(\tau + t) w_{nm}(\tau)}. \tag{10.68}
$$

If $M_{nm}(t)$ decays exponentially with time constant τ_c, evaluating the integrals in Eq. (10.67) is straightforward and yields

$$
\phi_{nm}(t) = \exp(-g_{nm}(t)) = \exp\left\{-\sigma_{nm}^2 \tau_c^2 [(t/\tau_c) - 1 + \exp(-t/\tau_c)]\right\}. \tag{10.69}
$$

Figure 10.9 shows the behaviour of this expression for several values of τ_c and σ_{nm}. At short times ($t \ll \tau_c$), Eq. (10.69) reduces to $\phi_{nm}(t) = \exp\left(-\sigma_{mn}^2 t^2/2\right)$. [You can see this by expanding the inner exponential in Eq. (10.69) as $\exp(-t/\tau_c) = 1 - t/\tau_c + (t/\tau_c)^2/2! - \ldots$]. In this limit the relaxation dynamics are independent of τ_c, and $\overline{\rho_{nm}}$ has a Gaussian dependence on time in accord with Eq. (B10.4.3). At the other extreme, when $t \gg \tau_c$, Eq. (10.69) goes to $\phi_{nm}(t) = \exp\left(-\sigma_{nm}^2 \tau_c t\right)$. $\overline{\rho_{nm}}$ then decays exponentially with a rate constant ($1/T_2^*$) of $\sigma_{\mathrm{nm}}^2 \tau_c$, in accord with the Redfield theory and Eqs. (10.29a and 10.29b). Kubo's function thus captures both the exponential dephasing that results from fast fluctuations and the Gaussian dephasing associated with slow fluctuations. It shows that the operational terms "fast" and "slow" relate to the ratio of the observation time (t) to τ_c, and it reproduces the reciprocal relationship between τ_c and the rate of pure dephasing that we discussed in the previous section.

If the off-diagonal matrix elements that describe the coherence between a ground state and an excited electronic state decay exponentially with time, the homogeneous absorption line should have a Lorentzian shape (Figs. 10.6 and 10.7; Eqs. (2.70) and (10.35)). More generally, as we discussed in Sect. 10.6, the spectral lineshape is the Fourier transform of the relaxation function:

$$
\frac{\varepsilon(\omega - \omega_o)}{\omega} \propto |\boldsymbol{\mu}_{ab}|^2 \int_{-\infty}^{\infty} e^{i(\omega - \omega_o)t} \phi(t)\, dt, \tag{10.70}
$$

where ω_o is the angular frequency corresponding to the 0–0 transition and we have omitted the subscript nm to generalize the relationship. Figure 10.10 shows spectra calculated by using this expression with the relaxation function $\phi(t)$ from

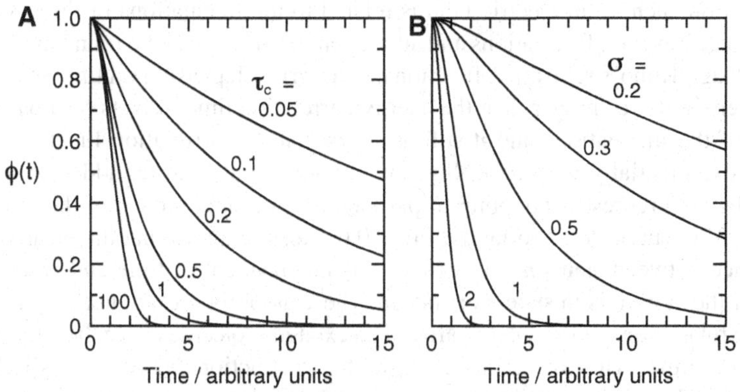

Fig. 10.9 The Kubo relaxation function $\phi(t)$ as given by Eq. (10.69) for an ensemble in which the correlation function decays exponentially with time constant τ_c. In (**A**), τ_c (indicated in arbitrary time units) is varied, while the rms amplitude of the fluctuations (σ) is fixed at 1 reciprocal time unit. In (**B**), τ_c is fixed at 1 time unit and σ is varied

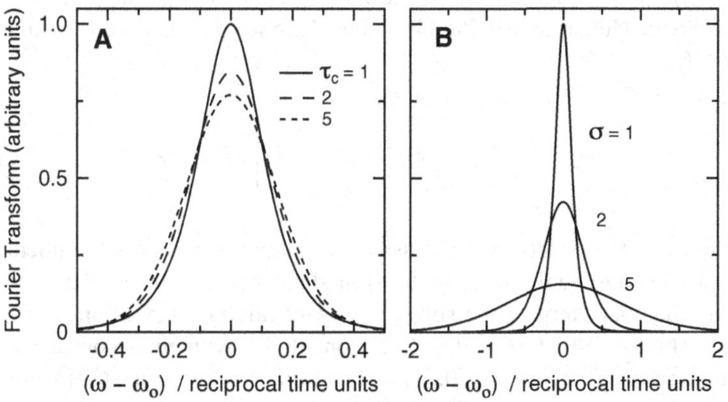

Fig. 10.10 Absorption spectral lineshapes calculated as the Fourier transform of the Kubo relaxation function $\phi(t)$. (**A**) τ_c (indicated in arbitrary time units) is varied, while σ is fixed at 1 reciprocal time unit; (**B**) τ_c is fixed at 1 time unit and σ is varied as indicated. To use the full Fourier transform (Eq. 10.70), $\phi(t)$ is treated as an even function of time (Fig. 10.11A)

Eq. (10.69) and several values of the underlying correlation time (τ_c) and variance (σ) of the fluctuations. As we saw in Fig. 10.9, $\phi(t)$ can vary from Gaussian to exponential depending on t/τ_c. Because the Fourier transform of a Gaussian is another Gaussian, whereas the Fourier transform of an exponential is a Lorentzian, the absorption lineshape can range from Gaussian to Lorentzian. A long correlation time gives a Gaussian absorption band, as we would expect for a spectrum that is inhomogeneously broadened; a short correlation time gives a Lorentzian band corresponding to the homogeneous lineshape.

Equation (10.70) is a manifestation of a general principle called the *Fluctuation-Dissipation theorem,* which describes how the response of a system to a small perturbation such as an electric field is related to the fluctuations of the system at thermal equilibrium. Derivations and additional discussion can be found in [25–31].

Although Kubo's relaxation function can describe dephasing on time scales that are either shorter or longer than the energy correlation time, it rests on a particular model of the fluctuations and it still assumes that the correlation function ($M(t)$) decays exponentially with time. Sinusoidal components can be added to $M(t)$ in Eq. (10.69) to represent coupling to particular vibrational modes of the molecule [29, 32, 33]. But more importantly, Eq. (10.69) also assumes that the mean energy difference between states n and $m\,(\hbar\overline{\omega_{nm}})$ is independent of time and is the same whether the system is in state n or state m. These assumptions are not very realistic because relaxations of the solvent around an excited molecule cause the emission to shift with time. This brings us back to the distinction between classical and semiclassical spectral density functions, which we discussed briefly in Sect. 10.6 in connection with the Redfield theory. We saw there that relaxations of a system to thermal equilibrium with the surroundings require a spectral density function that is not a purely even function of frequency (Eq. 10.54), or equivalently, a relaxation function that includes an imaginary component.

Noting that the relaxation and lineshape functions should, in principle, be complex, Shaul Mukamel [23, 29, 34] obtained the following more general expression for $g(t)$:

$$g(t) = \sigma^2 \int_0^t d\tau_1 \int_0^{\tau_1} M(\tau_2) d\tau_2 + i\Lambda_s \int_0^t M(\tau_1) d\tau_1. \tag{10.71}$$

Here as in Eq. (10.68), $M(t)$ is the classical correlation function of the fluctuations, which is assumed to be the same in the ground and excited states. The scale factor Λ_s for the imaginary term is the solvent reorganization energy in units of angular frequency (Fig. 4.28). If $M(t)$ decays exponentially with time constant τ_c, then evaluating the integrals in Eq. (10.71) and defining $\Gamma = 1 - \exp(-t/\tau_c)$ yields

$$g(t) = \sigma^2 \tau_c^2 [(t/\tau_c) - (1 - \exp(-t/\tau_c))] + i\Lambda_s \tau_c (1 - \exp(-t/\tau_c)), \tag{10.72}$$

and

$$\begin{aligned}\phi(t) &= \exp\{-\sigma^2 \tau_c^2 [(t/\tau_c) - \Gamma] - i\Lambda_s \tau_c \Gamma\} \\ &= \exp\{-\sigma^2 \tau_c^2 [(t/\tau_c) - \Gamma]\}[\cos(\Lambda_s \tau_c \Gamma) - i \sin(\Lambda_s \tau_c \Gamma)]\end{aligned} \tag{10.73}$$

Fourier transforms of $\phi(t)$ and $\phi^*(t)$ give the absorption and emission spectra, respectively [23, 29, 34, 35].

The reorganization energy Λ_s generally increases with σ^2. A large value of σ^2 means that the energy difference between the ground and excited state depends strongly on fluctuating interactions of the molecule with the solvent, which implies that relaxations of the solvent will cause a substantial decrease in the energy of the excited state. The expected relationship in the high-temperature limit is

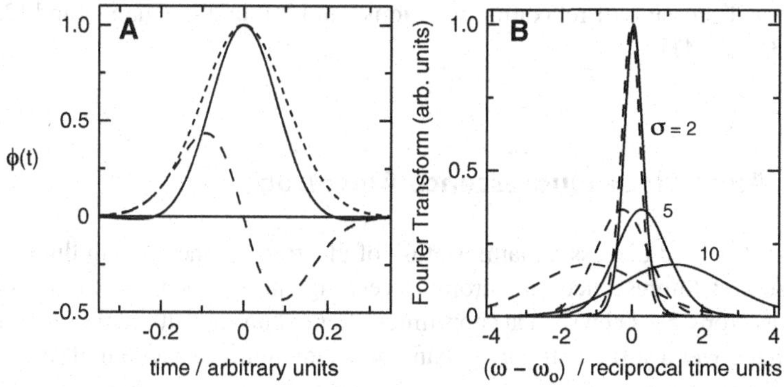

Fig. 10.11 (**A**) The real (*solid line*) and imaginary (*long dashes*) parts of the complex relaxation function $\phi(t)$ given by Eqs. (10.73) and (10.74) with $\tau_c = 1$ arbitrary time unit and $\sigma = 10$ reciprocal time units. The *short dashes* shows the Kubo relaxation function (Eq. 10.69), treated as an even function of time ($\phi(-t) = \phi(t)$). (**B**) Absorption (*solid lines*) and emission (*dashed lines*) spectra calculated as the Fourier transforms of, respectively, the complex relaxation function $\phi(t)$ (Eqs. 10.72 and 10.73) and its complex conjugate $\phi^*(t)$. The autocorrelation time constant τ_c was 1 arbitrary time unit, and σ was 2, 5 or 10 reciprocal time units, as indicated

$$\Lambda_s = \sigma^2\hbar/2k_BT \tag{10.74}$$

[23]. Evaluating the constants for 295 K gives $\Lambda_s = (1.295 \times 10^{-14}\ \mathrm{s})\sigma^2$.

Figure 10.11A shows the real and imaginary parts of the relaxation function ϕ given by Eqs. (10.73) and (10.74) when $\tau_c\sigma = 10$. Kubo's relaxation function (Eq. 10.69) is shown for comparison. Figure 10.11B shows the calculated absorption and emission spectra for $\tau_c\sigma = 2$, 5 and 10. Including the term involving Λ_s in the relaxation function shifts the absorption to higher energies and the fluorescence to lower energies. In agreement with the relationship we discussed in Chap. 5, the Stokes shift is $2\Lambda_s$. Increasing σ increases the Stokes shift as it broadens the spectra.

Relaxations of solvent-chromophore interactions can be studied experimentally by hole-burning spectroscopy, time-resolved pump-probe measurements, and photon-echo techniques that we discuss in the next chapter. If the temperature is low enough to freeze out pure dephasing, and a spectrally narrow laser is used to burn a hole in the absorption spectrum (Sect. 4.11), the zero-phonon hole should have the Lorentzian lineshape determined by the homogeneous lifetime of the excited state. The hole width increases with increasing temperature as the pure dephasing associated with σ^2 comes into play [36, 37].

The parameters σ^2 and τ_c also can be obtained by classical molecular-dynamics simulations in which one samples fluctuations of the energy difference between the ground and excited states during trajectories on the potential surfaces of ground and excited states. The variance and autocorrelation function of the fluctuations in the ground state provide σ^2 and τ_c for the absorption spectrum, and those in the excited state give the corresponding parameters for fluorescence. As we saw in Fig. 6.3, a Fourier transform of the autocorrelation function of the energy gap also can be used to identify the frequencies and dimensionless displacements of vibrational modes that are strongly coupled to an electronic transition [38, 39].

Further discussion of relaxation functions can be found in Chap. 8 and [27, 29, 30, 32–35, 40, 41].

10.8 Anomalous Fluorescence Anisotropy

In this section, we discuss a manifestation of electronic coherence in fluorescence spectroscopy: fluorescence anisotropy exceeding the classical maximum of 0.4. Such anisotropy has been seen at short times after excitation with femtosecond laser pulses that cover a sufficiently broad band of wavelengths to excite multiple optical transitions coherently.

To examine the effect of coherence on fluorescence anisotropy, consider a system with a ground state (state 1) and two electronically excited states (2 and 3), and assume as before that the interaction matrix elements H_{12}, H_{13} and H_{23} are zero in the absence of radiation. Suppose that an ensemble of systems in the ground state is excited with a weak pulse of light that is much shorter than the lifetime of the excited state (T_1). In the impulsive limit, the population of excited state 2 generated by the pulse is

$$\overline{\rho_{22}}(0) \approx \overline{|\hat{\mu}_{21} \cdot \hat{e}_i|^2} K_{22}, \tag{10.75a}$$

where

$$K_{22} = \Theta \int_{-\infty}^{0} dt \int_{0}^{\infty} \varepsilon_{21}(v) v^{-1} I(v,t) dv. \tag{10.75b}$$

Here $\hat{e}_i$ and $\hat{\mu}_{21}$ are unit vectors defining the polarization of the excitation light and the orientation of the transition dipole $\boldsymbol{\mu}_{21}$, respectively; $\Theta = (3000 \ \ln 10)/hN_{avo}$; $\varepsilon_{21}(v)$ is the molar extinction coefficient for excitation to state 2; and $I(v,t)$ describes the dependence of the excitation flash intensity on frequency and time. The bars indicate averaging over the ensemble as usual, but we assume here that $|\boldsymbol{\mu}_{21}|$ is the same for all the molecules, so that the averaging procedure needs to consider only the distribution of orientations relative to the polarization of the excitation light. To the same approximation,

$$\overline{\rho_{33}}(0) \approx \overline{|\hat{\mu}_{31} \cdot \hat{e}_i|^2} K_{33}, \tag{10.75c}$$

where

$$K_{33} = \Theta \int_{-\infty}^{0} dt \int_{0}^{\infty} \varepsilon_{31}(v) v^{-1} I(v,t) dv. \tag{10.75d}$$

The initial value of $\overline{\rho_{32}}$ created by the excitation flash can be estimated similarly:

$$\overline{\rho_{32}}(0) \approx \overline{|\hat{\mu}_{31} \cdot \hat{e}_i|^2 |\hat{\mu}_{21} \cdot \hat{e}_i|^2} K_{32}, \tag{10.76a}$$

with

$$K_{32} = \Theta \int dt \int [\varepsilon_{31}(\nu)\varepsilon_{21}(\nu)]^{1/2} \nu^{-1} I(\nu, t) d\nu. \tag{10.76b}$$

Here the square root can be taken arbitrarily to be positive. Equations (10.76a and 10.76b), which are a generalization of an expression suggested by Rahman et al. [42], arbitrarily assign $\overline{\rho_{32}}(0)$ a purely real value. In general, $\overline{\rho_{23}}(0)$ is complex and depends on the convolution of Eq. (B10.1.8) with the electric field in the excitation pulse, $\boldsymbol{E}(\nu, t)$. The main point is that a short excitation pulse can create coherence between states 2 and 3 if it overlaps the absorption bands for excitation to both states. Such overlap is a common feature of measurements made with femtosecond laser pulses, which inherently have large spectral widths.

Following the excitation pulse, $\overline{\rho_{32}}(t)$ will execute damped oscillations with a period of $h/|E_{32}|$. How does the transient coherence between states 2 and 3 affect the fluorescence intensity and anisotropy as a function of time after the excitation? Because fluorescence reflects radiative decay of $\overline{\rho_{22}}$ or $\overline{\rho_{33}}$, let's find steady-state expressions for $\partial\overline{\rho_{22}}/\partial t$ and $\partial\overline{\rho_{33}}/\partial t$ during a probe pulse of weak, broadband light. If the pump and probe pulses are well separated in time, these derivatives will tell us the rate of stimulated emission, which will be proportional to the intensity of spontaneous fluorescence. A more refined analysis that incorporates the detailed time courses of the pump and probe pulses requires evaluating the interaction of the probe field with the time-dependent third-order polarization as described in Chap. 11.

By proceeding as we did to obtain Eq. (10.39), but adding the term involving $\rho_{32}H_{13}$ from Eq. (B10.1.4), we obtain

$$\begin{aligned}\partial\rho_{12}(t)/\partial t = &(i/\hbar)(\rho_{22} - \rho_{11})\boldsymbol{\mu}_{12} \cdot \boldsymbol{E}_o \exp(2\pi i\nu t) + \{(i/\hbar)(E_{21}) - 1/T_2\}\rho_{12}(t) \\ &+ (i/\hbar)\boldsymbol{\mu}_{13} \cdot \boldsymbol{E}_o \exp(2\pi i\nu t)\rho_{32}(t).\end{aligned} \tag{10.77}$$

We have written this expression for the density matrix of an individual system rather than the averaged density matrix for an ensemble of systems. This is because, although the transition dipoles $\boldsymbol{\mu}_{12}$ and $\boldsymbol{\mu}_{13}$ for an individual system might have any angle with respect to the laboratory coordinates or the electric field vector of the pump pulse, we assume that their orientations do not change during the short time interval between the pump and probe pulses. We therefore need to average the product $\boldsymbol{\mu}_{13} \cdot \boldsymbol{E}_o \exp(2\pi i\nu t)\rho_{32}$ and the other similar products in Eq. (10.77) over all the orientations of the individual systems, rather than calculating the product of the ensemble-averages of the separate components. We will do the required averaging after we find $\partial\rho_{22}/\partial t$ and $\partial\rho_{33}/\partial t$ for an individual system.

Continuing as in Eqs. (10.39–10.42b) and making the steady-state approximation for ρ_{12} and ρ_{21} gives

$$\rho_{12}(\nu) \approx (-i/\hbar)\,[\boldsymbol{\mu}_{12} \cdot \boldsymbol{E}_o(\rho_{22} - \rho_{11}) + \boldsymbol{\mu}_{13} \cdot \boldsymbol{E}_o\rho_{32}]/[(i/\hbar)(E_{21} - h\nu) - 1/T_2] \tag{10.78a}$$

and

$$\rho_{21}(\nu) \approx (i/\hbar)\,[\boldsymbol{\mu}_{21} \cdot \boldsymbol{E}_o(\rho_{22} - \rho_{11}) + \boldsymbol{\mu}_{31} \cdot \boldsymbol{E}_o\rho_{23}]/[(i/\hbar)(E_{21} - h\nu) - 1/T_2]. \tag{10.78b}$$

And because H_{23} and H_{32} are assumed to be zero and we are considering only radiative transitions between states 1 and 2, the expression corresponding to Eq. (B10.1.2) becomes:

$$\begin{aligned}\partial\rho_{22}/\partial t &= (i/\hbar)(\rho_{21}(t)H_{12} - \rho_{12}(t)H_{21}) \\ &= (i/\hbar)(\exp(-2\pi i\nu t)\rho_{21}(\nu)H_{12} - \exp(2\pi i\nu t)\rho_{12}(\nu)H_{21}) \\ &= \Big\{(\rho_{11} - \rho_{22})|\boldsymbol{\mu}_{21} \cdot \boldsymbol{E}_o|^2 - (1/2)\,[(\boldsymbol{\mu}_{31} \cdot \boldsymbol{E}_o)(\boldsymbol{\mu}_{12} \cdot \boldsymbol{E}_o)\rho_{23} \\ &\quad + (\boldsymbol{\mu}_{13} \cdot \boldsymbol{E}_o)(\boldsymbol{\mu}_{21} \cdot \boldsymbol{E}_o)\rho_{32}]\Big\}\,W_{12}(\nu) \\ &= \Big\{(\rho_{11} - \rho_{22})|\boldsymbol{\mu}_{21} \cdot \boldsymbol{E}_o|^2 - \mathrm{Re}(\rho_{23})(\boldsymbol{\mu}_{21} \cdot \boldsymbol{E}_o)(\boldsymbol{\mu}_{31} \cdot \boldsymbol{E}_o)\Big\}\,W_{12}(\nu),\end{aligned} \tag{10.79}$$

where $W_{12}(\nu)$ is a Lorentzian lineshape function for fluorescence from state 2,

$$W_{12}(\nu) = (2/T_2)/\left[(E_{21} - h\nu)^2 + (\hbar/T_2)^2\right]. \tag{10.80}$$

Since $\mathrm{Re}(\rho_{23})$ oscillates and decays with time, Eq. (10.79) indicates that the amplitude of fluorescence from state 2 will execute similar damped oscillations, converging on a level that depends on the population difference $\rho_{11} - \rho_{22}$. The oscillatory term in Eq. (10.79) also applies to fluorescence from state 3, because $\mathrm{Re}(\rho_{23}) = \mathrm{Re}(\rho_{32})$.

To calculate the expected fluorescence anisotropy, we now evaluate the rate of fluorescence with frequency ν and polarization $\hat{e}_f$, following excitation with polarization $\hat{e}_i$. Because we are interested in the initial rate of radiative transitions from state 2 to state 1, we can set ρ_{11} to zero in Eq. (10.79). The density matrix elements ρ_{22} and ρ_{23} can be obtained from Eqs. (10.75a–10.75d, 10.76a, and 10.76b). Averaging over all orientations then gives the initial rate of emission from state 2:

$$\begin{aligned}F_{12}(\nu, \hat{e}_i, \hat{e}_f) = \Big[&\overline{(\hat{e}_i \cdot \hat{\mu}_{21})^2(\hat{e}_f \cdot \hat{\mu}_{21})^2}K_{22} \\ &+ \overline{(\hat{e}_i \cdot \hat{\mu}_{21})(\hat{e}_i \cdot \hat{\mu}_{31})(\hat{e}_f \cdot \hat{\mu}_{21})(\hat{e}_f \cdot \hat{\mu}_{31})}K_{23}\Big]W_{21}(\nu).\end{aligned} \tag{10.81}$$

In Chap. 5 we considered the fluorescence anisotropy of a molecule with only one excited state. In that case, or in the event that the excitation pulse does not overlap with both the $1 \rightarrow 2$ and $1 \rightarrow 3$ transitions, $K_{23} = 0$. The second term on the right side of Eq. (10.81) then drops out, leaving $F_{12} = \overline{(\hat{e}_i \cdot \hat{\mu}_{21})^2 (\hat{e}_f \cdot \hat{\mu}_{21})^2} K_{22} W_{21}(\nu)$. For an isotropic sample, the orientational average in this last expression evaluates to 1/5 if $\hat{e}_f$ is parallel to $\hat{e}_i$, and 1/15 if $\hat{e}_f$ and $\hat{e}_i$ are perpendicular (Box 10.5). The fluorescence polarized parallel to the excitation thus is three times the fluorescence with perpendicular polarization, which gives a fluorescence anisotropy of 0.4 in agreement with Eqs. (5.61) and (5.62). (Recall that the fluorescence anisotropy r is $(F_{\parallel} - F_{\perp})/(F_{\parallel} + 2F_{\perp})$.)

Box 10.5 Orientational Averages of Vector Dot Products

Orientational averages such as those in Eq. (10.81) can be evaluated by a general procedure involving the use of Euler angles [43]. Here are the results for several cases of interest.

Let the excitation polarization $\hat{e}_i$ be parallel to the laboratory x axis, and the fluorescence detection polarization $\hat{e}_f$ be parallel to x for a measurement of $F_{\parallel}$ or to y for a measurement of $F_{\perp}$; let ξ be the angle between transition dipoles $\boldsymbol{\mu}_{21}$ and $\boldsymbol{\mu}_{31}$. If the sample is isotropic, then

$$\overline{(\hat{x} \cdot \hat{\mu}_{21})^2} = 1/3, \tag{B10.5.1}$$

$$\overline{(\hat{x} \cdot \hat{\mu}_{21})^4} = 1/5, \tag{B10.5.2}$$

$$\overline{(\hat{x} \cdot \hat{\mu}_{21})(\hat{y} \cdot \hat{\mu}_{21})} = 1/15, \tag{B10.5.3}$$

$$\overline{(\hat{x} \cdot \hat{\mu}_{21})(\hat{x} \cdot \hat{\mu}_{31})} = (\cos \xi)/3, \tag{B10.5.4}$$

$$\overline{(\hat{x} \cdot \hat{\mu}_{21})^2 (\hat{x} \cdot \hat{\mu}_{31})^2} = (1 + 2\cos^2 \xi)/15, \tag{B10.5.5}$$

$$\overline{(\hat{x} \cdot \hat{\mu}_{21})^2 (\hat{y} \cdot \hat{\mu}_{31})^2} = (2 - \cos^2 \xi)/15, \tag{B10.5.6}$$

and

$$\overline{(\hat{x} \cdot \hat{\mu}_{21})(\hat{x} \cdot \hat{\mu}_{31})(\hat{y} \cdot \hat{\mu}_{21})(\hat{y} \cdot \hat{\mu}_{31})} = (3\cos^2 \xi - 1)/30. \tag{B10.5.7}$$

The values 3/5 and 1/5 given in Eqs. (5.61) and (5.62) refer to $\overline{(\hat{x} \cdot \hat{\mu})^4} / \overline{(\hat{x} \cdot \hat{\mu})^2}$ and $\overline{(\hat{x} \cdot \hat{\mu})^2 (\hat{y} \cdot \hat{\mu})^2} / \overline{(\hat{x} \cdot \hat{\mu})^2}$, respectively. Equations (B10.5.5) and (B10.5.6) account for the anisotropy of incoherent emission from a molecule with absorption and emission transition dipoles oriented at an angle ξ (Eq. 5.68).

(continued)

Box 10.5 (continued)

The text describes the use of the above expressions to calculate the fluorescence anisotropy of an ensemble of molecules in which states 2 and 3 are excited coherently. The same expressions also can be used to calculate the fluorescence anisotropy for an incoherent mixture of states 2 and 3. In the latter situation, $F_{\parallel}$ will be proportional to $\left[\overline{\rho_{22}}\overline{(\hat{x}\cdot\hat{\mu}_{21})^4}+\overline{\rho_{33}}\overline{(\hat{x}\cdot\hat{\mu}_{21})^2(\hat{x}\cdot\hat{\mu}_{31})^2}\right]$ $K_{22}+\left[\overline{\rho_{33}}\overline{(\hat{x}\cdot\hat{\mu}_{31})^4}+\overline{\rho_{22}}\overline{(\hat{x}\cdot\hat{\mu}_{31})^2(\hat{x}\cdot\hat{\mu}_{21})^2}\right]K_{33}$, where $\overline{\rho_{22}}$ and $\overline{\rho_{33}}$ are the populations of the two states and K_{22} and K_{33} are as defined in the text. $F_{\perp}$ will be proportional to $\left[\rho_{22}\overline{(\hat{x}\cdot\hat{\mu}_{21})^2(\hat{y}\cdot\hat{\mu}_{21})^2}+\rho_{33}\overline{(\hat{x}\cdot\hat{\mu}_{21})^2(\hat{y}\cdot\hat{\mu}_{31})^2}\right]$ $K_{22}+\left[\rho_{22}\overline{(\hat{x}\cdot\hat{\mu}_{31})^2(\hat{y}\cdot\hat{\mu}_{31})^2}+\rho_{33}\overline{(\hat{x}\cdot\hat{\mu}_{31})^2(\hat{y}\cdot\hat{\mu}_{21})^2}\right]K_{33}$. If $K_{22}=K_{33}$ and $\overline{\rho_{22}}=\overline{\rho_{33}}$, the anisotropy will be $(1+3\cos^2\xi)/10$, which is 0.4 for $\xi=0°$ and 0.1 for $\xi=90°$.

The second term on the right side of Eq. (10.81) represents the coherence between excited states 2 and 3. Because such coherences usually decay on a subpicosecond time scale, they are of little significance in conventional measurements of fluorescence amplitudes or anisotropy. We therefore neglected them in Chap. 5. However, coherence can have large effects on the fluorescence at short times [42–44]. Suppose for simplicity that $K_{23}\approx K_{22}$, and let the angle between the transition dipoles $\boldsymbol{\mu}_{21}$ and $\boldsymbol{\mu}_{31}$ be ξ. The initial anisotropy calculated from Eq. (10.81) and the orientational averages given in Box 10.5 then is

$$r=\frac{F-F_{\perp}}{F+2F_{\perp}}=\frac{F_{12}(\nu,\hat{x},\hat{x})-F_{12}(\nu,\hat{x},\hat{y})}{F_{12}(\nu,\hat{x},\hat{x})+2F_{12}(\nu,\hat{x},\hat{y})}$$
$$=\frac{[1/5+(1+2\cos^2\xi)/15]-[1/15+(3\cos^2\xi-1)/30]}{[1/5+(1+2\cos^2\xi)/15]+2[1/15+(3\cos^2\xi-1)/30]}=\frac{7+\cos^2\xi}{10+10\cos^2\xi}. \tag{10.82}$$

Figure 10.12 shows how the calculated anisotropy and the amplitude of the isotropic fluorescence $(F+2F_{\perp})$ depend on ξ. If $\xi=90°$, which as we discuss in Chap. 9 is the expected angle between the transition dipoles of the two exciton states of a dimer, coherence between states 2 and 3 does not affect the amplitude of the isotropic fluorescence. The initial anisotropy, however, will be 0.7 instead of 0.4. The predicted anisotropy will drop to 0.4 as the off-diagonal density matrix elements $\overline{\rho_{23}}$ and $\overline{\rho_{32}}$ decay to zero. For a three-state system with three orthogonal transition dipoles, the initial anisotropy is predicted to be 1.0, which means that the fluorescence is completely polarized parallel to the excitation [43]!

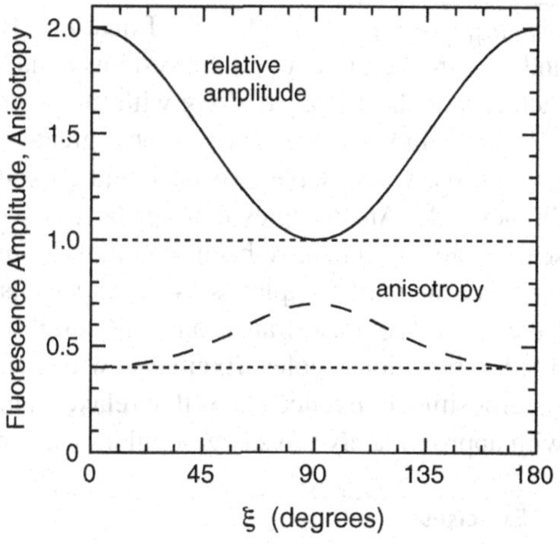

Fig. 10.12 Anisotropy and isotropic fluorescence amplitude calculated by Eqs. (10.84) and (10.85) (with $K_{23} = K_{22}$) for an ensemble with two coherently excited states, as functions of the angle (ξ) between the transition dipoles. The isotropic fluorescence is expressed relative to the amplitude expected for an incoherently excited ensemble of systems with the same value of ξ

To generalize Eq. (10.81), consider a system with M excited states. If we can evaluate the product of the density matrix and the fluorescence operator for an ensemble of systems, we can use Eq. (10.14) to calculate the fluorescence from any given excited state. Equation (10.74) suggests that the operator $\widetilde{\mathrm{F}}$ for the dipole strength of fluorescence with polarization $\hat{e}_f$ can be written symbolically as

$$\widetilde{\mathrm{F}} = \widetilde{\mu} \cdot \hat{e}_f |\psi_1\rangle\langle\psi_1|\widetilde{\mu} \cdot \hat{e}_f, \tag{10.83}$$

which has matrix elements $F_{mn} = \langle\psi_m|\widetilde{\mu} \cdot \hat{e}_f|\psi_1\rangle\langle\psi_1|\widetilde{\mu} \cdot \hat{e}_f|\psi_n\rangle$, or $(\boldsymbol{\mu}_{m1} \cdot \hat{e}_f)(\boldsymbol{\mu}_{n1} \cdot \hat{e}_f)$. Here ψ_m, ψ_n and ψ_1 are the wavefunctions of states m, n and the ground state, respectively. If we include the homogeneous emission spectrum $W_{1n}(\nu)$ as a generalization of $W_{12}(\nu)$ (Eq. 10.80), it follows from Eq. (10.14) that the flourescence from state n with polarization $\hat{e}_f$ and frequency ν is

$$F_{1n}(\nu, \hat{e}_i, \hat{e}_f) = \left[\overline{\rho_{nn}F_{nn}} + (1/2)\sum_{m \neq n}^{M} \left(\overline{\rho_{nm}F_{mn} + \rho_{mn}F_{nm}}\right)\right] W_{1n}(\nu), \tag{10.84}$$

where the $\overline{\rho_{nm}F_{jk}}$ depend on the excitation polarization. If the density matrix immediately after the excitation pulse is given by Eqs. (10.75a–10.75d, 10.76a, and 10.76b), the initial fluorescence from state n will be

$$F_{1n}(\nu, \hat{e}_i, \hat{e}_f) = \sum_m \overline{(\hat{e}_i \cdot \hat{\mu}_{n1})(\hat{e}_i \cdot \hat{\mu}_{m1})(\hat{e}_f \cdot \hat{\mu}_{n1})(\hat{e}_f \cdot \hat{\mu}_{m1})} K_{nm} W_{1n}(\nu). \tag{10.85}$$

Note that Eq. (10.85) describes the fluorescence immediately after the excitation pulse. As the off-diagonal elements decay to zero, the fluorescence becomes simply

$\overline{(\hat{e}_i \cdot \hat{\mu}_{n1})^2 (\hat{e}_i \cdot \hat{\mu}_{n1})^2} K_{nn} W_{1n}(\nu)$. Equation (10.85) also assumes that $W_{n1}(\nu)$, K_{nn} and K_{nm} are the same for all the systems in the ensemble; if this is not the case, these factors must be averaged along with the geometric parameters.

Initial fluorescence anisotropies greater than 0.4 have been measured in porphyrins, which have two degenerate excited states with orthogonal transition dipoles [44]. Anisotropies that appear to exceed 0.4 initially but decay on a time scale of about 30 fs have been seen in the stimulated emission from photosynthetic bacterial antenna complexes [45]. As we discussed in Chap. 8, these complexes have several allowed transitions with similar energies and approximately perpendicular transition dipoles. Excitation with a short pulse probably creates a coherent superposition of excited states that relaxes rapidly to an incoherent mixture of states with approximately the same populations, and then to a Boltzmann equilibrium.

Exercises

1. Suppose that molecule *a* of the dimer considered in exercise 8.1 is excited at time $t = 0$, and that the excitation then oscillates between the two molecules. (*a*) Calculate the frequency of the oscillation if the excitation energy of the two molecules is the same. (*b*) What would the oscillation frequency be if the excitation energy of molecule *a* is 1,000 cm^{-1} above that of molecule *b*? (*c*) Calculate the time-averaged probability of finding the excitation on molecule *a* for the second case ($E_{ba} = E_a - E_b = 1{,}000\ cm^{-1}$) in the absence of thermal relaxations, and compare your result with the probability of finding the excitation on this molecule at thermal equilibrium.
2. Consider the excited dimer as a system with two basis states, with $c_a(t)$ and $c_b(t)$ representing coefficients for the states in which the excitation is on, respectively, molecule *a* or molecule *b*. (*a*) Define the four elements of the density matrix ($\boldsymbol{\rho}$) for this system. (*b*) Using the notation H_{aa}, H_{ab}, etc., for the Hamiltonian matrix elements, write an expression for the time dependence of each element of $\boldsymbol{\rho}$ (e.g., $\partial \rho_{aa}/\partial t$) in the absence of stochastic relaxations. (*c*) What is the relationship between $\rho_{ab}(t)$ and $\rho_{ba}(t)$? (*d*) Suppose that interconversions of the two basis states are driven only by the quantum mechanical coupling element H_{ab}, but that stochastic fluctuations of the energies cause pure dephasing with a time constant T_2^*. What are the longitudinal (T_1) and transverse (T_2) relaxation times in this situation? (*e*) Write out the Stochastic Liouville expression for the time-dependence of each element of $\boldsymbol{\rho}$. (*f*) How would T_1 and T_2 be modified if the system also changes stochastically from state *a* to *b* with rate constant k_{ab} and from *b* to *a* with rate constant k_{ba}? (*g*) In what limit does the stochastic Liouville equation reduce to the golden-rule expression?
3. (*a*) Write an expression relating the stochastic rate constant for conversion of an ensemble of two-state quantum systems from state *a* to *b* ($-R_{aa,aa}$, where **R** is the Redfield relaxation matrix) to the spectral density of fluctuating electric fields from the surroundings. Your expression should indicate that the

rate constant depends on the fluctuations that occur at a particular frequency. (*b*) How does the important frequency depend on the energy difference between the two states (E_{ba})? (*c*) Relate the pertinent spectral density function to the autocorrelation function (memory function) of a quantum mechanical matrix element. (*d*) If the autocorrelation function decays exponentially with time constant τ_{c}, how does the rate constant depend on the value of τ_{c}? (*e*) In what limit of time does the Redfield theory apply? (*f*) Outline the modifications or extensions that are needed in order to account for a directional relaxation such as the Stokes shift of fluorescence relative to absorption.

References

1. Atkins, P.W.: Molecular Quantum Mechanics, 2nd edn. Oxford University Press, Oxford (1983)
2. Kubo, R.: The fluctuation-dissipation theorem. Rep. Progr. Theor. Phys. **29**, 255–284 (1966)
3. Kubo, R., Toda, M., Hashitsume, N.: Statistical Physics II: Nonequilibrium Statistical Mechanics. Springer, Berlin (1985)
4. Davidson, E.R.: Reduced Density Matrices in Quantum Chemistry. Academic Press, London (1976)
5. Lin, S.H., Alden, R.G., Islampour, R., Ma, H., Villaeys, A.A.: Density Matrix Method and Femtosecond Processes. World Scientific, Singapore (1991)
6. Blum, K.: Density Matrix Theory and Applications, 2nd edn. Plenum, New York (1996)
7. Haberkorn, R., Michel-Beyerle, M.E.: On the mechanism of magnetic field effects in bacterial photosynthesis. Biophys. J. **26**, 489–498 (1979)
8. Bray, A.J., Moore, M.A.: Influence of dissipation on quantum coherence. Phys. Rev. Lett. **49**, 1545–1549 (1982)
9. Reimers, J.R., Hush, N.S.: Electron transfer and energy transfer through bridged systems. I. Formalism. Chem. Phys. **134**, 323–354 (1989)
10. Kitano, M.: Quantum Zeno effect and adiabatic change. Phys. Rev. A **56**, 1138–1141 (1997)
11. Schulman, L.S.: Watching it boil: Continuous observation for the quantum Zeno effect. Found. Phys. **27**, 1623–1636 (1997)
12. Ashkenazi, G., Kosloff, R., Ratner, M.A.: Photoexcited electron transfer: Short-time dynamics and turnover control by dephasing, relaxation, and mixing. J. Am. Chem. Soc. **121**, 3386–3395 (1999)
13. Prezhdo, O.: Quantum anti-zeno acceleration of a chemical reaction. Phys. Rev. Lett. **85**, 4413–4417 (2000)
14. Facchi, P., Nakazato, H., Pascazio, S.: From the quantum zeno to the inverse quantum zeno effect. Phys. Rev. Lett. **86**, 2699–2703 (2001)
15. Kofman, A.G., Kurizki, G.: Frequent observations accelerate decay: The anti-Zeno effect. Zeit. Naturforsch. Sect. A **56**, 83–90 (2001)
16. Toschek, P.E., Wunderlich, C.: What does an observed quantum system reveal to its observer? Eur. Phys. J. D **14**, 387–396 (2001)
17. Parson, W.W., Warshel, A.: A density-matrix model of photosynthetic electron transfer with microscopically estimated vibrational relaxation times. Chem. Phys. **296**, 201–206 (2004)
18. Chiu, C.B., Sudarshan, E.C.G., Misra, B.: Time evolution of unstable quantum states and a resolution of Zeno's paradox. Phys. Rev. D **16**, 520–529 (1977)
19. Redfield, A.: The theory of relaxation processes. Adv. Magn. Res. **1**, 1–32 (1965)
20. Slichter, C.P.: Principles of Magnetic Resonance with Examples from Solid State Physics. Harper & Row, New York (1963)

21. Silbey, R.J.: Relaxation processes. In: Funfschilling, J.I. (ed.) Molecular Excited States, pp. 243–276. Kluwer Academic, Dordrecht (1989)
22. Oxtoby, D.W.: Picosecond phase relaxation experiments. A microscopic theory and a new interpretation. J. Chem. Phys. **74**, 5371–5376 (1981)
23. Yan, Y.J., Mukamel, S.: Photon echoes of polyatomic molecules in condensed phases. J. Chem. Phys. **94**, 179–190 (1991)
24. Mercer, I.P., Gould, I.R., Klug, D.R.: A quantum mechanical/molecular mechanical approach to relaxation dynamics: Calculation of the optical properties of solvated bacteriochlorophyll-a. J. Phys. Chem. B **103**, 7720–7727 (1999)
25. Callen, H.B., Greene, R.F.: On a theorem of irreversible thermodynamics. Phys. Rev. **86**, 702–710 (1952)
26. Greene, R.F., Callen, H.B.: On a theorem of irreversible thermodynamics. II. Phys. Rev. **88**, 1387–1391 (1952)
27. Berne, B.J., Harp, G.C.: On the calculation of time correlation functions. Adv. Chem. Phys. **17**, 63–227 (1970)
28. de Groot, S.R., Mazur, P.: Non-equilibrium Thermodynamics. Dover, Mineola, NY (1984)
29. Mukamel, S.: Principles of Nonlinear Optical Spectroscopy. Oxford University Press, Oxford (1995)
30. McHale, J.L.: Molecular Spectroscopy. Prentice Hall, Upper Saddle River, NJ (1999)
31. May, V., Kühn, O.: Charge and Energy Transfer Dynamics in Molecular Systems. Wiley-VCH, Berlin (2000)
32. Joo, T., Jia, Y., Yu, J.-Y., Lang, M.J., Fleming, G.R.: Third-order nonlinear time domain probes of solvation dynamics. J. Chem. Phys. **104**, 6089–6108 (1996)
33. de Boeij, W.P., Pshenichnikov, M.S., Wiersma, D.A.: System-bath correlation function probed by conventional and time-gated stimulated photon echo. J. Phys. Chem. **100**, 11806–11823 (1996)
34. Mukamel, S.: Femtosecond optical spectroscopy: A direct look at elementary chemical events. Annu. Rev. Phys. Chem. **41**, 647–681 (1990)
35. Fleming, G.R., Cho, M.: Chromophore-solvent dynamics. Annu. Rev. Phys. Chem. **47**, 109–134 (1996)
36. Volker, S.: Spectral hole-burning in crystalline and amorphous organic solids. Optical relaxation processes at low temperatures. In: Funfschilling, J.I. (ed.) Relaxation Processes in Molecular Excited States, pp. 113–242. Kluwer Academic, Dordrecht (1989)
37. Reddy, N.R.S., Lyle, P.A., Small, G.J.: Applications of spectral hole burning spectroscopies to antenna and reaction center complexes. Photosynth. Res. **31**, 167–194 (1992)
38. Warshel, A., Parson, W.W.: Computer simulations of electron-transfer reactions in solution and in photosynthetic reaction centers. Annu. Rev. Phys. Chem. **42**, 279–309 (1991)
39. Warshel, A., Parson, W.W.: Dynamics of biochemical and biophysical reactions: Insight from computer simulations. Q. Rev. Biophys. **34**, 563–679 (2001)
40. de Boeij, W.P., Pshenichnikov, M.S., Wiersma, D.A.: Ultrafast solvation dynamics explored by femtosecond photon echo spectroscopies. Annu. Rev. Phys. Chem. **49**, 99–123 (1998)
41. Myers, A.B.: Molecular electronic spectral broadening in liquids and glasses. Annu. Rev. Phys. Chem. **49**, 267–295 (1998)
42. Rahman, T.S., Knox, R., Kenkre, V.M.: Theory of depolarization of fluorescence in molecular pairs. Chem. Phys. **44**, 197–211 (1979)
43. van Amerongen, H., Struve, W.S.: Polarized optical spectroscopy of chromoproteins. Methods Enzymol. **246**, 259–283 (1995)
44. Wynne, K., Hochstrasser, R.M.: Coherence effects in the anisotropy of optical experiments. Chem. Phys. **171**, 179–188 (1993)
45. Nagarajan, V., Johnson, E., Williams, J.C., Parson, W.W.: Femtosecond pump-probe spectroscopy of the B850 antenna complex of *Rhodobacter sphaeroides* at room temperature. J. Phys. Chem. B **103**, 2297–2309 (1999)

Pump-Probe Spectroscopy, Photon Echoes and Vibrational Wavepackets 11

11.1 First-Order Optical Polarization

In the last chapter, we used a steady-state treatment to relate the shape of an absorption band to the dynamics of relaxations in the excited state. Because a period on the order of the electronic dephasing time will be required to establish a steady state, Eqs. (10.43) and (10.44) apply only on time scales longer than this. We need to escape this limitation if we hope to explore the relaxation dynamics themselves. Our first goal in this chapter is to develop a more general approach for analyzing spectroscopic experiments on femtosecond and picosecond time scales. This provides a platform for discussing how pump-probe and photon-echo experiments can be used to probe the dynamics of structural fluctuations and the transfer of energy or electrons on these short time scales.

To start, consider an ensemble of systems, each of which has two states (m and n) with energies E_m and E_n, respectively. In the presence of a weak radiation field $\boldsymbol{E}(t) = \boldsymbol{E}_o[\exp(i\omega t) + \exp(-i\omega t)]$, the Hamiltonian matrix element $\langle \psi_n | \widetilde{\mathrm{H}} | \psi_m \rangle$ can be written $H_{nm}(t) = H^0_{nm} + V_{nm}(t)$, where H^0_{nm} is the matrix element in the absence of the field, $V_{nm}(t) = -\boldsymbol{\mu}_{nm} \cdot \boldsymbol{E}_o\ [\exp(i\omega t) + \exp(-i\omega t)]$, and $\boldsymbol{\mu}_{nm}$ is the transition dipole. The commutator $[\overline{\boldsymbol{\rho}},\mathbf{H}]$, which determines the time dependence of the density matrix, then is the sum $[\overline{\boldsymbol{\rho}},\mathbf{H}^0] + [\overline{\boldsymbol{\rho}},\mathbf{V}]$. If the system is stationary in the absence of field (i.e., $H^0_{nm} = 0$ for $n \neq m$), the matrix elements of $[\overline{\boldsymbol{\rho}},\mathbf{H}^0]$ are $\left[\overline{\boldsymbol{\rho}},\mathbf{H}^0\right]_{nm} = E_{mn}\overline{\rho_{nm}} = -E_{nm}\overline{\rho_{nm}}$, where $E_{nm} = E_n - E_m$ (Eqs. 10.21–10.24).

We can simplify matters if we adjust all the elements of $\overline{\boldsymbol{\rho}}$ by subtracting the Boltzmann-equilibrium values in the absence of the radiation field (0 for off-diagonal elements, and $\overline{\rho^o_{nn}}$ as prescribed by Eq. (10.26) for diagonal elements). The rate constant for stochastic relaxations of the adjusted density matrix element $\overline{\rho_{nm}}$ then can be written as γ_{nm}, where $\gamma_{nm} = 1/T_1$ for $n = m$ and $1/T_2$ for $n \neq m$. To simplify the notation further, let $E_{nm}/\hbar = \omega_{nm}$.

W.W. Parson, *Modern Optical Spectroscopy*, DOI 10.1007/978-3-662-46777-0_11

With these definitions and adjustments, the stochastic Liouville equation (Eq. 10.30) for the ensemble of two-state systems in the presence of the radiation field takes the form

$$\partial\overline{\rho_{nm}}/\partial t = (i/\hbar)\left[\overline{\boldsymbol{\rho}}, \mathbf{H^0}\right]_{nm} + (i/\hbar)[\overline{\boldsymbol{\rho}}, \mathbf{V}]_{nm} - \gamma_{nm}\overline{\rho_{nm}} \tag{11.1a}$$

$$= (i/\hbar)[\overline{\boldsymbol{\rho}}, \mathbf{V}]_{nm} - (i\omega_{nm} + \gamma_{nm})\overline{\rho_{nm}}. \tag{11.1b}$$

The trick now is to expand $\overline{\rho_{nm}}(t)$ in a series of increasing orders of perturbation by the radiation field [1]:

$$\overline{\rho_{nm}} = \overline{\rho_{nm}^{(0)}} + \overline{\rho_{nm}^{(1)}} + \overline{\rho_{nm}^{(2)}} + \cdots, \tag{11.2}$$

where $\overline{\rho_{nm}^{(0)}}$ is the density matrix at equilibrium in the absence of the radiation field, $\overline{\rho_{nm}^{(1)}}$ is the perturbation to $\overline{\rho_{nm}^{(0)}}$ in the limit of a very weak field (a linear, or first-order perturbation), $\overline{\rho_{nm}^{(2)}}$ is a perturbation that is quadratic in the strength of the field, and so forth. If we view perturbations of progressively higher orders as developing sequentially in time, we can use Eq. (11.1b) to write their rates of change:

$$\frac{\partial\overline{\rho_{nm}^{(0)}}}{\partial t} = -(i\omega_{nm} + \gamma_{nm})\overline{\rho_{nm}^{(0)}}, \tag{11.3a}$$

$$\frac{\partial\overline{\rho_{nm}^{(1)}}}{\partial t} = (i/\hbar)\left[\overline{\boldsymbol{\rho}^{(0)}}, \mathbf{V}\right]_{nm} - (i\omega_{nm} + \gamma_{nm})\overline{\rho_{nm}^{(1)}}, \tag{11.3b}$$

$$\frac{\partial\overline{\rho_{nm}^{(2)}}}{\partial t} = (i/\hbar)\left[\overline{\boldsymbol{\rho}^{(1)}}, \mathbf{V}\right]_{nm} - (i\omega_{nm} + \gamma_{nm})\overline{\rho_{nm}^{(2)}}, \tag{11.3c}$$

and in general,

$$\partial\overline{\rho_{nm}^{(k)}}/\partial t = (i/\hbar)\left[\overline{\boldsymbol{\rho}^{(k-1)}}, \mathbf{V}\right]_{nm} - (i\omega_{nm} + \gamma_{nm})\overline{\rho_{nm}^{(k)}}. \tag{11.4}$$

The solution to Eq. (11.4) for $\overline{\rho_{nm}^{(k)}}$ at time τ is

$$\overline{\rho_{nm}^{(k)}}(\tau) = (i/\hbar)\int_0^\tau \left[\overline{\boldsymbol{\rho}^{(k-1)}}, \mathbf{V}\right]_{nm} \exp[-(i\omega_{nm} + \gamma_{nm})(\tau - t)]dt. \tag{11.5}$$

The terms in Eq. (11.2) thus represent the results of sequential interactions of the ensemble with the radiation field, convolved with oscillations and decay of the states and coherences generated by these interactions. If an ensemble of two-state systems starts out with all the systems in the ground state (a), a single interaction

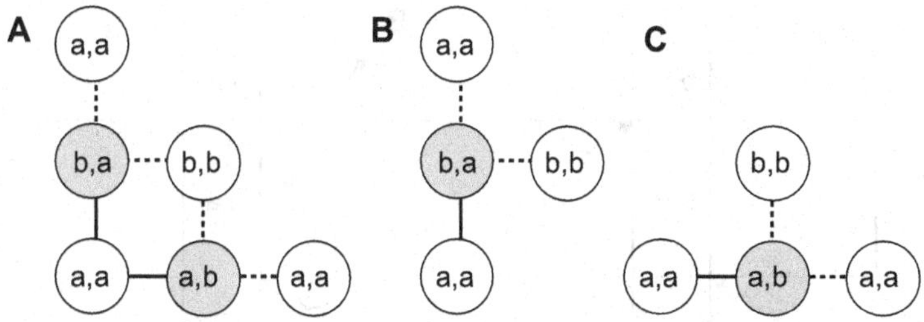

Fig. 11.1 Pathways in Liouville space. The circles labeled a,a and b,b represent the diagonal elements of the density matrix (populations) for a two-state system; those labeled a,b and b,a represent off-diagonal elements (coherences). *Lines* represent individual interactions with a radiation field, with *vertical lines* denoting interactions that change the left-hand (*bra*) index of the density matrix and *horizontal lines* those that change the right-hand (*ket*) index. In the convention used here, the zero-order density matrix ($\boldsymbol{\rho}^{(0)}$) is at the lower left, and time increases upwards and to the right; downward or leftward steps are not allowed. The coherences in the *shaded circles* are endpoints of the two one-step pathways [$\rho_{a,a} \to \rho_{b,a}$ (**B**) and $\rho_{a,a} \to \rho_{a,b}$ (**C**)] that contribute to the first-order density matrix ($\overline{\boldsymbol{\rho}^{(1)}}$) and the first-order optical polarization ($\boldsymbol{P}^{(1)}$). A second interaction with the radiation field (*dotted line*) can convert a coherence to the excited state (ρ_{bb}) or the ground (ρ_{aa}) state. The pathways in (**B**) and (**C**) are described as complex conjugates because one can be generated from the other by interchanging the two indices at each step

with the field creates one of the off-diagonal density matrix elements ($\overline{\rho_{ab}}$ or $\overline{\rho_{ba}}$), which represent coherences of state *a* with the excited state (*b*). A second interaction can either create a population in state *b* ($\overline{\rho_{bb}}$) or regenerate $\overline{\rho_{aa}}$. Such sequences of interactions are described as pathways in *Liouville space*, and can be represented schematically as shown in Fig. 11.1. A *Liouville-space diagram* consists of a square lattice, with each of the lattice points (circles in Fig. 11.1) labeled by the two indices of a density matrix element. A vertical line connecting two circles represents an interaction that changes the left index (*bra*); a horizontal line, an interaction that changes the right index (*ket*). The convention used here is that the density matrix of a resting ensemble begins at the lower left corner of the diagram and evolves upward or to the right on each interaction with an electromagnetic radiation field. The coherences or populations that are generated by a given number of interactions, and so make up that order of the density matrix, lie on an antidiagonal line. In Fig. 11.1, the coherences that contribute to $\overline{\boldsymbol{\rho}^{(1)}}$ are highlighted in gray.

Figure 11.2 shows another useful representation called a *double-sided Feynman diagram*. The two vertical lines in this diagram represent the left and right indices of the density matrix, and each interaction with the field is represented by a wavy arrow pointing to or away from one of the lines. Time increases upwards. Arrows pointing toward a vertical line are associated with absorption of a photon; arrows pointing away, with emission. The diagram also can convey additional information such as the wavevector and frequency of the radiation [2–11].

The power-series expansion of $\overline{\boldsymbol{\rho}}(t)$ can be used with Eq. (10.14) to find the expectation value of the macroscopic electric dipole for an ensemble of systems exposed to an electromagnetic field:

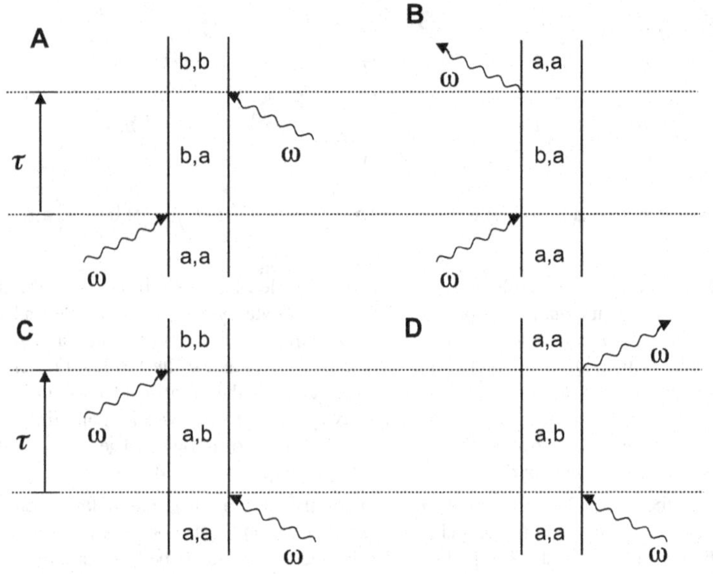

Fig. 11.2 Double-sided Feynman diagrams. The pairs of *vertical lines* represent the evolution of the left (*bra*) and right (*ket*) indices of the density matrix. Time increases from the bottom to the top of the diagram. In the two-state system considered here diagonal elements of ρ are labeled a,a and b,b for the ground and excited states, respectively, and a,b and b,a denote off-diagonal elements. *Wavy arrows labeled with a frequency* (ω) indicate interactions with an electromagnetic radiation field. Incoming arrows are associated with absorption of a photon (an increase in one of the indices of ρ); outgoing arrows, with emission (a decrease in one of the indices). A single interaction with the field at time zero generates a coherence that contributes to the first-order optical polarization. A second interaction at time τ converts a coherence to either the excited state (**A**, **C**) or the ground state (**B**, **D**). The sequences depicted in (**C**) and (**D**) are the complex conjugates of those in (**A**) and (**B**), respectively. (**A**) and (**B**) correspond to the paths shown in Fig. 11.1B; (**C**) and (**D**), to those in Fig. 11.1C. Double-sided Feynman diagrams also can be used to convey additional information, such as the wavevectors of incoming and outgoing radiation in experiments involving multiple pulses [3, 4, 7]

$$\langle \boldsymbol{\mu}(t) \rangle = \sum_k \boldsymbol{P}^{(k)}(t), \tag{11.6}$$

with

$$\boldsymbol{P}^{(k)}(t) = Tr\left(\overline{\boldsymbol{\rho}^{(k)}(t)\boldsymbol{\mu}}\right) = \overline{\sum_n \sum_m \rho_{nm}^{(k)}(t)\boldsymbol{\mu}_{mn}} \tag{11.7}$$

and $\boldsymbol{\mu}_{mn} = \langle \psi_m | \tilde{\mu} | \psi_n \rangle$. The terms $\boldsymbol{P}^{(k)}$ are referred to as various orders of the *optical polarization.* The first-order optical polarization is the quantum mechanical analog of the classical linear polarization of a dielectric by an oscillating electromagnetic field (Box B3.3); the higher-order polarizations correspond to classical components that depend on higher powers of the field. But note that, according to Eqs. (11.4–11.7), increasing orders of the optical polarization develop sequentially in time, whereas the classical polarizations with various dependences on the field were assumed to form instantaneously and simultaneously.

The benefit of expanding the optical polarization in this way is that various optical phenomena can be assigned to the terms with particular values of k [2, 6, 12–15]. The *first-order*, or *linear optical polarization* ($\boldsymbol{P}^{(1)}$) pertains to ordinary absorption of light; the *second-order optical polarization* ($\boldsymbol{P}^{(2)}$), to the generation of sum and difference frequencies; and the *third-order optical polarization* ($\boldsymbol{P}^{(3)}$), to "four-wave mixing" experiments that include pump-probe spectroscopy, transient-grating spectroscopy, and photon echoes. More precisely, each class of optical phenomena can be related to interaction of the electric field with a particular order of the polarization.

Classically, the optical polarization is viewed as a macroscopic oscillating dipole that can serve as either a source or an absorber of electromagnetic radiation. In a semiclassical picture in which we treat electromagnetic radiation classically, the energy of interaction of a radiation field $\boldsymbol{E}(t)$ with an optically polarized system is given by

$$\left\langle H'(t)\right\rangle = -\langle\boldsymbol{\mu}(t)\rangle\cdot\boldsymbol{E}(t) = -\sum_k \boldsymbol{P}^{(k)}(t)\cdot\boldsymbol{E}(t), \tag{11.8}$$

or using Eq. (10.14),

$$\left\langle H'(t)\right\rangle = \mathrm{Tr}\left(\overline{\boldsymbol{\rho}(t)}\,\mathbf{V}(t)\right), \tag{11.9}$$

where $\mathbf{V}$ again is the interaction matrix ($V_{nm} = -\boldsymbol{\mu}_{nm}\boldsymbol{P}(t)$). The rate of absorption of energy from the field is the derivative of this quantity with respect to time:

$$\frac{d\langle H'\rangle}{dt} = \frac{d}{dt}Tr(\overline{\boldsymbol{\rho}}\mathbf{V}) = \mathrm{Tr}\left(\frac{\partial(\overline{\boldsymbol{\rho}}\mathbf{V})}{\partial t}\right) = \mathrm{Tr}\left(\overline{\boldsymbol{\rho}}\,\frac{d\mathbf{V}}{dt}\right) + \mathrm{Tr}\left(\mathbf{V}\,\frac{d\overline{\boldsymbol{\rho}}}{dt}\right). \tag{11.10}$$

The last term on the right side of Eq. (11.10) is zero. You can show this by using the Von Neumann equation (Eq. 10.24) and noting that, because cyclic permutation of three matrices does not change the trace of the product of the matrices (Appendix A2), $\mathrm{Tr}(\mathbf{V}\,\overline{\boldsymbol{\rho}}\,\mathbf{V}) - \mathrm{Tr}(\mathbf{V}\,\mathbf{V}\,\overline{\boldsymbol{\rho}}) = 0$:

$$\begin{aligned}\mathrm{Tr}\left(\mathbf{V}\frac{d\boldsymbol{\rho}}{dt}\right) &= (i/\hbar)\mathrm{Tr}(\mathbf{V}\,[\boldsymbol{\rho},\mathbf{V}]) \\ &= (i/\hbar)\{\mathrm{Tr}(\mathbf{V}\,\boldsymbol{\rho}\,\mathbf{V}) - \mathrm{Tr}(\mathbf{V}\,\mathbf{V}\,\boldsymbol{\rho})\} = 0.\end{aligned} \tag{11.11}$$

If we drop this term, write the oscillating radiation field as in Eq. (10.37), and assume that the envelope of the field amplitude ($\boldsymbol{E}_o$) changes only slowly relative to $\exp(i\omega t)$, Eq. (11.10) gives

$$\frac{d\langle H' \rangle}{dt} = \mathrm{Tr}\left(\overline{\boldsymbol{\rho}}\frac{d\mathbf{V}}{dt}\right) = -\boldsymbol{P}(t) \cdot \frac{d}{dt}[\boldsymbol{E}_o\exp(i\omega t) + \boldsymbol{E}_o{}^*\exp(-i\omega t)]$$
$$\approx -i\omega \boldsymbol{P}(t) \cdot [\boldsymbol{E}_o\exp(i\omega t) + \boldsymbol{E}_o{}^*\exp(-i\omega t)]. \tag{11.12}$$

The instantaneous rate of excitation thus is proportional to the dot product of the optical polarization and the field, $\boldsymbol{P}(t)\cdot\boldsymbol{E}(t)$.

We saw in Eqs. (10.40)–(10.42) that the elements of the first-order density matrix contain factors that oscillate at the same frequency as the electromagnetic field that generates them. The same is true of the optical polarization. $\boldsymbol{P}$ therefore can be put in the form

$$\boldsymbol{P}(t) = \boldsymbol{P}_o(t)\exp(i\omega t) + \boldsymbol{P}_o{}^*(t)\exp(-i\omega t), \tag{11.13}$$

where $\boldsymbol{P}_o$ and $\boldsymbol{P}_o{}^*$, like $\boldsymbol{E}_o$, $\boldsymbol{E}_o{}^*$ and the factors $\overline{\overline{\rho_{ab}}}$ and $\overline{\overline{\rho_{ba}}}$ in Eqs. (10.40–10.42), change comparatively slowly with time. With $\boldsymbol{P}$ written in this way, Eq. (11.12) becomes

$$\frac{d\langle H' \rangle}{dt} = -i\omega[\boldsymbol{P}_o\exp(i\omega t) + \boldsymbol{P}_o{}^*\exp(-i\omega t)][\boldsymbol{E}_o\exp(i\omega t) - \boldsymbol{E}_o{}^*\exp(-i\omega t)]$$
$$= i\omega(\boldsymbol{P}_o\boldsymbol{E}_o{}^* - \boldsymbol{P}_o{}^*\boldsymbol{E}_o) + i\omega[\boldsymbol{P}_o\boldsymbol{E}_o\exp(2i\omega t) - \boldsymbol{P}_o{}^*\boldsymbol{E}_o{}^*\exp(-2i\omega t)]. \tag{11.14}$$

If we average this expression over a period of the oscillation, the factors containing $\exp(\pm 2i\omega t)$ drop out, leaving

$$\frac{d\langle H' \rangle}{dt} = i\omega(\boldsymbol{P}_o\boldsymbol{E}_o{}^* - \boldsymbol{P}_o{}^*\boldsymbol{E}_o) = 2\omega\mathrm{Im}(\boldsymbol{P}_o\boldsymbol{E}_o{}^*). \tag{11.15}$$

To illustrate the use of Eqs. (11.4–11.7) and (11.15), let's evaluate $\overline{\boldsymbol{\rho}^{(1)}}$ and $\boldsymbol{P}^{(1)}$ for an ensemble of two-state systems that are exposed to light. If the ensemble is at thermal equilibrium before the light is switched on, the initial density matrix is

$$\overline{\boldsymbol{\rho}^{(0)}} = \begin{bmatrix} \overline{\rho^o_{aa}} & 0 \\ 0 & \overline{\rho^o_{bb}} \end{bmatrix}. \tag{11.16}$$

Suppose that the amplitude of the oscillating electric field goes from zero to $\boldsymbol{E}_o$ abruptly at time zero and then remains constant at this level. Neglecting the initial rise, the perturbation matrix $\mathbf{V}$ for $t \geq 0$ then is given by $V_{ab} = -\boldsymbol{\mu}_{ab}\boldsymbol{E}(t)$. We assume here that the transition dipoles are the same for all members of the ensemble ($\overline{V_{ab}} = V_{ab}$), although V_{ab} usually must be averaged together with $\boldsymbol{\rho}$ over molecules with different orientations relative to the incident radiation. Let's also neglect any dependence of $\boldsymbol{E}_o$ and $\boldsymbol{\mu}_{ab}$ on ω, and assume further that the two basis states have no net charge or dipole moment, so that $V_{aa} = V_{bb} = 0$.

With these assumptions, the commutator $\left[\overline{\boldsymbol{\rho}^{(0)}}, \mathbf{V}\right]$ is zero for $t < 0$, and becomes

$$\left[\overline{\boldsymbol{\rho}^{(0)}}, \mathbf{V}\right] = \begin{bmatrix} 0 & \left(\overline{\rho_{aa}^{o}} - \overline{\rho_{bb}^{o}}\right)V_{ab} \\ \left(\overline{\rho_{bb}^{o}} - \overline{\rho_{aa}^{o}}\right)V_{ba} & 0 \end{bmatrix} \tag{11.17}$$

for $t \geq 0$. You can check these matrix elements by referring to Eqs. (10.21) and Box 10.1. For example, the entry for $[\overline{\boldsymbol{\rho}^{(0)}}, \mathbf{V}]_{ab}$, which pertains to the growth of $\overline{\rho_{ab}^{(1)}}$, is

$$\begin{aligned} \left[\overline{\boldsymbol{\rho}^{(0)}}, \mathbf{V}\right]_{ab} &= \sum_k \left(\overline{\rho_{ak}^{(0)}} V_{kb} - V_{ak}\overline{\rho_{kb}^{(0)}}\right) \\ &= \left(\overline{\rho_{aa}^{(0)}} - \overline{\rho_{bb}^{(0)}}\right)V_{ab} + \overline{\rho_{ba}^{(0)}}\left(V_{bb} - V_{aa}\right) = \left(\overline{\rho_{aa}^{(0)}} - \overline{\rho_{bb}^{(0)}}\right)V_{ab}. \end{aligned} \tag{11.18}$$

Similarly, $\left[\overline{\boldsymbol{\rho}^{(0)}}, \mathbf{V}\right]_{aa} = \overline{\rho_{ab}^{(0)}} V_{ba} - \overline{\rho_{ba}^{(0)}} V_{\mathrm{ab}} = 0$, because $\overline{\rho_{ab}^{(0)}}$ and $\overline{\rho_{ba}^{(0)}}$ are zero.

Using $[\overline{\boldsymbol{\rho}^{(0)}}, \mathbf{V}]_{ab}$ from Eq. (11.18) with V_{ab} from Eq. (10.37), Eq. (11.5) gives

$$\overline{\rho_{ab}^{(1)}}(\tau) = (i/\hbar)\left(\overline{\rho_{bb}^{o}} - \overline{\rho_{aa}^{o}}\right)\int_0^{\tau}\left\{\exp[-(i\omega_{ab} + \gamma_{ab})(\tau - t)]\,\boldsymbol{\mu}_{ab}\cdot\boldsymbol{E}_o[\exp(i\omega t) + \exp(-i\omega t)]\right\}dt \tag{11.19}$$

The integrand in Eq. (11.19) includes factors of the forms $\exp[i(\omega_{ab} + \omega)t]$ and $\exp[i(\omega_{ab} - \omega)t]$, where ω_{ab} again is $(H_{aa} - H_{bb})/\hbar$. Because $\omega_{ab} = -\omega_{ba}$, $\exp[i(\omega_{ab} + \omega)t]$ goes to 1 when $\omega \approx \omega_{ba}$. The factor $\exp[i(\omega_{ab} - \omega)t]$, on the other hand, becomes approximately $\exp(-2i\omega t)$, which oscillates rapidly between positive and negative values and contributes little to the integral at times greater than $1/\omega$. Neglecting the term $\exp(-2i\omega t)$ is essentially the same as the rotating-wave approximation we used in Sect. 10.6. Making this approximation, we have

$$\begin{aligned} \overline{\rho_{ab}^{(1)}}(\tau) &\approx \frac{i}{\hbar}\left(\overline{\rho_{bb}^{o}} - \overline{\rho_{aa}^{o}}\right)\boldsymbol{\mu}_{ab}\cdot\boldsymbol{E}_o\exp[-(i\omega_{ab} + \gamma_{ab})\tau]\int_0^{\tau}\left\{\exp[(i\omega_{ab} + i\omega + \gamma_{ab})t]\right\}dt \\ &= \frac{i}{\hbar}\left(\overline{\rho_{bb}^{o}} - \overline{\rho_{aa}^{o}}\right)\boldsymbol{\mu}_{ab}\cdot\boldsymbol{E}_o\exp[-(i\omega_{ab} + \gamma_{ab})\tau]\left(\frac{\exp[(i\omega_{ab} + i\omega + \gamma_{ab})\tau] - 1}{i(\omega_{ab} + \omega) + \gamma_{ab}}\right) \\ &= \frac{i}{\hbar}\left(\overline{\rho_{bb}^{o}} - \overline{\rho_{aa}^{o}}\right)\boldsymbol{\mu}_{ab}\cdot\boldsymbol{E}_o\frac{\exp(i\omega\tau) - \exp[-(i\omega_{ab} + \gamma_{ab})\tau]}{i(\omega_{ab} + \omega) + \gamma_{ab}}. \end{aligned} \tag{11.20}$$

For comparison of Eq. (11.20) with the corresponding steady-state expression, multiplying $\overline{\rho_{ab}^{(1)}}$ by $\exp(-i\omega\tau)$ to remove the rapid oscillation with time (Eq. 10.40), and setting $\omega \approx \omega_{ba} = -\omega_{ab}$, gives

$$\overline{\overline{\rho_{ab}^{(1)}}} = \overline{\rho_{ab}^{(1)}}\exp(-i\omega\tau) = \frac{i}{\hbar}\left(\overline{\rho_{bb}^{o}} - \overline{\rho_{aa}^{o}}\right)\boldsymbol{\mu}_{ab}\cdot\boldsymbol{E}_o\frac{1-\exp[-(i\omega_{ab}+i\omega+\gamma_{ab})\tau]}{i(\omega_{ab}+\omega)+\gamma_{ab}}$$
$$\approx\frac{i}{\hbar}\left(\overline{\rho_{bb}^{o}} - \overline{\rho_{aa}^{o}}\right)\boldsymbol{\mu}_{ab}\cdot\boldsymbol{E}_o\frac{1-\exp(-\gamma_{ab}\tau)}{i(\omega_{ab}+\omega)+\gamma_{ab}}. \tag{11.21}$$

Except for the additional factor $[1-\exp(-\gamma_{ab}\tau)]$ in the numerator, this is the same as Eq. (10.42a). As τ increases, the factor $[1-\exp(-\gamma_{ab}\tau)]$ takes $\overline{\rho_{ab}}\exp(-i\omega\tau)$ to its steady-state value with time constant $1/\gamma_{ab}$, which is T_2 in a two-state system.

Now that we have $\overline{\boldsymbol{\rho}^{(1)}}$, we can use Eq. (11.4) to find $\partial\overline{\rho_{bb}^{(2)}}/\partial t$, which will give us the time course of excitation to state b. The required element of the commutator $[\overline{\boldsymbol{\rho}^{(1)}},\mathbf{V}]$ is

$$\left[\overline{\boldsymbol{\rho}^{(1)}},\mathbf{V}\right]_{bb} = V_{ab}\overline{\rho_{ba}^{(1)}} - V_{ba}\overline{\rho_{ab}^{(1)}} = V_{ab}\left(\overline{\rho_{ab}^{(1)}}* - \overline{\rho_{ab}^{(1)}}\right), \tag{11.22}$$

and inserting $[\overline{\boldsymbol{\rho}^{(1)}},\mathbf{V}]_{bb}$ into Eq. (11.4) gives

$$\partial\overline{\rho_{bb}^{(2)}}/\partial t = (i/\hbar)V_{ab}\left\{\overline{\rho_{ab}^{(1)}}* - \overline{\rho_{ab}^{(1)}}\right\} - \gamma_{bb}\overline{\rho_{bb}^{(2)}}. \tag{11.23}$$

If we're interested in the rate of excitation at times that are short relative to the lifetime of the excited state ($t \ll T_1 = 1/\gamma_{bb}$), we can neglect the last term on the right side of Eq. (11.23). Using $\overline{\rho_{ab}^{(1)}}$ from Eq. (11.20), making the rotating-wave approximation of dropping terms that oscillate at frequencies greater than ω_{ba}, and setting $\omega+\omega_{ab}\approx 0$, Eq. (11.23) then gives

$$\partial\overline{\rho_{bb}^{(2)}}\Big/\partial t \approx -\left(\frac{i}{\hbar}\right)\left(\overline{\rho_{bb}^{(0)}} - \overline{\rho_{aa}^{(0)}}\right)\left|\boldsymbol{\mu}_{ab}\cdot\boldsymbol{E}_o\right|^2[\exp(i\omega\tau)+\exp(-i\omega\tau)] \quad\times$$
$$\left(\frac{i}{\hbar}\right)\left\{\frac{\exp(-i\omega\tau)-\exp[(i\omega_{ab}-\gamma_{ab})\tau]}{i(\omega_{ab}+\omega)-\gamma_{ab}} - \frac{\exp(i\omega\tau)-\exp[-(i\omega_{ab}+\gamma_{ab})\tau]}{i(\omega_{ab}+\omega)+\gamma_{ab}}\right\}$$
$$\approx\left(\overline{\rho_{bb}^{(0)}} - \overline{\rho_{aa}^{(0)}}\right)\frac{\left|\boldsymbol{\mu}_{ab}\cdot\boldsymbol{E}_o\right|^2}{\hbar^2}\left\{\frac{1-\exp(-\gamma_{ab}\tau)}{i(\omega_{ab}+\omega)-\gamma_{ab}} - \frac{1-\exp(-\gamma_{ab}\tau)}{i(\omega_{ab}+\omega)+\gamma_{ab}}\right\}$$
$$=\left(\overline{\rho_{aa}^{(0)}} - \overline{\rho_{bb}^{(0)}}\right)\frac{\left|\boldsymbol{\mu}_{ab}\cdot\boldsymbol{E}_o\right|^2}{\hbar^2}\left\{\frac{2\gamma_{ab}}{(\omega-\omega_{ba})^2+\gamma_{ab}{}^2}\right\}[1-\exp(-\gamma_{ab}\tau)]. \tag{11.24}$$

This is just the steady-state expression (Eq. 10.43) multiplied by the same factor of $[1-\exp(-\gamma_{ab}\tau)]$ that appears in Eq. (11.22). Eq. (11.24) indicates that the rate of formation of state 2 starts at zero, increases linearly with time at short times, and levels off at times exceeding $1/\gamma_{ab}$. The population of the excited state therefore will increase quadratically with time at short times, as we anticipated in Sect. 4.2. Figure 11.3 shows the predicted kinetics.

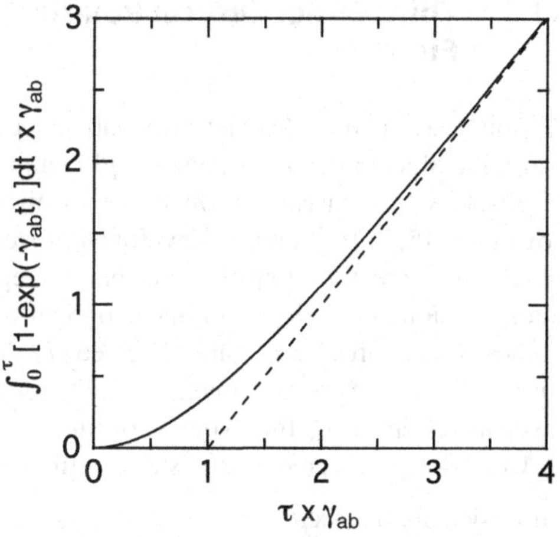

Fig. 11.3 The time dependence of $\overline{\rho_{bb}^{(2)}}$ according to Eq. (11.24). The *solid curve* is the integral of the function $[1-\exp(-\gamma_{ab}t)]$ from time $t=0$ to τ; the *dashed line* is the asymptote at long times. Time is scaled relative to $1/\gamma_{ab}$

Now let's see if we get the same result by evaluating the first-order polarization. Because $\overline{\rho_{aa}^{(1)}}$ and $\overline{\rho_{bb}^{(1)}}$ are zero, the first-order optical polarization at time t has only two terms:

$$\begin{aligned} \boldsymbol{P}^{(1)}(t) &= \sum_n \sum_m \overline{\rho_{nm}^{(1)}}(t)\boldsymbol{\mu}_{mn} = \overline{\rho_{ab}^{(1)}}(t)\boldsymbol{\mu}_{ba} + \overline{\rho_{ba}^{(1)}}(t)\boldsymbol{\mu}_{ab} \\ &= \overline{\overline{\rho_{ab}^{(1)}}}(t)\exp(i\omega t)\boldsymbol{\mu}_{ba} + \overline{\overline{\rho_{ba}^{(1)}}}(t)\exp(-i\omega t)\boldsymbol{\mu}_{ab} \\ &= \boldsymbol{P}_o^{(1)}(t)\exp(i\omega t) + \boldsymbol{P}_o^{(1)*}(t)\exp(-i\omega t), \end{aligned} \tag{11.25}$$

with $\overline{\overline{\rho_{ab}^{(1)}}}$ given by Eq. (11.21) and $\boldsymbol{P}_o^{(1)} \equiv \overline{\overline{\rho_{ab}^{(1)}}}\boldsymbol{\mu}_{ba}$.

Combining this result with Eqs. (11.15) and (11.21) gives the rate of absorption of energy:

$$\begin{aligned} \frac{d\langle H'\rangle}{dt} &= 2\omega \mathrm{Im}\left(\boldsymbol{P}_o^{(1)} \cdot \boldsymbol{E}_o^*\right) \\ &\approx 2\omega\ \mathrm{Im}\left[\frac{i}{\hbar}\left(\overline{\rho_{bb}^{o}} - \overline{\rho_{aa}^{o}}\right)(\boldsymbol{\mu}_{ab}\cdot\boldsymbol{E}_o)\frac{1-\exp(-\gamma_{ab}\tau)}{i(\omega_{ab}+\omega)+\gamma_{ab}}(\boldsymbol{\mu}_{ba}\cdot\boldsymbol{E}_o^*)\right] \\ &= \omega\left(\overline{\rho_{aa}^{(0)}} - \overline{\rho_{bb}^{(0)}}\right)\frac{|\boldsymbol{\mu}_{ab}\cdot\boldsymbol{E}_o|^2}{\hbar}\left(\frac{2\gamma_{ab}}{(\omega-\omega_{ba})^2+\gamma_{ab}{}^2}\right)[1-\exp(-\gamma_{ab}t)]. \end{aligned} \tag{11.26}$$

Except for a factor of $\hbar\omega$ (the energy absorbed per excitation), this is the same as Eq. (11.24). Using Eq. (11.15) thus gives the same result as working through the complete first-order density matrix.

11.2 Third-Order Optical Polarization and Non-linear Response Functions

Although evaluating the first-order optical polarization is reasonably straightforward, the algebra becomes increasingly cumbersome, and the interpretation of the results less transparent, as we move to higher-order terms of $\boldsymbol{P}$. Mukamel and coworkers [5, 7, 16, 17] have developed procedures based on pathways in Liouville space that greatly simplify the book-keeping and that help to clarify the interpretations of various nonlinear optical experiments. We'll give an overview of these procedures for $\boldsymbol{P}^{(1)}$ and $\boldsymbol{P}^{(3)}$. See [7] and [18] for additional details and more formal discussions of how operators that work in Liouville space relate to quantum mechanical operators that work in ordinary vector space (Hilbert space).

Consider an ensemble of systems with two electronic states (a and b), and an initial density matrix $\overline{\rho_{aa}^{(0)}} = 1$ and $\overline{\rho_{bb}^{(0)}} = \overline{\rho_{ab}^{(0)}} = \overline{\rho_{ba}^{(0)}} = 0$. Referring to the diagrams in Fig. 11.1, we see that each of the off-diagonal elements of $\boldsymbol{\rho}$ that contribute to $\boldsymbol{P}^{(1)}$ (ρ_{ab}, and ρ_{ba}) can be formed by a pathway in Liouville space that involves a single interaction with the radiation field ($\rho_{aa} \rightarrow \rho_{ab}$ or $\rho_{aa} \rightarrow \rho_{ba}$); a second interaction then is required to generate ρ_{bb}. Take the pathway through ρ_{ab}. Rewriting Eq. (11.20) for $\overline{\rho_{bb}^{(0)}} = 0$, $\overline{\rho_{ab}^{(1)}}$ at time τ is

$$\overline{\rho_{ab}^{(1)}}(\tau) = (-i/\hbar)\int_0^\tau \left\{\exp[-(i\omega_{ab} + \gamma_{ab})(\tau - t)]\,\boldsymbol{\mu}_{ab} \cdot \boldsymbol{E}(t)\overline{\rho_{aa}^{(0)}}\right\} dt, \qquad (11.27)$$

where $\boldsymbol{E}(t)$ is the electric field at time t. The right side of this expression can be dissected conceptually into three operations:

1. The electromagnetic radiation field acts on $\overline{\rho_{aa}^{(0)}}$ for a short interval of time (Δt) at time t, creating a small increment of an off-diagonal density matrix element representing coherence between states 1 and 2 ($\overline{\rho_{ab}^{(1)}}$). The magnitude of the increment is $\Delta\overline{\rho_{ab}^{(1)}}(t) = (-i/\hbar)\boldsymbol{\mu}_{ab} \cdot E(t)\overline{\rho_{aa}^{(0)}}\Delta t$.
2. The increment of $\overline{\rho_{ab}^{(1)}}$ evolves from time t until time τ, undergoing oscillations in the complex plane at frequency ω_{ab} and decaying with rate constant γ_{ab} as a result of fluctuating interactions with the surroundings. The fraction of the increment remaining at time τ is $\exp[-(i\omega_{ab}+\gamma_{ab})(\tau-t)]$.
3. Integrating processes 1 and 2 from $t=0$ to τ gives the total value of $\overline{\rho_{ab}^{(1)}}$ at time τ resulting from excitations at all earlier times. Steps 1–3 together comprise a convolution of the time-dependent excitation with a time-dependent response function.

To describe this process in a general way that can be extended to higher-order polarizations, imagine a Liouville-space operator $\widetilde{\mathrm{L}}_{mk,nk}$ that converts density matrix element ρ_{nk} into element ρ_{mk}. We also need a conjugate operator $\widetilde{\mathrm{L}}_{km,kn}$ that converts ρ_{kn} to ρ_{km}. Using these operators, we can write

$$-\boldsymbol{\mu}_{mn} \cdot \boldsymbol{E}(t)\widetilde{\mathrm{L}}_{mk,nk}\rho_{nk} = -\boldsymbol{\mu}_{mn} \cdot \boldsymbol{E}(t)\rho_{mk} \tag{11.28a}$$

and

$$-\boldsymbol{\mu}_{mn} \cdot \boldsymbol{E}(t)\widetilde{\mathrm{L}}_{km,kn}\rho_{kn} = -\boldsymbol{\mu}_{mn} \cdot \boldsymbol{E}(t)\rho_{km}. \tag{11.28b}$$

We next define a *time evolution operator* $\widetilde{\mathrm{G}}_{mn}(t_1)$ that acts on ρ_{mn} at a given time t and generates the value at time $t+\tau$, so that $\rho_{mn}(t+t_1) = \widetilde{\mathrm{G}}_{mn}(t_1)\rho_{mn}(t)$. Operators of this general type are called *Green functions* or *Green's functions* in recognition of the nineteenth century mathematician and physicist George Green. (The name is used broadly for solutions to linear partial differential equations. In problems concerning the transfer of heat, for example, a Green function might describe the heat at position x_2 and time $t+t_1$ after introduction of a small amount of heat at position x_1 and time t.) If ρ_{mn} decays exponentially with time constant γ_{nm}, the operator we need can be defined by its action on an arbitrary function A as

$$\widetilde{\mathrm{G}}_{mn}(t)A = \exp[-(i\omega_{mn} + \gamma_{mn})t]A. \tag{11.29}$$

With these definitions, we have

$$\boldsymbol{\mu}_{ba}\rho_{ab}^{(1)}(\tau) = |\boldsymbol{\mu}_{ba} \cdot \hat{e}\,|^2 \left(\frac{i}{\hbar}\right) \int_0^{\tau} R(\tau - t)\boldsymbol{E}(t)\,dt, \tag{11.30}$$

where

$$R(t) = \widetilde{\mathrm{G}}_{ba}(t)\widetilde{\mathrm{L}}_{ba,aa}\rho_{aa}, \tag{11.31}$$

and $\hat{e}$ is a unit vector indicating the polarization of the radiation field. Then, because $\rho_{ba} = \rho_{ab}{}^*$, the first-order optical polarization resulting from the paths through both ρ_{ba} and ρ_{ab} is

$$\begin{aligned} \boldsymbol{P}^{(1)}(\tau) &= \boldsymbol{\mu}_{ba}\rho_{ab}^{(1)}(\tau) + \boldsymbol{\mu}_{ab}\rho_{ab}^{(1)}{}^*(\tau) = |\boldsymbol{\mu}_{ba} \cdot \hat{e}\,|^2 \left(\frac{i}{\hbar}\right) \int_0^{\tau} \{R(\tau - t) - R^*(\tau - t)\}\boldsymbol{E}(t)\,dt \\ &= \int_0^{\tau} S^{(1)}(\tau - t)\boldsymbol{E}(t)\,dt, \end{aligned} \tag{11.32}$$

where the *linear response function* $S^{(1)}(t)$ is

$$S^{(1)}(t) = |\boldsymbol{\mu}_{ba} \cdot \hat{e}|^2 (i/\hbar)[R(t) - R^*(t)] = (i/2\hbar)|\boldsymbol{\mu}_{ba} \cdot \hat{e}|^2 \mathrm{Im}[R(t)]. \qquad (11.33)$$

This analysis rests on the basic assumption of *causality*, which means that the response of the system must follow the interaction with the field rather than preceeding it.

The Green function expressed in Eq. (11.29) assumes that off-diagonal elements of $\bar{\rho}$ decay by simple exponential kinetics. As we discussed in Sect. 10.7, actual relaxation functions generally are complex and can include Gaussian, and oscillatory components. To use a more realistic relaxation function, we can generalize Eq. (11.29) to

$$\widetilde{\mathrm{G}}_{mn}(t)A = \exp[-i\omega_{mn} - g(t)]A, \qquad (11.34)$$

with $g(t)$ given by Eq. (10.71) or (10.72) or another model such as a damped harmonic oscillator.

Let's now extend the Liouville-space approach to third-order optical polarization. Figure 11.4 shows four of the pertinent pathways for a two-state system and Fig. 11.5 shows double-sided Feynman diagrams for the same paths. As in Fig. 11.1, the first step starting from (a,a) in the lower-left corner in Fig. 11.4A generates a coherence that is represented diagrammatically by (b,a) or (a,b) depending on whether the perturbation acts on the left index (paths R_1 and R_4 in Figs. 11.4B and 11.5B) or on the right index (paths R_2 and R_3). The second interaction with the field, delayed by time t_1 after the first, converts the coherence to either (b,b) (paths R_1 and R_2) or (a,a) (paths R_3 and R_4). The third interaction occurs at time t_2 after the second and creates coherences again. There are eight (2^3) different three-step paths that end with a coherence represented by one of the shaded circles in Fig. 11.4A. The missing paths in Fig. 11.4 are the complex conjugates of the four paths that are shown (i.e., the paths obtained by replacing each vertical step by a horizontal step and vice versa, or simply interchanging the bra and ket subscripts of the density matrix element at each point). A fourth step from any of the coherences at time t (t_3 after the third interaction) gives either ρ_{aa} or ρ_{bb}.

By comparing the Feynman diagrams for $\boldsymbol{P}^{(3)}$ (Fig. 11.4) with those for $\boldsymbol{P}^{(1)}$ (Fig. 11.2), and referring to Eqs. (11.29–11.33) for the first-order polarization, you will find that the third-order polarization can be written

$$\boldsymbol{P}^{(3)}(t) = \int_0^t dt_3 \int_0^t dt_2 \int_0^t dt_1\, S_3(t_3,t_2,t_1)\, \boldsymbol{E}(t-t_3)\boldsymbol{E}(t-t_3-t_2)\boldsymbol{E}(t-t_3-t-t_1). \qquad (11.35)$$

Here $S^{(3)}(t_3, t_2, t_1)$ is the *third-order nonlinear response function*,

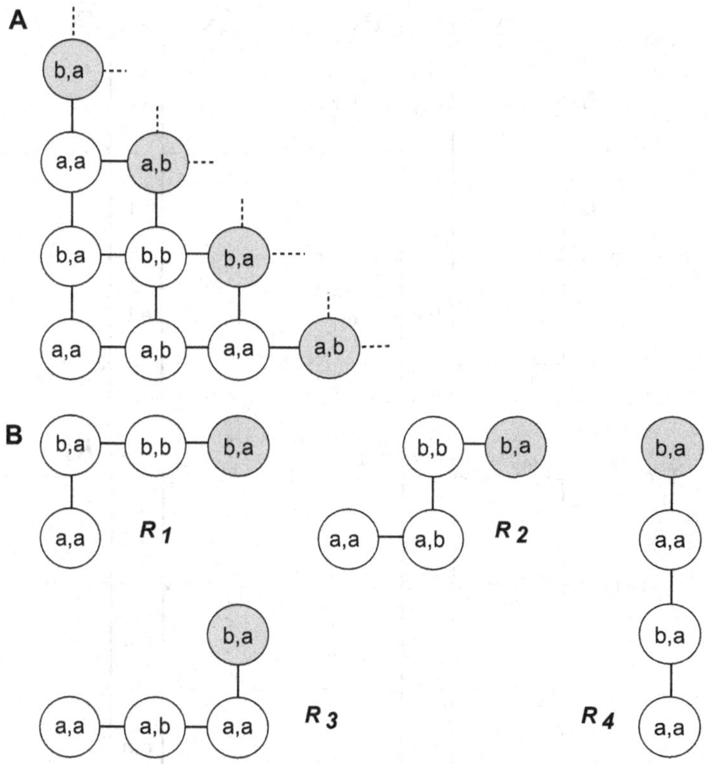

Fig. 11.4 Generation of the third-order polarization by pathways in Liouville space. (**A**) An extension of Fig. 11.1A, with the coherences that contribute to $\boldsymbol{\rho}^{(3)}$ and $\boldsymbol{P}^{(3)}$ denoted by *shaded circles*. There are eight three-step pathways that start at *a,a* (*lower-left corner*) and end at one of these ovals. (**B**) Four of these pathways are shown; the other four are the complex conjugates of these. A fourth interaction with the field (*vertical or horizontal dotted line* in **A**) generates either the excited state (*b,b*) or the ground state (*a,a*) (not shown). Pathways R_1, R_2, R_3 and R_4 correspond to the four individual response functions R_1 to R_4 (Eq. 11.37) that combine with their complex conjugates to make the third-order nonlinear response function S_3 (Eq. 11.36)

$$S^{(3)}(t_3,t_2,t_1) = |\boldsymbol{\mu}_{ba}\cdot\hat{e}\,|^4\left(\frac{i}{\hbar}\right)^3\sum_{\sigma=1}^{4}\{R_\sigma(t_3,t_2,t_1) - R_\sigma{}^*(t_3,t_2,t_1)\}, \tag{11.36}$$

in which the R_σ are response functions for the four pathways shown in Figs. 11.4 and 11.5:

$$R_1(t_3,t_2,t_1) = \widetilde{\mathrm{G}}_{ba}(t_3)\widetilde{\mathrm{L}}_{ba,bb}\times\widetilde{\mathrm{G}}_{bb}(t_2)\widetilde{\mathrm{L}}_{bb,ba}\times\widetilde{\mathrm{G}}_{ba}(t_1)\widetilde{\mathrm{L}}_{ba,aa}\overline{\rho_{aa}^{(0)}}, \tag{11.37a}$$

$$R_2(t_3,t_2,t_1) = \widetilde{\mathrm{G}}_{ba}(t_3)\widetilde{\mathrm{L}}_{ba,bb}\times\widetilde{\mathrm{G}}_{bb}(t_2)\widetilde{\mathrm{L}}_{bb,ab}\times\widetilde{\mathrm{G}}_{ab}(t_1)\widetilde{\mathrm{L}}_{ab,aa}\overline{\rho_{aa}^{(0)}}, \tag{11.37b}$$

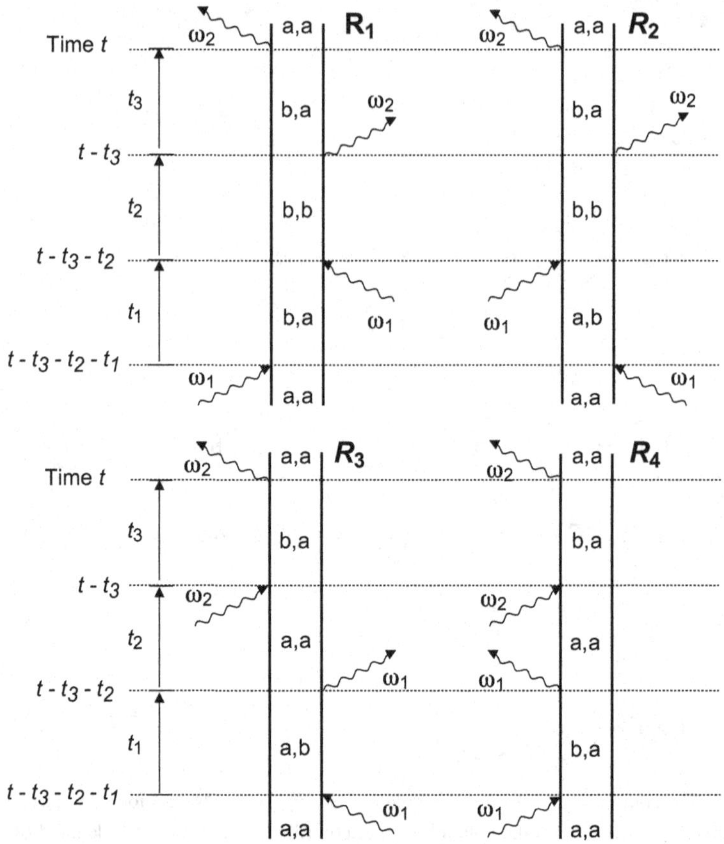

Fig. 11.5 Double-sided Feynman diagrams for third-order polarization in a two-state system. The symbols have the same meanings as in Fig. 11.2. The times between the interactions with the radiation field are indicated by τ_1, τ_2 and τ_3. Diagrams $\mathbf{R_1}$–$\mathbf{R_4}$ correspond to the Liouville-space paths in Fig. 11.4B and to third-order nonlinear response functions R_1 to R_4 (Eq. 11.37). However, only stimulated emission that repopulates the ground state (ρ_{aa}) is shown for the fourth interaction with the field. This interaction also could convert the last coherence or its complex conjugate to ρ_{bb}

$$R_3(t_3, t_2, t_1) = \tilde{\mathrm{G}}_{ba}(t_3)\tilde{\mathrm{L}}_{ba,aa} \times \tilde{\mathrm{G}}_{aa}(t_2)\tilde{\mathrm{L}}_{aa,ab} \times \tilde{\mathrm{G}}_{ab}(t_1)\tilde{\mathrm{L}}_{ab,aa}\overline{\rho_{aa}^{(0)}}, \tag{11.37c}$$

and

$$R_4(t_3, t_2, t_1) = \tilde{\mathrm{G}}_{ba}(t_3)\tilde{\mathrm{L}}_{ba,aa} \times \tilde{\mathrm{G}}_{aa}(t_2)\tilde{\mathrm{L}}_{aa,ba} \times \tilde{\mathrm{G}}_{ba}(t_1)\tilde{\mathrm{L}}_{ba,aa}\overline{\rho_{aa}^{(0)}}. \tag{11.37d}$$

The sequence of operations in these expressions proceeds from right to left and the symbol "×" denotes an ordinary scalar product.

Consider the path labeled R_1 in Figs. 11.4 and 11.5. In the notation used here, t_1, t_2, t_3 and t are the times between successive interactions with the radiation field. The first interaction occurs at $t–t_1–t_2–t_3$ and creates a coherence represented by ρ_{ba}. This coherence evolves for a period t_1, bringing the time to $t–t_2–t_3$, when the second interaction generates ρ_{bb}. The excited state evolves for period t_2, bringing us to $t–t_3$, when the third interaction regenerates ρ_{ba}. The third-order polarization created in this manner then evolves for period t_3, bringing us to t, when the final interaction yields either ρ_{aa} (as shown in Figs. 11.4 and 11.5) or ρ_{bb} (not shown). The complex conjugate of this pathway (also not shown) is $\rho_{aa} \rightarrow \rho_{ab} \rightarrow \rho_{bb} \rightarrow \rho_{ab} \rightarrow \rho_{aa}$ or ρ_{bb}.

Mukamel [7, 19] showed that the nonlinear response functions of Eqs. (11.37a–11.37d) can be calculated from a complex lineshape function $g(t)$ of the type we introduced in Eq. (10.71). The derivation uses a cumulant expansion parallel to that used to develop Eqs. (10.67) and (10.71): $\exp(-g(t))$ is expanded as a Taylor series in $g(t)$, and $g(t)$ is expanded in powers of the fluctuating electronic energy difference between the ground and excited states. Truncating the power series after the quadratic term gives the following results:

$$R_1 = \exp[-i\omega_{ba}(t_3 + t_1) - g^*(t_3) - g(t_1) - g^*(t_2) + g^*(t_2 + t_3) + g(t_1 + t_2) - g(t_1 + t_2 + t_3)], \tag{11.38a}$$

$$R_2 = \exp[-i\omega_{ba}(t_3 - t_1) - g^*(t_3) - g^*(t_1) + g(t_2) - g(t_2 + t_3) - g^*(t_1 + t_2) + g^*(t_1 + t_2 + t_3)], \tag{11.38b}$$

$$R_3 = \exp[-i\omega_{ba}(t_3 - t_1) - g(t_3) - g^*(t_1) + g^*(t_2) - g^*(t_2 + t_3) - g^*(t_1 + t_2) + g^*(t_1 + t_2 + t_3)], \tag{11.38c}$$

and

$$R_4 = \exp[-i\omega_{ba}(t_3 + t_1) - g(t_3) - g(t_1) - g(t_2) + g(t_2 + t_3) + g(t_1 + t_2) - g(t_1 + t_2 + t_3)]. \tag{11.38d}$$

These expressions apply to a variety of experiments that depend on the third-order polarization, including pump-probe and photon-echo experiments. Three-pulse photon echo experiments, for example, depend on R_2.

11.3 Pump-Probe Spectroscopy

In a typical pump-probe experiment, a sample is excited with a pulse with frequency ω_1 and wavevector $\boldsymbol{k}_1$, and is probed by a second pulse with frequency ω_2 and wavevector $\boldsymbol{k}_2$. The optical path of one of the pulses is varied to change the delay between the two pulses. The measured signal is the difference between the intensities of the transmitted probe pulses in the presence and absence of the excitation pulses, and usually is averaged over many pulses (Fig. 1.9). In a system with only two electronic states, the difference can reflect either stimulated emission from the excited state or bleaching of the absorption band of the ground state. The probe frequency often is selected by dispersing a spectrally broad probe beam after it passes through the sample, as illustrated schematically in Fig. 1.9 [20, 21]. Alternatively, part of the primary laser pulse can be split out before the sample and sent to a non-linear optical device to generate a probe pulse at a different frequency. The frequencies might, for example, be chosen to pump on the blue side of the absorption band and to measure bleaching or stimulated emission on the red side. Since the individual probe pulses are not resolved, the signal reflects an integral of the product $-\boldsymbol{\mu}_{ba}\boldsymbol{P}^{(3)}(t)\boldsymbol{E}(t)$ from time $t = -\infty$ to ∞.

The radiation fields that enter into Eq. (11.35) are combinations of the fields from the pump and probe pulses, and are given by

$$\begin{aligned} \boldsymbol{E}(t') = &\ \boldsymbol{E}_1(\boldsymbol{r}, t' + \tau)\{\exp(i\boldsymbol{k}_1 \cdot \boldsymbol{r} - i\omega_1 t') + \exp(-i\boldsymbol{k}_1 \cdot \boldsymbol{r} + i\omega_1 t')\} \\ &+ \boldsymbol{E}_2(t')\{\exp(i\boldsymbol{k}_2 \cdot \boldsymbol{r} - i\omega_2 t') + \exp(-i\boldsymbol{k}_2 \cdot \boldsymbol{r} + i\omega_2 t')\}, \end{aligned} \tag{11.39}$$

where $\boldsymbol{E}_j(t)$ is the temporal envelope and polarization of pulse j and τ is the delay between the peaks of the two pulses. In principle, the measured signal could leave the sample with any linear combination of the four individual wavevectors $(\pm\boldsymbol{k}_1 \pm \boldsymbol{k}_1 \pm \boldsymbol{k}_2 \pm \boldsymbol{k}_2)$ and the same combination of the four frequencies $(\pm\omega_1 \pm \omega_1 \pm \omega_2 \pm \omega_2)$, but only some of these combinations survive the rotating-wave approximation; others result in high-frequency oscillations that make little contribution to the measured signals [5, 7]. For example, if the pump and probe pulses are well separated in time, the exponents in the terms $\exp(\pm i\omega t)$ must have opposite signs for the first and second interactions with the field. Additional signals, often called "coherence artifacts" but actually predictable consequences of four-wave mixing, can be generated when the two pulses overlap.

In a three-state system, formation of the third state (c) can be probed by excited-state absorption (excitation from state b to state c) as well as by stimulated emission or ground-state bleaching. This simplifies the situation if the pump frequency (ω_1) is resonant for transitions between states a and b whereas the probe frequency (ω_2) is selective for transitions between states b and c. Figure 11.6 shows the pertinent double-sided Feynmann diagrams for four pathways that contribute to third-order optical polarization. Again, the third-order nonlinear response function ($S^{(3)}(t_3, t_2, t_1)$) also includes the complex conjugates of these pathways, which are not shown in the figure. Three interactions with the radiation field along pathway R_1, R_2 or R_3 lead to the coherence ρ_{bc}, which can generate either state b (ρ_{bb}) or state c (ρ_{cc}) when it is probed by the fourth interaction. Figure 11.6 shows only the branch leading to ρ_{cc}, and also omits the pathways shown in Fig. 11.5, which occur in both two- and three-state systems. Pathway R_4 reaches ρ_{ca} but then drops back to ρ_{ba}, which can give only ρ_{bb} or ρ_{aa} on the fourth interaction. It therefore does not contribute a measured signal except possibly when the pump and probe pulses overlap in time.

Probably the most interesting of the pathways shown in Fig. 11.6 is R_3. Although this pathway yields state c on the fourth interaction with the radiation field, state b is never actually populated: the route proceeds entirely through the coherences ρ_{ab}, ρ_{ac} and ρ_{bc}! The pathway should, therefore, be particularly sensitive to dephasing

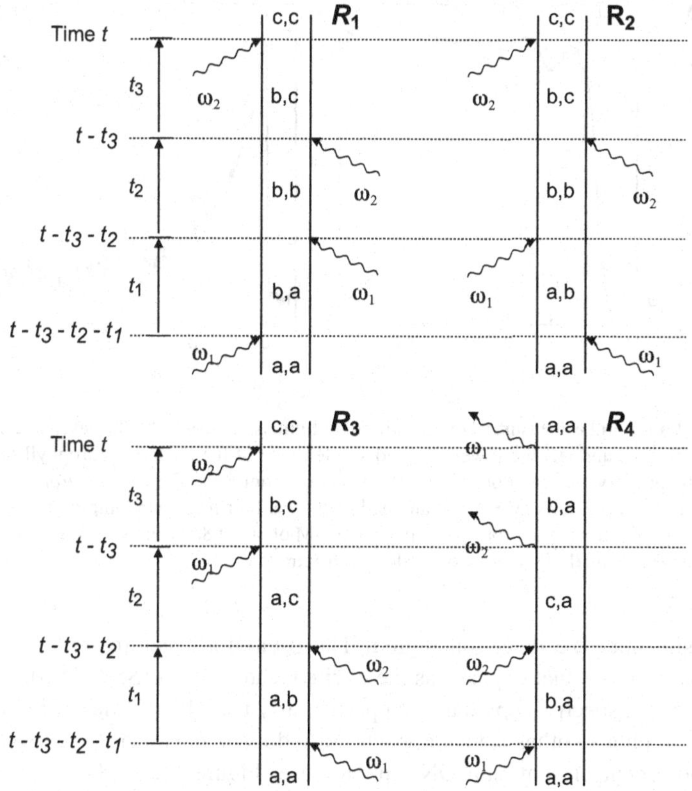

Fig. 11.6 Double-sided Feynman diagrams for the third-order polarization in a three-state system. Pathways that generate or pass through a coherence involving the third state are shown. Each of these has a complex conjugate that is not shown. Only formation of the third state is shown for the fourth interaction in pathways R_1 to R_3, and only regeneration of the ground state in R_4

and should occur only if delay t_2 is very short. This is similar to Raman scattering, which also involves coherences rather than "real" intermediate states (Chap. 12). Yan et al. [5] also relate this pathway to electron tunneling, in which electron transfer from a donor to an acceptor is facilitated by quantum mechanical coupling to a "virtual" intermediate state that lies higher in energy than either the initial or the final state.

Pump-probe techniques using picosecond and sub-picosecond laser pulses have made it possible to probe chemical processes on the time scale of nuclear motions [22]. Figure 11.7A shows a typical measurement of the early time course of stimulated emission from a dye molecule (IR132) in solution [23]. The dye was excited on the blue side of its absorption band (830 nm) and stimulated emission was measured at 900 nm. The signal includes a slow rise component with a time constant of several hundred femtoseconds that represents part of the Stokes shift of

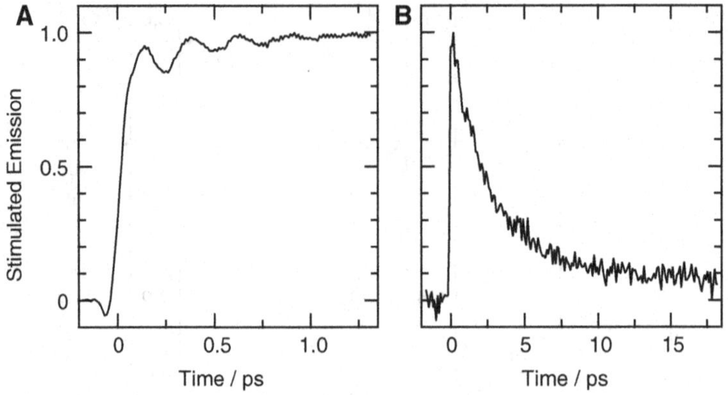

Fig. 11.7 Pump-probe measurements of stimulated emission from (**A**) the laser dye IR132 in dimethylsulfoxide, and (**B**) the lowest excited singlet state of the bacteriochlorophyll dimer that serves as the primary electron donor in reaction centers from *Rhodobacter sphaeroides* [23]. The excitation flash was centered at 830 nm and had a width at half-maximum amplitude (FWHM) of about 16 fs in both cases. The probe pulse had a FWHM of about 80 fs and was centered at 900 nm in (**A**) and 940 nm in (**B**). The ordinate scales are arbitrary

the emission into the detection region. The oscillatory features reflect coherent vibrational motions that dephase as the excited state relaxes (Sect. 11.5).

Pump-probe spectroscopy has been particularly useful in studies of light-driven processes in photosynthesis, vision, bacteriorhodopsin, photoactive yellow protein, green fluorescent protein, and DNA photolyase. Figure 11.7B shows a measurement of the kinetics of the initial electron-transfer step in photosynthetic bacterial reaction centers [23]. In this experiment, reaction centers were excited on the blue side of the long-wavelength absorption band of the bacteriochlorophyll dimer (P) that acts as the primary electron donor, and stimulated emission from the first excited singlet state of the dimer (P*) was measured at 940 nm. The decay of the stimulated emission reflects the dynamics of electron transfer from P* to the neighboring bacteriochlorophyll. Fitting the decay to a biexponential expression gives a major component with a time constant of 2.3 ps and a minor component with a time constant of 7.3 ps. For some other representative studies of bacterial reaction centers see [24–31].

11.4 Photon Echoes

We saw in Chap. 10 that static inhomogeneity in the energy of an excited electronic state causes an ensemble to lose coherence with Gaussian kinetics, whereas dynamic fluctuations cause an exponential decay. Photon-echo spectroscopy provides a way to measure the fluctuation dynamics without distortion from the effects of static inhomogeneity. We will focus on "three-pulse" photon echoes, which are generated by illuminating a sample with three short pulses separated by

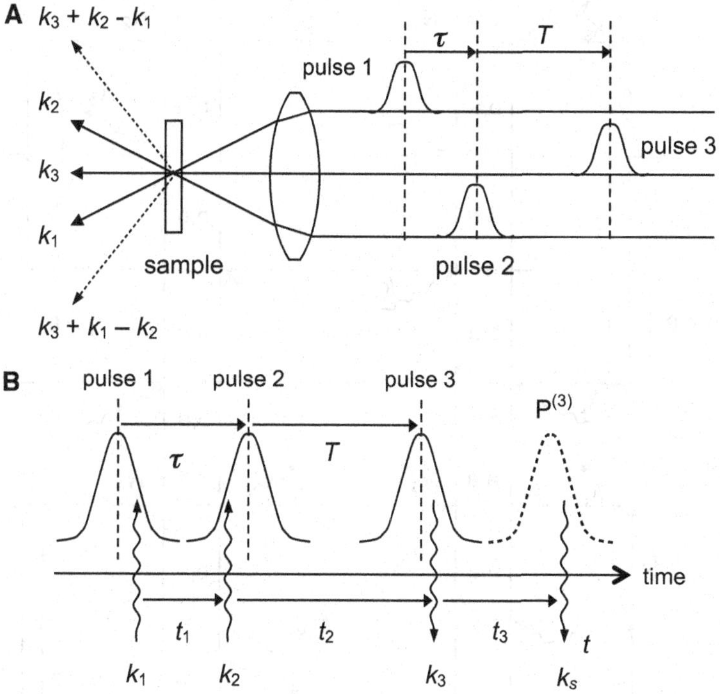

Fig. 11.8 Pulse sequence and field-matter interactions in a three-pulse photon-echo experiment. (**A**) Pulses 1, 2 and 3, traveling from *right to left* in this diagram, reach the sample with different wavevectors ($\boldsymbol{k}_1$, $\boldsymbol{k}_2$ and $\boldsymbol{k}_3$). Pulses 1 and 2 are separated by time interval τ; pulses 2 and 3, by interval T. Photon echos with wavevectors $\boldsymbol{k}_3 \pm (\boldsymbol{k}_2 - \boldsymbol{k}_1)$ are measured. (**B**) Interactions of the sample with the electromagnetic fields (*wavy arrows*) occur at times $t{-}t_3{-}t_2{-}t_1$ (sometime during pulse 1), $t{-}t_3{-}t_2$ (during pulse 2), and $t{-}t_3$ (during pulse 3), and the echo ($P^{(3)}$) is emitted at time t. The signal is integrated from $t_3 = 0$ to ∞. Figure 11.13 shows how movable mirrors can be used to control the pulse timing in such experiments

adjustable delays. Figure 11.8 shows the sequence of pulses and the optical layout in a typical experiment, and Fig. 11.9 shows the relevant double-sided Feynman diagrams. In both figures, interactions of the sample with the radiation field occur at times $t{-}t_3{-}t_2{-}t_1$, $t{-}t_3{-}t_2$, $t{-}t_3$, and t. The time between the centers of pulses 1 and 2 is τ, and T is the time between the centers of pulses 2 and 3.

The pulses in a three-pulse photon-echo experiment enter the sample at different angles of incidence, and the emitted light that forms the signal is collected at a particular angle with respect to the incident beams (Fig. 11.8A). As we discussed in the previous section, a signal can appear with any combination of the wavevectors of the three incident fields: $\boldsymbol{k}_s = \pm\boldsymbol{k}_1 \pm \boldsymbol{k}_2 \pm \boldsymbol{k}_3$, but only certain combinations conserve momentum and satisfy the rotating-wave conditions for a given pathway in Liouville space. The dominant signals usually are obtained at $\boldsymbol{k}_s = \boldsymbol{k}_3 + \boldsymbol{k}_2 - \boldsymbol{k}_1$, which meets the requirements for pathways R_2 and R_3 in Fig. 11.9, and at $\boldsymbol{k}_s{}' = \boldsymbol{k}_3 + \boldsymbol{k}_1 - \boldsymbol{k}_2$, which satisfies the conditions for R_2* and R_3* [9, 32, 33].

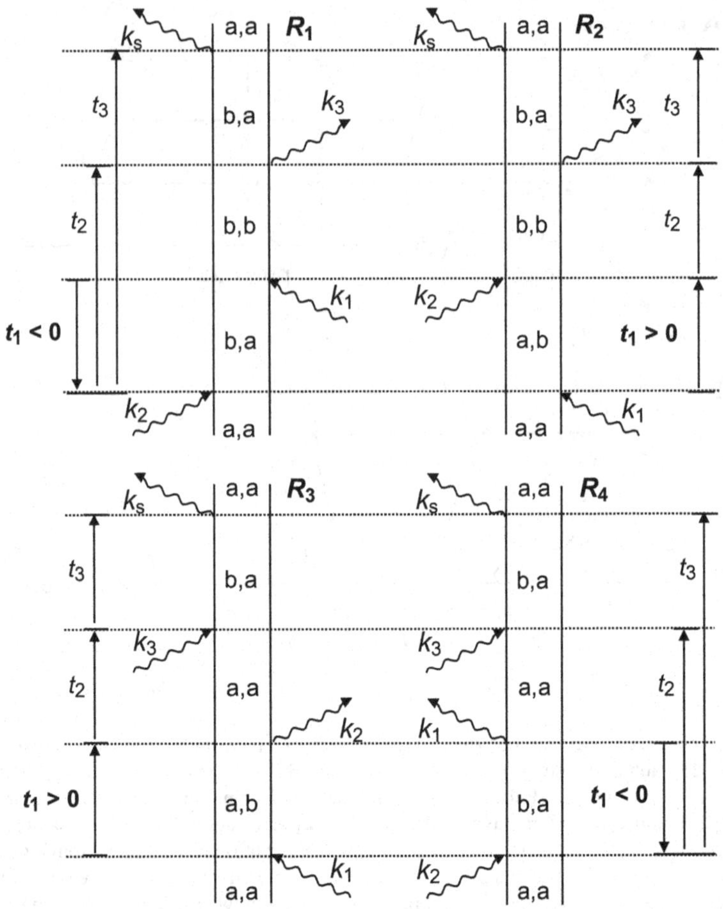

Fig. 11.9 Double-sided Feynman diagrams for the third-order polarization pathways that generate photon echoes within the rotating-wave approximation (paths R_2, R_3, and their complex conjugates), and the pathways that survive the rotating-wave approximation but do not generate photon echoes (paths R_1, R_4, and their complex conjugates). If the first three interactions with the field occur during three separate pulses as in Fig. 11.8, delaying pulse 1 so that it follows pulse 2 converts R_2 into R_1 and R_3 into R_4

Look first at path R_2, in which pulse 1 precedes pulse 2 ($t_1 > 0$) (Fig. 11.9). This path has two periods of coherence (delay periods t_1 and t_3) and the density matrix element in the second period (ρ_{ba}) is the complex conjugate of that in the first (ρ_{ab}). The two periods of coherence are separated by a "population" interval (t_2) when the systems are in the excited state (ρ_{bb}). During period t_1, both static and dynamic inhomogeneity in the energies of the molecules in the ensemble will cause the coherence to decay. But any truly static contribution to the energy of a given molecule will remain constant until period t_3, when it will have the opposite effect that it had during period t_1. This is because the Green function for ρ_{ba} is the complex conjugate (more precisely, the Hermitian conjugate) of that for ρ_{ab} (Eq. 11.37a–11.37d). Static inhomogeneity in the energy difference between states a and b creates a distribution of the factors $\exp(-i\omega_{ba}t)$ in the Green function for ρ_{ba}, which disappears when it is multiplied by the same distribution of the factors $\exp(-i\omega_{ab}t)$ in the Green function for ρ_{ab}. Thus, the dephasing caused by static inhomogeneity in period t_1 is reversed during period t_3, regenerating a coherence that will appear as a pulse of emitted light when $t_3 \approx t_1$. This is the photon echo.

Dephasing due to dynamic fluctuations does not undergo such a reversal, because it enters the Green functions for both ρ_{ba} and ρ_{ab} in an identical manner, for example as $\exp(-t/T_2)$ for both Green functions if the coherence has a simple exponential decay. The amplitude of the photon echo therefore will be largest when t_2 is short, and will decrease as t_2 is lengthened and fluctuations on a broader range of time scales invade the Hermitian relationship between the two Green functions. The dependence of the echo amplitude on t_2 thus will report on the fluctuation dynamics with relatively little disturbance by the effects of static inhomogeneity.

Now look at path R_1, which can be obtained from R_2 by simply advancing pulse 2 to that it precedes pulse 1 ($t_1 < 0$ in Figs. 11.8 and 11.9). R_1 and its complex conjugate each has two periods with the same coherence (ρ_{ba}) separated by a population interval. They do not result in echoes.

The complex conjugate of R_2 (not shown in Fig. 11.9) gives photon echoes that are the same as those of R_2 but, as mentioned above, appear with wavevector $\boldsymbol{k}_s = \boldsymbol{k}_1 - \boldsymbol{k}_2 - \boldsymbol{k}_3$. R_3 is similar to R_2 in having two periods with conjugate coherences separated by a population period. However, because the system spends the population period in the ground state (ρ_{aa}), this path and its complex conjugate usually do not contribute an observable signal [33]. R_4 and R_4* have corresponding relationships to R_1 and R_1*, and also give no echoes.

In what is probably the most informative type of three-pulse photon-echo experiment, the time between pulses 1 and 2 (τ) is varied while that between pulses 2 and 3 (T) is held constant (Figs. 11.8 and 11.9). The echo signal in direction $\boldsymbol{k}_s = \boldsymbol{k}_2 - \boldsymbol{k}_1 + \boldsymbol{k}_3$ is integrated over time t_3, and the measurements are repeated with different values of T. The signal reflects the third-order polarization with wavevector $\boldsymbol{k}_s$, $\boldsymbol{P}^{(3)}(\boldsymbol{k}_s)$, and can be detected either by collecting the emitted light directly (*homodyne detection*) or by mixing $\boldsymbol{P}^{(3)}(\boldsymbol{k}_s)$ with a separate, stronger radiation field (*heterodyne detection*). With homodyne detection, the signal depends on the squared amplitude (modulus) of the polarization:

$$I(\boldsymbol{k}_s) = \int_0^\infty \left|\boldsymbol{P}^{(3)}(\boldsymbol{k}_s)\right|^2 dt_3. \tag{11.40}$$

If we consider only path R_2, which dominates the signal if pulse 1 precedes pulse 2, the third-order polarization at time t is given by

$$\boldsymbol{P}^{(3)}(\boldsymbol{k}_s,t) = |\boldsymbol{\mu}_{ba}\cdot\hat{e}|^4\left(\frac{i}{\hbar}\right)^3\int_0^t dt_3\int_0^t dt_2\int_0^t dt_1 R_2(t_3,t_2,t_1)\left[\left|\boldsymbol{E}_3^o(t-t_3)\right|\exp(i\omega t_3)\right]$$
$$\times\left[\left|\boldsymbol{E}_2^o(t-t_3-t_2+T)\right|\exp(-i\omega t_2)\right]\left[\left|\boldsymbol{E}_1^o(t-t_3-t_2-t_1+\tau+T)\right|\exp(-i\omega t_1)\right], \tag{11.41}$$

where $|\boldsymbol{E}_1^o(t)|$, $|\boldsymbol{E}_2^o(t)|$ and $|\boldsymbol{E}_3^o(t)|$ are the envelopes of the field amplitudes in the three pulses and the response function $R_2(t_3, t_2, t_1)$ is related to the line-broadening function $g(t)$ by Eq. (11.38b) [7, 33, 34]. We have assumed that the three pulses have the same frequency (ω) and that the field envelopes are all real functions of time.

Now suppose that the light pulses are short relative to the time intervals t_1, t_2 and t_3, so that $t_1 \approx \tau$ and $t_2 \approx T$. In this limit, Eqs. (11.40) and (11.41) reduce to

$$I(\boldsymbol{k}_s,\tau>0) = |\boldsymbol{\mu}_{ba}\cdot\hat{e}|^8\left(\left|\boldsymbol{E}_3^o\right|^2\left|\boldsymbol{E}_2^o\right|^2\left|\boldsymbol{E}_1^o\right|^2/\hbar^6\right)\int_0^\infty |R_2|^2 dt_3, \tag{11.42}$$

where $|\boldsymbol{E}_1^o|$, $|\boldsymbol{E}_2^o|$ and $|\boldsymbol{E}_3^o|$ are averages of the field strengths over the pulses. The notation "$\tau > 0$" indicates that pulse 1 must precede pulse 2 to enforce the order of the first two interactions that make up path R_2. Using Eq. (11.38b) for R_2, we obtain

$$I(\boldsymbol{k}_s,\tau>0) \propto \int_0^\infty \left|\exp[-g^*(t_3)-g^*(t_1)+g(t_2)-g(t_2+t_3)-g^*(t_1+t_2)+g^*(t_1+t_2+t_3)]\right|^2 dt_3. \tag{11.43}$$

If pulse 1 is delayed so that it follows pulse 2 ($\tau<0$), $R_1(t_3, t_2, t_1)$ replaces $R_2(t_3, t_2, t_1)$, and using Eq. (11.38a) for R_1 gives

$$I(\boldsymbol{k}_s,\tau<0) = |\boldsymbol{\mu}_{ba}\cdot\hat{e}|^8\left(\left|\boldsymbol{E}_3^o\right|^2\left|\boldsymbol{E}_2^o\right|^2\left|\boldsymbol{E}_1^o\right|^2/\hbar^6\right)\int_0^\infty |R_1|^2 dt_3$$
$$\propto \int_0^\infty \left|\exp[-g^*(t_3)-g(t_1)-g^*(t_2)+g^*(t_2+t_3)+g(t_1+t_2)-g(t_1+t_2+t_3)]\right|^2 dt_3. \tag{11.44}$$

Figure 11.10 shows how the integrands in Eqs. (11.43) and (11.44) depend on time t_3 for three values of t_2 at a fixed value of t_1. The Kubo relaxation function (Eq. 10.69) was used for $\exp[g(t)]$. If pulse 1 precedes pulse 2, so that Eq. (11.43) applies, and if t_2 is close to zero, the signal peaks when $t_3 \approx t_1$, demonstrating the expected rephasing by path R_2 (Fig. 11.10A). The peak decreases in amplitude and

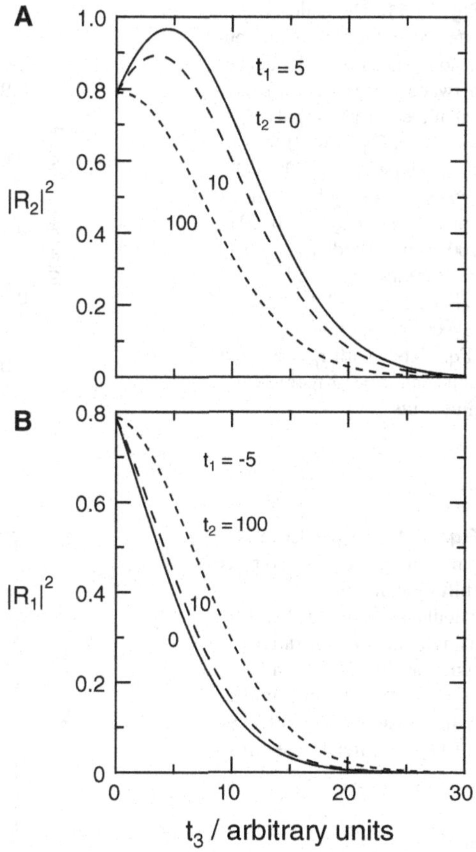

Fig. 11.10 Dependence of three-pulse photon-echo signals on delay time t_3, as calculated in the impulsive limit with Eq. (11.43) for $t_1 = \tau$ (the delay between pulses 1 and 2) = 5 (**A**) and with Eq. (11.44) for $t_1 = \tau = -5$ (**B**). The delay between pulses 2 and 3 ($t_2 = T$) was 0, 10 or 100, as indicated. The units of time are arbitrary. All calculations used the Kubo relaxation function (Eq. 10.69) with $\tau_c = 40$ time units and $\sigma = 0.1$ reciprocal time units

moves toward $t_3 = 0$ as t_2 is lengthened and the coherence created by pulse 1 decays. When pulse 1 follows pulse 2 so that Eq. (11.44) applies, the peak occurs at $t_3 = 0$ for all values of t_2 because path R_1 does not support rephasing (Fig. 11.10B). $|R_1|^2$ and $|R_2|^2$ become equivalent as t_2 goes to ∞ and the system loses all memory of the first pulse.

Figure 11.11 shows how the integrated signals depend on t_1 for the same three values of t_2. Equation (11.44) was used when $t_1 < 0$ and Eq. (11.43) when $t_1 < 0$. Again, positive values of t_1 result in an echo peak that collapses toward the origin as t_2 is increased. Fleming and coworkers [10, 33–38] have shown that the effect of t_2 on the shift of this peak away from zero on the t_1 axis (the *three-pulse photon-echo peak shift*) provides a particularly useful measure of the kinetics of dynamic dephasing. Plots of this effect are illustrated in Fig. 11.12 for several values of the correlation time constant τ_c in the Kubo relaxation function. As discussed in Sect. 10.7, the Kubo function includes both static (Gaussian) and dynamic (exponential) dephasing originating in energy fluctuations with a single correlation time

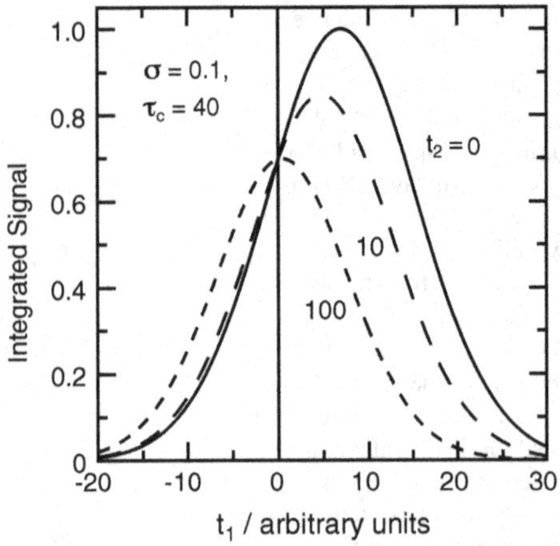

Fig. 11.11 Dependence of integrated three-pulse photon-echo signals on t_1 (the delay between pulses 1 and 2), as calculated in the impulsive limit with Eq. (11.43) for $t_1 < 0$ and with Eq. (11.44) for $t_1 > 0$. The delay between pulses 2 and 3 (t_2) was 0, 10 or 100, as indicated. The units of time are arbitrary. All calculations used the Kubo relaxation function (Eq. 10.69) with $\tau_c = 40$ time units and $\sigma = 0.1$ reciprocal time units

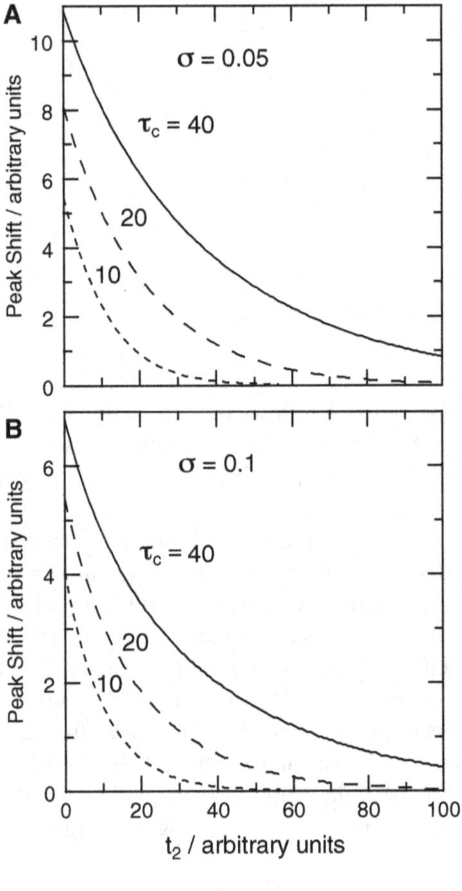

Fig. 11.12 Dependence of three-pulse photon-echo peak shift on delay time t_2, calculated as in Fig. 11.11 using the Kubo relaxation function (Eq. 10.69) with $\tau_c = 10$, 20 or 40 time units as indicated and $\sigma = 0.05$ (**A**) or 0.1 (**B**) reciprocal time units

constant. The three-pulse photon-echo peak shift recovers the dynamic component of this dephasing.

Photon echoes also can be obtained by using only two pulses instead of three. These can be described with the same theoretical formalism by letting the second and third pulses coincide and setting $\boldsymbol{k}_2 = \boldsymbol{k}_3$ and $\left|\boldsymbol{E}_2^o(t)\right| = \left|\boldsymbol{E}_3^o(t)\right|$ [39].

Since their first observation with ruby laser pulses lasting 10^{-8} s [40], photon echoes generated with a variety of pulsed lasers have been used to study solvation dynamics on time scales ranging from 10^{-14} to 10^{-1} s [10, 32–34, 36–38, 41–48]. Applications to the motions of ligand-binding sites in proteins have included studies of myoglobin [49–51], Zn cytochrome *c* [52], bacteriorhodopsin [53], antibodies [54], and calmodulin [55]. Proteins generally show multiphasic relaxation dynamics consistent with hierarchical, richly textured potential surfaces. Relaxations with effective correlation times of 1.3–4 ns were found to occur in Zn cytochrome *c* even at 1.8 K [52]. IR photon echoes have been used to study the vibrational dynamics of peptides and small molecules [56, 57], and have been combined with isotopic labeling to probe the dynamics of the amide I vibration at a specific residue in a transmembrane peptide [58].

11.5 Two-Dimensional Electronic and Vibrational Spectroscopy

A powerful extension of pump-probe and photon-echo experiments is to vary the frequency of the light that is detected independently from the frequency of the excitation. A two-dimensional spectrum then can display the signal amplitude with the excitation frequency plotted on one axis and the detection frequency on the other. Such experiments often employ heterodyne detection, in which third-order polarization of the sample following three short pulses of light is combined with the stronger field from a separate pulse called the "local oscillator," and the combined radiation is dispersed in a spectrometer equipped with a diode-array detector (Fig. 11.13). The intensity of light with frequency ω_s reaching the detector at time t is proportional to the modulus of the sum of the fields of the signal and the local oscillator ($|\boldsymbol{E}_{\mathrm{sig}}(t,\omega_s) + \boldsymbol{E}_{\mathrm{LO}}(t,\omega_s)|^2$). This quantity has three components: $|\boldsymbol{E}_{\mathrm{LO}}(t,\omega_s)|^2$, which can be measured separately and subtracted, $|\boldsymbol{E}_{\mathrm{sig}}(t,\omega_s)|^2$, which is negligible if the signal is much weaker than the local oscillator, and an interference term that depends on the product of the two fields. Mixing with the local oscillator thus can boost the signal substantially. If the local oscillator pulse is timed to be in phase with the third of the three pulses that impinge on the sample, cosine transforms of the interference term provide a one-dimensional spectrum of the signal amplitude versus ω_s for a particular choice of the delay periods τ and T in Fig. 11.8 [59]. This technique depends on the use of sub-picosecond pulses, which inherently contain a broad spectrum of frequencies. To obtain the dependence of the signal on the excitation frequency (ω_e), the measurements are repeated with different values of the delay between pulses 1 and 2 (τ in Fig. 11.8). A Fourier transform of the signal amplitude for a particular value of ω_s has positive and/or

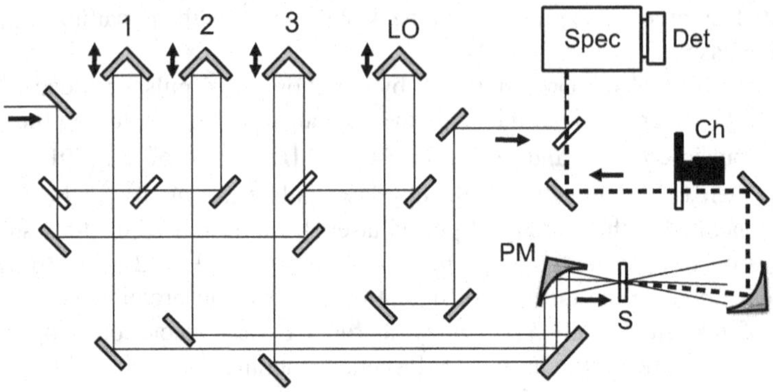

Fig. 11.13 An apparatus for 2-dimensional spectroscopy. Short pulses of light from a laser enter at the top left and are directed into multiple paths by beam splitters (*open rectangles*) and mirrors (*shaded rectangles*). A parabolic mirror (*PM*) focuses pulses with three different wavevectors on the sample (*S*). *Double arrows* indicate movable mirrors that control the timing of the pulses. Radiation leaving the sample with a chosen wavevector (*dashed line*) is sent through a chopper (*Ch*) and combined with a beam of stronger pulses (*LO*, local oscillator). The combined radiation is dispersed in a spectrometer (*Spec*) and sent to a detector array (*Det*). Signals from the local oscillator alone are measured when the chopper blocks light coming from the sample, and are subtracted. Polarizers and compensation plates are not shown. See [59, 72] for details, other optical schemes and information on data analysis

negative peaks corresponding to frequencies where the pump light excites the sample. The whole experiment then is repeated with various values of the delay between pulses 2 and 3 (T) to explore how the sample dephases or evolves in other ways after it is excited. The relative polarization of the pump and probe fields provides an additional independent variable.

A two-dimensional plot displaying the excitation frequency on the ordinate and the signal frequency on the abscissa typically has one or more peaks on the diagonal, representing ground-state bleaching or photon echoes at the excitation frequency. In addition, there can be off-diagonal signals that reflect excited states with different energies. These can be generated either by excited-state absorption, conformational changes, or energy transfer. For example, excitation from level 0 to level 1 of a vibrational mode gives a diagonal peak from ground-state bleaching of this absorption, and an off-diagonal peak with opposite sign representing excitation of the same mode from level 1 to level 2, as shown schematically in Fig. 11.14A. The difference between the signal frequencies for these peaks reveals the anharmonicity of the vibrational mode. Such measurements have been described for NO and CO bound to myoglobin and other hemoproteins, where the ligand stretching mode has a fundamental frequency in the range of 1,900–1,930 cm^{-1} and an anharmonicity of about 30 cm^{-1} [60–63].

Two-dimensional IR (2D-IR) spectra of CO-myoglobin exhibit additional off-diagonal peaks that are attributable to switching between different conformational states [61]. These peaks are not seen at short times after excitation, but

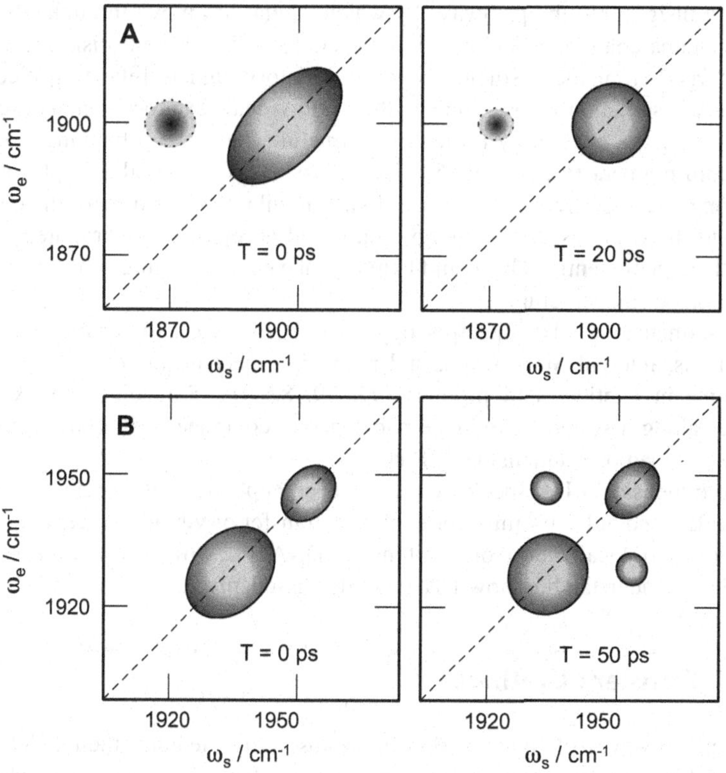

Fig. 11.14 Schematic depictions of 2-dimensional IR spectra of stretching modes of CO and NO ligands bound to hemoproteins. The excitation frequency is displayed on the ordinate and the signal frequency on the abscissa. (**A**) The N-O stretching mode gives a diagonal peak in the region of 1,900 cm^{-1} and an off-diagonal signal with opposite sign near 1,870 cm^{-1}. The main peak represents excitation from level $m=0$ to $m=1$; the lower-energy peak, excitation from $m=1$ to $m=2$. The off-diagonal peak is seen with very short values of the population time (T) as well as with longer delays. Structural disorder (inhomogeneous broadening) stretches the 0–1 peak along the diagonal at $T=0$, and rapid structural fluctuations (spectral diffusion) make the peak more symmetrical with time. (**B**) Spectra of the C-O stretching mode in CO-myoglobin have two diagonal peaks that represent distinct conformational states. Off-diagonal peaks develop as T increases, indicating that interconversions between the two states occur on the time scale of 50 ps. Spectral diffusion again causes the peaks to become more symmetrical with time

develop as the delay between pulses 2 and 3 (T) is increased, providing a direct measure of the conformational dynamics (Fig. 11.14B).

In a system with multiple chromophores, exciton interactions characteristically gives off-diagonal peaks at zero time, whereas transfer of excitations between pigments can give a ladder of off-diagonal signals that develop sequentially with time. Energy transfer between pigments with differently oriented transition dipoles also can give the off-diagonal signals a dependence on the relative polarization of the pump and probe fields. Measurements of such two-dimensional spectra have provided insight into the pathways by which excitations move through photosynthetic antenna complexes and reaction centers [64–75]. One surprising result that has emerged from these studies is that oscillatory signals reflecting electronic coherences sometimes last longer than individual excitons, indicating that excitations move coherently through the structure, rather than hopping stochastically from pigment to pigment [68, 71, 72, 76–81]. One possible explanation of these long-lived coherences is that the dominant vibrational modes of the complex modulate the energies of the energy donor and acceptor in a correlated manner rather than incoherently. This could help to make energy capture robust to thermal fluctuations of the structure.

Two-dimensional IR spectroscopy has been used to examine exciton interactions, internal Stark shifts, and relaxation mechanisms of amide I and II transitions in peptides and proteins [57, 59, 82–91]. Spreading the excitation energies along a second coordinate can separate components that are difficult to resolve in broad one-dimensional IR spectra.

Two-dimensional UV spectroscopy has been applied to nucleic acid bases and nucleotides, and holds promise for exploring transfer of vibrational energy and the migration and localization of excitons in DNA [92–96]. These processes are pertinent to understanding how DNA avoids photodamage.

11.6 Transient Gratings

If two plane waves of light overlap in an absorbing medium, their fields create sinusoidal interference patterns like those shown in Fig. 11.15. In regions where the interference is constructive, the absorbance of the medium can change as a result of ground-state bleaching or excited-state absorption; where it is destructive, little is changed. The spacing of the excitation bands is $\lambda_{band} = \lambda_{ex}/2\sin(\theta_{ex}/2)$, where θ_{ex} is the angle between the wavevectors of the two beams. The density of the medium also can change in the same pattern if the excitation causes local heating or volume changes, and this will affect the refractive index. Such bands of modified absorbance or refractive index can act as a diffraction grating to diffract a probe beam whose wavelength (λ_{probe}) and angle of incidence relative to the mean of the two wavevectors (θ_{probe}) satisfy the Bragg condition $j\lambda_{prrobe} = 2\lambda_{band}/\cos(\theta_{probe})$, where j is an integer. The grating, however, will disappear as the excited molecules return to their original state or diffuse away from their initial positions and as the heat generated by the excitation also diffuses away. The diffraction of the probe beam thus provides a way of measuring the decay dynamics of the excited state or of volume changes caused by the excitation [97].

Figure 11.16 shows a typical pump-probe arrangement for studying transient gratings on picosecond time scales [97–100]. In this scheme, a train of pulses from a laser is split into three beams, two of which converge in the sample to create the grating. The third beam is used to generate probe pulses at the same or a different frequency. The probe pulses strike the sample at an adjustable time after the pump pulses, and the portion of the probe beam that diffracts off the grating passes

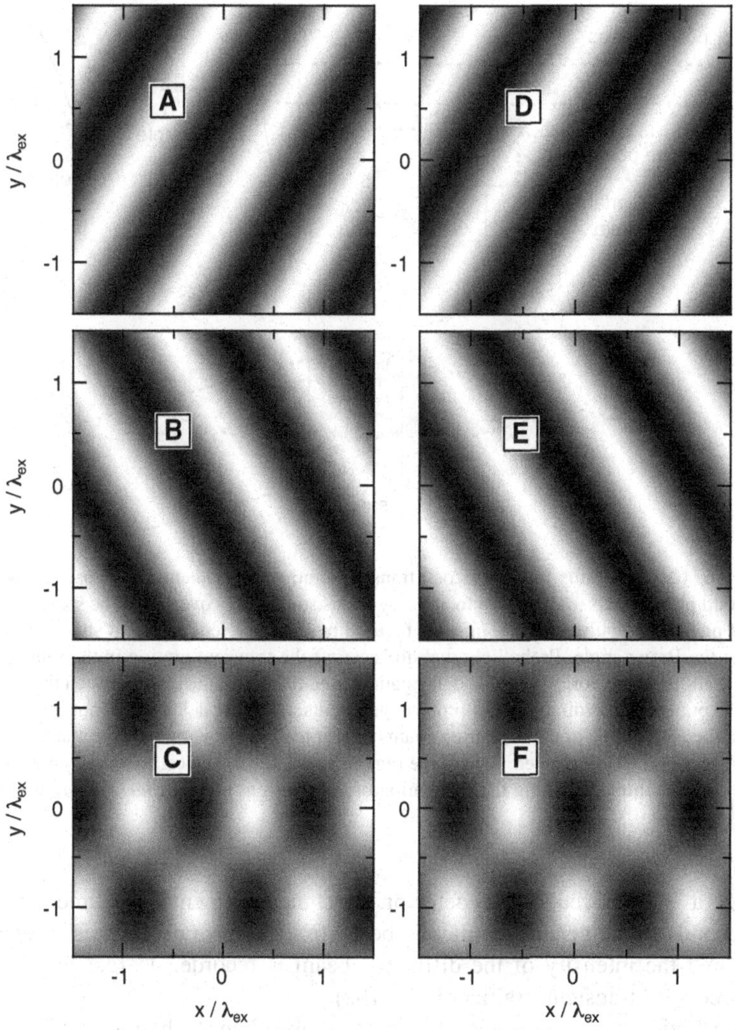

Fig. 11.15 Formation of a transient grating by overlapping plane waves. (**A**), (**B**) Image plots of the electric fields at zero time in two plane waves propagating to the right and downward (**A**) or upward (**B**) at an angle of 30° relative to the horizontal (x) axis ($\theta_{ex} = 60°$). The x and y coordinates are given relative to the wavelength of the radiation (λ_{ex}). Black indicates positive fields; white, negative. (**C**) The sum of the fields in (**A**) and (**B**). (**D–F**) The same as (**A–C**), respectively, at time $1/0.25\nu_{ex}$, where ν_{ex} is the frequency of the radiation. The vertical nodes in (**C**) and (**F**) move from left to right with time (note, e.g., the vertical gray stripe at $x/\lambda_{ex} = 0$ in (**C**) and at $x/\lambda_{ex} = 0.25$ in **F**), while the horizontal nodes (e.g., $y/\lambda_{ex} = \pm 0.5$) are stationary. The combined fields therefore will interact with the medium mainly in horizontal bands at $y/\lambda_{ex} = -1$, 0, +1, etc. In general, plane waves intersecting at angle θ_{ex} will give bands separated by $y/\lambda_{ex} = 1/2\sin(\theta_{ex}/2)$

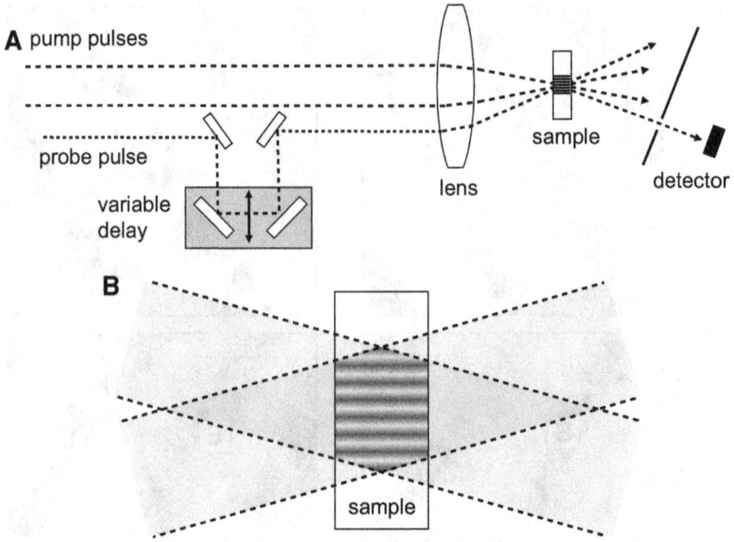

Fig. 11.16 (**A**) Apparatus for picosecond transient-grating experiments. Two parallel beams of picosecond pulses (*dashed lines*) are focused by a lens so that they overlap in the sample. A probe beam of picosecond pulses (*dotted lines*) is focused by the same lens so that it strikes the overlap region at the Bragg angle. Probe light that diffracts off the transient grating in the sample passes through a slit to a detector. A variable delay path is used to control the time between the excitation and probe pulses as in ordinary pump-probe experiments. (Another delay stage, not shown, is used to adjust the path of one of the excitation beams so that the two excitation pulses reach the sample at the same time.) (**B**) Expanded view of the region where the excitation beams (*light gray areas*) intersect in the sample, showing the orientation of the transient grating (*darker gray bands*)

through a slit to reach the detector. For measurements on time scales of nanoseconds or longer, a continuous beam from a separate laser is used as the probe, and the intensity of the diffracted beam is recorded in real time with an oscilloscope or transient-digitizer [101, 102].

Contributions to a transient grating from absorbance changes can be distinguished from effects involving the real part of the refractive index by their different dependence on the probe frequency [99]. Gratings created by thermal expansion can be recognized by their rapid decay, because thermal diffusion often is faster than other processes that follow the excitation [102, 103]. Transient-grating techniques thus offer a useful alternative to photoacoustic spectroscopy, in which thermal effects usually are identified by their dependence on the solvent's coefficient of thermal expansion (see, e.g., [104] and [105]). Applications have included studies of volume and enthalpy changes that follow photodissociation of carbon monoxide from myoglobin and hemoglobin [106–111], excitation of photoactive yellow protein [112] and rhodopsin [113], and folding of cytochrome *c* [114, 115]. Transient gratings also have been used to study exciton migration in photosynthetic antenna complexes [116].

Transient gratings also can be examined on a femtosecond time scale as a function of the time between the two pulses that create the grating [117]. As our discussion in Sect. 11.3 suggests, the two radiation fields do not actually need to be present in the sample simultaneously; the second field can interfere constructively or destructively with coherence generated by the first. This makes femtosecond transient-grating experiments potentially useful for exploring relaxations that destroy such coherence. However, photon-echo experiments provide a more thoroughly developed path to this end.

11.7 Vibrational Wavepackets

Fluorescence from molecules that are excited with short flashes can exhibit oscillations that reflect coherent excitation of multiple vibrational levels. Suppose that an ensemble of molecules occupy the lowest nuclear wavefunction of the ground electronic state. The probability that the flash will populate vibrational level k of an excited electronic state then depends on the electronic transition dipole, the spectrum and intensity of the flash, and the overlap integral $\langle \chi_{k(e)} | \chi_{0(g)} \rangle$, where $\chi_{0(g)}$ and $\chi_{k(e)}$ are the spatial parts of the ground and excited-state vibrational wavefunctions, respectively. If the flash includes a broad band of energies relative to the spacing of the vibrational eigenvalues, it will excite multiple vibronic levels ($j, k, l, \ldots$) coherently, and this coherence will be expressed in the off-diagonal elements of the vibrational density matrix. Consider, for example, an individual molecule that has a single vibrational mode with energy $h\upsilon$. If we neglect vibrational relaxations and decay of the excited electronic state following the flash, the off-diagonal density matrix elements will oscillate at frequencies that are various multiples of υ:

$$\rho_{jk}(t) = \rho_{jk}(0)\exp\left[-i\left(E_j - E_k\right)t/\hbar\right] = \rho_{jk}(0)\exp[-2\pi i(j-k)\upsilon t]. \qquad (11.45)$$

Constructive and destructive interference between these oscillations give rise to oscillatory features in the fluorescence.

An excited ensemble with vibrational coherence can be described by a linear combination of vibrational wavefunctions:

$$\mathrm{X}(u,t) = \sum_k C_k \chi_{k(e)}(u)\exp[-2\pi i(k+1/2)\upsilon t]. \qquad (11.46)$$

Here u represents a dimensionless nuclear coordinate and $\chi_{k(e)}(u)$ again denotes the spatial part of basis function k. Such a combination of wavefunctions is called a *wavepacket*. The coefficients C_k represent averages over the ensemble. Making the Born-Oppenheimer approximation and neglecting relaxations of the excited state and nonlinear effects in the excitation, they are given by

$$C_k \approx N^{-1} \sum_j \frac{\exp[-(j+1/2)h\upsilon/k_B T]}{Z} \left\langle \chi_{k(e)} \middle| \chi_{j(g)} \right\rangle I_{k,j} \,, \tag{11.47}$$

where $I_{k,j}$ is the spectral overlap of the excitation pulse with the homogeneous absorption band for the vibronic transition from level j of the ground state to level k of the excited state, k_B and T are the Boltzmann constant and temperature, Z is the vibrational partition function, and N^{-1} is a factor that depends on the electronic dipole strength and the intensity and width of the excitation flash. Figure 11.17B shows the probability function $|X(u,t)|^2$ for such a wavepacket immediately after the excitation and at several later times. For this illustration, we used one-dimensional harmonic oscillator wavefunctions as the basis and assumed that all the molecules

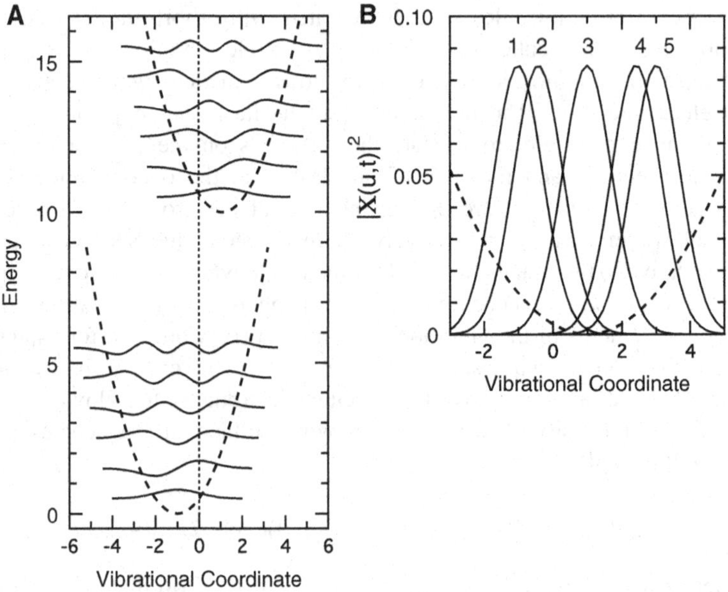

Fig. 11.17 (**A**) Wavefunctions and relative energies for a one-dimensional harmonic oscillator with a displacement (Δ) of 2.0 between the ground and excited electronic states. The abscissa is the dimensionless coordinate $u = (2\pi m_r \upsilon/h)^{-1/2} x$, where m_r and υ are the reduced mass and classical vibration frequency (Eq. 2.31); the origin is midway between the potential minima of the ground and excited states. The *dashed lines* show the potential energies. The 0–0 energy difference between the ground and excited states is arbitrary. (**B**) The probability function ($|X(u,t)|^2$) of a wavepacket at $t = 0$, $\tau/8$, $\tau/4$, $3\tau/8$, and $\tau/2$ (*curves* 1–5, respectively), where $\tau = 1/\upsilon$. The vibrational coordinate is expressed relative to the minimum of the ground-state potential as in **A**. Vibrational levels from $k = 0$ to 12 were included in the wavepacket with overlap integrals $\langle \chi_{k(e)} | \chi_{0(g)} \rangle$ calculated as described in Box 4.13. The excitation flash was centered at the Franck-Condon absorption maximum, which for $\Delta = 2$ is 1.5υ above the 0–0 energy. (The squares of the overlap integrals for $k = 0$ to 7 are 0.368, 0.520, 0.520, 0.425, 0.300, 0.190, 0.110 and 0.059.) The pulse included a broad band of energies (FWHM > > $h\upsilon$), so that $I_{k,0}/N \approx 1$ in Eq. (11.47) for all the vibrational levels of the excited state that overlap significantly with $\chi_{0(g)}$. Vibrational relaxations and dephasing were neglected

start in the lowest vibrational level of the ground state, which will be the case if $T \ll h\upsilon/k_B$. The potential energies for the ground and excited states are plotted in Fig. 11.17A along with the first few basis functions and energies.

At time zero, the probability function $|X(u,t)|^2$ of the wavefunction is centered at the potential minimum of the ground state. The wavepacket moves away from this point with time, passing through the potential minimum of the excited state when $t = \tau/4$, where τ is the vibrational period $1/\upsilon$ (*curve 3* in Fig. 11.17B). It reaches the other side of the potential well and turns around at $t = \tau/2$ (*curve 5*). All the individual wavefunctions come back into phase again at $t = \tau$, when the wavepacket arrives back at its original position. Note that Fig. 11.17B shows $|X(u,t)|^2$ rather than the wavepacket itself, which changes sign once each vibrational period and oscillates between purely real values when $t = 0, \tau, 2\tau, \ldots$ and purely imaginary values when $t = \tau/2, 3\tau/2, 5\tau/2, \ldots$.

The excitation pulse used in Fig. 11.17 was assumed to include a broad band of energies relative to the vibrational energy spacing $h\upsilon$. In this rather special situation, $|X(u,t)|^2$ has a Gaussian shape that remains constant indefinitely as the wavepacket oscillates. Such wavepackets provide a useful description of the radiation emitted from a continuous-wave laser. In this picture, the spatial oscillations of the wavepacket resemble the oscillating electric field associated with a continuous stream of photons with constant energy [118].

Anharmonic vibrational potential wells or excitation pulses with narrower spectral widths can create wavepackets with more complicated, time-dependent shapes [119–121]. Figure 11.18 shows an example in which the FWHM of the excitation

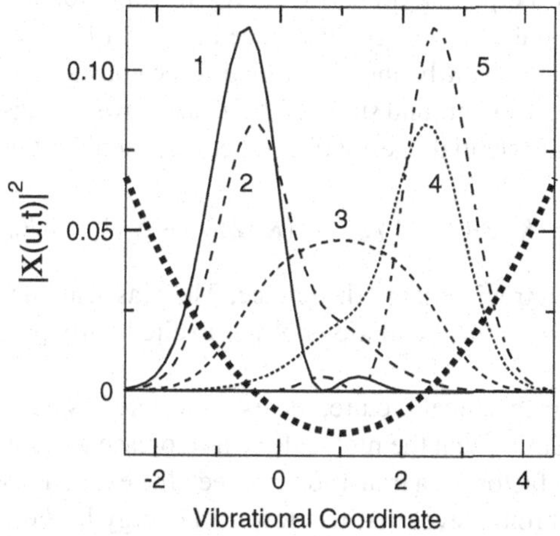

Fig. 11.18 Probability function for a wavepacket created with an excitation pulse width (FWHM) of $3h\upsilon$. *Curves 1–5* are $|X(u,t)|^2$ at times $t = 0, \tau/8, \tau/4, 3\tau/8$, and $\tau/2$, respectively; the potential energy of the excited state is shown in arbitrary units (*thick dashed curve*). $\Delta = 2.0$, and the abscissa gives u relative to a point halfway between the ground- and excited-state minima. Because the excitation spectrum biases the coefficients of the vibrational levels, $|X(u,t)|^2$ is less symmetrical than in Fig. 11.17 and initially peaks at $u \approx -0.5$ rather than -1.0

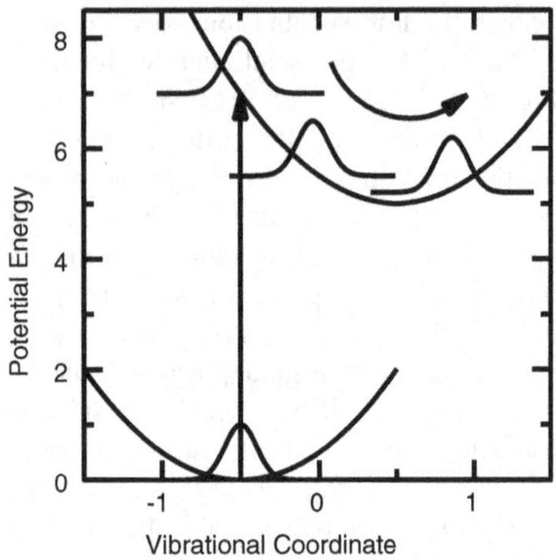

Fig. 11.19 A classical particle that is raised to an excited electronic state oscillates in the potential well of this state. If the potential well is displaced with respect to the ground-state well along one or more nuclear coordinates, the energy difference between the two states oscillates in phase with the motions of the particle

pulse is decreased to $3h\nu$. Here $|X_e(u,t)|^2$ broadens as it passes through the minimum of the potential well and becomes noticeably asymmetric as it moves away from the minimum. In addition, the wavepacket originates displaced from the ground-state minimum on the vibrational coordinate and its excursions are more restricted than they are with broader excitation. If the excitation spectrum is narrower than $h\nu$, $|X(u,t)|^2$ peaks at the Franck-Condon maximum (Fig. 4.21).

Now consider the spontaneous fluorescence from the excited system. In a classical picture, the frequency of the emission at any time depends on the vertical difference between the potential energy curves for the ground and excited states. For harmonic potentials, the energy difference is a linear function of the vibrational coordinate. To see this, let K be the vibrational force constant, and put the potential minima of the excited and ground states at $\pm\Delta/2$ as shown in Figs. 11.17, 11.18, and 11.19. The classical potential energy difference between the two states then is

$$V_e - V_g = E_{oo} + (K/2)(u - \Delta/2)^2 - (K/2)(u + \Delta/2)^2 = E_{oo} - uK\Delta, \quad (11.48)$$

where E_{oo} is the electronic energy difference. The classical energy difference thus oscillates in phase with the oscillations of the excited particle in its potential well (Fig. 11.19).

In the quantum mechanical picture, the excited state has a constant total energy. However, the probability that the molecule will fluoresce at frequency ν depends on the Franck-Condon factor for a transition between the excited-state wavepacket and the ground-state vibronic level that lies lower in energy by $h\nu$. As the wavepacket moves back and forth, its overlap with the various nuclear wavefunctions of the ground state changes, causing the fluorescence at a given frequency to oscillate. The fluorescence associated with transitions to vibrational level j of the ground state is

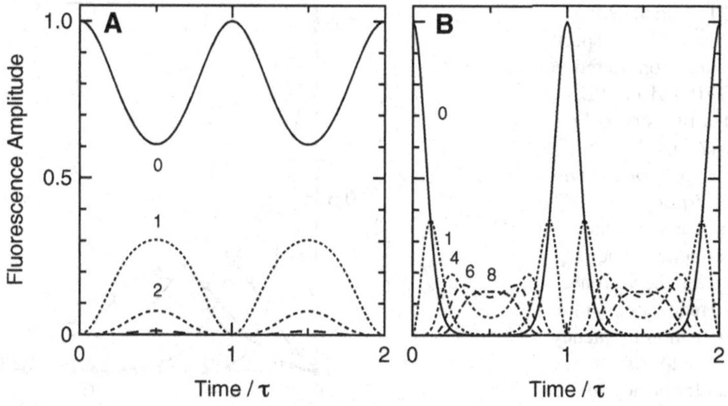

Fig. 11.20 Fluorescence at several frequencies, following formation of a vibrational wavepacket in an excited electronic state. A vibrational mode with period τ is coupled to the electronic transition with a displacement Δ of either 0.5 (**A**) or 2.0 (**B**). Wavepackets are created from the lowest vibrational level of the ground state by excitation with a spectrally broad pulse as in Fig. 11.17B. The *curves labeled 0* show the relative amplitude of fluorescence from transitions back to the lowest level of the ground state; these transitions contribute the high-energy edge of the emission spectrum. *Curves labeled 1, 2,* ... represent transitions to higher vibrational levels of the ground state, which give emission at progressively lower energies. (**A**) If Δ is small, transitions to levels 0, 1 and 2 account for essentially all the emission. (The *dashed curve* near the abscissa is for level 3.) (**B**) If Δ is larger, transitions to higher levels make significant contributions; five representative curves are shown. Vibrational dephasing and relaxations and the overall decay of the excited state are neglected, and the total fluorescence is normalized to 1.0 at all times

$$F_j(t) \propto \left| \sum_k C_k \left\langle \chi_{j(g)} \middle| \chi_{k(e)} \right\rangle \exp[-2\pi i(k+1/2)\upsilon t] \right|^2 , \qquad (11.49)$$

with C_k given by Eq. (11.47).

Figure 11.20 illustrates the calculated time course of the fluorescence at several different energies. Panel *A* is for a vibrational mode that is coupled weakly to the electronic transition ($\Delta = 0.5$), and *B* shows the results for a mode with the stronger coupling used in Fig. 11.17 ($\Delta = 2.0$). In both cases, the fluorescence resulting from transitions to the lowest vibrational level of the ground state (the curves labeled 0) peaks when $t = 0, \tau, 2\tau, \ldots$ (i.e., whenever the wavepacket is at its starting position), whereas fluorescence reflecting transitions to high vibrational levels of the ground state peaks at $t = \tau/2, 3\tau/2, 5\tau/2$ .... The oscillations of the fluorescence on the blue and red sides of the spectrum thus are 180° out of phase. For transitions with strong vibronic coupling, the fluorescence at intermediate wavelengths oscillates in a more complicated manner because the wavepacket passes through the middle of the potential well twice, moving in opposite directions, for each time it turns around at one of the edges (Fig. 11.20B curves 1, 4, 6).

Figure 11.21 shows the calculated fluorescence emission spectra at several different times between 0.1τ and 0.5τ for a system with $\Delta = 2.0$. As the fluorescence

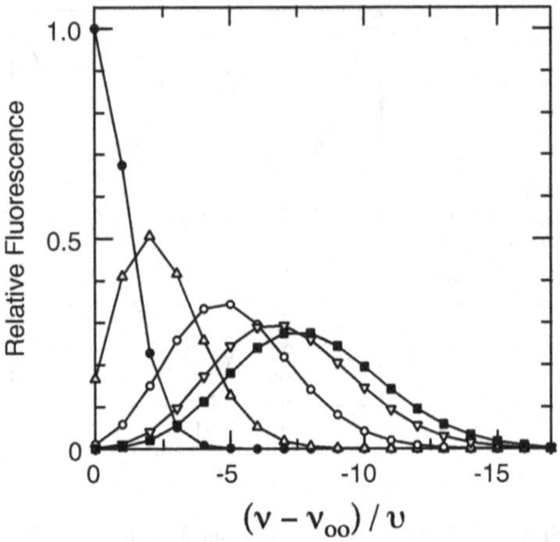

Fig. 11.21 Calculated fluorescence emission spectra for the system considered in Figs. 11.17B and 11.20B ($\Delta = 2.0$) at times $t = 0.1\tau$ (*filled circles*), 0.2τ (*up triangles*), 0.3τ (*open circles*), 0.4τ (*down triangles*) and 0.5τ (*squares*), where τ is the vibrational period. The abscissa gives the vibronic transition frequency (ν) minus the 0–0 transition frequency (ν_{oo}), relative to the vibrational frequency (υ)

oscillates between higher and lower frequencies, the width of the emission spectrum also oscillates. The spectrum is sharpest when when the wavepacket is at $u = 0$ and broadest at the opposite edge of the potential well. Again, anharmonic potential wells can give more complicated emission spectra that are broadest when the wavepacket is at intermediate positions. The time dependence of the emission spectrum thus can provide information on the shape of the potential surface [119–121], although it is unlikely to identify the shape uniquely in most cases.

It is instructive to compare these results with the fluorescence oscillations we discussed in Sect. 10.8 (Eq. 10.79). The electronic coherences we considered there caused the amplitudes of the fluorescence at different frequencies to oscillate in phase. Here the oscillations at different frequencies are out of phase. The difference is that in Sect. 10.8 we considered coherences between two excited electronic states that decay to a single ground state, whereas here we are discussing transitions from a single excited electronic state to various vibrational levels of the ground state.

Oscillations of fluorescence, stimulated emission and excited-state absorption have been studied by pump-probe techniques and fluorescence upconversion, and have been seen in numerous small molecules in solution (Fig. 11.7A; [120, 122–124]), and also in photosynthetic bacterial reaction centers [27, 125, 126]. They typically damp out over the course of several picoseconds as a result of vibrational relaxations and dephasing. Vibrational coherences generally decay more slowly than electronic coherences because the energies of vibrational states are not coupled as strongly to fluctuating interactions with the surroundings. Vibrational dephasing also tends to be less dependent on the temperature.

In multidimensional systems, coherently excited vibrational states of an ensemble of molecules probably are described best with the density-matrix formalism. Such a description has been used to rationalize oscillatory features in the dynamics of the initial electron-transfer step in photosynthetic bacterial reaction centers after excitation with sub-picosecond flashes [127, 128]. Vibrational modes that are coupled to electron transfer were identified by recording the fluctuating energy gap between the reactant and product states (P^*B and P^+B^-) during molecular-dynamics simulations (Fig. 6.3). The frequencies and displacements of the vibrational modes were obtained by taking a Fourier transform of the autocorrelation function of the fluctuations (Eqs. 10.51 and 10.52). A reduced density matrix then was constructed from approximately 650 vibronic states of five representative vibrational modes and the two electronic states. Initial values of the density matrix elements ($\rho_{jk}(0)$) were obtained for excitation of the reactive bacteriochlorophyll complex to P* with a Gaussian excitation pulse at a specified temperature by using the expressions

$$\rho_{jk}(0) = N^{-1}Z^{-1}\sum_{i \in P} \exp(-\varepsilon_i/k_BT)\, p_{i,j,k}, \tag{11.50a}$$

$$p_{i,j,k} = \left\{\prod_{m=1}^{5} \langle \chi_m^j | \chi_m^i \rangle\right\}\left\{\prod_{m=1}^{5} \langle \chi_m^k | \chi_m^i \rangle\right\} \exp\left\{-\left[(\varepsilon_j - \varepsilon_{ex})^2 + (\varepsilon_k - \varepsilon_{ex})^2\right]/2\Gamma_{ex}\right\}, \tag{11.50b}$$

$$N = Z^{-1}\sum_{i \in P} \exp(-\varepsilon_i/k_BT) \sum_{j \in P^*} p_{i,j,j}, \tag{11.50c}$$

and

$$Z = \sum_{i \in P} \exp(-\varepsilon_i/k_BT). \tag{11.50d}$$

Here i denotes a vibronic substate of the ground electronic state (P), and j and k denote substates of P*; χ_m^l is the harmonic-oscillator wavefunction for mode m in vibronic state l; $\hbar\upsilon_l$ is the energy of state l, ε_{ex} is the center of the excitation spectrum; and $\Gamma_{ex} = (W_{ex})^2/4\ln 2$, where W_{ex} is the full spectral width of the excitation spectrum at half-maximal amplitude [127]. The time dependence of the density matrix after the initial excitation was followed by integrating the stochastic Liouville equation (Eq. 10.30). Quantum mechanical interaction matrix elements for transitions between vibronic levels of different electronic states were written as products of the pertinent vibrational overlap integrals for all the modes and an assumed electronic coupling factor. The elements of the relaxation matrix **R** for thermal equilibration of two vibrational substates of a given electronic state were related to the numbers of vibrational quanta in the two substates, the energy difference between the substates, and a fundamental time constant for equilibration of the two lowest levels of each mode (T_1^o). Terms for pure dephasing and for transfer of coherence between different pairs of states also were included in **R**.

In the density-matrix model, the energy gap between P* and P^+B^- ($U(t)$) oscillates as the wavepacket evolves on the multidimensional potential surface of the excited state. The time dependence of the gap is given by

$$\langle U(t)\rangle = Tr\left\{\overline{\rho(t)\cdot \mathbf{U}(t)}\right\} = \sum_{k,j} \overline{\rho_{kj}(t) U_{jk}(t)}, \tag{11.51}$$

where $\mathbf{U}(t)$ is a matrix representation of $U(t)$ and the bars denote ensemble averages (Eq. 10.14). The matrix elements of $\mathbf{U}$ for a system in the state P* can be written

$$\mathbf{U}_{jk} = \sum_m \hbar\omega_m \Delta_m \left\{ \delta_{j,k}\frac{\Delta_m}{4} + (1-\delta_{j,k})\left\langle \chi_m^j \middle| \tilde{\mathrm{Q}}_m \middle| \chi_m^k \right\rangle \prod_{\mu\neq m} \left\langle \chi_\mu^j \middle| \chi_\mu^k \right\rangle \right\} \tag{11.52a}$$

$$= \sum_m \hbar\omega_m \Delta_m \left\{ \delta_{j,k}\frac{\Delta_m}{4} + (1-\delta_{j,k}) \left[\left(\frac{n_m^k+1}{2}\right)^{1/2} \delta_{n_m^j, n_m^k+1} + \left(\frac{n_m^k}{2}\right)^{1/2} \delta_{n_m^j, n_m^k-1} \right] \prod_{\mu\neq m} \delta_{n_\mu^j, n_\mu^k} \right\}. \tag{11.52b}$$

Here ω_m is the frequency of vibrational mode m, Δ_m is the dimensionless displacement along the coordinate for this mode in P^+B^- relative to P*, $\tilde{\mathrm{Q}}_m$ is the position operator for the mode (Eqs. 5.44 and 5.55), and n_m^k is the number of phonons of mode m in vibrational state k. The term $\delta_{j,k}\Delta_m/4$ for the diagonal elements of $\mathbf{U}$ assumes that the origin of coordinate m is midway between the potential minima of P* and P^+B, and changes sign for a system in the product state. The product of overlap integrals for all the modes (μ) other than m restricts the off-diagonal elements of $\mathbf{U}$ to those for gain or loss of a single phonon (($n_m^j = n_m^k \pm 1$ and $n_\mu^j = n_\mu^k$ for $\mu \neq m$). The matrix elements for gain or loss of more than one phonon would be much smaller.

Figure 11.22A shows the calculated dynamics of the energy gap when the fundamental time constant for vibrational equilbration (T_1^o) was taken to be 2 ps, a value that made the damping of the energy gap similar to that seen in the molecular-dynamics simulations [127, 128]. Electron transfer from P* to B was turned off for this simulation. The rapid decay of the energy gap during the first 0.1 ps results from dephasing of vibrational modes with different frequencies, whereas the slower damping reflects thermal equilibration. Figure 11.22B shows the calculated time dependence of the populations of P*, P^+B^-, and the next electron-transfer state (P^+H^-) when the electronic coupling is switched on. In this simulation, the probability of electron transfer peaks whenever the multidimensional wavepacket of P* approaches the potential energy surface of P^+B^-. Transient absorbance changes suggesting such step-wise electron transfer have been seen experimentally [29, 129–131]. The density-matrix model with T_1^o on the order of 1–10 ps also reproduced the unusual temperature dependence of the electron transfer rate [128].

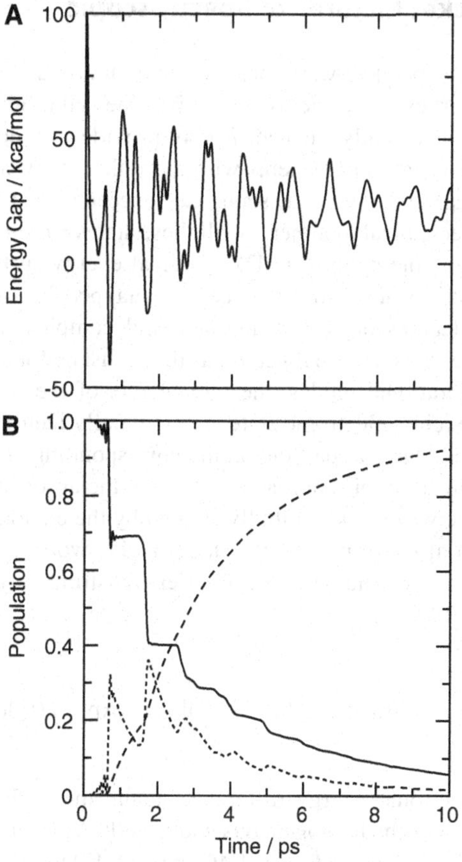

Fig. 11.22 (**A**) Calculated fluctuations of the energy gap between the excited state P* and the first electron-transfer state (P^+B^-) in a density-matrix model of photosynthetic reaction centers of *Rb. sphaeroides* [127, 128]. The model includes five effective vibrational modes. The simulation begins with excitation of P to P* by a short pulse of light, which generates a multidimensional wavepacket in the excited state. The energy gap depends on fluctuating electrostatic interactions of the protein and solvent with the electron carriers, the mean energy of the excitation flash, and a constant "gas-phase" energy difference that is adjusted arbitrarily for this figure. Electronic coupling between P* and P^+B^- is turned off in this simulation, so the oscillations are damped only by vibrational relaxations. (**B**) Calculated populations of P* (*solid curve*), P^+B^- (*short dashes*) and the next electronic state, P^+H^- (*long dashes*), in the same model when electronic coupling is turned on. Electron transfer occurs mainly at times when the energy gap between P* and P^+B^- is small

11.8 Wavepacket Pictures of Spectroscopic Transitions

The concept of time-dependent wavepackets is particularly useful for analyzing the spectroscopic properties of systems in which the vibrational states are too congested to treat individually. Indeed, it may provide the most realistic way to treat the spectral line shapes of systems with more than 5–10 atoms, in which the density of vibrational and rotational states can exceed 10^{10} per cm^{-1} [132]. An explicit quantum-mechanical treatment of all these states would be impossible with any presently conceivable computer. The classical energy surfaces of the ground and excited electronic states, however, can be mapped by molecular-dynamics simulations for molecules of almost any size and complexity (Box 6.1). If the vibrational states are spaced close together, as they must be for any molecule with a large number of vibrational modes, the movements of the quantum mechanical wavepacket in the excited electronic state are essentially identical to the dynamics of a Gaussian-shaped classical particle on the corresponding energy surface.

One fruitful application this approach is to relate the absorption spectrum to the spatial overlap of the wavepacket initially created by the excitation ($X(0)$) with the moving wavepacket at later times ($X(t)$). Heller and coworkers [132–137] showed that the spectrum can be obtained by a Fourier transform of the time-dependent overlap integral $\langle X(0)|X(t)\rangle$:

$$\frac{\varepsilon(\omega)}{\omega} \propto \int_{-\infty}^{\infty} \exp[i(\omega + E_{oo}/h)t]\langle \mathrm{X}(0)|\mathrm{X}(t)\rangle \exp(-t/\tau_c)dt. \tag{11.53}$$

Here E_{oo} is the electronic energy difference again and τ_c is a dephasing time constant that determines the homogeneous width of the vibronic lines. This expression can be rationalized by using Eqs. (11.46) and (11.47) to evaluate $\langle X(0)|X(t)\rangle$ for a one-dimensional system. If we factor out the dependence of $X(0)$ on the light source by setting $I_k/N = 1$ for all k, and assume for simplicity that only the lowest vibrational level of the ground state is occupied before the excitation, then $X(0) = \chi_{0(g)}$ and

$$\begin{aligned} \langle \mathrm{X}_g(0)|\mathrm{X}_e(t)\rangle &= \left\langle \chi_{0(g)} \middle| \sum_k C_k \chi_{k(e)} \exp[-2\pi i(k+1/2)\upsilon t] \right\rangle \\ &= \sum_k C_k \left\langle \chi_{0(g)} \middle| \chi_{k(e)} \right\rangle \exp[-2\pi i(k+1/2)\upsilon t] \\ &= \sum_k \left| \left\langle \chi_{0(g)} \middle| \chi_{k(e)} \right\rangle \right|^2 \exp[-2\pi i(k+1/2)\upsilon t]. \end{aligned} \tag{11.54}$$

A Fourier transform of this function gives a set of lines separated in frequency by υ, with each line weighted by the Franck-Condon factor $|\langle \chi_{k(e)}|\chi_{0(g)}\rangle|^2$. This is just the result we obtained in Chap. 4 for the shape of an absorption spectrum. If we multiply $\langle X(0)|X(t)\rangle$ by the damping factor $\exp(-t/\tau_c)$ that is included on the right side of Eq. (11.53), each vibronic line returned by the Fourier transform will have a Lorentzian shape with a width inversely proportional to τ_c, in agreement with the discussion in Chaps. 4 and 10.

Equation (11.54) suggests the picture that light of frequency ω acts continuously on the ground-state wavefunction, transforming it piecewise to $X_e(0)$ in the excited state with phase $\exp(i\omega t)$ [132, 135]. Meanwhile, pieces of the wavefunction that were promoted earlier move away from their birthplace (as $X_e(t)$) and then return to interfere constructively or destructively with the pieces that are just arriving. Constructive interferences give peaks in the absorption spectrum at the frequencies of the coupled vibrational modes. This picture may help to bring out the point that, although $\langle X(0)|X(t)\rangle$ and $\exp(-t/\tau)$ are functions of time, Eq. (11.53) is not intended to describe a time-dependent absorption spectrum; rather, it describes a continuous spectrum whose shape is determined by the vibrational dynamics of the excited state [138]. Lee and Heller [136] and Myers et al. [137] give more formal derivations of Eq. (11.54), along with explicit expressions for $\langle X(0)|X(t)\rangle$ in multi-dimensional harmonic systems.

Figure 11.23A shows the real and imaginary parts of the time-dependent overlap integral $\langle X(0)|X(t)\rangle$ as given by Eq. (11.54) for a single harmonic vibrational mode with frequency υ, $\Delta = 2.0$, and a relaxation time constant τ_c of $2\,\tau$. Panel *B* shows the normalized Fourier transform. As expected, the calculated spectrum has a vibronic line at the 0–0 transition frequency and a ladder of higher-frequency lines at intervals of υ. The vibrational lines are approximately Lorentzian, and although the figure does not demonstrate this, their widths depend inversely on the relaxation time constant τ_c. The overall width of the spectrum (dashed line in

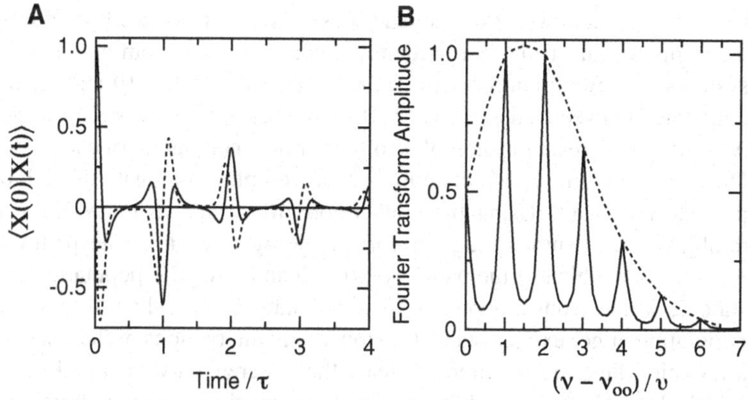

Fig. 11.23 (**A**) Real (*solid curve*) and imaginary (*dotted curve*) parts of the time-dependent overlap integral $\langle X(0)|X(t)\rangle$, as given by Eq. (11.54) for a molecule with a single harmonic vibrational mode with frequency υ, dimensionless displacement $\Delta = 2.0$ in the excited state, vibrational period $\tau = 1/\upsilon$, and vibrational relaxation time constant $\tau_c = 2\tau$. (**B**) Normalized magnitude of the Fourier transform of the overlap integral shown in **A**. The abscissa is the absorption frequency (ν) relative to the 0–0 transition frequency (ν_{oo}), in units of υ. The envelope of the calculated absorption spectrum (*dotted curve*) depends on the speed with which $\langle X(0)|X(t)\rangle$ decreases near $t = 0$

Fig. 11.23B) is determined by the fastest component in the time domain, which is the initial drop in $\langle X(0)|X(t)\rangle$ near $t=0$. In the semiclassical wavepacket model, the speed of this drop depends on the steepness of the excited-state energy surface at the point where the wavepacket is created [132].

The fluorescence emission spectrum can be obtained in the same way as the absorption spectrum by considering the time dependence of a wavepacket in the ground electronic state following a transition from the excited state. The weighting factor ω^{-1} on the left-hand side of Eq. (11.53) is simply replaced by ω^{-3} in accord with the Einstein coefficients (Chap. 5).

The main utility of the wavepacket approach lies in the evaluation of spectroscopic properties of molecules with congested vibrational modes. As we discussed in Chap. 4, the ground-state vibrational wavefunctions $X_{a(g)}$ for a molecule with multiple vibrational modes are products of the wavefunctions for all the individual modes: $X_{a(g)}=\chi_{a1(g)}\chi_{a2(g)}\chi_{a3(g)}\cdots$. If the normal coordinates do not change significantly in the excited electronic state, $X(t)$ will be a product of wavepackets for the individual modes ($X(t)=X_1(t)X_2(t)X_3(t)\ldots$) and the same product appears in the time-dependent overlap integral $\langle X(0)|X(t)\rangle$ [139]:

$$\langle \mathrm{X}(0)|\mathrm{X}(t)\rangle = \langle \mathrm{X}_1(0)|\mathrm{X}(t)\rangle\langle \mathrm{X}_2(0)|\mathrm{X}(t)\rangle\langle \mathrm{X}_3(0)|\mathrm{X}(t)\rangle \;\ldots\;. \tag{11.55}$$

The time dependences of the wavepackets for the individual modes can be predicted by molecular-dynamics simulations even for comparatively complex molecules. Applications of the wavepacket treatment to resonance Raman spectroscopy are discussed in Chap. 12.

Exercises

1. Consider a system with two excited states (1 and 2). (*a*) Draw Liouville-space diagrams for all of the independent pathways starting from an ensemble of systems at thermal equilibrium in the ground state (state 0) and leading to coherence between states 1 and 2. By "independent" here we mean that no two pathways are simply complex, or Hermitian, conjugates of each other. (*b*) Draw the Liouville-space diagram for the complex conjugate of each of the pathways in *a*. (*c*) Using the conventions illustrated in Fig. 11.2, draw a double-sided Feynman diagram corresponding to each of the pathways in *a* and *b*. (*d*) Which of the pathways in *a*, *b* and *c* require populating excited state 1; which require populating excited state 2; and which do not require populating either excited state? (*e*) How many interactions of the system with a radiation field are required to create the coherence by each pathway in *a*, *b* and *c*? (*f*) How many additional interactions with the radiation field would be required to convert the coherence into a population of state 2? (*f*) If the interactions in *e* and *f* occur sequentially during separate short flashes of light at times t_1, t_2, t_3, ..., which pathways will be most sensitive to the time interval t_3–t_2?
2. Suppose states 1 and 2 are different singlet electronic states with energies E_1 and E_2 and transition dipoles $\boldsymbol{\mu}_1$ and $\boldsymbol{\mu}_2$ for excitation from the ground state,

which also is a singlet state. (*a*) Write an approximate expression for the initial magnitude of the coherence relative to the population of state 1 ($|\rho_{12}(0)|/\rho_{11}(0)$) following excitation of the ensemble with a short flash of white light. Assume that "white" light has equal intensities at frequencies E_1/h and E_2/h, and that all the requisite interactions with the radiation field occur during the same flash. (*b*) How would $|\rho_{12}(0)|/\rho_{11}(0)$ differ from that in *a* if the intensity of the excitation flash at E_1/h is 100 times that at E_2/h? (*c*) How will the ratio $|\rho_{12}(t)|/\rho_{11}(t)$ depend on time (t) if the ensemble is isolated from stochastic interactions with its surroundings?

3. Suppose that states 1 and 2 are levels $m=1$ and $m=2$ of a harmonic vibrational mode of the same excited singlet electronic state. Assume that the ensemble starts entirely in vibrational level $m=0$ of the same mode in the ground electronic state, that the vibrational mode has an energy of $\hbar\omega=200\ \text{cm}^{-1}$, and that the coupling strength (Huang-Rhys factor) for the vibronic excitation is $S=0.2$. (*a*) Write an approximate expression for the initial value of the ratio $|\rho_{12}(0)|/\rho_{11}(0)$ following excitation of the ensemble with a short flash of white light. (*b*) How will the ratio $|\rho_{12}(t)|/\rho_{11}(t)$ depend on time in the absence of stochastic interactions with the surroundings? (*c*) Now assume that stochastic processes at temperature T cause transitions between levels 2 and 1 with rate constants $k_{21}=1\times10^{12}\ \text{s}^{-1}$ and $k_{12}=\exp(-\hbar\omega/k_BT)\times10^{12}\ \text{s}^{-1}$. How will $|\rho_{12}(t)|$ and $\rho_{11}(t)$ depend on time at 300 K in this situation?
4. Considering the vibronic system of exercise 3 again, suppose the excitation flash is centered 400 cm^{-1} above the 0–0 transition energy and has a Gaussian spectral envelope with a FWHM of 300 cm^{-1}. Though not truly "white," the flash then can populate multiple vibrational levels of the excited state. (*a*) Estimate the initial coefficients (C_m) for vibrational levels $m=0$ through 4. (*b*) Calculate the relative energy and time course of fluorescence from transitions from the excited system to vibrational level $m=0$ of the ground electronic state. (*c*) Calculate the relative energy and time course of fluorescence from transitions from the excited system to vibrational level $m=2$ of the ground electronic state.
5. In a three-pulse photon echo experiment, light pulses with wavevectors $\boldsymbol{k}_1$, $\boldsymbol{k}_2$ and $\boldsymbol{k}_3$ pass confocally through a sample at times τ_1, τ_2 and τ_3, respectively, and photon echos leaving the sample in direction $\boldsymbol{k}_3+\boldsymbol{k}_2-\boldsymbol{k}_1$ are measured. Explain why echos are seen only if $\tau_2>\tau_1$.
6. Explain why, in the semiclassical wavepacket picture, the width of an absorption band increases with the slope of the excited-state potential surface in the region of the Franck-Condon maximum, whereas the width of an emission band increases with the corresponding slope of the ground-state potential surface.

References

1. Slichter, C.P.: Principles of Magnetic Resonance with Examples from Solid State Physics. Harper & Row, New York (1963)
2. Ward, J.F.: Calculation of nonlinear optical susceptibilities using diagrammatic perturbation theory. Rev. Mod. Phys. **37**, 1–18 (1965)
3. Yee, T.K., Gustafson, W.G.: Diagrammatic analysis of the density operator for nonlinear optical calculations: pulsed and cw responses. Phys. Rev. A **18**, 1597–1617 (1978)
4. Druet, S.A.J., Taran, J.P.E.: CARS spectroscopy. Prog. Quantum Electron. **7**, 1–72 (1981)
5. Yan, Y.J., Fried, L.E., Mukamel, S.: Ultrafast pump-probe spectroscopy: femtosecond dynamics in Liouville space. J. Phys. Chem. **93**, 8149–8162 (1989)
6. Mukamel, S.: Femtosecond optical spectroscopy: a direct look at elementary chemical events. Annu. Rev. Phys. Chem. **41**, 647–681 (1990)
7. Mukamel, S.: Principles of Nonlinear Optical Spectroscopy. Oxford University Press, Oxford (1995)
8. Sepulveda, M.A., Mukamel, S.: Semiclassical theory of molecular nonlinear optical polarization. J. Chem. Phys. **102**, 9327–9344 (1995)
9. Joo, T., Albrecht, A.C.: Electronic dephasing studies of molecules in solution at room temperature by femtosecond degerate four wave mixing. Chem. Phys. **176**, 233–247 (1993)
10. Fleming, G.R., Cho, M.: Chromophore-solvent dynamics. Annu. Rev. Phys. Chem. **47**, 109–134 (1996)
11. Su, J.-J., Yu, I.A.: The study of coherence-induced phenomena using double-sided Feynman diagrams. Chin. J. Phys. **41**, 627–642 (2003)
12. Armstrong, J.A., Bloembergen, N., Ducuing, J., Pershan, P.S.: Interactions between light waves in a nonlinear dielectric. Phys. Rev. **127**, 1918–1939 (1962)
13. Bloembergen, N.: Nonlinear Optics. Benjamin, New York (1965)
14. Shen, Y.R.: The Principles of Nonlinear Optics. Wiley, New York (1984)
15. Butcher, P.N., Cotter, D.: The Elements of Nonlinear Optics. Cambridge University Press, Cambridge (1990)
16. Mukamel, S.: Collisional broadening of spectral line shapes in two-photon and multiphoton processes. Phys. Rep. **93**, 1–60 (1982)
17. Mukamel, S., Loring, R.F.: Nonlinear response function for time-domain and frequency-domain four-wave mixing. J. Opt. Soc. Am. B Opt. Phys. **3**, 595–606 (1986)
18. Schuler, B., Lipman, E.Å., Eaton, W.A.: Probing the free-energy surface for protein folding with single-molecule fluorescence spectroscopy. Nature **419**, 743–747 (2002)
19. Mukamel, S.: On the semiclassical calculation of molecular absorption and fluorescence spectra. J. Chem. Phys. **77**, 173–181 (1982)
20. Brito Cruz, C.H., Fork, R.L., Knox, W., Shank, C.V.: Spectral hole burning in large molecules probed with 10 fs optical pulses. Chem. Phys. Lett. **132**, 341–344 (1986)
21. Becker, P.C., Fork, R.L., Brito Cruz, C.H., Gordon, J.P., Shank, C.V.: Optical Stark effect in organic dyes probed with optical pulses of 6 fs duration. Phys. Rev. Lett. **60**, 2462–2464 (1988)
22. Zewail, A.H.: Laser femtochemistry. Science **242**, 1645–1653 (1988)
23. Nagarajan, V., Johnson, E., Schellenberg, P., Parson, W.: A compact, versatile femtosecond spectrometer. Rev. Sci. Instrum. **73**, 4145–4149 (2002)
24. Woodbury, N.W., Becker, M., Middendorf, D., Parson, W.W.: Picosecond kinetics of the initial photochemical electron-transfer reaction in bacterial photosynthetic reaction center. Biochemistry **24**, 7516–7521 (1985)
25. Fleming, G.R., Martin, J.-L., Breton, J.: Rates of primary electron transfer in photosynthetic reaction centers and their mechanistic implications. Nature **333**, 190–192 (1988)
26. Lauterwasser, C., Finkele, U., Scheer, H., Zinth, W.: Temperature dependence of the primary electron transfer in photosynthetic reaction centers from Rhodobacter sphaeroides. Chem. Phys. Lett. **183**, 471–477 (1991)

27. Vos, M.H., Jones, M.R., Hunter, C.N., Breton, J., Martin, J.-L.: Coherent nuclear dynamics at room temperature in bacterial reaction centers. Proc. Natl. Acad. Sci. U. S. A. **91**, 12701–12705 (1994)
28. Holzwarth, A.R., Muller, M.G.: Energetics and kinetics of radical pairs in reaction centers from Rhodobacter sphaeroides. A femtosecond transient absorption study. Biochemistry **35**, 11820–11831 (1996)
29. Streltsov, A.M., Aartsma, T.J., Hoff, A.J., Shuvalov, V.A.: Oscillations within the B_L absorption band of Rhodobacter sphaeroides reaction centers upon 30 femtosecond excitation at 865 nm. Chem. Phys. Lett. **266**, 347–352 (1997)
30. Kirmaier, C., Laible, P.D., Czarnecki, K., Hata, A.N., Hanson, D.K., et al.: Comparison of M-side electron transfer in Rb. sphaeroides and Rb. capsulatus reaction centers. J. Phys. Chem. B **106**, 1799–1808 (2002)
31. Haffa, A.L.M., Lin, S., Williams, J.C., Bowen, B.P., Taguchi, A.K.W., et al.: Controlling the pathway of photosynthetic charge separation in bacterial reaction centers. J. Phys. Chem. B **108**, 4–7 (2004)
32. Weiner, A.M., De Silvestri, S., Ippen, E.P.: Three-pulse scattering for femtosecond dephasing studies: theory and experiment. J. Opt. Soc. Am. B Opt. Phys. **2**, 654–662 (1985)
33. Joo, T., Jia, Y., Yu, J.-Y., Lang, M.J., Fleming, G.R.: Third-order nonlinear time domain probes of solvation dynamics. J. Chem. Phys. **104**, 6089–6108 (1996)
34. Cho, M., Yu, J.-Y., Joo, T., Nagasawa, Y., Passino, S.A., et al.: The integrated photon echo and solvation dynamics. J. Phys. Chem. **100**, 11944–11953 (1996)
35. Jimenez, R., van Mourik, F., Yu, J.-Y., Fleming, G.R.: Three-pulse photon echo measurements on LH1 and LH2 complexes of Rhodobacter sphaeroides: a nonlinear spectroscopic probe of energy transfer. J. Phys. Chem. B **101**, 7350–7359 (1997)
36. Passino, S.A., Nagasawa, Y., Fleming, G.R.: Three-pulse stimulated photon echo experiments as a probe of polar solvation dynamics: utility of harmonic bath models. J. Chem. Phys. **107**, 6094–6108 (1997)
37. Jordanides, X.J., Lang, M.J., Song, X., Fleming, G.R.: Solvation dynamics in protein environments studied by photon echo spectroscopy. J. Phys. Chem. B **103**, 7995–8005 (1999)
38. Lang, M.J., Jordanides, X.J., Song, X., Fleming, G.R.: Aqueous solvation dynamics studied by photon echo spectroscopy. J. Chem. Phys. **110**, 5884–5892 (1999)
39. Yan, Y.J., Mukamel, S.: Photon echoes of polyatomic molecules in condensed phases. J. Chem. Phys. **94**, 179–190 (1991)
40. Kurnit, N.A., Abella, I.D., Hartmann, S.R.: Observation of a photon echo. Phys. Rev. Lett. **13**, 567–568 (1964)
41. Hesselink, W.H., Wiersma, D.A.: Picosecond photon echoes stimulated from an accumulated grating. Phys. Rev. Lett. **43**, 1991–1994 (1979)
42. Becker, P.C., Fragnito, H.L., Bigot, J.Y., Brito Cruz, C.H., Fork, R.L., et al.: Femtosecond photon echoes from molecules in solution. Phys. Rev. Lett. **63**, 505–507 (1989)
43. Bigot, J.Y., Portella, M.T., Schoenlein, R.W., Bardeen, C.J., Migus, A., et al.: Non-Markovian dephasing of molecules in solution measured with 3-pulse femtosecond photon echoes. Phys. Rev. Lett. **66**, 1138–1141 (1991)
44. Nibbering, E.T.J., Fidder, H., Pines, E.: Ultrafast chemistry: using time-resolved vibrational spectroscopy for interrogation of structural dynamics. Annu. Rev. Phys. Chem. **56**, 337–367 (2005)
45. de Boeij, W.P., Pshenichnikov, M.S., Wiersma, D.A.: System-bath correlation function probed by conventional and time-gated stimulated photon echo. J. Phys. Chem. **100**, 11806–11823 (1996)
46. Fleming, G.R., Joo, T., Cho, M., Zewail, A.H., Letokhov, V.S., et al.: Femtosecond chemical dynamics in condensed phases. Adv. Chem. Phys. **101**, 141–183 (1997)
47. Momelle, B.J., Edington, M.D., Diffey, W.M., Beck, W.F.: Stimulated photon-echo and transient-grating studies of protein-matrix solvation dynamics and interexciton-state

radiationless decay in alpha phycocyanin and allophycocyanin. J. Phys. Chem. B **102**, 3044–3052 (1998)
48. Nagasawa, Y., Watanabe, Y., Takikawa, H., Okada, T.: Solute dependence of three pulse photon echo peak shift measurements in methanol solution. J. Phys. Chem. A **107**, 632–641 (2003)
49. Leeson, D.T., Wiersma, D.A.: Looking into the energy landscape of myoglobin. Nat. Struct. Biol. **2**, 848–851 (1995)
50. Leeson, D.T., Wiersma, D.A., Fritsch, K., Friedrich, J.: The energy landscape of myoglobin: an optical study. J. Phys. Chem. B **101**, 6331–6340 (1997)
51. Fayer, M.D.: Fast protein dynamics probed with infrared vibrational echo experiments. Annu. Rev. Phys. Chem. **52**, 315–356 (2001)
52. Leeson, D.T., Berg, O., Wiersma, D.A.: Low-temperature protein dynamics studied by the long-lived stimulated photon echo. J. Phys. Chem. **98**, 3913–3916 (1994)
53. Kennis, J.T.M., Larsen, D.S., Ohta, K., Facciotti, M.T., Glaeser, R.M., et al.: Ultrafast protein dynamics of bacteriorhodopsin probed by photon echo and transient absorption spectroscopy. J. Phys. Chem. B **106**, 6067–6080 (2002)
54. Jimenez, R., Case, D.A., Romesberg, F.E.: Flexibility of an antibody binding site measured with photon echo spectroscopy. J. Phys. Chem. B **106**, 1090–1103 (2002)
55. Changenet-Barret, P., Choma, C.T., Gooding, E.F., De Grado, W.F., Hochstrasser, R.M.: Ultrafast dielectric response of proteins from dynamic Stokes shifting of coumarin in calmodulin. J. Phys. Chem. B **104**, 9322–9329 (2000)
56. Ge, N.H., Hochstrasser, R.M.: Femtosecond two-dimensional infrared spectroscopy: IR-COSY and THIRSTY. PhysChemComm **5**, 17–26 (2002)
57. Park, J., Ha, J.-H., Hochstrasser, R.M.: Multidimensional infrared spectroscopy of the N-H bond motions in formamide. J. Chem. Phys. **121**, 7281–7292 (2004)
58. Mukherjee, P., Krummel, A.T., Fulmer, E.C., Kass, I., Arkin, I.T., et al.: Site-specific vibrational dynamics of the CD3z membrane peptide using heterodyned two-dimensional infrared photon echo spectroscopy. J. Chem. Phys. **120**, 10215–10224 (2004)
59. Khalil, M., Demirdöven, N., Tokmakoff, A.: Coherent 2D IR spectroscopy: molecular structure and dynamics in solution. J. Phys. Chem. A **107**, 5258–5279 (2003)
60. Golonzka, O., Khalil, M., Demirdöven, N., Tokmakoff, A.: Vibrational anharmonicities revealed by coherent two-dimensional infrared spectroscopy. Phys. Rev. Lett. **86**, 2154–2157 (2001)
61. Ishikawa, H., Kwac, K., Chung, J.K., Kim, S., Fayer, M.D.: Direct observation of fast protein conformational switching. Proc. Natl. Acad. Sci. U. S. A. **105**, 8619–8624 (2008)
62. Adamcyzk, K., Candelaresi, M., Kania, R., Robb, K., Bellota-Antón, C., et al.: The effect of point mutation on the equilibrium structural fluctuations of ferric myoglobin. Phys. Chem. Chem. Phys. **14**, 7411–7419 (2012)
63. Cheng, M., Brookes, J.E., Montfort, W.R., Khalil, M.: pH-dependent picosecond structural dynamics in the distal pocket of nitrophorin 4 investigated by 2D IR spectroscopy. J. Phys. Chem. B **117**, 15804–15811 (2013)
64. Brixner, T., Stenger, J., Vaswani, H.M., Cho, M., Blankenship, R.E., et al.: Two-dimensional spectroscopy of electronic couplings in photosynthesis. Nature **434**, 625–628 (2005)
65. Vaswani, H.M., Brixner, T., Stenger, J., Fleming, G.R.: Exciton analysis in 2D electronic spectroscopy. J. Phys. Chem. B **109**, 10542–10556 (2005)
66. Zigmantas, D., Read, E.L., Mancal, T., Brixner, T., Gardiner, A.T., et al.: Two-dimensional electronic spectroscopy of the B800–B820 light-harvesting complex. Proc. Natl. Acad. Sci. U. S. A. **103**, 12672–12677 (2006)
67. Read, E.L., Engel, G.S., Calhoun, T.R., Mancal, T., Ahn, T.K., et al.: Cross-peak-specific two-dimensional electronic spectroscopy. Proc. Natl. Acad. Sci. U. S. A. **104**, 14203–14208 (2007)

68. Engel, G.S., Calhoun, T.R., Read, E.L., Ahn, T.K., Mancal, T., et al.: Evidence for wavelike energy transfer through quantum coherence in photosynthetic systems. Nature **446**, 782–786 (2007)
69. Schlau-Cohen, G.S., Calhoun, T.R., Ginsberg, N.S., Read, E.L., Ballottari, M., et al.: Pathways of energy flow in LHCII from two-dimensional electronic spectroscopy. J. Phys. Chem. B **113**, 15352–15363 (2009)
70. Myers, J.A., Lewis, K.L.M., Fuller, F.D., Tekavec, P.F., Yocum, C.F., et al.: Two-dimensional electronic spectroscopy of the D1-D2-cyt b559 photosystem II reaction center complex. J. Phys. Chem. Lett. **1**, 2774–2780 (2010)
71. Collini, E., Wong, C.Y., Wilk, K.E., Curmi, P.M.G., Brumer, P., et al.: Coherently wired light-harvesting in photosynthetic marine algae at ambient temperature. Nature **463**, 644–648 (2010)
72. Schlau-Cohen, G.S., Ishizaki, A., Fleming, G.R.: Two-dimensional electronic spectroscopy and photosynthesis: fundamentals and applications to photosynthetic light-harvesting. Chem. Phys. **386**, 1–22 (2011)
73. Anna, J.M., Ostroumov, E.E., Maghlaoul, K., Barber, J., Scholes, G.D.: Two-dimensional electronic spectroscopy reveals ultrafast downhill energy transfer in photosystem I trimers of the cyanobacterium Thermosynechococcus elongatus. J. Phys. Chem. Lett. **3**, 3677–3684 (2012)
74. Fuller, F.D., Pan, J., Gelzinis, A., Butkus, V., Senlik, S.S., et al.: Vibronic coherence in oxygenic photosynthesis. Nat. Chem. **6**, 706–711 (2014)
75. Lewis, K.L.M., Ogilvie, J.P.: Probind photosynthetic energy and charge transfer with two-dimensional electronic spectroscopy. J. Phys. Chem. Lett. **3**, 503–510 (2013)
76. Lee, H., Cheng, Y.-C., Fleming, G.R.: Coherence dynamics in photosynthesis: protein protection of excitonic coherence. Science **316**, 1462–1465 (2007)
77. Ishizaki, A., Fleming, G.R.: Theoretical examination of quantum coherence in a photosynthetic system at physiological temperature. Proc. Natl. Acad. Sci. U. S. A. **106**, 17255–17260 (2009)
78. Calhoun, T.R., Ginsberg, N.S., Schlau-Cohen, G.S., Cheng, Y.-C., Ballottari, M., et al.: Quantum coherence enabled determination of the energy landscape in light-harvesting complex II. J. Phys. Chem. B **113**, 16291–16295 (2009)
79. Collini, E., Scholes, G.D.: Quantum coherent energy migration in a conjugated polymer at room temperature. Science **323**, 369–373 (2009)
80. Beljonne, D., Curutchet, C., Scholes, G.D., Silbey, R.: Beyond Förster resonance energy transfer in biological and nanoscale systems. J. Phys. Chem. B **113**, 6583–6599 (2009)
81. Panitchayangkoon, G., Hayes, D., Fransted, K.A., Caram, J.R., Harel, E., et al.: Long-lived quantum coherence in photosynthetic complexes at physiological temperature. Proc. Natl. Acad. Sci. U. S. A. **107**, 12766–12770 (2010)
82. Hamm, P., Lim, M., Hochstrasser, R.M.: Structure of the amide I band of peptides measured by femtosecond nonlinear-infrared spectroscopy. J. Phys. Chem. B **102**, 6123–6138 (1998)
83. Zanni, M.T., Hochstrasser, R.M.: Two-dimensional infrared spectroscopy: a promising new method for the time resolution of structures. Curr. Opin. Struct. Biol. **11**, 516–522 (2001)
84. Rubtsov, I.V., Wang, J., Hochstrasser, R.M.: Dual-frequency 2D-IR spectroscopy heterodyned photon echo of the peptide bond. Proc. Natl. Acad. Sci. U. S. A. **100**, 5601–5606 (2003)
85. Chung, H.S., Khalil, M., Smith, A.W., Ganim, Z., Tokmakoff, A.: Conformational changes during the nanosecond-to-millisecond unfolding of ubiquitin. Proc. Natl. Acad. Sci. U. S. A. **102**, 612–617 (2005)
86. DeChamp, M.F., DeFlores, L., McCracken, J.M., Tokmakoff, A., Kwac, K., et al.: Amide I vibrational dynamics of N-methylacetamide in polar solvents: the role of electrostatic interactions. J. Phys. Chem. B **109**, 11016–11026 (2005)
87. Hamm, P., Zanni, M.: Concepts and Methods of 2D Infrared Spectroscopy. Cambridge University Press, Cambridge (2011)

88. Baiz, C.R., Reppert, M., Tokmakoff, A.: An introduction to protein 2D IR spectroscopy. In: Fayer, M.D. (ed.) Ultrafast Infrared Vibrational Spectroscopy. Taylor & Francis, New York (2013)
89. Baiz, C.R., Reppert, M., Tokmakoff, A.: Amide I two-dimensional infrared spectroscopy: methods for visualizing the vibrational structure of large proteins. J. Phys. Chem. A **117**, 5955–5961 (2013)
90. Ganim, Z., Chung, H.S., Smith, A.W., DeFlores, L.P., Jones, K.C., et al.: Amide I two-dimensional infrared spectroscopy of proteins. Acc. Chem. Res. **41**, 432–441 (2008)
91. DeFlores, L., Ganim, Z., Nicodemus, R.A., Tokmakoff, A.: Amide I′-II′ 2D IR spectroscopy provides enhanced protein secondary structural sensitivity. J. Am. Chem. Soc. **131**, 3385–3391 (2009)
92. West, B.A., Womick, J.A., Moran, A.M.: Probing ultrafast dynamics in adenine with mid-UV four-wave mixing spectroscopies. J. Phys. Chem. A **115**, 8630–8637 (2011)
93. West, B.A., Womick, J.A., Moran, A.M.: Influence of temperature on thymine-to-solvent vibrational energy transfer. J. Chem. Phys. **135**, 114505 (2011)
94. Tseng, C.-H., Sándor, P., Kotur, M., Weinacht, T.C., Matsika, S.: Two-dimensional Fourier transform spectroscopy of adenine and uracil using shaped ultrafast laser pulses in the deep UV. J. Phys. Chem. A **116**, 2654–2661 (2012)
95. West, B.A., Womick, J.A., Moran, A.M.: Interplay between vibrational energy transfer and excited state deactivation in DNA components. J. Phys. Chem. A **117**, 5865–5874 (2013)
96. West, B.A., Molesky, B.P., Giokas, P.G., Moran, A.M.: Uncovering molecular relaxation processes with nonlinear spectroscopies in the deep UV. Chem. Phys. **423**, 92–104 (2013)
97. Salcedo, J.R., Siegman, A.E., Dlott, D.D., Fayer, M.D.: Dynamics of energy transport in molecular crystals: the picosecond transient-grating method. Phys. Rev. Lett. **41**, 131–134 (1978)
98. Nelson, K.A., Fayer, M.D.: Laser induced phonons: a probe of intermolecular interactions in molecular solids. J. Chem. Phys. **72**, 5202–5218 (1980)
99. Nelson, K.A., Caselegno, R., Miller, R.J.D., Fayer, M.D.: Laser-induced excited state and ultrasonic wave gratings: amplitude and phase grating contributions to diffraction. J. Chem. Phys. **77**, 1144–1152 (1982)
100. Genberg, L., Bao, Q., Bracewski, S., Miller, R.J.D.: Picosecond transient thermal phase grating spectroscopy: a new approach to the study of vibrational-energy relaxation processes in proteins. Chem. Phys. **131**, 81–97 (1989)
101. Terazima, M., Hirota, N.: Measurement of the quantum yield of triplet formation and short triplet lifetimes by the transient grating technique. J. Chem. Phys. **95**, 6490–6495 (1991)
102. Terazima, M., Hara, T., Hirota, N.: Reaction volume and enthalpy changes in photochemical reaction detected by the transient grating method: photodissociation of diphenylcyclopropenone. Chem. Phys. Lett. **246**, 577–582 (1995)
103. Hara, T., Hirota, N., Terazima, M.: New application of the transient grating method to a photochemical reaction: the enthalpy, reaction volume change, and partial molar volume measurements. J. Phys. Chem. **100**, 10194–10200 (1996)
104. Ort, D.R., Parson, W.W.: Flash-induced volume changes of bacteriorhodopsin-containing membrane fragments and their relationship to proton movements and absorbance transients. J. Biol. Chem. **253**, 6158–6164 (1978)
105. Arata, H., Parson, W.W.: Enthalpy and volume changes accompanying electron transfer from P-870 to quinones in Rhodopseudomonas sphaeroides reaction centers. Biochim. Biophys. Acta **636**, 70–81 (1981)
106. Richard, L., Genberg, L., Deak, J., Chiu, H.-L., Miller, R.J.D.: Picosecond phase grating spectroscopy of hemoglobin and myoglobin. Energetics and dynamics of global protein motion. Biochemistry **31**, 10703–10715 (1992)
107. Dadusc, G., Ogilvie, J.P., Schulenberg, P., Marvet, U., Miller, R.J.D.: Diffractive optics-based heterodyne-detected four-wave mixing signals of protein motion: from "protein quakes" to ligand escape for myoglobin. Proc. Natl. Acad. Sci. U. S. A. **98**, 6110–6115 (2001)

108. Sakakura, M., Morishima, I., Terazima, M.: The structural dynamics and ligand releasing process after the photodissociation of sperm whale carboxymyoglobin. J. Phys. Chem. B **105**, 10424–10434 (2001)
109. Sakakura, M., Yamaguchi, S., Hirota, N., Terazima, M.: Dynamics of structure and energy of horse carboxymyoglobin after photodissociation of carbon monoxide. J. Am. Chem. Soc. **123**, 4286–4294 (2001)
110. Choi, J., Terazima, M.: Denaturation of a protein monitored by diffusion coefficients: myoglobin. J. Phys. Chem. B **106**, 6587–6593 (2002)
111. Nishihara, Y., Sakakura, M., Kimura, Y., Terazima, M.: The escape process of carbon monoxide from myoglobin to solution at physiological temperature. J. Am. Chem. Soc. **126**, 11877–11888 (2004)
112. Takashita, K., Imamoto, Y., Kataoka, M., Mihara, K., Tokunaga, F., et al.: Structural change of site-directed mutants of PYP: new dynamics during pR state. Biophys. J. **83**, 1567–1577 (2002)
113. Nishioku, Y., Nakagawa, M., Tsuda, M., Terazima, M.: Energetics and volume changes of the intermediates in the photolysis of octopus rhodopsin at physiological temperature. Biophys. J. **83**, 1136–1146 (2002)
114. Nada, T., Terazima, M.: A novel method for study of protein folding kinetics by monitoring diffusion coefficient in time domain. Biophys. J. **85**, 1876–1881 (2003)
115. Nishida, S., Nada, T., Terazima, M.: Kinetics of intermolecular interaction during protein folding of reduced cytochrome c. Biophys. J. **87**, 2663–2675 (2004)
116. Salverda, J.M., Vengris, M., Krueger, B.P., Scholes, G.D., Czarnoleski, A.R., et al.: Energy transfer in light-harvesting complexes LHCII and CP29 of spinach studied with three pulse echo peak shift and transient grating. Biophys. J. **84**, 450–465 (2003)
117. Park, J.S., Joo, T.: Coherent interactions in femtosecond transient grating. J. Chem. Phys. **120**, 5269–5274 (2004)
118. Glauber, R.J.: Coherent and incoherent states of the radiation field. Phys. Rev. **131**, 2766 (1963)
119. Kowalczyk, P., Radzewicz, C., Mostowski, J., Walmsley, I.A.: Time-resolved luminescence from coherently excited molecules as a probe of molecular wave-packet dynamics. Phys. Rev. A **42**, 5622–5626 (1990)
120. Dunn, R.C., Xie, X.L., Simon, J.D.: Real-time spectroscopic techniques for probing conformational dynamics of heme-proteins. Methods Enzymol. **226**, 177–198 (1993)
121. Jonas, D.M., Bradforth, S.E., Passino, S.A., Fleming, G.R.: Femtosecond wavepacket spectroscopy. Influence of temperature, wavelength and pulse duration. J. Phys. Chem. **99**, 2594–2608 (1995)
122. Wise, F.W., Rosker, M.J., Tang, C.L.: Oscillatory femtosecond relaxation of photoexcited organic molecules. J. Chem. Phys. **86**, 2827–2832 (1987)
123. Mokhtari, A., Chesnoy, J., Laubereau, A.: Femtosecond time-resolved and frequency-resolved spectroscopy of a dye molecule. Chem. Phys. Lett. **155**, 593–598 (1989)
124. Zewail, A.H.: Femtochemistry: atomic-scale dynamics of the chemical bond. J. Phys. Chem. A **104**, 5560–5694 (2000)
125. Vos, M.H., Rappaport, F., Lambry, J.-H., Breton, J., Martin, J.-L.: Visualization of coherent nuclear motion in a membrane protein by femtosecond spectroscopy. Nature **363**, 320–325 (1993)
126. Stanley, R.J., Boxer, S.G.: Oscillations in spontaneous fluorescence from photosynthetic reaction centers. J. Phys. Chem. **99**, 859–863 (1995)
127. Parson, W.W., Warshel, A.: A density-matrix model of photosynthetic electron transfer with microscopically estimated vibrational relaxation times. Chem. Phys. **296**, 201–206 (2004)
128. Parson, W.W., Warshel, A.: Dependence of photosynthetic electron-transfer kinetics on temperature and energy in a density-matrix model. J. Phys. Chem. B **108**, 10474–10483 (2004)

129. Vos, M.H., Martin, J.-L.: Femtosecond processes in proteins. Biochim. Biophys. Acta **1411**, 1–20 (1999)
130. Yakovlev, A.G., Shkuropatov, A.Y., Shuvalov, V.A.: Nuclear wavepacket motion producing a reversible charge separation in bacterial reaction centers. FEBS Lett. **466**, 209–212 (2000)
131. Yakovlev, A.G., Shkuropatov, A.Y., Shuvalov, V.A.: Nuclear wavepacket motion between P* and $P^+B_A{}^-$ potential surfaces with subsequent electron transfer to H_A in bacterial reaction centers. 1. Room temperature. Biochemistry **41**, 2667–2674 (2002)
132. Heller, E.J.: The semiclassical way to molecular spectroscopy. Acc. Chem. Res. **14**, 368–375 (1981)
133. Heller, E.J.: Time-dependent approach to semiclassical dynamics. J. Chem. Phys. **62**, 1544–1555 (1975)
134. Heller, E.J.: Quantum corrections to classical photodissociation models. J. Chem. Phys. **68**, 2066–2075 (1978)
135. Kulander, K.C., Heller, E.J.: Time-dependent formulation of polyatomic photofragmentation. Application to $H_3{}^+$. J. Chem. Phys. **69**, 2439–2449 (1978)
136. Lee, S.Y., Heller, E.J.: Time-dependent theory of Raman scattering. J. Chem. Phys. **71**, 4777–4788 (1979)
137. Myers, A.B., Mathies, R.A., Tannor, D.J., Heller, E.J.: Excited-state geometry changes from pre-resonance Raman intensities. J. Chem. Phys. **77**, 3857–3866 (1982)
138. Myers, A.B., Mathies, R.A.: Resonance Raman intensities: a probe of excited state structure and dynamics. In: Spiro, T.G. (ed.) Biological Applications of Raman Spectroscopy, pp. 1–58. New York, Wiley (1987)
139. Myers, A.B.: "Time-dependent" resonance Raman theory. J. Raman Spectrosc. **28**, 389–401 (1997)

Raman Scattering and Other Multi-photon Processes 12

12.1 Types of Light Scattering

The vibrational transitions discussed in Chap. 6 occur by absorption of a photon whose energy matches a vibrational energy spacing, $h\upsilon$. Vibrational or rotational transitions also can occur when a molecule scatters light of higher frequencies; this is the phenomenon of *Raman scattering*. Raman scattering is one of a group of two-photon processes in which one photon is absorbed and another is emitted essentially simultaneously. Figure 12.1 illustrates the main possibilities. *Rayleigh scattering* is an *elastic* process, in which there is no net transfer of energy between the molecule and the radiation field: the incident and emitted photons have the same energy (Fig. 12.1, transition A). Raman scattering is an *inelastic* process in which the incident and departing photons differ in energy and the molecule is either promoted to a higher vibrational or rotational level of the ground electronic state, or demoted to a lower level. Raman transitions in which the molecule gains vibrational or rotational energy, called *Stokes* Raman scattering (Fig. 12.1, transition B), usually predominate over transitions in which energy is lost (*anti-Stokes* Raman scattering, Fig. 12.1, transition C) because resting molecules populate mainly the lowest levels of any vibrational modes with $h\upsilon > k_BT$. The strength of anti-Stokes scattering increases with temperature, and the ratio of anti-Stokes to Stokes scattering provides a way to measure the effective temperature of a molecule. Both Stokes and anti-Stokes Raman scattering increase greatly in strength if the incident light falls within a molecular absorption band (Fig. 12.1, transition D). The scattering then is termed *resonance Raman* scattering.

There are other types of light scattering that involve transfer of different forms of energy between the molecule and the radiation field. In *Brillouin scattering*, the energy difference between the absorbed and emitted photons creates acoustical waves in the sample; in *quasielastic* or *dynamic light scattering*, the energy goes into small changes in velocity or rotation. In *two-photon absorption*, the second photon is absorbed rather than emitted, leaving the molecule in a excited electronic state whose energy is the sum of the energies of the two photons.

W.W. Parson, *Modern Optical Spectroscopy*, DOI 10.1007/978-3-662-46777-0_12

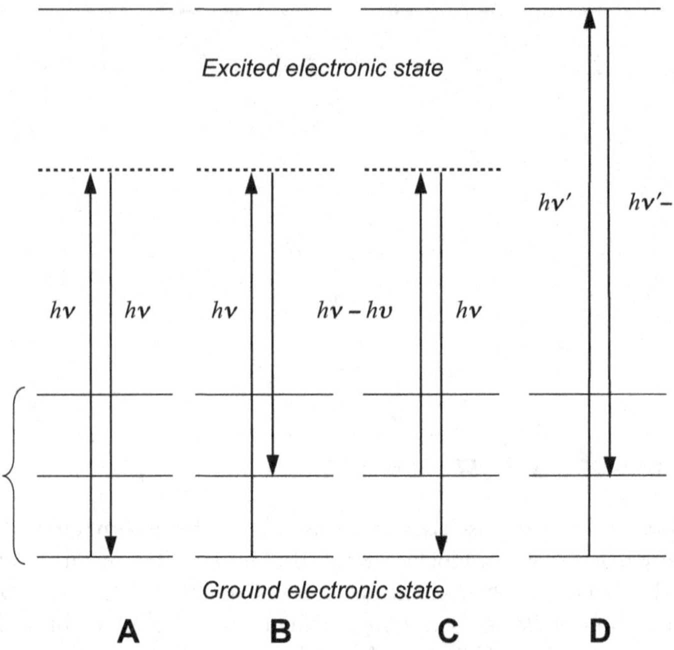

Fig. 12.1 Types of light scattering. The three *solid horizontal lines* at the *bottom* represent vibrational levels of the ground electronic state. In Rayleigh and Mie scattering (**A**) a molecule or particle absorbs a photon and emits a photon with the same energy. Rayleigh scattering refers to scattering by molecules or particles that are small relative to the wavelength of the scattered light; Mie scattering, to scattering by larger particles. In Raman scattering (**B–D**), a photon with a different energy is emitted, leaving the molecule in either a higher or lower vibrational level (Stokes (**B**) and anti-Stokes (**C**) Raman scattering, respectively). Resonance Raman scattering (**D**) occurs if the photon energies match transition energies to a higher electronic state

Raman scattering was discovered in 1928 by the Indian physicist C.V. Raman. It usually is measured by irradiating a sample with a narrow spectral line from a continuous laser, but time-resolved measurements also can be made by using a pulsed laser as the light source. Light scattered at 90° or another convenient angle from the axis of incidence is collected through a monochromator, and the intensity of the signal is plotted as a function of the diference in frequency or wavenumber between the excitation light and the scattered photons ($\nu_e - \nu_s$). The spectrum resembles an IR absorption spectrum (Fig. 12.2). However, the relative intensities of the Raman and IR lines corresponding to various vibrational modes generally differ, as we will discuss below. Resonance Raman spectra of macromolecules also differ from IR spectra and off-resonance Raman spectra in that signals from bound chromophores can be much stronger than the background signals from the protein. In Fig. 12.2D, for example, the resonance Raman spectrum of the chromophore in GFP is readily observable whereas absorption due to the protein would completely overwhelm the IR absorption spectrum of the chromophore. (Also note the higher

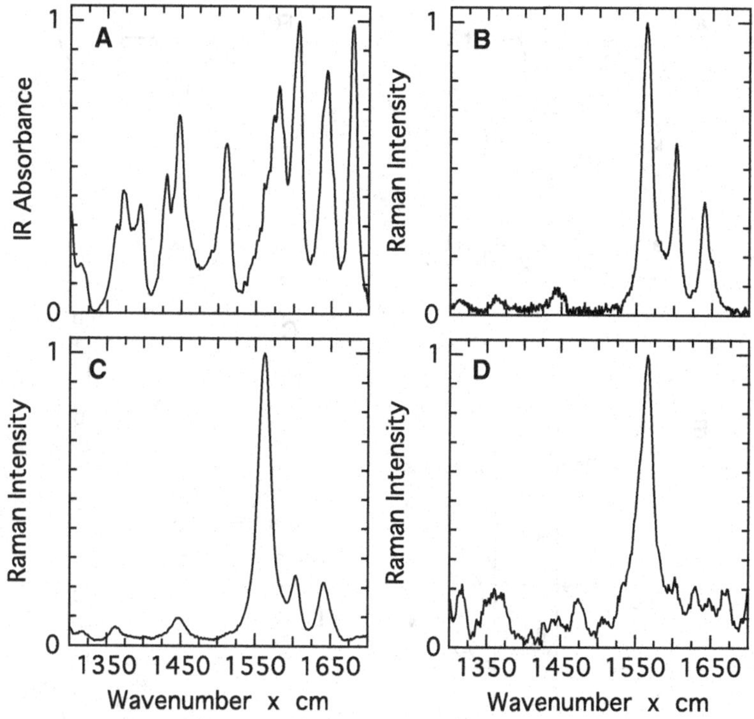

Fig. 12.2 Infrared absorption and Raman scattering spectra of the green fluorescent protein (GFP) and 4-hydroxybenzylidene-2,3-dimethyl-imidazolinone (HBDI, Fig. 5.10C), a model of the GFP chromophore [59, 60]. (**A**) FTIR absorption spectrum of HBDI in a KBr pellet. (**B**) Off-resonance Raman emission spectrum of HBDI in ethanol with excitation at 532.0 nm (18,800 cm^{-1}). (**C**) Resonance Raman emission spectrum of HBDI in ethanol with excitation at 368.9 nm (27,100 cm^{-1}). (**D**) Resonance Raman emission spectrum of GFP in aqueous solution with excitation at 368.9 nm. The abscissa for the Raman spectra is the difference between the wavenumbers of the signal and the excitation light. The amplitudes of the Raman spectra are normalized to the peak near 1,562 cm^{-1}, which is assigned to an in-plane stretching mode of the C=N bond in the imidazolinone ring and the C=C bond between the phenolic and imidazolinone rings [59, 60]. The FTIR spectrum is normalized to the peak at 1,605 cm^{-1}, which represents a mode that is localized mainly to the phenolic ring. HBDI in neutral ethanol has an absorption maximum at 372 nm and GFP has a corresponding absorption band at 398 nm

signal/noise ratio in the resonance Raman spectrum of HDBI in Fig. 12.2C compared to the off-resonance Raman spectrum in Fig. 12.2B.)

The Liouville-space diagrams in Fig. 12.3 help to clarify the main physical distinction between Raman scattering and ordinary fluorescence. Both processes require four interactions with a radiation field, and therefore four steps in Liouville space [1]. There are six possible pathways with four steps between the initial state whose population is indicated by a,a at the lower-left corner of Fig. 12.3A and the final state (b,b) at the upper right: the three paths shown in Fig. 12.3B–D and their complex conjugates. Ordinary fluorescence occurs by paths B and C, whereas

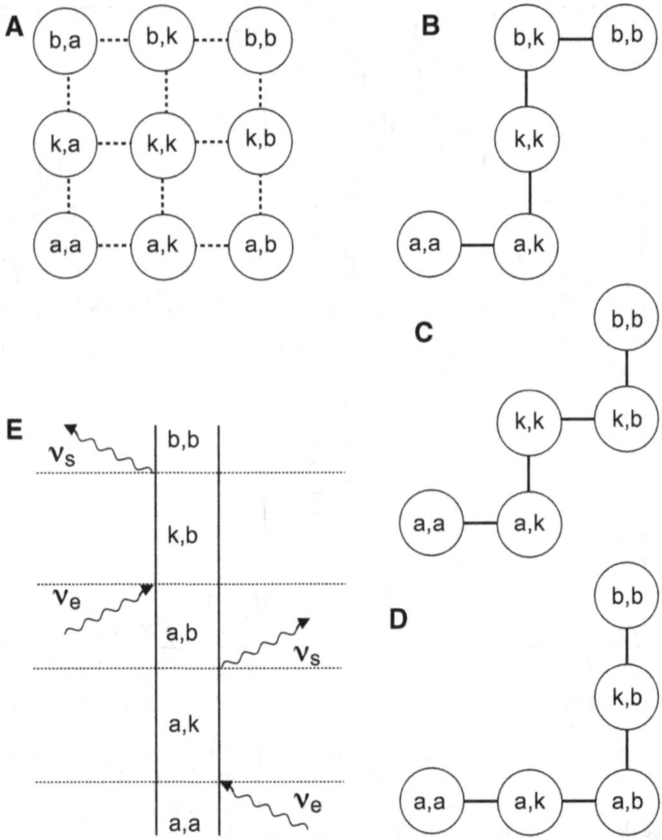

Fig. 12.3 Liouville-space diagrams for spontaneous fluorescence and Raman scattering. (**A**) Liouville-space pathways connecting an initial state (*a*), intermediate state (*k*) and a final state (*b*). (See Sect. 11.1, Figs. 11.1 and 11.4 for an explanation of these diagrams.) (**B–D**) Three of the six possible paths from *a* to *b* with four steps (four interactions with a radiation field). The other three paths are the complex conjugates of the ones shown. All six paths contribute to spontaneous fluorescence; Raman scattering involves only path (**D**) (and its complex conjugate), in which the intermediate state is never populated. (**E**) A double-sided Feynman diagram for path (**D**)

Raman scattering and the other two-photon processes discussed in this chapter occur by path D. Inspection of the Liouville diagrams shows that paths B and C proceed through an intermediate state (*k*) that is transiently populated and thus is, in principle, measureable. Path D passes through two coherences with this state (*a,k* and *k,b*) and a coherence between the initial and final states (*a,b*), but never generates a population in state *k*. Raman scattering thus differs from fluorescence in involving only a "virtual" intermediate state that is not directly measureable. A double-sided Feynman diagram for path D is shown in Fig. 12.3E.

Experimentally, Raman scattering differs from fluorescence in several ways. First, Raman emission lines are much narrower than fluorescence emission spectra.

Raman lines for small molecules in solution typically have widths on the order of 10 cm^{-1} as compared to several hundred cm^{-1} for fluorescence. Second, the emission spectra have very different dependences on the frequency of the excitation light. Fluorescence emission spectra of many molecules are essentially independent of the excitation frequency, whereas Raman lines shift linearly with ν_e so as to maintain a constant value of $|\nu_s - \nu_e|$. This reflects the requirement for overall conservation of energy during the Raman process ($\nu_s + \upsilon = \nu_e$ for Stokes Raman scattering, $\nu_s - \upsilon = \nu_e$ for anti-Stokes), in which the virtual intermediate has no opportunity to equilibrate thermally with the surroundings. Whereas spontaneous fluorescence typically has a lifetime of several ns, Raman scattering follows the time course of an excitation pulse with essentially no delay. Finally, the integrated strength of off-resonance Raman scattering usually is much lower than that of fluorescence.

In the classical explanation of Raman scattering, the incident electromagnetic radiation field $\mathbf{E}_o\cos\omega_e$ creates an oscillating induced dipole whose magnitude depends on the product of the field and the polarizability of the medium. The induced dipole constitutes the source of the radiation we detect as scattered light. If the polarizability (α) is modulated at a lower frequency by a molecular vibration ($\alpha = \alpha_o + \alpha_1\cos\omega_m$), the induced dipole will be proportional to the product $\mathbf{E}_o \cos\omega_e(\alpha_o + \alpha_1 \cos\omega_m)$, which is the same as $\mathbf{E}_o\alpha_o\cos(\omega_e) + (\mathbf{E}_o\alpha_1/2)[\cos(\omega_e + \omega_m) + \cos(\omega_e - \omega_m)]$. The scattering thus will have components at frequencies $\omega_e \pm \omega_m$ in addition to ω_e. The classical theory predicts correctly that the strength of Raman scattering depends on the extent to which a vibration changes the molecular polarizability as we discuss in Sect. 12.4, although it does not account readily for the difference between the strengths of Stokes and anti-Stokes scattering.

Kramers and Heisenberg [2], who predicted the phenomenon of Raman scattering several years before Raman discovered it experimentally, advanced a semiclassical theory in which they treated the scattering molecule quantum mechanically and the radiation field classically. Dirac [3] soon extended the theory to include quantization of the radiation field, and Placzec, Albrecht and others explored the selection rules for molecules with various symmetries [4, 5]. A theory of the resonance Raman effect based on vibrational wavepackets was developed by Heller, Mathies, Meyers and their colleagues [6–11]. Mukamel [1, 12] presented a comprehensive theory that considered the nonlinear response functions for pathways in Liouville space. Having briefly described the pertinent pathways in Liouville space above, we will first develop the Kramers-Heisenberg-Dirac theory by a second-order perturbation approach, and then turn to the wavepacket picture.

12.2 The Kramers-Heisenberg-Dirac Theory

Consider a molecule with ground-state wavefunction Ψ_a and excited-state wavefunction Ψ_k, and energies E_a and E_k. When a weak, continuous radiation field with frequency ν_e and amplitude $\mathbf{E}_e[\exp(2\pi i\nu_e t) + \exp(-2\pi i\nu_e t)]$ is introduced, the coefficient for state k (C_k) oscillates with time. We need an expression for C_k

that incorporates uncertainty in the energies caused by electronic dephasing or decay of state k. We can find C_k at a short time (τ) by evaluating the density matrix element $\overline{\rho}_{ka}(\tau)$ for an ensemble of molecules exposed to steady-state illumination. Recall that, in the Schrödinger representation,

$$\overline{\rho}_{ka}(\tau) = \overline{c_k(\tau)c_a^*(\tau)} = \overline{C_k(\tau)C_a^*(\tau)\exp[-i(E_k - E_a)\tau/\hbar]}, \qquad (12.1)$$

where the bars indicate averaging over the ensemble (Eqs. 10.8 and 10.12). If we replace $\overline{c_a^*(\tau)}$ by 1 on the assumption that virtually all the molecules are in the ground state, then

$$\overline{C}_k(\tau) = \overline{\rho}_{ka}(\tau)\exp[i(E_k - E_a)\tau/\hbar]. \qquad (12.2)$$

We found in Chap. 10 that the steady-state value of $\overline{\rho}_{ka}$ can be written

$$\overline{\rho}_{ka} \approx \overline{\overline{\rho_{ka}}}\exp(-i\omega_e t), \qquad (12.3)$$

with

$$\overline{\overline{\rho_{ka}}} = \frac{(i/\hbar)(\overline{\rho_{kk}} - \overline{\rho_{aa}})\boldsymbol{\mu}_{ka} \cdot \boldsymbol{E}_e}{i(\omega_e - \omega_{ak}) + 1/T_2} \qquad (12.4a)$$

$$\approx \frac{\boldsymbol{\mu}_{ka} \cdot \boldsymbol{E}_e}{E_k - E_a - h\nu_e - i\hbar/T_2}, \qquad (12.4b)$$

where $\boldsymbol{\mu}_{ka}$ is the transition dipole for absorption and T_2 is the time constant for decay of electronic coherence between states a and k (Eqs. 10.40 and 10.42b). We have dropped the term with $+h\nu_e$ in place of $-h\nu_e$, which is negligible when $E_k > E_a$, and in Eq. (12.4b) we have again set $\overline{\rho_{aa}} \approx 1$ and $\overline{\rho_{kk}} \approx 0$. Combining Eqs. (12.2–12.4b), and omitting the bar over C_k to simplify the notation gives

$$C_k(\tau) = \frac{\boldsymbol{\mu}_{ka} \cdot \boldsymbol{E}_e}{E_k - E_a - h\nu_e - i\hbar/T_2}\exp[-i(E_k - E_a - h\nu_e)\tau/\hbar]. \qquad (12.5)$$

Now suppose that a second radiation field, $\boldsymbol{E}_s[\exp(2\pi i\nu_s t) + \exp(-2\pi i\nu_s t)]$, couples state k to some other state, b. This could be either the radiation created by spontaneous fluorescence of state k, or an incident field that might cause stimulated emission (Sect. 12.6) or excite the molecule to a state with higher energy (Sect. 12.7). However, we assume that neither the first nor the second radiation field can, by itself, convert state a directly to b. Neglecting any population of state b that is present at zero time, we can find how the coefficient for state b (C_b) grows with time by continuing the same perturbation treatment that we used to find $C_k(\tau)$. But here we'll retain terms with either $+h\nu_s$ or $-h\nu_s$ in the denominator, so that E_b can be either greater or less than E_k. By applying Eq. (4.6b) to the transition from k to b, and considering a time t that is short enough so that C_b is small and C_a is still close to 1, we obtain

$$C_b(t) = \frac{i}{\hbar}(\boldsymbol{\mu}_{bk} \cdot \boldsymbol{E}_s) \int_0^t \{\exp[i(E_b - E_k + h\nu_s)\tau/\hbar] + \exp[i(E_b - E_k - h\nu_s)\tau/\hbar]\} C_k(\tau)\ d\tau$$
$$= \alpha_{ba} \left\{ \int_0^t \exp[i(E_b - E_a + h\nu_s - h\nu_e)\tau/\hbar] d\tau + \int_0^t \exp[i(E_b - E_a + h\nu_s - h\nu_e)\tau/\hbar] d\tau \right\}, \tag{12.6}$$

with

$$\alpha_{ba} = \frac{(\boldsymbol{\mu}_{bk} \cdot \boldsymbol{E}_s)(\boldsymbol{\mu}_{ka} \cdot \boldsymbol{E}_e)}{E_k - E_a - h\nu_e - i\hbar/T_2}. \tag{12.7}$$

Evaluating the integrals in Eq. (12.6) gives

$$C_b(t) = \alpha_{ba} \left[\frac{\exp[i(E_b - E_a + h\nu_s - h\nu_e)t/\hbar] - 1}{E_b - E_a + h\nu_s - h\nu_e} + \frac{\exp[i(E_b - E_a - h\nu_s - h\nu_e)t/\hbar] - 1}{E_b - E_a - h\nu_s - h\nu_e} \right]. \tag{12.8}$$

The first term in the brackets on the right-hand side of Eq. (12.8) accounts for Rayleigh and Raman scattering; the second accounts for two-photon absorption, which we'll discuss in Sect. 12.7. If $E_b = E_a$, as is the case for Rayleigh scattering, the first term goes to $it/\hbar$ when $\nu_s = \nu_e$ (Box 4.3), while the second term is negligible for any positive values of ν_e and ν_s. If states a and b are different, as they are in Raman scattering, the first term goes to $it/\hbar$ when $E_b - E_a = h\nu_e - h\nu_s$. Finally, if $E_b \gg E_a$, the first term in the braces usually is small but the second term goes to $it/\hbar$ when $E_b - E_a \approx h\nu_e + h\nu_s$.

The intensity of Rayleigh or Raman scattering of nearly monochromatic light should be proportional to the integral of $C_b^* C_b$ over a narrow band of the frequency difference $\nu_e - \nu_s$ (Eq. 4.8). If we perform this integration for the first term in the braces in Eq. (12.8), the rate of scattering becomes

$$S_{ba} = |\alpha_{ba}|^2 \frac{\rho_\nu(\nu)}{\hbar^2}, \tag{12.9}$$

where α_{ba} is given by Eq. (12.7), and $\rho_\nu(\nu)$ is the number of radiation modes that meet the condition $h\nu_e - h\nu_s = E_b - E_a$.

To this point, we have considered only a single intermediate state (k) between states a and b. A molecule generally will have many excited electronic states, each with many vibrational levels, and any of these vibronic states could serve as a virtual state for Rayleigh or Raman scattering. If we assume for simplicity that the dephasing time constant T_2 is approximately the same for all the important vibronic levels (clearly a significant approximation), then summing the contributions to $C_b(t)$ gives

$$C_b(t) \approx \left(\frac{\exp[i(E_b - E_a + h\Delta\nu)t/\hbar] - 1}{E_b - E_a + h\Delta\nu}\right)\sum_k \frac{(\boldsymbol{\mu}_{bk}\cdot\mathbf{E}_s)(\boldsymbol{\mu}_{ka}\cdot\mathbf{E}_e)}{E_k - E_a - h\nu_e - i\hbar/T_2}, \quad (12.10)$$

where $\Delta\nu = \nu_s - \nu_e$ and we have again retained only the first term from the braces in Eq. (12.8). The matrix element for light scattering thus becomes

$$\alpha_{ba} = \sum_k \frac{(\boldsymbol{\mu}_{bk}\cdot\mathbf{E}_s)(\boldsymbol{\mu}_{ka}\cdot\mathbf{E}_e)}{E_k - E_a - h\nu_e - i\hbar/T_2}, \quad (12.11)$$

or for the particular case of Rayleigh scattering (states a and b the same and $\boldsymbol{\mu}_{ka} = -\boldsymbol{\mu}_{ak}$),

$$\alpha_{aa} = \sum_k \frac{(\boldsymbol{\mu}_{ka}\cdot\mathbf{E}_s)(\boldsymbol{\mu}_{ka}\cdot\mathbf{E}_e)}{E_k - E_a - h\nu_e - i\hbar/T_2}. \quad (12.12)$$

Resonance Raman scattering occurs when the incident light falls within an absorption band, so that $E_{k(e)} - E_{a(g)} \approx h\nu_e$ for a set of vibronic levels (k) of the excited electronic state. Because the vibronic levels of this state will dominate the sum in Eq. (12.11), we can use the Born-Oppenheimer and Condon approximations (Sect. 4.10) to factor the transition dipole $\boldsymbol{\mu}_{ka}$ into a vibrational overlap integral ($\langle X_{k(e)}|X_{a(g)}\rangle$) and an electronic transition dipole that is averaged over the nuclear coordinates ($\boldsymbol{\mu}_{eg}$). Pulling the electronic transition dipoles out of the sum then yields

$$\alpha_{ba} \approx \left(\boldsymbol{\mu}_{ge}\cdot\mathbf{E}_s\right)\left(\boldsymbol{\mu}_{eg}\cdot\mathbf{E}_e\right)\sum_k \frac{\langle X_{b(g)}|X_{k(e)}\rangle\langle X_{k(e)}|X_{a(g)}\rangle}{E_{k(e)} - E_{a(g)} - h\nu_e - i\hbar/T_2}, \quad (12.13)$$

Where $\langle X_i|X_j\rangle$ is the vibrational overlap integral for states i and j.

Equations (12.9) and (12.12) indicate that the strength of Rayleigh scattering depends on $|\boldsymbol{\mu}_{ka}\cdot\mathbf{E}_e|^2|\boldsymbol{\mu}_{ka}\cdot\mathbf{E}_s|^2$, where $\mathbf{E}_e$ again is the excitation field and $\mathbf{E}_s$ is the field of spontaneous emission from the virtual excited state. From Chap. 4, we know that $|\boldsymbol{\mu}_{ka}\cdot\mathbf{E}_e|^2\rho_\nu(\nu) = (2\pi f^2/3cn)D_{ka}I$, where D_{ka}, I, n and f are, respectively, the dipole strength for excitation to state k, intensity of the incident light, refractive index and local-field correction factor (Eq. 4.12). From Eq. (5.12), $|\boldsymbol{\mu}_{ka}\cdot\mathbf{E}_s|^2 = (8\pi h n^3 n_s^3/c^3)|\boldsymbol{\mu}_{ka}\cdot\mathbf{E}_e|^2$. Because $\nu_e = \nu_s = \nu$ for Rayleigh scattering, the strength of Rayleigh scattering in photons/s is proportional to $I\nu^3$. Converting to units of energy/s gives an additional factor of $h\nu$. The Kramers-Heisenberg-Dirac theory thus reproduces the observed dependence of Rayleigh scattering on the fourth power of the frequency. The theory also predicts the polarization of the scattered light correctly. (See Sect. 12.9, [11] and [13] for further discussion of directional aspects of light scattering.)

The matrix elements given by Eqs. (12.11–12.13) also depend on a sum of products of weighted overlap integrals of vibrational level k of the excited electronic state with the initial and final levels of the ground state. Each term in the sum is weighted inversely by $\left(E_{k(e)} - E_{a(g)} - h\nu_e - i\hbar/T_2\right)$. But note that we have considered only a single vibrational level of the initial system. In a more complete description, $\langle X_{k(e)}|X_{a(g)}\rangle$ is replaced by a sum of thermally-weighted products of overlap integrals as described in Box 4.14.

The dephasing factor $i\hbar/T_2$ in Eqs. (12.11–12.13) makes the matrix elements for Rayleigh and Raman scattering complex quantities. The real and imaginary parts can be separated by multiplying each term in the sum by 1 in the form of $\left(E_{k(e)} - E_{a(g)} - h\nu_e - i\hbar/T_2\right)/\left(E_{k(e)} - E_{a(g)} - h\nu_e - i\hbar/T_2\right)$. Dissecting Eq. (12.13) in this way gives

$$\alpha_{ba} = \left|\boldsymbol{\mu}_{eg}\right|^2 \sum_k \left[\frac{\left(E_{k(e)} - E_{a(g)} - h\nu_e\right)\langle X_{b(g)}|X_{k(e)}\rangle\langle X_{k(e)}|X_{a(g)}\rangle}{\left(E_{k(e)} - E_{a(g)} - h\nu_e\right)^2 + (\hbar/T_2)^2} + i\frac{(\hbar/T_2)\langle X_{b(g)}|X_{k(e)}\rangle\langle X_{k(e)}|X_{a(g)}\rangle}{\left(E_{k(e)} - E_{a(g)} - h\nu_e\right)^2 + (\hbar/T_2)^2} \right]. \tag{12.14}$$

A comparison of Eq. (12.14) with Eq. (10.43) shows that the imaginary part of α_{aa} (the matrix element for resonance Rayleigh scattering) is proportional to the matrix element for ordinary absorption. The real part of α_{aa} can be related to the refractive index (Box 3.3 and [14]).

Figure 12.4 shows spectra of $|\alpha_{ba}|^2$ for a resonance Raman transition between vibrational levels 0 and 1 of the ground electronic state, as calculated by Eq. (12.13) for a molecule with a single harmonic vibrational mode. The spectra are plotted as functions of the excitation frequency (ν_e) for several values of T_2 and the displacement of the vibrational coordinate in the excited state (Δ). Note that these are excitation spectra for resonance Raman scattering, not plots of the emission intensity as a function of $\nu_e - \nu_s$ (*cf.* Fig. 12.2), and note also that they do not consider inhomogeneous broadening. For comparison, the figure also shows the homogeneous absorption spectra calculated as Im(α_{aa}) for the same systems.

Spectra of $|\alpha_{ba}|^2$ resemble a homogeneous absorption spectrum in having peaks at the 0-0 transition frequency and at integer multiples of υ above this, where the excitation energy matches the energy difference between the ground and excited vibronic states. However, the relative heights of the peaks differ. Note, for example, that the peak at $(\nu_e - \nu_{oo})/\upsilon = 2$ is missing entirely in the $|\alpha_{ba}|^2$ spectrum when $\Delta = 2$ (Fig. 12.4C, D), but not when $\Delta = 1$ (Fig. 12.4A, B). In addition, whereas changing Δ redistributes the ordinary absorption among the vibronic peaks without

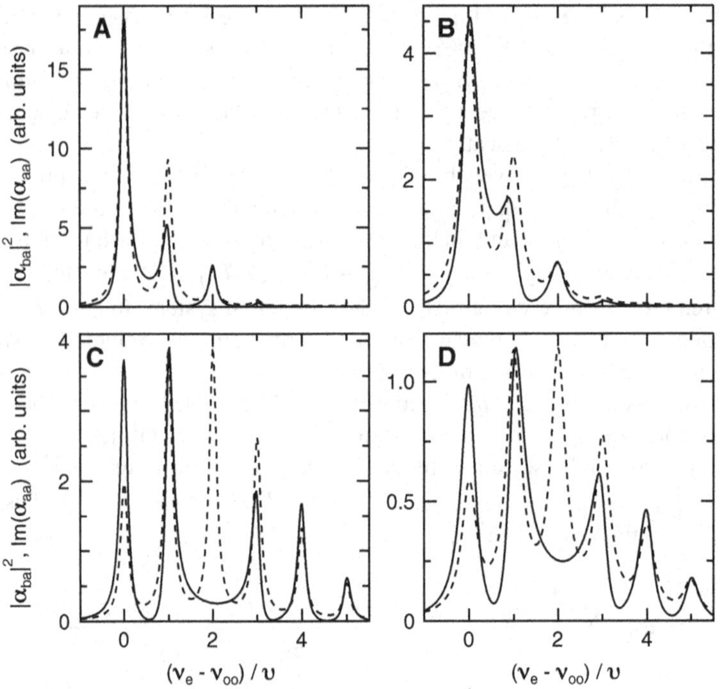

Fig. 12.4 Calculated resonance Raman excitation spectrum ($|\alpha_{ba}|^2$, *solid curves*) and homogeneous absorption spectrum (Im(α_{aa}), *dashed curves*) for a molecule with a single harmonic vibrational mode of frequency υ. α_{aa} is for scattering at the excitation frequency, and α_{ba} for scattering from the zero-point vibrational level into the first excited level ($h\nu_1 - h\nu_2 = h\upsilon$). The abscissa is the difference between the excitation frequency (ν_e) and the 0-0 transition frequency (ν_{oo}), in units of the vibrational frequency (υ). The absorption spectra are normalized to the Raman excitation spectra at the highest peaks in each panel. (**A**) $\hbar/T_2 = 0.1h\upsilon$ and Δ (dimensionless displacement of the vibrational coordinate in the excited electronic state) = 1.0. (**B**) $\hbar/T_2 = 0.2h\upsilon$, $\Delta = 1.0$. (**C**) $\hbar/T_2 = 0.1h\upsilon$, $\Delta = 2.0$. (**D**) $\hbar/T_2 = 0.2h\upsilon$, $\Delta = 2.0$. All vibrational levels of the excited state up to $k = 25$ were included in the sums. The overlap integrals were calculated as explained in Box 4.13

altering the integrated absorbance, it affects the integrated strength of Raman scattering. This point is illustrated in Fig. 12.5A, where the integrated Raman scattering cross section, $\int |\alpha_{ba}|^2 d\nu$, is plotted as a function of Δ. Scattering into the first excited vibrational level peaks at $\Delta \approx 0.9$.

The dephasing time constant T_2 also has different effects on Raman and absorption spectra. An integrated absorption spectrum is independent of $\hbar/T_2$, whereas the integrated strength of Raman scattering decreases as the dephasing becomes faster (Fig. 12.5B). Comparisons of the absolute cross sections for Raman scattering and ordinary absorption thus provide a way to measure the dynamics of dephasing [9, 10].

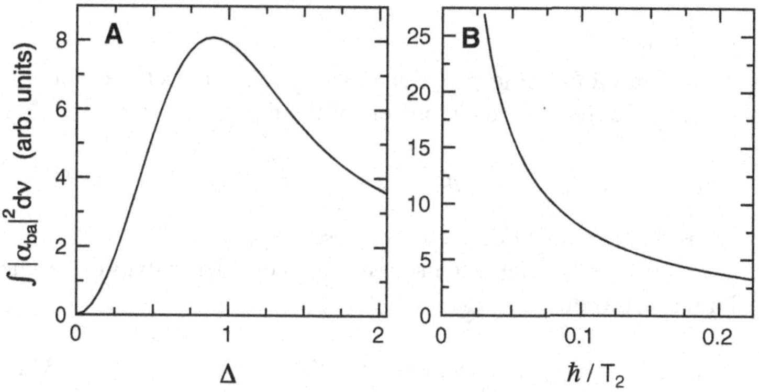

Fig. 12.5 Integrated Raman scattering excitation cross section, $\int |\alpha_{ba}|^2 dv$, for a molecule with a single harmonic vibrational mode with frequency υ. The model system is the same as in Fig. 12.4. (**A**) The integrated cross section as a function of Δ with $\hbar/T_2$ fixed at $0.1h\upsilon$. (**B**) The integrated cross section as a function of $\hbar/T_2$ with Δ fixed at 1.0

In the classical picture of light scattering, light passing through a polarizable medium generates oscillating induced electric dipoles that then can radiate light in various directions. The intensity of the scattering depends on the square of the induced dipole moment, and thus on the square of the polarizability. Box 12.1 describes a quantum mechanical treatment of electronic polarizability and shows that the matrix elements for light scattering are indeed proportional to the matrix elements for polarizability.

Box 12.1 Quantum Theory of Electronic Polarizability

Electronic polarizability generally is described by a second-rank tensor, $\boldsymbol{\alpha}$, which means that applying an electric field along a particular axis can generate an induced dipole with components perpendicular, as well as parallel, to this axis. However, it always is possible to choose a molecular coordinate system in which $\boldsymbol{\alpha}$ is diagonal, so that the polarizability can be described by a vector. Applying a field along the x, y or z axis of this coordinate system generates an induced dipole only along the same axis. These are called the "principal axes" of the polarizability. For an isotropic sample, the magnitude of the induced dipole is proportional to the scalar quantity $\alpha = (1/3)\mathrm{Tr}(\boldsymbol{\alpha})$, which does not depend on the choice of the coordinate system.

In a quantum mechanical picture, polarizability reflects mixing of the ground state with higher-energy states when a molecule is perturbed by an external electric field. We'll describe the theory first for a static field and then consider a time-dependent field.

(continued)

Box 12.1 (continued)

In the presence of a static field $\boldsymbol{E}$, a molecule's total electric dipole can be expressed as a Taylor's series in powers of the field:

$$\boldsymbol{\mu} = \boldsymbol{\mu}_0 + \boldsymbol{\mu}_{\text{ind}} = \boldsymbol{\mu}_0 + \boldsymbol{\alpha} \cdot \boldsymbol{E} + \ldots, \tag{B12.1.1}$$

where $\boldsymbol{\mu}_0$ is the permanent dipole in the absence of the field.

$\boldsymbol{\mu}_0$ and $\boldsymbol{\alpha}$ also can be related to the first and second derivatives of the energy (E) with respect to the field:

$$\boldsymbol{\mu}_0 = -\partial E / \partial \boldsymbol{E}, \tag{B12.1.2}$$

and

$$\boldsymbol{\alpha} = -\partial^2 E / \partial \boldsymbol{E}^2. \tag{B12.1.3}$$

Equation (B12.1.2) is the same as Eq. (4.2). Equations (B12.1.2) and (B12.1.3) follow from a general theorem called the *Hellman-Feynman theorem*, which says that the derivative of the energy with respect to any parameter is equal to the expectation value of the derivative of the Hamiltonian with respect to that parameter. In the present case, the Hellman-Feynman theorem informs us that, for a system with wavefunction Ψ and eigenvalue E, $\partial E / \partial \boldsymbol{E} = \langle \Psi | \partial H / \partial \boldsymbol{E} | \Psi \rangle = -\langle \Psi | \tilde{\mu} | \Psi \rangle = -\boldsymbol{\mu}$. To use this relationship, we first expand the energy as a Taylor's series in powers of $\boldsymbol{E}$:

$$E = E_0 + (\partial E / \partial \boldsymbol{E})_0 \boldsymbol{E} + \frac{1}{2} \left(\partial^2 E / \partial \boldsymbol{E}^2 \right)_0 \boldsymbol{E}^2 + \ldots, \tag{B12.1.4}$$

where all the derivatives are evaluated at $\boldsymbol{E} = (0,0,0)$. Differentiating Eq. (B12.1.4) with respect to $\boldsymbol{E}$ gives, according to the Hellman-Feynman theorem,

$$\boldsymbol{\mu} = -(\partial E / \partial \boldsymbol{E})_0 - \left(\partial^2 E / \partial \boldsymbol{E}^2 \right)_0 \boldsymbol{E} - \ldots, \tag{B12.1.5}$$

Equating terms with the same powers of $\boldsymbol{E}$ in Eqs. (B12.1.1) and (B12.1.5) then yields Eqs. (B12.1.2) and (B12.1.3).

Now consider a molecule with eigenfunctions Ψ_k in the absence of external fields. The wavefunction in the presence of a static field can be written as a linear combination of these basis functions:

(continued)

Box 12.1 (continued)

$$\Psi = \sum_k C_k \Psi_k. \tag{B12.1.6}$$

To find α_x by Eq. (B12.1.3), we must evaluate how the energy of this superposition state depends on the field. We can do this by following the procedure we have used to find how two states with diabatic energies E_a and E_k are mixed by a weak perturbation $\left(\widetilde{\mathrm{H}}'\right)$. Let state a be the ground state, and assume that the coefficient for any higher-energy state k is much smaller than that of state a ($0 < |C_k| \ll |C_a| \approx 1$). The energy of the system (E) then will be approximately the same as that of state a, so that from Eqs. (8.7a, b) we have

$$C_k = -C_a H_{ka}/(H_{kk} - E) \approx -H_{ka}/(H_{kk} - H_{aa}), \tag{B12.1.7}$$

and

$$\begin{aligned} E &= (C_a H_{aa} + C_k H_{ak})/C_a \approx H_{aa} - H_{ak} H_{ka}/(H_{kk} - H_{aa}) \\ &= E_a + H_{aa} - H_{ak} H_{ka}/(H_{kk} - H_{aa}), \end{aligned} \tag{B12.1.8}$$

where H_{aa}, H_{kk}, H_{ak} and H_{ka} are defined as in Eq. (8.8).

For the problem at hand, the perturbation operator $\widetilde{\mathrm{H}}'$ is the dipole operator $\tilde{\mu}$. Summing over all the higher-energy states, and letting $\boldsymbol{\mu}_{ij} = \langle \Psi_i | \tilde{\mu} | \Psi_j \rangle$ as usual, the energy of the system in the presence of the external field is

$$\begin{aligned} E &\approx E_a - \boldsymbol{\mu}_{aa} \cdot \boldsymbol{E} - \sum_{k \neq a} \frac{(\boldsymbol{\mu}_{ak} \cdot \boldsymbol{E})(\boldsymbol{\mu}_{ka} \cdot \boldsymbol{E})}{E_k - E_a} \\ &= E_a - \boldsymbol{\mu}_{aa} \cdot \boldsymbol{E} - \sum_{k \neq a} \frac{(\boldsymbol{\mu}_{ka} \cdot \boldsymbol{E})^2}{E_k - E_a}. \end{aligned} \tag{B12.1.9}$$

Taking the second derivative of the energy with respect to $\boldsymbol{E}$ gives, according to Eq. (B12.1.3),

$$\alpha \approx 2 \sum_{k \neq a} \frac{|\boldsymbol{\mu}_{ka}|^2}{E_k - E_a}. \tag{B12.1.10}$$

This derivation does not hold for a field that oscillates rapidly with time. In such a field, the induced dipole oscillates and the amplitude of these oscillations depends on the frequency. We can, however, define a frequency-dependent dynamic polarizability, or *molecular electric*

(continued)

Box 12.1 (continued)

susceptibility, as the ratio of the amplitudes of the two oscillations (Sect. 3.1.5). Suppose an ensemble of molecules in the ground state (a) is exposed to an oscillating field. If we write the density matrix $\overline{\boldsymbol{\rho}}$ with a basis of ground and excited-state wavefunctions, the expectation value of the total dipole is given by Eqs. (10.13) and (10.14):

$$\langle \boldsymbol{\mu}(t) \rangle = Tr(\overline{\boldsymbol{\rho}}(t)\boldsymbol{\mu}) = \sum_j \sum_k \overline{\rho}_{jk}(t)\boldsymbol{\mu}_{kj}$$
$$\approx \sum_k (\overline{\rho}_{ak}(t)\boldsymbol{\mu}_{ka} + \overline{\rho}_{ka}(t)\boldsymbol{\mu}_{ak}) = \boldsymbol{\mu}_{aa} + \sum_{k \neq a} (\overline{\rho}_{ak}(t)\boldsymbol{\mu}_{ka} + \overline{\rho}_{ka}(t)\boldsymbol{\mu}_{ak}). \tag{B12.1.11}$$

Here $\boldsymbol{\mu}_{aa}$ is the permanent dipole moment in the ground state, and the sum over k represents the induced dipole, $\boldsymbol{\mu}_{ind}$. If we use steady-state expressions for the density matrix elements (Eqs. 10.42a, b, 12.3, 12.4a, and 12.4b) in a field $\boldsymbol{E}_o[\exp(2\pi i\nu t) + \exp(-2\pi i\nu t)]$, and neglect terms that are more than first-order in the field strength, the induced dipole is

$$\langle \boldsymbol{\mu}_{ind}(t) \rangle = \sum_k \left[\frac{\boldsymbol{\mu}_{ak}\boldsymbol{\mu}_{ka} \cdot \boldsymbol{E}_0 \exp(-2\pi i\nu t)}{E_k - E_a - h\nu - i\hbar/T_2} + \frac{\boldsymbol{\mu}_{ka}\boldsymbol{\mu}_{ak} \cdot \boldsymbol{E}_0 \exp(2\pi i\nu t)}{E_k - E_a - h\nu + i\hbar/T_2} \right]. \tag{B12.1.12}$$

Because both the permanent and the induced components of $\langle \boldsymbol{\mu} \rangle$ are real, the polarizability operator must be Hermitian. This will be the case if we define the polarizability operator so that

$$\mu_{ind}(t) = a_{aa}\boldsymbol{E}_0 \exp(2\pi i\nu t) + a_{aa}{}^* \boldsymbol{E}_0 \exp(-2\pi i\nu t). \tag{B12.1.13}$$

A comparison of Eqs. (B12.1.1), (B12.1.12), and (B12.1.13) with Eq. (12.11) shows that the matrix elements of the dynamic polarizability defined in this way are the same as the matrix elements for Rayleigh scattering.

12.3 The Wavepacket Picture of Resonance Raman Scattering

Although evaluating the matrix element for resonance Raman scattering by Eq. (12.13) is staightforward for a molecule with only one or two vibrational modes, it rapidly becomes intractable for larger molecules, and a wavepacket treatment similar to the one described for absorption in Chap. 11 becomes increasingly useful. To recast Eq. (12.13) in a time-dependent form, we first note that the factor

$1/\left(E_{k(e)} - E_{a(g)} - h\nu_e - i\hbar/T_2\right)$ that appears in each term of the sum is $(2\pi)^{1/2}i$ times the half Fourier transform of the function $\exp\left[-i\left(E_{k(e)} - E_{a(g)}t/\hbar - \hbar t/T_2\right]\right.$:

$$\sqrt{2\pi}i\frac{1}{\sqrt{2\pi}}\int_0^\infty \left\{\exp\left[-i\left(E_{k(e)} - E_{a(g)}\right)t/\hbar - \hbar t/T_2\right]\right\}\exp(2\pi i\nu_e t)dt$$
$$= i\int_0^\infty \exp\left[-i\left(E_{k(e)} - E_{a(g)} - h\nu_e\right)/\hbar - \hbar t/T_2\right]dt = 1/\left(E_{k(e)} - E_{a(g)} - h\nu_e - i\hbar/T_2\right). \quad (12.15)$$

This function has the same form as the function we used in Sect. 2.5 to represent a wavefunction that decays exponentially with time (Eqs. 2.66–2.67 and Appendix A3). The decay time constant T in Eqs. (2.66) and (2.67) corresponds to $T_2/2\hbar$ in Eq. (12.15).

Let's now construct an excited-state wavepacket $X(t)$ as in Eqs. (11.46) and (11.47) but with a dephasing time constant of T_2, so that

$$\langle X_{b(g)}|X(t)\rangle = \left\langle X_{b(g)}\left|\sum_k C_k X_{k(e)}\exp\left(-iE_{k(e)}t/\hbar - \hbar t/T_2\right)\right.\right\rangle$$
$$= \sum_k \langle X_{b(g)}|X_{k(e)}\rangle\langle X_{k(e)}|X_{a(g)}\rangle\exp\left(-iE_{k(e)}t/\hbar - \hbar t/T_2\right). \quad (12.16)$$

Note that the products of vibrational overlap integrals in this expression are the same as those in Eq. (12.13).

Combining Eqs. (12.13), (12.15) and (12.16) gives

$$\alpha_{ba} = |\boldsymbol{\mu}_{eg}|^2\sum_k \frac{\langle X_{b(g)}|X_{k(e)}\rangle\langle X_{k(e)}|X_{a(g)}\rangle}{E_{k(e)} - E_{a(g)} - h\nu_e - i\hbar/T_2}$$
$$= i|\boldsymbol{\mu}_{eg}|^2\int_0^\infty \langle X_{b(g)}|X(t)\rangle\exp\left[i\left(E_{a(g)} + h\nu_e\right)t/\hbar\right]dt. \quad (12.17)$$

The matrix element for resonance Raman scattering thus is proportional to a half-Fourier transform of the overlap of the final vibrational wavefunction ($X_{b(g)}$) with the time-dependent wavepacket $X(t)$ created by exciting the molecule with white light in the ground state. See [6] and [8] for more complete proofs of this relationship, and [10] for a review of some of its extensions and applications. The resonance Raman excitation spectrum is proportional to $|\alpha_{ba}|^2$, as explained above.

Because the various vibrational wavefunctions of the ground electronic state are orthogonal, the overlap of $X_{b(g)}$ with the excited-state wavepacket $X(t)$ is zero at $t = 0$, when the wavepacket is simply a vertical projection of $X_{a(g)}$ onto the excited-state surface. The overlap grows with time as the wavepacket moves away from its origin (Fig. 12.6). This description provides a new perspective on the sensitivity of the

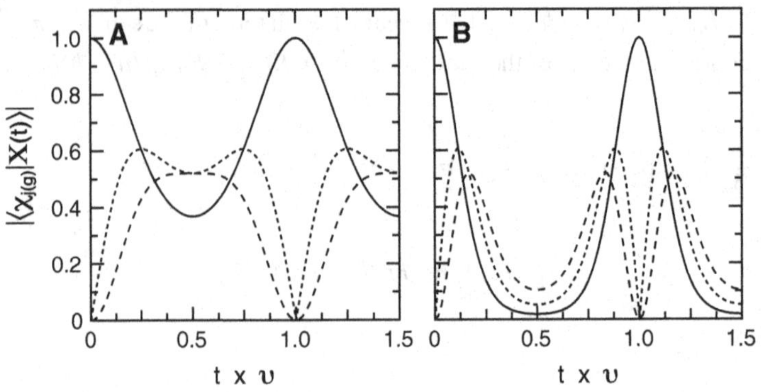

Fig. 12.6 Calculated magnitude of the overlap integral $|\langle\chi_{j(g)}|X(t)\rangle|$ as a function of time for $j=0$ (*solid curves*), 1 (*dotted curves*) or 2 (*dashed curves*), in a molecule with a single harmonic vibrational mode with frequency υ and $\Delta=1$ (**A**) or $\Delta=2$ (**B**). The dephasing time constant was assumed to be much longer than the vibrational period ($T_2=1{,}000/\upsilon$). Note that the overlap of the excited-state wavepacket ($X(t)$) with the zero-point wavefunction of the ground state ($\chi_{0(g)}$) is maximal at $t=0$, whereas the overlaps with higher levels begin at zero and peak at progressively later times. See Fig. 11.23 for plots of the real and imaginary parts of $|\langle\chi_{0(g)}|X(t)\rangle|$ when $\Delta=2$ and $T_2=2/\upsilon$

amplitude of resonance Raman scattering to the dephasing time constant T_2 (Fig. 12.5). If T_2 is short, the wavepacket decays before overlap of $X(t)$ with $X_{b(g)}$ has much opportunity to develop, with the result that Raman scattering is very weak. Ordinary absorption and Rayleigh scattering, by contrast, depend on the overlap of $X(t)$ with the initial wavefunction $X_{a(g)}$, which is maximal at zero time (Eq. 11.54).

If the normal coordinates do not change significantly upon excitation, the overall wavepacket $X(t)$ for a molecule with multiple vibrational modes is a product of wavepackets for the individual modes ($X(t)=X_1(t)X_2(t)X_3(t)\ldots$) and the time-dependent overlap integral $\langle X_{b(g)}|X(t)\rangle$ consists of a similar product [10]:

$$\langle X_{b(g)}|X(t)\rangle = \langle X_{b,i(g)}|X_i(t)\rangle\langle X_{b,j(g)}|X_j(t)\rangle\langle X_{b,k(g)}|X_k(t)\rangle\cdots. \tag{12.18}$$

Here $X_{b,i(g)}$ denotes a vibrational wavefunction for level b_i of mode i in the ground electronic state. As we discussed in connection with absorption spectra in Chap. 11, Eq. (12.18) makes the wavepacket formalism much more manageable than the Kramers-Heisenberg-Dirac expression for calculating Raman excitation spectra. It is not necessary to sum over all possible combinations of quantum numbers for the different modes.

12.4 Selection Rules for Raman Scattering

Inspection of Fig. 12.2 shows that some of the bands in an IR absorption spectrum also feature in the off-resonance Raman emission spectrum. The relative intensities of the bands are different, however, and there generally are bands that appear in only one spectrum or the other. The strong IR absorption band of HBDI at 1,677 cm^{-1} (Fig. 12.2A) has little or no Raman intensity (Fig. 12.2B), while the band at 1,562 cm^{-1} that dominates the Raman spectrum contributes only a weak shoulder on the side of the IR absorption band at 1,572 cm^{-1}. The relative intensities differ again in resonance Raman spectra (Fig. 12.2C).

We saw in Chap. 6 that there are two main selection rules for direct excitation of a harmonic oscillator from level n to level m: first, $m = n \pm 1$, and second, the vibration must change the permanent dipole moment of the molecule. Arguments parallel to those we used to find the selection rules for IR absorption can be used to predict qualitatively whether or not a particular vibrational mode will contribute to off-resonance Raman scattering. The difference is that for Raman scattering we relate the scattering matrix element α_{ba} to the molecular polarizability (α) rather than the permanent dipole moment. If the polarizability is expanded in a Taylor's series as a function of the normal coordinate (x) for the mode, the matrix element for Raman scattering becomes

$$\begin{aligned}\langle\chi_m|\alpha|\chi_n\rangle &= \alpha_0\langle\chi_m|\chi_n\rangle + (\partial\alpha/\partial x)_0\langle\chi_m|x|\chi_n\rangle + \frac{1}{2}(\partial^2\alpha/\partial x^2)_0\langle\chi_m|x^2|\chi_n\rangle + \ldots \\ &= (\partial\alpha/\partial x)_0\langle\chi_m|x|\chi_n\rangle + \frac{1}{2}(\partial^2\alpha/\partial x^2)_0\langle\chi_m|x^2|\chi_m\rangle + \ldots,\end{aligned} \tag{12.19}$$

where χ_n and χ_m are the initial and final vibrational wavefunctions for the mode, all the quantities refer to the ground electronic state, and the derivatives are evaluated at $x = 0$. This expression indicates that $\langle\chi_m|\alpha|\chi_n\rangle$ can be non-zero only if $m = n \pm 1$, so that $\langle\chi_m|x|\chi_n\rangle \neq 0$. In addition, $(\partial\alpha/\partial x)_0$ must be non-zero. The vibration therefore must change the molecule's polarizability, just as a vibration must change the molecule's permanent dipole in order to have an allowed IR transition.

As a rule of thumb, changes in α are associated with vibrations that increase the molecular size. Raman scattering thus is allowed in homonuclear diatomic molecules such as O_2, where IR transitions are forbidden by symmetry. Similarly, the symmetric stretching mode of a triatomic molecule gives an allowed Raman transition, whereas the antisymmetric stretching mode does not (Fig. 12.7). Note, however, that Eqs. (12.11–12.13) require summing the weighted products of $\langle X_{b(g)}|X_{k(e)}\rangle$ and $\langle X_{k(e)}|X_{a(g)}\rangle$ over all vibrational levels of the excited state (k) before we take the square of M_{ba} to find the Raman strength. This summation can lead to interferences that do not arise in IR transitions. Remember also that, as in electronic and vibrational absorption, the fact that a Raman transition is allowed by symmetry means only that the matrix element for the transition is not necessarily zero; it does not say anything more about the magnitude of the matrix element.

Resonance Raman scattering emphasizes the vibrational modes that are coupled most strongly to the resonant electronic transition (*i.e.*, modes with the largest displacement in the excited state). This tends to makes resonance Raman spectra much more selective than off-resonance Raman spectra, as we saw in Fig. 12.2. At the same time, the coupling to an electronic transition also relaxes the requirement

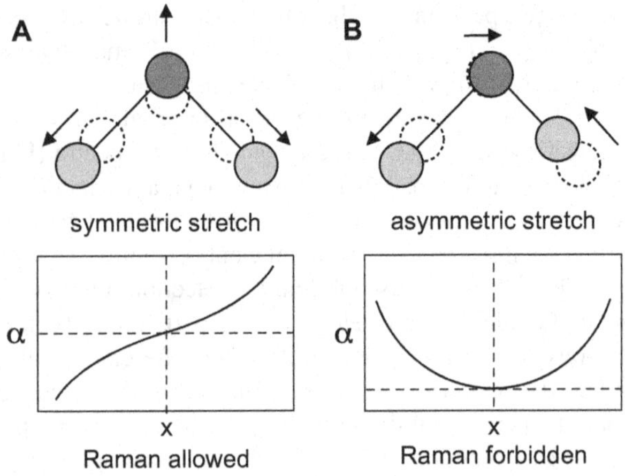

Fig. 12.7 Off-resonance Raman transitions for a vibrational mode with normal coordinate x are allowed when the molecular polarizability (α) has a non-zero slope at the mean value of the coordinate. This is the case for the symmetric stretching mode of a triatomic molecule (**A**), but not for the asymmetric stretching mode (**B**)

that $m = n \pm 1$. Consider the matrix element for resonance Raman scattering as given by Eq. (12.13). Whereas the orthogonality of the vibrational wavefunctions within a given electronic state makes $\langle X_{b(g)}|X_{a(g)}\rangle = 0$ unless $b = a$, both $\langle X_{b(g)}|X_{k(e)}\rangle$ and $\langle X_{k(e)}|X_{a(g)}\rangle$ can be non-zero because the two vibrational wavefunctions in each integral are for different electronic states. Resonance Raman scattering therefore has no formal selection rule that prevents scattering from a given vibrational level (m) to $m \pm 2$ or any other level. These transitions to higher-energy vibrational levels can make parts of the resonance Raman spectrum extremely rich, providing a unique "fingerprint" for a molecule. The *cis* and *trans* isomers of the retinylidine Schiff base chromophore in rhodopsin photoproducts were readily distinguishable in this way (Sect. 12.6).

In spite of the relaxation of the selection rule $m = n \pm 1$, the strongest resonance Raman scattering peaks usually reflect transitions in which the quantum number for a particular vibrational mode goes from 0 to 1 with no changes in other modes. We can rationalize this by using the wave-packet picture of Raman scattering. If all the vibrations are at their zero-point levels in the resting molecule, the overlap integral for Raman scattering from level 0 to level 1 of mode i (Eq. 12.18) reads

$$\langle X_{b(g)}|X(t)\rangle = \langle X_{1,i(g)}|X_i(t)\rangle \prod_{j \neq i} \langle X_{0,\,j(g)}|X_j(t)\rangle. \tag{12.20}$$

This usually is larger than the corresponding integral for scattering to $b_{i(g)} = 2$ because, in the wave-packet picture, $X_i(t)$ builds up overlap with $\chi_{1,i(g)}$ more quickly than it does with $\chi_{2,i(g)}$ (Fig. 12.6). Dephasing of the wavepacket frustrates the rise of $\langle \chi_{2,i(g)}|X_i(t)\rangle$. In addition, the build up of $\langle \chi_{2,i(g)}|X_i(t)\rangle$ must occur while the wavepackets for all the other modes ($j \neq i$) retain good overlap with the initial, ground-state wavefunction, which becomes progressively more difficult as the number of vibrational modes increases.

Because most of the molecules of interest in biophysics have very low symmetry, their vibrational modes cannot be described accurately as being either symmetric or antisymmetric, and the peaks in a Raman or resonance Raman spectrum cannot be assigned on the basis of simple selection rules. However, normal-mode analysis often can be used to identify the vibrational modes that are coupled most strongly to excitation of a chromophore. Isotopic labeling, chemical modifications of the chromophore, or site-directed mutagenesis can be used to shift a particular vibration to higher or lower frequency.

12.5 Surface-Enhanced Raman Scattering

The fortuitous discovery that Raman scattering by pyridine becomes much stronger when the pyridine is adsorbed on a roughened silver surface [15] initiated a continuing discussion of the physical basis for the enhancement [16–20]. Surface-enhanced Raman scattering (SERS) occurs with other noble metals including gold and copper, and with a variety of "nanoparticles" formed from colloidal metals or fabricated by lithography. Astonishing enhancements by a factor of 10^{14} have been obtained with single molecules adsorbed on colloidal silver particles [20–23], although factors between 10^8 and 10^{10} are more typical.

There is general agreement that the main factor leading to SERS is the strong radiation field created by surface plasmons in the metal (Box 3.2) [16, 17, 20, 24–27]. The highly curved surfaces of colloidal metal particles evidently lead to irregular localization of surface plasmons and to coupling of the fields generated by surface plasmons on neighboring particles, so that the fields acting on adsorbed molecules can vary enormously from site to site. Resonances with charge-transfer transitions in which an electron moves from the adsorbed molecule to the metal or *vice versa* probably contribute additional enhancement by factors of 10^2 to 10^3 [28–31]. Anions and cations affect the enhancement in ways that are still not well understood.

12.6 Biophysical Applications of Raman Spectroscopy

Raman spectroscopy has been particularly useful in studies of rhodopsin and bacteriorhodopsin. As discussed in Chap. 4, excitation of rhodopsin or bacteriorhodopsin by light causes isomerization of the retinyl chromophore. In rhodopsin, the chromophore changes from 11-*cis* to all-*trans*; in bacteriorhodopsin, it goes from all-*trans* to 13-*cis*. Resonance Raman measurements showed that the isomerization is essentially complete in metastable intermediate states that form within a few ps [32–38]. The conformations of these states were ascertained by comparisons of the resonance Raman spectra with those of model compounds.

Raman spectroscopy also has proved an effective way to study the ligation states and environments of hemes in proteins [39–49], and to examine the ligands and hydrogen bonding of the protein to the pigments in photosynthetic reaction centers and antenna complexes [50–56]. The bacteriochlorophyll ring has a characteristic vibrational mode near 1,615 cm^{-1} when the Mg has one axial ligand, and near 1,600 cm^{-1} when there are two ligands. Hydrogen bonding to the acetyl or keto group shifts the C=O stretching mode to lower frequency, and the magnitude of this shift depends approximately linearly on the strength of the H-bond [51, 54]. Shifts of the resonance Raman frequency upon site-directed mutagenesis have been used similarly to identify residues that form H-bonds to the formyl group of heme-*a* in cytochrome oxidase [39]. Measurements of resonance Raman scattering are advantageous in such studies because the incident light can be tuned to the absorption band of a particular subset of the pigments, such as the special pair of bacteriochlorophylls that serve as the initial electron donor in bacterial reaction centers. Comparisons of the resonance Raman excitation cross sections of the different pigments indicate that the dephasing time constant T_2 is very different for these pigments than for the other bacteriochlorophylls in the reaction center or in the antenna complexes [57, 58].

The experiments on GFP illustrated in Fig. 12.2 were aimed partly at the question of how excitation of the chromophore leads to dissociation of a proton from the phenolic -OH group (Fig. 5.9). Comparisons of resonance Raman spectra of GFP with spectra of the chromophore in ordinary and deuterated ethanol, together with normal-mode assignments of the Raman bands, indicated that stretching of the O-H bond is not strongly coupled to the initial excitation, and must develop later in the evolution of the excited state [59, 60].

Advancing techniques for obtaining laser light in the UV have opened the door to time-resolved resonance Raman studies of tyrosine, phenylalanine and tryptophan residues in proteins [41, 61–68]. UV resonance Raman appears to be a potentially powerful source of information on the secondary structure and the distribution of Ramachandran ψ angles in polypeptides [69, 70], and provides a way of probing the S-C-C-S dihedral angle of cystine residues in proteins [71].

Much of the interest in SERS to date has focused on analytical applications. The technique has been used in a sensitive biosensor of glucose [72] and to assay lysophosphatidic acid, a biomarker for ovarian cancer [73]. With the advent of single-molecule SERS, biophysical studies comparable to the single-molecule fluorescence experiments described in Sect. 5.9 should be interesting.

12.7 Coherent (Stimulated) Raman Scattering

"Coherent" or "stimulated" Raman scattering is a four-wave mixing technique in which Raman transitions are strongly enhanced [1, 74–77]. Although a detailed analysis reveals some interesting subtleties [1], the process is basically the same as ordinary off-resonance Raman scattering except that emission of Raman-shifted radiation is stimulated by radiation at the emission frequency. The magnitude of the resulting signal increases quadratically with the intensity of the stimulating radiation, and can be on the order of 10^4 as strong as the signal from unstimulated Raman transitions, in which the emission depends on the zero-photon radiation field. Either

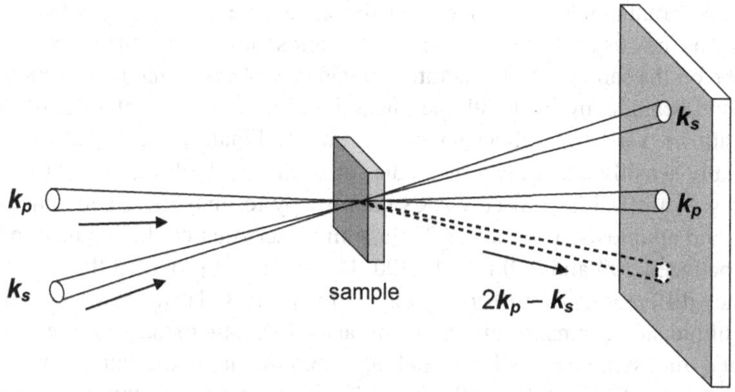

Fig. 12.8 In coherent Raman spectroscopy, a beam of electromagnetic radiation with frequency ν_p and wavevector $\boldsymbol{k}_p$, and a second beam with frequency ν_s and wavevector $\boldsymbol{k}_s$ are focused on the sample. Radiation emitted with frequency $\nu = 2\nu_p - \nu_s$ and wavevector $\boldsymbol{k}_f = 2\boldsymbol{k}_p - \boldsymbol{k}_s$ is collected. Stokes Raman transitions of the ground electronic state are stimulated when $\nu_p - \nu_s = \upsilon$, where $h\upsilon$ is a vibrational mode of the sample; anti-Stokes transitions are stimulated when $\nu_p - \nu_s = -\upsilon$

Stokes- or anti-Stokes Raman transitions can be stimulated, but anti-Stokes transitions (*coherent anti-Stokes Raman scattering*, or CARS) have the technical advantage of being shifted out of the region of background fluorescence. The term "coherent" in this context does not refer to coherences with a virtual intermediate state, which play the same role here as in spontaneous Raman scattering (Fig. 12.3D). It refers to the fact that emission from many molecules occurs coherently across a macroscopic region where the four waves overlap. In ordinary Raman scattering, the molecules in the illuminated region emit light independently and with random phases.

Figure 12.8 shows a typical arrangement of the excitation and signal beams for coherent Raman scattering. Two incident beams are required. The sample interacts twice with a field with frequency ν_p and wavevector $\boldsymbol{k}_p$ from the "pump" beam and twice a field with wavevector $\boldsymbol{k}_s$ and frequency ν_s from the "Stokes" beam. If the two fields overlap temporally and spatially, they can induce anti-Stokes Raman transitions of a vibrational mode with frequency υ when $\nu_s - \nu_p = \upsilon$. Stimulated Raman signals then propagate with wavevector $\boldsymbol{k} = 2\boldsymbol{k}_p - \boldsymbol{k}_s$. Coaxial arrangements of the excitation beams also can be used, and lend themselves well to confocal microscopy [78–82]. Signals propagating in the reverse direction can be greatly augmented by scattering of the forward-going signals in turbid materials such as living tissues [83].

Coherent Raman scattering has several features that make it particularly applicable to microscopy [83]. First, the excitation frequencies can be chosen to image structures with particular chemical compositions. Since the selectivity depends on the difference between the two frequencies rather than their absolute values, the wavelengths required are shorter than would be needed for imaging the same vibrational modes by IR absorption, and this provides greater spatial resolution.

Because external labels are not required, the signals are not limited by bleaching of the probe as they usually are in fluorescence microscopy, nor perturbed by effects of the probe on the sample. The quadratic dependence of the signal on the intensity of the second beam, combined with the linear dependence on the intensity of the first beam, allows very tight focusing of the image. Finally, the technique can be remarkably sensitive, as shown by studies of single lipid bilayers [84–86].

Cheng et al. [81] have used CARS microscopy to image live cells undergoing mitosis and apoptosis. Two pulsed Ti:Sapphire lasers that could be tuned independently between 700 and 900 nm (14,300–11,100 cm^{-1}) provided the radiation. A frequency difference of 2,870 cm^{-1} was used to probe C-H stretching vibrations of phospholipid side chains in cellular membranes. In another study, Cheng et al. [82] compared the symmetric CH_2 stretching vibration in multilamellar phosphatidylserine and phosphatidylcholine vesicles with the O-H stretching mode of water at the membrane surfaces (3,445 cm^{-1}). The dependence of the signals on the polarization of the radiation showed that the water molecules were oriented with their symmetry axis normal to the membrane surface.

12.8 Multi-photon Absorption

Two-photon absorption resembles Raman scattering in that alternating interactions of a molecule with two radiation fields create a series of coherences with a virtual state. However, the final state is an excited electronic state rather than the ground state. The Liouville-space diagrams for the process, one of which is shown in Figure 12.9A, are identical to those for Raman scattering (Fig. 12.3D). To indicate that all the steps are associated with absorption, the directions of two of the wavy arrows in the double-sided Feynman diagram (Fig. 12.9B) are reversed relative to those for Raman scattering (*cf.* Fig. 12.3E). As with Raman scattering, the essential feature of two-photon absorption is that it proceeds entirely through coherences. The same final state often can be attained by two discrete steps in which a lower excited state is populated as a real, if transient, intermediate. For comparison, Fig. 12.9C, D shows Liouville-space and double-sided Feynman diagrams for one of the pathways that contribute to the latter process.

The condition for energy conservation in two-photon absorption is $h\nu_1 + h\nu_2 = E_b - E_a$, where ν_1 and ν_2 are the frequencies of the two fields and E_a and E_b are the energies of the ground and excited states. Resonance with excitation to the virtual intermediate state is not required, although it would enhance the process just as it enhances Raman scattering. Experimentally, a single excitation beam is used in most cases, so $\nu_2 = \nu_1 = \nu$, $\mathbf{k}_2 = \mathbf{k}_1$, and $E_b - E_a = 2h\nu$. The product state can be measured by its spontaneous flourescence, stimulated emission, ground-state bleaching, excited-state absorption, or conversion of the excitation energy to heat [87]. If it is a higher excited singlet state, the product usually decays rapidly to the lowest such state by internal conversion so that, in accordance with Kasha's rule (Sect. 5.6), the fluorescence emission spectrum and lifetime are very similar to those obtained by single-photon excitation.

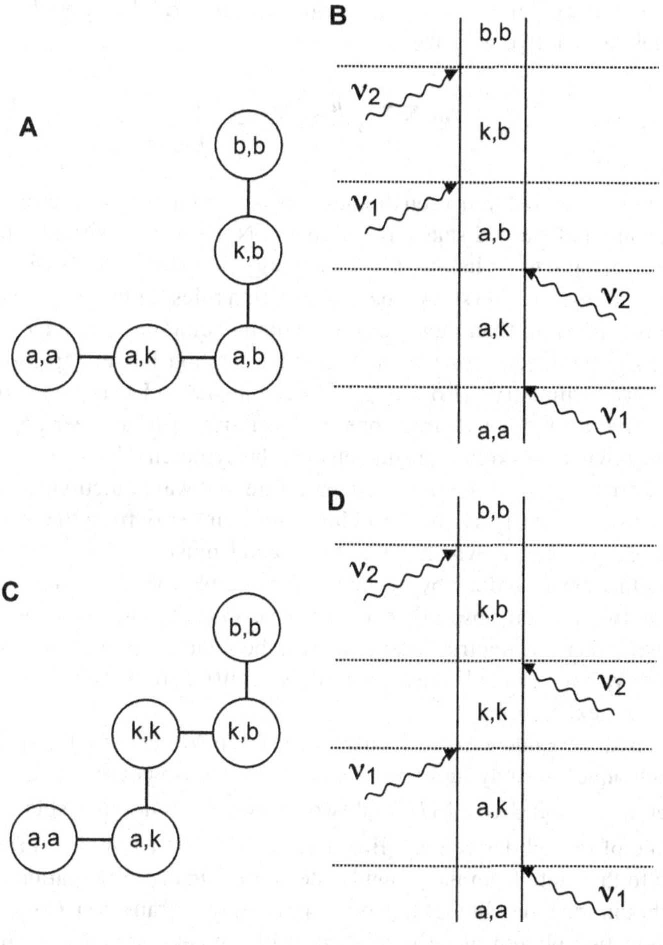

Fig. 12.9 Liouville-space and double-sided Feynman diagrams for two-photon absorption (**A, B**) and a representative pathway resulting in ordinary excited-state absorption (**C, D**). The ground state and the final excited state are labeled *a* and *b*. Excited-state absorption requires populating an intermediate state (*k*), whereas two-photon absorption proceeds entirely through coherences. Both processes also occur by the complex conjugates of the pathways shown. Excited-state absorption also can occur by the pathway shown in Fig. 12.3B and its complex conjugate

Two-photon excitation was predicted in her doctoral thesis by Maria Göppert-Mayer [88], who recognized that it was a corollary of the Kramers-Heisenberg-Dirac theory of light scattering. It was not observed experimentally until 30 years later, when pulsed ruby lasers finally provided the high photon flux that was required [89]. Göppert-Mayer received the physics Nobel prize in 1963 for unrelated work on nuclear structure.

The matrix element for two-photon absorption is essentially the same as that for off-resonance Raman scattering (Eq. 12.11). Assuming that the two photons have the same frequency ($\nu_1 = \nu_2 = \nu$) and field strength ($|\boldsymbol{E}|$) but possibly different polarizations ($\boldsymbol{e}_1$ and $\boldsymbol{e}_2$), we have

$$\alpha_{ba} = |\boldsymbol{E}|^2 \sum_k \frac{(\boldsymbol{\mu}_{bk} \cdot \boldsymbol{e}_2)(\boldsymbol{\mu}_{ka} \cdot \boldsymbol{e}_1)}{E_k - E_a - h\nu - i\hbar/T_2}, \tag{12.21}$$

where $\boldsymbol{\mu}_{ka}$ and $\boldsymbol{\mu}_{bk}$ are the transition dipoles connecting the virtual intermediate state k to the ground and excited states, respectively. Note that the virtual intermediate state k does not have to lie below state b in energy; in many cases it will be higher.

Inspection of Eq. (12.21) shows that the selection rules for two-photon excitation are somewhat different than those for one-photon excitation. Consider a molecule with inversion symmetry, for which each wavefunction has either gerade (g) or ungerade (u) symmetry (Box 4.8, Figs. 4.6–4.8). Excitations from one g wavefunction to another or from one u wavefunction to another are forbidden by symmetry, whereas excitations that change the symmetry ($u \rightarrow g$ or $g \rightarrow u$) are allowed. In order for α_{ba} to be non-zero, therefore, the wavefunction for the virtual intermediate state k must have different inversion symmetry from the wavefuctions for both state a and state b, which means that a and b must have the same symmetry. Excitations that are forbidden by symmetry for one-photon absorption thus can be allowed for two-photon absorption and *vice versa*. Although this selection rule breaks down in less symmetric chromophores, the relative strengths of two-photon excitation to various excited states generally will differ from the relative strengths for one-photon excitation.

The different dependence on orbital symmetries makes two-photon spectroscopy a useful technique for studying some excited states that are not readily accessible by one-photon excitation. Birge [87] used two-photon excitation to explore the $2^1A_g^-$ excited state of retinyl derivatives (Box 4.12) in solution and bound to rhodopsin. Excitation to this state from the ground state is, to a first approximation, forbidden in a one-photon transition but is allowed as a two-photon transition. Comparisons of the one- and two-photon absorption spectra of the unprotonated Schiff base of all-*trans*-retinal in solution showed that the $2^1A_g^-$ state lies below $1^1B_u^+$. Protonating the Schiff base moves $1^1B_u^+$ down in energy and inverts the order. In rhodopsin containing a "locked" 11-*cis*-retinyl derivative that was unable to undergo photoisomerization, $1^1B_u^+$ was found to lie below $2^1A_g^-$, in accord with other indications that the Schiff base is protonated [87, 90].

Comparisons of one- and two-photon absorption can be confusing because the two processes have different dependences on the light intensity. The rate of two-photon excitation depends more strongly on the spatial and temporal distribution of the excitation light [91]. The pertinent factor is the integral of $\langle I^2 \rangle$ over the illuminated volume, where I is the light intensity and $\langle I^2 \rangle$ is the average of I^2 over the period of the measurement. $\langle I^2 \rangle$ generally differs from $\langle I \rangle^2$, depending on the temporal coherence of the laser. Xu and Webb [91] have measured the two-photon absorption cross sections of a variety of fluorescent dyes, using wavelengths between 690 and 900 nm from a pulsed Ti:S laser. Typical values are between

10^{-50} and 10^{-48} $cm^4{\cdot}s$ per photon, or 1–100 in the unofficial but frequently used units of "Göppert-Mayers" (1 G.M. = 10^{-50} $cm^4{\cdot}s$). Psoralen has a two-photon absorption cross section of 20×10^{-50} $cm^4{\cdot}s$ [92], and quantum dots with cross sections as high as $47{,}000 \times 10^{-50}$ $cm^4{\cdot}s$ have been described [93]. To put these values in perspective, with the instrumentation described by Xu and Webb [91], a 1 mW continuous laser beam gave a pulse intensity of about 10^{28} photons/($cm^2{\cdot}s$), or $\langle I \rangle^2 \approx 10^{56}$ photon^2/($cm^4{\cdot}s^2$). A molecule with a two-photon absorption cross section of 10^{-50} $cm^4{\cdot}s$ per photon would be excited about 10^6 times per second.

The dependence on the polarization of the excitation light also differs for one- and two-photon absorption. For one-photon absorption in an isotropic system, the average of $(\boldsymbol{\mu}_{ba}{\cdot}\boldsymbol{e})^2$ over all possible orientations of the transition dipole ($\boldsymbol{\mu}_{ba}$) with respect to an axis of linear polarization ($\boldsymbol{e}$) is simply $|\boldsymbol{\mu}_{ba}|^2/3$ (Box 4.6). In two-photon absorption, the quantity α_{ba}^2 that is averaged over all orientations involves a tensor product rather than a simple dot product. As a result, the ratio (Ω) of the two-photon absorptivity with circularly polarized exciting light to that with linearly polarized light is particularly informative. Methods for relating this ratio to the symmetry of the initial and final states were developed by McClain and coworkers [94–97], and were extended to two-photon fluorescence anisotropy by Callis [98, 99]. In 3-methylindole, Ω is 1.4 for excitation to the ^1L_b state and 0.5 for the ^1L_a state [100]. The calculated cross sections for two-photon excitation with unpolarized light also differ for the ^1L_a and ^1L_b states, being 4–8 times larger for the ^1L_a state, although the overall absorption spectra measured with one- and two-photon excitation are coincidentally very similar [98–101].

Multi-photon excitation has proved to be especially useful in confocal fluorescence microscopy, where it has several distinct advantages over one-photon excitation [102–106]. Because multi-photon excitation has a quadratic or higher dependence on the light intentsity, it can be focused more tightly than one-photon excitation. This not only improves the spatial resolution, but also decreases the amount of absorption, photobleaching of the fluorophore, and photodamage to the specimen in regions that are out of the focal plane. UV optics are not required, and the red or near-IR excitation tends to penetrate tissues better than UV or blue light does. Finally, the emission spectrum of the excited fluorophore often is well to the blue of the excitation light, facilitating their separation.

Two-photon excitation has been used to visualize cell-cell interactions [107] and linear dichroism in cell membranes [108]. It has been combined with resonance energy transfer to image protein-protein interactions during T cell activation [109].

12.9 Quasielastic (Dynamic) Light Scattering (Photon Correlation Spectroscopy)

Our discussion of light scattering in Sects. 12.1 and 12.2 focused on scattering by an individual electron. We now consider how interference between light scattered from different regions of a macromolecule can provide information on molecular

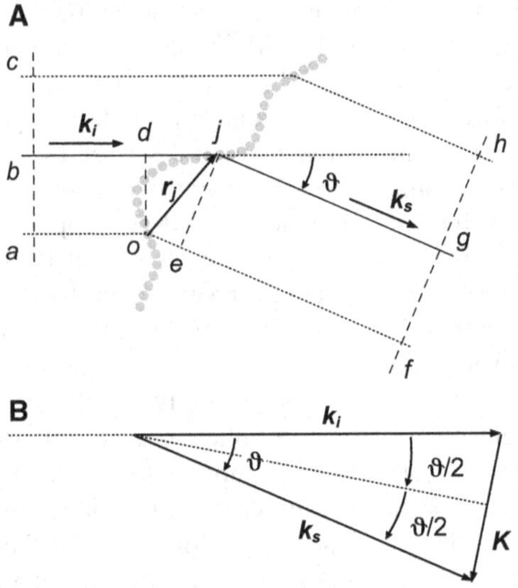

Fig. 12.10 (**A**) A plane wave of light with wavefront $\overline{ac}$ and wavevector $\boldsymbol{k}_i$ is scattered by various segments of a molecule (*gray circles*), and a detector collects a plane wave of scattered light with wavefront $\overline{fh}$ and wavevector $\boldsymbol{k}_s$. Segment j is located at $\boldsymbol{r}_j$ relative to the origin of an arbitrary coordinate system (o). *Lines* $\overline{od}$ and $\overline{ej}$ are perpendicular to $\boldsymbol{k}_i$ and $\boldsymbol{k}_s$, respectively. The field at the detector from light scattered by segment j experiences a phase shift of $(2\pi/\lambda)(\overline{oe} - \overline{dj}) = 2\pi \boldsymbol{r}_j \cdot (\boldsymbol{k}_s - \boldsymbol{k}_i)$ relative to light scattered at the origin. (**B**) A geometrical construction shows that the magnitude of the scattering vector $\boldsymbol{K}$ is $(2/\lambda)\sin(\vartheta/2)$, where ϑ is the angle between $\boldsymbol{k}_s$ and $\boldsymbol{k}_i$

structure and dynamics. Such interference occurs when the molecular dimensions are comparable to or greater than the wavelength of the light, so that the radiation field at any given time varies with the position in the molecule. We assume, however, that the molecule is still small enough, and the solution sufficiently dilute, so that the *velocity* of light is effectively constant throughout the solution under study. Scattering by large particles that delay light significantly relative to light passing by outside (called "Mie scattering" for spherical particles) is more strongly directional than the process we consider here.

Consider an idealized macromolecule consisting of N chemically identical "segments" that scatter light independently. Our first task is to relate the field of the radiation scattered by a given segment to the location of the segment relative to the incoming and outgoing light rays. Suppose the incident light is a plane wave propagating with wavevector $\boldsymbol{k}_i$ while the detector receives a plane wave of scattered light propagating with wavevector $\boldsymbol{k}_s$, the angle between $\boldsymbol{k}_s$ and $\boldsymbol{k}_i$ being ϑ (Fig. 12.10A). If the incident and scattered light have the same wavelength (λ) or very nearly so (*i.e.*, if the scattering is elastic or quasielastic), $\boldsymbol{k}_s$ and $\boldsymbol{k}_i$ have essentially the same magnitude ($1/\lambda$) and differ significantly only in direction. Referring to Fig. 12.10A, define point o as the origin of the coordinate system, so that the incident light reaches the origin along line $\overline{ao}$ and light scattered at the origin reaches the detector via line $\overline{of}$. Incident light reaches segment j of the molecule along $\overline{bj}$ and is scattered to the detector along $\overline{jg}$. Because these paths differ in length by $\overline{dj} - \overline{oe}$, the radiation field scattered by segment j is shifted in phase relative to the field scattered at the origin. The phase shift, δ_j, is

$$\delta_j = \frac{2\pi}{\lambda}\left(\overline{oe} - \overline{dj}\right) = 2\pi\left(\boldsymbol{r}_j \cdot \boldsymbol{k}_s - \boldsymbol{r}_j \cdot \boldsymbol{k}_i\right) = 2\pi \boldsymbol{r}_j \cdot \boldsymbol{K}, \tag{12.22}$$

where $\boldsymbol{r}_j$ is the position of segment j and $\boldsymbol{K} = \boldsymbol{k}_s - \boldsymbol{k}_i$. Figure 12.10B shows that the magnitude of $\boldsymbol{K}$, the *scattering vector*, is related to the scattering angle (ϑ) by

$$\begin{aligned} |\boldsymbol{K}| &= |\boldsymbol{k}_s - \boldsymbol{k}_i| = |\boldsymbol{k}_i| \sin(\vartheta/2) + |\boldsymbol{k}_s| \sin(\vartheta/2) \approx 2|\boldsymbol{k}_i| \sin(\vartheta/2) \\ &= \frac{2}{\lambda} \sin(\vartheta/2). \end{aligned} \tag{12.23}$$

From Eq. (12.22), we see that the phase δ_j can change with time if segment j moves, but that only the component of the movement parallel to $\boldsymbol{K}$ contributes to this change. Including the phase shift, the field of the light with polarization $\hat{e}_s$ reaching the detector at time t after scattering by segment j at time $\tau = 0$ can be written

$$\boldsymbol{E}_{s(j)}(t) = \boldsymbol{E}_{s(o)}(\tau=0)\left\{\exp\left[2\pi i\left(vt - \boldsymbol{k}_s \cdot \boldsymbol{R} + \boldsymbol{r}_j \cdot \boldsymbol{K}\right)\right] + \exp\left[-2\pi i\left(vt - \boldsymbol{k}_s \cdot \boldsymbol{R} + \boldsymbol{r}_j \cdot \boldsymbol{K}\right)\right]\right\}, \tag{12.24}$$

with

$$\boldsymbol{E}_{s(o)}(\tau = 0) = [\boldsymbol{E}_e(\tau = 0) \cdot \hat{e}_s]\alpha_{aa}\rho_v{}^{1/2} \sin(\theta)/\hbar|\boldsymbol{R}|. \tag{12.25}$$

In Eq. (12.25), θ is the angle between the polarization of the incident radiation ($\boldsymbol{E}_e$) and the direction of propagation of the scattered wave ($\boldsymbol{k}_s$), $\boldsymbol{R}$ is the position of the detector, α_{aa} is the dynamic polarizability of the segment, and $\rho_v(v)$ is the density of modes of the incident radiation at frequency v (Eqs. 12.9, 12.12, and B12.1.14). The factor $\sin(\theta)/|\boldsymbol{R}|$ is the same factor that determines the amplitude of the field from an oscillating electric dipole (Figs. 3.1 and 3.2), and the fluorescence from an excited molecule whose transition dipole is oriented along a fixed axis (Sect. 5.9). The polarizability α_{aa} can be obtained from the difference between the dielectric constant of the solution and that of the pure solvent.

Now suppose there are M macromolecules in the region where the incident and scattered beams overlap (the *scattering region*). We'll neglect photons that are scattered twice (*i.e.*, by more than one segment) before they reach the detector. To obtain the average intensity of the scattered light at the detector ($\bar{I}_s$), we must sum the fields from the individual segments of all the molecules in the scattering region, square the modulus of the total field, and integrate over the period of the radiation. Assuming that $|\boldsymbol{E}_{s(o)}|^2$ (the time-averaged square of the modulus of $\boldsymbol{E}_{s(o)}$) is the same for all the segments, the average intensity of the light reaching the detector with scattering vector $\boldsymbol{K}$ is

$$\overline{I_s}(\boldsymbol{K}) = \left|\boldsymbol{E}_{s(o)}\right|^2 \sum_{m_1=1}^{M} \sum_{m_2=1}^{M} \sum_{j=1}^{N} \sum_{k=1}^{N} \left\{ \exp\left[2\pi i \boldsymbol{K} \cdot (\boldsymbol{r}_j - \boldsymbol{r}_k)\right] + \exp\left[-2\pi i \boldsymbol{K} \cdot (\boldsymbol{r}_j - \boldsymbol{r}_k)\right] \right\}, \tag{12.26}$$

where the sum over j is for segments of molecule m_1 and the sum over k for segments of m_2. Terms of the form $\exp[4\pi i(\nu t - \boldsymbol{k}{\cdot}\boldsymbol{R} + \boldsymbol{r}_j{\cdot}\boldsymbol{K}/2 + \boldsymbol{r}_k{\cdot}\boldsymbol{K}/2)]$ have dropped out as a result of averaging over time.

In free solution, the positions of the segments in two different molecules usually will be uncorrelated. Assuming that $M \gg 1$, the terms of Eq. (12.26) for $m_1 \neq m_2$ therefore sum to zero, leaving us with

$$\overline{I_s}(\boldsymbol{K}) = \left|\boldsymbol{E}_{s(o)}\right|^2 M \sum_{j=1}^{N} \sum_{k=1}^{N} \left\{ \exp\left[2\pi i \boldsymbol{K} \cdot (\boldsymbol{r}_j - \boldsymbol{r}_k)\right] + \exp\left[-2\pi i \boldsymbol{K} \cdot (\boldsymbol{r}_j - \boldsymbol{r}_k)\right] \right\}, \tag{12.27}$$

with both sums pertaining to the same molecule. Expanding the exponentials in this expression gives, to second order in $|\boldsymbol{K}|$,

$$\overline{I_s}(\boldsymbol{K}) = \left|\boldsymbol{E}_{s(o)}\right|^2 M \sum_{j=1}^{N} \sum_{k=1}^{N} \left\{ 2 - \left[2\pi \boldsymbol{K} \cdot (\boldsymbol{r}_j - \boldsymbol{r}_k)\right]^2 - \cdots \right\}. \tag{12.28}$$

Inspection of Eq. (12.28) shows that the scattered irradiance at $|\boldsymbol{K}| = 0$ is $2MN^2 \left|\boldsymbol{E}_{s(o)}\right|^2$. The intensity of the scattering at small angles thus provides a way of determining the number of segments in the molecule (N), and from that, the molecular size [13, 110–112].

Since the segments are assumed to be identical, and so have the same mass, the sum of the quadratic terms in Eq. (12.28) is proportional to $-\left(|\boldsymbol{K}|^2 R_g^2\right)/3$, where R_g is the molecule's radius of gyration [13]. The factor of $(|\boldsymbol{K}|^2)/3$ comes from summing the square of the dot product over random orientations of the vector $(\boldsymbol{r}_j - \boldsymbol{r}_k)$ relative to $\boldsymbol{K}$. The slope of a plot of I_s versus $|\boldsymbol{K}|^2$ at small scattering angles therefore can be used to obtain R_g.

Brownian diffusion and internal motions of a macromolecule cause the intensity of quasielastically scattered light to fluctuate with time, and the autocorrelation function of the scattering provides information on the dynamics of these motions as we discussed in Sect. 5.11 for fluorescence fluctuations. Berne and Pecora [13], Schurr [113, 114], Chu [112] and Brown [115] give expressions for the autocorrelation functions that apply to various models for proteins and nucleic acids, along with further information on data collection and analysis. If the autocorrelation function decays with a single exponential time constant τ, the molecule's diffusion coefficient is $1/(2\tau|\boldsymbol{K}|^2)$. The autocorrelation function of dynamic light scattering by sample with a distribution of molecular sizes will have multiple components.

Exercises

1. The figure below shows an emission spectrum measured when a protein containing tryptophan was excited at 290 nm. (*a*) Suggest an assignment for the peak at 310 nm and point out the features of the spectrum support the assignment. (*b*) How could you test your explanation?

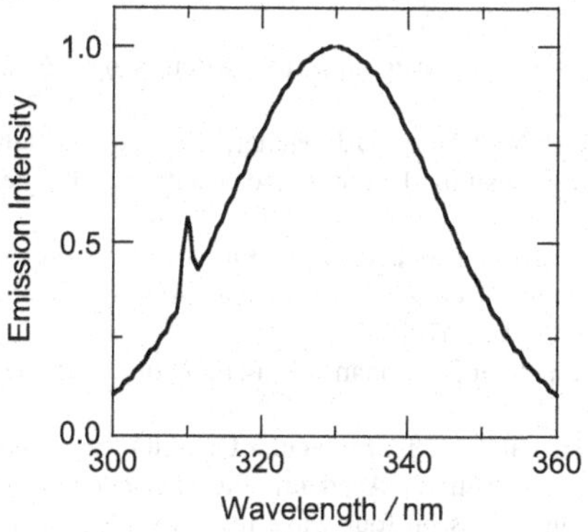

2. The strength of resonance Raman scattering by a molecule typically decreases strongly with increasing temperature, while the absorbance and fluorescence change very little. Explain this observation using (*a*) the Kramers-Heisenberg-Dirac theory and/or (*b*) the semiclassical wavepacket theory.
3. (*a*) Explain why a symmetric vibrational mode that makes little or no contribution to the IR absorption spectrum can contribute strongly to the Raman spectrum. (*b*) Does the formal selection rule $\Delta m = \pm 1$, where m is the vibrational quantum number, apply to both resonance and off-resonance Raman scattering?
4. How does coherent anti-Stokes Raman scattering resemble, but differ from (*a*) ordinary anti-Stokes Raman scattering, and (*b*) ordinary stimulated emission?
5. (*a*) What are the potentential advantages of two-photon excitation relative to one-photon excitation in fluorescence microscopy? (*b*) How do the selection rules for two-photon excitation differ from those for one-photon excitation?

References

1. Mukamel, S.: Principles of Nonlinear Optical Spectroscopy. Oxford University Press, Oxford (1995)
2. Kramers, H.A., Heisenberg, W.: On the dispersal of radiation by atoms. Z. Phys. **31**, 681–708 (1925) [Engl. transl. van der Waerden BL (ed.) Sources of Quantum Theory, Dover, 1967, pp. 1223–1252]
3. Dirac, P.A.M.: The quantum theory of dispersion. Proc. Roy. Soc. London **A114**, 710–728 (1927)
4. Albrecht, A.C.: On the theory of Raman intensities. J. Chem. Phys. **34**, 1476–1484 (1961)
5. Long, D.A.: The Raman Effect: A Unified Treatment of the Theory of Raman Scattering. Wiley, New York, NY (2002)
6. Lee, S.Y., Heller, E.J.: Time-dependent theory of Raman scattering. J. Chem. Phys. **71**, 4777–4788 (1979)
7. Heller, E.J.: The semiclassical way to molecular spectroscopy. Acc. Chem. Res. **14**, 368–375 (1981)
8. Myers, A.B., Mathies, R.A., Tannor, D.J., Heller, E.J.: Excited-state geometry changes from pre-resonance Raman intensities. Isoprene and hexatriene. J. Chem. Phys. **77**, 3857–3866 (1982)
9. Myers, A.B., Mathies, R.A.: Resonance Raman intensities: a probe of excited state structure and dynamics. In: Spiro, T.G. (ed.) Biological Applications of Raman Spectroscopy, pp. 1–58. Wiley, New York, NY (1987)
10. Myers, A.B.: "Time-dependent" resonance Raman theory. J. Raman Spectrosc. **28**, 389–401 (1997)
11. Craig, D.P., Thirunamachandran, T.: Molecular Quantum Electrodynamics: An Introduction to Radiation-Molecule Interactions. Academic Press, London (1984)
12. Mukamel, S.: Solvation effects on four-wave mixing and spontaneous Raman and fluorescence lineshapes of polyatomic molecules. Adv. Chem. Phys. **70 Part I**, 165–230 (1988)
13. Berne, B.J., Pecora, R.: Dynamic Light Scattering: With Applications to Chemistry, Biology, and Physics. Wiley, New York, NY (1976)
14. Yariv, A.: Quantum Electronics, 3rd edn. Wiley, New York, NY (1988)
15. Fleischmann, M., Hendra, P.J., McQuillan, A.J.: Raman spectra of pyridine adsorbed at a silver electrode. Chem. Phys. Lett. **26**, 163–166 (1974)
16. Moscovits, M.: Surface-enhanced spectroscopy. Rev. Mod. Phys. **57**, 783–826 (1985)
17. Moscovits, M., Tay, L.L., Yang, J., Haslett, T.: Optical properties of nanostructured random media. Topics Appl. Phys. **82**, 215–226 (2002)
18. Otto, A., Mrozek, I., Grabhorn, H., Akemann, W.: Surface-enhanced Raman scattering. J. Phys. Condensed Mat. **4**, 1143–1212 (1992)
19. Campion, A., Kambhampati, P.: Surface-enhanced Raman scattering. Chem. Soc. Revs. **27**, 241–250 (1998)
20. Wang, Z.J., Pan, S.L., Krauss, T.D., Du, H., Rothberg, L.J.: The structural basis for giant enhancement enabling single-molecule Raman scattering. Proc. Natl. Acad. Sci. U. S. A. **100**, 8638–8643 (2003)
21. Emory, S.R., Nie, S.: Near-field surface-enhanced Raman spectroscopy on single silver nanoparticles. Anal. Chem. **69**, 2631–2635 (1997)
22. Kneipp, K., Wang, Y., Kneipp, H., Perelman, L.T., Itzkan, I., et al.: Single molecule detection using surface-enhanced Raman scattering (SERS). Phys. Rev. Lett. **78**, 1667–1670 (1997)
23. Nie, S., Emory, S.R.: Probing single molecules and single nanoparticles by surface-enhanced Raman scattering. Science **275**, 1102–1106 (1997)
24. Jeanmaire, D.L., Van Duyne, R.P.: Surface Raman spectroelectrochemistry. Part I. Heterocyclic, aromatic, and aliphatic amines adsorbed on the anodized silver electrode. J. Electroanal. Chem. **84**, 1–20 (1977)

25. Moscovits, M.: Surface roughness and the enhanced intensity of Raman scattering by molecules adsorbed on metals. J. Chem. Phys. **69**, 4159–4161 (1978)
26. Haynes, C.L., Van Duyne, R.P.: Nanosphere lithography: a versatile nanofabrication tool for studies of size-dependent nanoparticle optics. J. Phys. Chem. B **105**, 5599–5611 (2001)
27. Futamata, M., Maruyama, Y., Ishikawa, M.: Metal nanostructures with single-molecule sensitivity in surface enhanced Raman scattering. Vibr. Spectrosc. **35**, 121–129 (2004)
28. Albrecht, M.G., Creighton, J.A.: Anomalously intense Raman spectra of pyridine at a silver electrode. J. Am. Chem. Soc. **99**, 5215–5217 (1977)
29. Lombardi, J.R., Birke, R.L., Lu, T., Xu, J.: Charge-transfer theory of surface enhanced Raman spectroscopy: Herzberg-Teller contributions. J. Chem. Phys. **84**, 4174–4180 (1986)
30. Doering, W.E., Nie, S.M.: Single-molecule and single-nanoparticle SERS: examining the roles of surface active sites and chemical enhancement. J. Phys. Chem. B **106**, 311–317 (2002)
31. Vosgrone, T., Meixner, A.J.: Surface- and resonance-enhanced micro-Raman spectroscopy of xanthene dyes: from the ensemble to single molecules. ChemPhysChem **6**, 154–163 (2005)
32. Aton, B., Doukas, A.G., Narva, D., Callender, R.H., Dinur, U., et al.: Resonance Raman studies of the primary photochemical event in visual pigments. Biophys. J. **29**, 79–94 (1980)
33. Eyring, G., Curry, B., Mathies, R.A., Fransen, R., Palings, I., et al.: Interpretation of the resonance Raman spectrum of bathorhodopsin based on visual pigment analogs. Biochemistry **19**, 2410–2418 (1980)
34. Pande, J., Callender, R.H., Ebrey, T.G.: Resonance Raman study of the primary photochemistry of bacteriorhodopsin. Proc. Natl. Acad. Sci. U. S. A. **78**, 7379–7382 (1981)
35. Loppnow, G.R., Mathies, R.A.: Excited-state structure and isomerization dynamics of the retinal chromophore in rhodopsin from resonance Raman intensities. Biophys. J. **54**, 35–43 (1988)
36. Doig, S.J., Reid, P.J., Mathies, R.A.: Picosecond time-resolved resonance Raman spectroscopy of bacteriorhodopsin-J, bacteriorhodopsin-K, bacteriorhodopsin-KL intermediates. J. Phys. Chem. **95**, 6372–6379 (1991)
37. Yan, M., Manor, D., Weng, G., Chao, H., Rothberg, L., et al.: Ultrafast spectroscopy of the visual pigment rhodopsin. Proc. Natl. Acad. Sci. U. S. A. **88**, 9809–9812 (1991)
38. Lin, S.W., Groesbeek, M., van der Hoef, I., Verdegem, P., Lugtenburg, J., et al.: Vibrational assignment of torsional normal modes of rhodopsin: probing excited-state isomerization dynamics along the reactive C-11=C-12 torsion coordinate. J. Phys. Chem. B **102**, 2787–2806 (1998)
39. Shapleigh, J.P., Hosler, J.P., Tecklenburg, M.M.J., Kim, Y., Babcock, G.T., et al.: Definition of the catalytic site of cytochrome-c oxidase: specific ligands of heme a and the heme a_3-Cu_B center. Proc. Natl. Acad. Sci. U. S. A. **89**, 4786–4790 (1992)
40. Varotsis, C., Zhang, Y., Appelman, E.H., Babcock, G.T.: Resolution of the reaction sequence during the reduction of O_2 by cytochrome oxidase. Proc. Natl. Acad. Sci. U. S. A. **90**, 237–241 (1993)
41. Hu, X.H., Spiro, T.G.: Tyrosine and tryptophan structure markers in hemoglobin ultraviolet resonance Raman spectra: mode assignments via subunit-specific isotope labeling of recombinant protein. Biochemistry **36**, 15701–15712 (1997)
42. Peterson, E.S., Friedman, J.M.: A possible allosteric communication pathway identified through a resonance Raman study of four b37 mutants of human hemoglobin A. Biochemistry **37**, 4346–4357 (1998)
43. Schelvis, J.P.M., Zhao, Y., Marletta, M., Babcock, G.T.: Resonance Raman characterization of the heme domain of soluble guanylate cyclase. Biochemistry **37**, 16289–16297 (1998)
44. Wang, D.J., Spiro, T.G.: Structure changes in hemoglobin upon deletion of C-terminal residues, monitored by resonance Raman spectroscopy. Biochemistry **37**, 9940–9951 (1998)
45. Hu, X.H., Rodgers, K.R., Mukerji, I., Spiro, T.G.: New light on allostery: dynamic resonance Raman spectroscopy of hemoglobin Kempsey. Biochemistry **38**, 3462–3467 (1999)

46. Huang, J., Juszczak, L.J., Peterson, E.S., Shannon, C.F., Yang, M., et al.: The conformational and dynamic basis for ligand binding reactivity in hemoglobin Ypsilanti (beta 99 Asp -> Tyr): origin of the quarternary enhancement effects. Biochemistry **38**, 4514–4525 (1999)
47. Lee, H., Das, T.K., Rousseau, D.L., Mills, D., Ferguson-Miller, S., et al.: Mutations in the putative H channel in the cytochrome c oxidase from Rhodobacter sphaeroides show that this channel is not important for proton conduction but reveal modulation of the properties of heme a. Biochemistry **39**, 2989–2996 (2000)
48. Maes, E.M., Walker, F.A., Montfort, W.R., Czernuszewicz, R.S.: Resonance Raman spectroscopic study of nitrophorin 1, a nitric oxide-binding heme protein from Rhodnius prolixus, and its nitrosyl and cyano adducts. J. Am. Chem. Soc. **123**, 11664–11672 (2001)
49. Smulevich, G., Feis, A., Howes, B.D.: Fifteen years of Raman spectroscopy of engineered heme containing peroxidases: what have we learned? Acc. Chem. Res. **38**, 433–440 (2005)
50. Mattioli, T.A., Hoffman, A., Robert, B., Schrader, B., Lutz, M.: Primary donor structure and interactions in bacterial reaction centers from near-infrared Fourier-transform resonance Raman spectroscopy. Biochemistry **30**, 4648–4654 (1991)
51. Mattioli, T.A., Lin, X., Allen, J.P., Williams, J.C.: Correlation between multiple hydrogen-bonding and alteration of the oxidation potential of the bacteriochlorophyll dimer of reaction centers from Rhodobacter sphaeroides. Biochemistry **34**, 6142–6152 (1995)
52. Goldsmith, J.O., King, B., Boxer, S.G.: Mg coordination by amino acid side chains is not required for assembly and function of the special pair in bacterial photosynthetic reaction centers. Biochemistry **35**, 2421–2428 (1996)
53. Olsen, J.D., Sturgis, J.N., Westerhuis, W.H., Fowler, G.J., Hunter, C.N., et al.: Site-directed modification of the ligands to the bacteriochlorophylls of the light-harvesting LH1 and LH2 complexes of Rhodobacter sphaeroides. Biochemistry **36**, 12625–12632 (1997)
54. Ivancich, A., Artz, K., Williams, J.C., Allen, J.P., Mattioli, T.A.: Effects of hydrogen bonds on the redox potential and electronic structure of the bacterial primary electron donor. Biochemistry **37**, 11812–11820 (1998)
55. Stewart, D.H., Cua, A., Chisolm, D.A., Diner, B.A., Bocian, D.F., et al.: Identification of histidine 118 in the D1 polypeptide of photosystem II as the axial ligand to chlorophyll Z. Biochemistry **37**, 10040–10046 (1998)
56. Lapouge, K., Naveke, A., Gall, A., Ivancich, A., Sequin, J., et al.: Conformation of bacteriochlorophyll molecules in photosynthetic proteins from purple bacteria. Biochemistry **38**, 11115–11121 (1999)
57. Cherepy, N.J., Shreve, A.P., Moore, L.P., Boxer, S.G., Mathies, R.A.: Temperature dependence of the Q_y resonance Raman spectra of bacteriochlorophylls, the primary electron donor, and bacteriopheophytins in the bacterial photosynthetic reaction center. Biochemistry **36**, 8559–8566 (1997)
58. Cherepy, N.J., Shreve, A.P., Moore, L.P., Boxer, S.G., Mathies, R.A.: Electronic and nuclear dynamics of the accessory bacteriochlorophylls in bacterial photosynthetic reaction centers from resonance Raman intensities. J. Phys. Chem. B **101**, 3250–3260 (1997)
59. Esposito, A.P., Schellenberg, P., Parson, W.W., Reid, P.J.: Vibrational spectroscopy and mode assignments for an analog of the green fluorescent protein chromophore. J. Mol. Struct. **569**, 25–41 (2001)
60. Schellenberg, P., Johnson, E.T., Esposito, A.P., Reid, P.J., Parson, W.W.: Resonance Raman scattering by the green fluorescent protein and an analog of its chromophore. J. Phys. Chem. B **105**, 5316–5322 (2001)
61. Deng, H., Callender, R.: Raman spectroscopic studies of the structures, energetics, and bond distortions of substrates bound to enzymes. Methods Enzymol. **308**, 176–201 (1999)
62. Balakrishnan, G., Case, M.A., Pevsner, A., Zhao, X., Tengroth, C., et al.: Time-resolved absorption and UV resonance Raman spectra reveal stepwise formation of T quarternary contacts in the allosteric pathway of hemoglobin. J. Mol. Biol. **340**, 843–856 (2004)
63. Balakrishnan, G., Tsai, C.H., Wu, Q., Case, M.A., Pevsner, A., et al.: Hemoglobin site-mutants reveal dynamical role of interhelical H-bonds in the allosteric pathway:

time-resolved UV resonance Raman evidence for intra-dimer coupling. J. Mol. Biol. **340**, 857–868 (2004)

64. Ahmed, Z., Beta, I.A., Mikhonin, A.V., Asher, S.A.: UV-resonance Raman thermal unfolding study of Trp-cage shows that it is not a simple two-state miniprotein. J. Am. Chem. Soc. **127**, 10943–10950 (2005)
65. Overman, S.A., Bondre, P., Maiti, N.C., Thomas Jr., G.J.: Structural characterization of the filamentous bacteriophage PH75 from Thermus thermophilus by Raman and UV-resonance Raman spectroscopy. Biochemistry **44**, 3091–3100 (2005)
66. Rodriguez-Mendieta, I.R., Spence, G.R., Gell, C., Radford, S.E., Smith, D.A.: Ultraviolet resonance Raman studies reveal the environment of tryptophan and tyrosine residues in the native and partially folded states of the E-colicin-binding immunity protein Im7. Biochemistry **44**, 3306–3315 (2005)
67. Sato, A., Mizutani, Y.: Picosecond structural dynamics of myoglobin following photodissociation of carbon monoxide as revealed by ultraviolet time-resolved resonance Raman spectroscopy. Biochemistry **44**, 14709–14714 (2005)
68. Balakrishnan, G., Hu, Y., Nielsen, S.B., Spiro, T.G.: Tunable kHz deep ultraviolet (193–210 nm) laser for Raman application. Appl. Spectrosc. **59**, 776–781 (2005)
69. Asher, S.A., Mikhonin, A.V., Bykov, S.V.: UV Raman demonstrates that a-helical polyalanine peptides melt to polyproline II conformations. J. Am. Chem. Soc. **126**, 8433–8440 (2004)
70. Mikhonin, A.V., Myshakina, N.S., Bykov, S.V., Asher, S.A.: UV resonance Raman determination of polyproline II, extended 2.5_1-helix, and b-sheet y angle energy landscape in poly-L-lysine and poly-L-glutamic acid. J. Am. Chem. Soc. **127**, 7712–7720 (2005)
71. van Wart, H.E., Lewis, A., Scheraga, H.A., Saeva, F.D.: Disulfide bond dihedral angles from Raman spectroscopy. Proc. Natl. Acad. Sci. U. S. A. **70**, 2619–2623 (1973)
72. Shafer-Peltier, K.E., Haynes, C.L., Glucksberg, M.R., Van Duyne, R.P.: Toward a glucose biosensor based on surface-enhanced Raman scattering. J. Am. Chem. Soc. **125**, 588–593 (2003)
73. Seballos, L., Zhang, J.Z., Sutphen, R.: Surface-enhanced Raman scattering detection of lysophosphatidic acid. Anal. Bioanal. Chem. **383**, 763–767 (2005)
74. Maker, P.D., Terhune, R.W.: Study of optical effects due to an induced polarization third order in the electric field strength. Phys. Rev. **137**, A801–A818 (1964)
75. Bloembergen, N.: The stimulated Raman effect. Am. J. Phys. **35**, 989–1023 (1967)
76. Druet, S.A.J., Taran, J.P.E.: CARS spectroscopy. Prog. Quant. Electr. **7**, 1–72 (1981)
77. Shen, Y.R.: The Principles of Nonlinear Optics. Wiley, New York, NY (1984)
78. Bjorklund, G.C.: Effects of focusing on third-order nonlinear processes in isotropic media. IEEE J. Quant. Electronics **11**, 287–296 (1975)
79. Zumbusch, A., Holtom, G.R., Xie, X.S.: Three-dimensional vibrational imaging by coherent anti-Stokes Raman scattering. Phys. Rev. Lett. **82**, 4142–4145 (1999)
80. Volkmer, A., Cheng, J.X., Xie, X.S.: Vibrational imaging with high sensitivity via epi-detected coherent anti-Stokes Raman scattering microscopy. Phys. Rev. Lett. **87**, 0239011–0239014 (2001)
81. Cheng, J.X., Jia, Y.K., Zheng, G., Xie, X.S.: Laser-scanning coherent anti-Stokes Raman scattering microscopy and applications to cell biology. Biophys. J. **83**, 502–509 (2002)
82. Cheng, J.X., Pautot, S., Weitz, D.A., Xie, X.S.: Ordering of water molecules between phospholipid bilayers visualized by coherent anti-Stokes Raman scattering microscopy. Proc. Natl. Acad. Sci. U. S. A. **100**, 9826–9830 (2003)
83. Evans, C.L., Potma, E.O., Puoris'haag, M., Cote, D., Lin, C.P., et al.: Chemical imaging of tissue in vivo with video-rate coherent anti-Stokes Raman scattering microscopy. Proc. Natl. Acad. Sci. U. S. A. **102**, 16807–16812 (2005)
84. Potma, E.O., Xie, X.S.: Detection of single lipid bilayers with coherent anti-Stokes Raman scattering (CARS) microscopy. J. Raman Spectrosc. **34**, 642–650 (2003)

85. Wurpel, G.W.H., Schins, J.M., Muller, M.: Direct measurement of chain order in single phospholipid mono- and bilayers with multiplex CARS. J. Phys. Chem. B **108**, 3400–3403 (2004)
86. Wurpel, G.W.H., Rinia, H.A., Muller, M.: Imaging orientational order and lipid density in multilamellar vesicles with multiplex CARS microscopy. J. Microscopy (Oxford) **218**, 37–45 (2005)
87. Birge, R.R.: 2-photon spectroscopy of protein-bound chromophores. Acc. Chem. Res. **19**, 138–146 (1986)
88. Göppert-Mayer, M.: Über Elementarakte mit zwei Quantensprüngen. Ann. Phys. **9**, 273–295 (1931)
89. Kaiser, W., Garrett, G.B.C.: Two-photon excitation in $CaF_2:Eu^{2+}$. Phys. Rev. Lett. **7**, 229–231 (1961)
90. Birge, R.R., Murray, L.P., Pierce, B.M., Akita, H., Balogh-Nair, V., et al.: Two-photon spectroscopy of locked-11-cis-rhodopsin: evidence for a protonated Schiff base in a neutral protein binding site. Proc. Natl. Acad. Sci. U. S. A. **82**, 4117–4121 (1985)
91. Xu, C., Webb, W.W.: Measurement of two-photon excitation cross sections of molecular fluorophores with data from 690 to 1050 nm. J. Opt. Soc. Am. B **13**, 481–491 (1996)
92. Oh, D.H., Stanley, R.J., Lin, M., Hoeffler, W.K., Boxer, S.G., et al.: Two-photon excitation of 4′-hydroxymethyl-4,5′,8-trimethylpsoralen. Photochem. Photobiol. **65**, 91–95 (1997)
93. Larson, D.R., Zipfel, W.R., Williams, R.M., Clark, S.W., Bruchez, M.P., et al.: Water-soluble quantum dots for multiphoton fluorescence imaging in vivo. Science **300**, 1434–1436 (2003)
94. Monson, P.R., McClain, W.M.: Polarization dependence of the two-photon absorption of tumbling molecules with application to liquid 1-chloronaphthalene and benzene. J. Chem. Phys. **53**, 29–37 (1970)
95. McClain, W.M.: Excited state symmetry assignment through polarized two-photon absorption studies of fluids. J. Chem. Phys. **55**, 2789–2796 (1971)
96. McClain, W.M.: Polarization of two-photon excited fluorescence. J. Chem. Phys. **58**, 324–326 (1972)
97. Drucker, R.P., McClain, W.M.: Polarized two-photon studies of biphenyl and several derivatives. J. Chem. Phys. **61**, 2609–2615 (1974)
98. Callis, P.R.: On the theory of two-photon induced fluorescence anisotropy with application to indoles. J. Chem. Phys. **99**, 27–37 (1993)
99. Callis, P.R.: Two-photon-induced fluorescence. Ann. Rev. Phys. Chem. **48**, 271–297 (1997)
100. Rehms, A.A., Callis, P.R.: Resolution of L_a and L_b bands in methyl indoles by two-photon spectroscopy. Chem. Phys. Lett. **140**, 83–89 (1987)
101. Callis, P.R.: Molecular orbital theory of the ^1L_b and ^1L_a states of indole. J. Chem. Phys. **95**, 4230–4240 (1991)
102. Denk, W., Strickler, J.H., Webb, W.W.: Two-photon laser scanning fluorescence microscopy. Science **248**, 73–76 (1990)
103. Helmchen, F., Denk, W.: Deep tissue two-photon microscopy. Nat. Methods **2**, 932–940 (2005)
104. Xu, C., Zipfel, W., Shear, J.B., Williams, R.M., Webb, W.W.: Multiphoton fluorescence excitation: new spectral windows for biological nonlinear microscopy. Proc. Natl. Acad. Sci. U. S. A. **93**, 10763–10768 (1996)
105. Yuste, R., Konnerth, A.: Imaging in Neuroscience and Development: A Laboratory Manual. Cold Spring Harbor Laboratory Press, Cold Spring Harbor, NY (2000)
106. Sanchez, S.A., Gratton, E.: Lipid-protein interactions revealed by two-photon microscopy and fluorescence correlation spectroscopy. Acc. Chem. Res. **38**, 469–477 (2005)
107. Buosso, P., Bhakta, N.R., Lewis, R.S., Robey, E.: Dynamics of thymocyte-stromal cell interactions visualized by two-photon microscopy. Science **296**, 1876–1880 (2002)
108. Benninger, R.K.P., Önfelt, B., Neil, M.A.A., Davis, D.M., French, P.M.W.: Fluorescence imaging of two-photon linear dichroism: cholesterol depletion disrupts molecular orientation in cell membranes. Biophys. J. **88**, 609–622 (2005)

109. Zal, T., Gascoigne, N.R.: Using live FRET imaging to reveal early protein-protein interactions during T cell activation. Curr. Opin. Immunol. **16**, 418–427 (2004)
110. Zimm, B.: Apparatus and methods for measurement and interpretation of angular variation of light scattering; preliminary results on polystyrene solutions. J. Chem. Phys. **16**, 1099–1116 (1948)
111. Kerker, M.: The Scattering of Light and Other Electromagnetic Radiation. Academic Press, New York, NY (1969)
112. Chu, B.: Laser Light Scattering: Basic Principles and Practice, 2nd edn. Academic Press, New York, NY (1991)
113. Schurr, J.M.: Dynamic light scattering of biopolymers and biocolloids. CRC Crit. Rev. Biochem. **4**, 371–431 (1977)
114. Schurr, J.M.: Rotational diffusion of deformable macromolecules with mean local cylindrical symmetry. Chem. Phys. **84**, 71–96 (1984)
115. Brown, W. (ed.): Dynamic Light Scattering: The Method and Some Applications. Clarendon Press, Oxford (1993)

Appendix A

A.1 Vectors

Vectors are used to represent properties that have both magnitude and direction. *Scalars* have a magnitude but no direction. Velocity, for example, is a vectorial property, while mass is a scalar. A vector in an N-dimensional coordinate space has N independent components (A_k), each parallel to one of the coordinate axes. In the text we denote a vector by a bold-face letter in italics or by enclosing a list of the individual components in parentheses:

$$\boldsymbol{A} = (A_1, A_2, A_3, \ldots). \tag{A1.1}$$

In a three-dimensional coordinate system, for example, $\boldsymbol{A} = (A_x, A_y, A_z)$ where A_x, A_y, and A_z are the components parallel to the x, y and z axes. The components can be arranged in either a row or a column. A vector with unit length parallel to the k axis is designated by a letter with a caret (^) on top $(\hat{k})$.

The *magnitude*, *modulus* or *length* of vector $\boldsymbol{A}$ is the square root of the sum of the squares of the individual components:

$$|\boldsymbol{A}| = \left(\sum_k A_k^2\right)^{1/2}. \tag{A1.2}$$

The sum or difference of two vectors is obtained by simply adding or subtracting the corresponding components. In three dimensions, for example,

$$\boldsymbol{A} \pm \boldsymbol{B} = (A_x \pm B_x,\ A_y \pm B_y,\ A_z \pm B_z). \tag{A1.3}$$

There are two types of vector products. The *dot product* or *scalar product* $\boldsymbol{A}\cdot\boldsymbol{B}$ of vectors $\boldsymbol{A}$ and $\boldsymbol{B}$ is a scalar whose magnitude is the sum of the products of the corresponding components:

W.W. Parson, *Modern Optical Spectroscopy*, DOI 10.1007/978-3-662-46777-0

$$A \cdot B = \sum_k A_k B_k. \tag{A1.4}$$

The magnitude of $\boldsymbol{A}$ therefore can be written as $|\boldsymbol{A}| = (\boldsymbol{A} \cdot \boldsymbol{A})^{1/2}$. In three dimensions,

$$\boldsymbol{A} \cdot \boldsymbol{B} = |\boldsymbol{A}||\boldsymbol{B}| \cos(\theta), \tag{A1.5}$$

where θ is the angle between the two vectors.

The *cross product* or *vector product* of two vectors, denoted $\boldsymbol{A} \times \boldsymbol{B}$ or $\boldsymbol{A} \wedge \boldsymbol{B}$, is a vector that is perpendicular to both $\boldsymbol{A}$ and $\boldsymbol{B}$ and has magnitude $|\boldsymbol{A}||\boldsymbol{B}| \sin(\theta)$. $\boldsymbol{A} \times \boldsymbol{B}$ is oriented in the direction in which a right-handed screw would advance if turning the screw rotates $\boldsymbol{A}$ onto $\boldsymbol{B}$. Thus $\boldsymbol{A} \times \boldsymbol{B}$ and $\boldsymbol{B} \times \boldsymbol{A}$ have the same magnitude but point in opposite directions. In vector notation,

$$\boldsymbol{A} \times \boldsymbol{B} = \left([A_y B_z - A_z B_y],\ -[A_x B_z - A_z B_x],\ [A_x B_y - A_y B_x] \right), \tag{A1.6}$$

which can be written in the form of a determinant:

$$\boldsymbol{A} \times \boldsymbol{B} = \begin{vmatrix} \hat{x} & \hat{y} & \hat{z} \\ A_x & A_y & A_z \\ B_x & B_y & B_z \end{vmatrix}. \tag{A1.7}$$

The *scalar triple product* of three vectors,

$$\boldsymbol{C} \cdot \boldsymbol{A} \times \boldsymbol{B} = \boldsymbol{C} \cdot (\boldsymbol{A} \times \boldsymbol{B}), \tag{A1.8}$$

is a scalar whose sign changes if the order of any two of the vectors is interchanged: $\boldsymbol{C} \cdot \boldsymbol{A} \times \boldsymbol{B} = -\boldsymbol{A} \cdot \boldsymbol{C} \times \boldsymbol{B} = \boldsymbol{B} \cdot \boldsymbol{C} \times \boldsymbol{A} = -\boldsymbol{B} \cdot \boldsymbol{A} \times \boldsymbol{C}$. The scalar triple product is zero if any two of the three vectors are parallel.

The *gradient of a scalar* quantitity A, which we will write $\tilde{\nabla} A$, is a vector whose components are derivatives of A with respect to the corresponding coordinates. In three dimensions, the gradient operator is

$$\tilde{\nabla} = (\partial A/\partial x,\ \partial A/\partial y,\ \partial A/\partial z). \tag{A1.9}$$

Several other functions of the derivatives of vectors occur frequently in discussions of electromagnetic fields. The *divergence* of a vector $\boldsymbol{A}$, written div$\boldsymbol{A}$, is defined as

$$\mathrm{div}\, \boldsymbol{A} = \tilde{\nabla} \cdot \boldsymbol{A} = \frac{\partial A_x}{\partial x} + \frac{\partial A_y}{\partial y} + \frac{\partial A_z}{\partial z}, \tag{A1.10}$$

and the curl (curl$\boldsymbol{A}$) is

$$\operatorname{curl} A = \tilde{\nabla} \times A = \begin{vmatrix} \hat{x} & \hat{y} & \hat{z} \\ \partial/\partial x & \partial/\partial y & \partial/\partial z \\ A_x & A_y & A_z \end{vmatrix}$$
$$= \hat{x}\left(\frac{\partial A_z}{\partial y} - \frac{\partial A_y}{\partial z}\right) + \hat{y}\left(\frac{\partial A_x}{\partial z} - \frac{\partial A_z}{\partial x}\right) + \hat{z}\left(\frac{\partial A_y}{\partial x} - \frac{\partial A_x}{\partial y}\right). \quad \text{(A1.11)}$$

Figure A1a illustrates a vector function of x and y that has a non-zero curl but a divergence of zero. A vector with a non-zero divergence but a curl of zero is shown in Fig. A1b.

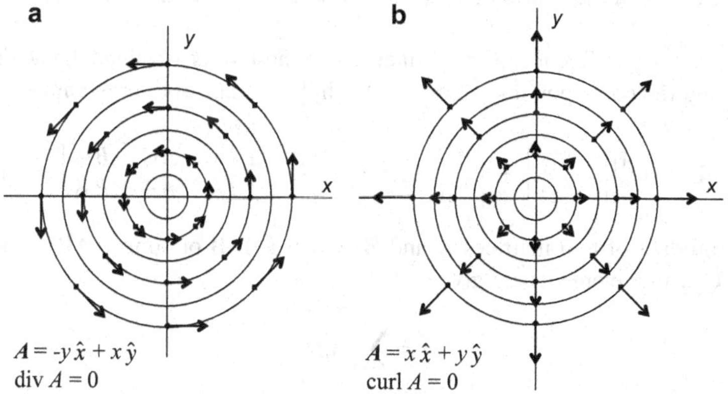

Fig. A1 (**a**) The vector function $-y\hat{x} + y\hat{x}$ has non-zero curl but zero divergence. (**b**) The function $x\hat{x} + y\hat{y}$ has non-zero divergence but zero curl

A.2 Matrices

A *matrix* is an ordered, two-dimensional array of *elements*, A_{ij}, with the first index (i) indicating the row in which the term is located in the array and the second (j) indicating the column. Matrices are denoted in the text by unitalicized letters in bold face or by enclosing the elements in square brackets. For example, if

$$\mathbf{A} = \begin{bmatrix} 41 & 73 \\ 9 & 12 \end{bmatrix}, \quad \text{(A2.1)}$$

then $A_{11} = 41$, $A_{12} = 73$, $A_{21} = 9$ and $A_{22} = 12$. We are concerned mainly with *square* matrices, which are matrices in which the number of rows is the same as the number of columns.

A *diagonal matrix* is a matrix in which non-zero elements occur only on the major diagonal, as for example,

$$\mathbf{A} = \begin{bmatrix} 17 & 0 \\ 0 & 3 \end{bmatrix}. \tag{A2.2}$$

The *trace*, or *character*, of matrix **A**, denoted Tr[**A**], is the sum of the diagonal elements:

$$Tr[\mathbf{A}] = \sum_k A_{kk}. \tag{A2.3}$$

For example, the trace of the matrix in Eq. (A2.1) is Tr[**A**] = 41 + 12 = 53. The trace of a matrix obeys the distributive law of arithmetic: if **C** = **A** + **B**, Tr[**C**] = Tr[**A**] + Tr[**B**].

The sum or difference of two matrices **A** and **B** is obtained by adding or subtracting the corresponding elements. With 2 × 2 matrices, for example,

$$\mathbf{A} \pm \mathbf{B} = \begin{bmatrix} A_{11} & A_{12} \\ A_{21} & A_{22} \end{bmatrix} \pm \begin{bmatrix} B_{11} & B_{12} \\ B_{21} & B_{22} \end{bmatrix} = \begin{bmatrix} A_{11} \pm B_{11} & A_{12} \pm B_{12} \\ A_{21} \pm B_{21} & A_{22} \pm B_{22} \end{bmatrix}. \tag{A2.4}$$

The product of two matrices **A** and **B** (written **A** · **B** or simply **AB**) is another matrix **C** whose elements are given by

$$C_{ij} = \sum_k A_{ik} B_{kj}. \tag{A2.5}$$

For example, the product of two 3 × 3 matrices is

$$\mathbf{A} \cdot \mathbf{B} = \begin{bmatrix} A_{11} & A_{12} & A_{13} \\ A_{21} & A_{22} & A_{23} \\ A_{31} & A_{32} & A_{33} \end{bmatrix} \cdot \begin{bmatrix} B_{11} & B_{12} & B_{13} \\ B_{21} & B_{22} & B_{23} \\ B_{31} & B_{32} & B_{33} \end{bmatrix}$$
$$= \begin{bmatrix} A_{11}B_{11} + A_{12}B_{21} + A_{13}B_{31} & A_{11}B_{12} + A_{12}B_{22} + A_{13}B_{32} & A_{11}B_{13} + A_{12}B_{23} + A_{13}B_{33} \\ A_{21}B_{11} + A_{22}B_{21} + A_{23}B_{31} & A_{21}B_{12} + A_{22}B_{22} + A_{23}B_{32} & A_{21}B_{13} + A_{22}B_{23} + A_{23}B_{33} \\ A_{31}B_{11} + A_{32}B_{21} + A_{33}B_{31} & A_{31}B_{12} + A_{32}B_{22} + A_{33}B_{32} & A_{31}B_{13} + A_{32}B_{23} + A_{33}B_{33} \end{bmatrix}. \tag{A2.6}$$

From Eqs. (A2.3) and (A2.5), the trace of the product **AB** is

$$\mathrm{Tr}[\mathbf{AB}] = \sum_i \sum_k A_{ik} B_{ki} = \sum_k \sum_i A_{ki} B_{ik} = \mathrm{Tr}[\mathbf{BA}]. \tag{A2.7}$$

And from this and the fact that **ABC** = **A** · (**BC**) = (**AB**) · **C** it follows that the trace of **ABC** is invariant to cyclic permutations:

$$\mathrm{Tr}[\mathbf{ABC}] = \mathrm{Tr}[\mathbf{CAB}] = \mathrm{Tr}[\mathbf{BCA}]. \tag{A2.8}$$

However, Tr[**ABC**] is not generally equal to Tr[**CBA**].

The product of a matrix **A** with a column vector ***B*** is a vector ***C*** with elements defined by

$$C_i = \sum_k A_{ik} B_k. \tag{A2.9}$$

The *transpose* ($\mathbf{A}^{\mathrm{T}}$) of matrix **A** is obtained by interchanging rows and columns, so that element A_{ij} becomes A_{ji}.

The *inverse* ($\mathbf{A}^{-1}$) of **A** is a matrix that, when multiplied by **A** gives a diagonal matrix with all the diagonal terms equal to 1. So, for a 2 × 2 matrix,

$$\mathbf{A}^{-1} \cdot \mathbf{A} = \begin{bmatrix} 1 & 0 \\ 0 & 1 \end{bmatrix}. \tag{A2.10}$$

Such a diagonal matrix of 1's is often denoted by a bold-face **1**. Finding the inverse of a square matrix (*inverting* the matrix) is a common procedure that provides the solutions to sets of linear algebraic equations. Press et al. [1] give efficient algorithms for doing this.

A matrix **A** is said to be *symmetric* if, for all its elements, $A_{ij} = A_{ji}$. It is *Hermitian* if, for all its elements, $A_{ij} = A^*_{ji}$, where A^*_{ji} is the complex conjugate of A_{ji}. All the matrices we discuss in the text are Hermitian. A matrix is called *orthogonal* if its transpose is the same as its inverse, so that

$$\mathbf{A}^{\mathrm{T}} \cdot \mathbf{A} = \mathbf{A}^{-1} \cdot \mathbf{A} = \mathbf{1}. \tag{A2.11}$$

The *gradient of a vector* function ***A***, which we write as $\tilde{\nabla} \boldsymbol{A}$, is a matrix in which element A_{ij} is the derivative of component *i* of the vector with respect to coordinate *j*. Thus if $\boldsymbol{A} = (A_x, A_y, A_z)$, its gradient is

$$\tilde{\nabla} \boldsymbol{A} = \begin{bmatrix} \partial A_x/\partial x & \partial A_y/\partial x & \partial A_z/\partial x \\ \partial A_x/\partial y & \partial A_y/\partial y & \partial A_z/\partial y \\ \partial A_x/\partial z & \partial A_y/\partial z & \partial A_z/\partial z \end{bmatrix}. \tag{A2.12}$$

The solutions to many problems in quantum mechanics and spectroscopy require *diagonalizing* matrices. Given a non-diagonal matrix **A**, the task is to find another matrix **C** and its inverse $\mathbf{C}^{-1}$ such that the product $\mathbf{C}^{-1} \cdot \mathbf{A} \cdot \mathbf{C}$ is diagonal. Computer algorithms are available for diagonalizing even large matrices rapidly [1].

A.3 Fourier Transforms

The *Fourier transform* of a function *f*(*t*) of time is the integral

$$F(\nu) = \int_{-\infty}^{\infty} f(t) \exp(2\pi i \nu t)\, dt. \tag{A3.1}$$

If $f(t)$ is defined everywhere in the interval $-\infty < t < \infty$, and the integral of $f(t)dt$ over this interval converges (i.e., is finite), the Fourier transform $F(\nu)$ also will converge. In addition, an *inverse Fourier transform* will regenerate the original function:

$$f(t) = \int_{-\infty}^{\infty} F(\nu)\exp(-2\pi i\nu t)\,d\nu. \tag{A3.2}$$

The pair of functions $f(t)$ and $F(\nu)$ can be viewed as two different representations of the same physical quantity. For example, if $f(t)$ expresses a quantity as a function of time (in seconds), $F(\nu)$ expresses the same quantity as a function of frequency (in cycles per second, or Hz). Sometimes it is convenient to use the angular frequency, $\omega = 2\pi\nu$, in units of rad/s; the transforms then must be scaled by a factor of $(2\pi)^{-1/2}$:

$$F(\omega) = \frac{1}{\sqrt{2\pi}} \int_{-\infty}^{\infty} f(t)\exp(i\omega t)\,dt \tag{A3.3}$$

and

$$f(t) = \frac{1}{\sqrt{2\pi}} \int_{-\infty}^{\infty} F(\omega)\exp(-i\omega t)\,d\omega. \tag{A3.4}$$

The same expressions can be used with other pairs of variables. The Fourier transform of a function of position (in units of, say, Å) gives a function of inverse length (cycles per Å). The Fourier transform of an interferogram obtained in an FTIR spectrometer thus gives the intensity of radiation as a function of the wavenumber $\bar{\nu}$.

In Chap. 2, we encountered a complex exponential function of the form $f(t) = \exp(-at - ibt)$ for $t > 0$ and $f(t) = 0$ for $t < 0$ (Eq. 2.65). The Fourier transform of this function is

$$F(\omega) = \left(\frac{1}{\hbar}\right)\left(\frac{1}{\sqrt{2\pi}}\right)\left(\frac{i}{\omega - b + ia}\right) \tag{A3.5a}$$

$$= \frac{1}{\hbar\sqrt{2\pi}}\left(\frac{i}{\omega - b + ia}\right)\left(\frac{\omega - b + ia}{\omega - b + ia}\right) = \frac{1}{\hbar\sqrt{2\pi}}\left(\frac{a + i(\omega - b)}{(\omega - b)^2 + a^2}\right), \tag{A3.5b}$$

where $\omega = E/\hbar$ and $d\omega = dE/\hbar$. Equation (2.71) is obtained by multiplying the real part of this expression by the normalization factor $(2/\pi)^{1/2}$.

The Fourier transform in Eqs. (A3.5a) and (A3.5b) includes both real and imaginary parts. This is because the original function $f(t)$ is not symmetrical around $t=0$. For functions that are symmetrical around zero in the sense that $|f(-t)| = |f(t)|$, the nature of the Fourier transform depends on whether the function has the same or opposite signs on either side of zero. A function is said to be *even* if $f(-t)=f(t)$, and *odd* if $f(-t)=-f(t)$. The Fourier transform of any real, even function is also real and even, whereas the transform of a real, odd function is purely imaginary and odd. The Fourier transform of $\exp(-|t/\tau|)$ (a real and even function) thus has only a real part, which turns out to be a Lorentzian:

$$f(t) = \exp(-|t/\tau|) \tag{A3.6}$$

and

$$F(\omega) = \sqrt{2/\pi}\left\{\frac{1/\tau}{(1/\tau)^2 + \omega^2}\right\} \tag{A3.7}$$

(Eq. (2.63) and Fig. 2.7).

If a function does not have either even or odd symmetry, its Fourier transform is complex. The real and imaginary parts of the transform consist of the cosine and sine Fourier transforms discussed in Appendix 4.

To illustrate these points, panels A and B of Fig. A2 show the even, two-sided decay function $f_2(t) = \exp(-|t|/\tau)$ and its Fourier transform ($F_2(\omega)$). Figure A2C shows the one-sided function $f_1(t) = \exp(-|t|/\tau)$ for $t \geq 0, f_1(t) = 0$ for $t < 0$, and Fig. A2D shows the Fourier transform of this function ($F_1(\omega)$) along with its real and imaginary parts. After scaling by a factor of 2 to compensate for the fact that it represents only positive values of t, the real part of $F_1(\omega)$ is the same as $F_2(\omega)$.

The Fourier transform of a Gaussian function centered at $t=$ zero (a real and even function of t) is another Gaussian:

$$f(t) = \exp(-at^2) \tag{A3.8}$$

and

$$F(v) = (2a)^{-1/2}\exp(-v^2/4a). \tag{A3.9}$$

If the Gaussian is centered at some value m other than zero, the Fourier transform is multiplied by $\exp(imv)$, or $\cos(mv) + i\ \sin(mv)$, and thus has an imaginary component.

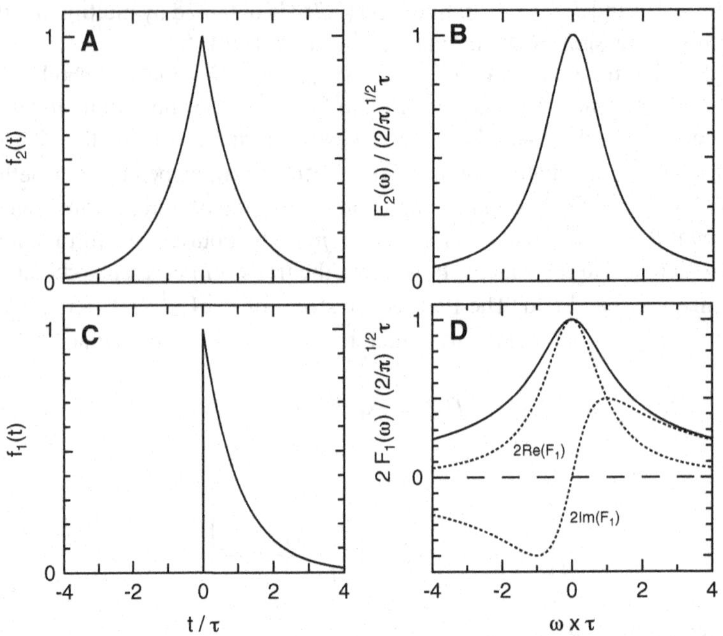

Fig. A2 (**A**) An even, two-sided decay function: $f_2(t) = \exp(-|t|/\tau)$. (**B**) $F_2(\omega)$, the Fourier transform of $f_2(t)$, is a purely real Lorentzian. (**C**) A one-sided decay function: $f_1(t) = \exp(-t/\tau)$ for $t \geq 0, f_1(t) = 0$ for $t < 0$. (**D**) $2\,F_1(\omega)$, the Fourier transform of $f_1(t)$, (*solid curve*) and its real and imaginary parts (*dotted curves*). $2Re(F_1(\omega))$ is identical to $F_2(\omega)$. Note that the ordinate scales are different in (**B**) and (**D**)

Fourier transforms provide a way of representing the *Dirac delta function*, $\delta(x)$, which is defined by the conditions $\delta(x) = 0$ if $x \neq 0$, and

$$\int_{-a}^{a} \delta(x)dx = 1 \tag{A3.10}$$

for $a > 0$. $\delta(x)$ is a function that peaks sharply at $x = 0$, in the limit that the width of the peak goes to zero while the height becomes infinite so that the area remains constant. It is useful for analyzing the dynamics of a process that occurs at a significant rate only when the energy difference between two states is close to zero. If a function $f(x)$ is defined over the region $x_1 < x < x_2$ and X is a particular value of x in this region, then

$$\int_{x_1}^{x_2} f(x)\delta(X - x)dx = f(X). \tag{A3.11}$$

$\delta(x)$ can be expressed as $(2\pi)^{-1/2}$ times the Fourier transform of the constant $f(x) = 1$:

$$\delta(x) = \frac{1}{2\pi}\int_{-\infty}^{\infty} \exp(ixy)dy. \tag{A3.12}$$

One way to look at this relationship is to note that $f(x)$ can be a constant only if its oscillation frequencies are distributed infinitely sharply around zero.

The Fourier transform of $\cos(\omega_o t)$ is $\delta(\omega \pm \omega_o)$, a pair of delta functions located at $\omega = \pm\omega_o$. The transform of $\sin(\omega_o t)$ is a similar pair of delta functions, but with imaginary amplitudes. The transform of an arbitrary, fluctuating function can be viewed as a superposition of many such delta functions with amplitudes reflecting the contributions that oscillations at particular frequencies make to the overall function.

Tables of Fourier transforms of many other functions are available [2], and there are rapid computational methods for finding the Fourier transform of an arbitrary function [1]. For additional information on Fourier transforms, see [3].

A.4 Phase Shift and Modulation Amplitude in Frequency-Domain Spectroscopy

To derive the expressions for the fluorescence phase shift (ϕ) and modulation amplitude (m) in Fig. 1.16, suppose the oscillatory part of the excitation light intensity is $I(t) = \sin(\omega t)$, and that the fluorescence ($F(t)$) generated by an instantaneous pulse of exciting light decays exponentially with time constant τ. The oscillatory part of the fluorescence signal observed at time t ($S(t)$) is obtained by integrating the fluorescence from the modulated excitation at all earlier times (t'):

$$\begin{aligned} S(t) &= \int_0^t I(t')F(t - t')dt' \Big/ \int_0^\infty F(t)dt \\ &= \int_0^t \sin(\omega t')\exp(-(t - t')/\tau)\,dt' \Big/ \int_0^\infty \exp(-t/\tau)\,dt. \end{aligned} \tag{A4.1}$$

The denominator in this expression is the total fluorescence generated by an instantaneous excitation pulse, which for a single-exponential decay is just τ. The *convolution integral* in the numerator can be evaluated straightforwardly:

$$\int_0^t \sin(\omega t')\exp(-(t-t')/\tau)\,dt' = \exp(-t/\tau)\int_0^t \sin(\omega t')\exp(t'/\tau)\,dt'$$
$$= \frac{(1/\tau)\sin(\omega t) - \omega\cos(\omega t) + \omega\exp(-t/\tau)}{(1/\tau)^2 + \omega^2}. \quad \text{(A4.2)}$$

The term $\omega\exp(-t/\tau)$ goes to zero at long times ($t \gg \tau$), giving

$$S(t) = \left(\frac{(1/\tau)\sin(\omega t) - \omega\cos(\omega t)}{(1/\tau)^2 + \omega^2}\right)\Big/\tau = \frac{\sin(\omega t) - (\omega\tau)\cos(\omega t)}{1 + (\omega\tau)^2}. \quad \text{(A4.3)}$$

Equating $S(t)$ to $m{\cdot}\sin(\omega t + \phi)$ and using the relationship $\sin(\omega t + \phi) = \sin(\omega t)\cos(\phi) + \cos(\omega t)\sin(\phi)$ gives the desired expressions:

$$m\cos(\phi) = \frac{1}{1 + (\omega\tau)^2}, \quad \text{(A4.4)}$$

$$m\sin(\phi) = \frac{\omega\tau}{1 + (\omega\tau)^2}, \quad \text{(A4.5)}$$

$$\tan(\phi) = m\sin(\phi)/m\cos(\phi) = \omega\tau, \quad \text{(A4.6)}$$

$$m^2 = m^2\cos^2(\phi) + m^2\sin^2(\phi) = \left(1 + (\omega\tau)^2\right)\Big/\left(1 + (\omega\tau)^2\right)^2, \quad \text{(A4.7)}$$

and

$$m = \left(1 + (\omega\tau)^2\right)^{-1/2}. \quad \text{(A4.8)}$$

If the fluorescence response to an instantaneous excitation pulse is multiexponential,

$$F(t) = \sum_k B_k\exp(-t/\tau_k), \quad \text{(A4.9)}$$

then Eq. (A4.3) becomes:

$$S(t) = \left(\sum_k B_k \frac{\tau_k\sin(\omega t) - \omega\tau_k^2\cos(\omega t)}{1 + (\omega\tau_k)^2}\right)\Big/\left(\sum_k B_k\tau_k\right). \quad \text{(A4.10)}$$

The cosine and sine terms in this expression can be viewed as, respectively, normalized *sine and cosine Fourier transforms* of the fluorescence decay function. If we define the normalized sine and cosine Fourier transforms of $F(t)$ as

$$S_{\sin}(\omega) = \int_0^\infty F(t)\sin(\omega t)\,dt \Big/ \int_0^\infty F(t)\,dt$$
$$= \left(\sum_k B_k \frac{\omega \tau_k^2}{1+(\omega\tau_k)^2}\right) \Big/ \left(\sum_k B_k \tau_k\right) \tag{A4.11}$$

and

$$S_{\cos}(\omega) = \int_0^\infty F(t)\cos(\omega t)\,dt \Big/ \int_0^\infty F(t)\,dt$$
$$= \left(\sum_k B_k \frac{\tau_k}{1+(\omega\tau_k)^2}\right) \Big/ \left(\sum_k B_k \tau_k\right), \tag{A4.12}$$

then Eqs. (A4.6) and (A4.8) take the forms

$$\tan(\phi) = S_{\sin}(\omega)/S_{\cos}(\omega), \tag{A4.13}$$

and

$$m = \left((S_{\sin}(\omega))^2 + (S_{\cos}(\omega))^2\right)^{-1/2}. \tag{A4.14}$$

More generally, the sine Fourier transform $G_{sin}(\omega)$ of a function $g(t)$ is defined as

$$G_{\sin}(\omega) = i(2\pi)^{-1/2} \int_{-\infty}^{\infty} g(t)\sin(\omega t)\,dt, \tag{A4.15}$$

which is zero if g is an even function of t. The cosine Fourier transform,

$$G_{\cos}(\omega) = (2\pi)^{-1/2} \int_{-\infty}^{\infty} g(t)\cos(\omega t)\,dt, \tag{A4.16}$$

is zero for odd functions of t. The continuous Fourier transform defined in Eq. (A3.3) is the sum of the sine and cosine Fourier transforms in Eqs. (A4.15) and (A4.16), as can be seen from the relationship $\exp(i\theta) = \cos(\theta) + i\sin(\theta)$. The factor i in Eq. (A4.15) is often omitted because only its product with the corresponding factor in the inverse transform ($-i$) is determined uniquely. Since the fluorescence decay function $F(t)$ is zero for $t < 0$, taking the integrals in Eqs. (A4.11) and (A4.12) from 0 to ∞ rather than from $-\infty$ to ∞ does not affect the results.

See [4, 5] for further information on data analysis and extensions to fluorescence anisotropy.

A.5 CGS and SI Units and Abbreviations

Physical quantity	CGS unit	SI (MKS) equivalent
Electric current	Abampere, biot (Bi)	10 amperes (A)
Energy	Calorie (cal)	4.1868 joule (J)
Dipole moment	Debye (D)	3.3356×10^{-30} coulomb meter (C m)
Force	Dyne (dyn)	10^{-5} newton (N)
Magnetic dipole moment	Emu	10^{-3} ampere·meter^2 (A m^2) 1.2566×10^{-3} tesla (T)
Energy, work	Erg	10^{-7} joule (J)
Electric charge	Esu, statcoulomb or franklin (Fr)	3.3356×10^{-10} coulomb
Magnetic flux density (magnetic induction)	Gauss (G)	10^{-4} tesla (T)
Wavenumber	Kayser (cm^{-1})	100 per meter
Luminance	Lambert (Lb)	3.1831×10^3 candela meter^{-2} (Cd m^2)
Magnetic flux	Maxwell (Mx)	10^{-8} weber (Wb)
Magnetic field strength	Oersted (Oe)	79.577 ampere-turns meter^{-1}
Illumination	Phot	10^4 lux
Dynamic viscosity	Poise (P)	0.1 pascal·second (Pa s)
Electric current	Statampere	3.3356×10^{-10} ampere (A)
Electric charge	Statcoulomb	3.3356×10^{-10} coulomb (C)
Potential	Statvolt	299.79 volts (V)
Magnetic flux	Unit pole	1.2564×10^{-7} weber (Wb)

References

1. Press, W.H., Flannery, B.P., Teukolsky, S.A., Vetterling, W.T.: Numerical Recipes in Fortran 77: The Art of Scientific Computing. Cambridge University Press, Cambridge (1989)
2. Beyer, W.H.: CRC Standard Mathematical Tables. CRC Press, Boca Raton, FL (1973)
3. Butkov, E.: Mathematical Physics. Addison-Wesley, Reading, MA (1968)
4. Weber, G.: Theory of differential phase fluorometry: detection of anisotropic molecular rotations. J. Chem. Phys. **66**, 4081–4091 (1977)
5. Lakowicz, J.R., Laczko, G., Cherek, H., Gratton, E., Limkeman, M.: Analysis of fluorescence decay kinetics from variable-frequency phase shift and modulation data. Biophys. J. **46**, 463–477 (1984)

W.W. Parson, *Modern Optical Spectroscopy*, DOI 10.1007/978-3-662-46777-0

Index

W.W. Parson, *Modern Optical Spectroscopy*, DOI 10.1007/978-3-662-46777-0